Withdrawn from Library Stock
Leeds Trinity

D1429116

LEEDS TRINITY LIBRARY

299038 0

Food Processing
Principles and Applications

Food Processing
Principles and Applications

Edited by
J. Scott Smith and Y. H. Hui

Blackwell Publishing

©2004 Blackwell Publishing
All rights reserved

Blackwell Publishing Professional
2121 State Avenue, Ames, Iowa 50014, USA

Orders: 1-800-862-6657
Office: 1-515-292-0140
Fax: 1-515-292-3348
Web site: www.blackwellprofessional.com

Blackwell Publishing Ltd
9600 Garsington Road, Oxford OX4 2DQ, UK
Tel.: +44 (0)1865 776868

Blackwell Publishing Asia
550 Swanston Street, Carlton, Victoria 3053, Australia
Tel.: +61 (0)3 8359 1011

Authorization to photocopy items for internal or personal use, or the internal or personal use of specific
clients, is granted by Blackwell Publishing, provided
that the base fee of $.10 per copy is paid directly to
the Copyright Clearance Center, 222 Rosewood Drive,
Danvers, MA 01923. For those organizations that
have been granted a photocopy license by CCC, a
separate system of payments has been arranged. The
fee code for users of the Transactional Reporting
Service is 0-8138-1942-3/2004 $.10.

First edition, 2004

Library of Congress Cataloging-in-Publication Data

Food processing : principles and applications / edited
    by J. Scott Smith and Y. H. Hui.—1st ed.
        p. cm.
    Includes index.
    ISBN 0-8138-1942-3 (acid-free paper)
    1. Food industry and trade.   I. Smith, J. Scott II.
    Hui, Y. H. (Yiu H.)

TP370.F626 2004
664—dc22
                                    2004007256

LEEDS TRINITY UNIVERSITY

The last digit is the print number: 9 8 7 6 5 4 3 2 1

Contents

Contributors

Sheryl A. Barringer, Ph.D. (Chapter 29)
Ohio State University
Department of Food Science and Technology
Room 110
2015 Fyffe Road
Columbus OH 43210-1007 USA
Phone: 614-688-3642
Fax: 614-292-0218
Email: barringer.11@osu.edu

Laura K. Basilio, M.S. (Chapter 19)
Sensory Evaluation Consultant
815 Josepi Drive
Knoxville, TN 37918 USA
Phone: 865-938-3017
E-mail: tbasilio2002@comcast.net

Samuel E. Beattie, Ph.D. (Chapter 14)
Extension Specialist
Department of Food Science and Human Nutrition
Iowa State University
Ames, IA 50011-1120 USA
Phone: 515-294-3357
Fax: 515-294-1040
E-mail: beatties@iastate.edu

Daniel W. Bena (Chapter 10)
Senior Fellow
PepsiCo International
700 Anderson Hill Road, 7/3-738
Purchase, NY 10577 USA
Phone: 914-253-3012
E-mail: dan.bena@pepsi.com

Gleyn E. Bledsoe, Ph.D., C.P.A. (Chapter 27)
Biological Systems Engineering Department
Washington State University
Pullman, WA 99164-6373 USA
Phone: 509-335-8167
Fax: 509-335-2722
E-mail: gleyn@wsu.edu

Lou Ann Carden, Ph.D. (Chapter 19)
Nutrition and Dietetics
Western Carolina University
126 Moore Hall
Cullowhee, NC 28723 USA
Phone: 828-227-3515
E-mail: lcarden@wcu.edu

Marijana Carić, Ph.D., P.E. (Chapter 17)
Professor
Faculty of Technology
University of Novi Sad
21000 NOVI SAD, Bulevar Cara Lazara 1
Serbia and Montenegro
Phone: 381 21 450-712
Fax: 381 21 450-413
E-mail: caricom@uns.ns.ac.yu

Ramesh C. Chandan, Ph.D. (Chapter 16)
Consultant
1364, 126th Avenue NW
Coon Rapids, MN 55448-4004 USA
Phone: 763-862-4768
Fax: 763-862-5049
E-mail address: chandanrc1@msn.com

Nanna Cross, Ph.D., R.D., L.D. (Chapter 8)
Consultant
1436 West Rosemont Avenue, Floor One
Chicago, IL 60660
Phone: 773-764-7749
E-mail: n.cross@sbcglobal.net

Jeff D. Culbertson, Ph.D. (Chapter 12)
Professor: Food Science and Toxicology
University of Idaho
202A Food Research Center
Moscow, Idaho 83844-1056 USA
Phone: 208-885-2572
Fax: 208-885-2567
E-mail: jeffc@uidaho.edu

James E. Dexter, Ph.D. (Chapter 13)
Canadian Grain Commission
Grain Research Laboratory
1404-303 Main Street
Winnipeg, Manitoba, Canada R3C 3G8
Phone: 204-983-6054
Fax: 204-983-0724
E-mail: jdexter@grainscanada.gc.ca

Robert Driscoll, Ph.D., P.E. (Chapter 2)
Department of Food Science and Technology
University of New South Wales
Sydney, NSW 2052 Australia
Phone: 0612-9385.4355
Fax: 0612-9385.5931
E-mail: R.Driscoll@unsw.edu.au

Susan E. Duncan, Ph.D., R.D. (Chapter 18)
Professor
Department of Food Science and Technology
Virginia Polytechnic Institute and State University
Blacksburg, VA 24061 USA
Phone: 540-231-8675
Fax: 540-231-9293
Email: duncans@vt.edu

Yi-Chung Fu, Ph.D., P.E. (Chapter 4)
Department of Food Science
National Chung Hsing University
P.O. Box 17-55, Taichung, Taiwan 40227, R.O.C.
Phone: 886-4-22853922
Fax: 886-4-22876211
E-mail: frank12@ms18.hinet.net

Ingolf U. Grün, Ph.D. (Chapter 20)
University of Missouri
Department of Food Science
256 William C. Stringer Wing
Columbia, MO 65211-5160 USA
Phone: 573-882-6746
Fax: 573-884-7964
Email: GruenI@missouri.edu

Y. H. Hui, Ph.D. (Chapters 1, 7, 21)
President
Science Technology System
P.O. Box 1374
West Sacramento, CA 95691 USA
Phone: 916-372-2655
Fax: 916-372-2690
Email: yhhui@aol.com

Ty Lawrence, Ph.D. (Chapter 22)
The Smithfield Packing Co.
15855 Hwy 87 West
Tar Heel, NC 28392 USA
Phone: 910-862-7675
Fax: 910-862-5249
E-mail: tylawrence@smithfieldpacking.com or
    shannty007@yahoo.com

Miang-Hoog Lim, Ph.D. (Chapter 1)
Univ. of Otago
Department of Food Science
PO Box 56
Dunedin, 9015 New Zealand
Phone: 64-3-4797953
Fax: 64-3-4797953
E-mail:
miang.lim@stonebow.otago.ac.nz

Maria de Lourdes Pérez-Chabela, Ph.D. (Chapters
    25, 26)
Departamento de Biotecnologia
Universidad Autonoma Metropolitana–Iztapalapa
Apartado Postal 55-535, C.P. 09340
Mexico D.F., Mexico
Phone: 52 5 724-4717/4726
Fax: 52 5 724-47 12
E-mail: lpch@xanum.uam.mx

Richard Mancini, M.S. (Chapter 22)
Department of Animal Sciences
Kansas State University
216 Weber Hall
Manhattan, KS 66502 USA
Phone: 785-532-1269
Fax: 785-532-7059
E-mail: rmancini@oznet.ksu.edu

Lisa J. Mauer, Ph.D. (Chapter 5)
Assistant Professor
Department of Food Science
Purdue University
745 Agriculture Mall Drive
West Lafayette, IN 47907-2009 USA
Phone: 765-494-9111
Fax: 765-494-7953
E-mail: mauer@purdue.edu.

Wai-Kit Nip, Ph.D. (Chapters 1, 3)
Department of Molecular Biosciences and
 Bioengineering
College of Tropical Agriculture and Human
 Resources
University of Hawaii at Manoa
1955 East-West Road
Honolulu, HI 96822 USA
Phone: 808-956-3852
Fax: 808-956-3542
E-mail: wknip@hawaii.edu

Sean Francis O'Keefe, Ph.D. (Chapter 11)
Associate Professor
Food Science and Technology Department
Virginia Polytechnic Institute and State University
Blacksburg VA 24061 USA
Phone: 540-231-4437
Fax: 540-231-9293
E-mail: okeefes@vt.edu

Banu F. Ozen, Ph.D. (Chapter 5)
Postdoctoral Associate
Department of Food Science
Purdue University
745 Agriculture Mall Drive
West Lafayette, IN 47907-2009 USA

Edith Ponce-Alquicira, Ph.D. (Chapter 24)
Departamento de Biotecnología, Universidad
 Autónoma Metropolitana-Iztapalapa
Av. San Rafael Atlixco 186, Col. Vicentina,
 Apartado postal 55-535, C.P. 09340.
México D.F., México
Phone: 5804-4717, 5804-4726
Fax: 5804-4712
Email: pae@xanum.uam.mx

Barbara A. Rasco, Ph.D., J.D. (Chapter 27)
Department of Food Science and Human Nutrition
Washington State University
Pullman, WA 99164-6376 USA
Phone: 509-335-1858
Fax: 509-335-4815
E-mail: Rasco@wsu.edu

Karen A. Schmidt, Ph.D. (Chapter 15)
Professor
Department of Animal Sciences and Industry
Kansas State University
Manhattan, KS 66506-1600 USA
Phone: 785-532-5654
Fax: 785-532-5681
E-mail: kschmidt@oznet.ksu.edu

J. Scott Smith, Ph.D. (Chapter 1)
Professor
Department of Animal Science and Industry
Kansas State University
Call Hall, Rm. 208
Manhattan, KS 66506, USA
Phone: 785-532-1219
Fax: 785-532-5681
E-mail: jsschem@ksu.edu

Peggy Stanfield, M.S., R.D. (Chapters 6, 28)
President
Dietetic Resources
167 Robbins Avenue W.
Twin Falls, ID 83301 USA
Voice/Fax: 208-733-8662
Email: pstandfld@pmt.org

Ruthann B. Swanson, Ph.D. (Chapter 9)
Associate Professor
Department of Foods and Nutrition
University of Georgia
174 Dawson Hall
Athens, GA 30602 USA
Phone: 706-542-4834
Fax: 706-542-5059
Email: rswanson@fcs.uga.edu

Fidel Toldrá, Ph.D. (Chapter 23)
Research Professor
Head of Laboratory of Meat Science
Department of Food Science
Instituto de Agroquimica y Tecnologia de
    Alimentos (CSIC)
P.O. Box 73
46100 Burjassot (Valencia)
Spain
Phone: 34 96 3900022
Fax: 34 96 3636301
E-mail: ftoldra@iata.csic.es

Alfonso Totosaus, Ph.D. (Chapters 25, 26)
Food Science Lab
Tecnológico de Estudios Superiores de Ecatepec
Av Tecnológico y Av. H. González
Ecatepec 55210, Edo. México, México
Phone: +52 55 5710 4560 ext. 307
Fax: +52 55 5710 4560 ext. 305
E-mail: totosaus@att.net.mx

P. H. F. Yu, Ph.D. (Chapter 1)
Department of Applied Biology and Chemical
    Technology
The Hong Kong Polytechnic University
Hung Hom, Kowloon
Hong Kong

Preface

In May 2002, the senior editor completed *Food Chemistry Workbook,* a student workbook to accompany his regular textbook, *Food Chemistry: Principles and Applications,* published in May 2000. In this workbook, he edited 30 chapters contributed by professionals in the United States and Mexico. Each chapter describes the manufacture of one kind of food product, with an emphasis on the principles of food chemistry presented in the textbook. Using some of these chapters as a foundation, but with a different emphasis, this book was born.

There are more than 60 undergraduate programs in food science and food technology in North America, with several programs offering food engineering or chemical engineering with an emphasis on food engineering. Most of them are in the approved list of programs under the leadership of the U.S. Institute of Food Technologists. As such, most of them also offer a course in the fundamentals of food processing. However, depending on a particular college or program, there are many variables in such a course for both teachers and students. The biggest ones are as follows:

- The placement of emphasis on three interrelated areas: food science, food technology, and food engineering.
- The establishment of several courses to cover the complex topics.
- The division of the course into components, each of which is taught in another course.

The structure and goal of our book combines the above approaches by grouping the 29 chapters into two sections. The first seven chapters cover some background information on food processing:

Principles of Food Processing
Food Dehydration

Food Fermentation
Microwave and Food Processing
Food Packaging
Food Regulations
Food Plant Sanitation and Quality Assurance

The remaining chapters discuss the details in the processing of individual food commodities such as

Beverages: Soft Drinks (Carbonated) and Beer.
Cereals: Muffins, Leavened Bread, Pasta, Noodles.
Dairy Products: Cheese, Dried Milk, Ice Cream, and Yogurt.
Fats and Oils: Mayonnaise, Shortening, and Processing Technology.
Fruits and Vegetables: Orange Juice and Tomatoes.
Meat: Hot Dogs, Fermented Meat.
Poultry Products: Poultry Ham, Poultry Nuggets, and Poultry Pâté.
Seafood: Frozen Aquatic Food Products and Seafood Processing Sanitation

There are many excellent books on the principles on food processing. This book is not designed to compete with these books. Rather, this book offers another option, both in the approach and the contents. The instructor can use this book by itself or use it to accompany another textbook in the market.

This book is the result of the combined effort of 30 plus authors from six countries who possess expertise in various aspects of food processing and manufacturing, led by two editors. The editors thank all the contributors for sharing their experiences in their fields of expertise. They are the people who made this book possible. We hope you enjoy and benefit from the fruits of their labor.

We know how hard it is to develop the contents of a book. However, we believe that the production of a professional book of this nature is even more difficult. We thank the production team at Blackwell Publishing, and express our appreciation to Ms. Lynne Bishop, coordinator of the entire project. You are the best judge of the quality of this book.

J. S. Smith
Y. H. Hui

Part I
Principles

1
Principles of Food Processing

Y. H. Hui, M.-H. Lim, W.-K. Nip, J. S. Smith, P. H. F. Yu

INTRODUCTION AND GOALS

This chapter provides an overview of the basic principles of food processing. The goals of modern food processing can be summarized as follows:

- *Formulation.* A logical basic sequence of steps to produce an acceptable and quality food product from raw materials.
- *Easy production procedures.* Develop methods that can facilitate the various steps of production.

The information in this chapter has been derived from documents copyrighted and published by Science Technology System, West Sacramento, California. ©2003. Used with permission.

- *Time economy.* A cohesive plan that combines the science of production and manual labor to reduce the time needed to produce the product.
- *Consistency.* Application of modern science and technology to assure the consistency of each batch of products.
- *Product and worker safety.* The government and the manufacturers work closely to make sure that the product is wholesome for public consumption, and the workers work in a safe environment.
- *Buyer friendliness.* Assuming the buyer likes the product, the manufacturer must do everything humanly possible to ensure that the product is user friendly (size, cooking instructions, keeping quality, convenience, etc.).

Obviously, to achieve all these goals is not a simple matter. This chapter is concerned mainly with the scientific principles of manufacturing safe food products. With this as a premise, the first question we can ask ourselves is: Why do we want to process food? At present, there are many *modern* reasons why foods are processed, for example, adding value to a food, improving visual appeal, and convenience. However, traditionally the single most important reason we wish to process food is to make it last longer without spoiling. Probably the oldest methods of achieving this goal are the salting of meat and fish, the fermenting of milk, and the pickling of vegetables. The next section discusses food spoilage and food-borne diseases.

FOOD SPOILAGE AND FOOD-BORNE DISEASES

FOOD SPOILAGE

Foods are made from natural materials and, like any living matter, will deteriorate in time. The deterioration of food, or food spoilage, is the natural way of recycling, restoring carbon, phosphorus, and nitrogenous matters to the earth. However, putrefaction (spoilage) will usually modify the quality of foods from good to bad, creating, for example, poor appearance (discoloration), offensive smell, and inferior taste. Food spoilage could be caused by a number of factors, chiefly by biological factors, but also by chemical and physical factors. Consumption of spoiled foods can cause sickness and even death. Thus, food safety is the major concern in spoiled foods.

Food Spoilage and Biological Factors

Processed and natural foods are composed mainly of carbohydrates, proteins, and fats. The major constituents in vegetables and fruits are carbohydrates, including sugars (sucrose, glucose, etc.), polymers of sugars (starch), and other complex carbohydrates such as fibers. Fats are the major components of milk and most cheeses, and proteins are the chief constituents of muscle foods. Under natural storage conditions, foods start to deteriorate once the living cells in the foods (plant and animal origins) are dead. Either when the cells are dead or if the tissues are damaged, deterioration begins with the secretion of internal proteases (such as chymotrypsin and trypsin to break up proteins at specific amino acid positions), lipases, and lyases from lyzosomes to disintegrate the cells, to hydrolyze proteins into amino acids and starch into simpler sugars (or monosaccharides), and to de-esterificate fats (triglycerides) into fatty acids. The exposure of foods and damaged cells to the environment attracts microorganisms (e.g., bacteria, molds, and virus) and insects, which in turn further accelerate the decomposition of the food. Foods contaminated with microorganisms lead to food-borne illnesses, which, as reported by the Centers for Disease Control and Prevention (CDC), cause approximately 76 million illnesses and 5000 deaths in the United States yearly (http//www.cdc.gov/foodsafety/). For most food poisoning, spoilage has not reached the stage where the sensory attributes (appearance, smell, taste, texture, etc.) of the food are abnormal.

Illness from food can be mainly classified as (1) food-borne infection caused by pathogenic bacteria (disease-causing microorganisms, such as *Salmonella* bacteria, multiplying in victim's digestive tract, causing diarrhea, vomiting and fever, etc.), and (2) food-borne intoxication (food poisoning resulting from toxin produced by pathogenic microorganisms, e.g., *Clostridium botulinum* and *Staphylococcus aureus*, in the digestive tract). Food-borne illness also has a major economic impact on society, costing billions of dollars each year in the form of medical bills, lost work time, and reduced productivity (McSwane et al. 2003). Some genera of bacteria found in certain food types are listed in Table 1.1, and some common types of microorganisms found in foods are listed in Table 1.2. Some major bacterial and viral diseases transmitted to humans through foods are listed in Table 1.3. The interactive behavior of microorganisms may contribute to their growth and/or spoilage activity (Gram et al. 2002).

Table 1.1. Most Common Bacteria Genera Found in Certain Food Types

Microorganisms	Foods
Corynebacterium, Leuconostoc	Dairy products
Achromobacter	Meat, poultry, seafoods
Bacteriodes, Proteus	Eggs and meats
Pseudomonas	Meats, poultry, eggs

Table 1.2. Most Common Pathogenic Bacteria and Viruses Found in Foods

Bacteria

Clostridium botulinum	*Listeria monocytogenes*
Salmonella spp.	*Staphlococcus aureus*
Clostridium perfringens	*Escherichia coli*
Botulinum spp.	*Campylobacter jejuni*
Streptococci spp.	*Bacillus cereus*
Lactobacillus spp.	*Proteus* spp.
Shigellas spp.	*Pseudomonas* spp.
Salnonella spp.	*Vibrio* spp.

Viruses

Hepatitis A virus	Echovirus
Rotavirus	Calcivirus

Table 1.3. Some Major Bacterial and Viral Diseases Transmitted to Humans through Food

Bacteria/Viruses	Disease
Bacteria	
Campylobacter jejuni	Campylobacteriosis
Listeria monocytogenes	Listeriosis
Salmonella spp.	Salmonellosis
Salmonella typhi	Typhoid fever
Shigella dysenteriae	Dysentery
Vibrio cholerae	Cholera
Yersinia enterocolitica	Diarrheal disease
Enterobacteriaceae	Enteric disease
Viruses	
ECHO virus	Gastroenteritis
HAV virus	Hepatitis type A
Norwalk agent	Viral diarrhea
Rotavirus	Infant diarrhea

Food Spoilage and Chemical Factors

In many cases, when foods are oxidized, they become less desirable or even rejected. The odor, taste, and color may change, and some nutrients may be destroyed. Examples are the darkening of the cut surface of a potato and the browning of tea color with time. Oxidative rancidity results from the liberation of odorous products during breakdown of unsaturated fatty acids. These products include aldehydes, ketones, and shorter-chain fatty acids.

Browning reactions in foods include three non-enzymatic reactions—Maillard, caramelization, and ascorbic acid oxidation—and one enzymatic reaction—phenolase browning (Fennema 1985). Heating conditions in the surface layers of food cause the Maillard browning reaction between sugars and amino acids, for example, the darkening of dried milk from long storage. The high temperatures and low moisture content in the surface layers also cause caramelization of sugars, and oxidation of fatty acids to other chemicals such as aldehydes, lactones, ketones, alcohols, and esters (Fellows 1992). The formation of ripening fruit flavor often results from Strecker degradation (the transamination and decarboxylation) of amino acids, such as the production of 3-methylbutyrate (apple-like flavor) from leucine (Drawert 1975). Further heating of the foods can break down some of the volatiles generated by Maillard reaction and Strecker degradation to produce burnt or smoky aromas. Enzymic browning occurs on cut surfaces of light-colored fruits (apples, bananas) and vegetables (potatoes) due to the enzymatic oxidation of phenols to orthoquinones, which in turn rapidly polymerize to form brown pigments known as melanins. Moisture and heat can also produce hydrolytic rancidity in fats; in this case, fats are split into free fatty acids, which may cause off odors and rancid flavors in fats and oils (Potter and Hotchkiss 1995).

Food Spoilage and Physical Factors

Food spoilage can also be caused by physical factors, such as temperature, moisture, and pressure acting upon the foods. Moisture and heat can also produce hydrolytic rancidity in fats; in this case, fats are split into free fatty acids, which may cause off odors and rancid flavors in fats and oils (Potter and Hotchkiss 1995). Excessive heat denatures proteins, breaks emulsions, removes moisture from food, and destroys nutrients such as vitamins. However, excessive coldness, such as freezing, also discolors fruits and vegetables, changes their texture and/or cracks their outer coatings to permit contamination by microorganisms. Foods under pressure will be squeezed and transformed into unnatural conformation. The compression will likely break up the surface structure, release degradative enzymes, and expose the damaged food to exterior microbial contamination.

Of course, many health officials consider physical factors to include such things as sand, glass, wood

chips, rat hair, animal urine, bird droppings, insect parts, and so on. These things may not spoil the food, but they do present hazards. Some of these foreign substances *do* lead to spoilage. Furthermore, insects and rodents can consume and damage stored foods, and insects can lay eggs and leave larvae in the foods, causing further damage later. Such foods are no longer reliable since they contain hidden contaminants. The attack of foods by insects and rodents can also contaminate foods further with microbial infections.

PREVENTION AND RETARDATION OF FOOD SPOILAGE

Food spoilage can be prevented by proper sanitary practices in food handling and processing, appropriate preservation techniques, and standardized storing conditions.

Food Handling and Processing

The entire process, from raw ingredients to a finished product ready for storage, must comply with a standard sanitation program. In the United States, the practice of HACCP (hazard analysis critical control points), though mandatory for several industries, may eventually become so for all food industries. At present, the application of HACCP is voluntary for most food processors. Similar sanitary programs apply to workers. It is important to realize that a food processing plant must have a basic sanitation system program before it can implement a HACCP program.

Food Preservation

There are many techniques used to preserve food such as legal food additives, varying levels of food ingredients or components, and new technology. Legal food additives, among other functions, can prevent oxidation and inhibit or destroy harmful microorganisms (molds and bacteria). Vitamin E or vitamin C can serve as an antioxidant in many food products, and benzoate in beverages can act as an anti-microbial agent. We can preserve food by manipulating the levels of food ingredients or components to inhibit the growth of microorganisms or destroy them. For example, keep the food low in moisture content (low water activity), high in sugar or salt content, or at a low pH (less than pH 5). Recently, new or alternative technologies are available to preserve food. Because they are new, their application is carefully monitored. Perhaps nothing in the last two decades has generated more publicity than the use of X rays in food processing. Although food irradiation has been permitted in the processing of several categories of food, its general application is still carefully regulated in the United States.

Food Packaging and Storage

Raw and processed foods should be packaged to prevent oxidation, microbial contamination, and loss of moisture. Storage of foods (when not contaminated) below $-20°C$ can keep food for several months or a year. Storing foods at $4°C$ can extend the shelf life to several days or a week (note that some bacteria such as *Listeria monocytogenes* can still grow and multiply even in foods at refrigerated temperatures).

Newly developed techniques to preserve foods include the incorporation of bacteriocin (so that it retains its activity) into plastic to inhibit the surface growth of bacteria on meat (Siragusa et al. 1999), and the application of an intelligent Shelf Life Decision System (SLDS) for quality optimization of the food chill chain (Giannakourou et al. 2001).

SOURCES OF INFORMATION

At present, all major western government authorities have established web sites to educate consumers and scientists on the safe processing of food products. Internationally, two major organizations have always been authoritative sources of information. They include The World Health Organization (WHO) and Food and Agriculture Organization (FAO). They also have comprehensive web sites.

In the United States, major federal authorities on food safety include, but are not limited to (1) the U.S. Department of Agriculture (USDA), (2) the Food and Drug Administration (FDA), (3) the Centers for Disease Control (CDC), (4) the Environmental Protection Agency (EPA), and (5) the National Institutes of Health (NIH).

Many trade associations in western countries have web sites that are devoted entirely to food safety. Some examples in the United States include (1) the American Society of Microbiologists, (2) the Institute of Food Technologists, (3) the International Association for Food Protection, (4) the National Food Processors Association, and (5) the National Restaurant Association.

All government or trade association web sites are easily accessible by entering the agency name into popular search engines.

PRODUCT FORMULATIONS AND FLOWCHARTS

As we have mentioned earlier, for many food products, processing is an important way to preserve the product. However, for some food products, many self-preserving factors, such as the ingredients and their natural properties, play a role. Three good examples are pickles, barbecue sauces, and hard candies. Preserving pickles is not difficult if the end product is very sour (acidic) or salty. Traditionally, barbecue sauces have a long shelf life because of the high content of sugar. Most unwrapped hard candies keep a long time, assuming the environment is at room temperature and not very humid. Most wrapped hard candies last even longer if the integrity of the wrappers is maintained. For baked products (cookies, bread), measures against spoilage take second place to consumer acceptance of freshness. So, the objectives of processing foods vary with the products. However, one aspect is essential to all manufacturers, as discussed below.

For a processed food product, it is assumed that the processor has a formula to manufacture the product. In countries all over the world, small family-owned food businesses usually start with home recipes for popular products instead of a scientific formula. Most of us are aware of the similar humble beginnings of major corporations manufacturing cola (carbonated), soft drinks, cheeses, breakfast cereals, and many others. When these family businesses started, there was not much science or technology involved. When a company becomes big and has many employees, it starts hiring food scientists, food technologists, and food engineers to study the "recipe" and refine every aspect of it until the entire manufacturing process is based on sound scientific, technical, and engineering principles. After that, all efforts are directed towards production. Even now, somewhere, a person will start making "barbecue sauce" in his garage and selling it to his neighbors. Although very few of these starters will succeed, this trend will continue, in view of the free enterprise spirit of the West.

Although any person can start manufacturing food using a home recipe, the federal government in the United States has partial or total control over certain aspects of the manufacturing processes for food and beverage products. This control automatically affects the recipes, formulas, or specifications of the products. Although the word "control" here refers mainly to safety, it is understood that it will affect the formulations to some extent, especially critical factors such as temperatures, pH, water activity, and so on.

Chapters in the second part of this book will provide formulations for manufacturing various food categories (bakery, dairy, fruits, etc). It also provides many operational flowcharts. Flowcharts differ from formulas in that they provide an overview of the manufacturing process. For illustration, Figures 1.1–1.8 provide examples of flowcharts for the manufacture of bakery (bread), dairy (yogurt), grain (flour), fruits (raisins), vegetables (pickles), and meat (frankfurters, frozen chicken parts), and seafood (canned tuna).

UNITS OF OPERATIONS

The processing of most food products involves raw materials; cleaning; separating; disintegrating; forming, raw; pumping; mixing; application methods (formulations, additives, heat, cold, evaporation, drying, fermenting, etc.); combined operations; and forming, finished product. We discuss some of these as units of operations. Certain items—heating, cooling, sanitation, quality control, packaging, and similar procedures—are discussed as separate topics rather than as units of operations.

According to the U.S. Department of Labor, there are hundreds of different categories of food products currently being manufactured. Correspondingly, there are hundreds of companies manufacturing each category of food products. In sum, there are literally thousands of food manufacturers. Two major reasons for this explosion of new companies are (1) the constant introduction of new products and (2) improvements in manufacturing methods and equipment.

To facilitate the technological processing of food at the educational and commercial levels, food-processing professionals have developed unifying principles and a systematic approach to the study of these operations. The involved processes of the food industry can be divided into a number of common operations, called unit operations. Depending on the processor, such unit operations vary in name and number. For ease of discussion, we use the following units of operations, in alphabetical order, for the most common ones: cleaning, coating, controlling,

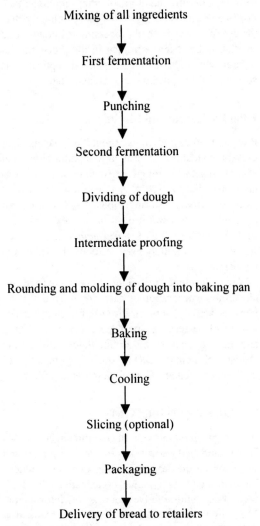

Figure 1.1. A general flowchart for the manufacture of bread.

decorating, disintegrating, drying, evaporating, forming, heating, mixing, packaging, pumping, raw materials handling, and separating.

During food processing, the manufacturer selects and combines unit operations into unit processes, which are then combined to produce more complex and comprehensive processes. We will now discuss these units in the order they appear in a food processing plant. Although emerging technology plays an important role in food processing as time progresses, this book is designed to provide students with the most basic approaches.

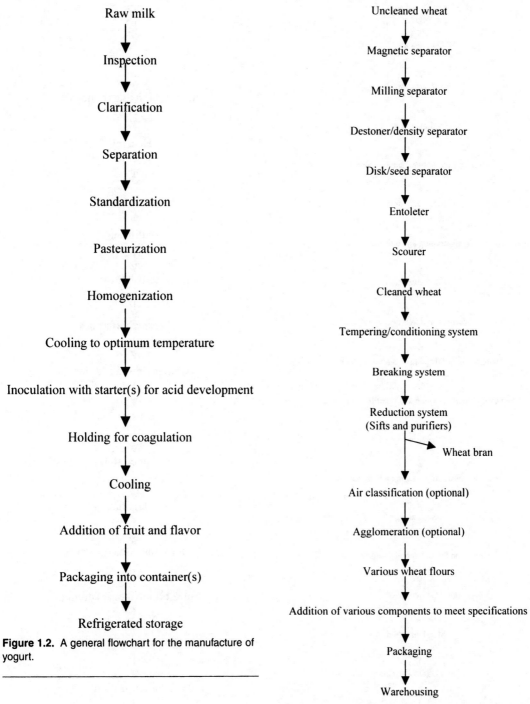

Raw milk
↓
Inspection
↓
Clarification
↓
Separation
↓
Standardization
↓
Pasteurization
↓
Homogenization
↓
Cooling to optimum temperature
↓
Inoculation with starter(s) for acid development
↓
Holding for coagulation
↓
Cooling
↓
Addition of fruit and flavor
↓
Packaging into container(s)
↓
Refrigerated storage

Figure 1.2. A general flowchart for the manufacture of yogurt.

Uncleaned wheat
↓
Magnetic separator
↓
Milling separator
↓
Destoner/density separator
↓
Disk/seed separator
↓
Entoleter
↓
Scourer
↓
Cleaned wheat
↓
Tempering/conditioning system
↓
Breaking system
↓
Reduction system
(Sifts and purifiers)
→ Wheat bran
↓
Air classification (optional)
↓
Agglomeration (optional)
↓
Various wheat flours
↓
Addition of various components to meet specifications
↓
Packaging
↓
Warehousing

Figure 1.3. A general flowchart for the production of flour from wheat.

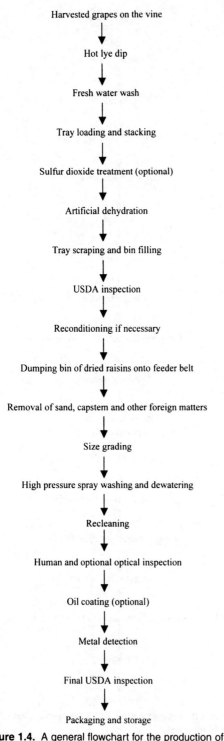

Harvested grapes on the vine

↓

Hot lye dip

↓

Fresh water wash

↓

Tray loading and stacking

↓

Sulfur dioxide treatment (optional)

↓

Artificial dehydration

↓

Tray scraping and bin filling

↓

USDA inspection

↓

Reconditioning if necessary

↓

Dumping bin of dried raisins onto feeder belt

↓

Removal of sand, capstem and other foreign matters

↓

Size grading

↓

High pressure spray washing and dewatering

↓

Recleaning

↓

Human and optional optical inspection

↓

Oil coating (optional)

↓

Metal detection

↓

Final USDA inspection

↓

Packaging and storage

Figure 1.4. A general flowchart for the production of raisins.

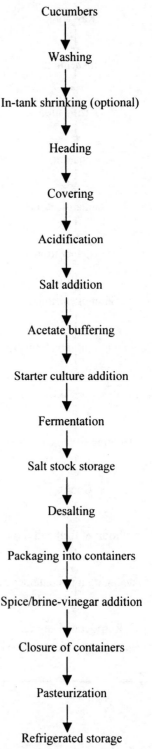

Cucumbers

↓

Washing

↓

In-tank shrinking (optional)

↓

Heading

↓

Covering

↓

Acidification

↓

Salt addition

↓

Acetate buffering

↓

Starter culture addition

↓

Fermentation

↓

Salt stock storage

↓

Desalting

↓

Packaging into containers

↓

Spice/brine-vinegar addition

↓

Closure of containers

↓

Pasteurization

↓

Refrigerated storage

Figure 1.5. A general flowchart for the production of pickles.

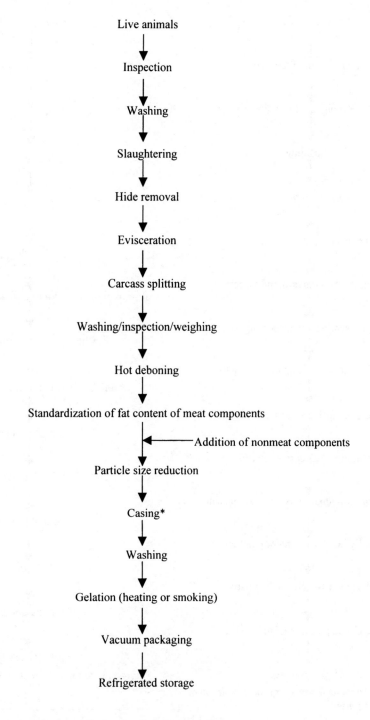

Live animals

↓

Inspection

↓

Washing

↓

Slaughtering

↓

Hide removal

↓

Evisceration

↓

Carcass splitting

↓

Washing/inspection/weighing

↓

Hot deboning

↓

Standardization of fat content of meat components

↓ ←——— Addition of nonmeat components

Particle size reduction

↓

Casing*

↓

Washing

↓

Gelation (heating or smoking)

↓

Vacuum packaging

↓

Refrigerated storage

* Use of inedible casing requires removal of casing before packaging.

Figure 1.6. A general flowchart for the production of Frankfurters.

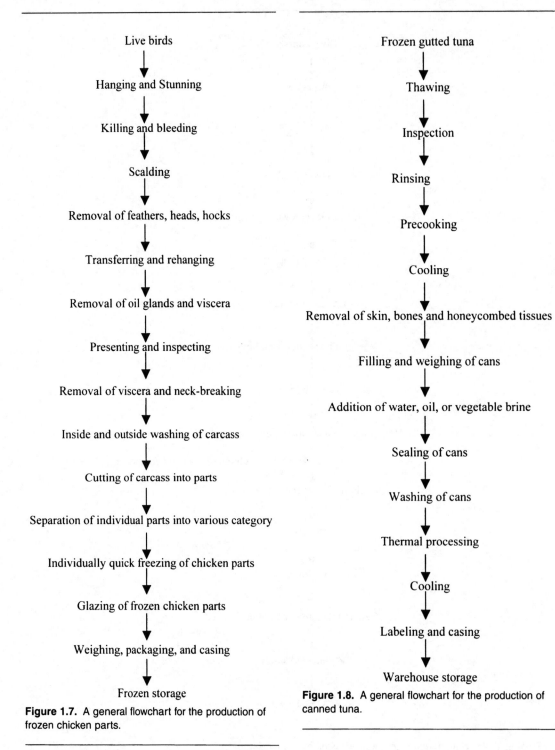

Live birds
↓
Hanging and Stunning
↓
Killing and bleeding
↓
Scalding
↓
Removal of feathers, heads, hocks
↓
Transferring and rehanging
↓
Removal of oil glands and viscera
↓
Presenting and inspecting
↓
Removal of viscera and neck-breaking
↓
Inside and outside washing of carcass
↓
Cutting of carcass into parts
↓
Separation of individual parts into various category
↓
Individually quick freezing of chicken parts
↓
Glazing of frozen chicken parts
↓
Weighing, packaging, and casing
↓
Frozen storage

Figure 1.7. A general flowchart for the production of frozen chicken parts.

Frozen gutted tuna
↓
Thawing
↓
Inspection
↓
Rinsing
↓
Precooking
↓
Cooling
↓
Removal of skin, bones and honeycombed tissues
↓
Filling and weighing of cans
↓
Addition of water, oil, or vegetable brine
↓
Sealing of cans
↓
Washing of cans
↓
Thermal processing
↓
Cooling
↓
Labeling and casing
↓
Warehouse storage

Figure 1.8. A general flowchart for the production of canned tuna.

RAW MATERIALS HANDLING

Raw materials are handled in various ways, including (1) hand and mechanical harvesting on the farm, (2) trucking (with or without refrigeration) of fruits and vegetables, (3) moving live cattle by rail, (4) conveying flour from transporting vehicle to storage bins.

For example:

- Oranges are picked on the farm by hand or mechanical devices, moved by truck trailers, usually refrigerated, to juice processing plants, where they are processed. Of course, the transport must take into account the size of the trucks, the length of time during transport, and temperature control. The major objective is to avoid spoilage. In recent years, the use of modified atmosphere packaging has increased the odds to favor the farmers and producers.
- Handling sugar and flour poses great challenges. When dry sugar reaches processing plants, via truck trailers or rail, it is transported to storage bins via a pneumatic lift system. The sugar will cake if the storage time, temperature, and humidity are not appropriate. Improper transfer of sugar may result in dusting and buildup of static electricity, which can cause an explosion, since sugar particles are highly combustible. The same applies to finely ground flour.

In handling raw materials, one wishes to achieve the following major objectives: (1) proper sanitation, (2) minimal loss of product, (3) acceptable product quality, (4) minimal bacterial growth, and (5) minimal holding time.

CLEANING

We all know what cleaning a raw product means. Before we eat a peach, we rinse it under the faucet. Before we make a salad, we wash the vegetables. Before we eat crabs, we clean them. Of course, the difference in cleaning between home kitchens and a food processing plant is volume. We clean one peach; they clean a thousand peaches.

Depending upon the product and the nature of the dirt, cleaning can be accomplished using the following methods or devices, individually or in combination: (1) air, high velocity; (2) brushes; (3) magnets; (4) steam; (5) ultraviolet light; (6) ultrasound; (7) vacuum; and (8) water. There are also other new technologies that will not be discussed here.

Water is probably the most common cleaning agent, and its application varies:

- Clams, oysters, crabs, and other shellfish commonly are hosed to remove mud, soil, and other foreign debris. If they are contaminated, they may have to be incubated in recirculating clean water.
- City water is not acceptable for manufacturing beverages. It must be further treated with chemical flocculation, sand filtration, carbon purification, microfiltration, deaeration, and so on. This is not considered a simple cleaning. Rather, it is a process in cleaning.
- Eviscerating poultry can be considered a cleaning operation if water is used, but the actual process of removing the entrails may involve vacuuming in addition to water.
- With a product like pineapples, the irregular surfaces are usually cleaned by the scrubbing action of high-pressure water jets.

Just as in a home kitchen where pots and pans require frequent cleaning, the equipment used in a food processing plant is required by state and federal regulations to be cleaned after each use. After dirt and mud is removed, some raw products require special sanitizing procedures. The use of sanitizers can be a complicated matter. It involves types of sanitizers, federal regulations, expertise, and so on.

SEPARATING

In food processing, separating may involve separating (1) a solid from a solid, as in peeling potatoes; (2) a solid from a liquid, as in filtration; (3) a liquid from a solid, as in pressing juice from a fruit; (4) a liquid from a liquid, as in centrifuging oil from water; and (5) a gas from a solid or a liquid, as in vacuum canning.

One time-honored technique in the separating operation is the hand sorting and grading of individual units (e.g., mushrooms, tomatoes, oranges). At present, many mechanical and electronic sorting devices have replaced human hands for various types of raw food products. An electronic eye can tell the difference in color as the products are going by on the conveyor belt. Built-in mechanisms can sort the products by color, "good" vs. "bad" color. The current invention of electronic noses shows promise.

Automatic separation according to size is easily accomplished by passing fruits or vegetables over different size screens, holes, or slits.

DISINTEGRATING

Disintegrating means subdividing large masses of foods into smaller units or particles. This may include cutting, grinding, pulping, homogenizing, and other methods. Examples include:

- Automatic dicing of vegetables,
- Mechanical deboning of meat,
- Manual and automatic cutting of meats into wholesale and retail sizes,
- Cutting bakery products with electric knives and water jets (high velocity and high pressure),
- Disintegrating various categories of food products with high-energy beams and laser beams, and
- Homogenization with commercial blenders, high pressure traveling through a valve with very small openings, ultrasonic energy, and so on.

Homogenization is probably one of the most important, if not *the* most important, stages in dairy processing. Homogenization produces disintegration of large globules and clusters of fat in milk or cream to minute globules. This is done by forcing the milk or cream under high pressure through a valve with very small openings.

FORMING

Forming is an important operation in many categories of the food industry: (1) meat and poultry patties, (2) confections (candies, jelly beans, fruit juice tablets), (3) breakfast cereals, (4) pasta, and (5) varieties (some cheese cubes, processed cheese slices, potato chips, etc.).

Meat and Poultry Patties

Patty-making machines are responsible for making ground meat and poultry patties by gently compacting the product into a disk shape. Uniform pressure is applied to produce patties with minimal variation in weight. Also, excessive pressure may result in tough cooked patties.

Pasta

Spaghetti is formed by forcing dough through extrusion dies of various forms and shapes before it is dried in an oven.

Confectionery

The shapes and forms in the confectionery industry (e.g., candies, jellies) are made in several ways. Two of the most popular methods are molds and special tableting machines. The traditional use of molds is responsible for confectionery such as fondants, chocolate, and jellies. The product is deposited into molds to cool and harden.

PUMPING

In food processing, pumping moves food (liquid, semisolid, paste, or solid) from one step to the next or from one location to another.

There are many types of pumps available, some with general, others specialized, applicability. The type of pump used depends on the food (texture, size, etc.). For example, broth, tomato pastes, ground meat, corn kernels, grapes, and other categories of food all require a "different" pump to do the job. Two important properties of pumps are (1) ability to break up foods and (2) ease of cleaning.

MIXING

The operation of mixing, for example, includes (1) kneading, (2) agitation, (3) blending, (4) emulsifying, (5) homogenizing, (6) diffusing, (7) dispersing, (8) stirring, (9) beating, (10) whipping, and (11) movements by hands and machines.

Examples of mixing include (1) homogenization to prevent fat separation in milk; (2) mixing and developing bread dough, which requires stretching and folding, referred to as kneading; (3) beating in air, as in making an egg-white foam; (4) blending dry ingredients, as in preparing a ton of dry cake mix, and (5) emulsifying, as in the case of mayonnaise.

Commercial mixers for food processing come in many shapes and forms, since many types of mixtures or mixings are possible. Two examples are provided as illustration.

1. *Mixing solids with solids (e.g., a dry cake mix).* The mixer must cut the shortening into the flour, sugar, and other dry ingredients in order to produce a fluffy, homogeneous dry mix. A ribbon blender is used.
2. *Beating air into a product while mixing, as when using a mixer-beater in an ice cream freezer.* The mixer turns in the bowl in which the ice cream mix is being frozen. This

particular operation permits the mixer to achieve several tasks or objectives: beat air into the ice cream to give the desired volume and overrun; keep the freezing mass moving to produce uniformity and facilitate freezing.

PROCESSING AND PRESERVATION TECHNIQUES

HEAT APPLICATION

Heat exchanging, or heating, is one of the most common procedures used in manufacture of processed foods. Examples include the pasteurizing of milk, bakery products, roasting peanuts, and canning. Foods may be heated or cooked using (1) direct injection of steam, (2) direct contact with flame, (3) toasters, (4) electronic energy as in microwave cookers, and (5) many forms of new technology.

Whatever the method, precise control of temperature is essential. Heating is used in (1) baking, (2) frying, (3) food concentration, (4) food dehydration, and (5) package closure.

Why are foods heated? All of us know why we cook food at home: to improve texture; to develop flavors; to facilitate mixing of water, oil, and starch; to permit caramelization; and so on. Commercially, the basic reasons for heating are simple and may include:

- Destruction of microorganisms and preservation of food. Food canning and milk pasteurization are common examples.
- Removal of moisture and development of flavors. Ready-to-eat breakfast cereals and coffee roasting are common examples.
- Inactivation of natural toxicants. Processing soybean meal is a good example.
- Improvement of the sensory attributes of the food such as color, texture, mouth-feel.
- Combination of ingredients to develop unique food attributes and attract consumer preferences.

Traditional thermal processing of foods uses the principles of transferring heat energy by conduction, convection, radiation, or a combination of these. At present, there are newer methods of heating food, such as electronic energy (microwave). Later in this chapter, other new technologies for heating foods will be discussed.

Foods are heated using various traditional equipments that were developed using basic principles of food engineering: heat exchangers, tank or kettle, re- torts, toasters. Other methods may include direct injection of steam, direct contact with flame, and of course, microwave.

Heat Exchangers for Liquid Foods

Since foods are sensitive to heat, special consideration is needed. Dark color, burned flavors, and loss of nutrients can result from heating, especially prolonged heat. Heat exchangers have special advantages. They permit (1) maximal contact of liquid food with the heat source and (2) rapid heating and cooling.

For example, a plate-type heat exchanger is used to pasteurize milk. This equipment is made up of many thin plates. When milk flows through one side of the plates, it is heated by hot water on the other side. This provides maximal contact between the heat source and the milk, resulting in rapid heating. The cooling is the reverse: after the milk has been heated, instead of hot water, cool water or brine is used.

Tanks or Kettles for Liquid Foods

During heating, hot water circulating in the jackets of the tanks or kettles heats the food; during cooling, circulating cool water or brine cools the food. This technique works for full liquid foods or partial liquid foods such as soups.

Pressure Cookers or Retorts for Packaged Foods

The most common method of sterilizing canned foods uses pressure cookers or retorts. Beginning with early seventies, the risk of botulism in canned food with low acidity prompted the U.S. Food and Drug Administration (FDA) to implement stringent regulations governing this group of foods. Although the name Hazards Analysis Critical Control Points (HACCP) did not have wide usage at the time, the regulations governing the production of low-acid canned foods can be considered the earliest form of the HACCP program. Large pressure cookers or retorts are used to ensure that the canned goods are heated above the boiling point of water. The high temperature is generated by steam under pressure in a large retort designed to withstand such temperature. In this case, convection and conduction of heat energy are achieved. Steam hits the outside of the cans, and energy is conducted into the can. Some form of moving or agitating device permits convection to occur inside the cans. Although there are other modern techniques for heating canned food products, many smaller companies still depend heavily on the traditional methods.

Roasters or Heated Vessels in Constant Rotation

Instead of one or two pieces of equipment, this system contains several units: loading containers, conveyor belts, hoppers, vats, or vessels. The vessels are usually cylindrical in shape with built-in heating devices. Heat is generated via one of the following methods:

- *Circulation of heated air.* This heats the food products inside the vessels.
- *Application of direct heat contacting outside of vessel such as steam, flame (gas), or air (hot).* Heat is radiated from the inside walls of the vessels to the food.

This unit system is best for roasting coffee beans or nuts.

Tunnel Ovens

Tunnel ovens can be used for a variety of food products. The product is placed on a conveyor belt that moves under a heat source. Sometimes, the product is vibrated so that heat distribution is even. Temperature control is essential, and products such as coffee beans or nuts can be roasted using this method.

HEAT REMOVAL OR COLD PRESERVATION

Cold preservation is achieved by the removal of heat. It is among the oldest methods of preservation. Since 1875, with the development of mechanical ammonia refrigeration systems, commercial refrigeration and freezing processes have become available. A reduction in the temperature of a food reduces the rate of quality changes during storage caused by the various factors. At low temperatures, microbial growth is retarded and microbial reproduction prevented. The rate of chemical reactions (e.g., oxidation, Maillard browning, formation of off flavors), biochemical reactions (e.g., glycolysis, proteolysis, enzymatic browning, and lipolysis), and physical changes resulting from interaction of food components with the environment (e.g., mois-

ture loss in drying out of vegetables) can also be reduced.

Most food spoilage organisms grow rapidly at temperatures above 10°C, although some grow at temperatures below 0°C, as long as there is unfrozen water available. Most pathogens, except some psychrophilic bacteria such as *Listeria monocytogenes* that commonly grows in dairy products, do not grow well at refrigeration temperatures. Below −9.5°C, there is no significant growth of spoilage or pathogenic microorganisms.

In general, the longer the storage period, the lower the temperature required. Pretreatment with intensive heat is not used in this process operation, but with adequate control over enzymatic and microbiological changes, the food maintains nutritional and sensory characteristics close to fresh status, resulting in a high quality product. In comparing chilled and frozen foods, chilled food has a higher quality but a shorter shelf life; frozen food has a much longer shelf life, but the presence of ice in the frozen product may create some undesirable changes in food quality.

Chilling and Refrigeration Process

Chilling process is the gentlest method of preservation with the least changes in taste, texture, nutritive value, and other attributes of foods. Generally it refers to storage temperature above freezing, about 16°C to −2°C. Most foods do not freeze until −2°C or slightly lower because of the presence of solutes such as sugars and salts. Commercial and household refrigerators usually operate at 4.5°C to 7°C.

In low-acid chilled foods, strict hygienic processing and packaging are required to ensure food safety. The chilling process is usually used in combination with other preservation methods such as fermentation, irradiation, pasteurization, mild heat treatment, chemicals (acids or antioxidants), and controlled atmosphere. The combination of these methods avoids extreme conditions that must be used to limit microbial growth, thus providing high quality product (e.g., marinated mussels and yogurt.)

Not all foods can be stored under chilled conditions. Tropical and subtropical fruits suffer chilling injury when stored below 13°C, resulting in abnormal physiological changes: skin blemishes (e.g., banana), browning in the flesh (e.g., mango), or failure to ripen (e.g., tomato). Some other foods should not be refrigerated; for example, breads stale faster at refrigeration temperature than at room temperature.

Starch in puddings also tends to retrograde at refrigeration temperatures, resulting in syneresis.

Important considerations in producing and maintaining high quality chilled foods include:

- *Quick removal of heat at the chilling stage.* Ideally, refrigeration of perishable foods starts at time of harvest or slaughter or at the finishing production line. Cooling can be accelerated by the following techniques:
  - *Evaporative cooling.* Spray water and then subject food to vacuum (e.g., leafy vegetables).
  - *Nitrogen gas (from evaporating liquid nitrogen on produce).* Of course, dry ice and liquid carbon dioxide are used to remove heat for different products.
  - *Heat exchangers.* (1) Thin stainless steel plates with enclosures, circulating on the outside by a chilled or "super-chilled" cooling fluid. (2) Coils with enclosures cooled by different means. Warm bulk liquid foods pass through the inside, and heat is transferred to the outside.
- *Maintaining low temperature during the chill storage.* This can be affected by:
  - *Refrigeration design (i.e., cooling capacity and insulation)* must be taken into account because the temperature can be affected by heat generated by lights and electric motors, people working in the area, the number of doors and how they are opened, and the kinds and amounts of food products stored.
  - *Refrigeration load.* The quantity of heat which must be removed from the product and the storage area in order to decrease from an initial temperature to the selected final temperature and to maintain this temperature for a specific time.
  - *Types of food.* (1) Specific heat of food: the quantity of heat that must be removed from a food to lower it from one temperature to another. The rate of heat removal is largely dependent on water content. (2) Respiration rate of food: Some foods (fruits and vegetables) respire and produce their own heat at varying rates. Products with relatively high respiration rates (snap beans, sweet corn, green peas, spinach, and strawberries) are particularly difficult to store.
- *Maintaining appropriate air circulation and humidity.* Proper air circulation helps to move heat away from the food surface toward refrigerator cooling coils and plates. Air velocity is especially important in commercial coolers or freezers for keeping the appropriate relative humidity because if the relative humidity is too high, condensation of moisture on the surface of cold food may occur, thus causing spoilage through microbial growth or clumping of the product. However, if relative humidity is too low, dehydration of food may occur instead. Therefore, it is important to control the RH (relative humidity) of the cooler and use proper packaging for the food.
- *Modification of gas atmosphere.* Chilled storage of fresh commodities is more effective if it is combined with control of the air composition of the storage atmosphere. A reduction in oxygen concentration and/or an increase in carbon dioxide concentration of the storage atmosphere reduces the rate of respiration (and thus maturation) of fresh fruits and vegetables and also inhibits the rate of oxidation, microbial growth, and insect growth. The atmospheric composition can be changed using three methods:
  - *Controlled atmosphere storage (CAS).* The concentrations of oxygen, carbon dioxide, and ethylene are monitored and regulated throughout storage. CAS is used to inhibit overripening of apples and other fruits in cold storage. Stored fruit and vegetables consume O_2 and give off CO_2 during respiration.
  - *Modified atmosphere storage (MAS).* The initially modified gas composition in sealed storage is allowed to change by normal respiration of the food, but little control is exercised. The O_2 is reduced but not eliminated, and CO_2 is increased (optimum differs for different fruits).
  - *Modified atmosphere packaging (MAP).* The fruit or vegetable is sealed in a package under flushed gas (N_2 or CO_2), and the air in the package is modified over time by the respiring product. Fresh meat (especially red meats) is packaged similarly.
- *Efficient distribution systems.* To supply high quality chilled foods to consumers, a reliable and efficient distribution system is also required. It involves chilled stores, refrigerated transportation, and chilled retail display cabinets. It requires careful control of the storage conditions as discussed above.

Freezing and Frozen Storage

Freezing is a unit operation in which the temperature of a food is reduced below the freezing point and a proportion of the water undergoes a phase change to form ice. Proper freezing preserves foods without causing major changes in their shape, texture, color and flavor. Good frozen storage requires temperatures of $-18°C$ or below, however, it is cost prohibitive to store lower than $-30°C$. Frozen foods have increased in their share of sales since the freezers and microwaves become more available.

The major commodities commonly frozen are (1) fruits (berries, citrus, and tropical fruit) either whole, pureed, or as juice concentrate; (2) vegetables (peas, green beans, sweet corn, spinach, broccoli, Brussels sprouts, and potatoes such as French fries and hash browns); (3) fish fillets and seafood, including fish fingers, fish cakes, and prepared dishes with sauces; (4) meats (beef, lamb, and poultry) as carcasses, boxed joints, or cubes, and meat products (sausages and beef burgers); (5) baked goods (bread, cakes, pastry dough, and pies); and (6) prepared foods (pizzas, desserts, ice cream, dinner meals).

Principles of Freezing. The freezing process implies two linked processes: (1) lowering of temperature by the removal of heat and (2) a change of phase from liquid to solid. The change of water into ice results in increase in concentration of unfrozen matrix and therefore leads to dehydration and lowering of water activity. Both the lowering of temperature and the lowering of water activity contribute to freezing as an important preservation method.

In order for a product to freeze, the product must be cooled below its freezing point. The freezing point of a food depends on its water content and the type of solutes present. The water component of a food freezes first and leaves the dissolved solids in a more concentrated solution, which requires a lower temperature to freeze. As a result, the freezing point decreases during freezing as concentration increases. Different solutes depress the freezing point to a different degree.

Rate of Freezing. Faster freezing produces small crystals, necessary for high quality products such as ice cream. There are two main opposing forces affecting the freezing rate: (1) The driving forces helping to freeze the product quickly include the difference in temperature between the freezing medium and the product (the bigger the difference, the faster the product will cool down), the high thermal conductivity of the freezing medium (the efficiency with which the refrigerating agent extracts heat), and di-

rect surface contact between the medium and the product. (2) On the other hand, the forces that resist freezing include product packed in large sizes, irregular product geometry that reduces direct contact of the product with the freezing agent, product composition that has a high heat capacity, and the thermal conductivity of food packages such as cardboard and plastics that may retard (by acting as an insulator) heat transfer and thus slow down freezing rate.

Quality Changes with Freezing and Frozen Storage. As a consequence of the formation of ice, some negative changes in the quality of food result. The two major causes are the freeze concentration effect and large ice crystal and recrystallization damage.

Freeze Concentration Effect. The quality of products will change if solutes in the frozen product precipitate out of solution (e.g., loss of consistency in reconstituted frozen orange juice because of aggregated pectic substances, and syneresis of starch pudding because of starch aggregation). The increase in ionic strength can lead to "salting out" of proteins, causing protein denaturation (reason for toughening of frozen fish). Increase in solute concentration may lead to the precipitation of some salts; the anion/cation ratio of colloidal suspensions is then disturbed, causing changes in pH. Such changes also cause precipitation of proteins and changes in color of anthocyanin in berries. The concentration of solutes in the extracellular fluid causes dehydration of adjacent tissues in fruit and vegetables, which are not able to rehydrate after thawing. Lastly, concentration of reactive compounds accelerates reactions such as lipid oxidation.

Large Ice Crystal and Recrystallization Damage. If the food is not stored under sufficiently cold and steady temperatures, the ice crystals will grow or recrystallize to large ice crystals that may cause consequential damage to the food texture. Damages such as physical rupture of cell walls and membranes and separation of plant and animal cells cause limp celery or green beans, drips in thawed berries and meat. Enlarged ice crystals also disrupt emulsions (butter and milk), frozen foams (ice cream), and gels (frozen pudding and pie fillings), thus making these frozen products less homogenous, creamy, and smooth.

Another quality damage relating to ice recrystallization is the freezer-burn problem. Freezer burn occurs when there is a headspace in the packaged food and the food is subjected to fluctuating storage temperatures. When the temperature increases, ice at the warmer surface will sublime into the headspace. As the temperature of the freezer or surroundings cools down, the water vapor recrystallizes on the inner surface of the package instead of going back into the product. This leads to dehydration of the surface of the product. If the frozen product is not packaged, the freezer-burn problem is more common and more severe.

Types of Common Freezers with Different Cooling Media.

Cold Air.

- *Blast/belt freezers* are large insulated tunnels in which air as cold as $-40°C$ is circulated to remove heat. The process is cheap and simple and is geared toward high-volume production. Rotating spiral tiers and multilayered belts are incorporated to move product through quickly and avoid "hot spots."
- *Fluidized bed freezers* are modified blast freezers in which cold air is passed at a high velocity through a bed of food, contained on a perforated conveyor belt. This produces a high freezing rate but it is restricted to particulate foods (peas, shrimp, and strawberries).

Cold Surface Freezers.

- *Plate freezers* work by increasing the amount of surface area that comes in direct contact with the product to be frozen. Typically, refrigerant runs in the coils that run through plates or drums on which products are laid out. Double-plated systems further increase the rate of heat transfer to obtain higher quality. This system is suitable for flat and uniform products such as fish fillets, beef burgers, and dinner meals.
- *Scraped-surface freezers*—the liquid or semisolid food (ice cream) is frozen on the surface of the freezer vessel, and the rotor scrapes the frozen portion from the wall. Typically, ice cream is only partially frozen in a scraped-surface freezer to about $-6°C$, and the final freezing is completed in a hardening room $(-30°C)$.

Cold Liquid Freezers.

- *Brine freezers* use super-saturated solutions for maximum surface contact by immersing the prod-uct into a liquid freezing agent, especially for irregular shapes such as crabs. Disadvantage—products are subject to absorption of salt as well as bacteria.

Cryogenic Freezing.

- *Liquid nitrogen and liquid carbon dioxide* (which vaporize at $-178°C$ and $-80°C$, respectively) freeze product extremely quickly. Suitable for premium products such as shrimp and crab legs because of the high cost of the nonrecoverable gas.

Tips for Obtaining Top Quality Frozen Product.

- Start with high quality product: freezing can maintain quality but not enhance it.
- Get the heat out quickly by removing any nonedible parts from the food.
- Maintain the integrity of the frozen product: proper cutting and packaging avoids drips.
- Store the product at the coldest temperature economically possible in a well-designed and maintained facility. Use proper inventory techniques to avoid deterioration.
- Avoid temperature fluctuations during storage and shipping.

EVAPORATION AND DEHYDRATION

Evaporation

During food processing, evaporation is used to achieve the following goals: (1) concentrate food by the removal of water, (2) remove undesirable food volatiles, and (3) recover desirable food volatiles.

Traditionally, evaporation is achieved via the following methods: (1) Use sun energy to evaporate water from seawater to recover the salts left behind. (2) Use a heated kettle or similar equipment to boil water from liquid or semisolid foods (e.g., sugar syrup). (3) An improved method is to evaporate under a vacuum. The term "vacuum evaporator" refers to a closed heated kettle or similar equipment connected to a vacuum pump. One principle to remember is that a major objective of vacuum evaporators is to remove water at temperatures low enough to avoid heat damage to the food.

There are, at present, many specialized pieces of equipment used for evaporating food products. But, overall, these three methods are most common.

Drying

Drying differs from evaporating in that the former takes the food to nearly total dryness or the equivalence of 97 or 98% solids. The oldest method of drying food is to put the food under a hot sun. This practice probably started thousands of years ago.

Although sun drying is still practiced, especially in many third world countries, modern food drying has been modified to a nearly exact science. Drying has multiple objectives: (1) to preserve the food from spoilage, (2) to reduce the weight and bulk of the food, (3) to make the food enjoy an availability and consumption pleasure similar to that of canned goods, and (4) to develop "new" or "novelty" items such as snacks.

Some well-known products prepared from drying include: (1) dried milk powder, (2) instant coffee, (3) fish and shellfish, (4) jerky, (5) dried fruits, and (6) dried potato flakes.

The central equipment in dehydrating food is dryers. There are many types of dryers: spray dryers, drum dryers, roller dryers, and so on. See Chapter 2, Food Dehydration, for additional information.

FOOD ADDITIVES

One popular method of food preservation uses chemicals, legally known as food additives in the United States. In January 1992, the U.S. Food and Drug Administration (FDA) and the International Food Information Council released a brochure that presented an overview of food additives. The information in this section has been derived from that document, with an update.

Perhaps, the main functional objectives of the use of food additives are (1) to keep bread mold free and salad dressings from separating, (2) to help cake batters rise reliably during baking and keep cured meats safe to eat, (3) to improve the nutritional value of biscuits and pasta and give gingerbread its distinctive flavor, (4) to give margarine its pleasing yellow color and prevent salt from becoming lumpy in its shaker, and (5) to allow many foods to be available year-round, in great quantity and the best quality.

Food additives play a vital role in today's bountiful and nutritious food supply. They allow our growing urban population to enjoy a variety of safe, wholesome, tasty foods year-round. And they make possible an array of convenience foods without the inconvenience of daily shopping.

Although salt, baking soda, vanilla, and yeast are commonly used in foods today, many people tend to think of any food additive as a complex chemical compound. All food additives are carefully regulated by federal authorities and various international organizations to ensure that foods are safe to eat and are accurately labeled. The purpose of this section is to provide helpful background information about food additives, why they are used in foods and how regulations govern their safe use in the food supply.

Why Are Additives Used in Foods?

Additives perform a variety of useful functions in foods that are often taken for granted. Since most people no longer live on farms, additives help keep food wholesome and appealing while en route to markets sometimes thousands of miles away from where it is grown or manufactured. Additives also improve the nutritional value of certain foods and can make them more appealing by improving their taste, texture, consistency, or color.

Some additives could be eliminated if we were willing to grow our own food, harvest and grind it, spend many hours cooking and canning, or accept increased risks of food spoilage. But most people today have come to rely on the many technological, aesthetic, and convenience benefits that additives provide in food.

Additives are used in foods for five main reasons:

1. *To maintain product consistency.* Emulsifiers give products a consistent texture and prevent them from separating. Stabilizers and thickeners give smooth uniform texture. Anticaking agents help substances such as salt to flow freely.
2. *To improve or maintain nutritional value.* Vitamins and minerals are added to many common foods such as milk, flour, cereal, and margarine to make up for those likely to be lacking in a person's diet or lost in processing. Such fortification and enrichment have helped reduce malnutrition in the U.S. population. All products containing added nutrients must be appropriately labeled.
3. *To maintain palatability and wholesomeness.* Preservatives retard product spoilage caused by mold, air, bacteria, fungi, or yeast. Bacterial contamination can cause food-borne illness, including life-threatening botulism. Antioxidants are preservatives that prevent fats and oils in baked goods and other foods from

becoming rancid or developing an off flavor. They also prevent cut fresh fruits such as apples from turning brown when exposed to air.

4. *To provide leavening or control acidity/ alkalinity.* Leavening agents that release acids when heated can react with baking soda to help cakes, biscuits, and other baked goods to rise during baking. Other additives help modify the acidity and alkalinity of foods for proper flavor, taste, and color.

5. *To enhance flavor or impart desired color.* Many spices and natural and synthetic flavors enhance the taste of foods. Colors, likewise, enhance the appearance of certain foods to meet consumer expectations. Examples of substances that perform each of these functions are provided in Table 1.4.

Many substances added to food may seem foreign when listed on the ingredient label, but they are actually quite familiar. For example, ascorbic acid is another name for vitamin C; alpha-tocopherol is another name for vitamin E; and beta-carotene is a source of vitamin A. Although there are no easy synonyms for all additives, it is helpful to remember that all food is made up of chemicals. Carbon, hydrogen, and other chemical elements provide the basic building blocks for everything in life.

What Is a Food Additive?

In its broadest sense, a food additive is any substance added to food. Legally, the term refers to "any substance the intended use of which results or may reasonably be expected to result, directly or indirectly, in its becoming a component or otherwise affecting the characteristics of any food." This definition includes any substance used in the production, processing, treatment, packaging, transportation, or storage of food.

If a substance is added to a food for a specific purpose in that food, it is referred to as a direct additive. For example, the low-calorie sweetener aspartame, which is used in beverages, puddings, yogurt, chewing gum and other foods, is considered a direct additive. Many direct additives are identified on the ingredient label of foods.

Indirect food additives are those that become part of the food in trace amounts due to its packaging, storage, or other handling. For instance, minute amounts of packaging substances may find their way into foods during storage. Food packaging manufacturers must prove to the FDA that all mate-

Table 1.4. Common Uses of Food Additives in Food Categories

Common Uses of Additives Additive Functions/Examples[a]	Foods Where Likely Used
Impart/maintain desired consistency Alginates, lecithin, mono- and diglycerides, methyl cellulose, carrageenan, glyceride, pectin, guar gum, sodium aluminosilicate	Baked goods, cake mixes, salad dressings, ice cream, processed cheese, coconut, table salt
Improve/maintain nutritive value Vitamins A and D, thiamine, niacin, riboflavin, pyridoxine, folic acid, ascorbic acid, calcium carbonate, zinc oxide, iron	Flour, bread, biscuits, breakfast cereals, pasta, margarine, milk, iodized salt, gelatin desserts
Maintain palatability and wholesomeness Propionic acid and its salts, ascorbic acid, butylated hydroxy anisole (BHA), butylated hydroxytoluene (BHT), benzoates, sodium nitrite, citric acid	Bread, cheese, crackers, frozen and dried fruit, margarine, lard, potato chips, cake mixes, meat
Produce light texture; control acidity/alkalinity Yeast, sodium bicarbonate, citric acid, fumaric acid, phosphoric acid, lactic acid, tartrates	Cakes, cookies, quick breads, crackers, butter, chocolates, soft drinks
Enhance flavor or impart desired color cloves, ginger, fructose, aspartame, saccharin, FD&C Red No.40, monosodium glutamate, caramel, annatto, limonene, turmeric	Spice cake, gingerbread, soft drinks, yogurt, soup, confections, baked goods, cheeses, jams, gum

[a]Includes GRAS and prior sanctioned substances as well as food additives.

rials coming in contact with food are safe, before they are permitted for use in such a manner.

What Is a Color Additive?

A color additive is any dye, pigment, or substance that can impart color when added or applied to a food, drug, or cosmetic, or to the human body. Color additives may be used in foods, drugs, cosmetics, and certain medical devices such as contact lenses. Color additives are used in foods for many reasons, including to offset color loss due to storage or processing of foods and to correct natural variations in food color.

Colors permitted for use in foods are classified as certified or exempt from certification. Certified colors are man-made, with each batch being tested by the manufacturer and the FDA to ensure that they meet strict specifications for purity. There are nine certified colors approved for use in the United States. One example is FD&C Yellow No.6, which is used in cereals, bakery goods, snack foods, and other foods.

Color additives that are exempt from certification include pigments derived from natural sources such as vegetables, minerals, or animals. For example, caramel color is produced commercially by heating sugar and other carbohydrates under strictly controlled conditions for use in sauces, gravies, soft drinks, baked goods, and other foods. Most colors exempt from certification also must meet certain legal criteria for specifications and purity.

How Are Additives Regulated?

Additives are not always byproducts of twentieth century technology or modern know-how. Our ancestors used salt to preserve meats and fish, added herbs and spices to improve the flavor of foods, preserved fruit with sugar, and pickled cucumbers in a vinegar solution.

Over the years, however, improvements have been made in increasing the efficiency and ensuring the safety of all additives. Today food and color additives are more strictly regulated than at any other time in history. The basis of modern food law is the Federal Food, Drug, and Cosmetic (FD&C) Act of 1938, which gives the Food and Drug Administration (FDA) authority over food and food ingredients and defines requirements for truthful labeling of ingredients.

The Food Additives Amendment to the FD&C

Act, passed in 1958, requires FDA approval for the use of an additive prior to its inclusion in food. It also requires the manufacturer to prove an additive's safety for the ways it will be used.

The Food Additives Amendment exempted two groups of substances from the food additive regulation process. All substances that FDA or the U.S. Department of Agriculture (USDA) had determined were safe for use in specific food prior to the 1958 amendment were designated as prior-sanctioned substances. Examples of prior-sanctioned substances are sodium nitrite and potassium nitrite used to preserve luncheon meats. However, at present, nitrites are called color-fixing agents for cured meats and not preservatives, according to the FDA.

A second category of substances excluded from the food additive regulation process is generally recognized as safe (GRAS) substances. GRAS substances are those whose use is generally recognized by experts as safe, based on their extensive history of use in food before 1958 or based on published scientific evidence. Salt, sugar, spices, vitamins, and monosodium glutamate are classified as GRAS substances, as are several hundred other substances. Manufacturers may also request that the FDA review the use of a substance to determine if it is GRAS.

Since 1958, FDA and USDA have continued to monitor all prior-sanctioned and GRAS substances in light of new scientific information. If new evidence suggests that a GRAS or prior-sanctioned substance may be unsafe, federal authorities can prohibit its use or require further studies to determine its safety.

In 1960, Congress passed similar legislation governing color additives. The Color Additives Amendments to the FD&C Act require dyes used in foods, drugs, cosmetics, and certain medical devices to be approved by the FDA prior to marketing.

In contrast to food additives, colors in use before the legislation were allowed continued use only if they underwent further testing to confirm their safety. Of the original 200 provisionally listed color additives, 90 have been listed as safe, and the remainder have either been removed from use by FDA or withdrawn by industry.

Both the Food Additives and Color Additives Amendments include a provision that prohibits the approval of an additive if it is found to cause cancer in humans or animals. This clause is often referred to as the Delaney Clause, named for its Congressional sponsor, Rep. James Delaney (D-NY).

Regulations known as good manufacturing practices (GMP) limit the amount of food and color additives used in foods. Manufacturers use only the amount of an additive necessary to achieve the desired effect.

How Are Additives Approved for Use in Foods?

To market a new food or color additive, a manufacturer must first petition the FDA for its approval. Approximately 100 new food and color additives petitions are submitted to the FDA annually. Most of these petitions are for indirect additives such as packaging materials.

A food or color additive petition must provide convincing evidence that the proposed additive performs as intended. Animal studies using large doses of the additive for long periods are often necessary to show that the substance will not cause harmful effects at expected levels of human consumption. Studies of the additive in humans also may be submitted to the FDA.

In deciding whether an additive should be approved, the agency considers the composition and properties of the substance, the amount likely to be consumed, its probable long-term effects, and various safety factors. Absolute safety of any substance can never be proven. Therefore, the FDA must determine if the additive is safe under the proposed conditions of use, based on the best scientific knowledge available.

If an additive is approved, the FDA issues regulations that may include the types of foods in which it can be used, the maximum amounts to be used, and how it should be identified on food labels. Additives proposed for use in meat and poultry products also must receive specific authorization by the USDA. Federal officials then carefully monitor the extent of Americans' consumption of the new additive and the results of any new research on its safety to assure that its use continues to be within safe limits.

In addition, the FDA operates an Adverse Reaction Monitoring System (ARMS) to help serve as an ongoing safety check of all additives. The system monitors and investigates all complaints by individuals or their physicians that are believed to be related to specific foods, food and color additives, or vitamin and mineral supplements. The ARMS computerized database helps officials decide whether reported adverse reactions represent a real public health hazard associated with food, so that appropriate action can be taken.

Summary

Additives have been used for many years to preserve, flavor, blend, thicken, and color foods, and they have played an important role in reducing serious nutritional deficiencies among Americans. Additives help assure the availability of wholesome, appetizing, and affordable foods that meet consumer demands from season to season.

Today, food and color additives are more strictly regulated than at any time in history. Federal regulations require evidence that each substance is safe at its intended levels of use before it may be added to foods. All additives are subject to ongoing safety review as scientific understanding and methods of testing continue to improve.

See Table 1.5 for additional information about food additives.

FERMENTATION

The availability of fermented foods has a long history among different cultures. Acceptability of fermented foods also differs among cultural habits. A product highly acceptable in one culture may not be so acceptable to consumers in another culture. The number of fermented food products is countless. Manufacturing processes for fermented products vary considerably due to variables such as food groups, form and characteristics of final products, kind of ingredients used, and cultural diversity. Fermented foods can be prepared from various products derived from dairy products, grains, legumes, fruits, vegetables, muscle foods, and so on. This book contains an entire chapter (Chapter 3, Fermented Product Manufacturing) devoted to the science and technology of food fermentation. Please refer to it for further information.

NEW TECHNOLOGY

At present, some alternative or new technologies in food processing are available.

On June 2, 2000, the United States Food and Drug Administration (FDA) released a report titled "Kinetics of Microbial Inactivation for Alternative Food Processing Technologies." This report evaluates the scientific information available on a variety of alternative food processing technologies. The purpose of the report is to help the Food and Drug Administration evaluate each technology's effectiveness in reducing and inactivating pathogens of public health concern.

Table 1.5. Answers to Some of the Most Popular Questions about Food Additives

Q What is the difference between "natural" and "artificial" additives?

A Some additives are manufactured from natural sources such as soybeans and corn, which provide lecithin to maintain product consistency, or beets, which provide beet powder used as food coloring. Other useful additives are not found in nature and must be man-made. Artificial additives can be produced more economically, with greater purity and more consistent quality than some of their natural counterparts. Whether an additive is natural or artificial has no bearing on its safety.

Q Is a natural additive safer because it is chemical-free?

A No. All foods, whether picked from your garden or your supermarket shelf, are made up of chemicals. For example, the vitamin C or ascorbic acid found in an orange is identical to that produced in a laboratory. Indeed, all things in the world consist of the chemical building blocks of carbon, hydrogen, nitrogen, oxygen and other elements. These elements are combined in various ways to produce starches, proteins, fats, water, and vitamins found in foods.

Q Are sulfites safe?

A Sulfites added to baked goods, condiments, snack foods, and other products are safe for most people. A small segment of the population, however, has been found to develop hives, nausea, diarrhea, shortness of breath, or even fatal shock after consuming sulfites. For that reason, in 1986 the FDA banned the use of sulfites on fresh fruits and vegetables intended to be sold or served raw to consumers. Sulfites added as a preservative in all other packaged and processed foods must be listed on the product label.

Q Does FD&C Yellow No.5 cause allergic reactions?

A FD&C Yellow No.5, or tartrazine, is used to color beverages, desert powders, candy, ice cream, custards, and other foods. The color additive may cause hives in fewer than one out of 10,000 people. By law, whenever the color is added to foods or taken internally, it must be listed on the label. This allows the small portion of people who may be sensitive to FD&C Yellow No.5 to avoid it. Actually, any certified color added to food is required to be listed on the label.

Q Does the low calorie sweetener aspartame carry adverse reactions?

A There is no scientific evidence that aspartame causes adverse reactions in people. All consumer complaints related to the sweetener have been investigated as thoroughly as possible by federal authorities for more than five years, in part under FDA's Adverse Reaction Monitoring System. In addition, scientific studies conducted during aspartame's preapproval phase failed to show that it causes any adverse reactions in adults or children. Individuals who have concerns about possible adverse reactions to aspartame or other substances should contact their physicians.

The information in this section has been completely derived from this report. For ease of reading, all references have been removed. Consult the original documents for unabridged data. The citation data for this document are the following: A report of the Institute of Food Technologists for the Food and Drug Administration of the U.S. Department of Health and Human Services, submitted March 29, 2000, revised June 2, 2000, IFT/FDA Contract No. 223-98-2333, Task Order 1, How to Quantify the Destruction Kinetics of Alternative Processing Technologies, http://www.cfsan.fda.gov/~comm/ift-pref.html.

This section will discuss briefly the following new technology: (1) microwave and radio frequency processing, (2) ohmic and inductive heating, (3) high-pressure processing, (4) pulse electric fields, (5) high-voltage arc discharge, (6) pulse light technology, (7) oscillating magnetic fields, (8) ultraviolet light, (9) ultrasound, and (10) pulse X rays.

Table 1.5. Answers to Some of the Most Popular Questions about Food Additives *(continued)*

Q Do additives cause childhood hyperactivity?

A No. Although this theory was popularized in the 1970s, well-controlled studies conducted since that time have produced no evidence that food additives cause hyperactivity or learning disabilities in children. A Consensus Development Panel of the National Institutes of Health concluded in 1982 that there was no scientific evidence to support the claim that additives or colorings cause hyperactivity.

Q Why decisions sometimes are changed about the safety of food ingredients?

A Since absolute safety of any substance can never be proven, decisions about the safety of food ingredients are made on the best scientific evidence available. Scientific knowledge is constantly evolving. Therefore, federal officials often review earlier decisions to assure that the safety assessment of a food substance remains up to date. Any change made in previous clearances should be recognized as an assurance that the latest and best scientific knowledge is being applied to enhance the safety of the food supply.

Q What are some other food additives that may be used in the future?

A Among other petitions, FDA is carefully evaluating requests to use ingredients that would replace either sugar or fat in food. In 1990, FDA confirmed the GRAS status of Simplesse®, a fat replacement made from milk or egg white protein, for use in frozen desserts. The agency has also confirmed the use of the food additive Olestra, which will partially replace the fat in oils and shortenings.

Q What is the role of modern technology in producing food additives?

A Many new techniques are being researched that will allow the production of additives in ways not previously possible. One approach, known as biotechnology, uses simple organisms to produce additives that are the same food components found in nature. In 1990, FDA approved the first bioengineered enzyme, rennin, which traditionally has been extracted from calves' stomachs for use in making cheese.

Microwave and Radio Frequency Processing

Microwave and radio frequency heating refer to the use of electromagnetic waves of certain frequencies to generate heat in a material through two mechanisms—dielectric and ionic. Microwave and radio frequency heating for pasteurization and sterilization are preferred to conventional heating because they require less time to reach the desired process temperature, particularly for solid and semisolid foods. Industrial microwave pasteurization and sterilization systems have been reported on and off for over 30 years, but commercial radio frequency heating systems for the purpose of food pasteurization or sterilization are not known to be in use.

For a microwave sterilization process, unlike conventional heating, the design of the equipment can dramatically influence the critical process parameter—the location and temperature of the coldest point. This uncertainty makes it more difficult to make general conclusions about processes, process deviations, and how to handle deviations.

Many techniques have been tried to improve the uniformity of heating. The critical process factor when combining conventional heating and microwave or any other novel process will most likely remain the temperature of the food at the cold point, primarily due to the complexity of the energy absorption and heat transfer processes.

Since the thermal effect is presumably the sole lethal mechanism, time-temperature history at the coldest location will determine the safety of the process and is a function of the composition, shape, and size of the food; the microwave frequency; and the applicator (oven) design. Time is also a factor in the sense that, as the food heats up, its microwave absorption properties can change significantly, and the location of cold points can shift.

For further information, please refer to Chapter 4

in this book, Fundamentals and Industrial Applications of Microwave and Radio Frequency in Food Processing.

Ohmic and Inductive Heating

Ohmic heating (sometimes also referred to as Joule heating, electrical resistance heating, direct electrical resistance heating, electroheating, or electroconductive heating) is defined as the process of passing electric currents through foods or other materials to heat them. Ohmic heating is distinguished from other electrical heating methods by the presence of electrodes contacting the food, by frequency, or by waveform.

Inductive heating is a process wherein electric currents are induced within the food due to oscillating electromagnetic fields generated by electric coils. No data about microbial death kinetics under inductive heating have been published.

A large number of potential future applications exist for ohmic heating, including its use in blanching, evaporation, dehydration, fermentation, and extraction. The principal advantage claimed for ohmic heating is its ability to heat materials rapidly and uniformly, including products containing particulates. The principal mechanisms of microbial inactivation in ohmic heating are thermal. While some evidence exists for nonthermal effects of ohmic heating, for most ohmic processes, which rely on heat, it may be unnecessary for processors to claim this effect in their process filings.

High-Pressure Processing (HPP)

High-pressure processing (HPP), also described as high hydrostatic pressure (HHP) or ultra high-pressure (UHP) processing, subjects liquid and solid foods, with or without packaging, to pressures between 100 and 800 MPa. Process temperature during pressure treatment can be specified from below 0°C to above 100°C. Commercial exposure times can range from a millisecond pulse to over 20 minutes. Chemical changes in the food generally will be a function of the process temperature and treatment time.

HPP acts instantaneously and uniformly throughout a mass of food independent of size, shape, and food composition. Compression will uniformly increase the temperature of foods approximately 3°C per 100 MPa. The temperature of a homogenous food will increase uniformly due to compression.

Compression of foods may shift the pH of the food as a function of imposed pressure and must be determined for each food treatment process. Water activity and pH are critical process factors in the inactivation of microbes by HPP. An increase in food temperature above room temperature and, to a lesser extent, a decrease below room temperature increase the inactivation rate of microorganisms during HPP treatment. Temperatures in the range of 45–50°C appear to increase the rate of inactivation of food pathogens and spoilage microbes. Temperatures ranging from 90 to 110°C in conjunction with pressures of 500–700 MPa have been used to inactivate spore-forming bacteria such as *Clostridium botulinum.* Current pressure processes include batch and semicontinuous systems, but no commercial continuous HPP systems are operating.

The critical process factors in HPP include pressure, time at pressure, time to achieve treatment pressure, decompression time, treatment temperature (including adiabatic heating), product initial temperature, vessel temperature distribution at pressure, product pH, product composition, product water activity, packaging material integrity, and concurrent processing aids. Other processing factors present in the process line before or after the pressure treatment were not included.

Because some types of spores of *C. botulinum* are capable of surviving even the most extreme pressures and temperatures of HPP, there is no absolute microbial indicator for sterility by HPP. For vegetative bacteria, nonpathogenic *L. innocua* is a useful surrogate for the food-borne pathogen, *L. monocytogenes.* A nonpathogenic strain of *Bacillus* may be useful as a surrogate for HPP-resistant *E. coli* O157:H7 isolates.

Pulsed Electric Fields (PEFs)

High intensity pulsed electric field (PEF) processing involves the application of pulses of high voltage (typically 20–80 kV/cm) to foods placed between two electrodes. PEF may be applied in the form of exponentially decaying, square wave, bipolar, or oscillatory pulses at ambient, subambient, or slightly above ambient temperature for less than one second. Energy loss due to heating of foods is minimized, reducing the detrimental changes of the sensory and physical properties of foods.

Some important aspects of pulsed electric field technology are the generation of high electric field intensities, the design of chambers that impart uniform treatment to foods with minimum increase in temperature, and the design of electrodes that minimize the effect of electrolysis.

Although different laboratory- and pilot-scale treatment chambers have been designed and used for PEF treatment of foods, only two industrial-scale PEF systems are available. The systems (including treatment chambers and power supply equipment) need to be scaled up to commercial systems.

To date, PEF processing has been applied mainly to improve the quality of foods. Application of PEF processing is restricted to food products that can withstand high electric fields, have low electrical conductivity, and do not contain or form bubbles. The particle size of the liquid food in both static and flow treatment modes is a limitation.

Several theories have been proposed to explain microbial inactivation by PEFs. The most studied theories are electrical breakdown and electroporation.

Factors that affect the microbial inactivation with PEFs are process factors (electric field intensity, pulse width, treatment time and temperature, and pulse wave shapes), microbial entity factors (type, concentration, and growth stage of microorganism), and media factors (pH, antimicrobials and ionic compounds, conductivity, and medium ionic strength).

Although PEF processing has potential as a technology for food preservation, existing PEF systems and experimental conditions are diverse, and conclusions about the effects of critical process factors on pathogens of concern and the kinetics of inactivation need to be further studied.

High Voltage Arc Discharge

Arc discharge is an early application of electricity to pasteurize fluids by applying rapid discharge voltages through an electrode gap below the surface of aqueous suspensions of microorganisms. A multitude of physical effects (intense wave) and chemical compounds (created from electrolysis) are generated, inactivating the microorganisms. The use of arc discharge for liquid foods may be unsuitable largely because electrolysis and the resulting formation of highly reactive chemicals occur during the discharge. More recent designs may show some promise for use in food preservation, although the results reported should be confirmed by independent researchers.

Pulsed Light Technology

Pulsed light as a method of food preservation involves the use of intense, short-duration pulses of broad spectrum "white light," (ultraviolet to the near infrared region). For most applications, a few flashes applied in a fraction of a second provide a high level of microbial inactivation.

This technology is applicable mainly in sterilizing or reducing the microbial population on packaging or food surfaces. Extensive independent research on the inactivation kinetics across a full spectrum of representative variables of food systems and surfaces is needed.

Oscillating Magnetic Fields

Static and oscillating magnetic fields have been explored for their potential to inactivate microorganisms. For static magnetic fields (SMFs), the magnetic field intensity is constant with time, while an oscillating magnetic field (OMF) is applied in the form of constant amplitude or decaying amplitude sinusoidal waves. OMFs applied in the form of pulses reverse the charge for each pulse. The intensity of each pulse decreases with time to about 10% of the initial intensity. Preservation of foods with an OMF involves sealing food in a plastic bag and subjecting it to 1–100 pulses in an OMF with a frequency between 5 and 500 kHz at a temperature of 0–50°C for a total exposure time ranging from 25 to 100 ms.

The effects of magnetic fields on microbial populations have produced controversial results. Consistent results concerning the efficacy of this method are needed before considering this technology for food preservation purposes.

Ultraviolet Light

There is a particular interest in using ultraviolet (UV) light to treat fruit juices, especially apple juice and cider. Other applications include disinfection of water supplies and food contact surfaces. Ultraviolet processing involves the use of radiation from the UV region of the electromagnetic spectrum. The germicidal properties of UV irradiation (UVC 200–280 nm) are due to DNA mutations induced by DNA absorption of the UV light. This mechanism of inactivation results in a sigmoidal curve of microbial population reduction.

To achieve microbial inactivation, the UV radiant

exposure must be at least 400 J/m^2 in all parts of the product. Critical factors include the transmissivity of the product; the geometric configuration of the reactor; the power, wavelength, and physical arrangement of the UV source(s); the product flow profile; and the radiation path length. UV radiation may be used in combination with other alternative process technologies, including various powerful oxidizing agents such as ozone and hydrogen peroxide, among others.

Ultrasound

Ultrasound is energy generated by sound waves of 20,000 or more vibrations per second. Although ultrasound technology has a wide range of current and future applications in the food industry, including inactivation of microorganisms and enzymes, most current developments for food applications are nonmicrobial.

Data on inactivation of food microorganisms by ultrasound in the food industry are scarce, and most applications are used in combination with other preservation methods. The bactericidal effect of ultrasound is attributed to intracellular cavitations, that is, micromechanical shocks that disrupt cellular structural and functional components up to the point of cell lysis. The heterogeneous and protective nature of food that includes particulates and other interfering substances severely curtails the singular use of ultrasound as a preservation method. Although these limitations make the current probability of commercial development low, combination of ultrasound with other preservation processes (e.g., heat and mild pressure) appears to have the greatest potential for industrial applications.

Critical processing factors are assumed to be the amplitude of the ultrasonic waves, the exposure/contact time with the microorganisms, the type of microorganism, the volume of food to be processed, the composition of the food, and the temperature during treatment.

Pulsed X rays

It is important to realize that pulsed X ray is one form of irradiation that has been applied to the preservation of several categories of food in the United States. Electrons have a limited penetration depth of about 5 cm in food, while X rays have significantly higher penetration depths (60–400 cm), depending upon the energy used.

Pulsed X ray is a new alternative technology that utilizes a solid-state opening switch to generate electron beam x-ray pulses of high intensity (opening times from 30 ns down to a few nanoseconds; repetition rates up to 1000 pulses/second in burst mode operation). The specific effect of pulses in contrast to nonpulsed X rays has yet to be investigated.

The practical application of food irradiation by X rays in conjunction with existing food processing equipment is further facilitated by: (1) the possibility of controlling the direction of the electrically produced radiation, (2) the possibility of shaping the geometry of the radiation field to accommodate different package sizes, and (3) its high reproducibility and versatility.

Potentially, the negative effects of irradiation on the food quality can be reduced.

PACKAGING

The obvious reason for packaging a food product is to protect the food so it will not be exposed to the elements until it is ready to be prepared and consumed. In the world of food manufacturing, this is not a small matter because the FDA has rigid control over the materials used in food packaging. As far as the FDA is concerned, any packaging material is considered a food additive. All packaging materials used to contain food must comply with rigid regulations for the use of a food additive.

The term "food additive" refers to any substance whose intended use results or may reasonably be expected to result, directly or indirectly, in its becoming a component or otherwise affecting the characteristics of any food (including any substance intended for use in producing, manufacturing, packing, processing, preparing, treating, packaging, transporting, or holding food—if such substance is not generally recognized as safe).

Recently, the FDA has established the Food Contact Notification Program. It issues administrative guidance and regulations for the use of packaging materials, among others. FDA's website (www.FDA.gov) provides details for this program.

Different materials are used as packaging containers, including, but not limited to glass, plastic, laminates (paper based), and metal cans.

See Chapter 5, Food Packaging, for further information.

GLOSSARY

ARMS—Adverse Reaction Monitoring System.
CDC—Centers for Disease Control.
EPA—U.S. Environmental Protection Agency.
FAO—Food and Agriculture Organization.
GMP—good manufacturing practice.
HHP—high hydrostatic pressure processing.
HPP—high-pressure processing.
NIH—National Institutes of Health.
OMF—oscillating magnetic field.
PEF—pulsed electric fields.
SMF—static magnetic field.
UHP—ultra high-pressure processing.
USDA—U.S. Department of Agriculture.
UV—ultraviolet.
WHO—World Health Organization.

GENERAL REFERENCES

CV Barbosa-Canovas (editor), H. Zhang (editor), Gustavo V. Barbosa-Canovas. 2000. Innovations in Food Processing. CRC Press, Boca Raton, Fla.

ST Beckett (editor). 1996. Physico-Chemical Aspects of Food Processing. Kluwer Academic Publishers, New York.

J Bettison, JAG Rees 1995. Processing and Packaging of Heat Preserved Foods. Kluwer Academic Publishers, New York.

PJ Fellows. 2000. Food Processing Technology: Principles and Practice, 2nd edition. CRC Press, Boca Raton, Fla.

DR Heldman, RW Hartel (contributor). 1997. Principles of Food Processing, 3rd edition. Kluwer Academic Publishers, New York.

YH Hui et al. (editors). 2003. Food Plant Sanitation. Marcel Dekker, New York.

___. 2001. Meat Science and Applications. Marcel Dekker, New York.

___. 2004. Handbook of Vegetable Preservation and Processing. Marcel Dekker, New York.

___. 2004. Handbook of Frozen Foods. Marcel Dekker, New York.

___. 2004. Handbook of Food and Beverage Fermentation Technology. Marcel Dekker, New York.

MJ Lewis, NJ Heppell. 2000. Continuous Thermal Processing of Foods: Pasteurization and UHT Sterilization. Kluwer Academic Publishers, New York.

TC Robberts. 2002. Food Plant Engineering Systems. CRC Press, Boca Raton, Fla.

GD Saravacos, AE Kosaropoulos, AF Harvey. 2003. Handbook of Food Processing Equipment. Kluwer Academic Publishers, New York.

SPECIFIC REFERENCES

Drawert F. 1975. In Proceedings of the International Symposium on Aroma Research, 13–39. Center for Agricultural Publications and Documents, PUDOC, Wageningen.

Fellows PJ. 1992. Food Processing Technology (new edition), 323–324. Ellis Horwood, New York.

Fennema OR. 1985. Food Chemistry (new edition), 445–446. Marcel Dekker, Inc. New York.

Giannakourou MC, K Koutsoumanis, GJE Nychas, PS Taoukis. 2001. J. Food Protection, 64(7): 1051–1057.

Gram L, L Ravn, M Rasch, JB Bruhn, AB Christensen, M Givskov. 2002. Intl. J. Food Microbiology, 78(1–2): 79–97.

McSwane D, N Rue, R Linton. 2003. Essentials of Food Safety and Sanitation, 4. Prentice Hall, N.J.

Potter NN, JH Hotchkiss. 1995. Food Science (new edition), 377–378. Chapman and Hall, New York.

Siragusa GR, CN Cutter, JL Willett. 1999. Food Microbiology, 16(3): 229–235.

2
Food Dehydration

R. Driscoll

INTRODUCTION

Dehydration is the removal of water from a product. Our purpose in dehydrating (drying) is usually to improve the shelf life of the product, and thus dehydration is a unit operation of great importance to the food industry.

The effect on shelf life is due to the link between moisture content and a property called water activity, a measure of the availability of water to take part in chemical reactions. As moisture is reduced, the water activity of the product is also reduced. Once the water activity has dropped to about 0.6, the product is generally considered to be shelf stable.

Products may be dried for other reasons; for example, to control texture properties such as crispness (biscuits), to standardize composition, and to reduce weight for transport. The most important reason, however, is control of water activity.

Drying is expensive, since the energy required to remove water is high. Heat recovery systems (for example, heat pumps) may be used to reduce this cost, but they have higher capital costs and add complexity.

Drying is the single most common unit operation in the food industry. Dryers are often designed for specific products, and the range of dryer types is large. In this chapter we will examine the effects of drying on quality, the theory of drying, and drying equipment.

DRYING AND QUALITY

Dehydration changes food products in several ways, affecting the organoleptic qualities of the product. Dehydration normally requires high temperatures, which can cause chemical reactions such as nonenzymatic browning, caramelization, and denaturation of proteins in the product. Drying also affects the physical parameters of the product, as removal of water causes shrinkage (plums become prunes, grapes become sultanas). Due to these changes, rehydration after drying may not restore the original product.

WATER ACTIVITY

We often assume that the purpose of drying is to control moisture, and in a sense this is true. More important from a preservation point of view, however, is control of water activity.

Water activity (a_w) is the relation between fugacity of water vapor in the food and that of pure water vapor, expressed as fugacity of water vapor in food over fugacity of pure water vapor. If the conditions are such that the water vapor is an ideal gas, water activity is numerically equivalent to the relative vapor pressure, $RVP = p_v/p_s$, where p_v is the vapor pressure of water in the food and p_s is the vapor pressure of pure water at the same temperature. Relative vapor pressure is commonly called relative humidity (RH).

Water activity is a measure of the availability of water for chemical reaction, and varies between 0 and 1. Water has reduced availability if it is bound to the nonsoluble solids in a food matrix. The presence of soluble solids in the product's liquid phase will also reduce water availability.

Water activity (a_w) is not the same as moisture content. Chemical and biological processes correlate well with a_w at higher moistures, but not with moisture concentration. Physical effects such as shrinkage are, conversely, better explained in terms of moisture.

Air in contact with a product at water activity a_w comes to an equilibrium relative humidity (ERH) given by

$$a_w \approx \frac{p_v}{p_s} = ERH / 100 \qquad 2.1$$

where p_v is the vapor pressure of moisture in the air, and p_s is the saturation vapor pressure at the same temperature (see notes later on RH). The factor of 100 is required because, by convention, relative humidity is normally expressed as a percentage. Thus, we can measure a product's water activity by measuring the RH of air adjacent to the product, for example, by placing the product in a sealed jar and measuring the headspace RH.

Water activity can be controlled by other methods than dehydration. A common method in the food industry is to add humectants, chemicals such as sugars, salts, and glycerol, which bind available water.

DETERIORATION REACTIONS IN FOODS

Microbial Stability

The limits for microbial growth are determined by water activity. For example, most bacteria need $a_w >$ 0.91, and most molds need $a_w > 0.80$. The exact water activity limit for a specific organism depends on other factors such as pH, oxygen availability, the nature of the solutes present, nutrient availability, and temperature. Generally, the less favorable the factors, the higher the value of a_w required for growth.

The effect of microbial action on quality may simply be *economic loss,* for example, discoloration, physical damage, off flavors, and off odors (spoilage microbes); or it may be a *health issue,* for example, pathogens. Reduction in moisture will increase the microbiological stability of the product, increasing shelf life.

Chemical Stability

Water may take part in chemical reactions as a solvent, a reactant, a product (for example, in nonenzymic browning reactions) and/or a modifier (in catalytic or inhibitory activities). Reactions that depend on moisture to bring reactants together will become increasingly limited by drying, due to the reduced molecular mobility of the reactants. At low moistures, a further preservation mechanism becomes significant. As the moisture content of a food is reduced during drying, solutes become more concentrated, and solution viscosity rises. Drying temperature also has an important effect on reaction rates, and hence quality. Some examples of important food chemical reactions are:

- *Enzymic reactions.* These reactions, which are not completely understood, are very slow at low a_w values due to the lack of mobility of the substrate to diffuse to the active site of the enzyme.
- *Nonenzymic browning (NEB).* Water-dependent reaction with maximum reaction rates around a_w = 0.6–0.7. Water is also a reaction product. Too much water inhibits reaction by dilution, and too little gives inadequate mobility.
- *Lipid oxidation.* Reaction that is fast at both low and high values of a_w.

- *Loss of nutrients.* For example, vitamin B or C losses due to breakdown at high temperatures.
- *Loss of volatiles.* For example, loss of flavors and aromas from the product.
- *Release of structural water.* Changes food texture.

Physical Stability

Physical deterioration *does* correlate with moisture content. Some examples of physical effects are:

- *Softening/hardening of texture.* Texture softens at high moisture and hardens at low moisture (water acts as a plasticizer of the food material).
- *Differential shrinkage.* Outer layers shrink relative to inner layers, leading to either surface cracks or radial cracks.
- *Surface wetting effects.* Moisture works on the product surface to expand pores and capillaries.
- *Case hardening.* A hydrophobic layer may be formed in an oil-rich product during rapid drying of outer layers, which traps moisture inside the product.
- *Cell collapse.* Cells may collapse if internal moisture is removed, leading to the product wrinkling (e.g., prunes, sultanas).

WATER AND AIR

How Do We Dry?

The main method for drying is to pass dry air over the product. The boundary layer of air adjacent to the product equilibrates with the product, and moisture diffusing through this boundary escapes to the airstream to be carried out of the dryer, the driving force being the difference in vapor pressure between the boundary layer and the upstream air. As moisture evaporates to replenish the boundary layer, the heat of evaporation must be replaced by heat transferred to the product surface. Thus hot air will dry faster than cold air at the same humidity.

Product Equilibrium

If we place some product in a jar (Figure 2.1) and then seal the jar, the product and air will come to equilibrium over time. At equilibrium, the rate of evaporation from the surface of the product matches the rate of condensation, and the air moisture content is determined by the product moisture and temperature only. This air relative humidity is the equilibrium relative humidity (ERH), and the moisture

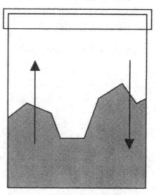

Figure 2.1. Product in a sealed glass jar.

content is the equilibrium moisture content (EMC). By measuring the ERH at different moisture contents, but the same temperature, a product *isotherm* can be constructed.

How Does Water Evaporate?

Energies in liquid water have a Gaussian distribution. Only high-energy molecules have sufficient energy (in excess of the intermolecular bond energy) to escape. As they leave, the average remaining energy per molecule is reduced, and hence the product cools. This is called evaporative cooling.

Wet Basis and Dry Basis Measurement

Industry conventionally measures the moisture content of a product as the ratio of water mass to total product mass. For example, for a product containing 40 kg of water for every 100 kg, the moisture content is expressed as 40%. This basis of measurement is called *wet basis* (wb).

For dryer analysis, however, it is more convenient to measure moisture content on a *dry basis* (db) (the ratio of water mass to dry solids mass). Since in the example above there are 40 kg of water to every 60 kg of dry solids, the dry basis moisture content is 40/60 or 67% db. The product moisture is thus the water concentration in the product.

Using the symbol W for wet basis moisture and M for dry basis moisture,

$$M = \frac{m_w}{m_s} \qquad W = \frac{m_w}{m_w + m_s} \qquad 2.2$$

where m_w is the mass of water and m_s is the dry solids mass in a sample. For example:

- 20% wb is equal to 25% db.
- 75% wb is 300% db (which means three parts water to one part dry solids).

To convert from wet basis to dry basis,

$$M = \frac{100W}{100 - W} \qquad 2.3$$

Exercise 1

Using the conversion formulae above, complete the following table:

Total Mass Product (kg)	Dry Product Mass (kg)	Dry Basis Moisture Content (%)	Wet Basis Moisture Content (%)
100	50		
100	90		
100		20	
100			20
	50	40	
	20		85

Can wet basis moisture be over 100%?
Can dry basis moisture be over 100%?

Exercise 2

How much moisture would I need to remove to dry 100 kg of wet product from:

(a) 50% wet basis (wb) to 20% wb?
(b) 50% dry basis (db) to 20% db?
(c) 50% wb to 20% db?

Do I get the correct answer by simply subtracting moisture contents in any of these cases?

IMPORTANT PSYCHROMETRIC EQUATIONS

The measurement of the properties of a gas/vapor mix is called psychrometry, an important branch of physical chemistry. The moisture content of the air is called absolute humidity, H:

$$H = \frac{m_w}{m_a} \qquad 2.4$$

where m_w is the mass of water, m_a is the mass of the dry (non-water) air components, and so the units of H may be written as kg/kg dry air, since by convention air humidity is measured on a dry basis.

The equation for an ideal gas is

$$PV = nRT \qquad 2.5$$

where P is absolute pressure (Pa), V is the volume occupied by the gas (m^3), n is number of moles, R is the universal gas constant (8.314 kJ/ kmol·K), and T is absolute temperature. The total pressure is the sum of the partial pressures exerted by each component in the gas mix:

$$p_T = \sum_{i=1}^{n} p_i \qquad 2.6$$

where subscript T indicates total pressure and subscript i refers to the ith component of a mix of n components.

Air enthalpy measured relative to the natural state of its components at 0°C is

$$h = m_a[c_a T + H (c_v T + \lambda_0)] \qquad 2.7$$

where c_a is the dry air specific heat, c_v is the specific heat of water vapor (both in kJ/kg·K), T is air temperature (°C) and λ_0 is the latent heat of evaporation of pure water at 0°C. In this equation, $m_a c_a T$ is the sensible heat of the dry air components, $m_a H c_v T$ is the sensible heat of pure water, and $m_a H \lambda_0$ is the latent heat of converting $m_a H$ kilograms of water to water vapor.

Exercise 3

The specific heat of dry air is 1.01 kJ/kg·K, of water vapor 1.83 kJ/kg·K, and of liquid water 4.19 kJ/kg·K, and the latent heat of water at 0°C is 2501 kJ/kg (values at 20°C). From this information, calculate the heat per kilogram of each term in Equation 2.7 at a temperature of 20°C and a humidity of 10 g/kg dry air.

Air can carry very little water vapor, so typically, H is small, of the order of 10 or 20 g water/kg dry air. Thus the second term makes little contribution to the total enthalpy. However, since water has a very high latent heat of evaporation, the last term is substantial. The enthalpy h is measured in kJ, c_a is 1.007 kJ/kg·K, c_v is 1.876 kJ/kg·K, and λ_0 is 2500 kJ/kg. Lines of constant enthalpy are drawn on charts of air/water properties, called psychrometric charts, and are important in drying.

Since air rapidly saturates with water, there is a maximum vapor partial pressure for water at any given temperature. The ratio of the actual vapor

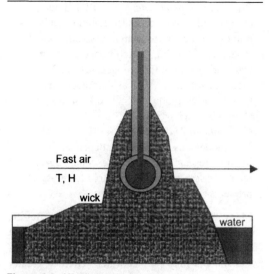

Figure 2.2. Wet bulb temperature.

pressure p_v to this maximum p_s at the same temperature is the relative humidity (RH):

$$RH = 100 \frac{p_v}{p_s} \qquad 2.8$$

Saturated air has a relative humidity of 100%.

WET BULB TEMPERATURE

If we place a glass thermometer in an airstream, the temperature measured is called the dry bulb temperature (Fig. 2.2). If the wick is kept wet, evaporation of water cools the wick, and the indicated temperature, T_{wb}, is the wet bulb temperature.

The rate of evaporation (m_e, kg/s) is

$$m_e = k_y A \left(H_s - H\right) \qquad 2.9$$

where A is the area (in m^2) for evaporation of moisture from the wick, k_y is the mass transfer coefficient in kg/m^2·s, and H_s is the saturation humidity of the air at the temperature of the air.

The rate of heat flow (q, kJ/s) into the wick is

$$q = hA \left(T - T_{wb}\right) \qquad 2.10$$

where h is the heat transfer coefficient (W/m^2·K). At equilibrium,

$$q = m_e \cdot \lambda_T \qquad 2.11$$

where λ_T is the latent heat of free water at the evaporation temperature T (°C). Thus,

$$T - T_{wb} = (k_y \lambda_T / h)(H_s - H) \qquad 2.12$$

Equation 2.12 represents a straight line (linear function of H vs. T) that can be plotted on a psychrometric chart. In practice, small changes in the value of λ_T causes small variations from linearity.

Drying a thin layer of product under constant conditions, air leaves at close to the same enthalpy (assuming negligible heat losses) as the inlet air. Lines of constant enthalpy are very close to the wet bulb temperature lines given by Equation 2.12.

DRYING THEORY

MOISTURE DEFINITIONS

Equilibrium moisture content is the moisture level at which product is in equilibrium with the moisture of its surroundings (air). (See Fig. 2.3, below.) *Bound moisture* is the moisture that exerts a vapor pressure lower than 100% RH, so that product water activity

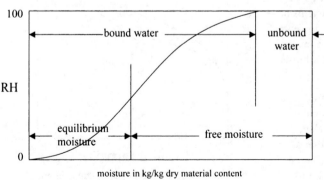

Figure 2.3. Schematic description of the state of water in a food product.

is less than 1.0. *Unbound moisture* is moisture in excess of the minimum required to give a water activity of 1.0. *Free moisture* is moisture in excess of the equilibrium moisture content, so it can be removed by drying. Further moisture cannot be removed without reducing the relative humidity of the air, and hence changing the product equilibrium moisture content.

Problem: Air at 25°C, 40% RH is being used to dry a 5 ton bed of rice at 20% wb moisture. Estimate how much air is required to dry the rice to 14% db, if the air leaves at 84% RH, in equilibrium with the grain.

Solution: From the psychrometric chart:

T_{wb} = 16.2°C

From chart, grain cools to 18°C.

H_{out} = 0.011 kg/kg dry air
H_{in} = 0.008 kg/kg dry air

Dry mass = 5000 kg × 80% = 4000.0 kg
Initial moisture = 5000 kg × 10% = 1000.0 kg
Final moisture = 14% × 4000 kg = 560.0 kg

∴ Weight of moisture to remove = 440.0 kg
Air required = 440/(0.011 − 0.008) = 1470 kg

Note use of *wet basis* (wb) when total mass is used as the reference quantity and *dry basis* (db) when solids mass is used.

VAPOR ADSORPTION THEORIES

Moisture moves through the product to its outer surface and then evaporates to the boundary layer. The mechanism of moisture movement is still not completely understood, but is thought to be a combination of several phenomena at once. Several models have been proposed to explain this mechanism, leading to models that predict the water content/water activity relation of the product. At a constant temperature, this product-dependent relationship is called an isotherm.

- *Langmuir* (1918). Langmuir studied the first bonding of water molecules condensing onto a product surface (monolayer adsorption). Hence Langmuir was able to predict that the rate of adsorption was proportional to $(M - M_e) \cdot p_v$ at the surface. This model describes isotherms at low moistures only.

- *Brunnauer, Emmett and Teller* (BET model) (1938). This model extended Langmuir adsorption to multilayer absorption, assuming Van der Waal's H-H bonding as additional layers of water are added to the product surface. This model results in the typical sigmoidal curves found in food products, and works well up to about 40% RH.

- *Guggenheim-Anderson-deBoer* (GAB model). This model was discovered independently by three researchers. It gives excellent agreement over the full isotherm curve for most products.

Many other models exist, most empirical. But the above three models have a theoretical basis, and the last model (GAB) describes product behavior well. Some models may predict water activities greater than unity, which is physically impossible, and this limitation may affect the correct choice of model for a given situation. Generally, the constants required for the equation are determined experimentally by measuring the product isotherm under a few reference conditions.

HYSTERESIS

As moisture adsorbs to the product surface, the product structure may be modified by work done on the surface. This results in an effect called hysteresis, in which desorption isotherms differ from adsorption. This effect decreases after successive cycling of the product (see Fig. 2.4).

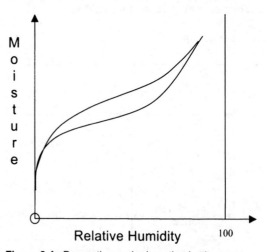

Figure 2.4. Desorption and adsorption isotherms.

The Four Drying Rate Periods

For the purpose of studying the drying rate of a product, it is convenient to define a thin layer of product as follows: the air leaving the thin layer is not detectably different from the inlet air. Thus, all of the product is effectively in contact with the same air. This does not necessarily imply that the product itself is thin. The opposite of a thin-layer drying situation is deep-bed drying. Mathematically, deep-bed drying is modeled as if it consists of multiple thin layers.

Thus we can study the drying properties of a thin layer, and based on this information, we can predict how the product will dry in any situation. Plotting moisture against time for a thin layer gives a product-drying curve. For a high moisture product, this has four identifiable regions:

- *Initial transient.* Thermal equilibration to the inlet air wet bulb temperature, accompanied by a small moisture change as evaporative cooling or condensation occurs to match air to product enthalpies at the product surface.
- *Constant rate period (CRP)* is

$$m_s = \frac{dM}{dt} = k_y A (H_s - H_a) \qquad 2.13$$

where t is drying time, M is the sample moisture content (as a fraction, db), m_s is the dry solids weight of the sample (kg), and other symbols are as defined for the dew point equation (2.9).

- *Transitional region.* Surface dry spots appear at the critical moisture content M_c, so that the product no longer behaves like a free water surface, and diffusion starts to limit moisture loss.
- *Falling rate period (FRP).* Evaporation is determined by diffusion:

$$\frac{\partial M}{\partial t} = D(T) \frac{\partial^2 M}{\partial x^2} \qquad 2.14$$

Equation 2.14 is called Fick's Law, and M is a function of position and time. Variation in moisture content within the product causes moisture gradients resulting in moisture movement.

Shown in Figure 2.5 is a typical thin-layer drying curve. M_o is the initial moisture, M_c is the critical moisture and M_e is the final equilibrium moisture (measured when the product has come to complete equilibrium with the air). From this diagram, it can be seen that the initial transient region (near the initial moisture, M_o) is only significant for a short time

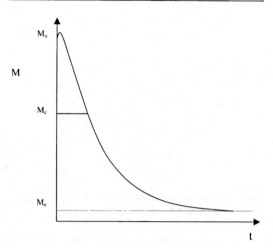

Figure 2.5. Thin-layer drying curve: moisture versus time.

near the start of drying. Despite this, it is an important region, as we will see presently. Note also that the constant rate period (from M_o to M_c) is a straight line (when plotted using dry basis moistures). The transition region occurs around M_c, but is usually ignored because it is difficult to see in practice. The remainder of the curve represents the falling rate period.

Thin-layer drying may also be shown on a psychrometric chart (Fig. 2.6). This helps to show the drying process from the perspectives of both the air and the product. The product entering the dryer is positioned on the chart according to its temperature and equilibrium relative humidity (point P in Fig. 2.6). For example, if the product is at 20°C and has a water activity of 0.95, then P is the intersection of the 20°C temperature line and the 95% relative humidity line.

The product will dry more quickly if the inlet air is chosen with high temperature and low relative humidity (point I in Fig. 2.6). During drying the total heat content remains constant (see discussion later), so a line can be drawn through I at constant enthalpy towards the air saturation line to represent the drying process.

Point B is the intersection of the equilibrium relative humidity line for the product with the enthalpy line that passes through I. The initial transient for the product (compare Fig. 2.5) is represented by the curve PB, where equilibration between the enthalpy of the product and the enthalpy of the drying air takes place.

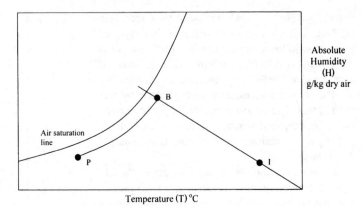

Figure 2.6. Simplified representation of drying on a psychrometric chart.

Only products with a water activity of 1 will exhibit a constant rate period (CRP). For these products, point B will be on the saturation line at the wet bulb temperature of the air, and the product will remain at point B until the critical moisture content is reached.

Many studies of drying use the falling rate period to model the complete drying curve.

MODELS OF THE FALLING RATE PERIOD

Fick's diffusion equation can be solved for simple regular shapes, homogeneous product, and constant drying conditions (Crank 1956), resulting in equations of the form:

$$M(x,t) = a_0 + \sum_{n=1}^{\infty} d_n \exp a_n x \cdot \exp b_n t \qquad 2.15$$

where a_n, b_n, and d_n are constants. Generally, we need only the average moisture content of the product:

$$M = c_0 + c_1 \exp - k_1 t + c_2 \exp - k_2 t + \dots \qquad 2.16$$

where c_n and k_n are constants dependent on product shape and inlet air properties, and k_n is related to the mass diffusivity of the product. This model is called a multi-compartment *model*.

The simplest useful form of this equation is to keep the first two terms only:

$$M = c_0 + c_1 \exp - k_1 t$$

Substituting $M = M_o$ at $t = 0$ and $M = M_e$ at $t = \infty$ gives

$$MR = \frac{(M - M_e)}{(M_o - M_e)} = e^{-k_1 t} \qquad 2.17$$

where MR is the moisture ratio, and varies from 1 to 0 during drying. If a constant rate period exists, then time t refers to time *after* the critical moisture is reached, and M_o is replaced with M_c.

Differentiating the equation gives

$$\frac{dM}{dt} = -k_1 (M - M_e) \qquad 2.18$$

(where k_1 is now called the drying rate constant), showing that the drying rate is proportional to the difference between the present product moisture and its final equilibrium moisture. This model, although based on several simplifying assumptions, has been found to represent product drying adequately for most commercial situations.

Problem

A 4 m² tray of liquid product (density 1000 kg/m³) has an initial solids content of 8% and dries at a rate of 440 g/min. At a moisture content of 133% db, the product drying rate starts to reduce. If the product depth is 10 mm on the tray, find the drying rate in the constant rate period.

Solution

Initial dry weight of product is

4 m² × (7 × 10⁻³ m) × 1000 kg/m³ × 0.08 = 2.24 kg.

Mass drying rate is 420/1000 = 0.42 kg/s.

Thus, drying rate is $0.42 \times 100/2.24 = 18.8\%$ db/min.

Problem

A product in the falling rate period (below the critical moisture) has a drying rate constant of 1/120 min^{-1}. Calculate the time to dry from 60% to 10% db if the equilibrium moisture content is 4% db.

Solution

Note that all moistures must be expressed on the same basis, which must be consistent with the drying rate constant, k. So if k was determined from dry basis moisture measurements, we should use dry basis in solving the problem.

For the falling rate period,

$dM/dt = k(M - M_e)$

Integrating (compare Eq. 2.16),

$(M_f - M_e) / (M_i - M_e) = \exp(-kt)$

Substituting known values:

$t = -120 \ln[(10 - 4)/(60 - 4)] = 268$ minutes
= 4 hours 28 minutes

THEORIES FOR THE FALLING RATE PERIOD

There are several possible diffusion mechanisms:

- *Liquid diffusion.* Moisture moves through the product in proportion to the liquid water gradient at any point. This model gives poor prediction of moisture profiles.
- *Vapor diffusion.* This assumes that the product is porous, so that vapor can diffuse through the material. This model predicts moisture profiles well.
- *Capillary movement.* Liquid water moves by capillary action through pores. This is a good model when the water activity is 1, and the model works well in combination with the liquid diffusion model.

A COMPLETE DRYING MODEL

The drying rate for the constant rate period can be represented by a constant (k_o). Combining models gives a simple overall model:

For $M \geq M_c$:

$$\frac{dM}{dt} = -k_o$$

For $M < M_c$:

$$\frac{dM}{dt} = -k_1(M - M_e)$$

2.19

EFFECT OF AIRFLOW

At low airspeeds, there is insufficient air to remove water from the product surface, and the air leaves saturated. Above the minimum airspeed required to remove this moisture, airspeed has little effect on the rate of drying for a thin layer, as drying cannot occur faster than the heat required for evaporation is supplied from the air. This is true for both the constant rate and falling rate periods. Experimentally, the rate of drying for a thin layer increases roughly with the cube root of airspeed. This small effect is due to an increase in convective heat transfer coefficient from the air to the product surface as airspeed increases.

If the product is being dried in a deep bed (or has been spread out on trays in the direction of airflow), then airspeed has a major effect on the total drying rate, as the air spends a greater time in contact with the product. Under these conditions, drying capacity is affected strongly by the rate of air supply to the dryer.

DRYING EQUIPMENT

There is a large range of dryer types. Dryer designs depend largely on the particular needs of the enormous variety of food products that require drying. Examples are tray, cabinet, vacuum, osmotic, column, recirculating, and freeze dryers.

Dryers may be categorized by:

- **Mode of operation.** *Batch* dryers (for example, a kiln dryer) are loaded and operated, and then dried product is unloaded. *Continuous* dryers (for example, a rotary dryer) are loaded and unloaded while the dryer operates.
- **Method of heating.** *Direct* heating means that the flue gases from combustion come in contact with the product; *indirect* means that a heat exchanger is used to transfer heat from the flue gases to the drying air, thus protecting the product from possible contaminants. Electrical forms of heating (ohmic, microwave, and radio frequency) could be considered as a special case.

- **Nature of product.** Product might be loaded into the dryer as *solid, liquid, slurry,* or *granules,* each requiring a different form of dryer. Liquids can be dried in spray dryers, solids on meshes, granular materials in deep or fluidized beds, slurries in trays.
- **Direction of airflow.** The drying air may be *concurrent, countercurrent* or *cross-flow*. A dryer may have zones with different airflow directions (combination dryers).

In the following section, dryers are classified as batch or continuous, and the continuous dryers are further classified by the direction of airflow.

BATCH DRYERS

Batch dryers handle a single batch of product at once. The product is first loaded into the dryer, then the dryer runs through its drying cycle and switches off, then the dried product is unloaded. Batch dryers are often simpler in design than continuous dryers but do not interface well with continuous processing lines. In addition, the time required for loading/unloading will reduce the effective time of utilization of the dryer. Batch dryers tend to be used for small-scale production such as rapidly changing product lines, pilot-plant processing, rural production, and high value products.

Kiln Dryers

Kiln dryers are a simple, universal form of dryer used for drying thin layers of product. They consist of a drying tray over some form of heat source, for example, a biomass combustor or a furnace, and so are usually direct fired. They are inefficient, as the hot air, after passing through a single layer of product, must be vented to the atmosphere.

Kiln dryers are commonly used for drying fruits, vegetables, and cocoa beans.

In-Store Dryers

In-store dryers (bin dryers) are dryers that can both dry and store product (similarly to the way a cool room both cools and stores). The granular product is placed in bulk on a mesh supporting screen, and air is pumped into a plenum chamber below the product, passing through the screen and then through the product. These dryers have high thermal efficiency, as they operate at near ambient temperatures, with the drying front submerged in the product mass for a large proportion of the drying time, so that the air leaves close to saturated. These dryers are suitable for granular products such as grains, nuts, and berries.

Tray Dryers

Tray dryers (cabinet dryers) have product on trays in a closed cabinet, where air enters the dryer, is mixed with recirculated air, heated, and then passed across the trays. A proportion of the exit air is vented from the dryer. They are suited to small-scale operations and rapid changes in product line. The benefit of recirculation is that the heat requirements of tray dryers are much less than those of kiln dryers, because the heat content of the air is used more efficiently.

Freeze Dryers

Freeze dryers use sublimation rather than evaporation of moisture. They are expensive, but are suitable for high-value, heat-labile products. The product is placed on heated shelves and the drying chamber evacuated. The effect of reducing the air pressure is to cause the product to cool by sublimation of moisture until the temperature is about -20 to $-40°C$. Evaporative drying is slow at low temperatures, but sublimation drying is relatively fast. The low temperatures protect the product from changes due to heating and reduce the loss of volatiles such as aroma and flavor. In addition, freeze-drying generally better preserves the structure of the product.

CONTINUOUS DRYERS

With continuous dryers, food product enters the dryer by conveyor, passes through the required drying treatment (which may consist of multiple sections at different temperatures), and then exits the dryer without stopping. Continuous dryers are suited to running for long periods of time with the same product and are usually fitted with feedback control to maintain drying conditions and/or product exit conditions.

Rotary Dryers

A rotary dryer consists of a long cylinder supported on girth rings used to rotate the dryer. Product is fed in at one end, and heated air comes in contact with the material as it passes down the cylinder, which is usually slightly inclined to allow the material to flow down. As the cylinder rotates, product may be picked up by flights mounted along the inside surface of the barrel, carried, and then dropped, allowing good air/product contact and product mixing.

This dryer is suited to granular products. It may be directly or indirectly fired, and is run con- or countercurrent.

Drum Dryers

A large range of drum dryer designs exists, but the essential feature is a steam-heated drum being coated with liquid product. As the drum rotates, a thin film of liquid is picked up on the surface of the drum. The thickness of this layer can be controlled by blades close to the drum surface or by a second rotating drum. In the time that the drum rotates, the thin product layer is dried to a solid and scraped off the surface as flakes.

Spray Dryers

Spray dryers create a fine spray of liquid product in a hot air environment. The liquid is pumped into a nozzle (preferably using a positive displacement pump to ensure a uniform flow of product), which forces the liquid through an atomizer, a device that imposes high shear stresses in the liquid. Examples of atomizers are:

- *Paired disks* with one rotating, the other stationary. The liquid is forced between the two plates from an axial feed.
- *Perforated plate.* The liquid is forced under high pressure through small holes in the plate.

This process breaks the liquid into fine droplets, which assume a spherical shape owing to surface tension. As the liquid leaves the high shear region, it enters a hot air region, which may be designed con- or countercurrently to the product flow. At high temperatures the outside of the droplet dries quickly, forming a hard shell. Water inside the droplet boils, rupturing the hard shell to create distinctive partial shells. For many products, the product temperature is kept below boiling, producing honeycombed patterns, depending on the product properties. This open structure leads to efficient reconstitution (the addition of water to rehydrate the product).

The product must be dried within a short distance of the nozzle exit, because wet product contacting the inside dryer surfaces is a major cleaning problem. Thus, the air and product flow rates and the drying air temperature must be chosen carefully to ensure good final product quality.

The resulting mixture of dried product powder and air exits the dryer and is separated in a cyclone. Note that a mixture of air and powdered food may be an explosion hazard. In some cases, the product may be passed to a fluidized bed dryer to complete the drying process.

Fluidized Bed Dryers

Fluidized bed dryers are designed for granular solids. The material to be dried is placed on perforated screen, and air is blown through the screen at sufficient speed (typically over 2 m/s) until the resulting pressure drop across the bed matches the total product weight. At this point the bed starts to fluidize (act like a fluid) and flow. Further increases in airspeed have little effect on the pressure drop across the bed of product. The product is successfully fluidizing when aeration cells, in which product and air mix uniformly, form in the bed.

The benefit of fluidization is that the product dries from all sides, creating a more uniform moisture distribution, reducing the thickness of the vapor boundary layer around the product, and reducing drying time. The final product dries quickly in a small space to a uniform moisture content.

Care must be taken to ensure that the dryer design is suited to the specific product. The size, shape, and cohesiveness of the product affect drying. Also, fluidized bed dryers are susceptible to dead pockets, areas where insufficient air is supplied to fluidize. The stagnant product collects, heats, and becomes a contamination or fire hazard.

Large particles can often be successfully fluidized by entrainment with finer particles, reducing the average particle size to a range where fluidization works effectively.

Fluidized bed dryers are increasingly finding application as high technology dryers in the food industry. They can be operated as batch or continuous dryers. Although more expensive than conventional dryers, these dryers are high throughput units with excellent final product uniformity.

Spouted Bed Dryers

These dryers operate on the same principle as a fluidized bed dryer, except that only a central core of product is fluidized; the remaining product forms an annular region around the central air spout. Product entrained in the spout is heated rapidly before falling (the fountain region) back into the annular region. This region is also aerated, so the heated product continues to dry as it falls through the annulus, before being guided back to the air spout.

Spouted bed dryers are potentially more energy efficient than fluidized bed dryers, but in practice problems with scale-up from design prototypes to full-scale units have limited their application to the food industry.

A simple modification of the unit allows liquids to be dried. The drying chamber is first filled with inert spheres (for example nylon) and the liquid product sprayed into the chamber. The liquid coats the spheres, dries rapidly, and then breaks off due to friction between the spheres. The resulting powder leaves with the air spout and can be collected in a cyclone separator.

Flash Dryers

Technically, the phrase *flash dryer* refers to the use of pressure reduction (partial vacuum), creating an evaporative drying effect owing to the reduced partial pressure of water vapor around the product.

Multistage Dryers

Drying is essentially a slow process, limited by the rate of moisture diffusion from the center of the product to the outside. For this reason, many belt dryers are arranged in multiple stages, the air conditions at each stage being chosen to give the best drying effect and least quality degradation. Many vegetable dryers are multistage, with high temperatures used in the early stages and low temperatures used to finish the drying process. The product may take three to nine hours to pass through the dryer, depending on the type of product and the degree of drying required. The stages may be arranged in series, stretching out to a long continuous dryer, or to save floor space, the belts may be positioned above each other so that product tumbles from one belt to the next. As it falls to the next belt, fresh product surfaces are exposed, accelerating the drying process.

Column Dryers

Column dryers are suited to granular products. The central drying chamber is fed continuously with wet product from elevators or buffer bins. The product falls slowly through the dryer as dried product is re-moved from the base. Hot air is introduced into the product mass through vents or a central air column. The purpose of the column arrangement is to save floor space.

ANALYSIS OF DRYERS

MOISTURE AND HEAT BALANCES

Analysis of a kiln dryer provides a suitable starting point for dryer analysis, as the simplest dryer configuration is to have air enter the drying chamber, interact with the product, then exit the chamber.

Assume that the inlet air conditions remain constant. Then, equating the moisture change of the product to the difference between the inlet and outlet air moisture content gives

$$\dot{m}_a(H_I - H_E) = m_p \frac{dM}{dt} \qquad 2.20$$

where m_a is the airflow rate, H_I and H_E are the inlet and outlet absolute humidities, m_p is the mass of product in the dryer, and dM/dt is the drying rate (dry basis), which is negative for drying.

The psychrometric chart (Fig. 2.7) shows how to measure the required absolute humidities graphically.

Exercise 4

Using the psychrometric chart (Fig. 2.7) below,

1. Describe what happens to the product.
2. Describe what happens to the air.

Note that H_E will be somewhere between H_I and H_B, the exact position depending on factors such as the airspeed and drying time.

In this diagram, line *AI* represents heating ambient air in the furnace, and line *IE* represents the air picking up moisture from the product and exiting the dryer. Line *PB* is the thermal equilibration of the product with the inlet air, and *BI* is the product drying on the tray. As it dries, points *B* and *E* will move towards the inlet air.

Secondly, we can conduct a heat balance across the kiln dryer.

$$\dot{m}_a(h_I - h_E) = m_p c_w \frac{dM}{dt} \qquad 2.21$$

where h_I and h_E are the inlet and exit air enthalpies (see Eq. 2.22), and c_w is the specific heat of water in

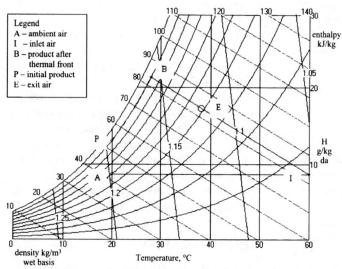

Figure 2.7. Representation of a kiln dryer on a psychrometric chart.

the product (about 4.2 kJ/kg·K). This equation is valid for the drying regions only, as it assumes the product temperature changes slowly. Substituting from Equation 2.19:

$$h_E - h_I = c_w(H_E - H_I) \qquad 2.22$$

Since absolute humidities are of the order of 10^{-2} for moisture in air, the enthalpy difference must also be very small. For this reason, drying can be considered a constant enthalpy process. This can be proven more generally as follows.

From the first law of thermodynamics,

$$dQ = dU + PdV$$

for the drying air, where dQ is the heat loss from the dryer, dU is the change in internal energy of the air, P is air pressure and dV is volume change of the air. Assuming (as before) that heat losses from the dryer are negligible, $dQ = 0$:

$$dU + PdV = 0$$

From the thermodynamic definition of enthalpy of a gas:

$$h = U + PV$$

Differentiating:

$$dh = dU + PdV + VdP$$

Combining equations and noting that since a dryer is open to the atmosphere, pressure differences through a conventional dryer are small ($dP = 0$):

$$dh = 0$$

Thus, conventional dryers are isoenthalpic. For this reason, enthalpy lines on the psychrometric chart may be used to represent drying.

Problem

A cabinet dryer circulates 30 kg/min. dry air across a stack of 10 drying trays. Each tray holds 1 kg of product at 95% wb. If the ambient air is at 20°C and 8 g/kg dry air, and the exit air is at 50°C and 20 g/kg dry air, estimate

1. The rate of moisture removal from the trays if the rate of internal air recirculation is 90%,
2. The heater size for the unit, and
3. The time to dry in the constant rate period to 45% wb.

Solution

1. The amount of air exiting the dryer is

10% of 30 kg/min.= 3 kg/min.

Thus, the rate of moisture removal is

3 (kg/min.) × (20 − 8)/1000 × 60 (min/hr) = 2.16 kg/h

2. To find the size of heater required, solve the air mixture equations to find the thermal difference between the air before the heater and the air leaving the heater.

If I mix 90% of the exit air with 10% of the inlet air, then:

$$T_m = 90\% \times 50 + 10\% \times 20 = 47°C$$

$$H_m = 90\% \times 20 + 10\% \times 8 = 18.8 \text{ g/kg}$$

where T_m and H_m are the mixture point temperature and absolute humidity. The humidity term neglects a small mass associated with cross product terms in H and T.

From a psychrometric chart, the enthalpy of air at 47°C and 18.8 g/kg is 96 kJ/kg dry air. Drawing an enthalpy line through the exit air conditions, locate the inlet conditions with the same enthalpy as the exit air (102.3 kJ/kg) and same absolute humidity as the mixture air (18.8 g/kg). The difference in enthalpy between these two points is $(102.3 - 96) = 6.3$ kJ/kg, and this heat must come from the air heater. Thus, the required heater size is:

$$6.3 \text{ (kJ/kg)} \times 30 \text{ (kg/min.)} \times (1/60) \text{ (min/sec)} = 3.15 \text{ kW}$$

(This assumes dryer heat losses are negligible.)

Compare the resulting drying path on a psychrometric chart with that of the kiln dryer.

3. The rate of moisture removal is

$$30 \text{ kg/min.} \times (20 - 18.8)/1000 = 0.036 \text{ kg/min.}$$

Amount of moisture to remove from each tray is

$$1 \text{ kg} \times (1 - 0.95) \times (0.95/0.05 - 0.45/0.55) = 0.91 \text{ kg}$$

The time required to remove 0.91 kg moisture from 10 trays is

$$t = 10 \times 0.91/(0.036) = 252.5 \text{ minutes} = 4 \text{ hour } 12 \text{ minutes}$$

So after about four hours, the product will enter the falling rate period.

In conclusion, many dryers have been developed for the food industry, using a range of physical principles and technologies, more than can be adequately covered in an introductory text. Using heat and mass balances and the chemistry of air/vapor mixes, many dryers can be analyzed, resulting in better prediction of drying times and a better understanding of the basic processes involved.

BIBLIOGRAPHY

Crank J. 1956. The Mathematics of Diffusion. Oxford University Press.

Fellows PJ. 2000. Chapter 15, Food Processing Technology Principles and Practise, 2nd edition. CRC/Woodhead.

Fennema OR. 1996. Chapter 2, Food Chemistry, 3rd edition. Marcel Dekker.

Heldman DR. 1975. Chapter 6, Food Process Engineering. Avi .

Holman JP. 1992. Chapter 6, Heat Transfer, 7th edition. McGraw Hill.

Perry RH, D Green. 1984. Chapters 12 and 20, Perry's Chemical Engineers' Handbook, 6th (50th anniversary) edition. McGraw Hill.

Singh RP, DR Heldman. 1984. Chapters 8 and 9, Introduction to Food Engineering. Academic Press.

Toledo RT. 1991. Chapters 3, 4, 5, and 12. Fundamentals of Food Process Engineering, 2nd edition. Van Nostrand Reinhold/Avi.

3
Fermented Product Manufacturing

W.-K. Nip

INTRODUCTION

The availability of fermented foods has a long history among the different cultures. Acceptability of fermented foods also differs because of cultural habits. A product highly acceptable in one culture may not be so acceptable by consumers in another culture. The number of fermented food products is countless. Manufacturing processes of fermented products vary considerably owing to variables such as food group, form, and characteristics of final products; kind of ingredients used; and cultural diversity. It is beyond the scope of this chapter to address all the manufacturing processes used to produce fermented foods. Instead, this chapter is organized to address fermented food products based on food groups such as dairy, meat, cereal, soy, and vegetables. Within each food group, manufacturing processes of typical products are addressed. This chapter is only an introduction to manufacturing processes for selected fermented food products. Readers should consult the references below and other available literature for detailed information.

The information in this chapter has been derived from documents copyrighted and published by Science Technology System, West Sacramento, California, ©2003. Used with permission.

FERMENTED DAIRY PRODUCTS

INGREDIENTS AND KINDS OF PRODUCTS

Fermented dairy products are commonly produced in milk-producing countries and by nomadic peoples. These products are highly acceptable in these cultures. They have been gradually accepted by other cultures because of cultural exchange. It is generally accepted that most fermented dairy products were first discovered and developed by nomadic peoples. The production of a fermented dairy product nowadays can be a highly sophisticated process. However, the production of another fermented dairy product can still be conducted in a fairly primitive manner in another location. The quality of a fermented dairy product varies due to the milk, microorganisms, and other ingredients used in the manufacturing process. Many factors affect the gross composition of milk (Early 1998, Jenness 1988, Kosikowski and Mistry 1997, Spreer 1998, Waltsra et al. 1999). The factors most significant to the processing of milk products are breed, feed, season, region, and herb health. Reviews of animal milks are available in the literature. Table 3.1 lists the approximate composition of cow's milk (Early 1998, Jenness 1988, Kosikowski and Mistry 1997, Robinson 1986, Spreer 1998, Waltsra et al. 1999). In industrial countries, milk composition is standardized to meet a country's requirements. However, it is understood that the requirements in one country may not be the same as those in another; thus, the composition may vary for the same product. International agreements to standardize some products are now available. However, products produced in different locations still can vary because of microorganisms and culturing practices used in their production.

Fermented dairy products can be grossly divided into three big categories: cheeses, yogurts, and fermented liquid milks. Within each of these categories, there are subcategories. Table 3.2 presents examples for each of these categories (Early 1998, Jenness 1988, Kosikowski and Mistry 1997, Robinson 1986, Spreer 1998, Waltsra et al. 1999).

In the manufacturing of fermented dairy products, various ingredients such as the milk itself, microorganism(s), coagulants, salt, sugar, vitamins, buffering salts, bleaching (decolorizing) agents, dyes (coloring agents), flavoring compounds, stabilizers, and emulsifiers may be used. The use of these ingredients in fermented liquid milks, yogurts, and natural and processed cheeses are summarized in Table 3.3

(Early 1998, Jenness 1988, Kosikowski and Mistry 1997, Robinson 1986, Spreer 1998, Waltsra et al. 1999).

Various microorganisms such as lactic acid bacteria, yeasts, and molds are used in the manufacturing of fermented dairy products to produce the various characteristics in these products. Table 3.4 lists some of the more common dairy microorganisms and their uses in fermented liquid dairy products, yogurts, and cheeses (Davies and Law 1984, Jay 1996, Robinson 1990).

Cultures of the different microorganisms are available in various forms, such as liquid, frozen, or freeze-dried. Examples of their usage in the manufacturing of fermented dairy products are listed in Table 3.5 (Early 1998, Jenness 1988, Kosikowski and Mistry 1997, Robinson 1986, Spreer 1998, Waltsra et al. 1999).

Because the starter cultures are available in various forms, the preparation steps for these cultures, before inoculation, are different. Table 3.6 lists some of the preparation procedures used in the industry for different forms of starter cultures (Early 1998, Jenness 1988, Kosikowski and Mistry 1997, Robinson 1986, Spreer 1998, Waltsra et al. 1999).

Different microorganisms have different temperature requirements for their optimum growth and functioning. Some fermented dairy products, such as mold-ripened cheeses, may require more than one microorganism to complete the manufacturing process. These molds function best during the long ripening period and therefore have standard incubation temperatures in the refrigerated range. This is also true for some cheeses that require long ripening periods. Microorganisms requiring higher incubation temperatures are used in the production of fermented liquid milks that require only a short incubation time. Table 3.7 lists some of the dairy microorganisms used in some products and their incubation temperatures (Davies and Law 1984, Emmons 2000, Jay 1996, Law 1997, Nath 1993, Robinson 1990, Scott et al. 1998, Specialist Cheesemakers Association 1997).

CHEESES

Cheeses can be classified into different categories based on their moisture, the way the milk is processed, and the types of microorganisms used for the ripening process (Table 3.8) (Early 1998, Jenness 1988, Robinson 1986).

In the processing of cheese, the amount of curd

used for each block of cheese to be made differs considerably, resulting in different block weights (Table 3.9) (Early 1998, Jenness 1988, Kosikowski and Mistry 1997, Law 1997, Nath 1993, Robinson 1986, Scott et al. 1998, Specialist Cheesemakers Association 1997). Harder cheeses have much larger blocks than the soft cheeses. This may be due to the ease of handling after ripening.

Cheeses are packaged in different forms to satisfy consumer consumption patterns and, to some extent, to be compatible with the way the cheese is ripened and for marketing purposes. The various packaging materials are selected to protect the cheeses in a sanitary condition, extend shelf life, and delay the deterioration of the final products. Table 3.10 lists some of the requirements of cheese packaging materials (Early 1998, Emmons 2000, Kosikowski and Mistry 1997, Nath 1993, Robinson 1986, Scott et al. 1998, Specialist Cheesemakers Association 1997, Spreer 1998, Waltsra et al. 1999).

All cheeses produced must be coagulated from acceptable milk to form curd, followed by removal of the whey. Most cheeses are made from standardized and pasteurized milk. Nonpasteurized milk is also used in some exceptional cases, provided they do not carry pathogens. The majority of cheeses are made from cow's milk. Milks from other animals are also used for specialty products. The coagulation process is conducted through the addition of coagulant (rennin or chymosin) and incubation of appropriate lactic acid bacteria in milk to produce enough acid and appropriate pH for curdling the milk. After the casein is recovered, it is salted and subjected to fermentation, with or without inoculation with other microorganisms to produce the desirable characteristics of the various cheeses. Variations in the different manufacturing steps thus produce a wide variety of cheeses with various characteristics. Table 3.11 summarizes the basic steps in the cheese manufacturing process (Davies and Law 1984; Early 1998; Jay 1996; Jenness 1988; Kosikowski and Mistry 1997; Nath 1993; Robinson 1986, 1990; Scott et al. 1998; Specialist Cheesemakers Association 1997; Spreer 1998; Waltsra et al. 1999). Table 3.12 summarizes the ripening conditions for various cheeses. Selected examples are introduced below to provide an overview of the complexity of cheese manufacturing (Davies and Law 1984; Early 1998; Jay 1996; Jenness 1988; Kosikowski and Mistry 1997; Nath 1993; Robinson 1986, 1990; Robinson and Tamime 1991; Scott et al. 1998; Specialist Cheesemakers Association 1997; Spreer 1998; Waltsra et al. 1999).

Cottage Cheese Manufacturing

Cottage cheese is a product with very mild fermentation treatment. It is produced by incubating (fermenting) the standardized and pasteurized skim milk with the starter lactic acid bacteria to produce enough acid and appropriate pH for the curdling of milk. The curd is then recovered and washed, followed by optional salting and creaming. The product is then packed and ready for marketing. No further ripening is required for this product. This is different from most fermented cheeses that require a ripening process. Table 3.13 lists the various steps involved in the production of cottage cheese (Early 1998, Kosikowski and Mistry 1997, Nath 1993, Robinson 1986, Scott et al. 1998, Spreer 1998, Waltsra et al. 1999).

Cheddar Cheese Manufacturing

Cheddar cheese is a common hard cheese without eyes used in the fast-food industry and in the household. Its production process is characterized by a requirement for milling and cheddaring of the curd. This cheese can be ripened with a wax rind or rindless (sealed under vacuum in plastic bags.) It is also categorized into regular, mild, or sharp based on the aging period (45–360 days). The longer the aging period, the sharper the flavor. It is packaged as a large block or in slices. Table 3.14 lists the basic steps in the manufacturing of cheddar cheese (Early 1998, Kosikowski and Mistry 1997, Nath 1993, Robinson 1986, Scott et al. 1998, Spreer 1998, Waltsra et al. 1999).

Swiss Cheese Manufacturing

Swiss cheese is also a common cheese used in the fast-food industry and in the household. It is characterized by having irregular eyes inside the cheese. These eyes are produced by *Propionicbacterium freudenreichii* subsp. *shermanii,* which produces gases trapped inside the block of cheese during fermentation and ripening. A cheese with eyes like Swiss cheese has become the icon for cheese in graphics. Swiss cheese is also characterized by its propionic acid odor. The salting process for Swiss cheese utilizes both the dry- and brine-salting processes. Like cheddar cheese, it can be categorized into regular, mild, and sharp, depending on the length of the curing process. Table 3.15 lists the basic steps in the manufacture of Swiss cheese

(Early 1998, Kosikowski and Mistry 1997, Nath 1993, Robinson 1986, Scott et al. 1998, Spreer 1998, Waltsra et al. 1999).

Blue Cheese

Blue cheese is characterized by its strong flavor and by blue mold filaments from *Penicillium roqueforti* inside the cheese. It is commonly consumed as cheese or made into a salad dressing. In the manufacturing of blue cheese, as in that of Swiss cheese, salting is accomplished by the application of dry-salting and brining processes. It is characterized by a cream-bleaching step to show off the blue mold filament with a lighter background and by needling the block of curd so that the mold can spread its filaments inside the block. It also has a soft and crummy texture due to the needling process and to the gravity draining procedure used to drain the curd. The curing period of two to four months is shorter than for hard cheeses. Its shelf life of two months is also shorter than that of its harder counterparts. Table 3.16 lists the basic steps in the manufacture of blue cheese (Early 1998, Kosikowski and Mistry 1997, Nath 1993, Robinson 1986, Scott et al. 1998, Spreer 1998, Waltsra et al. 1999).

American Style Camembert Cheese

American style Camembert cheese is categorized as a soft cheese. It is characterized by a shell of mold filament on the surface produced by *Penicillium camembertii*. Brie cheese is a similar product. Addition of annatto color is optional. Like blue cheese, it is gravity drained. Therefore it has a soft, smooth texture. This cheese is surface salted and has a total curing period of three weeks before distribution. It is usually cut into wedges and wrapped individually for direct consumption. Table 3.17 lists the basic steps in the manufacture of American style Camembert cheese (Early 1998, Kosikowski and Mistry 1997, Nath 1993, Robinson 1986, Scott et al. 1998, Spreer 1998, Waltsra et al. 1999).

Feta Cheese Manufacturing

Feta cheese is a common cheese in the Mediterranean countries. It is a soft cheese characterized by its brine curing (maturation) process, which is not common in cheese making. Instead, it has a similarity to the manufacture of sufu (Chinese fermented tofu, see below in this chapter). Like other soft cheese,

the curing period is only two to three months. Table 3.18 lists the basic steps in the manufacture of Feta cheese (Robinson and Tamime 1991).

YOGURT

Yogurt can be considered as a curdled milk product. Plain yogurt is yogurt without added flavor, stabilizer, or coagulant. Its acceptance is limited to those who really enjoy eating it. With the development of technology, other forms of yogurt, such as flavored and sweetened yogurt, stirred yogurt, yogurt drinks, and frozen yogurt, are now available. Its popularity varies from location to location. It is considered as a health food when active or live cultures are added to the final product. Table 3.19 lists the basic steps involved in the manufacture of yogurt. Table 3.3, presented earlier, should also be consulted for reference to other ingredients (Chandan and Shahani 1993, Tamime and Robinson 1999).

Most commercially produced yogurt and its products contain sweeteners, stabilizers, or gums (Table 3.20); fruit pieces; natural and synthetic flavors (Table 3.21); and coloring compound (Table 3.22) (Chandan and Shahani 1993, Tamime and Robinson 1999).

Different countries also have different standards on the percent fat and percent solids-not-fat (SNF) contents in their yogurt products (Table 3.23) (Chandan and Shahani 1993, Tamime and Robinson 1999).

The different variables described above make the situation complicated. The term "yogurt" in one country may not have the same meaning in another country. This creates difficulties for international trade. Consensus or agreement among countries, and proper labeling are needed to identify the products properly.

FERMENTED LIQUID MILKS

In milk-producing countries, it is common to have fermented milk products. These products were first discovered or developed by accident. Later, the process was modified for commercial production. Fermented liquid milks are similar to plain yogurt drinks. It is basically milk that has gone through an acid and or alcoholic fermentation. The final product is maintained in the liquid form rather than in the usual soft-gel form of yogurt. There are different fermented liquid milks available, but only sour milk, kefir, and acidophilus milk are discussed below.

Readers should refer to the references listed below and other available literature on related products.

Sour Milk Manufacturing

Table 3.24 presents the basic steps in the manufacturing of the most basic fermented liquid milk, sour milk. The milk is standardized, pasteurized, inoculated, incubated, homogenized, and packaged. It is a very straightforward procedure compared to those for the other two products, kefir and acidophilus milk (Davies and Law 1984, Early 1998, Jay 1996, Jenness 1988, Kosikowski and Mistry 1997, Robinson 1990, Spreer 1998, Waltsra et al. 1999).

Kefir Manufacturing

Kefir is a fermented liquid milk product characterized by the small amount of alcohol it contains and its inoculant, the kefir grains. It is a common product in the Eastern European countries and is considered to have health benefits. Among all the fermented dairy products, only this and similar products contain small amounts of alcohol. Also, in all the other fermented dairy products, pure cultures of bacteria, yeasts, and/or molds are used, but in kefir, the kefir grains are used and recycled. Kefir grains are masses of bacteria, yeasts, polysaccharides, and other products of bacterial metabolism, together with curds of milk protein. Production of kefir is a two-step process: (1) the production of mother kefir and (2) the production of the kefir drink. Table 3.25 lists the basic steps in kefir manufacturing (Davies and Law 1984; Early 1998; Farnworth 1999; Jay 1996; Jenness 1988; Kosikowski and Mistry 1997; Robinson 1986, 1990; Spreer 1998; Waltsra et al. 1999).

ACIDOPHILUS MILK

Acidophilus milk is considered to have probiotic benefits. Like yogurt, it is advertised as having live cultures of *Lactobacillus acidophilus* and *Bifidobacterium bifidum* (optional). These live cultures are claimed to provide the benefit of maintaining a healthy intestinal microflora. Traditional acidophilus milk has a considerable amount of lactic acid and is considered to be too sour for the regular consumers in some locations. Therefore, a small amount of sugar is added to the final product to make it more palatable. This later product is called sweet acidophilus milk. Table 3.26 lists the basic

steps in the manufacture of acidophilus milk (Davies and Law 1984; Early 1998; Jay 1996; Jenness 1988; Kosikowski and Mistry 1997; Robinson 1986, 1990; Spreer 1998; Waltsra et al. 1999).

MEAT PRODUCTS

INGREDIENTS AND TYPES

Fermented meat products such as ham and sausages have been available to different cultures for centuries. It is interesting to learn that the ways these products are produced are basically very similar in different cultures. Besides the meat, nitrite and salt, and sugar (optional), pure cultures are sometimes used, especially in fermented sausages. Microorganisms do not merely provide the characteristic flavor for the products; the lactic acid bacteria also produce lactic and other acids that can lower the pH of the products. Pure cultures are sometimes used in hams to lower the pH and thus inhibit the growth of *Clostridium botulinum.* The raw meat for ham manufacturing is basically a large chunk of meat, and it is difficult for microorganisms to penetrate into the center, unless they are injected into the interior. Microbial growth is mainly on the surface, and the microbial enzymes are gradually diffused into the center. By contrast, in sausages the cultures, if used, are mixed with the ingredients at the beginning, and the fermentation is carried out without difficulty. Besides, sausages are much smaller than hams. Table 3.27 lists some of the ingredients used in the manufacture of hams and sausages (Cassens 1990, Hammes et al. 1990, Huang and Nip 2001, Incze 1998, Roca and Incze 1990, Skrokki 1998, Toldra et al. 2001, Townsend and Olsen 1987, Xiong et al. 1999).

HAMS

Hams, as indicated earlier, are made from large chunks of meat. Western cultures manufacture ham using either a dry cure and/or a brine cure process, sometimes followed by a smoking process. Tables 3.28 and 3.29 list the basic steps involved with the dry cure and brine cure of hams, respectively. These two processes are similar except for the salting step (Cassens 1990, Townsend and Olsen 1987).

Chinese hams are basically manufactured using a dry curing process. Procedures differ slightly, depending on the regions where the hams are made. The most famous Chinese ham is the Jinghua ham made in central China. Yunan ham, from southern China, also has a good reputation. In the old days, without refrigeration facilities during processing, transportation, and storage, it is believed that the ham completed its aging process during the transportation and storage stages. Today, with controlled temperature and relative humidity rooms, the hams are produced under controlled conditions. Table 3.30 lists the current process used in China for Jinghua ham (Huang and Nip 2001, Xiong et al. 1999).

SAUSAGES

Many European-type sausages are manufactured using a fermentation process. These sausages have their own characteristic flavors due to the formulations and curing processes used. It is not the intent of this chapter to list the various formulations. Readers should consult the references in this chapter and other references available elsewhere. Commercial inocula are available. Bacteria and some yeasts grow inside the sausage during the ripening period, producing the characteristic flavor. Molds can grow on the surface during storage if sausages are not properly packaged and stored in the refrigerator. Because these sausages are not sterilized, fermentation is an on-going process, and the aged sausages carry a stronger flavor. Table 3.31 lists the basic steps in the manufacture of dry fermented sausages (Hammes et al. 1990, Incze 1998, Roca and Incze 1990, Toldra et al. 2001).

FERMENTED CEREAL PRODUCTS (BREADS AND RELATED PRODUCTS)

KINDS OF PRODUCTS AND INGREDIENTS

In wheat-producing countries or areas, baked yeast bread is a major staple in people's diets. This is common in the major developed countries. In other countries, other forms of bread may be the major staple. Baked bread may come in different forms such as regular yeast breads, flat breads, and specialty breads. Today, even retarded (chilled or frozen) doughs are available to meet consumers' preference for a semblance of home-cooked food. For countries or areas with less available energy, other forms of bread such as steamed bread and boiled breads are available. Fried breads are consumed mainly as breakfast or snack items. Table

3.32 lists some examples of different types of breads (Cauvain and Young 1998, Groff and Steinbaecher 1995, Huang 1999, Pyler 1988, Quail 1998, Qarooni 1996).

Today, as a result of centuries of breeding selection, there are different types of wheat available to suit production environments in various regions with diverse climatic conditions. Wheat used for making bread is hard wheat, soft wheat, or a combination of both to meet product specifications. Wheat kernels are milled with removal of the bran and germ and further processed into wheat flour. Traditionally, this flour is the major ingredient for baking bread. For some health conscious consumers, whole wheat flour is the flour of choice for making bread nowadays. Wheat bran is also added to increase the fiber content of the product. Table 3.33 lists the proximate composition of wheat and some of their common wheat products (Cauvain and Young 1998, Groff and Steinbaecher 1995, Pyler 1988).

In the manufacture of various wheat-based breads and related products, the major ingredients are wheat flour, yeast, sourdough bacteria (optional), salt, and water. Other ingredients vary considerably with the types of products produced. These may be grossly classified as optional ingredients, additives, or processing aids. Each country has its own regulations and requirements. Table 3.34 lists basic ingredients, optional ingredients, additives, and processing aids used in the manufacturing of bread and related products (Cauvain and Young 1998, Groff and Steinbaecher 1995, Pyler 1988).

REGULAR BREAD

Table 3.35 lists the basic steps in bread manufacturing (Cauvain and Young 1998, Groff and Steinbaecher 1995, Pyler 1988,).

There are three basic processes in commercial bread making: straight dough process, sponge-and-dough process, and continuous-baking process. The process to be used is determined by the manufacturer and the equipment available in the baking plant. Table 3.36 lists the basic steps in the different processes. The major difference is in the way the dough is prepared and handled (Cauvain and Young 1998, Groff and Steinbaecher 1995, Pyler 1988).

Because the dough may be prepared in various ways, the amounts of ingredients used differ accordingly. Table 3.37 lists two formulations, comparing the differences in ingredients that arise from differences in the dough preparation processes (Cauvain and Young 1998, Groff and Steinbaecher 1995, Pyler 1988).

RETARDED DOUGH

As indicated earlier, retarded dough is also available to some consumers. This type of dough is more accessible where refrigerators and freezers are more common. Dough is prepared so that the fermentation is carefully controlled, and the dough is packed inside the container. Storage of this package is also carefully controlled. When the package is open, consumers can just follow the instructions on the package to bake their own bread. The technology is proprietary to the manufacturers, but there are some guidelines available (Table 3.38) (Cauvain and Young 1998, Groff and Steinbaecher 1995, Pyler 1988).

FLAT (LAYERED) BREAD

Flat bread is a general term for bread products that do not rise to the same extent as regular bread. Flat breads are common commodities in Middle Eastern countries and in countries or areas with less accessible energy. In developed countries, flat breads are considered specialty breads. The making of the dough is similar to that of regular bread. But, the dough is flattened and sometimes layered before it is baked directly inside the hearth or in an oven. Table 3.39 lists the general production scheme for flat breads (Qarooni 1996, Quail 1998).

CROISSANTS AND DANISH PASTRIES

Croissants and Danish pastries can be considered as products that result from modifications of the basic bread making process. The dough preparation steps are similar, but the ingredients are different. Table 3.40 compares the ingredients used in making croissants and Danish pastries. From this table, it is clear that even within each group, the ingredient formulation can vary considerably, producing a wide variety of products available in the market (Cauvain and Young 1998, Groff and Steinbaecher 1995, Pyler 1988).

STEAMED BREAD (MANTOU)

Steamed bread is common in the Chinese community. Plain steamed bread is consumed as the major staple in the northern provinces of China. However, stuffed steamed breads are consumed as specialty items in various parts of China. Manufacture of steamed bread differs from that of regular bread mainly in the dough solidification process. Regular bread uses a baking process, whereas in steamed bread, steaming is used instead of baking. Consequently, in steamed bread, there is no brown crust on the bread surface because the temperature used is not high enough to cause the browning reaction. Steamed bread is always consumed hot or held in a steamer because the bread is soft at this temperature. Sometimes the bread is deep-fried before consumption. Steamed bread hardens when it cools down, making it less palatable. Various procedures are available for the production of steamed bread. Table 3.41 lists the basic steps in steamed bread processing in China (Huang 1999).

FERMENTED SOY PRODUCTS

KINDS OF PRODUCTS AND INGREDIENTS

Soybeans have been available to the Chinese for centuries, and various fermented soy products were developed and spread to neighboring countries. These countries further developed their own fermented soy products. Soy sauce originating in China probably is the most famous and widely accepted fermented soy product. The credit for this wide acceptance also goes to the Kikkoman Company from Japan, which has helped spread soy sauce worldwide through their marketing strategy. Fermented whole soybeans such as ordinary natto, salted soybeans (e.g., Japanese Hama-natto and Chinese douchi), and tempe (Indonesia); fermented soy pastes (e.g., Japanese miso and Chinese dou-pan-chiang); and fermented tofus (e.g., sufu and stinky tofu or chao-tofu of Chinese origin) are more acceptable to ethnic groups. Consumers worldwide are gradually accepting these products through cultural exchange activities. The manufacturing of these products varies widely. Table 3.42 summarizes the ingredients needed for the manufacture of common fermented soy products (Ebine 1986; FK Liu 1986; KS Liu 1997, 1999; Steinkraus 1996; Sugiyama 1986; Teng et al. 2004; Winarno 1986; Yoneya 2003).

SOY SAUCE

There are many types of soy sauce, depending on the ratio of ingredients (wheat and soybeans), the fermentation and extraction procedures, and the flavoring ingredients (caramel and others) used. However, the procedures for manufacturing are similar. Basically, soy sauce is made by fermenting cooked soybeans in salt or brine under controlled conditions to hydrolyze the soy proteins and starches into smaller flavoring components. The soy sauce is then extracted from the fermented soybeans for standardization and packaging. Table 3.43 lists a generalized scheme for the manufacture of soy sauce. More detailed information is presented in references listed in this chapter and available literature elsewhere (Ebine 1986; FK Liu 1986; KS Liu 1997, 1999; Sugiyama 1986; Yoneya 2003).

FERMENTED WHOLE SOYBEANS

Ordinary (Itohiki) Natto

Ordinary natto is a typical Japanese fermented whole soybean product. The sticky mucilaginous substance on the surface of soybeans is its characteristic. It is produced by a brief fermentation of cooked soybeans with *Bacillus natto,* and it has a short shelf life. Table 3.44 lists the basic steps in the manufacture of ordinary natto. For detailed information on ordinary natto, please refer to the references in this chapter (KS Liu 1997, 1999; Yoneya 2003).

Hama-natto and Dou-chi

Hama-natto is fermented whole soybeans produced in the Hama-matsu area of Japan. Similar products are produced in Japan, prefixed with different names taken from the production location. A very similar product in the Chinese culture is "tou-chi" or "dou-chi." It is produced by fermenting the cooked soybeans in salt, brine, or soy sauce and then drying them as individual beans. Hama-natto includes ginger in its flavoring, whereas the inclusion of ginger flavoring is optional in dou-chi. Table 3.45 lists the basic steps in the production of Hama-natto and dou-chi. For further information, readers should refer to the references in this chapter and other available literature (FK Liu 1986; KS Liu 1997, 1999; Yoneya 2003).

FERMENTED SOY PASTES

Both the Chinese and Japanese have fermented soy pastes available in their cultures, and they are made in a similar manner. However, the usage of these two products is quite different. The Japanese use their fermented soy paste, miso, in making miso soup, and to a lesser extent, for example, in marinating/flavoring of fish. Miso soup is common in traditional Japanese meals. The Chinese use their fermented soy paste, dou-pan-chiang, mainly as condiment in food preparation. Dou-pan-chiang can also be made from wing beans, and this is beyond the scope of this chapter. Table 3.46 lists the basic steps in the manufacture of miso. For detailed information on miso and dou-pan-chiang, readers should consult the references for this chapter and other literature available elsewhere (Ebine 1986; FK Liu 1986; KS Liu 1997, 1999; Steinkraus 1996; Sugiyama 1986; Yoneya 2003).

FERMENTED TOFU

Sufu (Fermented Soy Cheese)

Sufu, or fermented soy cheese, is made by fermenting tofu that is made by coagulating the soy protein in soy milk with calcium and/or magnesium sulfate. It is similar to Feta cheese in its fermentation process. Both products are matured in brine in sealed containers. Some packed sufu contains flavoring ingredients. Table 3.47 lists the basic steps in the manufacture of sufu. For detailed information, readers should refer to the list of references in this chapter and the other available literature (FK Liu 1986; KS Liu 1997, 1999; Teng et al. 2004).

Stinky Tofu

Stinky tofu is a traditional Chinese food made by fermenting tofu briefly in "stinky brine." The tofu is hydrolyzed slightly during this brief fermentation and develops its characteristic flavoring compounds. When this raw stinky tofu is deep-fried, these compounds volatilize and produce the characteristic stinky odor, thus the name "stinky tofu." It is usually consumed with chili and soy sauces. Stinky tofu is also steamed with condiments for consumption. Table 3.48 lists the basic steps in the manufacture of stinky tofu. Readers should consult the references in this chapter for further reading (FK Liu 1986; KS Liu 1997, 1999; Teng et al. 2004).

TEMPE (TEMPEH)

Tempe is a traditional Indonesian food consumed commonly by its people. It is made by fermenting cooked soybeans wrapped in wilted banana leaves or plastic wraps. The mold *Rhizopus oligosporus* produces its mycelia, and these mycelia penetrate into the block of soybeans. The mold mycelia also surround the block. This kind of fermentation is similar to molded cheese fermentation. Tempe is gradually being accepted by vegetarians in the West as a nutritious and healthy food. It is generally consumed as a deep-fried product. Table 3.49 lists the basic steps in the production of tempe (KS Liu 1997, 1999; Winarno 1986; Yoneya 2003).

FERMENTED VEGETABLES

KINDS OF PRODUCTS AND INGREDIENTS

Fermented vegetables were produced in different cultures in the old days to preserve the harvested vegetables when they are not available or due to climatic limitations. Some of these products started as traditional cultural foods but became widely accepted in other cultures. It is interesting that most of these processes are similar. Salt is used in the production of the product or the salt stock. Natural lactic acid fermentation, to produce enough lactic acid to lower product pH, is the major microbial activity in these processes. With the amount of salt added and lactic acid produced, these two ingredients create an environment that can inhibit the growth of spoilage microorganisms. Available leafy vegetables, fruits (commonly used as vegetables), and roots are used as the raw materials. Starter cultures are used occasionally. Vinegar is used in some products. Chili pepper and other spices are used in many products. Preservatives may also be used to extend shelf life after the package is opened. Table 3.50 compares the ingredients used in different fermented vegetable products (Anonymous 1991, Beck 1991, Brady 1994, Chiou 2003, Desroiser 1977, Duncan 1987, Fleming et al. 1984, Hang 2003, Lee 2003, Park and Cheigh 2003).

SAUERKRAUT

The term sauerkraut literally means sour *(sauer)* cabbage *(kraut)*. It is a traditional German fermented vegetable product that has spread to other cultures; it is used on its own or in food preparations. Its sequential growth of lactic acid bacteria has long been recognized. Each lactic acid bacterium dominates the fermentation until its end product becomes inhibitory for its own development and creates another environment suitable for an-

other lactic acid bacterium to take over. The fermentation continues until most of the available fermentable sugars are exhausted. The production of sauerkraut is not risk-free and sanitary: precautions must be taken to avoid spoilage. Table 3.51 presents the basic steps in sauerkraut processing (Anonymous 1991, Desrosier 1977, Fleming et al. 1984, Hang 2003).

PICKLES

Western-style pickles are produced by salting the pickling cucumbers in vats in salt stocks for long-term storage, followed by desalting, and bottling in sugar and vinegar, with or without spices. The fermentation is still lactic acid fermentation. However, it is more susceptible to spoilage because air may be trapped inside the slightly wax-coated cucumbers. In the salt curing of cucumbers, spoilage can occur, and precautions should be taken to avoid its occurrence. Because of their high acidity and low pH as well as their high salt content, the products are generally mildly heat-treated to sterilize or pasteurize them. Table 3.52 lists the basic steps in the production of Western-style pickles (Anonymous 1991, Beck 1991, Brady 1994, Desrosier 1977, Duncan 1987, Fleming et al. 1984).

KIMCHI

Kimchi is a traditional Korean fermented vegetable. Most kimchi is characterized by its hot taste because of the fairly high amount of chili pepper used in the product and its visibility. However, some kimchis are made without chili pepper, but with garlic and ginger as well as other vegetables and ingredients. Vegetables used in making kimchi vary with its formulation: Chinese cabbage, cucumber, and large turnip are more common. Either chili pepper, or garlic and ginger can be used to provide a hot sensation. Other ingredients may also be added to provide a typical flavor. The fermentation is still lactic acid fermentation. Traditionally, kimchi was made in every household in rural areas in Korea to provide vegetables for the winter, when other fresh vegetables were not readily available. Today, it is a big industry in Korea, and kimchi is available year-round. Even small kimchi refrigerators are now available to meet the demands of consumers living in cities. In other parts of the world where Koreans are residents, kimchi is available either as a household item or as a commercial product. Kimchi is usually not heat sterilized after packaging in jars. Pasteurization is optional. Kimchi is considered perishable and is stored refrigerated. Table 3.53 lists the basic steps in the manufacture of kimchi (Lee 2003, Park and Chiegh 2003).

CHINESE PICKLED VEGETABLES

The Chinese also manufacture a wide range of pickled vegetables. Various kinds of vegetables are used as raw materials. The fermentation can be either a dry-salting or a brining process, depending on the product to be manufactured. However, the fermentation is still lactic acid fermentation. The major difference between Chinese-style pickled vegetable products and Western-style pickles is that desalting is usually not practiced in the manufacture of Chinese-style pickled vegetables. The desalting process is left to the consumers, if needed. Also, some Chinese-style vegetables are made into intermediate moisture products that are not produced in their Western-style counterparts. Table 3.54 lists some of the basic steps in the manufacturing of selected Chinese pickled vegetables (Chiou 2003, Lee 2003).

APPLICATION OF BIOTECHNOLOGY IN THE MANUFACTURING OF FERMENTED FOODS

With the advances in biotechnology, microorganisms with special characteristics for the manufacturing of fermented foods have become available. The most significant example is the approval by the FDA of Chy-Max (chymosin produced by genetic manipulation) used in the production of cheese. Its availability greatly reduces the reliance on chymosin from young calves and produces economic savings.

Other products with similar or other properties are also available in the market. Genetically modified lactic acid bacteria and yeasts used in fermented food production are also available nowadays to reduce production costs. Gradual acceptance by consumers is the key to the further development and success of biotechnology (Barrett et al. 1999, Early 1998, Geisen and Holzapfel 1996, Henriksen et al. 1999, Jay 1996, Kosikowski and Mistry 1997, Scott et al. 1998, Spreer 1998, Walstra et al. 1999). Readers should refer to the references in this chapter and other references available for further information.

PROCESS MECHANIZATION IN THE MANUFACTURE OF FERMENTED FOODS

Fermented foods produced by traditional methods are labor intensive and rely a great deal on the experience of the manufacturers. The main drawback is product inconsistency. In most developed countries, products such as many cheeses, yogurts, breads, sausages, and soy sauce are now made by highly mechanized processes to standardize the products (Prasad 1989, Hamada et al. 1991, Iwasaki et al. 1992, Caudill 1993, Dairy and Food Industries Supply Association 1993, Gilmore and Shell 1993, Muramatsu et al. 1993, Luh 1995, Kamel and Stauffer 1993, Belderok 2000). This not only provides product consistency, but also reduces production costs. Consumers benefit from these developments. However, some consumers, even in developed countries, still prefer the traditional products, even at an increased cost, because of their unique product characteristics. There are also fermented products that are still made by traditional or semimechanized processes because mechanization processes have not been developed for them.

Table 3.1. Approximate Composition of Cow's Milk

Components	Average Content in Milk (% w/w)	Range (% w/w)	Average Content in Dry Matter (% w/w)
Water	87.1	85.3–88.7	
Solid-not-fat	8.9	7.9–10.0	69
Fat in dry matter	31	22–38	31
Lactose	4.6	3.8–5.3	36
Fat	4	2.5–5.5	31
Protein	3.25	2.3–4.4	25
Caesin	2.6	1.7–3.5	20
Mineral substances	0.7	0.57–0.83	5.4
Organic acids	0.17	0.12–0.21	1.3
Miscellaneous	0.15		1.2

Sources: Early 1998, Jenness 1988, Koshikowski and Mistry 1997, Robinson 1986, Spreer 1998, Walstra et al. 1999.

Table 3.2. Kinds of Fermented Dairy Products with Examples

Kinds	Examples
Fermented liquid milks	
Lactic fermentation	Buttermilk, Acidophillus,
With alcohol and lactic acid	Kefir, Koumiss
With mold and lactic acid	Villi
Concentrated	Ymer, Skyr, Chakka
Yogurts	
Viscous/liquid	Yogurt
Semisolid	Strained yogurt
Solid	Soft/hard frozen yogurt
Powder	Dried yogurt
Cheeses	
Extra hard	Parmesan, Romano, Sbrinz
Hard with eyes	Emmeental, Gruyere, Swiss
Hard without eyes	Cheddar, Chester, Provolone
Semi-hard	Gouda, Edam, Caerphilly
Semi-hard, internally mold ripened	Rouquefort, Blue, Gorgonzola
Semisoft, surface ripened with bacteria	Limburger, Brick, Munster
Soft, surface mold ripened	Brie, Camembert, Neufchatel
Soft, unripened	Cream, Mozzarella, USA–Cottage

Sources: Early 1998, Jenness 1988, Kosikowsi and Mistry 1997, Robinson 1986, Spreer 1998, Walstra et al. 1999.

Table 3.3. Ingredients for Fermented Dairy Food Production

Ingredients	Fermented Liquid Milk Products	Yogurt	Processed Natural Cheese	Cheese Products
Milk				
Raw	Optional	Optional	Optional	Optional
Standardized (fat and milk solids)	Preferred	Preferred	Preferred	Preferred
Milk powders	Optional	Optional	Optional	Optional
Microorganisms				
Starter bacteria	Required	Required	Required	Required
Mold	Optional	Optional	Optional	Optional
Yeast	Optional	Optional	Optional	Optional
Genetically modified microorganisms	Optional	Optional	Optional	Optional
Coagulant				
Rennet	Preferred	Preferred	Preferred	Preferred
Acid	Optional	Optional	Optional	Optional
Microbial protease(s)	Optional	Optional	Optional	Optional
Common salt (sodium chloride)	No	No	Required	Required
Sugar	Optional	Optional	No	No
Vitamins	Preferred	Preferred	Preferred	Preferred
Buffering salts (calcium chloride hydroxide phosphates, sodium or potassium phosphates)	Optional	Optional	Optional	Optional
Bleaching (decolorizing) agents	No	No	Optional	Optional
Antimicrobial agents	Optional	Optional	No	Preferred
Dyes (coloring agents)	No	No	Optional	Optional
Flavoring compounds (fruits, spices spice oils, fruits, fruit flavors, artificial smoke)	Optional	Optional	Optional	Optional
Stabilizers	No	Preferred	No	Preferred
Emulsifiers	Optional	Optional	No	Preferred

Sources: Early 1998, Jenness 1988, Kosikowski and Mistry 1997, Robinson 1986, Spreer 1998, Walstra et al. 1999.

Table 3.4. Some Common Organisms Used in Fermented Milk Products

Microorganisms	Buttermilk	Cream	Fermented Milk	Yogurt	Kefir	Cheese
Bifidobacterium bifidum			X	X		X
Enterococcus durans						X
Enterococcus faecalis						X
Geotrichum candidum						X
Lactobacillus acidophilus			X			
Lactobacillus casei						X
Lactobacillus delbrueckii subsp. *bulgaricus*	X	X				X
Lactobacillus heleveticus						X
Lactobacillus kefir					X	
Lactobacillus lactis						X
Lactobacillus lactis biovar.*diacetylactis*		X				X
Lactobacillus lactis subsp. *cremoris*	X	X				X
Lactobacillus lactis subsp. *lactis*						X
Lactobacillus lactis var. *hollandicus*						X
Leuconostoc mesenteroidis subsp. *cremoris*						X
Leuconostoc mesenteroides subsp. *dextranicum*						X
Propionibacterium freudenreichii subsp. *shermanii*						X
Penicillium camberberti						X
Penicillium glaucum						X
Penicillium roqueforti						X
Streptococcus thermophilus				X		X

Sources: Davies and Law 1984, Jay 1996, Robinson 1990.

Table 3.5. Dairy Starter Cultures

Physical Form	Usage
Liquid cultures in skim milk or whole milk (antibiotic free)	For inoculation of intermediate cultures
Liquid culture—frozen	For inoculation of intermediate cultures
	For inoculation into bulk cultures
Dried culture—from normal liquid culture	For inoculation of intermediate culture
Spray dried cultures	For inoculation into bulk cultures
	For direct-to-vat inoculation.
Frozen cultures in special media (frozen at $-40°C$)	For inoculation into bulk cultures
	For direct-to-vat inoculation
Frozen concentrated culture (in sealed containers at $-196°C$)	For inoculation into bulk cultures
	For direct-to-vat inoculation
Single strain lypholized cultures (in foil sachets with known activity)	For inoculation into bulk cultures
	For direct-to-vat inoculation

Sources: Early 1998, Jenness 1988, Kosikowski and Mistry 1997, Robinson 1986, Spreer 1998, Walstra et al. 1999.

Table 3.6. Types of Starter Cultures and Their Preparation Prior to Usage

Kinds	Preparation Steps	Timing
Regular starter culture	Preparation of starter culture blanks	8:00 a.m
	Storing milk blanks	11:00 a.m.
	Activating lypholized culture powder	3:00 p.m.
	Daily mother culture preparation	3:00 p.m.
	Semibulk and bulk starter preparation	3:00 p.m.
Frozen culture and bulk starter application	Store frozen culture at −40°C or less	
	Warm to 31°C and use directly	
Reconsitituted milk or whey-based starter	Reconstitution	8:00 a.m.
	Heating and tempering	8:30 a.m
	Inoculating and incubating	10:00 a.m.
Bulk starter from ultrafiltrated milk	Ultrafiltration	1:00 p.m.
	Heating and tempering	3:30 p.m.
	Inoculating and incubating	5:00 p.m.

Sources: Early 1998, Jenness 1988, Kosikowski and Mistry 1997, Robinson 1986, Spreer 1998, Walstra et al. 1999.

Table 3.7. Temperature Requirements and Acid Production for Some Dairy Microbes

Microorganisms	Product Group[a]	Standard Temperature for Incubation, °C	General Maximun Titratable Acidity Produced in Milk, %
Bacteria			
Bifidobacterium bifidum	1, 2	36–38	0.9–1.0
Lactobacillus acidophilus	1, 2	38–44	1.2–2.0
Lactobacillus delbrueckii subsp. *bulgarius*	1	43–47	2.0–4.0
Lactobacillus lactics subsp. *cremoris*	2	22	0.9–1.0
Lactobacillus subsp. *lactis*	2	22	0.9–1.0
Leuconostoc mesenteroides subsp. *cremoris*	2	20	0.1–0.3
Streptococcus durans	2	31	0.9–1.1
Streptococcus thermophilus	2	38–44	0.9–1.1
Molds			
Penicillium roqueforti	3	11–16	NA
Penicillium camerberti	3	10–22	NA

Sources: Davies and Law 1984, Emmons 2000, Jay 1996, Law 1997, Nath 1993, Robinson 1990, Scott et al. 1998, Specialist Cheesemakers Association 1997.

[a]Product group: 1 = yogurt, 2 = fermented liquid milk, 3 = cheese.

Table 3.8. Classification of Cheese According to Moisture Content, Scald Temperature, and Method of Ripening

Hard cheese (moisture 20–42%; fat in dry matter, 32–50%, minimum)

Low Scald, Lactic Starter	Medium Scald, Lactic Starter	High Scald, Propionic Eyes	Plastic Curd, Lactic Starter or Propionic Eyes
Gouda	Cheddar	Parmesan	Provolone
Cheshire	Svecia	Beaufort	Mozzarella

Semi-hard cheese (moisture 45–55%; fat in dry matter, 40–50%, minimium)

Lactic Starter	Smear Coat	Blue-veined Mold
St. Paul	Limburg	Roquefort
Lanchester	Munster	Danablue

Soft cheese (moisture >55%; fat in dry matter, 4–51%, minimum)

Acid-Coagulated	Smear Coat or Surface Mold	Surface Mold	Normal Lactic Starter	Unripened Fresh
Cottage cheese (USA)	Brie	Camembert	Quarg	Cottage (UK)
Quesco-Blanco	Bel Paese	Neufchatel	Petit Suisse	York

Sources: Early 1998, Jenness 1988, Robinson 1986.

Table 3.9. Approximate Weight of Cheese Block for Various Cheese Varieties

Cheese Variety	Approximate Weight (kg)
Hard to semi-hard or semisoft	
Wensleydale	3–5
Caerphilly	3–6
White Stilton	4–8
Single Gloucester	10–12
Leichester	13–18
Derby	14–16
Sage Derby	14–16
Cheddar	18–28
Cheshire	20–22
Dunlap	20–27
Double Gloucester	22–28
Lancashire	22
Internally mold-ripened (blue-veined) cheese	
Blue Wensleydale	3–5
Blue Vinney	5–7
Blue Stilton	6–8
Blue Cheshire	10–20
Soft cheese	
Colwich	0.25–0.50
Cambridge	0.25–1.0
Melbury	2.5

Sources: Early 1998, Jenness 1988, Kosikowski and Mistry 1997, Law 1997, Nath 1993, Robinson 1986, Scott et al. 1998, Specialist Cheesemakers Association 1997, .

Table 3.10. Requirements of Cheese Packaging Materials

Low permeability to oxygen, carbon dioxide, and water vapor
Strength and thickness of film
Stability under cold or warm conditions
Stability to fats and lactic acid
Resistance to light, especially ultraviolet
Ease of application, stiffness, elasticity
Ability to seal and accept adhesives
Laminated films to retain laminated
Low shrinkage or aging unless shrinkage is a requisite
Ability to take printed matter
Should not impart odors to the cheese
Suitability for mechanization of packaging
Hygienic considerations in storage and use
Cost effectiveness as a protective wrapping

Sources: Early 1998, Emmons 2000, Nath 1993, Kosikowski and Mistry 1997, Robinson 1986, Scott et al. 1998, Specialist Cheesemakers Association 1997, Spreer 1998, Walstra et al. 1999.

Table 3.11. Basic Cheese Making Steps

Standardize cheese milks.
Homogenize cheese milks.
Heat-treat or pasteurize cheese milks.
Add starter.
Add color and additives.
Coagulation/curdling:
  Cut coagulum/curd.
  Stir and scald.
  Wash curd cheese.
Salt cheese.
Press cheese.
Coat, bandage, and wrap cheese.
Let cheese ripen.
Package for retail.
Store.

Sources: Davies and Law 1984; Early 1998; Jay 1996; Jenness 1988; Kosikowski and Mistry 1997; Nath 1993; Robinson 1986, 1990; Scott et al. 1998; Specialist Cheesemakers Association 1997; Spreer 1998; Walstra et al. 1999.

Table 3.12. Cheese Ripening Conditions

Types of Cheese	Storage Period (days)	Temperature (°C)	Relative Humidity (%)
Soft	12–30	10–14	90–95
Mold ripened	15–60	4–12	85–95
Cooked, e.g., Emmental			
Cold room	7–25	10–15	80–85
Warm room	25–60	18–25	80–85
Hard, e.g., Cheddar	45–360	5–12	87–95

Sources: Davies and Law 1984, Robinson 1986, Jenness 1988, Robinson 1990, Robinson and Tamimie 1991, Nath 1993, Jay 1996, Kosikowski and Mistry 1997, Specialist Cheesemakers Assocciation 1997, Early 1998, Scott et al. 1998, Spreer 1998, Walstra et al. 1999.

Table 3.13. Basic Steps in Making Cottage Cheese

Standardize skim milk.

Pasteurize milk with standard procedure and cool to 32°C.

Inoculate with active lactic starter, add rennet, and set curd:

Rennet addition—at 2 ml single strength (prediluted, 1:40) per 1000 kg milk within 30 minutes of starter addition

Specifications	Short Set	Medium Set	Long Set
Starter concentration	5%	3%	0.5%
Temperature of milk set	32°C	27°C	22°C
Time from setting to cutting	5 hr	8 hr	14–16 hr

Final pH and whey titratable acidity—4.6 and 0.52%, respectively.

Cut curd with 1.3, 1.6, or 1.9 cm wire cheese knife.

Cook curd:

Let curd cubes stand for 15–30 minutes and cook to 51–54°C at 1.7°C per 10 minutes.

Roll the curds gently every 10 minutes after initial 15–30 minute wait.

Test curd firmness and hold 10–30 minutes longer to obtain proper firmness.

Wash curd:

First wash with 29°C water temperature

Second wash with 16°C water temperature

Third wash with 4°C water temperature

Drain washed curd (by gravity) for about 2.5 hours.

Salt and cream at 152 kg creaming mixture per 454 kg with final 0.5–0.75% salt content and 4% fat content (varies with products and optional).

Package in containers.

Store at refrigerated temperature.

Sources: Early 1998, Kosikowski and Mistry 1997, Nath 1993, Robinson 1986, Scott et al. 1998, Spreer 1998, Walstra et al. 1999.

Table 3.14. Basic Steps in Making Cheddar Cheese

Standardize cheese milk.

Homogenize milk.

Pasteurization and additional heating of milk.

Cool milk to 31°C.

Inoculate milk with lactic starter (0.5–2% active mesophilic lactic starter).

Add rennet or other protease(s)—198 ml single strength (1:15,000) rennet per 1000 kg milk. Dilute the measured rennet 1:40 before use. Agitate at medium speed.

Set the milk to proper acidity—25 minutes.

Cut the curd using 0.64 cm or wider wire knife. Stir for 5 minutes at slow speed.

Cook the curd at 38°C for 30 minutes with 1°C for every 5 minute increment. Maintain temperature for another 4–5 minutes and agitate periodically at medium speed.

Drain the curd at 38°C.

Cheddar the curd at pH 5.2–5.3.

Mill the curd slabs.

Salt the curd at 2.3–3.5 kg salt per 100 kg curd in three portions in 30 minutes.

Waxed cheddar cheese:

Hoop and press at 172 kPa for 30–60 seconds then 172–344 kPa overnight.

Dry the cheese at 13°C at 70% RH for 2–3 days.

Paraffin the whole cheese at 118°C for 6 seconds.

Rindless cheddar cheese:

Press at 276 kPa for 6–18 hours.

Prepress for 1 minute, followed by 45 minutes under 686 mm vacuum.

Remove and press at 345 kPa for 60 minutes.

Remove and vacuum seal in bags with hot water shrinkage at 93°C for 2 seconds.

Ripen at 85% RH at 4°C for 60 days or longer, up to 9–12 months, or at 3°C for 2 months then 10°C for 4–7 months, up to 6–9 months.

Sources: Early 1998, Kosikowski and Mistry 1997, Nath 1993, Robinson 1986, Scott et al. 1998, Spreer 1998, Walstra et al. 1999.

Table 3.15. Basic Steps in Making Swiss Cheese

Standardize cheese milk to 3% milk fat—treatment with H_2O_2-catalase optional.

Pasteurize the milk.

Inoculate with starters:

Streptococcus thermophilus, 330 ml per 1000 kg milk

Lactobacillus delbruechii subsp. *bulgaricus,* 330 ml per 1000 kg milk

Propionibacterium freudenreichii subsp. *shermanii,* 55 ml per 1000 kg milk

Add rennet, 10–20 minutes after inoculation—154 ml single-strength (1:15,000) rennet extract per 1000 kg milk, prediluted 1:40 with tap water before addition. Stir for 3 minutes.

Let milk set (coagulate) for 25–30 minutes.

Cut the curd with 0.64 wire knife; let stand undisturbed for 5 minutes; stir at medium speed for 40 minutes.

Cook the curd slowly to 50–53°C for about 30 minutes and stir at medium speed, then turn off steam and continue stirring for 30–60 minutes with pH reaching 6.3–6.4.

Allow the curd to drip for 30 minutes.

Press the curd—with preliminary pressing, then at 69 kPa overnight.

Salt the curd:

First salting—in 23% salt brine for 2–3 days at 10°C

Second salting—at 10–16°C, 90% RH. Wipe the cheese surface from the brine soaking, then sprinkle salt over cheese surface daily for 10–14 days

Third salting—at 20–24°C, 80–85% RH. Wash cheese surface with salt water and sprinkle with dry salt 2–3 times weekly for 2–3 weeks

Rinded block Swiss cheese:

Cure—at 7°C or lower (USA) or 10–25°C (Europe) for 4–12 months.

Package in container and store at cool temperature.

Rindless block Swiss cheese:

Wrap or vacuum pack the blocks.

Cure stacked cheese at 3–4°C for 3–6 weeks.

Store at cool temperature.

Sources: Early 1998, Kosikowski and Mistry 1997, Nath 1993, Robinson 1986, Scott et al. 1998, Spreer 1998, Walstra et al. 1999.

Table 3.16. Basic Steps in Making Blue Cheese

Milk preparation:
  Separate cream and skim milk.
  Pasteurize skim milk by HTST, cool to 30°C.
  Bleach cream with benzoyl peroxide (optional) and heat to 63°C for 30 seconds.
  Homogenize hot cream at 6–9 mPa and then 3.5 mPa, cool, and mix with pasteurized skim milk.
Inoculate milk at 30°C with 0.5% active lactic starter. Let stand for 1 hour.
Add rennet—158 ml single strength (prediluted 1:40) per 1000 kg milk. Mix well.
Let coagulate or set, 30 minutes.
Cut curd with 1.6 cm standard wire knife.
Cook curd at 30°C, let stand 5 minutes, and then agitate every 5 minutes for 1 hour. Whey should have 0.11 to 0.14 titratable acidity.
Drain whey by gravity for 15 minutes.
Inoculate with *Penicillium roqueforti* spores—2 kg coarse salt and 28 g *P. requeforti* spore powder per 100 kg curd followed by thorough mixing. Add food grade lipase (optional).
Salting:
  First salting—dip the curd in 23% brine for 15 minutes, then press or mold at 22°C, turning every 15 minutes for 2 hours and every 90 minutes for rest of day.
  Second salting—salt cheese surface every day for 5 days at 16°C, 85% RH.
  Final dry salting or brine salting in 23% brine for 24–48 hours. Final salt concentration about 4%.
Incubate for 6 days at 16°C, 95% RH. Wax and needle air holes or vacuum pack and needle air holes.
Mold filament development in air holes at 16°C for 6–8 days.
Cure at 11°C and 95% RH for 60–120 days.
Cleaning and storing:
  Strip off the wax or vacuum packaging bag.
  Clean cheese, dry, and repack in aluminum foil or vacuum packaging bags.
  Store at 2°C.
Product shelf life—2 months.

Sources: Early 1998, Kosikowski and Mistry 1997, Nath 1993, Robinson 1986, Scott et al. 1998, Spreer 1998, Walstra et al. 1999.

Table 3.17. Basic Steps in Making American Style Camembert Cheese

Standardize milk.
Homogenize milk.
Pasteurize milk at 72°C for 6 seconds.
Cool milk to 32°C.
Inoculate with 2% active lactic starter followed by 15–30 minutes acid ripening to 0.22% titratable acidity.
Add annatto color at 15.4 ml per 1000 kg milk (optional).
Add rennet —220 ml single-strength (prediluted 1:40) rennet per 1000 ml, then mix for 3 minutes and let stand for 45 minutes.
Cut curd with 1.6 cm standard wire knife.
Cook curd at 32°C for 15 minutes with medium speed stirring.
Drain curd at 22°C for 6 hours with occasional turning.
Inoculate with *Penicillium camerberti* spores by spray gun on both sides of cheese once.
Press and mold curd by pressing for 5–6 hours at 22°C without any weight on surface.
Surface salt cheese; let cheese stand for about 9 hours.
Cure—at 10°C, 95% RH for 5 days undisturbed, then turn once and continue curing for 14 days.
Packaging, storage, and distribution:
  Wrap cheese and store at 10°C, 95–98% RH for another 7 days.
  Move to cold room at 4°C and cut into wedges, if required, and rewrap.
  Distribute immediately.

Sources: Early 1998, Kosikowski and Mistry 1997, Nath 1993, Robinson 1986, Scott et al. 1998, Spreer 1998, Walstra et al. 1999.

Table 3.18. Basic Steps in Making Feta Cheese

Standardize milk with 5% fat, enzyme treated and decolorized.

Homogenize milk.

Pasteurize by standard procedure and cool to 32°C.

Inoculate with 2% active lactic starter as cheddar cheese and allow to ripen for 1 hour.

Add rennet at 198 ml single-strength rennet (prediluted, 1:40) per 1000 kg milk and let set for 30–40 minutes.

Cut the curd with 1.6 cm standard wire knife and let stand 15–20 minutes.

Allow curd to drip for 18–20 hour at 12–18 kg on 2000 cm^2, with pH and titratable acidity developed to 4.6 and 0.55%, respectively.

Prepare cheese blocks of 13 × 13 × 10 cm each.

Salt in 23% salt brine for 1 day at 10°C.

Can and box cheese blocks in 14% salt brine (sealed container).

Cure for 2–3 months at 10°C.

Soak cured cheese in skim milk for 1–2 days before consumption to reduce salt.

Yield—15 kg/100 kg of 5% fat milk.

Source: Robinson and Tamime 1991.

Table 3.19. Basic Steps in the Production of Yogurt

Standardize liquid milk.

Homogenize liquid milk.

Heat-treat or pasteurize liquid milk at 90°C for 5 minutes or equivalent.

Cool pasteurized milk to 1–2°C above inoculation temperature.

Add starter (inoculation), 1–3% operational culture.

Add flavor, sweetener, gums, and/or color (optional).

Incubate at 40–45°C for 2.5–3.0 hours for standard cultures.

Break curd (optional).

Cool to 15–20°C in 1–1.5 hours.

Add live culture (optional).

Package.

Store at ≤ 10°C.

Sources: Chandan and Shahani 1993, Tamime and Robinson 1999.

Table 3.20. Some Common Gums that Could Be Used in Yogurt Manufacturing

Kind	Name of Gum
Natural	Agar
	Alginates
	Carageeenan
	Carob gum
	Corn starch
	Casein
	Furcelleran
	Gelatin
	Gum arabic
	Guar gum
	Karaya gum
	Pectins
	Soy protein
	Tragacanth gum
	Wheat starch
Modified gums	Cellulose derivatives
	Dextran
	Low-methoxy pectin
	Modified starches
	Pregelatinized starches
	Propylene glycole alginate
	Xanthin
Synthetic gums	Polyethylene derivatives
	Polyvinyl derivatives

Sources: Chandan and Shahani 1993, Tamime and Robinson 1999.

Table 3.21. Some Common Flavors for Yogurt

Retail Flavor	Natural Characteristic— Impact Compound	Synthetic Flavoring Compound Available
Apricot	NA	g-Undecalactone
Banana	3-Methylbutyl acetate	NA
Bilberry	NA	NA
Blackcurrant	NA	*trans-* and *cis- p-*Methane-8-thiol-3-one
Grape, Concord	Methyl antranilate	NA
Lemon	Citral	15 compounds
Peach	g-Decalactone	g-Undecalactone
Pineapple	NA	Allyl hexanoate
Raspberry	1-p-Hydroxyphenyl-3-butanone	NA
Strawberry	NA	Ethyl-3-methyl-3-phenylglycidate

Sources: Chandan and Shahani 1993, Tamime and Robinson 1999.

Table 3.22. Permitted Yogurt Colorings

Name of Color	Maximum Level (mg /kg)
Intigotine	6
Brilliant black PN	12
Sunset yellow FCF	12
Tartrazine	18
Cochineal	20
Carminic acid	20
Erythrosine	27
Red 2G	30
Ponceau	48
Caramel	150

Sources: Chandan and Shahani 1993, Tamime and Robinson 1999.

Table 3.23. Existing or Proposed Standards for Commercial Yogurt Composition [% Fat and % Solid-not-fat (SNF)] in Selected Countries

Country	% Fat			% SNF
	Low	Medium	Normal	
Australia	NA	0.5–1.5	3	NA
France	0.5	NA	3	NA
Italy	1	NA	3	NA
Netherlands	1	NA	3	NA
New Zealand	0.3	NA	3.2	NA
UK	0.3	1.0–2.0	3.5	8.5
USA	0.5–1.0	2	3.25	8.5
West Germany	0.5	1.5–1.8	3.5	8.25–8.5
FAO/WHO	0.5	0.5–3.0	3	8.2
Range	0.3–1.0	0.5–3.0	3–3.5	8.2–8.5

Sources: Chandan and Shahani 1993, Tamime and Robinson 1999.

Table 3.24. Basic Steps in Sour Milk Processing

Standardize milk.

Heat milk to 85–95°C, then homogenize.

Cool milk to 19–25°C and transfer to fermentation tank.

Add 1–2% start culture (inoculation).

Allow shock-free fermentation to pH 4.65–4.55.

Homogenize gel.

Cool to 4–6°C.

Fill bottles, jars, or one-way packs or wholesale packs.

Sources: Davies and Law 1984, Early 1998, Jay 1996, Jenness 1988, Kosikowski and Mistry 1997, Robinson 1990, Spreer 1998, Walstra et al. 1999.

Table 3.25. Basic Steps in Kefir Processing

Preparation of mother "kefir"

Standardize milk for preparation of mother "kefir."

Pasteurize milk at 90–95°C for 15 minutes and cool to 18–22°C.

Spread kefir grains at the bottom of a container (5–10 cm thick) and add pasteurized milk (20–30 times the amount of kefir grains).

Ferment for 18–24 hours, mixing 2–3 times. Kefir grains float to the surface.

Filter out the kefir grains with a fine sieve, wash the grains with water, and save for the next fermentation.

Save the fermented milk for the next-step inoculation.

Preparation of drinkable kefir

Blend fermented milk from above with 8–10 times fresh, pasteurized, untreated milk.

Pour into bottles, then close the bottles and ferment mixture for 1–3 days at 18–22°C. [Another option is to mix the fermented milk with fresh milk at 1–5% and ferment at 20–25°C for 12–15 hours (until pH 4.4–4.5 is reached), then ripen in storage tanks 1–3 days at 10°C. Product is not as traditional but is acceptable.]

Cool to refrigerated temperature.

Store and distribute.

Sources: Davies and Law 1984; Early 1998; Farnworth 1999; Jay 1996; Jenness 1988; Kosikowsiki and Mistry 1997; Robinson 1986, 1990; Spreer 1998; Walstra et al. 1999.

Table 3.26. Basic Steps for Sweet Acidophilus Milk Processing

Procedure 1:

Standardize milk.

Heat milk to 95°C for 60 minutes, cool to 37°C, and hold for 3–4 hours; reheat to 95°C for 10–15 minutes, then cool to 37°C.

Inoculate with 2–5% bulk starter.

Incubate for up to 24 hours or to 1% lactic acid.

Cool to 5°C.

Pack and distribute.

Procedure 2:

Standardize milk.

Homogenize milk at 14.5 mPa.

Heat to 95°C for 60 minutes.

Cool to 37°C.

Inoculate with direct vat inoculation (DVI) starter.

Incubate for 12–16 hours or to about 0.65% lactic acid.

Heat at ultra high temperature (UHT), 140–145°C for 2–3 seconds to eliminate undesirable contaminants.

Cool to 10°C or lower.

Package and distribute.

Sources: Davies and Law 1984; Early 1998; Jay 1996; Jenness 1988; Kosikowski and Mistry 1997; Robinson 1986, 1990; Spreer 1998; Walstra et al. 1999.

Table 3.27. Raw Ingredients for Fermented Meat Products

Ingredient	Ham	Sausage
Meat		
Pork	Yes	Optional
Beef	No	Optional
Casing	No	Yes
Salt	Yes	Yes
Sugar	Optional	Optional
Starter microorganisms	Optional	Optional
Lactobacillus sakei, L. curvatus, L. plantarum, L. pentosus,		
L. pentoaceus		
Pediococcus pentosaceus, P. acidilactic		
Staplyococcus xylosus, S. carnosus		
Kocuria varians		
Debaryomyces hansenit		
Candida famata		
Penicillium nagiovense, P. chrysogenum		
Spices	Optional	Optional
Other flavoring compounds	Optional	Optional
Moisture retention salts	Optional	Optional
Preservatives	No	No

Sources: Cassens 1990, Hammes et al. 1990, Huang and Nip 2001, Incze 1998, Roca and Incze 1990, Skrokki 1998, Toldra et al. 2001, Townsend and Olsen 1987, Xiong et al. 1999.

Table 3.28. Basic Steps in Dry Cured Ham Processing

Prepare pork for dry curing.

Mix the proper ratio of ingredients [salt, sugar, nitrite, and inocula (optional)].

Rub the curing mixture into the meat.

Stack the green ham for initial dry curing at 36–40°C.

Rerub the green ham and stack for additional curing at 36–40°C. [The ham should be left in the cure for the equivalent of 3 days per pound of meat.]

Soak the cured ham for 2–3 hours, then thoroughly scrub.

Place green ham in tight-fitting stockinette and hang in smokehouse to dry overnight.

Smoke at about 60 or 80°C with 60% RH for 12–36 hours.

Cool.

Vacuum pack and place in cool storage.

Sources: Cassens 1990, Townsend and Olsen 1987.

Table 3.29. Basic Steps in Brine Cured Ham Processing

Prepare pork for brine curing.

Mix the proper ratio of ingredients (salt, sugar, and nitrite with inocula optional): 5 gallons of brine for 100 pounds meat.

Soak the meat in the prepared brine, or stitch pump the brine into the meat (10% of the original weight of the meat) followed by soaking in the brine for 3–7 days vacuum tumbling or massaging (optional).

Remove the meat from the cover brine and wash.

Place green ham in tight-fitting stockinette and hang in smokehouse to dry overnight.

Smoke at about 60 or 80°C and 60% RH for 12–36 hours.

Cool.

Vacuum pack and place in cool storage.

Sources: Cassens 1990, Towsend and Olsen 1987.

Table 3.30. Basic Steps in Chinese Jinghua Ham Processing

Select pork hind leg, 5–7.5 kg.
Trim.
Salt, 7–8 kg salt per 10 kg ham.
Stack and overhaul at 0–10°C for 33–40 days.
Wash with cold water and brush.
Dry in the sun for 5–6 days.
Ferment (cure) for 2–3 months at 0–10°C (harmless green mold will develop on surface).
Brush off the mold and trim.
Age for 3–4 months, maximum 9 months; alternate aging process in temperature-programmable room with 60% RH for 1–2 months.
Grade.
Package and distribute.
(Yield: about 55–60%.)

Sources: Huang and Nip 2001, Xiong et al. 1999.

Table 3.31. Basic Steps in Dry (Fermented) Sausage Processing

Select meat for processing.
Chop and mix chopped meat with spices, seasonings, and inocula at temperature of about 10°C.
Stuff the mixture in suitable casings.
Make links.
Cure or dry for 1–3 months in rooms with temperature, relative humidity, and air circulation regulated according to the type of sausage being produced.
Package and place in cool storage.

Sources: Hammes et al. 1990, Incze 1998, Roca and Incze 1990, Toldra et al. 2001.

Table 3.32. Types of Bread and Related Products

Type	Examples
Baked Breads	
Regular yeast breads	Bread (white, whole wheat or multi-grain)
Flat (layered) breads	Pocket bread, croissants
Specialty breads	Sourdough bread, rye bread, hamburger bun, part-baked bread, Danish pastry, stuffed bun
Chilled or frozen doughs	Ready-to-bake doughs, retarded pizza doughs, frozen proved dough
Steamed breads	Chinese steamed bread (mantou), steamed stuffed buns
Fried breads	Doughnuts
Boiled breads	Pretzels

Sources: Cauvain and Young 1998, Groff and Steinbaecher 1995, Huang 1999, Pyler 1988, Qaroni 1996, Quail 1998.

Table 3.33. Composition of Wheat, Flour, and Germ

Material	Moisture %	Protein %	Fat %	Total CHO %	Fiber %	Ash %
Wheat						
Hard red spring	13	14	2.2	69.1	2.3	1.7
Hard red winter	12.5	12.3	1.8	71.7	2.3	1.7
Soft red winter	14	10.2	2	72.1	2.3	1.7
White	11.5	9.4	2	75.4	1.9	1.7
Durum	13	12.7	2.5	70.1	1.8	1.7
Flour, straight						
Hard wheat	12	11.8	1.2	74.5	0.4	0.46
Soft wheat	12	9.7	1	76.9	0.4	0.42
Flour, patent						
Bread	12	11.8	1.1	74.7	0.3	0.44
Germ	11	25.2	10	49.5	2.5	4.3

Sources: Cauvain and Young 1998, Groff and Steinbaecher 1995, Pyler 1988.

Table 3.34. Bread Making—Functional Ingredients

Kind	Examples
Basic ingredient	
Wheat flour	Bread flour, whole wheat flour
Yeast	Compressed yeast, granular yeast, cream yeast, dried yeast, instant yeast, encapsulated yeast, frozen yeast, pizza yeast, deactivated yeast
	Saccharomyces cervisiae, S. carlsburgenis, S. exisguus
Salt	
Water	
Optional ingredients	Whole wheat flour, gluten, soya flour, wheat bran, other cereals or seeds, milk powder, fat, malt flour, egg, dried fruit, vitamins
	Sourdough bacteria:
	Lactobacillus plantarum, L. brevis, L. fermentum, L. sanfrancisco
	Other yeasts
Additives	
Emulsifier	Diacetylated tartaric acid esters of mono- and diglycerides of fatty acids (DATA esters), Sodium stearyl-2-lactylate (SSL), distlled monoglyceride, lecithin
Flour treatment agents	Ascorbic acid, L-cysteine, potassium bromate, potassium iodate, azodicarbonamide
Preservatives	Acetic acid, potassium acetate, sodium diacetate, sorbic acid, potassium sorbate, calcium sorbate, propionic acid, sodium propionate, calcium propionate, potassium propionate
Processing aids	Alpha-amylase, hemicellulose, proteinase, novel enzyme systems (lipases, oxidases, peroxidases)

Sources: Cauvain and Young 1998, Groff and Steinbaecher 1995, Pyler 1988.

Table 3.35. Basic Steps in Regular or Common Bread Making

Prepare basic and optional ingredients.
Prepare yeast or sourdough for inoculation.
Mix proper ingredients to make dough.
Allow to ferment.
Remix dough (optional).
Sheet.
Mold and pan.
Proof in a temperature and relative humidity controlled chamber.
Decoratively cut dough surface (optional).
Bake, steam, fry, or boil.
Cool.
Package.
Store.

Sources: Cauvain and Young 1998, Groff and Steinbaecher 1995, Pyler 1988.

Table 3.36. Various Bread Making Processes

Straight dough baking process:
  Weigh out all ingredients.
  Add all ingredients to mixing bowl.
  Mix to optimum development.
  Allow first fermentation, 100 minutes, room temperature, or at 27°C for 1.5 hours.
  Punch.
  Allow second fermentation, 55 minutes, room temperature, or at 27°C for 1.5 hours.
  Divide.
  Allow intermediate proofing, 25 minutes, 30–35°C, 85% RH
  Mold and pan.
  Allow final proofing, 55 minutes at 30–35°C, 85% RH.
  Bake at 191–232°C for 18–35 minutes to approximately 100°C internal temperature.

Sponge-and-dough baking process:
  Weigh out all ingredients.
  Mix part of flour, part of water, yeast, and yeast food to a loose dough (not developed).
  Ferment 3–5 hours at room temperature, or at 21°C for 12–16 hours.
  Add other ingredients and mix to optimum development.
  Allow fermentation (floor time), 40 minutes.
  Divide.
  Allow intermediate proofing, 20 minutes, 30–35°C, 85% RH, or 27°C for 30 minutes.
  Mold and pan.
  Allow final proofing, 55 minutes, 30–35°C, 85% RH
  Bake at 191–232°C for 18–35 minutes to approximately 100°C internal temperature.

Continuous-baking process:
  Weigh out all ingredients.
  Mix yeast, water, and maybe part of flour to form liquid sponge.
  Add remaining flour and other dry ingredients.
  Mix in dough incorporator.
  Allow fermentation, 2–4 hours, 27°C.
  Pump dough to development chamber.
  Allow dough development under pressure at 80 psi.
  Extrude within 1 minute at 14.5°C and pan.
  Proof for 90 minutes.
  Bake at 191–232°C for 18–35 minutes to approximately 100°C internal temperature.

Sources: Cauvain and Young 1998, Groff and Steinbaecher 1995, Pyler 1988.

Table 3.37. Sample Bread Recipes

White pan bread (bulk fermentation or straight dough process):

Ingredients	Percent of flour weight
Flour	100.0
Yeast	1.0
Salt	2.0
Water	57.0

Optional dough improving ingredients	
Fat	0.7
Soya flour	0.7
Malt flour	0.2

White pan bread (sponge and dough process):

Sponge ingredient	Percent of total flour weight
Flour	25.0
Yeast	0.7
Salt	0.5
Water	14.0

Dough ingredients	Percent of total flour weight
Flour	75.0
Yeast	2.0
Salt	1.5
Water	44.0

Optional improving ingredients	
Fat	0.7
Soya flour	0.7
Malt flour	0.2

Sources: Cauvain and Young 1998, Groff and Steinbaecher 1995, Pyler 1988.

Table 3.38. General Guidelines for Retarded Dough Production

Reduce yeast levels as storage times increase.

Keep yeast levels constant when using separate retarders and provers.

Reduce yeast levels as the dough radius increases.

Reduce yeast levels with higher storage temperatures.

The lower the yeast level used, the longer the proof time will be to a given dough piece volume.

Yeast levels should not normally be less than 50% of the level used in scratch production.

For dough stored below −5°C, the yeast level may need to be increased.

Reduce the storage temperature to reduce expansion and weight loss from all dough pieces.

Lower the yeast levels to reduce expansion and weight losses at all storage temperatures.

Dough pieces of large radius are more susceptible to the effects of storage temperatures.

The lower freezing rate achieved in most retarder-provers, combined with the poor thermal conductivity of dough, can cause quality losses.

Proof dough pieces of large radius at a lower temperature than those of small radius.

Lower the yeast level in the dough to lengthen the final proof time and to help minimize temperature differentials.

Maintain a high relative humidity in proofing to prevent skinning.

Sources: Cauvain and Young 1998, Groff and Steinbaecher 1995, Pyler 1988.

Table 3.39. General Production Scheme for Flat Bread

Ingredient preparation.

Mixing of ingredients (dough formation).

Fermentation.

Dough cutting and rounding.

Extrusion and sheeting (optional).

First proofing.

Flattening and layering.

Second proofing.

Second pressing (optional).

Baking or steaming.

Cooling.

Packaging and distribution.

Sources: Qarooni 1996, Quali 1998.

Table 3.40. Formulations for Croissant and Danish Pastries

Ingredients	Croissant (%)	Danish Pastries (%)
Flour	100	100
Salt	1.8–2.0	1.1–1.56
Water	52–55.4	43.6–52
Yeast (compressed)	4–5.5	6–7.6
Shortening	2–9.7	6.3–12.5
Sugar	2–10	9.2–25
Egg	0–24	5–25
Skimmed milk powder	3–6.5	4–6.25
Laminating margarine/butter	32–57	50–64

Sources: Cauvain and Young 1998, Groff and Steinbaecher 1995, Pyler 1988.

Table 3.41. Basic Steps in Steamed Bread Processing

Selecting flour and ingredients such as milk powder and sugar (optional).

Mixing dough.

Fermentation:

Full fermentation—1–3 hours

Partial fermentation—0.5–1.5 hours

No-time fermentation—0 hours

Remixed fermentation dough—remixing of fully fermented dough with up to 40% of flour by weight.

Neutralizing with 40% sodium bicarbonate and remixing.

Molding.

Proofing at 40°C for 30–40 minutes (no-time dough).

Steaming for about 20 minutes.

Steamed bread is maintained at least warm to preserve quality.

Source: Huang 1999.

Table 3.42. Raw Ingredients for Fermented Soy Products

Ingredient	Soy Sauce	Natto	Soy Nuggets	Soy Paste	Tempe	Soy Cheese	Stinky Tofu
Major ingredients:							
Soy							
Soybean	Yes	Yes	Yes	Optional	Yes	Yes	Yes
Soybean flour	Optional	No	No	Yes	No	Optional	Optional
Salt	Yes	Yes	Yes	Yes	No	Yes	No
Wheat	Optional	No	No	No	No	No	No
Rice flour	No	No	No	Optional	No	No	No
Major microorganism(s):							
Mold							
Aspergillus oryzae	Yes	No	Yes	Yes	No	Optional	No
Aspergillus sojae	No	No	No	Optional	No	No	No
Mucor hiemalis, M. silivaticus	No	No	No	No	No	Yes	No
M. piaini	No	No	No	No	No	Yes	No
Actinomucor elegans	No	No	No	No	No	Yes	No
A. repens, A. taiwanensis	No	No	No	No	No	Yes	No
Rhizopus oligosporus	No	No	No	No	Yes	No	No
R. chinesis var. *chungyuen*	No	No	No	No	No	Yes	No
Bacteria							
Bacillus natto	No	Yes	No	No	No	No	No
Klebsiella pneumoniae	No	No	No	No	Yes	No	No
Bacillus sp.	No	No	No	No	No	No	Yes
Streptococcus sp.	No	No	No	No	No	No	Yes
Enterococcus sp.	No	No	No	No	No	No	Yes
Lactobacillus sp.	No	No	No	No	No	No	Yes
Halophlic yeasts							
Saccharomyces rouxii	Yes	No	Yes	Yes	No	No	No
Torulopsis versatlis	Yes	No	Yes	Yes	No	No	No
Halophilic lactic bacteria							
Pediococcus halophilus	Yes	No	Yes	Yes	No	No	No
Bacillus subtilus	Yes	No	Yes	Yes	No	No	No
Other ingredients:							
Additional flavor added	Optional	No	No	No	No	Optional	No
Preservative added	Optional	No	No	No	No	No	No

Sources: Ebine 1986; FK Liu 1986; KS Liu 1997, 1999; Steinkraus 1996; Sugiyama 1986; Teng et al. 2004; Winarno 1986; Yoneya 2003.

Table 3.43. Production Scheme for Soy Sauce

Select and soak beans.
Cook clean or defatted soybean under pressurized steam at 1.8 kg/cm² for 5 minutes.
Cool cooked bean to 40°C.
Roast and crush wheat.
Mix prepared soybeans and wheat.
Inoculate with *Aspergillus oryzae* or *sojae*.
Incubate mixture to make starter koji at 28–40°C.
Add brine (23% saltwater) to make moromi (mash).
Inoculate with halophilic yeasts and lactic acid bacteria (optional).
Brine fermentation at 15–28°C.
Add saccharified rice koji (optional).
Age moromi (optional).
Separate raw soy sauce by pressing or natural gravity.
Refine soy sauce.
Add preservative and caramel (option).
Package and store.

Sources: Elbine 1986; FK Liu 1986; KS Liu 1997, 1999; Sugiyama 1986; Yoneya 2003.

Table 3.44. Production Scheme for Itohiki (Ordinary) Natto

Start with clean, whole soybeans.
Wash and soak at 21–25°C for 10–30 hours.
Cook soybean under pressurized steam at 1–1.5 kg/cm² for 20–30 minutes.
Drain and cool soybean at 80°C.
Inoculate with *Bacillus natto.*
Mix and package in small packages.
Incubation:
    40–43°C for 12–20 hours, or 38°C for 20 hours plus 5°C for 24 hours.
Final product.
Refrigerate to prolong shelf life.

Sources: KS Liu 1997, 1999; 2003.

Table 3.45. Production Scheme for Soy Nuggets (Hama-natto and Dou-chi)

Start with clean, whole soybeans.
Wash and soak for 3–4 hours at 20°C.
Steam cook soybean at ambient pressure for 5–6 hours or at 0.81.0 kg/cm² for 30–40 minutes.
Drain and cool soybean to 40°C.
Add alum (optional for dou-chi).
Mix with wheat flour (optional for Hama-natto).
Inoculate with *Aspergillus oryzae.*

Procedure 1 (Hama-natto):
    Incubate for 50 hours at 30–33°C.
    Soak inoculated soybean in flavoring solution for 8 months.
    Incubate under slight pressure in closed containers.

Procedure 2 (dou-chi):
    Incubate at 35–40°C for 5 days.
    Wash.
    Incubate for 5–6 days at 35°C.

Remove beans from liquid for drying.
Mix with ginger soaked in soy sauce (Hama-natto only).
Package final product (soy nuggets).
Refrigerate to prolong shelf life (optional).

Sources: FK Liu 1986; KS Liu 1997, 1999; Yoneya 2003.

Table 3.46. Production Scheme of Fermented Soybean Pastes (Miso)

Start with whole, clean soybeans.
Wash and soak at 15°C for 8 hours.
Cook at 121°C for 45–50 minutes or equivalent.
Cool and mash the soybeans.
Prepare soaked, cooked, and cooled rice (optional).
Prepare parched barley (optional).
Inoculate rice or barley with *Aspergillus oryzae* (tane-koji, optional).
Mix koji and rice or barley mixture.
Add salt to koji and rice or barley mixture and mix.
Inoculate halophilic yeasts and lactic acid bacteria (optional).
Pack mixture (mashed soybean and koji) into fermenting vat with 20–21% salt brine.
Ferment at 25–30°C for 50–70 days.
Blend and crush ripened miso.
Add preservative and colorant (optional).
Pasteurize (optional).
Package and store.

Sources: Ebine 1986; FK Liu 1986; Liu 1997, 1999; Steinkraus 1996; Sugiyama 1986; Yoneya 2003.

Table 3.47. Production Scheme for Sufu (Chinese Soy Cheese)

Clean whole soybeans.
Soak.
Grind with water.
Strain through cheesecloth to recover soymilk.
Heat to boiling and then cool.
Coagulate soymilk with calcium and/or magnesium sulfate.
Cool to 50°C.
Press to remove water (formation of tofu).
Sterilize at 100°C for 10 minutes in hot-air oven.
Inoculate with *Mucor, Actinomucor,* and/or *Rhizopus* sp.

Procedure 1:
Incubate in dry form for 2–7 days, depending on inocula.
Incubate (ferment in 25–30% salt brine) for 1 month or longer.
Brine and age in small containers with or without addition of alcohol or other flavoring ingredients.

Procedure 2:
Incubate at 35°C for 7 days until covered with yellow mold.
Pack in closed container with 8% brine and 3% alcohol.
Ferment at room temperature for 6–12 months.

Final product (sufu or Chinese soy cheese).

Sources: FK Liu 1986; KS Liu 1997, 1999; Teng et al. 2004.

Table 3.48. Production Scheme for Stinky Tofu

Clean whole soybeans.
Soak.
Grind with water.
Strain through cheesecloth to recover soymilk.
Heat to boiling and then cool.
Coagulate soymilk with calcium and/or magnesium sulfate.
Cool to 50°C.
Press to remove water (formation of tofu).
Press to remove additional water.
Soak in fermentation liquid for 4–20 hours at 5–30°C.
Fresh stinky tofu, ready for frying or steaming.
Refrigerate to prolong shelf life.

Sources: FK Liu 1986; KS Liu 1997, 1999, Teng et al. 2004.

Table 3.49. Production Scheme for Tempe

Start with whole, clean soybeans.
Rehydrate in hot water at 93°C for 10 minutes.
Dehull.
Soak with or without lactic acid overnight.
Boil for 68 minutes.
Drain and cool to 38°C.
Inoculate with *Rhizopus oligosporus* w/o *Klebsiella pneumonia.*
Incubate on trays at 35–38°C, 75–78 % RH for 18 hours.
Dehydrate.
Wrap.

Sources: KS Liu 1997, 1999; Winarno 1986; Yoneya 2003.

Table 3.50. Raw Ingredients for Fermented Vegetables

Ingredient	Sauerkraut	Western Pickles	Jalapeño Peppers	Kimchi	Oriental Vegetables
Vegetable					
Head cabbage	Yes	No	No	Optional	Optional
Chinese cabbage	No	No	No	Major	Optional
Mustard green	No	No	No	Optional	Optional
Turnip	No	No	No	Optional	Optional
Jalapeño Pepper	No	No	Yes	Optional	Optional
Chili pepper	No	No	No	Yes	Optional
Pickle/cucumber	No	Yes	No	Optional	Optional
Salt	Yes	Yes	Yes	Yes	Yes
Starter culture (lactic acid bacteria)	Optional	Optional	Optional	No	No
Added vinegar	No	Yes	Yes	No	Optional
Added spices	No	Optional	Optional	Optional	Optional
Other added flavors	No	Yes	No	Optional	Optional
Preservative(s)	No	Optional	Optional	Optional	Optional

Sources: Anonymous 1991, Beck 1991, Brady 1994, Chiou 2003, Desroiser 1977, Duncan 1987, Fleming et al. 1984, Hang 2003, Lee 2003, Park and Park 2003.

Table 3.51. Basic Steps in Sauerkraut Processing

Select and trim white head cabbage.
Core and shred head cabbage to 1/8 inch thick.
Salt with 2.25–2.50% salt by weight with thorough mixing.
Store salted cabbage in vats with plastic cover, weighed with water to exclude air in the cabbage.
Ferment at 7–23°C for 2–3 months or longer to achieve an acidity of 2.0% (lactic).
Heat kraut to 73.9°C before filling the cans or jars, then exhaust, seal, and cool.
Store and distribute.

Sources: Brady 1994, Desrosier 1977, Fleming et al. 1984, Hang 2003.

Table 3.52. Basic Steps in Fermented Pickles Processing

Size and clean cucumbers.
Prepare 5 (low salt) or 10% brine (salt stock).
Cure (ferment) cucumbers in brine for 1–6 weeks to 0.7–1.0% acidity (lactic) and pH of 3.4–3.6, dependent on temperature, with salinity maintained at a desirable level (15% for salt stock). Addition of sugar, starter culture, and spices is optional.
Recover pickles from brine, then rinse or desalt (salt stock).
Grade.
Pack pickles into jars filled with vinegar, sugar, spices, and alum, depending on formulation.
Pasteurize at 74°C for 15 minutes, followed by refrigerated storage; exhaust to 74°C at cold point, then seal and cool; or vacuum pack and heat at 74°C (cold point) for 15 minutes, then cool.
Store and distribute.

Sources: Anonymous 1991, Beck 1991, Brady 1994, Desrosier 1977, Duncan 1987, Fleming et al. 1984.

Table 3.53. Basic Steps in Kimchi Processing

Select vegetables (Chinese cabbage, radish, cucumber, or others).
Wash vegetables.
Cut vegetables, if necessary.
Prepare 8–15% brine.
Immerse vegetables in brine for 2–7 hours to achieve 2–4% salt in vegetable.
Rinse and drain briefly.
Add seasoning.
Ferment at 0°C to room temperature for about 3 days.
Package (can also be done before fermentation).
Store at 3–4°C.

Sources: Lee 2003, Park and Cheigh 2003.

Table 3.54. Basic Steps in Fermented Chinese Vegetables

Select and clean vegetables.
Cut vegetables (optional).

Procedure 1:
   Wilt vegetables for 1–2 days to remove moisture.
   Dry salt vegetables in layers with weights on top (5–7.5% salt).
   Ferment for 3–10 days.
   Wash.
   Dry or press fermented vegetables (optional).
   Add spices and flavoring compounds.
   Package.
   Sterilize (optional).

Procedure 2:
   Wilt cut vegetables.
   Rinse fermentation container in hot water.
   Fill the container with cut vegetables.
   Add 2–3% brine and other flavoring compounds (optional).
   Ferment at 20–25°C for 2–3 days.
   Ready for direct consumption or packaging and cool storage.

Sources: Chiou 2003, Lee 2003.

REFERENCES

Anonymous. 1991. Let's Preserve Pickles. Cooperative Extension Service, Purdue University, West Lafayette, Ind.

Barrett FM, AL Kelly, PLH McSweeney, PF Fox. 1999. Use of exogenous urokinase to accelerate proteolysis in Cheddar cheese during ripening. International Dairy Journal 9(7): 421–427.

Beck P. 1991. Making Pickled Products. Extension Service, North Dakota State University, Fargo.

Belderok B. 2000. Developments in bread making processes. Plant Foods for Human Nutrition 55(1): 1–86.

Brady PL. 1994. Making Brined Pickles and Sauerkraut. Cooperative Extension Service, University of Arkansas, Little Rock.

Cassens RG. 1990. Nitrite-cured meat: A food safety issue in perspective. Trumbull, Conn.: Food and Nutrition Press.

Caudill V. 1993. Engineering: Plant design, processing, and packaging. In: YH Hui, editor. Dairy Science and Technology Handbook, vol. 3, 295–329. New York: VCH Publishers, Inc.

Cauvain SP, LS Young. 1998. Technology of Breadmaking. London: Blackie Academic and Professional.

Chandan RC , KM Shahani. 1993. Yogurt. In: YH Hui, editor. Dairy Science and Technology Handbook, vol. 2, 1–56. New York: VCH Publishers, Inc.

Chiou YY. 2003. Leaf mustard pickles and derived products. In: YH Hui, S Ghazala S, Murrell KD, Graham DM, Nip WK, editors. Handbook of Vegetable Preservation and Processing, 169–178. New York: Marcel Dekker, Inc.

Dairy and Food Industries Supply Association. 1993. Directory of Membership Products and Services.

Davies FL, BA Law, editors. 1984. Advances in the Microbiology and Biochemistry of Cheese and Fermented Milk. New York: Elsevier Applied Science Publishers.

Desroiser NW. 1977. Elements of Food Technology. Westport, Conn.: AVI Publishing Co.

Duncan A. 1987. Perfecting the pickle. Oregon's Agricultural Progress 33(2/3): 6–9.

Early R. (editor). 1998. The Technology of Dairy Products. London: Blackie Academic.

Ebine H. 1986. Miso preparation and use (also hot and sweet pastes). In: EW Lusas, DR Erickson, WK Nip, editors. Food Uses of Whole Oil and Protein Seeds, 131–147. Champaign, Ill.: American Oil Chemists Society.

Emmons DB. 2000. Practical Guide for Control of Cheese Yield. Brussels: International Dairy Federation.

Farnworth ER. 1999. From folklore to regulatory approval. J. Nutraceuticals, Functional, and Medical Foods 1(4): 57–68.

Fleming HP, RF McFeeters, JL Ethchells, TA Bell. 1984. Pickled vegetables. In: ML Speck, editor. Compendium of Methods for the Microbiological Examination of Foods, 663–681. APHA Technical Committee on Microbiological Methods of Foods.

Geisen R, WH Holzapfel. 1996. Genetically modified starter and protective cultures. International Journal of Food Microbiology 30(3): 315–334.

Gilmore T, J Shell. 1993. Dairy equipment and supplies. In: YH Hui, editor. Dairy Science and

Technology Handbook, Vol. 3, 155–294. New York: VCH Publishers, Inc.

Groff ET, MA Steinbaecher. 1995. Baking technology. Cereal Foods World 40(8): 524–526.

Hamada T, M Sugishita, Y Fukusima, T Fukase, M Motal. 1991. Continuous production of soy sauce by a bioreactor system. Process Biochemistry 26(1): 39–45.

Hammes WP, Bantleon, S Min. 1990. Lactic acid bacteria in meat fermentation. FEMS Microbiology Letters 87(1/2): 165–174.

Hang DY. 2003. Western fermented vegetables: Sauerkraut. In: YH Hui, S Ghazala, KD Murrell, DM Graham, WK Nip, editors. Handbook of Vegetable Preservation and Processing, 223–230. New York: Marcel Dekker, Inc.

Henriksen CM, Nilsson D, Hansen S, Johasen E. 1999. Industrial applications of genetically modified microorganisms: Gene technology at Chr. Hansen A.S. Proceedings 9th European Congress on Biotechnology. July 11–15, 1999, Brussels.

Huang SD. 1999. Wheat products: 2. Breads, cakes, cookies, pastries and dumplings. In: CYW Ang, KS Liu, YW Huang, editors. Asian Foods: Science and Technology, 71–109. Lancaster, Pa.: Technomics Publishing Co., Inc.

Huang TC, WK Nip. 2001. Intermediate-moisture meat and dehydrated meat. In: YH Hui, WK Nip, RW Rogers, OA Young, editors. Meat Science and Applications, 403–442. New York: Marcel Dekker, Inc.

Incze K. 1998. Dry fermented sausages. Meat Science. 49(Supplement): S169–177.

Iwasaki K, M Nakajima, H Sasahara. 1992. Rapid continuous lactic acid fermentation by immobilized lactic acid bacteria for soy sauce production. Process Biochemistry 28(1): 39–45.

Jay JM. 1996. Modern Food Microbiology. New York: Chapman and Hall.

Jenness R. 1988. Composition of milk. In: NP Wang, R Jenness, RM Keeney, EH Marth, editors. Fundamentals of Dairy Chemistry, 3rd edition, 1–38. New York: Klumer Academic Publishers (VanNostrand Reihold Co.).

Kamel BS, CE Stauffer, editors. 1993. Advances in Baking Technology. New York: John Willey and Sons (VCH Publishers, Inc.)

Kosikowski FV, VV Mistry. 1997. Cheese and Fermented Milk Foods, 3rd edition, vols. 1 and 2. Wesport, Conn.: F.V. Kosikowski.

Law BA, editor. 1997. Microbiology and Biochemistry of Cheese and Fermented Milk. London: Blackie Academic and Professional.

Lee KY. 2003. Fermentation: Principles and microorganisms. In: YH Hui, S Ghazala, KD Murrell,

DM Graham, WK Nip, editors. Handbook of Vegetable Preservation and Processing, 155–168. New York: Marcel Dekker, Inc.

Liu FK. 1986. Food uses of soybeans. In: EW Lusas, DR Erickson, WK Nip, editors. Food Uses of Whole Oil and Protein Seeds, 148–158. Champaign Ill.: American Oil Chemists Society.

Liu KS. 1997. Soybean: Chemistry, Technology and Utilization. New York: Chapman and Hall.

___. 1999. Oriental soyfoods. In: CYW Ang, KS Liu, YW Huang, editors. Asian Foods: Science and Technology, 139–199. Lancaster, Pa.: Technomic Publishing Co., Inc.

Luh BS. 1995. Industrial production of soy sauce. Journal of Industrial Microbiology 14(6): 467–471.

Muramatsu S, Y Sano, Y Uzuka. 1993. Rapid fermentation of soy sauce. Application of preparation of soy sauce low in sodium chloride. In: AM Spanier, H Okai, M Tamura, editors. Food Flavor and Safety: Molecular Analysis and Design. (ACS Symposium 528), 200–210. Columbus, Ohio: American Chemical Society.

Nath KR. 1993. Cheese. In: YH Hui, editor. Dairy Science and Technology Handbook, vol. 2, 161–255. New York: VCH Publishers, Inc.

Park KY, HS Cheigh. 2003. Fermented Korean vegetables (kimchi). In: YH Hui, S Ghazala, KD Murrell, DM Graham, WK Nip, editors. Handbook of Vegetable Preservation and Processing, 189–222. New York: Marcel Dekker, Inc.

Prasad KSK. 1989. Dairy Plant. Secunderabad, India: KSC Prasad.

Pyler EJ. 1988. Baking Science and Technology, 3rd edition, vols. 1 and 2. Merriam, Kans.: Sosland Publishing Co.

Qarooni J. 1996. Flat Bread Technology. New York: Chapman and Hall.

Quail KJ. 1998. Arabic Bread Production. St. Paul, Minn.: American Association of Cereal Chemists.

Robinson RK, editor. 1986. Modern Dairy Technology, vol. 2. New York: Elsevier Applied Science Publishers.

___, editor. 1990. Dairy Microbiology, vols. 1 and 2. London: Applied Science.

Robinson RK, AY Tamime, editors. 1991. Feta and Related Cheeses. New York: Chapman and Hall (Ellis Horwood, Ltd.)

Roca M, K Incze. 1990. Fermented sausages. Food Review International 6(1): 91–118.

Scott R, RK Robinson, RA Wilbey. 1998. Cheese Making Practice. New York: Chapman and Hall.

Skrokki A. 1998. Additives in Finnish sausages and other meat products. Meat Science 39(2): 311–315.

Specialist Cheesemakers Association. 1997. The Specialist Cheesemakers: Code of Best Practice. Staffordshire, Great Britain: Specialist Cheesemakers Association.

Spreer E. (A Mixa, translator). 1998. Milk and Dairy Technology. New York: Marcel Dekker, Inc.

Steinkraus KH. 1996. Handbook of Indigenous Fermented Foods, 2nd edition, revised and expanded. New York: Marcel Dekker, Inc.

Sugiyama S. 1986. Production and uses of soybean sauces. In: EW Lusas, DR Erickson, WK Nip, editors. Food Uses of Whole Oil and Protein Seeds, 118–130. Champaign, Ill.: American Oil Chemists Society. Pp. 118-130.

Tamime AY, RK Robinson. 1999. Yogurt: Science and Technology. Boca Raton, Fla.: CRC Press.

Teng DF, CS Lin, PC Hsieh. 2004. Fermented tofu: Sufu and stinky tofu. In: YH Hui, LM Goddik, AS Hansen, J Josephsen, WK Nip, PS Stanfield, F Toldra, editors. Handbook of Food and Beverage Fermentation Technology. New York: Marcel Dekker, Inc. (Forthcoming.)

Toldra F, Y Sanz, M Flores. 2001. Meat fermentation technology. In: YH Hui, WK Nip, RW Rogers, OA Young, editors. Meat Science and Applications, 538–591. New York: Marcel Dekker, Inc.

Townsend WE, DG Olsen. 1987. Cured meat and meat products processing. In: JF Price, BS Scheweigert , editors. The Science of Meat and Meat Products, 431–456. Westport, Conn.: Food and Nutrition Press.

Walstra P, TJ Geurts, A Noomen, A Jellema, MAJS van Boekel. 1999. Dairy Technology: Principles of Milk Properties and Processes. New York: Marcel Dekker, Inc.

Winarno FG. 1986. Production and uses of soybean tempe. In: EW Lusas, DR Erickson, WK Nip, editors. Food Uses of Whole Oil and Protein Seeds, 102–130. Champaign, Ill.: American Oil Chemists Society.

Xiong YL, FQ Yang, XQ Lou. 1999. Chinese meat products. In: CYW Ang, KS Liu, YW Huang, editors. Asian Food Products: Science and Technology, 201–213. Lancester, Pa.: Technomic Publishing Co., Inc.

Yoneya T. 2003. Fermented soy products: Tempe, nattos, miso and soy sauce. In: YH Hui, S Ghazala, DM Graham, KD Murrell, WK Nip, editors. Handbook of Vegetable Preservation and Processing, 251–272. New York: Marcel Dekker, Inc.

4

Fundamentals and Industrial Applications of Microwave and Radio Frequency in Food Processing

Y.-C. Fu

INTRODUCTION

Microwave heating of food has existed since 1949. Growth in the number of homes with microwave ovens, combined with the industrial use of microwaves, has created a large market for microwave-processed foods, and consequently, has changed food preferences and preparation methods. Using microwaves as a source of heat in the processing (heating, thawing, drying, etc.) of food materials is advantageous because it offers a potential for rapid heat penetration, reduced processing times and, hence, increased production rates, more uniform heating, and improved nutrient retention. It is indisputable that microwave heating has many advantages over conventional heating, but the process itself is extremely complicated.

The use of microwaves represents the use of sophisticated technology in the food industry. Lack of sufficient and unified knowledge of this complex and radically different heating process has been the primary contributor to its unpredictability. Although there has been a lack of predictive models to understand how microwave energy fields interact with the product to produce heat (Mudgett 1986), significant progress has been made in the last 17 years. A lack of understanding often leads to undesirable temperature and moisture content distributions in microwave-heated products. Dubious empirical research or the "black box" approach employed by the food industry to develop commercial applications should be avoided. Emphasis should be on basic research to better understand the interaction between microwave energy and product. This chapter will present the fundamentals of microwaves and a description of microwave processes in the food industry.

INDUCTION, DIELECTRIC, AND MICROWAVE HEATING

The temperature of a material can be increased either directly or indirectly. Indirect methods are those in which heat is generated outside the product and is transferred to the product by conduction, convection, or radiation. Direct methods are those in which heat is generated within the material itself. These include induction, dielectric, and microwave techniques. Direct heating methods enable (1) high concentration of heat energy, (2) selectivity in the location of heat application, and (3) accurate control of heat duration (Anonymous 1980). These factors are important advantages because they lead to increased output, improved quality, and reduction of production costs.

Induction (ohmic) heating is used with materials (usually metallic) that are conductors of electricity. The material to be heated is placed inside the coil or inductor, which is energized with an alternating current. Frequencies utilized range from 50 Hz to 2 MHz. An overview of ohmic and inductive heating is given by Sastry (1994) and the Food and Drug Administration (2000). Dielectric heating is used with insulating materials. The material to be heated is placed between two electrodes and forms the dielectric component of a capacitor. Excitation is by means of a high frequency voltage (2 to 100 MHz) applied to the condenser plates. Radio frequency heating, which is at a much lower frequency, has thrived as an industry alongside microwaves over the decades. Radio frequency heating in the United States can be performed at any of the three frequencies: 13.56, 27.12, and 40.68 MHz. Microwave heating is a special field of dielectric heating in which very high frequencies (300 MHz to 30 GHz) are applied.

Domestic microwave ovens operate at 2450 MHz, and industrial processing systems generally use either 2450 MHz or 915 MHz (896 MHz in the United Kingdom). Two types of applicators are commonly used. With a multimode applicator, the microwaves are discharged in a random configuration, using the walls of the applicator to cause random reflections of the waves. The disadvantage of this method is that there can be a concentration of microwave energy at various points in the material, which results in localized overheating. In a domestic microwave oven, this effect is overcome by rotating the foodstuff on a turntable or by using a metal stirrer, which alters the electromagnetic field resonance pattern. With a single mode applicator, horns used to discharge the microwaves are designed to give a constant electromagnetic field strength along the length of the applicator to which the foodstuff is exposed, perpendicular to the length of travel. This significantly reduces local overheating because the waves pass directly into the foodstuff, with minimal reflection from the surrounding surfaces.

As electrically nonconducting (dielectric) materials are poor heat conductors, heat applied from the outside by convection, radiation, or conduction is inefficient. In some cases, the heat applied causes a skin or crust, which is in itself a thermal barrier, to form on the outside. The single most important thing about microwave heating is the unique opportunity to create heat within a material—the volumetric heating effect—which is not achievable by any other conventional means. No temperature differential is required to force heat into the center of the material. Generation of heat within food products by microwave energy is primarily caused by molecular friction attributed to the breaking of hydrogen bonds associated with water molecules and ionic migration of free salts in an electric field of rapidly changing polarity. Substances that respond to, and therefore can be processed by, microwave energy are composed of polar (e.g., water), ionic, or conductive (e.g., carbon black) compounds. Nonpolar substances, for example, polyethylene and paraffin, are unaffected.

PROPAGATION OF ELECTROMAGNETIC WAVES

A briefly theoretical explanation of what microwaves are and how they interact with food matter is needed in order to understand their general behavior. These equations are fundamental and have resulted in the derivation of all basic equations and terminology for microwave heating, such as wave propagation, power dissipation, Lambert's law, penetration depth, and so on.

This section begins with the four fundamental equations of electromagnetism that bear the name of James Clerk Maxwell (1831–1879), who developed the classical theory of electromagnetism and correctly predicted that an electromagnetic wave has associated electric field, E, and magnetic field, H, properties. The set of four fundamental equations of electromagnetism in differential form are (Cheng 1990)

$$\nabla \times E = \frac{\partial B}{\partial t} \qquad\qquad 4.1$$

$$\nabla \times H = J + \frac{\partial D}{\partial t} \qquad\qquad 4.2$$

$$\nabla \cdot D = \rho_e \qquad\qquad 4.3$$

$$\nabla \cdot B = 0 \qquad\qquad 4.4$$

where E and H are the electric and magnetic field intensities, J is total current density, ρ_e is total electric charge density, D is electric displacement (electric flux density), and B is the magnetic flux density. They are known as Maxwell's equations. The above equations are general in that the media can be nonhomogeneous, nonlinear, and nonisotropic. The constitutive relations relating J, D, B, E, and H are $D = \varepsilon E$, $B = \mu H$, and $J = \sigma E$, which describe the macroscopic properties of the medium in terms of permittivity, ε; permeability, μ; and conductivity, σ. In problems of wave propagation, we are concerned with the behavior of an electromagnetic wave in a source-free region where ρ_e and J are zero. In other words, we are often interested not so much in how an electromagnetic wave originates, but in how it propagates. If the wave is in a simple (linear, isotropic, and homogeneous) nonconducting medium characterized by ε and $\mu (\sigma = 0)$, Maxwell's equations (Eqs 4.1–4.4) reduce to

$$\nabla \times E = -\mu \frac{\partial H}{\partial t} \qquad 4.5$$

$$\nabla \times H = \varepsilon \frac{\partial E}{\partial t} \qquad 4.6$$

$$\nabla \cdot E = 0 \qquad 4.7$$

$$\nabla \cdot H = 0 \qquad 4.8$$

Equations 4.5–4.8 can be combined to give a second-order homogeneous vector wave equation in E and H alone.

$$\nabla^2 E - \mu\varepsilon \frac{\partial^2 E}{\partial t^2} = 0 \qquad 4.09$$

or, since $u = 1 / \sqrt{\mu\varepsilon}$,

$$\nabla^2 E - \frac{1}{u^2} \frac{\partial^2 E}{\partial t^2} = 0 \qquad 4.10$$

In free space, the source-free wave equation for E is

$$\nabla^2 E - \frac{1}{c_0^2} \frac{\partial^2 E}{\partial t^2} = 0 \qquad 4.11$$

where c_0 is the phase velocity in free space

$$c_0 = 1 / \sqrt{\varepsilon_0 \mu_0} = 3 \times 10^8 (m/s) \qquad 4.12$$

Field vectors that vary with space coordinates and are sinusoidal functions of time can similarly be represented by vector phasors that depend on space coordinates but not on time. As an example, we can write a time-harmonic E field (referring to $\cos\omega t$) as

$$E(x,y,z,t) = \mathrm{Re}\left[E(x,y,z) e^{j\omega t} \right] \qquad 4.13$$

$\mathrm{Re}[E(x,y,z)e^{j\omega t}]$ is the real part of $[E(x,y,z)e^{j\omega t}]$. Alternatively $e^{j\omega t}$ can be used to express the time dependence. If $E(x,y,z,t)$ is to be represented by the vector phasor, $E(x,y,z)$, then $\partial E(x,y,z,t)/\partial t$ and $\int E(x,y,z,t)dt$ would be represented by vector phasors $j\omega E(x,y,z)$ and $E(x,y,z)/j\omega$, respectively. In a simple, nonconducting, source-free medium characterized by $\rho_e = 0$, $J = 0$, $\sigma = 0$, the time-harmonic Maxwell's equations (Eqs 4.5–4.8) become

$$\nabla \times E = -j\omega\mu H \qquad 4.14$$

$$\nabla \times H = j\omega\varepsilon E \qquad 4.15$$

$$\nabla \cdot E = 0 \qquad 4.16$$

$$\nabla \cdot H = 0 \qquad 4.17$$

From Equation 4.9 we obtain

$$\nabla^2 E + k^2 E = 0 \qquad 4.18$$

where $k = \omega\sqrt{\mu\varepsilon}$ is called the wave number.

If the simple medium is conducting ($\sigma \neq 0$), a current $J = \sigma E$ will flow, and Equation 4.15 should be changed to:

$$\nabla \times H = (\sigma + j\omega\varepsilon)E = j\omega\left(\varepsilon + \frac{\sigma}{j\omega} \right)E = j\omega\varepsilon_c E \qquad 4.19$$

with

$$\varepsilon_c = \varepsilon - j\frac{\sigma}{\omega} = \varepsilon' - j\varepsilon'' = \varepsilon_0(\varepsilon_r' - j\varepsilon_r'') \qquad 4.20$$

$$\varepsilon_r' \equiv \varepsilon'/\varepsilon_0; \quad \varepsilon_r'' \equiv \varepsilon''/\varepsilon_0 \qquad 4.21$$

where ε_0 is the permittivity of free space (8.8542E-12 Farad/m). Hence, all the previous equations for nonconducting media will apply to conducting media if ε is replaced by the complex permittivity ε_c. The material's ability to store electrical energy is represented by, ε' and ε'' accounts for losses through energy dissipation. ε'_r is often called "relative dielectric constant." This is somewhat inappropriate, as the term "constant" should be used only for true constants. ε'_r varies significantly both with temperature and frequency for many typical workload substances. ε''_r is called the relative dielectric loss incorporating all of the energy losses due to dielectric relaxation and ionic conduction. The ratio $\varepsilon''/\varepsilon'$ is called a loss tangent because it is a measure of the power loss in the medium:

$$\tan\delta_c = \frac{\varepsilon''}{\varepsilon'} \cong \frac{\sigma}{\omega\varepsilon} \qquad 4.22$$

The quantity δ_c may be called the loss angle. Alternatively, we may define an equivalent conductivity representing all losses and write $\sigma = \omega\varepsilon''$. On the basis of Equation 4.19, a medium is referred to as a good conductor if $\sigma \gg \omega\varepsilon$, and a good insulator if $\omega\varepsilon \gg \sigma$. Thus, a material may be a good conductor at low frequencies but may have the properties of a lossy dielectric at very high frequencies. In a lossy dielectric medium, the real wave number k should be changed to a complex wave number:

$$k_c = \omega\sqrt{\mu\varepsilon_c} = \omega\sqrt{\mu(\varepsilon' - j\varepsilon'')} \qquad 4.23$$

A uniform plane wave characterized by $E = a_x E_x$ propagating in a lossy medium in the +z-direction has associated with it a magnetic field $H = a_y H_y$. Thus E and H are perpendicular to each other, and both are transverse to the direction of propagation (a particular case of a transverse electromagnetic, TEM, wave). The solution to be considered here is that of a plane wave, which for the electric field attains the form

$$E(x) = E_{max} e^{j\omega t - \gamma x} \qquad 4.24$$

A propagation constant, γ, is defined as

$$\gamma = jk_c = j\omega\sqrt{\mu\varepsilon_c} \qquad 4.25$$

The propagation factor $e^{-\gamma x}$ can be written as a product of two factors:

$$E(x) - E_{max} e^{-\alpha z} e^{j(\omega t - \beta z)} \qquad 4.26$$

where α and β are the real and imaginary parts of γ, respectively. Since γ is complex, we write, with the help of Equation 4.20,

$$\gamma = \alpha + j\beta = j\omega\sqrt{\mu\varepsilon}\left(1 + \frac{\sigma}{j\omega\varepsilon}\right)^{1/2}$$
$$= j\omega\sqrt{\mu\varepsilon'}\left(1 - j\frac{\varepsilon''}{\varepsilon'}\right)^{1/2} \qquad 4.27$$

$$\alpha = \frac{\sqrt{2}\pi f}{c}\sqrt{\varepsilon_r'\left(\sqrt{1 + \tan^2\delta} - 1\right)} \qquad 4.28$$

$$\beta = \frac{\sqrt{2}\pi f}{c}\sqrt{\varepsilon_r'\left(\sqrt{1 + \tan^2\delta} + 1\right)} \qquad 4.29$$

As we shall see, both α and β are positive quantities. The first factor, $e^{-\alpha z}$, decreases as z increases and thus is an attenuation factor, and α is called an attenuation constant. The second factor, $e^{-j\beta z}$, is a phase factor; β is called a phase constant, which expresses the shift of phase of the propagating wave and is related to the wavelength of radiation in the medium (λ_m) by $\lambda_m = 2\pi/\beta$ which, in free space, reduces to $\lambda_0 = 2\pi/\beta = c_0/f$.

From Equation 4.26, the first exponential term gives the attenuation of the electric field, and therefore, the distribution of the dissipated or absorbed power in the homogeneous lossy material follows the exponential law (Lambert's Law):

$$P_{diss} = P_{trans} e^{-2\alpha z} \qquad 4.30$$

where P_{trans} is the power through the surface in the z direction. Theoretically, the power penetration depth, D_p, is defined as the depth below a large plane surface of the substance where the power density of a perpendicularly impinging, forward propagating, plane electromagnetic wave has decayed by $1/e$ from the surface value, $1/e \approx 37\%$ (Risman 1991). The absorbed power in the top layer of this thickness in relation to the totally absorbed power (per surface area) is then 63%.

$$D_p = \frac{1}{2\alpha} \qquad 4.31$$

Substitution of Equation 4.28 into Equation 4.31 yields the general expression for the penetration depth:

$$D_p = \frac{c}{4\pi f}\sqrt{\frac{2}{\varepsilon_r'\left(\sqrt{1 + \tan^2\delta} - 1\right)}}$$
$$= \frac{\lambda_0}{2\pi\sqrt{2\varepsilon_r'}\left(\sqrt{1 + \left(\varepsilon_r''/\varepsilon_r'\right)^2} - 1\right)^{1/2}} \qquad 4.32$$

The skin depth D_s, where the electric field strength is reduced to $1/e$ [and the power density thus to $(1/e)^2$] is twice the power penetration depth, $D_s = 2D_p$. There are many texts where it is not clear whether D_s or D_p is referred to. Even worse, there are instances where a stated formula or numerical values do not correspond to the terminology used (Risman 1991). Some authors in the United States have tried to avoid the confusion by using half-power depth $D_{1/2}$ or D_{50}. This relates to D_p by

$$D_{1/2} = D_p \cdot \ln2 \approx 0.69 D_p \qquad 4.33$$

MICROWAVE POWER DISTRIBUTION

Most practical materials treated by microwave power are nonhomogeneous and very frequently anisotopic; the permittivity of these materials changes with temperature and moisture content (drying process). Thermal losses from the material surface and heat transfer in the bulk of material produce additional complications. The generation of heat in food materials is also accompanied by significant moisture migration that, in turn, affects the energy absorption characteristics of food, creating a coupling of heat and mass transport that complicates mathematical analysis. From the physical point of view, microwave heating is a combination of at least four different processes: distribution of power, absorption of power, heat transfer, and mass transfer. The magnitude and uniformity of temperature distribution are affected by both food and oven factors such as (1) the magnitude and distribution of microwave power (i.e., external electric field) where the food is placed, (2) the reflection of waves from the food surface and penetration depth, as characterized by the food geometry and properties, (3) the distribution of absorbed power as well as power dissipated at a particular point (i.e., internal electric field) as functions of the material parameters, temperature, and time (due to drying), and (4) simultaneous heat and mass transfer.

INTERNAL ELECTRIC FIELD INTENSITY

Electromagnetic waves transport energy through space. The amount of microwave energy absorbed is, in turn, determined by the electric field inside the microwave applicator. It offers an intangible link between the electromagnetic energy and the material to be treated.

For microwave heating, the governing energy equation includes volumetric heat generation that results in a temperature rise in the material:

$$\frac{\partial T}{\partial t} = \alpha \nabla^2 T + \frac{Q_{abs}}{\rho C_p} \qquad 4.34$$

In this equation, Q_{abs} (watts/cm^3) corresponds to the volumetric rate of internal energy generation due to

dissipation of microwave energy. Basically, the apparatus is placed in the oven at the position of interest, and the rate of temperature rise, $\partial T/\partial t$ is measured. C_p (cal/g-°C) is the heat capacity of the material, and ρ (g/cm^3) is the density of the material. Assuming no temperature gradients in a small mass of dielectric medium, the energy balance can be obtained by simplifying Equation 4.34:

$$Q_{abs} = \frac{P_{abs}}{V} = \rho C_p \frac{\partial T}{\partial t} \qquad 4.35$$

where P_{abs} is the total power absorbed by the dielectric medium (watts). Its relationship to the E-field at the location can be derived from Maxwell's equations of electromagnetic waves (Metaxas and Meredith 1983).

$$Q_{abs} = 2\pi f \varepsilon_0 \varepsilon''_{eff} E^2_{rms} \qquad 4.36$$

where f is the microwave frequency (2450 MHz), ε''_{eff} is the dielectric loss factor for the dielectric material being heated, and E_{rms} is the root mean square value of the electric field intensity. Since the rate of temperature rise is known, the heat generation, Q_{abs}, can be determined and equated to the "internal" electric field, $E_{rms, internal}$, using Equation 4.36.

$$E_{rms, internal} = \sqrt{\frac{\rho C_p}{2\pi f \varepsilon_0 \varepsilon''_{eff}} \frac{\partial T}{\partial t}} \qquad 4.37$$

EXTERNAL ELECTRIC FIELD INTENSITY

The Poynting theorem simply states that there is conservation of energy in electromagnetic fields. The power flow through a closed surface can be calculated from the integration of the Poynting vector, $P = E \times H$ (W/m^2) , over the surface (Eq. 4.38).

$$P_{av} = -\int_A P \cdot dA = -\frac{1}{2} \int_A \mathbf{Re}(E \times H) \cdot dA \qquad 4.38$$

where Re$(E \times H)$ means the real part of $(E \times H)$. The negative sign represents the rate at which electromagnetic energy flows "into" the closed surface. The time-averaged power density, P_{av}, is equal to the time-averaged energy density, $\varepsilon_0 \cdot E^2_{rms, ext}$, multiplied by the phase velocity, c (Eq. 4.39).

$$P_{av} = \varepsilon_0 \cdot E^2_{rms, ext} \cdot c \qquad 4.39$$

where $E_{rms, ext}$ is root-mean-square of external electric field intensity (Lorrain et al. 1988). In a medium where there is no wave reflection at an interface and 100% of the wave energy is absorbed by the dielectric material, then

$$Q_{abs,interal} V = P_{abs} = \rho C_p \frac{\partial T}{\partial t} V$$

$$= \varepsilon_0 \cdot E^2_{rms,external} \cdot c \cdot A \qquad 4.40$$

where A is the surface area (m^2). In the case for heating a dielectric medium, part of the wave that strikes the dielectric medium will be reflected and part will enter the dielectric droplet, where it is partially absorbed. Two parameters are introduced to solve the "*external*" E-field value when reflection and absorption are taken into consideration (Eq. 4.41).

$$Q_{abs,internal} V = P_{abs} = \rho C_p \frac{\partial T}{\partial t} V$$

$$= \varepsilon_0 \cdot E^2_{rms,external} \cdot c \cdot A \cdot \sigma \cdot \Gamma \qquad 4.41$$

Γ, the transmission coefficient, indicates the fraction of power that is transmitted to the dielectric medium. σ, the absorption coefficient, indicates the fraction of power that is absorbed and produces heating. Since absorption and transmission coefficients are known, Equation 4.42 may be used to calculate the external electric field intensity (White 1970):

$$E_{external} = \sqrt{\rho C_p \frac{1}{3} a \frac{\partial T}{\partial t} \frac{1}{c \cdot \varepsilon_0} \frac{1}{\sigma \cdot \Gamma}} \qquad 4.42$$

LAMBERT'S LAW

In several computational studies of microwave heating, heat generation has been modeled by Lambert's law, according to which microwave power attenuates exponentially as a function of distance of penetration into the sample (Ayappa et al. 1991a,b; Nykvist and Decareau 1976; Ohlsson and Bengtsson 1971; Stuchly and Hamid 1972; Taoukis et al. 1987). It must be emphasized that these penetration depth calculations are valid only for materials undergoing plane wave incidence and for semi-infinite media; this will be referred to henceforth as Lambert's law limit (Ayappa et al. 1991a, Stuchly and Hamid 1972). Although Lambert's law is valid for samples thick enough to be treated as infinitely thick, it is a poor approximation in many practical situations and often does not describe accurately the microwave heating of food in a cavity.

To determine the conditions of the approximate

applicability of Lambert's law for finite slabs, Ayappa and others (1991a) compared it with the microwave heating predicted by Maxwell's equation. The critical slab thickness, L_{crit} (cm) above which the Lambert's law limit is valid can be estimated from $L_{crit} = 2.7/D_p - 0.08$. Fu and Metaxas (1992) proposed a new definition for the power penetration depth, Δ_p, which is the depth at which the power absorbed by the material is reduced to $(1 - e^{-1})$ of the total power absorbed. This definition allows a unique value of Δ_p to be found for all thicknesses and also gives an indication of the validity of assuming exponential decay within the slab. Another approach is used where a spherical dielectric load is assumed to absorb energy from a surrounding radiation field (MacLatchy and Clements 1980).

The power absorption inside a dielectric medium can be estimated in the following way. Assume that the power flux (power per unit area) entering through the surface of the dielectric medium is uniform and that all the waves are transmitted into the medium (i.e., there is no wave reflection). Then power decays exponentially, $P(x) = P_0 \cdot \exp(-x/D_p)$, where P_0 is the incident power at the surface. From the Poynting theorem, the field energy that dissipates as heat in the enclosed volume is equal to the total power flowing into a closed surface minus the total power flowing out of the same closed surface (Cheng 1990).

$$\sum P_{eff} = \int \frac{P_0 e^{-x/D_p}}{D_p} dx \qquad 4.43$$

$$= -P_0 \left[\int_0^a \frac{e^{-(a-r)/D_p}}{D_p} d(a-r) - \int_0^{-a} \frac{e^{-(a-r)/D_p}}{D_p} d(a-r) \right]$$

$$P_{abs} = \sum P_{eff} = P_0 \left[1 - e^{-2a/D_p} \right] \qquad 4.44$$

where a is the radius of the spherical dielectric load, P_{eff} is the effective magnitude of the Poynting vector, and P_{abs} is the total power absorption by the dielectric medium. So, the absorption coefficient, σ, used in Equation 4.37 is

$$\sigma = P_{abs} / P_0 = 1 - e^{-2a/D_p} \qquad 4.45$$

The use of Lambert's law requires an estimate of the transmitted power intensity, P_{trans} (Eq. 4.30), which is obtained from calorimetric measurements

(Ohlsson and Bengtsson 1971, Taoukis et al. 1987) or used as an adjustable parameter to match experimental temperature profiles with model predictions (Nykvist and Decareau 1976). Thus, P_{trans} measured by the above methods represents the intensity of transmitted radiation, the accuracy of the estimate depending on the method used. Alternately, if P_{trans} is the incident power flux, Lambert's law must be modified to account for the decrease in power due to reflection at the surface of the sample. Since Lambert's law does not yield a comprehensive approach, a more accurate estimate of the heating rate based on predicting or measuring the fundamentally nonuniform electric field intensity in a cavity should be the most important subject of current research. How the shape and volume (relative to the microwave oven) of a food material change the rate of heating must be investigated further. The interior electric field, the moisture movement in solid foods, and changes in the dielectric and other properties combine to make designing microwave processes a difficult task.

MEASUREMENT OF ELECTRIC FIELD INTENSITY

Lack of information on electric field intensity or power density distribution surrounding an object (load) during heating is a major concern to the food industry because it can be used as input for the purpose of developing mathematical models to predict heating patterns in microwave-heated foods and for computer simulation of food processing. Measurement of microwave E-field or power density distribution inside an oven is needed. E-field distribution is complex and is beyond the scope of simple calculations. Currently, there is comparatively little literature on measuring electric field intensity in food systems during microwave heating (Goedeken 1994, Mullin and Bows 1993). Thus far, measuring the distribution of the electric field in a microwave oven has proven most difficult.

For many years electric fields have been measured in air and in material media (Bassen and Smith 1983), but not for food applications during microwave heating. The previously developed techniques either do not give a quantitative value of the field, or perturb the field, or both (Bosisio et al. 1974, MacLatchy and Clements 1980, Washisu and Fukai 1980). Indirect measurements of field intensity are often accomplished using the temperature rise in small amounts of liquids placed in various lo-

cations inside the cavity (MacLatchy and Clements 1980,Watanabe et al. 1978, White 1970). The method of using a large load of water in the cavity yields a measurement of the power absorbed from the field and may be used to estimate electric field intensity (White 1970). However, when traditional methods of computation are used, an erroneously high value of the field is obtained (MacLatchy and Clements 1980). Luxtron® Corp. developed an E-field probe based on a fiber-optic temperature sensor that measures the temperature of a resistive element when it is exposed to an electromagnetic field. A second sensor is used to measure the ambient temperature, and the difference between the two measurements is the temperature rise of the resistive element (Randa 1990, Wickersheim and Sun 1987, Wickersheim et al. 1990). Advantages of the design are that it is small and nonperturbing, and it can be used in high electromagnetic fields. Very few studies of the use of this probe to measure the electric field inside a food sample heated in a microwave oven are reported. However, Luxtron® stopped making this electric-field-strength probe in 1997.

INTERACTION OF MICROWAVES WITH FOOD

Food shape, volume, surface area, and composition are critical factors in microwave heating. These factors can affect the amount and spatial pattern of absorbed energy, leading to effects such as corner and edge overheating, focusing, and resonance. Food composition, in particular moisture and salt percentages, has a much greater influence on microwave processing than on conventional processing, due to its influence on dielectric properties. Interference from side effects like surface cooling, interior burning, steam distillation of volatiles, and short cooking time alter the extent of interactions.

DIELECTRIC PROPERTIES

The dielectric properties of foods are very important in describing the way foods are heated by microwaves. The most comprehensive collection of dielectric property data to date is that of von Hippel (1954). The important properties are the dielectric constant (ε'), which relates to the ability of a food to store microwave energy, and the dielectric loss constant (ε''), which relates to the ability of the food to dissipate microwave energy as heat. The dielectric properties of foods vary considerably with composition, changing with variations in water, fat, carbohydrate, protein, and mineral content (Kent 1987). Dielectric properties also vary with temperature. As indicated earlier, the dielectric properties affect the depth to which microwave energy penetrates into the food to be dissipated as heat. The magnitude of the penetration depth, defined as the depth at which 63% of the energy is dissipated, can be used to quantitatively describe how microwave energy interacts with the food. A large penetration depth indicates that energy is poorly absorbed, while a small penetration depth indicates predominantly surface heating.

Dielectric property data for agricultural products, biological substances, and various materials for microwave processing are widely dispersed in the technical literature (Datta et al. 1995, Nelson 1973, Stuchly and Stuchly 1980, Tinga and Nelson 1973). Unfortunately, most of the literature values on these properties are only available at room temperature to 60°C and are not readily available at sterilization temperature. Those literature data can provide guidelines, but the variability of food product composition and other specific conditions for particular applications often require carefully conducted measurements.

GEOMETRICAL HEATING EFFECTS— CORNER, EDGE, AND FOCUSING EFFECTS

With conventional cooking methods, heat is transferred from outside to the food product by conduction, convection, or infrared radiation. There is a temperature gradient from the outside to the inside. It is often said that with microwaves, heating takes place from the inside to the outside. This is not true; heating occurs throughout the whole food simultaneously, although it may not be evenly distributed. Probably this misinterpretation is due to the fact that surface temperatures tend to be lower than temperatures inside the food (because of evaporative cooling and a geometrical heating effect). For foods with a high loss factor, most of the microwave energy of a wave impinging on the food will be absorbed near the surface, and penetration and in-depth heating will be limited. In general, the surface will heat more rapidly than the interior, but there are exceptions. Refraction and reflection at interfaces will cause reinforcement of the field pattern near corners and edges of rectangular-shaped foods, resulting in overheating. Core heating effects of the same nature occur in foods of spherical or cylindrical shape at

certain dimensions, causing energy concentration and overheating of the central part.

The concentration heating effect means maximum heating occurs in the center for certain spherical and cylindrical geometries (Ohlsson and Risman 1978). The well-known explosion of eggs during microwave heating is one of the most significant demonstrations of core heating effects. This occurs because center heating causes formation of steam, which induces an energy impulse with such high power as to move the surrounding masses away from each other. This kind of thermal behavior has already been observed by Mudgett (1986), Nykvist and Decareau (1976), Ohlsson and Risman (1978), and Whitney and Porterfield (1968) for cylindrically and spherically shaped foods. The maximum heating regions also move slowly from the center towards the surface when the diameter increases. If the diameter is much greater than penetration depth, the temperature profile will be similar to that observed for a "semi-infinite" body. That is, the temperature will decrease exponentially from the surface in accordance to Lambert's law, which governs the absorption of microwave power. If the diameter is much less than penetration depth, the heating profile will be flat. In between these extremes the focusing effect occurs.

Moreover, Mudgett (1986) pointed out the effect of salt on drying behavior. With addition of sodium chloride, penetration depth decreases significantly; therefore, the heating profile could shift from that of focusing and center heating to one of surface heating. Parent and others (1992) showed the temperature and moisture profile of a cylindrical sample with a diameter of 3.5 cm during microwave heating. Without salt, the center heated and dried faster than the surface. However, for a sample with 4% salt, the surface heated and dried faster than the center.

Another reason for uneven heating in lossy products can be traced to the electromagnetic boundary conditions at edges and corners (Pearce et al. 1988). These are the so-called edge and corner effects. In an electric field, where the wavelength is larger than the dimensions of the heated object, field bending will give rise to concentrations at some locations. The convergence of two or more waves at a corner results in a higher volumetric power density than on the flat surface. Higher heating rates will thus be obtained at the corners. If the electric field is strong enough, an arc may emanate from there when the air ionizes (Yang and Pearce 1989). Square containers can cause burning in the corners of the product due to a greater surface area/volume ratio, resulting in more microwave energy absorption. Circular or oval containers help reduce the strong edge and corner effects because energy absorption occurs evenly around the edge, but core heating effects may then be introduced.

MICROWAVE BUMPING

Another phenomenon during microwave heating is the "bumping" that may occur in microwave cooking. The term, "microwave bumping," also known as microwave popping or microwave splattering, is descriptive of an explosion phenomenon characterized by a jostling or shaking of the container, usually accompanied by an audible explosion. When microwave bumping occurs, the explosive sounds, which can be heard some distance away, are annoying, and are an unexpected surprise to consumers. Microwave bumping is due to the explosion of food particulates, not localized boiling of the liquid. In studies by Fu and others (1994), microwave bumping was characterized. Increasing the viscosity of the liquid did not result in a significant difference in intensity or frequency of bumping. The degree of microwave bumping is believed to be directly related to local superheating effects. The higher the electric field intensity is, the greater the incidence of bumping. Due to edge, corner, and focus heating effects by microwave, container shape influences the heating pattern of a food product and the location of bumping in the container. Sterilizing vegetable particulates (which causes excessive softening) and salting food particulates (which causes a high microwave heating rate) are two conditions that are indispensable to producing microwave bumping (Fu et al. 1994).

EVAPORATIVE COOLING AND STEAM DISTILLATION

During the heating of foods containing water, the resulting evaporation at the surface causes a depression of the temperature, known as evaporative cooling. The surface of food is seen to be cooler than the region just below the surface and warmer than the surrounding air. This phenomenon is readily seen during the cooking of a meat roast (Nykvist 1977, Nykvist and Decareau 1976). At the same time, this surface evaporation can cause steam distillation of certain flavor components. Flavor release in microwave cooking is increased by steam distillation.

In microwave heating, water vapor (steam) is one of the most important transport mechanisms contributing to movement of flavor compounds within a food matrix (Fu et al. 2003b,c,d). Individual compounds that make up a flavor, which are of particularly low molecular weight and water soluble, may be driven off or steam distilled out of the product during microwave heating. Fruit and other "sweet" flavorings are more of a problem. They evaporate easily in foods with high initial water content because they contain a great number of short-chain, volatile flavoring substances. Moreover, they are often of a more hydrophilic character: therefore, a great part of the flavoring substance migrates to the aqueous phase of the food, which selectively absorbs the greater part of microwave energy (Van Eijk 1992). The percent loss may range from less than 10% for high-boiling compounds to 95% for very volatile compounds (Risch 1989). It is the very volatile compounds that create a strong aroma, which is necessary when the flavor is designed to impart a balanced aroma profile in the room during microwave heating (Steinke et al. 1989). In this case, the flavor is added solely for aroma generation and contributes very little to the flavor profile of the microwave product itself. However, this phenomenon, flash off, often leads to an imbalance of flavor concentration in a finished product that has a different character from the flavor that was added before cooking. Formulations that compensate for flash off may require a highly imbalanced flavor character prior to microwaving. The specific loss is dependent on the types of flavor components used and the food system in which it is incorporated. Moreover, the amount of flash off can be highly variable within a product because temperature at any given moment can be quite local. As the outward migration of water vapor is the most important factor influencing flavor retention in the food product, the flavorings used for microwave application should have low water vapor volatility unless the flavorings are intended either to create the "oven aroma" of conventional cooking methods or to cover undesirable off notes released during microwave cooking.

LACK OF CRISPNESS (TEXTURE) AND BROWNING (COLOR, FLAVOR) OF MICROWAVE FOODS

With conventional cooking methods, we have high temperature ambient air (e.g., 180°C) at a rather low relative humidity. Heat permeates the surface, and there is a temperature gradient and a corresponding water vapor pressure gradient directed towards the center. Since water vapor density is highest near the surface, this results in a pressure gradient that creates a driving force from the surface toward the center (Wei et al. 1985a,b), thus helping to retain volatiles within the product. Due to the high ambient temperature, surface dehydration, protein denaturation, starch gelatinization, caramelization, and so on take place, and result in the formation of a crust (Van Eijk 1992). When collapse of the surface occurs, a sealing surface layer surrounds the food product and prevents or delays further evaporation of the water vapor and the associated flavoring substances into the ambient air. The very subtle but mouthwatering flavor nuances found in oven baked products are largely due to flavors generated from the Maillard reaction. The Maillard reaction, which encompasses a complex series of reactions that start with the condensation of amino acids and reducing sugars, has long been used as a tool for reproducing, enhancing, and improving mother nature's handiwork in a whole variety of food products. For the nonenzymatic browning or Maillard reaction to occur, the moisture content of the food product's surface must be greatly reduced (water activity levels between 0.6 and 0.8), and the surrounding air cannot be saturated with moisture (Risch 1989). There is no distinct temperature that must be attained for browning to occur; however, the higher the temperature, the greater the extent of browning.

The texture of a microwaveable food may directly affect its acceptance. Toughness or lack of crispness in bread slightly overcooked in a microwave oven may not directly change its flavor, but it does influence the consumer's perception of the product. Microwave toughening is most probably related to moisture migration and loss in these reheated baked products, which also can lead to other undesirable protein-protein interactions.

The lack of conventional-style browning and crisping in microwave ovens is due to the microwave frequency used. At 2450 MHz, the wavelength, 12.2 cm, is too long to create the intense surface heat that occurs at the higher infrared frequencies, limiting the food item to a temperature of approximately 100°C. This is ideal for wet foods like vegetables and stews, but unacceptable for pastry, breaded or batter-coated items, and roast meat. In contrast to the convectively heated food, in most cases we have relatively low temperature ambient air (60–75°C) with a rather high relative humidity

during microwave heating. The level of maximum temperature and consequently of maximum water vapor pressure generally lies farther below the surface. The main driving force, therefore, is directed towards the surface instead of towards the center (Wei et al. 1985a,b). Water vapor generated inside the food continuously migrates to the surface, drawing flavoring substances with it on the way out: the evaporation rate of water is not high enough to dry out the surface, and the evaporated water is continuously replaced by migration of water from the inside (Van Eijk 1992). For foods that require a long heating time, for example, meat joints, the effect can be significant, and the resulting moisture loss from the surface of the product can be appreciable. An electromagnetic phenomenon creating "hot" and "cold" spots is inherent in all microwave ovens and is responsible for much of the uneven cooking associated with them. Liquid products quickly dissipate the microwave energy, resulting in a more uniform product. Solid food products, multiphase systems, or frozen products develop hot and cold spots during heating, which further complicates flavor delivery in these systems (Steinke et al. 1989).

During microwave heating the low surface temperature, its much higher water activity (approximately 1.0), and the lack of prolonged baking time have the following consequences: (1) no crust is formed because the necessary physical changes (protein denaturation, starch gelatinization, etc.) are inhibited, and (2) the formation of many flavor compounds and/or pigments (Maillard browning reactions) do not occur to the required extent. Thus, some flavors that typically develop in a conventionally cooked product will not necessarily work in a microwaved product. Van Eijk (1992) states that the differences in flavor generation and the performance of flavoring substances in microwave foods can be explained satisfactorily by the differences in heating pattern, the corresponding differences in water vapor migration, and the resulting physical changes, particularly at the surface of the food. No athermal effects have been observed.

FOOD INGREDIENTS

The dielectric and thermal properties of foods can be modified by adjusting food ingredients and formulations and are manageable within certain limits. Ingredients in foods such as water, ionized salts, and fats and oils, in particular, and the distribution of these ingredients in the food product exert a strong influence on temperature level and distribution. These ingredients interact physically and chemically to an extent dictated by numerous factors, including mode of heating. Frozen pure water has no microwave dipole relaxation and is therefore microwave transparent. Frozen foods, however, are not microwave transparent since some of the water is still in free liquid form. So when deep-frozen foods are defrosted by microwave energy, particularly difficult problems arise once both ice and water are present. Hot spots and run-away heating may be the consequence in this case. Fats have a low dielectric loss and consequently do not generate as much heat directly from the microwave field. Once heat has been generated, conduction and convection become the main mechanisms of heat transfer. Fats reach very high temperatures due to their high boiling points, whereas water is limited to a maximum temperature of 100°C plus the effects of boiling point elevation exercised by dissolved substances. However, since the heat capacity of fats is about half that of water, they heat more quickly in the microwave.

Factors that affect dielectric properties of water, including the presence of other interactive constituents such as hydrogen bonding due to the presence of glycerol and propylene glycol, and sugar and carbohydrate-like polyhydroxy materials, will also impact microwave heating (Shukla and Anantheswaran 2001). Salts and sugars can be used to modify the browning and crisping of food surfaces. Heating a sample with higher salt content can change the microwave heating pattern from center heating to surface heating (Parent et al. 1992). In addition to direct microwave interactions, lipids, salts, sugar, and polyhydroxy alcohols can raise the boiling point of water. This allows the food to reach the higher temperatures needed for the development of reaction flavors and Maillard browning reactions.

Linking the formation of roast or baked flavor notes only to Maillard reactions is an oversimplification. The reactions of fats with other food constituents (e.g., in meat) are also of great importance for the ultimate flavor profile. Because reactions of this type are also lacking in microwave cooking, an incomplete flavor profile may result. The ability to simulate a specific flavor in a food is significantly influenced by the flavor-binding capacity of the protein used. Denaturing the protein can enhance flavor absorption. This probably reflects the greater exposure of hydrophobic segments of the protein, since hydrophobic interaction is the principal force in the random coil folding of proteins and accounts for binding of nonpolar flavor compounds. The extent to which different proteins bind flavors cannot always be predicted in complicated food systems since the presence of other factors (salts, lipids) will influence flavor behavior. Indeed, the moisture content of the system can influence the extent of aroma released.

To obtain useful and meaningful information on the contributions of rates of flavor migration and kinetics of degradation under various conditions, Fu and others (2003a) designed an apparatus for on-line measurement of flavor concentration, to formulate a thermally stable flavor-dough system and to accomplish isothermal heating. A photoionization detection method (Fu et al. 2001) and a cold-trap, on-line sampling method (Fu et al. 2003b) were used to investigate the migration of flavor compounds in a solid food matrix subjected to microwave heating. As the moisture concentration decreased below 0.1g water/g solid during microwave heating of gelatinized flour dough, a type of encapsulation occurred that prevented flavor from being released. The results of microwave reheating of limonene-formulated dough showed that limonene is very stable, with no significant limonene concentration profile in the sample and less than 1% overall change in total limonene concentration (Fu et al. 2003c).

MICROWAVE PROCESSING

In the quest for better quality shelf stable, low-acid foods, a number of emerging technologies have been considered (Food and Drug Administration 2000).Food engineering will continue to evolve, but more and more food engineering research will be shifted to nontraditional processing and nonthermal processing, such as microwave and radio frequency processing, ohmic heating, high-pressure processing, pulsed electric fields, and so on. Ohmic heating had a lot of potential in 1989–91 and went through some testing, but except for a liquid egg processor, nobody is using it for particulates (Mermelstein 2001). Pulsed electric fields (PEFs) are considered a form of pasteurization, suitable for high-acid foods such as fruit juices (Clark 2002a). High-pressure processing extends shelf life, but the product still requires refrigeration, since the pressure does not inactive spores (Clark 2002b). To achieve shelf stable foods, high-pressure processing must be combined with a mild heat treatment. Although alternative processes have been developed over the years, thermally processed food products maintain a clear dominance in the marketplace, primarily as a result of the wealth of theoretical and empirical knowledge that has been developed regarding thermal inactivation of pathogenic microorganisms and their spores (Mermelstein 2001). Microwave sterilization is a nontraditional, but solely thermal, process and so can be regarded by technologists and regulators as another terminal thermal sterilization technique.

Microwave heating offers numerous advantages in productivity over conventional heating methods such as hot air, steam, and so on. These advantages include high speed, selective energy absorption, excellent energy penetration, instantaneous electronic control, high efficiency and speed, and environmentally clean processing (Cober Electronics, Inc. 2003). Unfortunately, although for the last 35 years expectations have been high that radio frequency and microwave processing of foods might find a niche in the industry, there has been only modest growth in sales of microwave processing equipment over this period. Currently, both microwave and radio frequency are laboratory or pilot scale, and there are no known large microwave systems operating in the food industry, except for bacon precooking or tempering (Schiffmann 2001). But it remains a very exciting processing tool, unmatched by any other technology if attention is paid to its selection. The following sections examine a number of microwave food processes that are interesting from an academic point of view.

DRYING AND DEHYDRATION

Microwave drying is more rapid, more uniform, and more energy efficient than conventional hot air drying, and sometimes it results in improved product quality. But it is highly unlikely that an economic advantage will be demonstrated if only bulk water removal by microwave heating, such as occurs in the constant-rate region is desired (Buffler 1993). During the falling-rate period, because of low thermal conductivity and the evaporative cooling effect, high product temperatures are not easily obtained using convective drying. Surface hardening and thermal gradients again provide further resistance to moisture transfer. Actually, it has been suggested that microwave energy should be applied in the falling-rate period or at low moisture content for finish drying (Funebo and Ohlsson 1998, Kostaropoulos and Saravacos 1995, Maskan 1999, Prabhanjan et al. 1995). Correspondingly, sensory and nutritional damage caused by long drying times or high surface

temperatures can be prevented. It is important to understand the dielectric properties of the material as a function of moisture content during microwave drying. The ability of dielectric heating to selectively heat areas with higher dielectric loss factors and the potential for automatic moisture leveling afford a major advantage even for drying of these types of materials (Buffler 1993).

Because internal microwave heating facilitates a more predominant vapor migration from the interior of the material than occurs during conventional drying, microwave dried products have been reported to show a higher porosity because of the puffing effect caused by internal vapor generation (Fu 1996, Tong et al. 1990, Torringa et al. 1996). Similar results are also found for pasta drying (Buffler 1993). Microwave drying produces a slightly puffed, porous noodle that rehydrates in half the time required for noodles dried by conventional methods (MicroGas Corporation 2003). Using miniature fiber-optic temperature and pressure probes, Tong and others (1990) investigated temperature and pressure distribution during microwave heating in a dough system with porosity ranging from 0.01 to 0.7. Pressure build-up to approximately 14 kPa occurred during the initial stages of the heating process when the initial porosity was less than 0.15 and disappeared when the pressure exceeded the rupture strength of the dough. Volume expansion was observed up to the point where the dough sample ruptured, producing visible cracks in the structure. So microwaves produce a pressure gradient that pumps out the moisture. This property can be used to advantage to speed up the drying process. The results might be positive or negative to the dried product. If the rupture strength of the sample is smaller than the pressure build-up, the solid matrix might be damaged, and visible cracks in the structure would be seen. In an experiment on microwave finish drying of starch pearls, significant visible cracks developed on the outside at microwave powers greater than 200 W, which created unacceptable product (Fu et al. 2003e). Alternatively, if the pressure build-up does not exceed the rupture strength of the structure, the result may be an enhanced porous structure of the samples. So, it is a difficult task to reduce drying time and increase quality at the same time. Careful studies need to be done to establish the correct amount of microwave energy to be used in the process.

Nonuniformities in the microwave electric field and associated heating patterns can lead to high temperatures in various previously dried regions, causing product degradation (Lu et al. 1999). To achieve improvement, fluidized bed dryers or spouted bed dryers can be used to average the uneven electric field (Feng and Tang 1998, Kudra 1989). The combination of microwave and vacuum drying (Boehm et al. 2002, Durance et al. 2001, Gunasekaran 1999, Langer 2000, Sunderland 1980, Whalen 1992) or freeze-drying (Barrett et al. 1997; Litvin et al. 1998; Ma and Peltre 1975a,b; Wang and Shi 1999) also has potential. The vacuum process opens the cell structures (puffing) due to fast evaporation, resulting in an open pore structure. Reduced drying time is the primary advantage of using microwaves in the freeze-drying process, but no commercial industrial application can be found, due to high costs and a small market for freeze-dried food products.

Pasta and potato chips have been dried successfully. Freeze-drying and vacuum drying, in conjunction with microwave energy, have also shown promise, and although the process is interesting from an academic point of view, it does not meet economic criteria. A new technology from Battelle Ingenieurtechnik GmbH of Germany for drying fruits and vegetables has been developed, wherein air belt drying is followed by microwave-vacuum puffing, then further air belt drying or vacuum drying, before sorting and packaging. Effects of this procedure on physicochemical properties, sensory properties, and the ultrastructure of fruits and vegetables are considered together with the avoidance of microwave hot spots and other products that would be suitable for processing by this method (Langer 2000, Räuber 1998). Recently, a relatively new and successful combination of microwave energy and frying is used to produce fried goods, such as chips, noodles, and chickens, with 60% reduced time, 50% reduced fat content and 33–60% energy saving (FIRDI 2003).

PASTEURIZATION AND STERILIZATION

Pasteurization inactivates pathogenic vegetative cells of bacteria, yeast, or molds. Pasteurized products generally have to be refrigerated. Sterilization processes are designed to inactivate microorganisms or their spores. Thermal sterilization is usually done at temperatures in excess of 100°C, which means they are usually done under pressure. Industrial microwave pasteurization and sterilization systems have been reported on and off for over 30 years. Studies with implications for commercial pasteurization and sterilization have also appeared for many years (Burfoot et al. 1988,1996; Cassanovas et al.

1994; Hamid et al. 1969; Knutson et al. 1988; Kudra et al. 1991; Proctor and Goldblith 1951; Villamiel et al. 1997; Zhang et al. 2001). Early operational systems include batch processing of yogurt in cups (Anonymous 1980) and continuous processing of milk (Sale 1976). A very significant body of knowledge has been developed related to these processes. As of this writing, two commercial systems worldwide can be found that currently perform microwave pasteurization and/or sterilization of foods (Akiyama 2000, Tops 2000). As a specific example, Tops Foods (Belgium) (Tops 2000) produced over 13 million ready-to-eat meals in 1998 and installed a newly designed system in 1999. Although continuous microwave heating in a tube flow arrangement has been studied at the research level, no commercial system is known to exist for food processing.

Microwave pasteurization can reduce the come-up time, which can be shortened to a small fraction of the time used in the conventional process. After pasteurization, the microwave-heated meals pass into a nonmicrowave hot air tunnel for the hold time period, and then to the cooler. With microwaves it is difficult to hold a constant temperature, and they should not be used at this stage. Especially in Europe, food pasteurization by microwave processing has been successfully accomplished for decades. The major advantage of the microwave process is that the product may be pasteurized within a package. A wrapped product goes through the line continually, package by package, pallet by pallet. Shelf life can be extended from days to over a month without preservatives. For example, due to higher moisture content, the usability of untreated toast bread is quite short—approximately six days. Distinctive pasteurization effects can be achieved by fast microwave heating (< 35 seconds) and a 15-minute pause at a temperature higher than 50°C (ROMill®). The condition of durability can be optimally fulfilled, even from the microbiological point of view, at an output temperature of 77°C after only 20 seconds of exposure from the initial temperature of 22°C, which is considerably faster than any other method of heating. If slow cooling follows, the tests of shelf life show a usability time of longer than 45 days.

For commercial sterilization, temperatures in the product may be 121–129°C (250–265°F), with hold times of 20–40 minutes. Come-up time may be significantly reduced by use of microwaves, and reduced come-up time would provide greater product quality since quality attributes normally have an activation energy much lower (10–40 kcal/mol) than that of microbial spores (50–95 kcal/mol). Microwave sterilization is more flexible than ohmic heating and aseptic processing. Liquid, semisolid, and solid prepackaged food products can also be sterilized. CAPPS and Industrial Microwave Systems manufacture a flow-through cylindrical microwave reactor to eliminate the heat-up time of thermal processing. In the cylindrical reactor, microwaves are focused to provide uniform exposure of product to energy within the reactor cavity. The uniform energy exposure region of the reactor is approximately 1.5 inches in diameter and 6 inches long. This reactor also allows for integration with existing continuous processing lines (Mermelstein 2001). In Europe, microwave-sterilized foods, primarily pasta dishes such as lasagna and ravioli, are on many grocery shelves, with no reported difficulties. Safety regulations are less stringent in Europe. For example, in one implementation (Tops 2000) the process design consists of microwave tunnels with several launchers for each different type of product (ready meals). Microwave-transparent and heat-resistant trays with shapes adapted for microwave heating are used. Exact positioning of the package is made within the tunnel, and the package receives a precalculated, spatially varying microwave power profile optimized for that package. The process consists of heating, holding, and cooling in pressurized tunnels. The entire operation is highly automated.

Use of microwaves for food sterilization has not been approved by the Food and Drug Administration in the United States. There are several practical concerns and problems that must be addressed before microwave sterilization can be applied at the industrial level. The main issue has been the regulation of process parameters so that commercial sterility can be achieved. For conventional retort processes, monitoring the time-temperature history at the cold point with a thermocouple thermometer is reasonably easy and accurate for determining microbial lethality through mathematical calculations. But, determining the microbial lethality for a microwave sterilization process is not straightforward. The cold point during microwave sterilization is not always located on the central axis. The difficulty of providing a uniformly heated product makes it extremely time consuming and costly to adjust the microwave pattern to produce the quality advantage that is theoretically possible with the use of microwaves. Each product could require custom adjustment. The presence of uneven heating (hot and cold spots) makes it very difficult to ensure that all portions of a meal have reached a kill temperature. Microbiological safety is the major reason for the slow acceptance of microwave sterilization. In addition, the technical ability to accurately measure the temperature distribution throughout an entire microwave-sterilized product has not been demonstrated. From the engineering point of view, no computer simulation models are available for investigating the feasibility of microwave sterilization. These computer simulation models are not only required by the Food and Drug Administration for regulating and approving microwave sterilization processes, but also are in high demand by the food industry for performing cost/benefit analyses. Without reliable inputs of dielectric properties, thermophysical properties, and boundary conditions, a computer model is completely useless. Unfortunately, literature values on these properties are only available at room temperature to 60°C and are not readily available for sterilization temperatures.

TEMPERING AND THAWING

Thawing and tempering of biological products used to be a slow process. For many production processes, incoming raw material is frozen in thick blocks and stored at −23 to −10°C until ready to use. The first operation on this material usually is to dice, slice, or separate individual sections into smaller pieces. This mechanical operation requires that the blocks be "tempered" from their solid frozen state to a point just below freezing (−7 to −1°C), at which point cutting or separation can be done without damage to the product. Thawing and tempering of frozen food materials is an important part of some food processes, especially in the meat industry and in food service. Reduced thawing time results in a decrease in product quality, such as more drip loss and surface drying, as well as increased risk of microbial growth.

Frozen foods can be considered a mixture containing two components: (1) a fixed structure of ice and biological material surrounded by a monomolecular layer of strongly bound water and (2) liquid water saturated with dissolved salts. The dielectric activity of this mixture is much higher than that of pure ice, but much lower than that of the same material at temperatures above 0°C. The loss factor (ε'') of water is approximately 12, while that of ice is approximately 0.003. The penetration depth in water (1.4 cm) is much lower than in ice (1160 cm) (von Hippel 1954).If the thickness value is much greater than the penetration depth, the temperature profile will be similar to that observed for a "semi-infinite" body. That is, the temperature will decrease exponentially from the surface in accordance with Lambert's law. Surface layers thus absorb more energy and heat up a little faster than the inside of the product. But for thickness values smaller than a certain value, resonance cannot be avoided, and the inside of the slab can be heated directly at a high intensity, resulting in quick thawing. As the loss factor increases with the temperature, the surface heats up faster and faster, and the penetration depth continually decreases. Spots of free water and spots that have reached temperature > 0 °C absorb more energy than ice crystals, which leads to further acceleration of heating. Microwave energy penetrates a food material and produces heat internally. The main advantage of microwave energy consists in speed: tempering by microwaves takes minutes instead of hours or even dozens of hours. For example, a 20 cm thick piece of beef, frozen to −16°C, thaws in more than 10 hours at the surrounding temperature of +4 °C. On the other hand, the whole cycle of microwave tempering following slicing, modification, and repeated freezing takes only 30 minutes (ROMill® 2003). In another example, from Microdry Corporation, cartons of frozen food in solid blocks weighing up to 100 pounds are raised in temperature to just below freezing using conventional tempering (Microdry, Inc. 2003). Most plants dunk the blocks into warm water. Others use hot air. Many use floor tempering alone, without any heat aid, which may take 48–72 hours. By contrast, microwave tempering is applied on a moving belt to food still in cartons and generally takes less than five minutes. Thus, without doubt, this is a major successful application of microwave heating in industry. There are at least 400 tempering systems operating in the United States alone. Food is heated to just under freezing temperatures, allowing easy chopping, cutting, processing, and so on. In the United Kingdom there are several large systems, up to 200 kW, utilized for tempering frozen beef, as well as butter. The lower frequencies, for example, the 915 MHz band, are used to advantage for microwave thawing and tempering of larger blocks of food. As a general rule, microwave energy at 915 MHz has three times the penetration depth of 2450 MHz, thus allowing for greater bed depths and processing of larger product geometries. For example, when tempering 18 cm thick blocks at 915 MHz, the temperature gradient is half that of the gradient for 2450 MHz (ROMill® 2003). 915 MHz tempering systems, batch and continuous, are sold worldwide.

Although microwaves have been successfully applied to tempering frozen products, microwave thawing remains a major problem. A main difficulty is formation of large temperature gradients (runaway heating) within the product. The preferential absorption of microwaves by liquid water over ice is a major cause for runaway heating. Maximum homogeneity is achieved with temperatures slightly above zero. After that the nonhomogeneity rises again. Therefore, it is advantageous to reduce the thawing process to plain tempering, that is, to stop the heating at temperatures of −5 to −2°C. Another reason to prefer tempering is the progress of energy consumption as a function of temperature. With most biological materials and water, energy consumption starts to rise sharply at temperatures above −5°C; the less fat the product contains, the higher the microwave absorption. Since thawed material has a much higher dielectric loss, microwave penetration depth at the surface is significantly reduced, in effect developing a "shield." Surface cooling helps reduce the gradient in a frozen food, thus enabling the microwave power to remain on longer, further decreasing the thawing time. Temperature uniformity during microwave thawing can be improved when appropriate sample thickness, microwave power level, frequency, and/or surface cooling are applied (Bengtsson 1963, Bialod et al. 1978; Decareau 1985, Virtanen et al. 1997). Today, there continues to be a great deal of interest and some research and development activity in thawing and tempering by microwaves (Chamchong and Datta 1999a,b; George 1997; Li and Sun 2002).

BAKING

Baking, in all cases except unleavened products, involves the creation, expansion, and setting of edible foams through the use of heat. Proofing is the step of allowing the dough to rise and precedes the final baking, or frying in the case of doughnuts. During baking of raw bread dough, significant volume change occurs, and the dough is converted from a viscoelastic material containing airtight gas cells with the ability to expand to a rigid structure that is highly permeable to gas flow. The cell walls are elastic but strong, and the increasing gas pressure must cease while the cell walls set. Baking is a complex physicochemical reaction in which all the events must be carefully timed and must occur in a well-defined sequence. All baked products form some sort of crust, which acts as a shield, making it even harder for heat to reach the inside. The heat transfer problems encountered in conventional heating can be easily overcome by microwave heating. Pei (1982) reviewed heat and mass transfer in the bread baking process and discussed the application of microwave energy. Goedeken (1994) investigated microwave baking of bread dough with simultaneous heat and mass transfer. Highly porous products, such as bread, lend themselves well to the use of microwave energy because of the greater penetration of microwave energy, which results in more uniform energy distribution within the product. However, the microwave application must be carefully controlled or heating and expansion will occur too quickly, and while the product may look fully expanded and baked, it will collapse to a pancake when the microwave energy is removed.

There are four broad classes of products that have been studied for microwave applications: yeast-raised (bread, Danish pastry), chemically leavened (doughnuts, cake), steam-leavened (angle food cake, Chinese-style steamed bread), and unleavened products (cookies, crackers, matzos) (Schiffmann 2001). Yeast-raised dough has a well-defined structure and shape prior to the final heat-setting treatment: baking or frying. Chemically leavened batter is flowable and amorphous in shape and therefore requires some sort of shape defining structure (e.g., a cake pan or the rapidly formed crust of a doughnut) to be present.

Bread baking by means of microwave energy was first reported in the literature by Fetty (1966). Decareau (1967) noted the possibility of combining microwave energy and hot air to produce typically brown and crusted loaves of bread in a shorter time than by conventional baking methods. One microwave baking process that was quite successful for several years was microwave frying of doughnuts. Frying times of approximately two-thirds normal time are possible with 20% larger volumes, or 20% less doughnut mix required for standard volume. Fat absorption can be 25% lower than in conventional frying. This proofing system was developed by DCA Food Industries; it operated at 2450 MHz and varied in output from 2.5 to 10 kW for production rates of 400–1500 dozen doughnuts per hour (Schiffmann 1971; Schiffmann et al. 1971, 1979).

One difficulty in the microwave baking process was to find a microwavable baking pan that was sufficiently heat resistant and not too expensive for commercial use. A patent by Schiffmann and others (1981) describes microwave proofing and baking of bread in metal pans. This technique utilizes partial proofing in a conventional proofing system followed by proofing in a microwave proofer utilizing warm, humidity-controlled air. This process reduces proofing time by 30–40%. This was then followed by microwave baking in a separate oven. Four patents by Schiffmann and others (1979, 1981, 1982, 1983) describe procedures for baking bread utilizing metal pans and, in some cases, also provided for partial proofing of the bread in the pans.

In the procedure described in the aforementioned patents, the microwave baking process involved the simultaneous application of microwave energy and hot air to both bake and brown the bread, producing thoroughly browned and crusted loaves of comparable volume, gain structure, and organoleptic properties. It was found that the use of either 915 MHz or combinations of 915 and 2450 MHz were quite effective in baking a loaf of bread. The system for microwave frying of doughnuts was very successful for quite some time during the 1970s. These doughnuts have longer shelf life, better sugar stability, and excellent eating quality. The larger volume and lower fat absorption provided high profits for the bakery. However, the microwave frying system disappeared after several years. The reasons are quite complex and have little or nothing to do with their performance or the quality of the doughnuts. Generally, the baking industry is extremely slow to adopt new technologies because baking ovens are expensive and represent major capital investments. Furthermore, it is almost impossible to retrofit an existing baking oven with microwaves, primarily because of problems of microwave leakage, so it is only possible to install a microwave baking oven or proofer when a new line is installed.

To date, some very sophisticated packaging, coupled with an advanced susceptor technology, has been the predominant solution to the lack of conventional-styled browning and crisping. Susceptors rapidly heat to temperatures where browning readily occurs and thus help produce flavor in the product. However, susceptors solve the flavor-related problems only on the product surface. Another possible solution to the lack of browning during microwave cooking is the addition of compounds that give a roasted or toasted reaction flavor.

RADIO FREQUENCY PROCESSING

Radio frequency and microwave heating refers to the use of electromagnetic waves of certain frequencies to generate heat in a material (Metaxas 1996, Metaxas and Meredith 1983, Roussy and Pearce 1995). Radio frequency heating, which is at a much lower frequency than microwave heating, has thrived as an industry alongside microwaves over the decades. Radio frequency heating in the United States can be performed at any of the three frequencies, 13.56, 27.12, and 40.68 MHz. The heating mechanism of radio frequencies is simply resistance heating, which is similar to ohmic heating. This lossy dielectric arises from the electrical conductivity of the food and is different from the resonant dipolar rotation of microwave frequencies.

Unlike microwave sources, one cannot purchase a high-power radio frequency source. Due to the high impedance nature of radio frequency coupling, the radio frequency source and applicator normally need to be designed and built together. Manufacturers of radio frequency equipment develop the whole system, rather than only the power source. Therefore, developments in radio frequency processing must involve the commercial radio frequency manufacturers. Radio frequency equipment is available commercially at much higher power levels than microwave sources. While commercial microwave sources are available only below 75kW, radio frequency equipment at hundreds of kilowatts is very common. At these high levels, the price per watt of radio frequency equipment is much cheaper than that of microwaves. In addition to higher power and lower cost, another advantage of radio frequency equipment over microwaves is in the control area. In high-power radio frequency systems, the source and the load are commonly locked together in a feedback circuit. Therefore, variations in the load can be followed by the source without external controls (Mehdizadeh 1994).

The question is when to use microwave and when to use radio frequency? For the same electric field, the higher the frequency, the higher the amount of power transferred into the material. This is why microwaves are conceptually a more effective means of heating. However, radio frequency equipment has several advantages that workers in processing may find more suitable for scale-up of some processes. Microwave fields attenuate within the bulk of conductive materials and in materials with high dielectric loss. Furthermore, the penetration depth of microwaves is much lower. This is particularly troublesome for larger scale processes. But this type of nonuniformity is frequency dependent and becomes less severe as frequency is lowered. Because of the much longer wavelengths of radio frequencies, they have better uniformity. Also, the depth of penetration is much higher. So, in cases where uniformity of heating is a critical issue, use of radio frequencies and 915 MHz microwave frequency may have potential for the future (Lau et al. 1999, Wig et al. 1999).

Using radio frequencies allows processing of a large range of materials from thin, wide webs of paper to large three-dimensional objects like textile packages. In general terms, microwave is better for irregular shapes and small dimensions, and radio frequency is better for regular shapes and large dimensions. Microwave is more suitable for hard-to-heat dielectrics. Actually, many applications are suitable for either microwave or radio frequency, but radio frequency is cheaper if it fits. Radio frequency equipment is easier to engineer into process lines and can be made to match the physical dimensions of the up- and downstream plant. In the case of microwaves, in a continuous process, complex arrangements may be necessary to allow the product to move in and out of the enclosure without giving rise to excessive leakage of energy (Jones and Rowley 1997). This is because the wavelengths at microwave frequencies (e.g., 12.54 cm at 2450 MHz) are very much shorter than those at radio frequencies (e.g., 1100 cm at 27.12 MHz).

An overview of food and chemical processes using radio frequency can be seen in Minett and Witt (1976) and Kasevich (1998). Industrial applications using radio frequency include textiles (drying of yarn packages, webs, and fabrics), foods (bulk drying of grains; moisture removal and moisture leveling in finished food products), pharmaceuticals (moisture removal in tablet and capsule production processes), and woodworking (adhesive curing for wood joinery). Radio frequency heating has been used in the food processing industry for many decades. The postbaking of biscuits, crackers, and snack foods is one of the most accepted and widely used applications of radio frequency heating in the food processing industry. A relatively small radio frequency unit can be incorporated directly into a new or existing oven line (a hot air oven or conventional baking line) to increase the line's productivity and its ability to process a greater range of products.

The benefits of radio frequency–assisted baking are precise moisture control, reduced checking, improved color control, and increased oven-line throughput (Radio Frequency Co., Inc. 2003). Radio frequency drying is intrinsically self-leveling, with more energy being dissipated in wetter regions than in drier ones (Jones and Rowley 1997). This radio frequency leveling leads to improvements in product quality and more consistent final products. On the Goldfish line, Pepperidge Farm has added radio frequency drying equipment that reduces the moisture of the snack cracker by half without impacting color, size, or other baking characteristics. Today, the plant is able to double its production capacity. Another application is the drying of products such as expanded cereals and potato strips. Recently, radio frequency cooking equipment for pumpable foods has been developed. These devices involve pumping a food through a plastic tube placed between two electrodes, shaped to give uniform heating (Ohlsson 1999). The primary advantage of improved uniformity of heating was also shown for in-package sterilization of foods in large packages using radio frequency at 27.12 MHz, although enhanced edge heating continued to be an issue (Wig et al. 1999). Commercial radio frequency heating systems for the purpose of food pasteurization or sterilization are not known to be in use, although it has been researched over the years (Bengtsson and Green 1970, Houben et al. 1991, Wig et al. 1999). Defrosting frozen food using radio frequency was a major application, but problems of uniformity with foods of mixed composition limited the actual use. The interest in radio frequency defrosting has increased again in recent of years (Ohlsson 1999).

Today, the use of a more recent 50Ω radio frequency heating device that allows the radio frequency generator to be placed at a convenient location away from the radio frequency applicator offers the possibility of advanced process control (Rowley 2001). Whether conventional or 50Ω dielectric heating systems are used, the radio frequency applicator must be designed for the particular product being heated or dried. Radio frequency postbaking, radio frequency–assisted baking, and radio frequency meat and fish defrosting systems will continue to benefit both existing and emerging food applications, and the availability of low cost radio frequency power sources could lead to major growth in use of radio frequency heating in the commercial food sectors. Radio frequency heating is well established in industry, and for many applications, it is the standard method. Its equipment is well proven and reliable. It is an excellent choice where it fits.

FUTURE OF MICROWAVE/RADIO FREQUENCY HEATING IN FOOD INDUSTRY

- The fundamentals of microwave heating should be studied in depth before spending a great deal of time and effort on trial and error. As to the future, successful development of the microwave-assisted food industry can be achieved as a result of greater scientific and technological understanding of microwave-food interactions and continued cooperation between scientists, food technologists, food process engineers, and electrical engineers in this area.

- Microwave and radio frequency heating provide a product that is potentially superior in quality to a product produced by conventional techniques. This point is key to almost all industrial processes. Commercial success of a microwave process is possible if the products are of high intrinsic economic value and can carry the extra cost burden put on them. Economic considerations usually eliminate commodity products from consideration.

- The term "hybrid energy" refers to a microwave/radio frequency processing in conjunction with hot air and steam. The potential synergistic effects of microwaves combined with steam, forced-air convection, and/or infrared will probably lead future expansion of microwave processing technology. Internally, the foods will heat rapidly by microwave; at the surface, the traditional heat processes will provide the desired texture, color, and appearance.

ACKNOWLEDGMENT

The author thanks Professor Daryl B. Lund, Executive Director, North Central Regional Association of State Agricultural Experiment Station Directors, University of Wisconsin, Madison, and Professor An-I Yeh, Graduate Institute of Food Science and Technology, National Taiwan University, for reviewing the first draft of this chapter. Any remaining deficiencies belong solely to the author.

REFERENCES

Akiyama H. 2000. Otsuka Chemical Co., Ltd., 463 Kagasuno Kawauchi, CHO, Tokushima 771-0193 Japan. Phone: 0886-65-6672. Email: hakiyama@otsukac.co.jp

Anonymous. 1980. The potential of Bach. Food Manufacture 55(10): 53.

Ayappa KG, HT Davis, G Crapiste, EA Davis, J Gordon. 1991a. Microwave heating: An evaluation of power formulations. Chemical Engineering Science 46(4): 1005–1016.

Ayappa KG, HT Davis, EA Davis, J Gordon. 1991b. Analysis of microwave heating of materials with temperature-dependent properties. AIChE Journal 37(3): 313–322.

Barrett AH, AV Cardello, A Prakash, L Mair, IA Taub, LL Lesher. 1997. Optimization of dehydrated egg quality by microwave assisted freeze-drying and hydrocolloid incorporation. Journal of Food Processing and Preservation 21(3): 225–244.

Bassen HI, GS Smith. 1983. Electric field probes–A review. IEEE Transactions on Antennas and Propagation, Ap-31(5): 710–718.

Bengtsson NE. 1963. Electronic defrosting of meat and fish at 35 and 2450 MHz. A laboratory comparison. Food technology 17(10): 97.

Bengtsson NE, W Green. 1970. Radio-frequency pasteurization of cured hams. Journal of Food Science 35:681–687.

Bialod D, M Jolion, R LeGoff. 1978. Microwave thawing of food products using associated surface cooling. Journal of Microwave Power 13:269.

Bosisio RG, M Nachman, R Nobert. 1974. A simple method for determining the electric field distribution along a microwave applicator. Journal of Microwave Power 10(2): 223–231

Boehm M, M Bade, B Kunz. 2002. Quality stabilization of fresh herbs using a combined vacuum-microwave drying process. Advances in Food Sciences 24(2): 55–61.

Buffler CR. 1993. Microwave Cooking and Processing: Engineering Fundamentals for the Food Scientist, 128–141. New York: Van Nostrand Reinhold.

Burfoot D, WJ Griffin, SJ James. 1988. Microwave pasteurization of prepared meals. Journal of Food Engineering. 8: 145–156.

Burfoot D, CJ Railton, AM Foster, R Reavell. 1996. Modeling the pasteurization of prepared meal with microwaves at 896 MHz. Journal of Food Engineering 30:117–133.

Casasnovas J, RC Anantheswaran, J Shenk, VM Puri. 1994. Thermal processing of food packaging waste using microwave heating. Journal of Microwave Power Electromagnetic Energy 29:171.

Chamchong M, AK Datta. 1999a. Thawing of foods in a microwave oven. I. Effect of power levels and power cycling. Journal Microwave Power and Electromagnetic Energy 34(1): 9–21.

___. 1999b. Thawing of foods in a microwave oven. II. Effect of load geometry and dielectric properties. Journal Microwave Power and Electromagnetic Energy 34(1): 22–32.

Cheng DK. 1990. Field and wave electromagnetics, 2nd edition, 321–343. New York: Addison-Wesley Publishing Company.

Clark JP. 2002a. Thermal and nonthermal processing. Food Technology 56(12): 63–64.

___. 2002b. Extending the shelf life of fruits and vegetables. Food Technology 56(4): 98–100.

Cober Electronics, Inc. 2003. 151 Woodward Avenue, Norwalk, CT 06854. Voice: 203-855-8755. Fax: 203-855-7511. E-mail: sales@cober.com

Datta AK, E Sun, A Solis. 1995. Food dielectric property data and its composition-based prediction. In: MA Rao, SSH Rizvi, editors. Engineering properties of Food, 457–494. New York: Marcel Dekker.

Decareau RV. 1967. Application of high frequency energy in the baking field. Baker's Digest 41(6): 52–53.

___. 1985. Microwaves in the food processing Industry. New York: Academic Press, Inc.

Durance TD, Z Vaghri, CH Scaman, DD Kitts, JH Wang, C Hu. 2001. Process for Dehydration of Berries. United States Patent US 6,312,745 B1

Fetty H. 1966. Microwave baking of partially baked products. Proceedings of the American Society of Bakery Engineers, 145–166. Chicago, Ill.

Feng H, Tang J. 1998. Microwave finish drying of diced apples in a spouted bed. Journal of Food Science 63(4): 679–683.

FIRDI. 2003. Food Industry Research and Development Institute, 331 Shih-Pin Road, Hsinchu, Taiwan 300, R.O.C. Voice: 886-3-5223191 (ext 307). Fax: 886-3-5214016. E-mail: bby@firdi.org.tw

Food and Drug Administration. 2000. Kinetics of Microbial Inactivation for Alternative Food Processing Technologies. FDA Center for Food Safety and Applied Nutrition report–A report of the IFT for the FDA of the U.S. Department of Health and Human Services. June 2.

Fu W, AC Metaxas. 1992. A mathematical derivation of power penetration depth for thin lossy materials. Journal Microwave Power and Electromagnetic Energy 27(4): 217–222.

Fu YC. 1996. Microwave-assisted Heat and Mass Transfer in Food. Ph.D. dissertation, Rutgers–The State University of New Jersey, New Brunswick, N.J.

Fu YC, CH Tong, DB Lund. 1994. Microwave bumping: Quantifying explosions in foods during microwave heating. Journal of Food Science 59(4): 899–904.

___. 2001. Photoionization detection (PID) method for on-line measurement of flavor concentration. Food Science and Agricultural Chemistry 3:97–101.

___. 2003a. Flavor migration out of food matrices: I. System development for on-line measurement of flavor concentration. Journal of Food Science 68(3): 775–783.

___. 2003b. Flavor migration out of food matrices: II. Quantifying flavor migration from dough undergoing isothermal heating. Journal of Food Science 68(3): 923–930.

___. 2003c. Flavor migration out of food matrices: III. Migration of limonene and pyrazine in formulated dough undergoing microwave reheating. Journal of Food Science 68(3): 931–936.

___. 2003d. Moisture migration in solid food matrices. Journal of Food Science 68(8): 2497-2503.

Fu YC, L Dai, BB Yang. 2003e. Microwave finish drying of (tapioca) starch pearls. International Journal of Food Science and Technology (Accepted for publication.).

Funebo T, T Ohlsson. 1998. Microwave-assisted air dehydration of apple and mushroom. Journal of Food Engineering 38: 353–367.

George M. 1997. Industrial microwave food processing. Food Review 24(7): 11–13.

Goedeken DL. 1994. Microwave Baking of Bread Dough with Simultaneous Heat and Mass Transfer. PhD. Dissertation, Rutgers–The State University of New Jersey, New Brunswick, N.J.

Gunasekaran S. 1999. Pulsed microwave-vacuum drying of food materials. Drying Technology 17(3): 395–412.

Hamid MAK, RJ Boulanger, SC Tong, RA Gallop, RR Pereira. 1969. Microwave pasteurization of raw milk. Journal of Microwave Power. 4(4): 272–275.

Houben J, L Schoenmakers, E van Putten, P van Roon, B Krol. 1991. Radio-frequency pasteurization of sausage emulsions as a continuous process. Journal of Microwave Power Electromagnetic Energy. 26(4): 202–205.

Jones PL, AT Rowley. 1997. Chapter 8. Dielectric dryers. In: CJ Baker, editor. Industrial Drying of Foods. London:Chapman and Hall.

Kasevich RS. 1998. Understand the potential of radiofrequency energy. Chemical Engineering Progress, 75–81.

Kent M. 1987. Electrical and Dielectric Properties of Food Materials. Essex, England: Science and Technology Publishers, Ltd.

Knutson KM, EH Marth, MK Wagner. 1988. Use of microwave ovens to pasteurize milk. Journal of Food Protection 51(9): 715–719.

Kostaropoulos AE, GD Saravacos. 1995. Microwave pretreatment for sun-died raisins. Journal of Food Science 60: 344–347.

Kudra T. 1989. Dielectric drying of particulate materials in a fluidized state. Drying Technology 7(1): 17–34.

Kudra T, FR Van De Voort, SV Raghavan, HS Ramaswamy. 1991. Heating characteristics of milk constituents in a microwave pasteurization system. Journal of Food Science 56(4): 931–934.

Lau MH, J Tang, IA Taub, TCS Yang, CG Edwards, FL Younce. 1999. HTST processing of food in microwave pouch using 915 MHz microwaves. AIChE Annual Meeting.

Langer G. 2000. Microwave on high then quickly dry. Food Technology in New Zealand 35(4): 8.

Li B, DW Sun. 2002. Novel methods for rapid freezing and thawing of foods–A review. Journal of Food Engineering 54(3): 175–182.

Litvin S, CH Mannheim, J Miltz. 1998. Dehydration of carrots by a combination of freeze drying, microwave heating and air or vacuum drying. Journal of Food Engineering 36(1): 103–111.

Lorrain P, DP Corson, F Lorrain. 1988. Electromagnetic Fields and Waves, 3rd edition, 522–529. New York: W.H. Freeman and Company.

Lu L, J Tang, X Ran. 1999. Temperature and moisture changes during microwave drying of sliced food. Drying Technology 17(3): 413–432.

Ma YH, PR Peltre. 1975a. Freeze dehydration by microwave energy. Part I. Theoretical investigation. AIChE Journal 21(2): 335–344.

___. 1975b. Freeze dehydration by microwave energy. Part II. Experimental study. AIChE Journal 21(2): 344–350.

Maskan M. 1999. Microwave/air and microwave finish drying of banana. Journal of Food Engineering 44:71–78.

MacLatchy CS, Clements RM. 1980. A simple technique for measuring high microwave electric field strengths. Journal of Microwave Power 15(1): 7–14

Mehdizadeh M. 1994. Engineering and scale-up considerations for microwave induced reactions. Res. Chem. Intermed. 20(1): 79–84.

Mermelstein NH. 2001. High-temperature, short-time processing. Food Technology 55(6): 65–66.

Metaxas AC, RJ Meredith. 1983. Industrial microwave heating. IEE Power Engineering Series, No. 4, 70–83. London: Peter Peregrinus Ltd.

Metaxas R. 1996. Foundations of Electroheat: A Unified Approach. John Wiley and Sons. Chichester, U.K.

Microdry Incorporated. 2003. 7450 Highway 329, Crestwood, KY 40014. E-mail: engineering@microdry.com

MicroGas Corporation. 2003. 127 Bellevue Way SE, Bellevue, WA 98004-6229. Phone: 425-453-9223. Fax: 425-455-0981. E-mail: rtidball@energyint.com

Minett PJ, JA Witt. 1976. Radio frequency and microwaves. Food Processing Industry, 36–37.

Mudgett RE. 1986. Microwave properties and heating characteristics of foods. Food Technology, 40(6): 84–93.

Mullin J, J Bows. 1993. Temperature measurement during microwave cooking. Food Additives and Contaminants 10(6): 663–672.

Nelson SO. 1973. Electrical properties of agricultural products–A critical review. Transaction ASAE 16(2): 384–400.

Nykvist WE. 1977. Microwave Meat Roasting–A Computer Analysis for Cylindrical Roasts. Technical Report NATICK TR-77/022. United States Army, Natick Research and Development Command, Natick, Mass.

Nykvist WE, RV Decareau. 1976. Microwave meat roasting. Journal of Microwave Power 11(1): 3–24.

Ohlsson T. 1999. Chapter 6. Minimal processing of foods with electric heating methods. In: FAR Oliveira, JC Oliveira, editors. Processing Foods–Quality Optimization and Process Assessment. New York: CRC Press LLC.

Ohlsson T, N Bengtsson. 1971. Microwave heating profiles in foods–A comparison between heating experiments and computer simulation. Microwave Energy Applications Newsletter 4(6): 3–8.

Ohlsson T, PO Risman. 1978. Temperature distribution of microwave heating–Spheres and cylinders. Journal Microwave Power 13(4): 303–310.

Parent A, CH Tong, DB Lund. 1992. Temperature and moisture distributions in porous food materials during microwave heating. Personal Communication.

Pearce JA, SI Yang, PS Schmidt. 1988. A research program for dielectric heating and drying of industrial materials. In: WH Sutton, MH Brooks, IJ Chabinsky, editors. Microwave Processing of Materials. Materials Research Society Symp. Proc. 124:329–334.

Pei DCT. 1982. Microwave baking, new developments. Bakers Digest 2:8–9.

Prabhanjan DG, HS Ramaswamy, GSV Raghavan. 1995. Microwave-assisted convective air drying of thin layer carrots. Journal of Food Engineering 25: 283–293.

Proctor BE, SA Goldblith. 1951. Electromagnetic radiation fundamentals and their applications in food technology. Advances in Food Research 3:120–196.

Radio Frequency Co., Inc. 2003. 150 Dover Road, Millis, MA 02054. Phone:508-376-9555. Fax: 508-376-9944. Email: rfc@radiofrequency.com

Randa J. 1990. Theoretical considerations for a thermo-optic microwave electric-field-strength probe. Journal of Microwave Power and Electromagnetic Energy 25(3): 133–140

Räuber H. 1998. Instant-Gemüse aus dem östlichen Dreiländereck. Gemüse, 10'98.

Risch SJ. 1989. Flavors for microwavable foods. Cereal Foods World 34(2): 226.

Risman O. 1991. Terminology and notation of microwave power and electromagnetic energy. Journal Microwave Power and Electromagnetic Energy 26(4): 243–250.

ROMill®. 2003. spol. s r.o. Kotlarska 53, Brno CZ-658 92, Czech Republic. E-mail: info@romill.cz

Roussy G, Pearce J. 1995. Foundations and Industrial Applications of Microwaves and Radio Frequency Fields. New York: Wiley.

Rowley AT. 2001. Chapter 9. Radio frequency heating. In: P Richardson, editor. Thermal Technologies in Food Processing. Boca Raton: CRC Press LLC.

Sale AJH. 1976. A review of microwave for food processing. Journal of Food Technology 11:319–329.

Sastry SK. 1994. Ohmic heating. In: RP Singh, FAR Oliveira, editors. Minimal processing of foods and process optimization: An interface, 17–33. Boca Raton, Fla.: CRC Press, Inc.

Schiffmann RF. 1971. Applications of microwave energy to doughnut production. Food Technology 25:718–722.

___. 2001. Microwave Processes for the Food Industry. In: AK Datta, RC Anantheswaran, editors. Handbook of Microwave Technology for Food Applications, 299–337. New York: Marcel Dekker, Inc.

Schiffmann RF, EW Stein, HB Kaufman, Jr. 1971. Dough Proofing. U.S. Patent 3,630,755.

Schiffmann RF, AH Mirman, RJ Grillo, SA Wouda. 1979. Microwave Baking of Brown and Serve Products. U.S. Patent 4,157,403.

Schiffmann RF, AH Mirman, RJ Grillo. 1981. Microwave Proofing and Baking Bread Utilizing Metal Pans. U.S. Patent 4,271,203.

___. 1982. Method of Baking Firm Bread, U.S. Patent 4,318,931.

Schiffmann RF, AH Mirman, RJ Grillo, RW Batey. 1983. Microwave Baking with Metal Pans. U.S. Patent 4,388,335.

Shukla TP, RC Anantheswaran. 2001. Chapter 11. Ingredient Interactions and Product Development for Microwave Heating. In: AK Datta, RC Anantheswaran. editors. Handbook of Microwave

LEEDS TRINITY UNIVERSITY

Technology for Food Applications. New York: Marcel Dekker, Inc.

Stuchly SS, MAK Hamid. 1972. Physical parameters in microwave heating processes. Journal Microwave Power 7(2): 117–137.

Stuchly MA, SS Stuchly. 1980. Dielectric properties of biological substances–Tabulated. Journal of Microwave Power 15(1): 19–26.

Steinke JA, CM Frick, JA Gallagher, KJ Strassburger. 1989. Chapter 49. Influence of microwave heating on flavor. In: TH Parliament, RJ McGorrin, CT Ho, editors. Thermal Generation of Aromas, 519–525. American Chemistry Society.

Sunderland JE. 1980. Microwave freeze drying. Journal of Food Process Engineering 4:195–212.

Taoukis P, EA Davis, HT Davis, J Gordon, Y Talmon. 1987. Mathematical modeling of microwave thawing by the modified isotherm migration method. Journal of Food Science 52:455–463.

Tinga WR, SO Nelson. 1973. Dielectric properties of materials for microwave processing– Tabulated. Journal of Microwave Power 8(1): 23–65.

Tong CH, YC Fu, DB Lund. 1990. Temperature, pressure, and moisture content of porous and nonporous food materials during microwave heating. Presented at the IDS'90 Seventh International Drying Symposium, August 26–30, Prague, Czechoslovakia.

Tops R. 2000. Tops Foods N.V., Lammerdries 26, B-2250 Olen, Belgium. Phone: 0032/14 28 55 60. Fax: 0032/14 28 55 80. E-mail : info@topsfoods.com

Torringa EM, EJ van Dijk, PS Bartels. 1996. Microwave puffing of vegetables: Modeling and measurements. In: Proceedings of 31st Microwave Power Symposium, Int. Microwave Power Inst., Manassas, Va.

Van Eijk T. 1992. Flavorings in microwave foods. Paper presented at the ASC Symposium. Washington D.C., August 27.

Villamiel M, R LopezFandino, A Olano. 1997. Microwave pasteurization of milk in a continuous flow unit. Effects on the cheese-making properties of goat's milk. Milchwissenschaft. 52(1): 29–32.

Virtanen AJ, DL Goedeken, CH Tong. 1997. Microwave assisted thawing of model frozen foods using feed-back temperature control and surface cooling. Journal of Food Science 62(1): 150–154.

von Hippel AR. 1954. Dielectric Properties and Applications. New York: The Technology Press of M.I.T. and John Wiley and Sons, Inc.

Wang ZH, MH Shi. 1999. Microwave freeze drying characteristics of beef. Drying Technology 17(3): 433–447.

Washisu S, Fukai I. 1980. A simple method for indicating the electric field distribution in a microwave oven. Journal of Microwave Power 15(1): 59–61.

Watanabe M, M Suzuki, S Ohbrawa. 1978. Analysis of power density distribution in microwave ovens. Journal of Microwave Power 13(2): 173–181.

Wei CK, HT Davis, EA Davis, J Gordon. 1985a. Heat and mass transfer in water-laden sandstone: Convective heating. AIChE Journal 31(8): 1338–1348.

___. 1985b. Heat and mass transfer in water-laden sandstone: Microwave heating. AIChE Journal 31(5): 842–848.

Whalen P. 1992. Half products for microwave puffing of expanded food product. U.S. Patent 5,102,679.

White JR. 1970. Measuring the strength of the microwave field in a cavity. Journal of Microwave Power 5(2): 145–147.

Whitney JD, JG Porterfield. 1968. Moisture movement in a porous, hygroscopic solid. Transaction of the ASAE. 11(5): 716–719.

Wickersheim KA, MH Sun. 1987. Fiberoptic thermometry and its applications. Journal of Microwave Power 22:85–94.

Wickersheim KA, MH Sun, A Kamal. 1990. A small microwave E-field probe utilizing fiberoptic thermometry. Journal of Microwave Power 25(3): 141–148.

Wig T, J Tang, F Younce, L Hallberg, CP Dunne, T Koral. 1999. Radio frequency sterilization of military group rations. AIChE Annual Meeting.

Yang S-I, JA Pearce. 1989. Boundary condition effects on microwave spatial power deposition patterns. Center for Energy Studies, Balcones Research Center, The University of Texas at Austin.

Zhang H, AK Datta, I Taub, C Doona. 2001. Experimental and numerical investigation of microwave sterilization of solid foods. AIChE Journal. 47(9): 1957–1968.

5

Food Packaging

L. J. Mauer, B. F. Ozen

BACKGROUND INFORMATION

INTRODUCTION

Food packaging is an integral and essential part of modern food processing and will play an increasingly significant role in the food industry as the use of new and alternative food processing operations

---

L. J. Mauer is corresponding author.

expands. Food packaging is defined as a coordinated industrial and marketing system for enclosing products in a container to meet the following needs: containment, protection, preservation, distribution, identification, communication, and convenience (Robertson 1993, Soroka 1999). Historically, developments in food packages have coincided with developments in society. The mass production of products and an increasingly urban society during the industrial revolution in the 1700s created the need for distribution packaging to move large quantities of products out of factories and bring large quantities of foods into cities. In the late 1800s, developments in supermarkets and refrigeration increased national distribution of products, and brand marks were introduced to package labeling. The development of fast-food restaurants in the 1950s created the demand for disposable single-service packages. Currently, an increasingly urban, international society with consumer demands for convenience and a wider variety of food products has increased the need for packaging to extend the shelf life of foods. Single events, such as the need for tamper-evident packages created after the Tylenol tampering incident in 1982, also may change the dynamics of the packaging industry (Soroka 1999). The role of packaging in the food industry will continue to evolve to meet the demands created as new societal and consumer expectations develop.

FUNCTIONS OF PACKAGING

The general functions of food packaging are referred to in the definition of packaging: containment, protection, preservation, distribution (transportation), identification, communication, and convenience (Robertson 1993, Soroka 1999). An ideal package will enable a safe, quality food product to reach the consumer at minimum cost. The importance of each function of packaging will depend on the type of food product, the location of the packaged product in the distribution chain, and the intended destination end point. Often, package functions are interdependent. It is important to note that if a package fails to function properly, much of the expense and energy put into the production and processing of the food product will be wasted.

Containment

This basic function of packaging is a key factor for all other packaging functions. A food product must be contained before a package can protect, preserve,

and identify it and before it can be moved from one location to another.

Protection

A good package will protect its contents from the environment (water vapor, oxygen, light, microorganisms, other contaminants, vibration, shock, etc.) while protecting the environment from its contents. Protection often occurs simultaneously with containment and/or preservation. If a product is not contained during distribution, the environment will not be protected, and the product will not be preserved.

Preservation

Packaging can function to preserve and/or extend the shelf life of food products. A can or pouch preserves thermostabilized foods by providing a barrier between the processed, shelf-stable foods and the environment (most significantly microorganisms). Other ways packaging can preserve foods include (1) acting as a barrier to water vapor, oxygen, carbon dioxide, other volatiles and contaminants, light, and microorganisms and (2) interacting with the product to extend shelf life (active packaging).

Distribution (Transportaion)

A good package design will facilitate effective movement of products through the entire distribution chain. The ability to efficiently unitize individual packages into larger containers is desirable for shipping and handling. Food products are often placed in industrial distribution packages (such as corrugated boxes and unitized pallet loads) designed to protect the product from stresses encountered during transportation, including vibration, shock, and compression forces.

Identification and Communication

By law, a package must display the specific name of the product, the quantity contained, the address of the responsible company, and often, nutritional information. The package design also will communicate by the material, shape and size, color, recognizable symbols or brands, and illustrations (Soroka 1999). Universal Product Code (UPC) symbols are used to facilitate both rapid retail checkout and tracking and inventory at warehouses and distribution centers (Robertson 1993). The basic structure of a bar code contains high-contrast rectangular bars and spaces, and the ratio of bar and space widths contains manufacturer and product information (Soroka 1999). The ultimate goal of identification and communication via labeling and design of packaging is to sell the product.

Convenience

Consumers demand products that fit into their lifestyles; therefore, packaging must be designed to be convenient and user friendly. Convenient design of packaging will make a package easy to open, hold, and use. Convenient design features include apportionment, ergonomic design, and the capacities to be reused, resealed, easily opened, microwaved, recycled, and easily recognized.

LEVELS OF PACKAGING

There are four basic levels of packaging (primary package, secondary package, distribution or tertiary package, and unit load or quaternary package) and two package types defined by destination (consumer package and industrial package), as shown in Figure 5.1. A primary package is in direct contact with the food product and is responsible for many, if not all, of the general packaging functions. A secondary package contains the primary package and often provides physical protection for the food product and the primary package. A distribution package contains secondary packages and functions to protect their contents and enable handling. Finally, a

unit load is comprised of distribution packages that are bound together to facilitate handling and storage throughout the distribution chain. In addition to these levels, packaging can be classified by end destinations: consumer packages are sold in a grocery store and ultimately reach the consumer (usually primary and secondary packages), and industrial packages are used for warehousing and distribution of grocery store products (distribution and unit load packages) or transporting products for further processing from one manufacturer to another.

A consumer package for breakfast cereals or crackers consists of a primary package (the plastic bag containing the food that extends its shelf life by protecting it from unwanted moisture migration into the package) and a secondary package (the paperboard box that protects the primary package and food, facilitates handling, and carries branding and labeling information to communicate with the consumer) (Soroka 1999). In contrast, a consumer package for potato chips consists only of a primary package, the laminate bag [often made of layers of oriented polypropylene (OPP), polyethylene (PE), and metallized, heat-sealable polypropylene] that contains the chips, extends their shelf life by protecting them from moisture, oxygen, and light migration into the package, facilitates handling, and carries branding and labeling information. These consumer packages are placed into corrugated shipping containers, which are then palletized or otherwise assembled into a unit load to protect the products during distribution. This chapter focuses on primary packages commonly used for foods.

RAW MATERIALS PREPARATION

This section addresses preparation of materials used for food packaging, formation of food packages using these materials, and the advantages and disadvantages of each package material. A later section in the chapter, Properties of Plastics, will provide a more detailed description of the properties of plastics, sample calculations related to permeability and shelf life, and the uses of individual thermoplastics and laminates.

PAPER AND PAPERBOARD

Types of Paper and Paperboard

Paper is made of plant fiber that is matted or felted into a sheet. The difference between paper and pa-

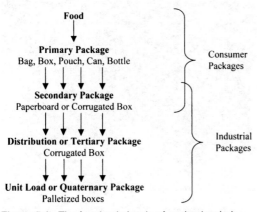

Figure 5.1. The four basic levels of packaging (primary, secondary, tertiary, unit load) and two package types defined by destination (consumer package, industrial package).

perboard is related to thickness (caliper) and/or weight (grammage). Paperboard is thicker (> 300µm) and/or weighs more (> 250g/m^2) than paper (Hanlon et al. 1998). To make paper and paperboard, wood is made into a pulp (by mechanical, chemical, or combination methods); the pulp may be bleached and additives and sizings added to control desired functional or visible properties; and the resulting "furnish" is put into a papermaking machine (four-drinier or cylinder) (Soroka 1999). Recycled paper products are generally not used for direct food contact applications. Paper is typically made on a four-drinier machine, where the furnish is put on a moving wire screen that allows the water to drain before the wet paper is moved around a series of heated drying drums; then the dried paper is wound into mill rolls. Paperboard is made on cylinder machines, which have a series of six to eight wire-mesh cylinders. Each cylinder rotates in an individual vat of furnish and deposits a layer of pulp furnish onto a moving felt blanket. Since the cylinders are in series, six to eight layers of pulp furnish are deposited onto the felt, and the resulting product is thicker than that from the fourdrinier machine. After the paper or paperboard is dried, it may be calendered (passed through a series of heavy rolls) to produce a more dense paper with a glossy, smooth surface.

Types of paper used in food packaging include natural kraft paper, bleached paper, greaseproof paper, glassine paper, parchment paper, waxed paper, tissue paper, label paper, and linerboards. Of these, natural kraft paper is the strongest and most often used. Greaseproof, glassine, parchment, and waxed papers are resistant to grease and oil and are used for baked, greasy, and sometimes wet foods. Linerboard is a kraft paper used for the liners of corrugated paperboard. Types of paperboard used in food packaging include chipboard, white-lined paperboard, clay-coated newsback, solid bleached sulfate, food board, and solid unbleached sulfate. Chipboard is the lowest cost, lowest quality paperboard and is made from 100% recycled fiber. White-lined paperboard is lined with a white pulp on one or both sides to improve appearance and printability. Clay coatings also are applied to some paperboards to improve appearance and printability. Solid bleached and unbleached sulfate (kraft) paperboards and food board are stronger than other types and are therefore used in high-speed machines and for carrying containers (carrying baskets for colas and beers in glass bottles). When coated, solid bleached kraft paperboard is used for food contact applica-

tions such as frozen food boxes and butter cartons. Hanlon and others (1998) provide a good description of corrugated paperboard, which is most often used for secondary and/or distribution packaging.

Formation of Paper and Paperboard Packages

Two basic types of food packages are made with paper and paperboard: paper bags and folding cartons (Fig. 5.2.). Paper bag types include single- and multiwall bags (flat, satchel bottom, square bottom). Seams and bases of paper bags are sealed with glue. Flat bags have a lengthwise seam and a folded, glued base; square-bottomed bags have additional bellows folds along the sides; and satchel-bottomed bags have a base folded to provide a flat bottom when opened. Folding cartons are made from paperboard that has been printed, creased, scored, cut, folded, and glued. Common designs include vertical end-filled, horizontal end-filled, and top-filled cartons. Cartons are often delivered in a collapsed form and set up at the location where they are filled. Paper

Figure 5.2. Examples of paper and paperboard packages.

and paperboard used for bags and cartons may be coated prior to forming (with polyethylene, wax, glassine, etc.) to improve the functional properties of the end package.

Characteristics of Paper and Paperboard Packages

Paper is one of the most widely used package types, especially for distribution packaging (corrugated shipping boxes, etc.). Advantages for using paper and paperboard for food packages include relative low cost and lightweight. Often, inexpensive products, such as flour and sugar, are packaged in paper bags. Products with short shelf life or those rapidly consumed, such as donuts or fast food items, also are given to the consumer in paper or paperboard packages. Paperboard boxes are commonly used for cereals, cake mixes, and many other foods where the food product is contained in a plastic pouch that is placed in the box. Paper also serves as a printable layer in laminate juice box structures. Disadvantages of paper include its hygroscopic and hygroexpansive nature, its viscoelasticity, its poor moisture and gas barrier properties, its lack of resistance to pests, and its limited formability. Paper will absorb and lose moisture as environmental conditions vary; thus, it will expand in humid summers and possibly create problems in laminated structures, warp, or distort printed graphics. Paper is viscoelastic, meaning that paper will distort over time as compressive forces are applied (as in stacked boxes). Unless coated, paper cannot be used for greasy or moist products, and products packaged in paper or paperboard may absorb aromas from the environment. Pests such as bugs and rodents are easily able to penetrate through paper packages. The sizes and shapes of packages made from paper are more limited than for versatile plastic package designs. Due to these limitations of paper for food packages, products such as cereals and crackers often are packaged in plastic or laminate bags that are placed into paperboard cartons.

METAL

Types of Metal

Four metals are commonly used in food packaging: steel, aluminum, tin, and chromium. Of these, tinplate (a composite of tin and steel), electrolytic chromium-coated steel, and aluminum are most widely used. Bare steel, or black plate, corrodes

when exposed to moisture and is therefore not commonly used for food products. Tinplate is made by electrolytically coating bare steel sheets with a thin layer of tin, which protects the steel from rust and corrosion. A layer of oil is added over the tin to add additional protection against corrosion and to protect the tin during formation and handling. Electrolytic chromium-coated steel is a bare steel sheet coated with chromium, chromium oxide, and oil to protect the steel, and it is more heat resistant and cheaper than tinplate.

Formation of Metal Packages

Common types of metal packages used for foods include three-piece cans, two-piece cans, and foil pouches. Metallized films are also used in many flexible laminate packages.

Three-Piece Cans. Three-piece cans are made from tinplate or electrolytic chromium-coated steel sheets and two end pieces, as shown in Figure 5.3. A slitter cuts the steel sheets to can-width strips, which then are curled, side seamed, and welded. The cans are transferred to a flanger, which flares the end edges to receive can ends. The body of the can is ribbed or beaded to increase strength and provide resistance to collapse due to the external processing temperatures and internal vacuum pressures encountered during thermal processing. The ends of the cans also have concentric beads for the same purpose. The ends of three-piece cans are double-seamed onto the can body. A double seam forms a hermetic seal by interlocking the cover and body of a can as shown in Figure 5.4.

Two-Piece Cans. Two-piece cans are often made from electrolytic chromium-coated steel or aluminum sheets and one end piece, as shown in Figure 5.5. Draw-and-redraw and draw-and-iron processes are used to make two-piece cans. In the draw-and-redraw process, a metal blank, often of electrolytic chromium-coated steel, is stamped (drawn) through a die to form a shallow can shape. The shape from the first draw is forced through additional dies (second and third draws) to increase the height of the can without decreasing the thickness of the can. For thermal process applications, the cans are beaded, and the end is double-seamed onto the can body. In the draw-and-iron process, a metal blank, often of aluminum, is drawn into a wide cup, which then is

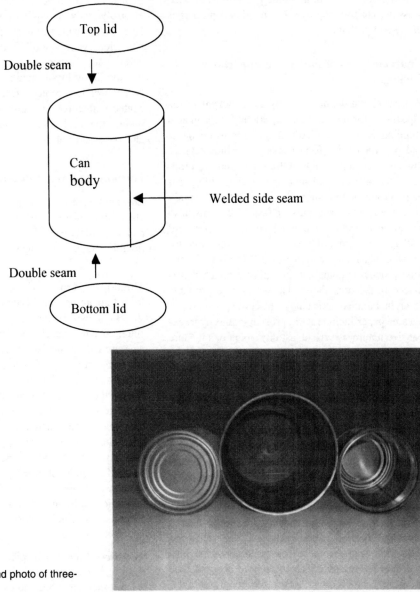

Figure 5.3. Diagram and photo of three-piece cans.

redrawn to the finished can diameter and ironed to reduce sidewall thickness (Fig. 5.6.). Drawn-and-ironed cans are widely used for carbonated beverages that create internal pressure to keep the sidewalls from denting. The ends of drawn-and-ironed cans often are necked or narrowed to reduce the size of the ends, thereby requiring a smaller end-piece. This size decrease in the closure results in reduced packaging costs.

Metal Foils. Metal foils are most commonly made from aluminum and are defined as rolled sections of metal that are less than 0.006 inches thick (Hanlon et al. 1998). Metal foils are most commonly used in multilayer or laminate flexible packages, such as retort, meals-ready-to-eat (MRE), and aseptic pouches and juice boxes. An adhesive, often an ionomer, is used to bond a metal foil to other layers in a laminate material. The layers of a retort pouch, from out to in,

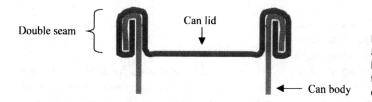

Double seam {

Can lid →

← Can body

Figure 5.4. A model cross section of a double seam commonly used to hermetically seal lids onto two- and three-piece cans. A cross section of a double seam will contain five layers.

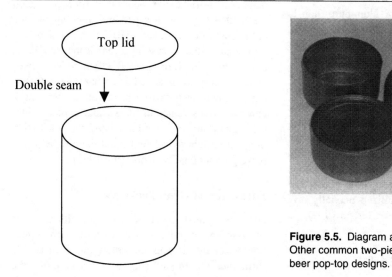

Top lid

Double seam

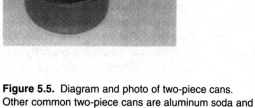

Figure 5.5. Diagram and photo of two-piece cans. Other common two-piece cans are aluminum soda and beer pop-top designs.

Figure 5.6. Stages of formation (right to left) of drawn and ironed two-piece cans.

are polyethylene terephthalate (PET)/adhesive/foil/ polyolefin/food product. The layers of an aseptic juice box are low-density polyethylene (LDPE)/printed polyethylene (PE)/paper/ LDPE/foil/ionomer/LDPE/ juice product. The metal foil adds barrier properties (to moisture, gases, oils, and light) to the package.

Metallized Films. Metallized films are made by vapor-depositing a thin layer of aluminum onto a plastic film in a high-vacuum chamber. These metallized plastic films are less expensive than aluminum foil and also provide a barrier to moisture, gases, oils, and light. Oriented polypropylene (OPP) is the most widely used metallized film, but metallized PET and nylon are also used. Metallized films are used in laminate structures for snack foods (chips) and coffee. The layers of a snack food pouch are often reverse-printed biaxially oriented polypropylene (BOPP)/adhesive/metallized BOPP/sealing polymer/food product. The layers of a pouch for vacuum-sealed coffee are metallized biaxially oriented nylon (BON)/LDPE/coffee (Soroka 1999).

Characteristics of Metal Packages

Advantages of using metals for food packaging include thermal stability, mechanical strength and rigidity, ease of processing on high-speed lines, recyclability, excellent barrier properties, and consumer acceptance. Specific advantages of aluminum include the highest heat conductivity of any food packaging material, resistance to corrosion, and capacity for being rolled thinner than other metals for use in multilayer film packages (such as aseptic juice boxes)(Hanlon et al. 1998). The introduction of pop-top or easy-open ends to metal cans has increased the convenience of using metal packages because the consumer no longer needs a can opener. Disadvantages of metals include the weight of the cans, cost, corrosion, and reactivity with foods (tin will react with acids in foods). Metals in both three- and two-piece cans are coated with an enamel to improve corrosion resistance, package performance and compatibility with a variety of food products, and appearance. Cans are most commonly used for thermally processed, shelf-stable food products such as soups, fruits, vegetables, and canned meats. Foils and metallized films are used in pouches and laminates, such as retort pouches and snack chip bags, as a barrier layer to moisture, gases, and light.

GLASS

Types of Glass

The most widely used glass for food packaging is soda-lime glass, a rigid, amorphous, inorganic product. Soda-lime glass contains mostly silica sand (~73%), limestone (~12%), soda ash (~13%), and aluminum oxide (~1.5%), with small amounts of magnesia, ferric oxide, and sulfur trioxide, which are melted together in a gas-fired melting furnace until fusion occurs (near 1510°C) and cooled to a rigid state without crystallization (Robertson 1993, Sacharow 1976, Soroka 1999). Often cullet, broken or recycled glass, is added as an ingredient. Coloring additives such as iron or sulfur (amber glass), chrome oxides (emerald glass), and cobalt oxides (blue glass) may be added to control the penetration of specific light wavelengths and thereby help preserve product quality (Robertson 1993).

Formation of Glass Packages

Glass is formed into food packages either by the blow-and-blow process (used to form narrow-neck bottles) or by the press-and-blow process (used to form wide-mouth jars) (Fig. 5.7). For both of these processes, a gob (lump) of molten glass is transferred from the furnace to a blank mold (or parison mold). A plunger in the base of the mold is used to form the finish (the threaded part that will receive the closure) and the neck ring of the package. For the blow-and-blow process, air is then blown through the finish to expand the glass into the mold and form the parison. For the press-and-blow process, a metal plunger rather than air pushes the gob into the mold. A completed parison resembles a test tube with a threaded top, as shown in Figure 5.8. For both processes, completed parisons are transferred into blow molds, where air forces the glass to conform to the shape of the blow mold. The blow mold is the size and shape of the finished package. Once a glass package is formed, it is transferred to an annealing oven, or lehr, which gradually cools the glass to minimize internal stresses and possible cracking created by uneven cooling of package surfaces and inner sections. Coatings may be applied to the glass to strengthen the surface and minimize scratching: hot-end coatings (tin or titanium chloride) are applied prior to the annealing oven, and cold-end coatings (waxes, silicones, and polyethylenes) are applied at the end of the annealing process (Soroka 1999).

Figure 5.7. Narrow-neck glass bottles formed by the blow-and-blow process and wide-mouth jars formed by the press-and-blow process.

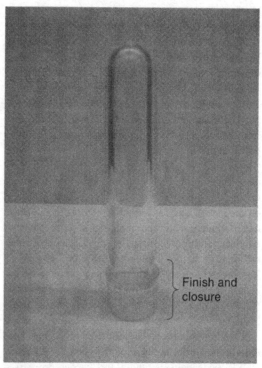

Finish and closure

Figure 5.8. Photo of an injection-molded parison (preform) used for producing bottles. The bit of plastic on the bottom of the parison indicates this was made using an injection process in which a bit of plastic remains at the gate point.

Characteristics of Glass Packages

Advantages of using glass for food packaging include chemical inertness, nonpermeability, strength, resistance to high internal pressure, optical properties, and surface smoothness (Sacharow 1976). Applesauce is often hot-filled into glass bottles that withstand high temperatures but allow the consumer to see the product. Disadvantages of glass packages include fragility, brittleness, and heavy weight (Sacharow 1976). The heavy weight of glass and/or safety concerns related to broken or chipped glass in foods have decreased the use of glass for many food products, such as carbonated cola beverages, which are now packaged in aluminum cans and PET bottles. However, the nonpermeability trait of glass outweighs its disadvantages for applications such as beer bottling.

PLASTICS

Types of Plastic

Plastics are a group of synthetic and modified natural polymers that can be formed into a wide variety of shapes using heat and pressure. Most polymers used for food packaging originate from the petrochemical industry. The type and arrangement of monomer units in a polymer and the processing conditions are used to identify types of plastics. There are two basic classes of plastic polymers: thermoset and thermoplastic. Thermoset plastics are formed by irreversible polymerization of monomers into highly cross-linked three-dimensional structures. Thermoplastic plastics can be reversibly solidified and melted; therefore, thermoplastic materials are recyclable, and scrap can be recovered. These thermoplastic materials are the most widely used in food packaging.

Types of thermoplastics used in food packaging include *polyolefins* (polyethylenes, polypropylene); *substituted olefins* (polystyrene, polyvinyl alcohol, polyvinyl chloride, polyvinylidene chloride, polytetrafluoroethylene); *copolymers of ethylene* (ethylene-vinyl acetate, ethylene-vinyl alcohol); *polyesters* (polyethylene terephthalate); *polycarbonates; polyamides* (nylons); and *acrylonitriles* (styrenes) (Robertson 1993). Refer to Table 5.1 for properties of select polymers. The chemical structures comprising these polymers have a significant influence on the barrier and functional properties of the plastics. Polymers with more nonpolar structures (polyethylene and polypropylene) interact with water differently than polymers with polar structures. Varying the density of the polymers alters the properties the polymer: low-density polyethylene (LDPE) is more flexible but has poorer barrier properties than high-density polyethylene (HDPE). Orientation processes also influence properties: oriented and bioriented polypropylene (OPP and BOPP) have better strength and barrier properties than polypropylene (PP). Refer to the plastics structural and mechanical property sections, below, for further discussion of these topics.

Formation of Plastic Packages

Plastics are formed into packages by various methods: compression molding, extrusion, thermoforming, injection molding, and blow molding (extrusion blow molding, injection blow molding, and injection stretch blow molding).

Compression Molding. Compression molding is used to mold thermoset resins into closures by placing a set weight of resin into a heated mold, closing the mold, and allowing the pressure and heat of the mold to cure and set the resin into the desired shape. Compression molding also is commonly used to mold thermoplastics into closures, with the largest application being the screw caps for plastic soda bottles. No visible markings on the final package are produced by the compression molding process.

Extrusion. All of the plastic-forming techniques, except compression molding, require an extruder. Thermoplastics are formed into sheets, films, or tubes using screw extrusion. A powdered plastic resin is fed into a screw extruder and passed through a die to form a sheet or tube. The flat film (cast film) process is used to make thin films of plastic, while the tubular or blown film process is used to make thin tubes of plastic that are commonly slit into a film. Due to the faster rate of cooling, the cast film process produces films with better surface thickness and uniformity and a more amorphous structure than films produced by blown film extrusion (Soroka 1999). The blown film process has lower equipment costs, orients films biaxially, and can produce wider films once the tube is slit into a film. Most plastic bags are made using the blown film process. Extrusion also is used to produce continuous parisons (continuous plastic tubes) that are used in extrusion blow molding processes.

Thermoforming. In thermoforming processes, a sheet of thermoplastic plastic is placed over a mold, heated to its softening temperature, and formed into the desired shape by vacuum forming, positive air-pressure forming, or matched mold forming. Packages made by thermoforming include clamshell display packages, blister packages, and some tubs. The process produces no visible markings on the final package. Packages with undercuts and narrow necks cannot be made by the thermoforming process.

Injection Molding. In injection molding, a precise amount of thermoplastic resin is injected into a fully closed mold, cooled, and ejected. Packages made by injection molding will have a bit of plastic at the gate point and faint parting lines on the sides. Injection molding produces the most dimensionally accurate parts of any process and is used for making closures, wide-mouth tubs with snap-on lids, complex shapes, and parisons for injection blow molding processes.

Blow Molding. In blow molding processes, a parison is placed into a mold, and air is blown into the parison to force it to contour with the mold and cool. The mold is then opened and the bottle ejected. There are three basic types of blow molding: extrusion blow molding, injection blow molding, and injection stretch blow molding.

In *extrusion blow molding,* the parison is continuously extruded and must be cut, usually by trapping between the two halves of a mold, prior to blow molding. Bottles produced by extrusion blow mold-

Table 5.1. Properties of Select, Commercially Important Polymers

Polymer	Typical Molecular Mass (g·mol⁻¹)[a]	Relative Density[b]	T_g (°C)[c]	T_m (°C)[c]	Maximum Use Temperature (°C)[d]	$P_{O_2}\times10^{-11}$ (cm³·cm· cm⁻²·s⁻¹· cmHg⁻¹, 30°C)[c]	$P_{\text{water vapor}}\times10^{-11}$ (cm³·cm· cm⁻²·s⁻¹· cmHg⁻¹, 25°C and 90% RH)[c]	Tensile Strength (mPa)[a,d]	Impact Strength (J·m⁻¹)[a]	Elongation at Break (%)[a,d]
Low-density polyethylene (LDPE)	$3\text{–}40\times10^4$	0.91	−25	98	66	55	800	10–60	No break	100–800
High-density polyethylene (HDPE)	1×10^3 to 8×10^6	0.941	−125	137	100	10.6	130	102–310	30–200	18–400
Polypropylene (PP)	$1\text{–}6\times10^6$	0.88	−18	176	116	23	680		27–no break	600
Polyethylene terephthalate (PET)	$3\text{–}8\times10^5$	1.36	69	267	204	0.22	1300	50	90	50
Polyvinyl chloride (PVC)		1.23	87	212	93	1.2	1560	55–57		85
Polyvinylidene chloride (PVDC)	$1\text{–}100\times10^4$	1.64	−35	198		0.053	14	73–110	21–54	35–55
Polyhexamethylene adipamide (nylon 6,6)	$12\text{–}20\times10^5$	1.13	50	265	177	0.38	7000	42–129	53–64	30–300

Note: Many properties of polymers vary with polymer composition (molecular weight, copolymers, additives, etc.) and testing conditions (method, temperature, humidity, etc.).

[a]Modified from Mark 1999.
[b]Density of 0.001 in thick plastic film relative to density of water (Soroka 1999).
[c]Modified from Robertson 1993.
[d]Modified from Soroka 1999.

ing will have a pinch-off line across the bottom and parting lines on the sides. This method is used for forming bottles, including milk gallon containers that have handles.

In *injection blow molding,* the parison is formed by injection molding and then transferred into another mold for blow molding. This process produces more dimensionally accurate bottles with less scrap than extrusion blow molding, and it is used to make wide-mouth bottles. Injection blow molded packages will have a bit of plastic at the gate point and faint parting lines on the sides.

In *injection stretch blow molding,* the parison is stretched prior to blow molding. This stretching step orients the polymers and produces bottles with better tensile and impact strengths, better gas and water vapor barrier properties, and improved visible gloss and transparency (Soroka 1999). The injection stretch blow molding process is used to produce PET bottles for carbonated beverages, sports drinks, and juices, and for hot-fill and aseptic processes (Fig. 5.9). Injection stretch blow molded bottles can

Figure 5.9. Photo of an injection-molded parison (center) and injection stretch blow molded bottles for hot fill processes (left) and aseptic process (right).

be recognized by a circular bull's-eye pattern on the bottom and faint parting lines on the sides.

Characteristics of Plastic Packages

Advantages of using plastics for food packages include ease and versatility of shaping, light weight, resistance to breakage, brilliant colors, transparency, and high function-to-cost ratio. Plastic packaging materials are lightweight alternatives to glass and metal containers used in food packaging. Plastic packages can be flexible or rigid and are produced in many sizes, shapes, and designs. Plastic polymer materials have variable properties that enable custom design of packages for specific products and product needs (including modified atmosphere and active packaging). The use of plastics in food packages is widespread and increasing due to the variety of advantageous functions and versatility in design. However, plastics are not absolute barriers, and the transfer of gases and odors from/to the plastic package might result in shortened shelf life or decreased quality of the food (although permeability traits can be exploited in modified atmosphere packaging applications described in a later section). Flavor scalping may occur when flavor and aroma compounds in foods migrate through plastics. Polyolefins are known to absorb oil-based flavors, such as d-limonene in orange juice, thereby decreasing product quality (Jenkins and Harrington 1991). Therefore, it is important to design packages with an inner layer that minimizes solubilization of important flavor components [such as ethylene-vinyl alcohol (EVOH) for orange juice]. Also, additives in plastics that are used to enhance the properties of the plastics can migrate into the packaged food and possibly cause off flavors to develop. Plastics are recyclable, although direct contact between recycled plastics and foods is a concern due to possible contaminants; however, many plastics are not biodegradable. An example of a biodegradable plastic is PLA (polylactide resin). PLA is made from plants (corn stalks, wheat straw, grasses) and is used for some bakery, deli, meat, and dairy package applications.

PROPERTIES OF PLASTICS

This section will provide a summary of structural and mechanical properties of plastics, discuss permeability theory and equations used for calculating permeability and shelf life, and describe common traits and uses of thermoplastics for food packaging.

STRUCTURAL PROPERTIES

The structural and mechanical properties of plastics discussed in this chapter are among those most commonly considered for the design of food packages. The structural properties described are molecular weight, glass transition temperature, and crystalline melting temperature. More detailed information on these topics can be found in basic polymer textbooks such as Billmeyer (1971) and Sperling (1992).

Molecular Weight

Plastic polymers are comprised of repeating units of a variety of monomer structures (ethylene, propylene, etc.). The degree of polymerization (DP) is used to describe the average number of each of these monomers in a polymer. For example, the DP for polyvinylidene chloride (PVDC) is 100–10,000 (Andrady 1999). During the formation of polymers, chains with many different lengths and DPs are produced; therefore, polymers have average molecular weights rather than molecular weights (shown in Table 5.1). Depending on how molecular weight is measured, different types of averages are obtained (number average or weight average). Number averaged molecular weight ($\overline{M}_n$) is the total weight of molecules divided by the total number of molecules, shown in Equation 5.1.

$$\overline{M}_n = \frac{\sum\limits_{i=1}^{\infty} N_i \, M_i}{\sum\limits_{i=1}^{\infty} N_i} \qquad 5.1$$

$\overline{M}_n$ is independent of molecular size and is sensitive to small molecules in the mixture (Robertson 1993). Most of the thermodynamic properties of polymers (colligative properties, osmotic pressure, freezing point depression) depend on $\overline{M}_n$ (Billmeyer 1971).

Weight averaged molecular weight ($\overline{M}_w$) is more complex and is calculated as shown in Equation 5.2.

$$\overline{M}_w = \frac{\sum\limits_{i=1}^{\infty} N_i \, M_i^2}{\sum\limits_{i=1}^{\infty} N_i M_i} \qquad 5.2$$

Heavier molecules become more important in the calculation of $\overline{M}_w$, and most of the bulk properties of polymers (viscosity, strength) depend on $\overline{M}_w$.

Above their critical molecular weight, polymers begin to show strength and toughness, which drastically influence processing capability. Critical molecular weight depends on polymer chain entanglement. Increasing entanglement increases a polymer's molecular weight and brings it closer to the critical molecular weight (Sperling 1992). Tensile strength of polymers first increases with increasing molecular weight but then reaches a maximum at the critical molecular weight. Viscosity increases continuously with increasing molecular weight. This increase in viscosity makes polymers nonprocessible above their critical molecular weight.

Glass Transition and Crystalline Melting Temperatures

The temperature at which a polymer changes from a glassy state to a rubbery state is called the glass transition temperature (T_g). T_g is a characteristic of amorphous polymers, which do not form regular structures due to the interference of chain and pendant groups. The crystalline melting temperature (T_m), on the other hand, is the temperature at which a crystalline polymer undergoes a transition from a crystalline solid to a liquid. Above T_m, a polymer is in a liquid (melted) state. Since most plastic polymers used in food packaging are semicrystalline (they contain both amorphous and crystalline regions), they have both T_g and T_m (as shown in Table 5.1). At temperatures below T_g, amorphous regions of polymers are in the glassy state. In the glassy state, molecules have no segmental motion but vibrate slightly. Structures in the glassy state do not have the regularity of crystalline structures, but the physical properties of glassy and crystalline structures (such as hardness and brittleness) are similar. At temperatures above T_g, amorphous structures are in the rubbery state, and the polymer becomes soft and flexible as molecular movement increases. In the section on package filling, the importance of selecting plastics with T_g above temperatures encountered during food processing is described.

Crystalline and glassy regions in a plastic polymer provide barriers to permeants, and a plastic is more permeable above its T_g due to the decrease in glassy regions. Therefore, knowledge of the T_g and T_m of a plastic is essential for designing good food packages with the desired barrier properties. The mobility of the polymer chain is the determining factor for T_g; therefore, factors that restrict the rotational motion of the molecules cause an increase in

T_g (e.g., increasing intermolecular forces and increasing numbers of bulky pendant groups), and factors that increase molecular movement and flexibility cause a decrease in T_g (e.g., plasticizers and flexible pendant groups) (Sperling 1992). Increasing the number of polar side groups will form stronger intermolecular forces and lead to higher T_g. Bulky pendant groups such as benzene rings can restrict the rotational freedom of neighboring chains and increase T_g; however, flexible pendant groups such as aliphatic chains can limit chain packing and decrease T_g. Increasing the cross-linking of polymers will decrease free volume, restrict molecular rotational motion, and raise T_g. Addition of low molecular weight plasticizers to a plastic will increase the flexibility of the polymers, weaken the intermolecular forces between the polymer chains, and lower T_g.

MECHANICAL PROPERTIES

Mechanical properties of polymers describe their behavior (strength, stiffness, brittleness, and hardness) under stress. Tensile properties, tear strength, and impact strength are measures used to describe mechanical properties. Understanding these properties is important for designing proper packages to withstand stresses and forces encountered during processing, shipping, distribution, warehousing, and consumer use.

Tensile properties include tensile strength, yield strength, elongation, and Young's modulus. Tensile properties are determined from stress-strain curves that are constructed by plotting the change in the length (strain) of the polymer with respect to tensile stress applied to the polymer (as shown in Fig. 5.10). Tensile strength is the maximum stress a polymer can sustain at its break point. This property is quite important for polymers that need to be stretched and is an indication of the resistance of the polymer to continuous stress (as in a screw cap on a bottle) (Hanlon et al. 1998). Polymers with low tensile strength can be used for packaging dry soup, coffee, or confectionery products; however, high tensile strength is needed for packaging bulk products (Soroka 1999). The strain at the break point of a polymer is called elongation at break, and it is expressed as the percent change of the original length of the polymer. Elongation is a good measure of toughness and the ability of a plastic material to conform to an irregular surface (Hanlon et al. 1998). Polymers with low elongation are used for packaging heavy products. Yield strength is the stress at the

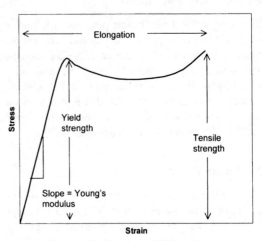

Figure 5.10. An example of a typical stress-strain curve.

point of a nonelastic deformation of a polymer. The slope of the stress-strain curve over the range for which this ratio is constant (the initial slope of the stress-strain curve) is called Young's modulus. Young's modulus is a good measure of the intrinsic stiffness of a polymer.

The shape of the stress-strain curve also provides information about other mechanical properties of the polymer. Toughness is measured from the area under the stress-strain curve, and is a measure of the energy a polymer can absorb before it breaks. Overall toughness is also related to the impact strength of the polymer. Impact strength is the resistance to breakage or rupture as a result of a sudden stress, while tensile strength is a measure of the resistance to breaking as a result of a slowly applied stress.

PERMEABILITY

Since plastic packaging materials, unlike glass and metal, are not absolute barriers, they allow the transport of gases and odors to and from the package. This exchange of gases, or permeability, has a drastic impact on the shelf life, quality, and safety of food products. Therefore, permeability characteristics are one of the most important properties of plastics for the design of food packages. This section provides a brief theoretical background for permeability and discusses factors affecting the permeability of gases through plastics. The next section provides sample permeability and shelf-life calculations based on the equations presented here.

The permeability coefficient is described by the following equation (Crank 1975):

$$P = D \times S \qquad 5.3$$

where P is the permeability coefficient that describes the total mass transport at a steady state through a film; D is the diffusion coefficient, which is a measure of how fast the permeant molecules are moving in the plastic polymer; and S is the solubility coefficient that measures how many permeant molecules are moving in the plastic polymer. A polymer with low permeability will have low diffusion and solubility coefficients. Permeation of molecules through polymers involves the following stages (Ashley 1985): (1) absorption of the permeant onto the surface of the polymer, (2) solubilization of the permeant in the polymer matrix, (3) diffusion of the permeant through the polymer along a concentration gradient, and (4) desorption of the permeant from the other polymer surface as shown in Figure 5.11. These stages, and therefore the permeability of plastic packaging materials, are influenced by the properties of the plastic polymers, the properties of the permeating molecules, the degree of interaction between the polymer and the permeating molecules, and the environmental conditions (temperature and pressure). The properties of polymers that affect permeability include crystallinity; polarity; chain-to-chain packing ability; glass transition temperature; size, shape, and polarity of the permeant; temperature; and pressure (Pascat 1986, Robertson 1993, Sperling 1992).

Crystallinity. Because diffusion of molecules occurs in the amorphous regions of a polymer, the permeability of highly crystalline polymers is significantly less than the permeability of highly amorphous polymers, as shown in Figure 5.12. For example, the oxygen (O_2) permeability of high-density polyethylene with 80% crystallinity is about 4.5 times lower than the O_2 permeability of low-density polyethylene with 50% crystallinity (Pascat 1986).

Polarity. Highly polar polymers are excellent barriers to nonpolar permeant molecules (such as oxygen) but poor barriers to polar permeant molecules (such as water vapor). An increase in relative humidity will cause an increase in the permeability of polar polymers. The two nonpolar polymers commonly used in food packaging are polyethylene and polypropylene; most other polymers for food packaging are polar.

Chain-to-Chain Packing Ability. Linear polymers with simple molecular structures have higher (more dense) chain packing and lower gas permeability than more complex and branched polymers. Poly-

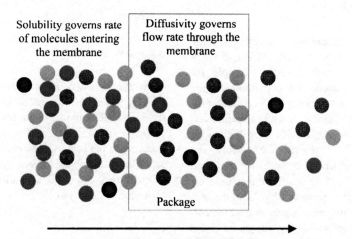

Permeability = Diffusivity x Solubility

Figure 5.11. Diagram of how solubility and diffusivity relate to permeability. As described by Ashley (1985), a permeant molecule must absorb at the surface of the polymer, solubilize in the polymer matrix, diffuse through the polymer along a concentration gradient, and desorb at the opposite polymer surface.

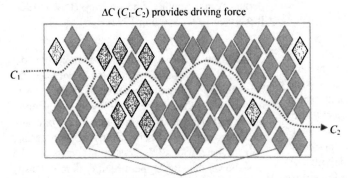

ΔC (C_1-C_2) provides driving force

C_1

C_2

Crystalline domains provide resistance (obstacles) to
diffusion and therefore decrease permeability

Figure 5.12. Diagram of the influence
of crystalline regions on permeation of a
molecule through a package.

mers with bulky side chains have poor packing ability and higher permeability. HDPE has a more linear structure than LDPE, and the permeability of HDPE is lower than that of LDPE.

Glass Transition Temperature (T_g). The free volume and mobility of polymer molecules below their T_g are reduced. Therefore, at temperatures below the T_g, a polymer has fewer voids, permeating molecules have a more tortuous path to travel through the polymer, and permeability is reduced. Polymers with T_gs higher than their end-use temperature for food packaging have improved barrier properties. Table 5.1 shows T_g, T_m, O_2 permeability, and H_2O permeability values for select polymers.

Size, Shape, and Polarity of the Permeating Species. Smaller molecules more readily diffuse through polymers than larger molecules. For example, for LDPE the diffusion coefficient of carbon dioxide (CO_2), which has a 3.4 angstrom (Å) molecular diameter, is 0.37×10^{-6} cm^2·s^{-1}, while the diffusion coefficient for the smaller O_2 with a 3.1 Å diameter is 0.46×10^{-6} cm^2·s^{-1} (Pascat 1986). However, since permeability is affected by both diffusivity and solubility (refer to Eq. 5.3), smaller molecules may not always have higher permeability. Also, the permeability of linear molecules is greater than the permeability of molecules with bulky side chains. If the polarities of both the permeating molecule and the polymer are the same, the permeating molecule may easily diffuse through the polymer. However, when the polarity of the permeating molecule is opposite that of the polymer, interaction will occur between

the permeant and the polymer, and permeability will decrease.

Temperature and Pressure. Permeability (P) is independent of pressure if there is no interaction between the polymer and the permeant. However, P becomes pressure dependent and increases with increasing pressure for polymers having an interaction with the permeant. Permeability, diffusion, and solubility coefficients vary exponentially with temperature according to the Arrhenius law:

$$D = D_0 \exp(-E_D / RT) \qquad 5.4$$

$$P = P_0 \exp(-E_p / RT) \qquad 5.5$$

$$S = S_0 \exp(-E_s / RT) \qquad 5.6$$

where P_0, D_0, and S_0 are pre-exponential constants; E_P, E_D, and E_s are activation energies for permeation, diffusion, and sorption, respectively; R is the universal gas constant; and T is the absolute temperature. Since the permeability coefficient is the product of the diffusion coefficient and the solubility coefficient (Eq. 5.3), the activation energy for the permeation is equal to

$$E_p = E_D + E_s \qquad 5.7$$

Calculations for permeability of food packages are based on Fick's first law. Fick's first law is used to describe the permeation of a molecule, called the permeant, through a plastic film at a steady state.

For unidirectional diffusion, Fick's first law is given by

$$J = -D\frac{\partial c}{\partial x} \qquad 5.8$$

where J is the flux or the amount of permeant diffusing per unit area per unit time, D is the diffusion coefficient or diffusivity, c is the concentration of the permeant in the film, and x is the distance across which the permeant travels (package thickness). If (1) steady state mass transport, (2) negligible convective transport, and (3) a constant diffusion coefficient are assumed, Equation 5.8 can be integrated across the total thickness of the package (l) to give Equation 5.9:

$$J = D\frac{c_1 - c_2}{l} \qquad 5.9$$

where c_1 and c_2 are permeant concentrations at the package surfaces and l is the package thickness. The flux, J, of a permeant in a film can be defined as the amount of permeant (Q) passing through a surface of unit area (A) in one direction of flow during unit time (t). The equation for calculating flux is

$$J = \frac{Q}{At} \qquad 5.10$$

where Q is the total amount of permeant passing through per unit area per unit time.

Equation 5.10 can be substituted into Equation 5.9 to give Equation 5.11. Equation 5.11 enables the calculation of the total amount of permeant passing through a film with an area A in a period of time t:

$$Q = \frac{D(c_1 - c_2)At}{l} \qquad 5.11$$

When measuring gas permeation, it is more convenient to measure the partial pressure of the permeant rather than its concentration. According to Henry's law, the concentration of the permeant in the film (c) is expressed as:

$$c = Sp \qquad 5.12$$

where S is the solubility coefficient and p is the partial pressure of the permeant in the gas phase.

By combining Equation 5.11 with Equation 5.12, Equation 5.13 is formed:

$$Q = \frac{DS(p_1 - p_2)At}{l} \qquad 5.13$$

Since the product of D and S is the permeability coefficient, P (as shown in Eq. 5.3), Equation 5.13 can be rewritten as:

$$P = \frac{Ql}{At(p_1 - p_2)} \qquad 5.14$$

According to the SI system, the units of P are

$$P = \frac{cm^3 (STP) \times cm}{cm^2 \times s \times Pa}$$

As a molecule permeates through a package, an unsteady-state diffusion precedes the steady-state diffusion of the permeant through the polymer (Fig. 5.13.). Mass transfer during unsteady-state diffusion can be described by Fick's second law. The solution of Fick's second law yields Equation 5.15 for a system with (1) a concentration-independent diffusion constant, (2) a polymer that is initially free from permeant, and (3) only one surface of the polymer exposed to the permeant gas at pressure p_1 (Comyn 1985):

$$Q = \frac{Dc_1}{l}\left(t - \frac{l^2}{6D}\right) \qquad 5.15$$

If the linear portion of steady-state line in Figure 5.13 is extrapolated to $Q = 0$, then the intercept on the x-axis, which is known as time lag (τ), can be expressed as

$$\tau = \frac{l^2}{6D} \qquad 5.16$$

Equation 5.16 provides the basis for calculating diffusivity, D.

Mutilayer or laminate films are composed of several layers of different types of polymers in order to maximize functional properties while minimizing cost. For calculating the permeability coefficient (P) for a multilayer film that consists of n layers of different types of plastics (Fig. 5.14.), the following series of equations can be used. If it is assumed that the flux of the permeant molecules is at a steady state and the areas where permeation takes place are equal, the following equation can be used to express Q of the layered package:

$$Q_T = Q_1 = Q_2 \ldots\ldots = Q_n \qquad 5.17$$

For multilayer films, Equation 5.14 can be written as:

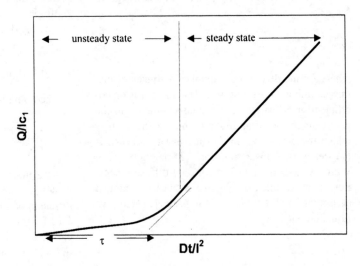

Figure 5.13. An example of a typical permeation curve.

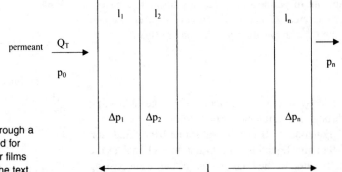

Figure 5.14. A diagram of permeation through a multilayer plastic film. This diagram is used for calculating permeability through multilayer films as described by Equations 5.17–5.20 in the text.

$$Q_T = \frac{P_1 A (p_0 - p_1)}{l_1} = \frac{P_2 A (p_1 - p_2)}{l_2} = \ldots\ldots$$
$$= \frac{P_n A (p_{n-1} - p_n)}{l_n}$$

5.18

$$\Delta p = (p_0 - p_n) = (p_1 - p_2) + (p_2 - p_3)$$
$$+ \ldots\ldots + (p_{n-1} - p_n)$$

5.19

And for multilayer films, the permeability coefficient P_T can be calculated from the following equation:

$$P_T = \frac{l}{(l_1 / P_1) + (l_2 / P_2) + \ldots\ldots + (l_n / P_n)}$$

5.20

In addition to passing through plastics by permeation, gases also may pass through plastics via pores, pinholes, cracks, defective seals, or other defects. Packages must be intact for the permeability equations described above to be valid. If the packages have pores, holes, or defects, the permeability calculations will underestimate permeability, and the shelf-life calculations will overestimate shelf life. Package testing procedures (described in the section Finished Product, below) are designed to detect leaks or defects that could limit package performance beyond the shelf life and permeability calculated using the equations described above.

SAMPLE PERMEABILITY CALCULATIONS

Example 1: Calculation of Permeability through a Monolayer Film

How much oxygen would permeate through a 20 cm × 20 cm plastic bag made of linear low-density polyethylene (P_{O_2} = 4.18×10^{-8} cm³·cm·cm^{-2}·s^{-1}·atm^{-1}) or PET (P = 1.67×10^{-10} cm³·cm·cm^{-2}·s^{-1}·atm^{-1}) per second? The thickness of the plastic is 0.2 cm, and the partial pressure of oxygen across the film is 0.21 atm.

Using Equation 5.14:

$$\frac{Q}{t} = \frac{PA(p_1 - p_2)}{l}$$

for linear low-density polyethylene (LLDPE)

$$\frac{Q}{t} = \frac{4.18 \times 10^{-8} (\text{cm}^3 \cdot \text{cm} \cdot \text{cm}^{-2} \cdot \text{s}^{-1} \cdot \text{atm}^{-1}) \times (20 \times 20)(\text{cm}^2) \times 0.21(\text{atm})}{0.2(\text{cm})}$$

$$= 1.76 \times 10^{-5} \text{cm}^3 \cdot \text{s}^{-1}$$

for PET

$$\frac{Q}{t} = \frac{1.67 \times 10^{-10} (\text{cm}^3 \cdot \text{cm} \cdot \text{cm}^{-2} \cdot \text{s}^{-1} \cdot \text{atm}^{-1}) \times (20 \times 20)(\text{cm}^2) \times 0.21(\text{atm})}{0.2(\text{cm})}$$

$$= 7.014 \times 10^{-8} \text{cm}^3 \cdot \text{s}^{-1}$$

Example 2: Calculation of Permeability through a Multilayer Film

How much oxygen would permeate through a 20 cm × 20 cm multilayer plastic bag made of polyethylene (P_{O_2} = 4.18×10^{-8} cm³·cm·cm^{-2}·s^{-1}·atm^{-1}) and PET (P_{O_2} = 1.67×10^{-10} cm³·cm·cm^{-2}·s^{-1}·atm^{-1}) per second? The thickness of each plastic layer (PE and PET) is 0.1 cm, and the partial pressure of oxygen across the film is 0.21 atm.

P_T can be calculated from Equation 5.20:

$$\frac{l}{P_T} = \frac{l_1}{P_1} + \frac{l_2}{P_2}$$

$$\frac{0.2}{P_T} = \frac{0.1}{4.18 \times 10^{-8}} + \frac{0.1}{1.67 \times 10^{-10}}$$

$$P_T = 3.327 \times 10^{-10} \text{cm}^3 \cdot \text{cm} \cdot \text{cm}^{-2} \cdot \text{s}^{-1} \cdot \text{atm}^{-1}$$

and Q_T is

$$\frac{Q_T}{t} = \frac{P_T A(p_0 - p_2)}{l}$$

$$\frac{Q_T}{t} = \frac{3.326 \times 10^{-10} (\text{cm}^3 \cdot \text{cm} \cdot \text{cm}^{-2} \cdot \text{s}^{-1} \cdot \text{atm}^{-1}) \times (20 \times 20)(\text{cm}^2) \times 0.21(\text{atm})}{0.2(\text{cm})}$$

$$\frac{Q_T}{t} = 1.397 \times 10^{-7} \text{cm}^3 \cdot \text{s}^{-1}$$

Example 3: Calculation of the Shelf Life of a Food Product Packaged in a Monolayer Film

A food product becomes rancid when it absorbs 2.1 ml of O_2. What is the shelf life of this product if it is packaged with LDPE (P_{O_2} = 4.18×10^{-8} cm³·cm·cm^{-2}·s^{-1}·atm^{-1})? What is the shelf life of the product if it is packaged with a PET film (P_{O_2} = 1.67×10^{-10} cm³·cm·cm^{-2}·s^{-1}·atm^{-1})? The surface area of the package is 400 cm², and the package thickness is 0.1 cm. The partial pressure of oxygen across the package is 0.21 atm.

This problem can be solved using Equation 5.14:

$$\frac{Q}{t} = \frac{PA(p_1 - p_2)}{l}$$

The t in Equation 5.14 is the shelf life of the product (t_s) for this example. Equation 5.14 can be rewritten as

$$t_s = \frac{Ql}{PA\Delta p}$$

For LDPE:

$$t_s = \frac{2.1(cm^3) \times 0.1(cm)}{4.18 \times 10^{-8} (\text{cm}^3 \cdot \text{cm} \cdot \text{cm}^{-2} \cdot \text{s}^{-1} \cdot \text{atm}^{-1}) \times 400(\text{cm}^2) \times 0.21(\text{atm})}$$

$$t_s = 59.8 \times 10^3 s = 0.69 \text{days}$$

Thus, LDPE will not provide the needed O_2 barrier properties for this product.

For PET:

$$t_s = \frac{2.1(cm^3) \times 0.1(cm)}{1.67 \times 10^{-10} (\text{cm}^3 \cdot \text{cm} \cdot \text{cm}^{-2} \cdot \text{s}^{-1} \cdot \text{atm}^{-1}) \times 400(\text{cm}^2) \times 0.21(\text{atm})}$$

$$t_s = 14.97 \times 10^6 \cdot s = 173 \text{days}$$

Thus, PET will provide a much better shelf life for this product than LDPE.

USES OF THERMOPLASTICS FOR FOOD PACKAGING

Types of thermoplastics used in food packaging include *polyolefins* (polyethylenes, polypropylene), *substituted olefins* (polystyrene, polyvinyl alcohol, polyvinyl chloride, polyvinylidene chloride, polyte-trafluoroethylene), *copolymers of ethylene* (ethylene-vinyl acetate, ethylene-vinyl alcohol), *polyesters* (polyethylene terephthalate), *polycarbonates; poly-amides* (nylons), and *acrylonitriles* (styrenes) (Robertson 1993). The chemical composition of each thermoplastic will influence its performance in processing, forming, and use of packages as well as its performance and interactions with a variety of foods. Therefore, specific food applications generally use select thermoplastics that will function well in the parameters of the application. When an individual thermoplastic cannot provide all the functions necessary or is too expensive for a certain product, a laminate system is often used. A laminate is made by bonding two or more layers of different package material types (plastic, paper, metal) to optimize package performance for a specific product. A summary of common traits and uses for selected individual thermoplastics and laminates is provided below.

Individual Thermoplastic Applications

Polyolefins (Polyethylene, Polypropylene). The polyolefin class of plastics includes the major nonpolar plastics used in food packaging: polyethylene and polypropylene. Since these plastics are nonpolar, they generally act as good water vapor barriers and poor oil and oxygen barriers. Low-density polyethylene (LDPE) is the most widely used and least expensive plastic packaging material. LDPE is a good water vapor barrier and a poor oxygen barrier, can scalp flavors from foods (such as D-limonene from citrus juices), and can form heat seals. Films made of LDPE are soft, flexible, and easily stretched; they are used for packaging produce and baked products (bread bags). LDPE is used as an adhesive in multilayer package structures, including aseptic juice boxes, and as a water- and grease-resistant coating for paperboard packages (Jenkins and Harrington 1991). LDPE packages will not hold a vacuum due to their high gas permeability (Soroka 1999). Linear low-density polyethylene (LLDPE) is stronger, more stable at high and low temperatures, more resistant to chemicals, and more resistant to stress cracking than LDPE (Robertson 1993). High-density polyethylene (HDPE) is stronger, more dense, a better barrier, more crystalline, and more difficult to heat seal than LDPE. HDPE is used for grocery bags, blow molded bottles, and injection molded tubs for butter, yogurt, and ice cream. Metallocene polyethylene (mPE)

was introduced to the market in the 1990s, and metallocene-catalyzed LLDPE bag film is used for fresh-cut produce and salad packaging, including modified atmosphere package applications. Using metallocene as a catalyst for polymer synthesis allows for better structural uniformity, better control of branching and molecular weight distribution, better melt characteristics, increased impact strength and toughness, and improved film clarity (Prasad 1999).

Polypropylene (PP) is the least dense polymer used for food packaging, is less crystalline than PE, is a good water vapor barrier, and has good live hinge properties (Jenkins and Harrington 1991). PP is more heat resistant than PE and can be used for hot-filled bottles; however, PP can become brittle at temperatures below 0°F and is therefore not used for packaging frozen foods. Antioxidants are commonly added to PP because it is susceptible to oxidative degradation (Robertson 1993). Injection molded closures and containers for margarine and yogurt are made from PP. Oriented PP (OPP) has better strength and barrier properties than PP but will not heat seal. Biaxially oriented polypropylene (BOPP) is up to four times stronger than PP and is clearer than PP and OPP due to layering of crystalline structures (Jenkins and Harrington 1991).

Substituted Olefins (Polyvinyl Chloride, Polyvinylidene Chloride). Polyvinyl chloride (PVC) is a brittle, rigid plastic that requires high concentrations of plasticizers (up to 50%) to make it useful for food packaging; however, the plasticizers can migrate out of the plastic over time, which can decrease PVC functionality and potentially cause food quality and safety issues (Jenkins and Harrington 1991). The poor water vapor barrier characteristic of PVC limits moisture condensation on the inside of a film, which is useful for the stretch wrap films used to cover trays of meat (Robertson 1993). Polyvinylidene chloride (PVDC) copolymerized with PVC forms soft, clear, strong, excellent oxygen and water vapor barrier films with excellent cling characteristics (Andrady 1999, Jenkins and Harrington 1991). These films are commonly known as Saran®. Pure PVDC is often not used for food packaging due to its stiff nature (Robertson 1993).

Copolymers of Ethylene (Ethylene Vinyl Alcohol). Ethylene vinyl alcohol (EVOH or EVAL) is an excellent oxygen and flavor barrier but is expensive and sensitive to moisture. When EVOH is exposed to moisture, its oxygen permeability greatly increases; therefore, EVOH is often placed between two polyolefin layers that provide the needed moisture vapor barrier. EVOH is used as a barrier layer in laminate packages used for packaging aseptically processed juice drinks, ketchup, and citrus juices, and in hot-fill and retort applications.

Polyesters (Polyethylene Terephthalate). Polyethylene terephthalate (PET) provides excellent strength, toughness, barrier, and clarity characteristics. PET has a high T_g but cannot be heat sealed. PET is used for injection stretch blow molded bottles for carbonated sodas and beer, and for hot-fill bottles for juices and sports drinks. CPET (crystallized PET) is a rapidly crystallized form of PET; thermoformed CPET trays are used for dual-oven applications and are able to function in the freezer, microwave, and conventional oven. APET (amorphous PET) is an amorphous form of PET with higher molecular weight and is used for thermoformed sheets and snap-on overlids for CPET trays. PETG (PET glycol) is a copolyester used in some thermoforming applications for clamshell displays and in injection molded heavy-wall jars (Hanlon et al. 1998).

Polycarbonates. Polycarbonates are lightweight and shatter resistant package alternatives to glass but are more expensive than many plastic polymers. Polycarbonates are used in microwaveable packaging, for shatterproof, refillable five-gallon water bottles, and for baby bottles (Hanlon et al. 1998).

Polyamides (Nylons). Nylons are formed by a condensation reaction of a diamine and a dibasic acid or by polymerization of select amino acids; nylon 6 and nylon 6/6 are used for food packaging (Soroka 1999). The 6 in nylon 6 indicates the number of carbon atoms in the basic amino acid. In nylon 6/6, the first number indicates the number of carbon atoms in the reacting amine, and the second number indicates the number of carbon atoms in the reacting dibasic acid (Soroka 1999). Nylons are clear, are good barriers to gases and aromatics, and are strong, tough, and resistant to cracking; however, they are poor water vapor barriers and are not heat sealable. As nylon is exposed to water or high humidity, its barrier properties are negatively affected. Nylon 6 is used for vacuum packaged meat and cheese products, and oriented nylons (BON, biaxially oriented nylon) are

commonly used as layers in laminate structures for vacuum coffee packaging (Soroka 1999).

Polystyrene. Polystyrene (PS) films have poor water vapor and gas barrier properties that facilitate their use in "breathable" wraps for fresh produce (Soroka 1999). PS is clear, has a low impact strength, and has a low T_m (190°F) that negates its use for hot foods (Hanlon et al. 1998). Expanded PS (EPS) is used for egg cartons and trays for meats and to provide insulation for cold-temperature distribution package applications. High-impact PS (HIPS) is used for trays for vegetable and potato products.

Laminate Applications

When an individual thermoplastic is unable to meet all functional packaging needs for a food product, a laminate structure is often used in order to combine desirable traits of more than one package material while minimizing cost. A laminate is made by bonding together at least two layers of paper, plastic, aluminum foil, and metallized film to form a final package with the desired structural, performance, barrier, and aesthetic properties (Soroka 1999). The layers are bound together by wet bonding, dry bonding, hot-melt bonding, or extrusion or coextrusion methods (Soroka 1999). Common adhesives used to bond layers in flexible packages include ionomers (which have excellent adhesion to foils) and LDPE (Soroka 1999). The structural, barrier, cost, and other properties of a laminate can be modified by changing the types and/or thicknesses of its layers. For example, increasing the thickness of the PET layer in Example 2 (above) would decrease the permeability of the laminate, but substituting a layer of PP for the PET layer would increase the permeability.

Examples of laminate structures have been given in previous sections, for example, the layers in an aseptic juice box: LDPE, printed PE, paper, LDPE, foil, ionomer, LDPE, juice product. The function of each layer contributes to the overall functionality of the package. In the juice box, the LDPE on the exterior of the package acts as a moisture barrier and protects the printed materials underneath. The printed PE carries the required labeling information as well as other communication functions (branding, UPC codes, etc.). The paper layer provides stiffness. The LDPE bonds the paper to the foil layer. The metal foil adds barrier properties (to moisture, gases, oils, and light) to the package. The ionomer layer bonds the foil to the LDPE layer. The LDPE layer on the interior of the package enables heat-sealing.

Many other types of laminate structures are used in food packaging. For crackers, packages with moisture, oxygen, and flavor barriers are needed. Laminate structures designed for crackers include OPP/PVDC/OPP, HDPE/nylon/ionomer, polyolefin/ EVOH/polyolefin, OPP/metallized PET, and OPP/ metallized OPP (Jenkins and Harrington 1991). Chocolate candy bars may be packaged in lacquer/ ink/white OPP/PVDC/ cold-seal laminates where the lacquer protects the ink, the white OPP is a light barrier and adds strength, the PVDC is an oxygen and moisture barrier, and the cold seal enables package sealing without melting the chocolate (Jenkins and Harrington 1991). The combinations of layers reflect the requirements of the product, the distribution, the market, the cost, and the consumer; therefore, there are numerous types of laminates, which change as new products are introduced to the market.

PACKAGE FILLING

Filling systems are designed to place a food product into a primary package. There are two basic categories of filling systems based on product type: (1) liquids and viscous products and (2) dry products. Volumetric and constant-level fillers are used for liquid and viscous products. Volumetric fillers deliver a precise volume of product and are generally used for expensive or viscous products; however, differences in bottle dimensions can result in the appearance of unequal amounts of product in transparent bottles or jars. Constant-level fillers deliver a specific level of a product to a package, which is desirable when the consumer can see the level of the product (Jenkins and Harrington 1991). There are three basic categories of filling systems for dry products: (1) count, (2) weight, and (3) volume. Count-based dry filling systems are used to deliver an exact number of products (such as cookies) to a package (Jenkins and Harrington 1991). Weight-based filling systems are used to deliver a gross or net weight of a dry product to a package. Net-weight filling is the most accurate, but gross-weight filling is used for fragile and sticky products (Hanlon et al. 1998). Volume-based dry filling systems include vacuum, auger, cup, and constant-stream systems. Vacuum dry filling is similar to liquid volumetric filling and is used to minimize loss of product dust (Jenkins and Harrington 1991). Auger fillers deliver a precise volume of product via a set number of turns of the auger; cup fillers deliver a precise volume of product via a cup with a preset volume; and constant stream fillers deliver a constant volume of product to packages moving at a set speed under the filler (Jenkins and Harrington 1991).

The filling method chosen depends on the type of product and the type of package. In addition to the filler categories described above, fillers can be classified by processes that occur in them. These classifications include (1) fill and seal, (2) form, fill, and seal, (3) thermoform, fill, and seal, and (4) blow mold, fill, and seal:

1. In a *fill and seal process,* a preformed container is filled and then sealed. *Example:* A preformed bottle is filled with a product (such as a juice or sports drink) and capped prior to exiting the filler. Glass, paper, and metal packages are generally preformed.
2. In a *form, fill, and seal process,* a package material enters the filler, is formed into the end package shape, is filled, and then is sealed. *Example:* A laminate film is formed into a juice box shape for an aseptically processed juice drink, filled, and sealed prior to exiting the filler.
3. In a *thermoform, fill, and seal process,* the packaging material enters the filler as a roll stock, is heated and thermoformed (usually into cups), is filled, and then is sealed with a lid material (often aluminum foil coated with LDPE for heat-sealing).
4. In a *blow mold, fill, and seal process,* an extrudable material (PET, PP, PE) is blow molded into a container that is filled and sealed in place before the mold is opened.

For plastic packages, the temperature of filling is extremely important. If a bottle is hot filled at temperatures above the T_g of the plastic material (which is possible for hot-filled juice and sports drink beverages and applesauce), the bottle and finish could distort. Thus, plastics chosen for hot-fill processes have relatively high T_gs (refer to Table 5.1). LDPE is not used for hot-filled products, while PET often is used (although careful attention to T_g, processing temperatures, and rapid cooling after filling is necessary). A wider variety of plastics and containers with thinner sidewalls can be used for cold or aseptically filled processes due to the low temperature of filling (below the T_g) and the lack of vacuum forces generated during cooling of hot products.

PACKAGE CLOSURES

A closure is often the most critical part of a package and must provide all the basic packaging functions in addition to being easy to open. The process of adding a closure to a package must not create defects or damage the package, and the area of the package that will receive the closure (often the finish of the package) must be free of food particles, which could prevent hermetic sealing. Selection criteria for package closures include the compatibility of the closure, the package, and the food; the barrier and sealing properties of the closure; the processing and handling requirements for the closure; the need for multiuse closures; cost; convenience; the targeted consumers; and the need for tamper-evident systems (Soroka 1999).

There are four basic categories of closures: (1) plugs, (2) caps, (3) cap liners, and (4) seals. A common plug is the natural or plastic "cork" used to close wine bottles. Natural cork is dense, provides a good barrier to oxygen and water, and is elastic and compressible. Common caps include screw caps, lug caps, and crown closures. Screw caps are threaded, twist onto a similarly threaded package finish, and seal the package with the contact created between the threads and between the top edge of the finish and the interior top of the closure. Screw caps often are made from polypropylene, polyethylene, and metal and are used on bottled beverages. Lug caps are often made from metal and are used for closing glass wide-mouth jars for hot-filled applesauce, pickles, and salsa (Hanlon et al. 1998). Crown systems are often made from metal lined with polyethylene or PET and are used on beer bottles. Traditional crown closures require a bottle opener for removal; however, twist-off crown closures are now widely used on bottled beer and can be removed by hand. Cap liners contain a resilient backing and a facing material and are made from a variety of package materials (pulpboard, PE, PP, PVC, PET, aluminum foil, wax coating) (Hanlon et al. 1998). Cap liners are often used under caps to provide additional protection (sometimes tamper-evident) and a hermetic seal with the bottle rim, but cap liners also may be used alone, as in yogurt packages with peel-off foil laminate lids (Jenkins and Harrington 1991). Seals are generally formed via heat-sealing. Common types of polymers that form seals when heated are PE and PP. In laminate structures, PE and PP are included as the inner layer to enable heat-sealing of the package, pouch, or box. Induction sealing enables the use of aluminum foil coated with a hot-melt type adhesive (Soroka 1999).

ADDITIONAL PACKAGING TYPES

ASEPTIC PACKAGING

Commercially sterilized food products are filled into commercially sterile containers under aseptic conditions and sealed hermetically during aseptic processing (Fig. 5.15). The terms "commercially sterile" and "sterile" are often interchanged in discussions of aseptic processing. Aseptic systems allow the use of high-temperature short-time (HTST) and ultra high-temperature (UHT) processes because the product and the package are sterilized separately. Due to the shortened exposure to high temperatures in comparison with more traditional canning/retorting processes, aseptically processed products have excellent sensory qualities and better retention of nutritional components (heat labile vitamins). Aseptic processing also provides flexibility in the selection of containers to be used in packaging the product since the packaging materials do not need to withstand the harsh temperature and pressure conditions of conventional thermal processes.

Properties of the product, desired shelf life, and storage temperature determine the required reduc-

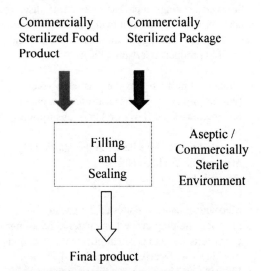

Figure 5.15. Diagram of an aseptic process in which a commercially sterilized food is filled and sealed into a commercially sterilized package in a commercially sterile environment..

tion in microbial count for the sterilization of food-contact packaging materials. While a minimum 4D reduction in bacterial spores is required for packages used with nonsterile acidic products (pH < 4.5), a 6D reduction is necessary for packages used with sterile, neutral, low-acid (pH > 4.5) food products (Robertson, 1993). The main sterilization techniques used for the sterilization of packaging materials for aseptic processes include *irradiation* (ultraviolet rays, infrared rays, and ionizing radiation), *heat* (saturated steam, superheated steam, hot air, hot air plus steam, and extrusion), and *chemical treatments* (hydrogen peroxide, peracetic acid, ethylene oxide, ozone, and chlorine) (Floros 1993). These techniques are used individually as well as in combination. For the verification of a sterilization process, the contact surfaces of the package are inoculated with an indicator organism and passed through the package sterilization operation on an aseptic processing line. The package is then filled with growth medium and incubated, and a microbial count is obtained to determine the D value for the process.

Due to the separate package and food product sterilization, a wider variety of package designs and materials can be used than in traditional thermal processes (canning/retorting, hot-filling). The thickness and amount of PET used in a bottle for aseptic products is significantly less than the amount of PET necessary for a hot-filled product; thus there is significant cost savings in packaging materials used for aseptic processes. The following factors influence the choice of packaging material for aseptically processed products (Carlson 1996):

- Functional properties of the plastic polymer [gas and water vapor barrier properties, chemical inertness, and flavor and odor absorption (scalping)],
- Potential interactions between the plastic polymer and the food product,
- Desired shelf life,
- Cost,
- Mechanical characteristics of the packaging material [molding properties, material handling characteristics, and compatibility with packaging machinery and sterilization methods],
- Shipping and handling conditions [toughness, type of overwrap or cases required, vibration, and compression],
- Compliance with regulations, and
- Targeted consumer group.

MODIFIED AND CONTROLLED ATMOSPHERE PACKAGING

Both modified atmosphere packaging (MAP) and controlled atmosphere packaging (CAP) are designed to extend the shelf life of foods held at ambient and refrigerated temperatures by modifying the gaseous environment in which the foods are stored. MAP is accomplished by modifying the gaseous environment in a package by gas flush packaging or vacuum packaging when the food is placed into the package, and no further control is exercised (Brody 1989). In gas flush packaging, air is replaced with a controlled mixture of gases (usually O_2, CO_2, and N_2); in vacuum packaging all air is removed. CAP systems, on the other hand, first alter and then selectively control the gaseous environment in a package in order to maintain a precisely defined gaseous atmosphere. True CAP systems are impractical, however, due to the chemical and microbial nature of foods and the physical characteristics, including permeability, of packages (Ooraikul and Stiles 1991). Fresh food products (such as lettuce, carrots, and apples) and microorganisms continue to respire after they are packaged, and the CO_2 produced and O_2 consumed change the gas concentration inside the package. In theory, a CAP system would respond to these changes by scavenging excess CO_2 and releasing O_2 to replace what was consumed in order to maintain the desired gaseous environment.

The important factors for CAP/MAP preservation of foods include the composition of the gas atmosphere in the package; the type of food; the type of microorganisms present; and the temperature, moisture, and pressure. Control of the concentration of the gases, particularly CO_2, inside the package is the fundamental concept of MAP and CAP food preservation. An increase in CO_2 or a decrease in O_2 concentration can slow the respiration rate of foods and the growth of microorganisms, thereby extending the shelf life of the foods. The gas atmosphere in a package is a function of the gas transmission rate of the packaging material, the respiration rate of food and bacteria in the package, the initial atmospheric composition in the package, and any control mechanisms added to the package to respond to changes in gas concentrations (Stiles 1991a). Optimum MAP/CAP gaseous atmospheric conditions depend on the type of food and the type of microorganism. For fresh produce, an increase in CO_2 concentration may have beneficial effects, but the total absence of O_2 will result in the development of off flavors. Different types of fruits and vegetables have specific gas concentration requirements for optimum storage life. For example, recommended MAP conditions for apples are 0–5°C, 2–3% O_2, 1–2% CO_2, and 95–98% N_2; for bananas, 12–15°C, 2–5% O_2, 2–5% CO_2, and 90–96% N_2; and for lettuce, 0–5°C, 2–5% O_2, 0% CO_2, and 95–98% N_2 (Kader et al. 1989). Optimum MAP conditions will reduce the respiration rate, decrease ethylene production, delay initiation of ripening, retard senescence, inhibit microbial growth and spoilage, and reduce some physiological disorders such as chilling injury (Powrie and Skura 1991).

Although increased CO_2 levels will retard the growth of some microorganisms (including *Pseudomonas*, which causes off flavor to develop in meats), elevated levels of CO_2 have less effect on other microorganisms, for example, fermentative bacteria such as lactic acid bacteria. The minimum effective CO_2 concentration for extending the shelf life of meat is 20–30% (Stiles 1991b). Packaging low-acid foods (pH > 4.6) in anaerobic conditions could allow *Clostridium botulinum* growth and toxin production. Therefore, understanding the microorganisms present in the packaged food is extremely important for designing appropriate MAP/CAP systems. In addition to modifying gas concentrations, a decrease in temperature will slow the respiration rate and spoilage of foods, and low pressure can be used to remove ethylene (a ripening hormone) to extend the shelf life of fresh produce (Stiles 1991a).

ACTIVE PACKAGING

Active packaging, also known as interactive or smart packaging, involves an interaction between the packaging components and the food product (Labuza and Breene 1989). Active packages respond to changes in the internal or external environment by changing their own properties or attributes to enhance the preservation of food products while maintaining nutritional quality (Brody et al. 2001). Active substances are contained in sachets or incorporated directly into the packaging component. The major active packaging technologies include oxygen scavengers, ethylene scavengers, moisture regulators, and antimicrobial agents (Rooney 1995, Vermeiren et al. 1999).

Oxygen Scavengers

The majority of commercially available O_2 scavengers work on the principle of oxidation of iron powder by chemical means or enzymes to prevent the deterioration of food constituents by oxidation or spoilage. In the first case, iron kept in a small sachet that is highly permeable to O_2 is placed inside a food package and is oxidized to iron oxide. This oxidation of the iron removes oxygen from the package and limits O_2 interaction with the food product. In enzyme systems, an enzyme such as glucose oxidase reacts with a substrate to scavenge oxygen.

Ethylene Scavengers

Ethylene acts as a growth hormone and accelerates ripening and senescence of fruits. Removing ethylene from the environment surrounding a fruit can extend the shelf life of the fruit. Most ethylene scavengers are based on potassium permanganate ($KMnO_4$), which oxidizes ethylene to acetate and ethanol. Charcoal or finely dispersed minerals such as zeolites are also used as ethylene scavengers, but they are less effective than the $KMnO_4$ scavengers.

Moisture Regulators

Several desiccants such as silicates and humidity-controlling substances are used in food packaging to control the moisture content inside the package of very dry foods or of respiring, wet, and high relative humidity fresh/minimally processed foods.

Antimicrobial Agents

Antimicrobial agents such as sorbates, benzoates, ethanol, and bacteriocins are incorporated into or onto polymeric packaging materials to reduce the microbial growth on the surface of food products. In some packaging systems, these antimicrobial agents are released from the packaging film into the food product over time. In other systems, the antimicrobial agent is immobilized in the packaging material.

EDIBLE COATINGS AND FILMS

Edible films and coatings have the same functions as other packaging materials (e.g., preventing moisture loss, acting as a barrier to oxygen, and reducing flavor and aroma loss). In addition, they provide the further benefits of (1) being formed from natural substances and reducing waste and environmental pollution; (2) enhancing the organoleptic, physical, and nutritional properties of the foods; (3) continuing to offer protection after the package (often a plastic) has been opened; and (4) providing protection for small pieces (such as raisins, nuts, etc.) (Labuza and Breene 1989).

Edible films and coatings can be classified as polysaccharide-based, protein-based, lipid-based, and multiconstituent films and coatings (Krochta et al. 1994). Polysaccharide-based films generally are poor water vapor barriers due to their hydrophilic nature. They also have poor oxygen barrier properties at high relative humidities. Polysaccharide-based films are used to retard the ripening of climacteric fruits without creating severe anaerobic conditions. Protein-based films and coatings are generally formed from gelatin, whey protein, casein, corn zein, wheat gluten, and soy protein. They have much better oxygen barrier properties than polysaccharide-based films and also add nutritional value to the product. Lipid-based films are generally used to prevent weight loss in fruit and vegetables; however, anaerobic respiration and off-flavor development are possible. Since most lipid-based films lack sufficient structural integrity and durability to form freestanding films, they are used in combination with polysaccharide- and protein-based films. These multiconstituent films are formed to combine the desirable properties of each component (barrier properties) while minimizing individual component weaknesses (structural integrity). Antimicrobial agents, antioxidant vitamins, and flavors can be added to modify the functionality of the films and coatings. Films and coatings are applied to food products by dipping the product into the film solution, spraying the film solution onto the surface of the product, or casting freestanding films and applying these to the product.

FINISHED PRODUCT

As described at the beginning of this chapter, the general functions of a food package are containment, protection, preservation, distribution (transportation), identification, communication, and convenience (Robertson 1993, Soroka 1999). If a package fails to function properly, much of the expense and energy put into the production and processing of the food product will be wasted. Therefore, food packages are subjected to a variety of tests to ensure package performance through distribution and to provide the consumer with a safe product. Several package-testing procedures are described below. A section on recycling is also added. Specific packaging regulations are published in the *Federal Register* and the *Code of Federal Regula-*

tions. These regulations relate to (1) weights and measures, (2) adulteration, (3) public safety, (4) information, and (5) the environment.

PACKAGE TESTING

One of the most important functions of a food package is to protect the contents of the package from microbial spoilage. In most cases, loss of package integrity due to defects in seals and in sensitive points on the packages (pinholes or cracks in corners and folds) is the major cause of spoilage. The minimum defect dimension that can lead to microbial contamination in the packaged food is determined by variables such as pressure differential, microorganism type and concentration, depth and shape of the defect, and viscosity of the packaged foods (Blakistone and Harper 1995, Floros and Gnanasekharan 1992). Reported values of minimum defect sizes for bacterial penetration vary between 0.2 and 80 mm (Harper et al. 1995); however, package-testing methods are usually designed to detect the presence of leaks or defects rather than the size or location of the defects. Methods used for package evaluation include visual examination, field trials, observations of actual performance, nonreproducible testing, and reproducible testing (Floros and Gnanasekharan 1992).

Package and seal integrity tests are classified as destructive or nondestructive, and many of these methods are described in American Society for Testing and Materials (ASTM 1998–2002) documents and in the Food and Drug Administration's Bacteriological Analytical Manual (FDA 2001).

Destructive Package Tests

Destructive methods involve tests that partially or completely destroy the packages. Destructive integrity tests are the simplest way to effectively evaluate how a package will behave in "real world" conditions. Destructive methods are costly because the packages used for testing are no longer fit for sale; however, these tests reveal the true behavior of the package and can provide important information about conditions required to induce package failure. The following are common examples of destructive package testing methods: bubble test, electrolytic test, dye test, burst test, and microbial challenge test.

Bubble Test. The bubble test is performed by submerging a package in a liquid and applying pressure or pulling a vacuum. Any leak in the package will result in the formation of bubbles. Although this test is rapid and inexpensive, viscous food materials can easily clog the leaks, causing a defective package to appear to be intact (no bubbles formed in this test when leaks are clogged by food materials); therefore, the type of food must be considered. Results from the bubble test are qualitative not quantitative.

Electrolytic Test. This test is based on the principle that a container with no leak is an electrical insulator. When an electrical potential is applied through a defective package that is partially filled with a brine solution, a current flow can be observed with a voltmeter. This test is only applicable for packages that have at least one nonconducting layer. The electrolytic test is qualitative and does not give information about the position of the defect; therefore, a dye test is often used following this test.

Dye Test. In the dye test, a dye solution is applied to one side of a package. The other side of the package is visually inspected for the presence of penetrant dye after an adequate time is allowed for dye penetration. The dye test is qualitative and visually shows the location of holes or defects, but the size of the defect is not determined by this test.

Burst Test. The burst test determines the strength of a flexible package when internal pressure is applied at a uniform flow rate, as described in ASTM designations F 1140 and F 2045 (ASTM 2000). Once enough pressure is applied to burst the package, seal strength and the location of weak seal positions can be determined.

Microbial Challenge Test. For microbial challenge tests, packages are filled with a microbiological media, sealed or closed, and either immersed in a bacterial liquid suspension or sprayed with an aerosol bacterial suspension. Presence of microbial growth in the package will indicate the presence of a leak or other defect. Microbial challenge tests do not provide quantitative information about the size or location of the leak. Detailed procedures are explained in the FDA's Bacteriological Analytical Manual (2001).

Nondestructive Package Tests

Nondestructive tests are designed to not damage the package or its contents. Unlike destructive tests, nondestructive tests can be off-line or on-line. Off-line procedures are performed in laboratories, off of the production lines, and do not interfere with the production line. On-line tests are designed to test up to 100% of the packages in a process without interrupting the rate of production. Nondestructive tests include visual inspection, pressure difference, capacitance, ultrasonic, and infrared thermography methods.

Visual Inspection. Visual inspection is the simplest of the nondestructive testing methods, and it includes inspection of seals for the presence of defects such as voids, wrinkles and pleats, or product contamination. Dimensional checks are also part of the visual inspection. Computer-aided video inspection is also another way of checking the defects in seal areas. Characterization of the defects under investigation is the main problem with this technique since it requires the examination of a large number of representative samples. For on-line testing, two- and three-dimensional images of the packages can be produced with magnetic resonance imaging using magnetic fields and radio waves. Defects are detected due to differences in the signal intensities of defective and nondefective seals (Blakistone and Harper 1995).

Pressure Difference. Leak detection methods based on pressure difference principles are also commonly used in nondestructive testing of the packages and can be categorized as pressure or vacuum decay methods and trace gas detection methods (Floros and Gnanasekharan 1992). Pressure or vacuum decay tests are performed by monitoring the change in the pressure outside the package that is located in a pressurized chamber. This type of testing produces quantitative results. The trace gas detection method involves measurement of the presence or absence of a preselected trace gas in a package such as O_2 or CO_2. The sensitivity of this test depends on the pressure differential used to force the tracer gas out of the package and the sensitivity of the trace gas detection system (Floros and Gnanasekharan 1992).

Capacitance Test. A capacitance test is conducted by passing a package between conducting plates and measuring capacitance. An increase in dielectric constant across the seal indicates the presence of defects (Blakistone and Harper 1995).

Ultrasonics/Acoustics. In ultrasound systems, sound waves are transmitted through a package and medium (such as water) and then measured by a laser vibrometer. Defects such as low or high fill level, low vacuum, missing lids, and fill density can be predicted from changes in vibrational characteristics of the package (Rodriguez 1995).

Infrared Thermography. Infrared thermography is best used for predictive maintenance and process control for heat-sealed applications (pouches, lid stock, etc.). Infrared cameras placed in the process line immediately after heat-sealing are able to detect hot and cold spots in seal areas, both of which can affect the seal integrity of the packages.

Distribution and Storage Package Tests

In addition to destructive and nondestructive package integrity tests, there are transportation, distribution, and storage simulation tests to evaluate the ability of package systems (primary, secondary, and tertiary) to protect products through handling, distribution, and storage environments. These tests are described in ASTM documents (ASTM 1998–2002), including ASTM D4169, and are performed by subjecting a package system to a simulated distribution environment that includes shock, drop, vibration, and compression forces. A list of ASTM standards for package testing can be found in Hanlon and others (1998) and in ASTM documents (ASTM 1998–2002). Following these tests, destructive and nondestructive evaluation of the primary package can be performed to determine package integrity, as described above.

The *free-fall drop test*, as described by ASTM D 5276 (ASTM 1998), evaluates the ability of a package to withstand handling by people and machinery at loading and unloading points. A package is repeatedly dropped on flat sides, corners, and edges and the amount of damage caused by each drop is recorded. This enables observation of the progressive failure of a package system and the respective damage to the package contents, information that is useful for developing appropriate distribution package designs.

The *compression test* is used to determine the ultimate compression strength a single package or a unit load can withstand during shipping and long-term stacking/warehousing practices. As described in ASTM D642 (ASTM 2000), the compressive re- sistance of a package can be determined by applying either a constant rate compressive force or constant force compression to package faces, edges, and corners. Factors for humidity, temperature, and duration of stacking [found in ASTM D4169 (2001)] are used to predict package performance in the "real world" from simulated laboratory tests.

The *vibration test,* described in ASTM D999 and D3580 documents (ASTM 2001), is designed to simulate vibrations encountered during shipping, from 0.08 Hz in slow moving trucks up to 1100 Hz in moving freight cars and ships, with the most problematic at 30 Hz and below (Soroka 1999). A vibration table is used to create vertical linear motion at a desired range of frequencies and amplitudes and to simulate vibration forces encountered during standard shipping (several days) into a much shorter time period (an hour or more).

RECYCLING

Increased environmental concerns have created a need for recycling of packaging materials. Recycling reduces the volume of package materials entering the waste stream and saves materials and energy as long as the energy to ship and reprocess the recycled materials does not exceed that of virgin materials (Marsh 1991). A concern for using recycled package materials for food contact uses (primary packages) is that contaminants could jeopardize the safety or quality of the food. Generally, recycled glass and metal containers are acceptable for food contact use, but recycled plastic and paper are not. The heat used during the melting and forming of glass and metals during recycling is sufficient to pyrolyze organic compounds and kill any microorganisms that might be present (Marsh 1991). However, shipping costs for heavy glass and metal may be prohibitive. Most types of paper are recycled; however, recycled paper might not be suitable for food contact use since recycling processes may allow contaminants to be present in the recycled paper product. The linerboard used in cereal boxes is an example of the use of recycled paper in packaging that does not come into contact with food since the cereal is contained in a plastic pouch placed in the linerboard box.

Unlike glass, plastics are not inert to foods, and components of plastic can migrate into packaged foods, or food components can interact with the plastic. There also is the possibility that the plastic package was used for a second purpose before entering the recycling stream (i.e., a plastic milk gallon was filled with motor oil). Therefore, recovered plastic materials may have more chemical contaminants than virgin plastics. In addition, all recycled plastics cannot be mixed together due to their varying composition (PET is not compatible with LDPE). Therefore, the recovery of plastic packaging wastes is more difficult and costly than recovery of glass or metal. Recycling techniques for plastics include reuse, physical (mechanical), and chemical techniques (Castle 1994, Crockett and Sumar 1996).

Reuse Recycling

The reuse technique involves refilling rigid containers after washing. This approach is common for glass bottles and has been used for rigid plastic milk containers. However, safety concerns related to this type of recycling are due to the possible presence of wash-resistant contaminants. This concern is more substantial for plastic packages than for glass packages due to possible interactions between the plastic and the product.

Physical/Mechanical Recycling

Physical recycling is the remelting and reextrusion or molding of plastic packages into films or containers. Sources of recycled plastics could include scraps from manufacturers or previously used plastic packages. Scrap materials are comparable with virgin materials and could be appropriate for direct food contact if the manufacturer has total control over the source. Controlling previously used plastics is difficult because the composition of the plastic can change due to migration of components to or from plastics, chemical transformations, and accumulation of additives (Castle 1994). The FDA does not encourage the use of recycled plastics for food-contact use; however, recycled plastics could be used for secondary packages or internal layers of a multilayer laminate package.

Chemical Recycling

Waste materials are depolymerized back to monomers or very short molecules in chemical recycling. Fresh plastic is produced by purification of monomers followed by polymerization. The safety of chemically recycled plastics depends on the monomer purification process. Chemically recycled plastics may be the safest among the recycled plas-

tics and can be suitable for food-contact use. Regenerated PET (RPET) is one example for this type of recycling.

ACKNOWLEDGMENT

The authors would like to thank K.D. Hayes, J. Marcy, K.P. Sandeep, M. El-Abiad, D. Granizo, B. Prado, and I. Weiss for their support and suggestions.

GLOSSARY

A—area
APET—amorphous polyethylene terephthalate.
ASTM—American Society for Testing and Materials.
BON—biaxially oriented nylon.
BOPP—biaxially oriented polypropylene.
C—concentration of permeant in a film.
Caliper—term used to describe the thickness of paper and paperboard.
CAP—controlled atmosphere packaging.
CFR—Code of Federal Regulations.
CPET—crystallized polyethylene terephthalate.
D—diffusion coefficient.
Diffusion—the net movement of molecules from an area of high concentration/pressure to an area of low concentration/pressure.
DP—degree of polymerization.
EPS—expanded polystyrene.
EVAL—ethylene-vinyl alcohol (also abbreviated EVOH).
EVOH—ethylene-vinyl alcohol (also abbreviated EVAL).
FDA—U.S. Food and Drug Administration.
Finish—the part of a package, usually threaded, that receives a closure.
Flux—the amount of permeant passing through an area of package during a unit of time.
Furnish—mixture of water, wood pulp, and additives that is fed into papermaking machines to make paper and paperboard.
Grammage—term used to describe the weight of paper and paperboard.
HDPE—high-density polyethylene.
HIPS—high-impact polystyrene.
HTST—high temperature short time.
J—flux.
l—thickness.
Laminate—a package material made by bonding together two or more layers of paper, plastic, foil, and/or metallized film.
LDPE—low-density polyethylene.
LLDPE—linear low-density polyethylene.
MAP—modified atmosphere packaging.

mPE—metallocene polyethylene.

OPP—oriented polypropylene.

P—permeability coefficient.

p—partial pressure of permeant.

Parison—the plastic or glass test-tube–like threaded preform of a package that is later blown into the final package shape.

PE—polyethylene.

Permeability—the movement of molecules through a package material via activated diffusion or solution-diffusion processes.

Permeant—gases, vapors, and other molecules that can solubilize in a package material and then diffuse or permeate through the package.

PET—polyethylene terephthalate.

PETG—polyethylene terephthalate glycol.

PLA—polylactide resin.

Plasticizer—a substance added to a plastic polymer to increase its flexibility.

PP—polypropylene.

PS—polystyrene.

PVC—polyvinyl chloride.

PVDC—polyvinylidene chloride (Saran®).

Q—amount of permeant

RPET—regenerated polyethylene terephthalate.

S—solubility coefficient.

τ—time lag.

T_g—glass transition temperature.

T_m—crystalline melting temperature.

UHT—ultra high temperature.

UPC—Universal Product Code .

REFERENCES

American Society for Testing and Materials. 1998–2002. Annual book of ASTM standards. ASTM International.

Andrady AW. 1999. Poly(vinylidene chloride). In: JE Mark, editor. Polymer Data Handbook, 945–948. Oxford University Press.

Ashley RJ. 1985. Permeability and plastic packaging. In: J Comyn, editor. Polymer permeability, 269–308. London: Elsevier.

Billmeyer FW. 1971. Textbook of Polymer Science, 2nd edition. New York:Wiley-Interscience.

Blakistone BA, CL Harper. 1995. New developments in seal integrity testing. In: B Blakistone and C Harper, editors. Plastic Package Integrity Testing, 1–10. Herndon, Va.: Institute of Packaging Professionals; Washington, D.C.: Food Processors Institute.

Brody AL. 1989. Introduction. In: AL Brody, editor. Controlled/Modified Atmosphere/Vacuum Packaging of Foods, 1–16. Trumbull: Food and Nutrition Press.

Brody AL, ER Strupinsky, LR Kline. 2001. Active Packaging for Food Applications. Lancaster: Technomic Publishing Co.

Carlson RV. 1996. Food-packaging equipment. In: JRD David, RH Graves, VR Carlson, editors. Aseptic processing and packaging of food, 127–146. Boca Raton, Fla.: CRC Press.

Castle L. 1994. Recycled and re-used plastics for food packaging? Packag Technol Sci 7:291–7.

Comyn J. 1985. Introduction to polymer permeability and the mathematics of diffusion. In: J Comyn, editor. Polymer Permeability, 1–10. London: Elsevier.

Crank J. 1975. The Mathematics of Diffusion, 2nd edition. London: Oxford University Press.

Crockett C, S Sumar. 1996. The safe use of recycled and reused plastics in food contact materials—Part I. Nutr Food Sci 3:32–37.

Floros JD. 1993. Aseptic packaging technology. In: JV Chambers, PE Nelson, editors. Principles of Aseptic Processing And Packaging, 2nd edition, 115–148. Washington, D.C.: The Food Processors Institute.

Floros, JD, V Gnanasekharan. 1992. Principles, technology and applications of destructive and nondestructive package integrity testing. In: RK Singh, PE Nelson, editors. Advances in Aseptic Processing Technologies, 157–189. London: Elsevier Applied Sci.

Food and Drug Administration (United States), Center for Food Safety and Applied Nutrition. 2001. Chapter 22, Bacteriological Analytical Manual. www.cfsan.fda.gov/~ebam/bam-22c.html.

Hanlon JH, RJ Kelsey, HE Forcinio. 1998. Handbook of Package Engineering. Lancaster: Technomic Publ. Co.

Harper CL, BA Blakistone, JB Litchfield, SA Morris. 1995. Developments in food packaging integrity testing. Trends Food Sci Tech 6:336–340.

Jenkins WA, JP Harrington. 1991. Packaging Foods with Plastic. Lancaster: Technomic Publishing Co., Inc.

Kader AA, D Zagory, EL Kerbel. 1989. Modified atmosphere packaging of fruits and vegetables. CRC CR Rev Food Sci 28:1–30.

Krochta JM, EA Baldwin, M Nisperos-Carriedo. 1994. Edible Coatings and Films to Improve Food Quality. Lancaster: Technomic Publ. Co.

Labuza TP, WM Breene. 1989. Applications of "active packaging" for improvement of shelf-life and nutritional quality of fresh and extended shelf-life of foods. J Food Process Pres 13:1–69.

Mark JE. 1999. Polymer Data Handbook. Oxford University Press.

Marsh KS. 1991. Effective management of food packaging: From production to disposal. Food Technol 45:225–234.

Ooraikul B, ME Stiles. 1991. Introduction: Review of the development of modified atmosphere packaging. In: B Ooraikul, ME Stiles, editors. Modified Atmosphere Packaging of Food, 1–17. New York: Ellis Horwood.

Pascat B. 1986. Study of some factors affecting permeability. In: M Mathlouthi, editor. Food Packaging and Preservation, 7–24. London: Elsevier Applied Science.

Powrie WD, BJ Skura. 1991. Modified atmosphere packaging of fruits and vegetables. In: B Ooraikul, ME Stiles, editors. Modified Atmosphere Packaging of Food, 169–245. New York: Ellis Horwood.

Prasad, A. 1999. Polyethylene, metallocene linear low-density. In: J Mark, editor. Polymer Data Handbook, 529–539. Oxford University Press, Inc.

Robertson GL. 1993. Food Packaging. New York: Marcel Dekker.

Rodriguez JG. 1995. Noncontacting acoustic ultrasonic analysis development. In: B Blakistone and C Harper, editors. Plastic Package Integrity Testing, 107–111. Herndon, Va.: Institute of Packaging Professionals; Washington, D.C.: Food Processors Institute.

Rooney ML. 1995. Active Food Packaging. London: Blackie.

Sacharow S. 1976. Handbook of Package Materials. Westport, Conn.: The AVI Publishing Company, Inc.

Soroka W. 1999. Fundamentals of Packaging Technology, 2nd edition. Institute of Packaging Professionals.

Sperling LH. 1992. Introduction to Physical Polymer Science. New York: Wiley.

Stiles ME. 1991a. Scientific principles of controlled/modified atmosphere packaging. In: B Ooraikul, ME Stiles, editors. Modified Atmosphere Packaging of Food, 18–25. New York: Ellis Horwood.

6

Food Regulations
in the United States

P. Stanfield

AN OVERVIEW

This section provides a summary of the legal requirements affecting manufacture and distribution of food products produced within and imported into the United States. The U.S. Food and Drug Administration (FDA) has provided a description of these requirements to the public at large. The information has been translated into several languages, and it is reproduced below with some minor updating by the author.

The FDA regulates all food and food-related products, except commercially processed egg products and meat and poultry products, including combination products (e.g., stew, pizza) containing 2% or more poultry or poultry products or 3% or more red meat or red meat products, which are regulated by the U.S. Department of Agriculture's Food Safety and Inspection Service (FSIS). Fruits, vegetables, and other plants are regulated by the USDA's Animal and Plant Health Inspection Service (APHIS), to prevent the introduction of plant diseases and pests into the United States. The voluntary grading of fruits and vegetables is carried out by the USDA's Agricultural Marketing Service (AMS).

All nonalcoholic beverages and wine beverages containing < 7% alcohol are the responsibility of the FDA. All alcoholic beverages, except wine beverages (i.e., fermented fruit juices) containing < 7% alcohol, are regulated by the Bureau of Alcohol, Tobacco, and Firearms of the U.S. Department of the Treasury.

In addition, the Environmental Protection Agency (EPA) regulates pesticides. The EPA determines the safety of pesticide products, sets tolerance levels for

---

The data provided in this chapter have been modified from a document published and copyrighted by Science Technology System, West Sacramento, California, ©2002. Used with permission.

pesticide residues in food under a section of the Federal Food, Drug, and Cosmetic Act (FD&C Act), and publishes directions for the safe use of pesticides. It is the responsibility of the FDA to enforce the tolerances established by the EPA

Within the United States, compliance with the FD&C Act is secured through periodic inspections of facilities and products, analysis of samples, educational activities, and legal proceedings. A number of regulatory procedures or actions are available to the FDA to enforce the FD&C Act and thus help protect the public's health, safety, and well-being.

Adulterated or misbranded food products may be voluntarily destroyed or recalled from the market by the shipper or may be seized by U.S. Marshals on orders obtained by the FDA from federal district courts. Persons or firms responsible for violation may be prosecuted in the federal courts and if found guilty may be fined and/or imprisoned. Continued violations may be prohibited by federal court injunctions. The violation of an injunction is punishable as contempt of court. Any or all types of regulatory procedures may be employed, depending upon the circumstances.

A recall may be initiated either voluntarily by the manufacturer or shipper of the food commodity or at the request of the FDA. Special provisions on recalls of infant formulas are in the FD&C Act. While the cooperation of the producer or shipper with the FDA in a recall may make court proceedings unnecessary, it does not relieve the person or firm from liability for violations.

It is the responsibility of the owner of the food in interstate commerce to ensure that the article complies with the provisions of the FD&C Act, the Fair Packaging and Labeling Act (FPLA), and their implementing regulations. In general, these acts require that the food product be a safe, clean, wholesome product and that its labeling be honest and informative.

The FD&C Act gives the FDA the authority to establish and impose reasonable sanitation standards on the production of food. The enclosed copy of Title 21, Code of Federal Regulations, Part 110 (21 CFR Part 110) contains the current good manufacturing practice (GMP) regulations concerning personnel, buildings and facilities, equipment, and product process controls for manufacturing, packing, and holding human food; if scrupulously followed, these regulations may give manufacturers some assurance that their food is safe and sanitary. In 21 CFR 110.110, the FDA recognizes that it is not possible to grow, harvest, and process crops that are

totally free of natural defects. Therefore, the agency has published the defect actions for certain food products. These defect action levels are set on the basis of no hazard to health. In the absence of a defect action level, regulatory decisions concerning defects are made on a case-by-case basis.

The alternative to establishing natural defect levels in food would be to insist on increased utilization of chemical substances to control insects, rodents, and other natural contaminants. The FDA has published "action levels" for poisonous or deleterious substances to control levels of contaminants in human food and animal feed. However, a court in the United States invalidated the FDA's "action levels" for poisonous or deleterious substances on procedural grounds. In the interim, the agency is using Action Levels for Poisonous or Deleterious Substances in Human Food and Animal Feed as guidelines, which do not have the "force and effect" of law. The agency has made it clear that action levels are procedural guidelines rather than substantive rules.

The FDA does not approve, license, or issue permits for domestic products shipped in interstate commerce. However, all commercial processors, whether foreign or domestic, of thermally processed low-acid canned foods (LACF) packaged in hermetically sealed containers or of acidified foods (AF) are required by regulations to register each processing plant. In addition, each process for a LACF or an AF must be submitted to the FDA and accepted for filing by the FDA before the product can be distributed in interstate commerce.

A low-acid food is defined as any food, other than alcoholic beverages, with a finished equilibrium pH > 4.6 and a water activity > 0.85—many canned food products are LACF products, and packers are therefore subject to the registration and process filing requirements. The only exceptions are tomatoes and tomato products that have a finished equilibrium pH < 4.7. An acidified food is a low-acid food to which acid(s) or acid food(s) are added, resulting in a product having a finished equilibrium pH of ≤ 4.6.

The FDA's LACF regulations require that each hermetically sealed container of a low-acid processed food be marked with an identifying code that must be permanently visible to the naked eye. The required identification must identify, in code, the establishment where the product is packed, the product contained therein, the year and day of the pack, and the period during the day when the product was packed [21 CFR 113.60(c)]. There is no requirement that a product be shipped from the United States within a stipulated period of time from the date of manufacture. If a LACF or an AF is properly processed, it does not require any special shipping or storage conditions.

FDA regulations require that scheduled processes for LACF be established by qualified persons having expert knowledge of thermal processing requirements for low-acid foods in hermetically sealed containers and having adequate facilities for making such determinations (21 CFR 113.83). All factors critical to the process are required to be specified by the processing authority in the scheduled process. The processor of the food is required to control all critical factors within the limits specified in the scheduled process.

The FDA has the responsibility to establish U.S. identity, quality, and fill of container standards for a number of food commodities. Food standards, which essentially are definitions of food content and quality, are established under provisions of the FD&C Act. Standards have been established for a wide variety of products. These standards give consumers some guarantee of the kind and amount of major ingredients in these products. A food that purports to be a product for which a food standard has been promulgated must meet that standard, or it may be deemed to be out of compliance and therefore be subject to regulatory action.

Amendments to the FD&C Act establish nutrient requirements for infant formulas and provide the FDA authority to establish good manufacturing practices and requirements for nutrient quantity, nutrient quality control, record keeping, and reporting. Under these amendments, the FDA factory inspection authority was expanded to manufacturer's records, quality control records, and test results necessary to determine compliance with the FD&C Act.

The FDA has mandated hazard analysis critical control point (HACCP) procedures for several food categories including seafood and selected fruit and vegetable products. Such procedures assure safe processing, packaging, storage, and distribution of both domestic and imported fish and fishery products and fruit and vegetable products. HACCP is a system by which food processors evaluate the kinds of hazards that could affect their products, institute controls necessary to keep hazards from occurring, monitor the performance of the controls, and maintain records of this monitoring as a matter of routine practice. The purpose is to establish mandatory preventative controls to ensure the safety of the products sold commercially in the United States and exported abroad. The FDA will review the adequacy of HACCP controls in addition to its traditional inspection activities.

The food labeling regulations found in 21 CFR 101 and 105 contain the requirements that, when followed, result in honest and informative labeling of food. Mandatory labeling of food includes a statement of identity (common or usual name of the product—21 CFR 101.3); a declaration of net quantity of contents (21 CFR 101.105); the name and place of business of the manufacturer, packer, or distributor (21 CFR 101.5); and if fabricated from two or more ingredients, a list of ingredients in descending order of predominance by their common or usual names (21 CFR 101.4 and 101.6). Spices, flavorings, and some coloring, other than those sold as such, may be designated as spices, flavoring, and coloring without naming each item. However, food containing a color additive that is subject to certification by the FDA must be declared, in the ingredients statement, to contain that color.

On January 6, 1993, the FDA issued final rules concerning food labeling as mandated by the Nutrition Labeling and Education Act (NLEA). These rules significantly revise many aspects of the existing food labeling regulations, mainly nutrition labeling and related claims for food. The NLEA regulations apply only to domestic food shipped in interstate commerce and to food products offered for import into the United States. The labeling of food products exported to a foreign country must comply with the requirements of that country.

If the label on a food product fails to make all the statements required by the FD&C Act, the FPLA, and the regulations promulgated under these acts, or if the label makes unwarranted claims for the product, the food is deemed misbranded. The FD&C Act provides for both civil and criminal actions for misbranding. The FPLA provides for seizure and injunction. The legal responsibility for full compliance with the terms of each of these acts and their regulations, as applied to labels, rests with the manufacturer, packer, or distributor when the goods are entered into interstate commerce. The label of a food product may include the Universal Product Code (UPC) as well as a number of symbols which signify that the trademark is registered with the U.S. Patent Office; the literary and artistic content of the label is protected against infringement under the copyright laws of the United States; and the food has been prepared and/or complies with dietary laws of certain religious groups. It is important to note that neither the UPC nor any of the symbols mentioned above are required by, or are under the authority of, any of the acts enforced by the U.S. Food and Drug Administration.

The FD&C Act requires premarket approval for food additives (substances the intended use of which results or may reasonably be expected to result, directly or indirectly, in their either becoming a component of food or otherwise affecting the characteristics of food). The approval process involves a very careful review of the additive's safety for its intended use. Following the approval of a food additive, a regulation describing its use is published in the Code of Federal Regulations. As defined in the CFR, the term safe or safety, ". . . means there is a reasonable certainty in the minds of competent scientists that the substance is not harmful under the intended conditions of use." It is impossible in the present state of scientific knowledge to establish with complete certainty the absolute harmlessness of the use of any substance. Premarket clearance under the FD&C Act does assure that the risk of adverse effects occurring due to a food additive is at an acceptably small level.

The FDA's regulation of dietary supplements is under the authority of the Dietary Supplements Health and Education Act of 1994. It ensures that the products are safe and properly labeled and that any disease- or health-related claims are scientifically supported. The legal provisions governing the safety of dietary supplements depend on whether the product is legally a food or a drug. In either instance, the manufacturer is obligated to produce a safe product. Premarket safety review by the FDA is required for new drugs.

The label of a dietary supplement must state what the product contains, how much it contains, how it should be used, and what precautions are necessary to assure safe use, and all other information provided must be truthful and not misleading. If the dietary supplement is a food, a review of any disease- or health-related claim is conducted under the NLEA health claim provisions.

CURRENT GOOD MANUFACTURING PRACTICE REGULATIONS (CGMPR), HAZARDS ANALYSIS CRITICAL CONTROL POINTS REGULATIONS (HACCPR), AND THE FOOD CODE

Nearly 25 years ago, the U.S. Food and Drug Administration (FDA) started using umbrella regulations to help food industries produce wholesome food as required by the Federal Food, Drug, Cosmetic Act (the Act). In 1986, the FDA promulgated the first umbrella regulations under the title of good manufacturing practice (GMP) regulations (GMPR). Since then, many aspects of the regulations have been revised. Traditionally, industry and regulators have depended on spot checks of manufacturing conditions and random sampling of final products to ensure safe food. The current good manufacturing practice regulations (CGMPR) forms the basis on which the FDA will inform the food manufacturer about deficiencies in its operations. This approach, however, tends to be reactive rather than preventive and can definitely be improved.

For more than 30 years, the FDA has been regulating the low-acid canned food (LACF) industries with a special set of regulations, many of which are preventive in nature. This action aims at preventing botulism. In the last 30 years, threats from other biological pathogens have increased tremendously. Between 1980 and 1995, the FDA studied use of the hazard analysis and critical control points (HACCP) approach. For this approach, the FDA uses the LACF regulations as a partial guide. Since 1995, the FDA has issued HACCP regulations (HACCPR) for the manufacture or production of several types of food products. These include the processing of seafood and fruit/vegetable juices.

Since 1938, when the Act was first passed by Congress, the FDA and state regulatory agencies have worked hard to reach a uniform set of codes for the national regulation of food manufacturing industries and state regulation of retail industries associated with food (e.g., groceries, restaurants, catering, etc.). In 1993, the first document titled Food Code was issued jointly by the FDA and state agencies. It has been revised twice since then. This chapter discusses CGMPR, HACCPR, and the Food Code.

CURRENT GOOD MANUFACTURING PRACTICE REGULATIONS (CGMPR)

The current good manufacturing practice regulations (CGMPR) cover the topics listed in Table 6.1. These regulations are discussed in detail here. Please note that the word "shall" in a legal document means mandatory and is used routinely in the FDA regulations published in the U.S. Code of Federal Regulations (CFR). In this chapter, the words "should" and "must" are used to make for smoother reading. However, this in no way diminishes the legal impact of the original regulations.

Table 6.1. Contents of the Current Good Manufacturing Regulations (CGMPR)

21 CFR 110.3	Definitions.
21 CFR 110.5	Current good manufacturing practice.
21 CFR 110.10	Personnel.
21 CFR 110.19	Exclusions.
21 CFR 110.20	Plant and grounds.
21 CFR 110.35	Sanitary operations.
21 CFR 110.37	Sanitary facilities and controls.
21 CFR 110.40	Equipment and utensils.
21 CFR 110.80	Processes and controls.
21 CFR 110.93	Warehousing and distribution.

DEFINITIONS (21 CFR 110.3)

FDA has provided the following definitions and interpretations for several important terms.

1. *Acid food* or *acidified food* means foods that have an equilibrium pH $\leq$ 4.6.
2. *Batter* means a semifluid substance, usually composed of flour and other ingredients, into which the principal components of a food are dipped, with which they are coated, or which may be used directly to form bakery foods.
3. *Blanching,* except for tree nuts and peanuts, means a prepackaging heat treatment of foodstuffs for a sufficient time and at a sufficient temperature to partially or completely inactivate the naturally occurring enzymes and to affect other physical or biochemical changes in the food.
4. *Critical control point* means a point in a food process where there is a high probability that improper control may cause a hazard or filth in the final food or decomposition of the final food.
5. *Food* includes raw materials and ingredients.
6. *Food-contact surfaces* are those surfaces that contact human food and those surfaces from which drainage onto the food or onto surfaces that contact the food ordinarily occurs during the normal course of operations. *Food-contact surfaces* include utensils and the food-contact surfaces of equipment.
7. *Lot* means the food produced during a period of time indicated by a specific code.
8. *Microorganisms* are yeasts, molds, bacteria, and viruses and include, but are not limited to, species having public health significance. The term *undesirable microorganisms* includes those microorganisms that are of public health significance, that promote decomposition of food, or that indicate that food is contaminated with filth.
9. *Pest* refers to any objectionable animals or insects including, but not limited to, birds, rodents, flies, and insect larvae.
10. *Plant* means the building or facility used for the manufacturing, packaging, labeling, or holding of human food.
11. *Quality control operation* means a planned and systematic procedure for taking all actions necessary to prevent food from being adulterated.
12. *Rework* means clean, unadulterated food that has been removed from processing for reasons other than unsanitary conditions or that has been successfully reconditioned by reprocessing and that is suitable for use as food.
13. *Safe moisture level* is a level of moisture low enough to prevent the growth of undesirable microorganisms in the finished product under the intended conditions of manufacturing, storage, and distribution. The maximum safe moisture level for a food is based on its water activity, a_w. A particular a_w will be considered safe for a food if adequate data are available that demonstrate that the food at or below the given a_w will not support the growth of undesirable microorganisms.
14. *Sanitize* means to adequately treat food-contact surfaces by a process that is effective in destroying vegetative cells of microorganisms that are of public health significance and in substantially reducing numbers of other undesirable microorganisms without adversely affecting the product or its safety for the consumer.
15. *Water activity (a_w)* is a measure of the free moisture in a food and is the quotient of the water vapor pressure of the substance divided by the vapor pressure of pure water at the same temperature.

PERSONNEL (SECTION 110.10)

Plant management should take all reasonable measures and precautions to ensure compliance with the following regulations.

1. *Disease control.* Any person who, by medical examination or supervisory observation, is

shown to have an illness or open lesion, including boils, sores, or infected wounds, by which there is a reasonable possibility of food, food-contact surfaces, or food-packaging materials becoming contaminated, should be excluded from any operations which may be expected to result in such contamination until the condition is corrected. Personnel should be instructed to report such health conditions to their supervisors.

2. *Cleanliness.* All persons working in direct contact with food, food-contact surfaces, and food-packaging materials should conform to hygienic practices while on duty. The methods for maintaining cleanliness include, but are not limited to, the following:

   a. Wearing outer garments suitable to the operation to protect against the contamination of food, food-contact surfaces, or food-packaging materials.

   b. Maintaining adequate personal cleanliness.

   c. Washing hands thoroughly (and sanitizing if necessary to protect against contamination with undesirable microorganisms) in an adequate hand-washing facility before starting work, after each absence from the work station, and at any other time when the hands may have become soiled or contaminated.

   d. Removing all unsecured jewelry and other objects that might fall into food, equipment, or containers and removing hand jewelry that cannot be adequately sanitized during periods in which food is manipulated by hand. If such hand jewelry cannot be removed, it may be covered by material which can be maintained in an intact, clean, and sanitary condition and which effectively protects against their contamination of the food, food-contact surfaces, or food-packaging materials.

   e. Maintaining gloves, if they are used in food handling, in an intact, clean, and sanitary condition. The gloves should be of an impermeable material.

   f. Wearing, where appropriate, hairnets, headbands, caps, beard covers, or other effective hair restraints.

   g. Storing clothing or other personal belongings in areas other than where food is exposed or where equipment or utensils are washed.

   h. Confining the following personal practices to areas other than where food may be exposed or where equipment or utensils are washed: eating food, chewing gum, drinking beverages, or using tobacco.

   i. Taking any other necessary precautions to protect against contamination of food, food-contact surfaces, or food-packaging materials with microorganisms or foreign substances including, but not limited to, perspiration, hair, cosmetics, tobacco, chemicals, and medicines applied to the skin.

3. *Education and training.* Personnel responsible for identifying sanitation failures or food contamination should have a background of education or experience sufficient to provide the level of competency necessary for production of clean and safe food. Food handlers and supervisors should receive appropriate training in proper food handling techniques and food-protection principles and should be informed of the danger of poor personal hygiene and unsanitary practices.

4. *Supervision.* Responsibility for assuring compliance by all personnel with all legal requirements should be clearly assigned to competent supervisory personnel.

PLANT AND GROUNDS (SECTION 110.20)

1. *Grounds.* The grounds surrounding a food plant that are under the control of the plant manager should be kept in a condition that will protect against the contamination of food. The methods for adequate maintenance of grounds include, but are not limited to, the following:

   a. Properly storing equipment, removing litter and waste, and cutting weeds or grass within the immediate vicinity of the plant buildings or structures that may constitute an attractant, breeding place, or harborage for pests.

   b. Maintaining roads, yards, and parking lots so that they do not constitute a source of contamination in areas where food is exposed.

   c. Adequately draining areas that may contribute contamination to food by seepage or foot-borne filth, or by providing a breeding place for pests.

   d. Operating systems for waste treatment and disposal in an adequate manner so that they do not constitute a source of contamination in areas where food is exposed. If the plant grounds are bordered by grounds not under the operator's control and not maintained in an acceptable manner, steps must be taken to exclude pests, dirt, and filth that may be a source of food contamination. Implement inspection, extermination, or other counter-measures.

2. *Plant construction and design.* Plant buildings and structures should be suitable in size, construction, and design to facilitate maintenance and sanitary operations for food manufacturing purposes. The plant and facilities should:

   a. Provide sufficient space for such placement of equipment and storage of materials as necessary for the maintenance of sanitary operations and the production of safe food.

   b. Take proper precautions to reduce the potential for contamination of food, food-contact surfaces, or food-packaging materials with microorganisms, chemicals, filth, or other extraneous material. The potential for contamination may be reduced by adequate food safety controls and operating practices or effective design, including the separation of operations in which contamination is likely to occur, by one or more of the following means: location, time, partition, air flow, enclosed systems, or other effective means.

   c. Take proper precautions to protect food in outdoor bulk fermentation vessels by any effective means, including (1) using protective coverings, (2) controlling areas over and around the vessels to eliminate harborages for pests, (3) checking on a regular basis for pests and pest infestation, and (4) skimming the fermentation vessels, as necessary.

   d. Be constructed in such a manner that floors, walls, and ceilings may be adequately cleaned and kept clean and in good repair; that drip or condensate from fixtures, ducts, and pipes does not contaminate food, food-contact surfaces, or food-packaging materials; and that aisles or working spaces are provided between equipment and walls and are adequately unobstructed and of adequate width to permit employees to perform their duties and to protect against contaminating food or food-contact surfaces with clothing or personal contact.

   e. Provide adequate lighting in hand-washing areas, dressing and locker rooms, and toilet rooms; in all areas where food is examined, processed, or stored; and where equipment or utensils are cleaned. Also provide safety-type light bulbs, fixtures, skylights, or other glass where such items are suspended over exposed food in any step of preparation, or otherwise protect against food contamination in case of glass breakage.

   f. Provide adequate ventilation or control equipment to minimize odors and vapors (including steam and noxious fumes) in areas where they may contaminate food; and locate and operate fans and other air-blowing equipment in a manner that minimizes the potential for contaminating food, food-packaging materials, and food-contact surfaces.

   g. Provide, where necessary, adequate screening or other protection against pests.

SANITARY OPERATIONS (SECTION 110.35)

1. *General maintenance.* Buildings, fixtures, and other physical facilities of the plant should be maintained in a sanitary condition and should be kept in repair sufficient to prevent food from becoming adulterated within the meaning of the Act. Cleaning and sanitizing of utensils and equipment should be conducted in a manner that protects against contamination of food, food-contact surfaces, or food-packaging materials.

2. *Substances used in cleaning and sanitizing; storage of toxic materials.*

   a. Cleaning compounds and sanitizing agents used in cleaning and sanitizing procedures should be free from undesirable microorganisms and should be safe and adequate under the conditions of use. Compliance with this requirement may be verified by any effective means including purchase of these substances under a supplier's guarantee or certification, or examination of these substances for contamination. Only the following toxic materials may be used or stored in a plant where food is processed or exposed: (1) those required to maintain clean and sanitary conditions, (2) those necessary for use in laboratory testing procedures, (3) those necessary for plant and equipment maintenance and operation, and (4) those necessary for use in the plant's operations.

b. Toxic cleaning compounds, sanitizing agents, and pesticide chemicals should be identified, held, and stored in a manner that protects against contamination of food, food-contact surfaces, or food-packaging materials.

3. *Pest control.* No pests should be allowed in any area of a food plant. Guard or guide dogs may be allowed in some areas of a plant if the presence of the dogs is unlikely to result in contamination of food, food-contact surfaces, or food-packaging materials. Effective measures should be taken to exclude pests from the processing areas and to protect against the contamination of food on the premises by pests. The use of insecticides or rodenticides is permitted only under precautions and restrictions that will protect against the contamination of food, food-contact surfaces, and food-packaging materials.

4. *Sanitation of food-contact surfaces.* All food-contact surfaces, including utensils and food-contact surfaces of equipment, should be cleaned as frequently as necessary to protect against contamination of food.

a. Food-contact surfaces used for manufacturing or holding low-moisture food should be in a dry, sanitary condition at the time of use. When the surfaces are wet-cleaned, they should, when necessary, be sanitized and thoroughly dried before subsequent use.

b. In wet processing, when cleaning is necessary to protect against the introduction of microorganisms into food, all food-contact surfaces should be cleaned and sanitized before use and after any interruption during which the food-contact surfaces may have become contaminated. Where equipment and utensils are used in a continuous production operation, the utensils and food-contact surfaces of the equipment should be cleaned and sanitized as necessary.

c. Non-food-contact surfaces of equipment used in the operation of food plants should be cleaned as frequently as necessary to protect against contamination of food.

d. Single-service articles (such as utensils intended for one-time use, paper cups, and paper towels) should be stored in appropri-

ate containers and should be handled, dispensed, used, and disposed of in a manner that protects against contamination of food or food-contact surfaces.

e. Sanitizing agents should be adequate and safe under conditions of use. Any facility, procedure, or machine is acceptable for cleaning and sanitizing equipment and utensils if it is established that the facility, procedure, or machine will routinely render equipment and utensils clean and provide adequate cleaning and sanitizing treatment.

5. *Storage and handling of cleaned portable equipment and utensils.* Cleaned and sanitized portable equipment with food-contact surfaces and utensils should be stored in a location and manner that protects food-contact surfaces from contamination.

SANITARY FACILITIES AND CONTROLS (SECTION 110.37)

Each plant should be equipped with adequate sanitary facilities and accommodations including, but not limited to:

1. *Water supply.* The water supply should be sufficient for the operations intended and should be derived from an adequate source. Any water that contacts food or food-contact surfaces should be safe and of adequate sanitary quality. Running water at a suitable temperature, and under pressure as needed, should be provided in all areas where required for the processing of food; for the cleaning of equipment, utensils, and food-packaging materials; or for employee sanitary facilities.

2. *Plumbing.* Plumbing should be of adequate size and design and adequately installed and maintained to:

a. Carry sufficient quantities of water to required locations throughout the plant.

b. Properly convey sewage and liquid disposable waste from the plant.

c. Avoid constituting a source of contamination to food, water supplies, equipment, or utensils or creating an unsanitary condition.

d. Provide adequate floor drainage in all areas where floors are subject to flooding-type cleaning or where normal operations release or discharge water or other liquid waste on the floor.

e. Provide that there is no backflow from, or cross-connection between, piping systems that discharge wastewater or sewage and piping systems that carry water for food or food manufacturing.

3. *Sewage disposal.* Sewage disposal should be made into an adequate sewerage system or through other adequate means.

4. *Toilet facilities.* Each plant should provide its employees with adequate, readily accessible toilet facilities. Compliance with this requirement may be accomplished by:

   a. Maintaining the facilities in a sanitary condition.
   b. Keeping the facilities in good repair at all times.
   c. Providing self-closing doors.
   d. Providing doors that do not open into areas where food is exposed to airborne contamination, except where alternate means have been taken to protect against such contamination (such as double doors or positive airflow systems).

5. *Hand-washing facilities.* Hand-washing facilities should be adequate and convenient and be furnished with running water at a suitable temperature. Compliance with this requirement may be accomplished by providing:

   a. Hand-washing and, where appropriate, hand-sanitizing facilities at each location in the plant where good sanitary practices require employees to wash and/or sanitize their hands.
   b. Effective hand cleaning and sanitizing preparations.
   c. Sanitary towel service or suitable drying devices.
   d. Devices or fixtures, such as water control valves, designed and constructed so as to protect against recontamination of clean, sanitized hands.
   e. Readily understandable signs directing employees handling unprotected food, unprotected food-packaging materials, or food-contact surfaces to wash and, where appropriate, sanitize their hands before they start work, after each absence from post of duty, and when their hands may have become soiled or contaminated. These signs may be posted in the processing room(s) and in all other areas where employees may handle such food, materials, or surfaces.

f. Refuse receptacles that are constructed and maintained in a manner that protects against contamination of food.

6. *Rubbish and offal disposal.* Rubbish and any offal should be so conveyed, stored, and disposed of as to minimize the development of odor, minimize the potential for the waste becoming an attractant and harborage or breeding place for pests, and protect against contamination of food, food-contact surfaces, water supplies, and ground surfaces.

EQUIPMENT AND UTENSILS (SECTION 110.40)

1. All plant equipment and utensils should be so designed and of such material and workmanship as to be adequately cleanable and should be properly maintained. The design, construction, and use of equipment and utensils should preclude the adulteration of food with lubricants, fuel, metal fragments, contaminated water, or any other contaminants. All equipment should be so installed and maintained as to facilitate the cleaning of the equipment and of all adjacent spaces. Food-contact surfaces should be corrosion resistant when in contact with food. They should be made of nontoxic materials and designed to withstand the environment of their intended use and the action of food, and, if applicable, cleaning compounds and sanitizing agents. Food-contact surfaces should be maintained to protect food from being contaminated by any source, including unlawful indirect food additives.

2. Seams on food-contact surfaces should be smoothly bonded or maintained so as to minimize accumulation of food particles, dirt, and organic matter and thus minimize the opportunity for growth of microorganisms.

3. Equipment that is in the manufacturing or food-handling area and that does not come into contact with food should be so constructed that it can be kept in a clean condition.

4. Holding, conveying, and manufacturing systems, including gravimetric, pneumatic, closed, and automated systems, should be of a design and construction that enables them to be maintained in an appropriate sanitary condition.

5. Each freezer and cold storage compartment used to store and hold food capable of supporting growth of microorganisms should be fitted

with an indicating thermometer, temperature-measuring device, or temperature-recording device so installed as to show the temperature accurately within the compartment, and with an automatic control for regulating temperature or an automatic alarm system to indicate a significant temperature change in a manual operation.

6. Instruments and controls used for measuring, regulating, or recording temperatures, pH, acidity, water activity, or other conditions that control or prevent the growth of undesirable microorganisms in food should be accurate and adequately maintained, and adequate in number for their designated uses.

7. Compressed air or other gases mechanically introduced into food or used to clean food-contact surfaces or equipment should be treated in such a way that food is not contaminated with unlawful indirect food additives.

PROCESSES AND CONTROLS (SECTION 110.80)

All operations in the receiving, inspecting, transporting, segregating, preparing, manufacturing, packaging, and storing of food should be conducted in accordance with adequate sanitation principles. Appropriate quality control operations should be employed to ensure that food is suitable for human consumption and that food-packaging materials are safe and suitable. Overall sanitation of the plant should be under the supervision of one or more competent individuals assigned responsibility for this function. All reasonable precautions should be taken to ensure that production procedures do not contribute contamination from any source. Chemical, microbial, or extraneous material testing procedures should be used where necessary to identify sanitation failures or possible food contamination. All food that has become contaminated to the extent that it is adulterated within the meaning of the Act should be rejected, or if permissible, treated or processed to eliminate the contamination.

1. *Raw materials and other ingredients.*
   a. Raw materials and other ingredients should be inspected and segregated or otherwise handled as necessary to ascertain that they are clean and suitable for processing into food and should be stored under conditions that will protect against contamination and minimize deterioration. Raw materials should be washed or cleaned as necessary to remove soil or other contamination. Water used for washing, rinsing, or conveying food should be safe and of adequate sanitary quality. Water may be reused for washing, rinsing, or conveying food if it does not increase the level of contamination of the food. Containers and carriers of raw materials should be inspected on receipt to ensure that their condition has not contributed to the contamination or deterioration of food.

   b. Raw materials and other ingredients should either not contain levels of microorganisms that may produce food poisoning or other disease in humans, or they should be pasteurized or otherwise treated during manufacturing operations so that they no longer contain levels that would cause the product to be adulterated within the meaning of the Act. Compliance with this requirement may be verified by any effective means, including purchasing raw materials and other ingredients under a supplier's guarantee or certification.

   c. Raw materials and other ingredients susceptible to contamination with aflatoxin or other natural toxins should comply with current FDA regulations, guidelines, and action levels for poisonous or deleterious substances before these materials or ingredients are incorporated into finished food. Compliance with this requirement may be accomplished by purchasing raw materials and other ingredients under a supplier's guarantee or certification, or may be verified by analyzing these materials and ingredients for aflatoxins and other natural toxins.

   d. Raw materials, other ingredients, and rework susceptible to contamination with pests, undesirable microorganisms, or extraneous material should comply with applicable FDA regulations, guidelines, and defect action levels for natural or unavoidable defects if a manufacturer wishes to use the materials in manufacturing food. Compliance with this requirement may be verified by any effective means, including purchasing the materials under a supplier's guarantee or certification, or examination of these materials for contamination.

e. Raw materials, other ingredients, and rework should be held in bulk, or in containers designed and constructed so as to protect against contamination, and should be held at such temperature and relative humidity as to prevent the food from becoming adulterated. Material scheduled for rework should be identified as such.

f. Frozen raw materials and other ingredients should be kept frozen. If thawing is required prior to use, it should be done in a manner that prevents the raw materials and other ingredients from becoming adulterated.

g. Liquid or dry raw materials and other ingredients received and stored in bulk form should be held in a manner that protects against contamination.

2. *Manufacturing operations.*

a. Equipment and utensils and finished food containers should be maintained in an acceptable condition through appropriate cleaning and sanitizing, as necessary. Insofar as necessary, equipment should be taken apart for thorough cleaning.

b. All food manufacturing, including packaging and storage, should be conducted under such conditions and controls as are necessary to minimize the potential for the growth of microorganisms or for the contamination of food. One way to comply with this requirement is careful monitoring of physical factors (such as time, temperature, humidity, a_w, pH, pressure, and flow rate) and manufacturing operations (such as freezing, dehydration, heat processing, acidification, and refrigeration) to ensure that mechanical breakdowns, time delays, temperature fluctuations, and other factors do not contribute to the decomposition or contamination of food.

c. Food that can support the rapid growth of undesirable microorganisms, particularly those of public health significance, should be held in a manner that prevents the food from becoming spoiled. Compliance with this requirement may be accomplished by any effective means, including (1) maintaining refrigerated foods at 45°F (7.2°C) or below as appropriate for the particular food involved, (2) maintaining frozen foods in a frozen state, (3) maintaining hot foods at 140°F (60°C) or above, and (4) heat treating acid or acidified foods to destroy mesophilic microorganisms when those foods are to be held in hermetically sealed containers at ambient temperatures.

d. Measures such as sterilizing, irradiating, pasteurizing, freezing, refrigerating, controlling pH, or controlling a_w that are taken to destroy or prevent the growth of undesirable microorganisms, particularly those of public health significance, should be adequate under the conditions of manufacture, handling, and distribution to prevent food from being adulterated.

e. Work-in-process should be handled in a manner that protects against contamination.

f. Effective measures should be taken to protect finished food from contamination by raw materials, other ingredients, or refuse. When raw materials, other ingredients, or refuse are unprotected, they should not be handled simultaneously in a receiving, loading, or shipping area if that handling could result in contaminated food. Food transported by conveyor should be protected against contamination as necessary.

g. Equipment, containers, and utensils used to convey, hold, or store raw materials, work-in-process, rework, or food should be constructed, handled, and maintained during manufacturing or storage in a manner that protects against contamination.

h. Effective measures should be taken to protect against the inclusion of metal or other extraneous material in food. Compliance with this requirement may be accomplished by using sieves, traps, magnets, electronic metal detectors, or other suitable effective means.

i. Food, raw materials, and other ingredients that are adulterated should be disposed of in a manner that protects against the contamination of other food. If the adulterated food is capable of being reconditioned, it should be reconditioned using a method that has been proven to be effective, or it should be reexamined and found to be unadulterated before being incorporated into other food.

j. Mechanical manufacturing steps such as washing, peeling, trimming, cutting, sorting and inspecting, mashing, dewatering, cooling, shredding, extruding, drying, whipping, defatting, and forming should be performed so as to protect food against contamination. Compliance with this requirement may be accomplished by providing adequate physical protection of food from contaminants that may drip, drain, or be drawn into the food. Protection may be provided by adequate cleaning and sanitizing of all food-contact surfaces and by using time and temperature controls at and between each manufacturing step.

k. Heat blanching, when required in the preparation of food, should be effected by heating the food to the required temperature, holding it at this temperature for the required time, and then either rapidly cooling the food or passing it to subsequent manufacturing without delay. Thermophilic growth and contamination in blanchers should be minimized by the use of adequate operating temperatures and by periodic cleaning. Where the blanched food is washed prior to filling, water used should be safe and of adequate sanitary quality.

l. Batters, breading, sauces, gravies, dressings, and other similar preparations should be treated or maintained in such a manner that they are protected against contamination. Compliance with this requirement may be accomplished by any effective means, including one or more of the following: (1) using ingredients free of contamination, (2) employing adequate heat processes where applicable, (3) using adequate time and temperature controls, (4) providing adequate physical protection of components from contaminants that may drip, drain, or be drawn into them, (5) cooling to an adequate temperature during manufacturing, and (6) disposing of batters at appropriate intervals to protect against the growth of microorganisms.

m. Filling, assembling, packaging, and other operations should be performed in such a way that the food is protected against contamination. Compliance with this requirement may be accomplished by any effective means, including (1) use of a quality control operation in which the critical control points are identified and controlled during manufacturing, (2) adequate cleaning and sanitizing of all food-contact surfaces and food containers, (3) using materials for food containers and food-packaging materials that are safe and suitable, (4) providing physical protection from contamination, particularly airborne contamination, and (5) using sanitary handling procedures.

n. Food such as, but not limited to, dry mixes, nuts, intermediate-moisture food, and dehydrated food that relies on the control of a_w for preventing the growth of undesirable microorganisms should be processed to and maintained at a safe moisture level. Compliance with this requirement may be accomplished by any effective means, including employment of one or more of the following practices: (1) monitoring the a_w of the food, (2) controlling the soluble solids/water ratio in finished food, and (3) protecting finished food from moisture pickup by use of a moisture barrier or other means, so that the a_w of the food does not increase to an unsafe level.

o. Food such as, but not limited to, acid and acidified food that relies principally on the control of pH for preventing the growth of undesirable microorganisms should be monitored and maintained at a pH ≤ 4.6. Compliance with this requirement may be accomplished by any effective means, including employment of one or more of the following practices: (1) monitoring the pH of raw materials, food-in-process, and finished food and (2) controlling the amount of acid or acidified food added to low-acid food.

p. When ice is used in contact with food, it should be made from water that is safe and of adequate sanitary quality and should be used only if it has been manufactured in accordance with current good manufacturing practice.

q. Food manufacturing areas and equipment used for manufacturing human food should not be used to manufacture nonhuman-food-grade animal feed or inedible products, unless there is no reasonable possibility for the contamination of the human food.

WAREHOUSING AND DISTRIBUTION (SECTION 110.93)

Storage and transportation of finished food should be under conditions that will protect food against physical, chemical, and microbial contamination as well as against deterioration of the food and the container.

NATURAL OR UNAVOIDABLE DEFECTS IN FOOD FOR HUMAN USE THAT PRESENT NO HEALTH HAZARD (SECTION 110.110)

1. Some foods, even when produced under current good manufacturing practice, contain natural or unavoidable defects that at low levels are not hazardous to health. The FDA establishes maximum levels for these defects in foods produced under current good manufacturing practice and uses these levels in deciding whether to recommend regulatory action.
2. Defect action levels are established for foods whenever it is necessary and feasible to do so. These levels are subject to change upon the development of new technology or the availability of new information.
3. The mixing of a food containing defects above the current defect action level with another lot of food is not permitted and renders the final food adulterated within the meaning of the Act, regardless of the defect level of the final food.
4. A compilation of the current defect action levels for natural or unavoidable defects in food for human use that present no health hazard may be obtained from the FDA in printed or electronic versions.

HAZARD ANALYSIS CRITICAL CONTROL POINTS REGULATIONS (HACCPR)

In 1997, the FDA adopted a food safety program that was developed nearly 30 years ago for astronauts and is now applying it to seafood and to fruit and vegetable juices. The agency intends to eventually use it for much of the U.S. food supply. The program for the astronauts focuses on preventing hazards that could cause food-borne illnesses by applying science-based controls, from raw material to finished products. The FDA's new system will do the same.

Many principles of this new system, now called hazard analysis and critical control points (HACCP), are already in place in the FDA-regulated low-acid canned food industry. Since 1997, FDA has mandated HACCP for the processing of seafood, fruit juices, and vegetable juices. Also, FDA has incorporated HACCP into its Food Code, which gives guidance to and serves as model legislation for state and territorial agencies that license and inspect food-service establishments, retail food stores, and food vending operations in the United States.

The FDA now is considering developing regulations that would establish HACCP as the food safety standard throughout other areas of the food industry, including both domestic and imported food products. HACCP has been endorsed by the National Academy of Sciences, the Codex Alimentarius Commission (an international, standard-setting organization), and the National Advisory Committee on Microbiological Criteria for Foods. Several U.S. food companies already use the system in their manufacturing processes, and it is also in use in other countries, including Canada.

WHAT IS HACCP?

HACCP involves seven principles.

1. *Analyze hazards.* Potential hazards associated with a food and measures to control those hazards are identified. The hazard could be biological (e.g., a microbe), chemical (e.g., a toxin), or physical (e.g., ground glass or metal fragments).
2. *Identify critical control points.* These are points in a food's production—from its raw state through processing and shipping to consumption by the consumer—at which the potential hazard can be controlled or eliminated. Examples are cooking, cooling, packaging, and metal detection.
3. *Establish preventive measures* with critical limits for each control point. For a cooked food, for example, this might include setting the minimum cooking temperature and time required to ensure the elimination of any harmful microbes.
4. *Establish procedures to monitor the critical control points.* Such procedures might include determining how and by whom cooking time and temperature should be monitored.
5. *Establish corrective actions* to be taken when monitoring shows that a critical limit has not been met—for example, reprocessing or disposing of food if the minimum cooking temperature is not met.
6. *Establish procedures to verify that the system is working properly*—for example, testing time and temperature recording devices to verify that a cooking unit is working properly.
7. *Establish effective record keeping* to document the HACCP system. This would include

records of hazards and their control methods, the monitoring of safety requirements, and action taken to correct potential problems.

Each of these principles must be backed by sound scientific knowledge such as published microbiological studies on time and temperature factors for controlling food-borne pathogens.

NEED FOR HACCP

New challenges to the U.S. food supply have prompted the FDA to consider adopting a HACCP-based food safety system on a wider basis. One of the most important challenges is the increasing number of new food pathogens. For example, between 1973 and 1988, bacteria not previously recognized as important causes of food-borne illness—such as *Escherichia coli* O157:H7 and *Salmonella enteritidis*—became more widespread. There also is increasing public health concern about chemical contamination of food: for example, the effects of lead in food on the nervous system.

Another important factor is that the size of the food industry and the diversity of products and processes have grown tremendously—in the amount of domestic food manufactured and the number and kinds of foods imported. At the same time, the FDA and state and local agencies have the same limited level of resources to ensure food safety. The need for HACCP in the United States, particularly in the seafood industry, is further fueled by the growing trend in international trade for worldwide equivalence of food products and the Codex Alimentarius Commission's adoption of HACCP as the international standard for food safety.

ADVANTAGES AND PLANS

HACCP offers a number of advantages over previous systems. Most importantly, HACCP (1) focuses on identifying and preventing hazards from contaminating food, (2) is based on sound science, (3) permits more efficient and effective government oversight, primarily because the record keeping allows investigators to see how well a firm is complying with food safety laws over a period rather than how well it is doing on any given day, (4) places responsibility for ensuring food safety appropriately on the food manufacturer or distributor, (5) helps food companies compete more effectively in the world

market, and (6) reduces barriers to international trade.

Here are the seven steps used in HACCP plan development.

1. Preliminary steps: (a) General information. (b) Describe the food. (c) Describe the method of distribution and storage. (d) Identify the intended use and consumer. (e) Develop a flow diagram.
2. Hazard Analysis Worksheet: (a) Set up the Hazard Analysis Worksheet. (b) Identify the potential species-related hazards. (c) Identify the potential process-related hazards. (d) Complete the Hazard Analysis Worksheet. (e) Understand the potential hazard. (f) Determine if the potential hazard is significant. (g) Identify the critical control points (CCPs).
3. HACCP Plan Form: (a) Complete the HACCP Plan Form. (b) Set the critical limits (CLs).
4. Establish monitoring procedures: (a) What. (b) How. (c) How often. (d) Who.
5. Establish corrective action procedures.
6. Establish a record-keeping system.
7. Establish verification procedures.

It is important to remember that apart from HACPR promulgated for seafood and juices, the implementation of HACCP in other categories of food processing is voluntary. However, the FDA and various types of food processors are working together so that eventually HACCPR will become available for many other food processing systems under the FDA jurisdiction. Using the HACCPR for seafood processing as a guide, the following discussion for a HACCP plan applies to all categories of food products being processed in United States.

HAZARD ANALYSIS

Every processor should conduct a hazard analysis to determine whether there are food safety hazards that are reasonably likely to occur for each kind of product processed by that processor and to identify the preventive measures that the processor can apply to control those hazards. Such food safety hazards can be introduced both within and outside the processing plant environment, including food safety hazards that can occur before, during, and after harvest. A food safety hazard that is reasonably likely to occur is one for which a prudent processor would establish controls because experience, illness data, scientific reports, or other information provide a basis to conclude that there is a reasonable possibility that it will occur in the particular type of product being processed in the absence of those controls.

THE HACCP PLAN

Every processor should have and implement a written HACCP plan whenever a hazard analysis reveals one or more food safety hazards that are reasonably likely to occur. A HACCP plan should be specific to (1) each location where products are processed by that processor and (2) each kind of product processed by the processor.

The plan may group kinds of products or kinds of production methods together, if the food safety hazards, critical control points, critical limits, and procedures that must be identified and performed are identical for all products so grouped or for all production methods so grouped.

Contents of the HACCP Plan

The HACCP plan should, at a minimum:

- *List the food safety hazards* that are reasonably likely to occur, as identified, and that thus must be controlled for each product. Consideration should be given to whether any food safety hazards are reasonably likely to occur as a result of natural toxins; microbiological contamination; chemical contamination; pesticides; drug residues; decomposition in products where a food safety hazard has been associated with decomposition; parasites, where the processor has knowledge that the parasite-containing product will be consumed without a process sufficient to kill the parasites; unapproved use of direct or indirect food or color additives; and physical hazards.
- *List the critical control points* for each of the identified food safety hazards including, as appropriate, (1) critical control points designed to control food safety hazards that could be introduced in the processing plant environment and (2) critical control points designed to control food safety hazards introduced outside the processing plant environment, including food safety hazards that occur before, during, and after harvest.

- *List the critical limits* that must be met at each of the critical control points.
- *List the procedures,* and frequency thereof, that will be used to monitor each of the critical control points to ensure compliance with the critical limits.
- *Include any corrective action plans* that are to be followed in response to deviations from critical limits at critical control points.
- *List the verification procedures,* and frequency thereof, that the processor will use.
- *Provide for a record-keeping system* that documents the monitoring of the critical control points. The records should contain the actual values and observations obtained during monitoring.

Signing and Dating the HACCP Plan

The HACCP plan should be signed and dated (1) upon initial acceptance, (2) upon any modification, and (3) upon verification of the plan. The plan should be signed and dated either by the most responsible individual on site at the processing facility or by a higher level official of the processor. This signature should signify that the HACCP plan has been accepted for implementation by the firm.

SANITATION

Sanitation controls may be included in the HACCP plan. However, to the extent that they are otherwise monitored, they need not be included in the HACCP plan.

IMPLEMENTATION

This book is not the proper forum to discuss in detail the implementation of HACCPR. Readers interested in additional information on HACCP should visit the FDA HACCP website http://vm.cfsan.fda.gov/ that lists all the currently available documents on the subject.

FDA FOOD CODE

The FDA Food Code (the Code) is an essential reference that provides guidelines on how to prevent food-borne illness to retail outlets such as restaurants and grocery stores and institutions such as nursing homes. Local, state and federal regulators use the Code as a model in developing or updating their own food safety rules and to be consistent with national food regulatory policy. Also, many of the over one million retail food establishments apply Food Code provisions to their own operations. The Code is updated every two years, to coincide with the biennial meeting of the Conference for Food Protection. The conference is a group of representatives from regulatory agencies at all levels of government, the food industry, academia, and consumer organizations that works to improve food safety at the retail level. A brief discussion of the Code is provided below. Further information, including access to the Code, may be obtained from the Food Safety Training and Education Alliance (www.fstea.org).

The Code establishes definitions; sets standards for management and personnel, food operations, and equipment and facilities; and provides for food establishment plan review, permit issuance, inspection, employee restriction, and permit suspension. The Code discusses GMP for equipment, utensils, linens, water, plumbing, waste, physical facilities, poisonous or toxic materials, compliance, and enforcement. The Code also provides guidelines on food establishment inspection, HACCP guidelines, food-processing criteria, model forms, guides, and other aids.

Although this guide is designed for retail food protection, more than half of the data included are directly applicable to food processing plants, for example, equipment design (cleanability), clean-in-place (CIP) system, detergents and sanitizers, refrigeration and freezing storage parameters, water requirements, precautions against backflow (air, valve, etc.), personnel health and hygiene, rest rooms and accessories, pest control, storage of toxic chemicals, inspection forms, inspection procedures, and many more. Some of the data in the present book can be readily traced to the Code.

The Code consists of eight chapters and seven annexes. The annex that covers inspection of a food establishment applies equally well to both retail food protection and sanitation in food processing. According to the Code, the components of an inspection would usually include the following elements: (1) introduction, (2) program planning, (3) staff training, (4) conducting the inspection, (5) inspection documentation, (6) inspection report, (7) administrative procedures by the state/local authorities, (8) temperature measuring devices, (9) calibration procedures, (10) HAACCP Inspection Data Form, (11) food establishment inspection report, (12) FDA electronic inspection system, and (13) establishment scoring.

Details of these items will not be discussed here; some are further explored in various chapters in this book (please consult the index for specific topics). Instead, the next two sections trace the history and practices of food establishment inspection and how basic sanitation controls are slowly evolving into the prerequisites for HACCP plans in both retail food protection *and* food processing plants.

PURPOSE

A principal goal of food establishment inspection is to prevent food-borne disease. Inspection is the primary tool a regulatory agency has for detecting procedures and practices that may be hazardous and for taking action to correct deficiencies. Code-based laws and ordinances provide inspectors with science-based rules for food safety. The Code provides regulatory agencies with guidance on planning, scheduling, conducting, and evaluating inspections. It supports programs by providing recommendations for training and equipping the inspection staff and attempts to enhance the effectiveness of inspections by stressing the importance of communication and information exchange during regulatory visits. Inspections aid the food-service industry by:

- Serving as educational sessions on specific Code requirements as they apply to an establishment and its operation,
- Conveying new food safety information to establishment management and providing an opportunity for management to ask questions about general food safety matters, and
- Providing a written report to the establishment's permit holder or person in charge so that the responsible person can bring the establishment into conformance with the Code.

CURRENT APPLICATIONS OF HACCP

Inspections have been a part of food safety regulatory activities since the earliest days of public health. Traditionally, inspections have focused primarily on sanitation. Each inspection is unique in terms of the establishment's management, personnel, menu, recipes, operations, size, population served, and many other considerations.

Changes to the traditional inspection process were first suggested in the 1970s. The terms "traditional" or "routine" inspection have been used to describe periodic inspections conducted as part of an ongoing regulatory scheme. A full range of approaches was tried, and many were successful in managing a transition to a new inspection philosophy and format. During the 1980s, many progressive jurisdictions started employing the HACCP approach to refocus their inspections. The term "HACCP approach" is used to describe an inspection using the Hazard Analysis Critical Control Point concept. Food safety is the primary focus of a HACCP approach inspection. One lesson learned was that good communication skills on the part of the person conducting an inspection are essential.

The FDA has taught thousands of state and local inspectors the principles and applications of HACCP since the 1980s. The State Training Branch and FDA's Regional Food Specialists have provided two-day to week-long courses on the scientific principles on which HACCP is based, the practical application of these principles including field exercises, and reviews of case studies. State and local jurisdictions have also offered many training opportunities for HACCP.

A recent review of state and local retail food protection agencies shows that HACCP is being applied in the following ways:

- *Formal studies.* Inspector is trained in HACCP and is using the concepts to study food hazards in establishments. These studies actually follow foods from delivery to service and involve the write-up of data obtained (flow charts, cooling curves, etc.).
- *Routine use.* State has personnel trained in HACCP and is using the hazard analysis concepts to more effectively discover hazards during routine inspections.
- *Consultation.* HACCP-trained personnel are consulting with industry and assisting them in designing and implementing internal HACCP systems and plans.
- *Alternative use.* Jurisdiction used HACCP to change inspection forms or regulations.
- *Risk-based.* Jurisdiction prioritized inventory of establishments and set inspection frequency using a hazard assessment.
- *Training.* Jurisdiction is in the active process of training inspectors in the HACCP concepts.

Personnel in every sort of food establishment should have one or several copies of the Food Code readily available for frequent consultation.

APPLICATIONS

The sanitary requirements in the CGMPR and the Food Code serve as the framework for the chapters in this book. The HACCPR will be touched on when they help to clarify the discussion. Essentially, this book shows how to implement the umbrella regulations provided under the CGMPR. Each chapter handles one aspect of these complicated regulations. Most chapters discuss the regulations applicable to all types of food products being processed. Several chapters concentrate on the sanitary requirements from the perspectives of the processing of a specific category of food.

GLOSSARY

AF—acidified foods.
AMS—Agricultural Marketing Service, USDA.
APHIS—Animal and Plant Health Inspection Service, USDA.
a_w—water activity.
CFR—U.S. Code of Federal Regulations.
CGMPR—current good manufacturing practice regulations.
CIP—clean-in-place.
EPA—Enviornmental Protection Agency.
FD&C—Federal Food, Drug, and Cosmetic Act.
FDA—U.S. Food and Drug Administration.
FPLA—Fair Packaging and Labeling Act .
FSIS—Food Safety and Inspection Service, USDA.
GMP—good manufacturing practice.
GMPR—good manufacturing practice regulations.
HACCP—hazard analysis critical control point.
HACCPR—hazards analysis critical control points regulations.
LACF—low-acid canned foods.
NLEA—Nutrition Labeling and Education Act.
UPC—Universal Product Code.
USDA—U.S. Department of Agriculture.

REFERENCES

Food and Drug Administration. 2001a. Current Good Manufacturing Practice in Manufacturing, Packing, or Holding Human Food. 21 CFR 110. U.S. Government Printing Office, Washington, D.C.
___. Food Code. 2001b. U. S. Department of Health and Human Services, Washington, D.C.
___. 2001c. Hazard Analysis and Critical Control Point (HACCP) Systems. 21 CFR 120. U.S. Government Printing Office, Washington, D.C.

7

Food Plant Sanitation and Quality Assurance

Y. H. Hui

SANITATION

The sanitation in a food processing plant is to assure that the food product the company manufactures is wholesome and safe to eat. This usually means that the food does not contain, among other potential undesirable substances, any biological toxins, chemical toxicants, environmental contaminants, or extraneous substances.

To achieve this goal, a food processor with products sold through interstate commerce in the United States uses the following approaches:

- Implementation of a basic food plant sanitation program.
- Compliance with the good manufacturing practice (GMP) regulations issued by the United States Food and Drug Administration.
- Long-term plan to developing a food hazards analysis and critical control points (HACCP) program for those food industries that are not currently mandated to have a HACCP program.

FOOD PLANT SANITATION PROGRAM

Most food processors have a sanitation program to make sure that their products are safe. Most programs have the following components, among others: (1) the product and its ingredients, (2) cleaning, (3) housekeeping, (4) personnel hygiene and safety,

---

The information in this chapter has been derived from documents copyrighted and published by Science Technology System, West Sacramento, California, ©2002. Used with permission.

(5) warehousing, (6) distribution and transportation, and (7) sanitation inspections.

Let us use the manufacture of bakery products as an example to study the above factors: Bakery goods include bread, cakes, pies, cookies, rolls, crackers, and pastries. Ingredients consisting of flour, baking powder, sugar, salt, yeast, milk, eggs, cream, butter, lard shortening, extracts, jellies, syrups, nuts, artificial coloring, and dried or fresh fruits are blended in a vertical or horizontal mixer after being brought from storage, measured, weighed, sifted, and mixed. After mixing, the dough is raised, divided, formed, and proofed. Fruit or flavored fillings are cooked and poured into dough shells. The final product is then baked in electric or gas-fired ovens, processed, wrapped, and shipped. Loaves of bread are also sliced and wrapped.

RAW INGREDIENTS AND THE FINAL PRODUCT

Sanitation considerations apply to every stage of the processing operation: raw materials and operations.

Critical Factors in the Evaluation of Raw Materials

- Raw materials must come from warehouses that comply with local, county, state, and federal requirements for food warehouse sanitation.
- For certain ingredients such as egg and milk products, their sources, types, and so on should be ascertained. If frozen eggs are used, are they pasteurized and received under a *Salmonella*-free guarantee? Some food plants require routine testing of critical raw materials for bacterial load including *Salmonella* and other pathogens.
- Are raw materials requiring refrigeration (or freezing) or refrigerated (or frozen)?
- Is there any "blend off," mixing contaminated raw materials with clean raw materials?

Critical Factors in Evaluating the Sanitation of Operations

- *Room temperature, bottleneck, and bacterial contamination.* During a certain stage of an assembly line operation, always check sites where "bottleneck" frequently occurs. Room temperature and periods of bottlenecking are related to chances of bacterial contamination.

- *Metal detection.* During a production operation, always check metal detection or removal devices to make sure that they are working properly.
- *Time and temperature.* Identify stages in the operation where time and temperature are major and/or critical variables. Intense education must assure that any abuses that may allow growth of, and possible toxin formation by, microbial contaminants are strictly forbidden.
- *Equipment design.* Be alert for poorly designed conveyors or equipment that might add to bacterial load through product delay or "seeding."

CLEANING

Imagine your kitchen. We have to wash the kitchen floor because water, oil, and other cooking ingredients are dropped on them accidentally or intentionally. Then there are the dishes and pots and pans. They have to be cleaned and put away.

Of course the same problems exist in a bakery processing plant but on a much bigger scale.

Almost all bakery processing plants have a written plan on plant sanitation:

- Is there water on the floor?
- Has all flour dust been removed?
- Are different and clearly identified containers used for salvaged material and returned goods? Any noncompliance may be the cause of contaminating newly produced products.
- Is the distance between garbage disposal containers and stations where food ingredients are processed acceptable to avoid any potential contamination?

These few examples are among hundreds of details that a good sanitation program will carefully identify and for which it will establish who is responsible and what the responsibility is.

Similarly, food processing equipment requires a highly structured sanitation program.

Major components or equipment in the process flow are flour bins, elevator boots, conveyor systems, sifters, dump scale apparatus, production line flouring devices, dough proofers, overhead supports and ledges, and transport vehicles. All of them have removal inspection ports. Scheduled checks should make sure that any accumulation of insect and/or rodent infestations, for example, urine, hair, parts, is removed. The amount accumulated will vary, depending on equipment, age of building, and so on. The only course of action is to remove them as soon as possible for some old equipment or buildings. If the accumulation is excessive, an investigation is required to identify the actual sources of contamination or routes of entry. Scheduled checks should also remove accumulated ingredients such as flour, sugar, seeds, crumbs, and other debris.

Most companies require the inspection of equipment prior to production to determine the adequacy of clean-up and sanitizing operations.

It is doubtful that a food processing company will survive long if it does not have a comprehensive and workable program for cleaning the equipment used to manufacture its products. The objective of cleaning any equipment that has been used in food processing is to remove any residue or dirt from the surfaces that may or may not touch any food or ingredient.

Some of the equipment may be subjected to further sanitization and sterilization. Such attempts will be questioned if there is still visible dirt or debris attached to any surface of the equipment.

The wet-cleaning process, used by all food processors, has three components: prerinse, cleaning, postrinse. This can be done manually or by circulation.

1. *Prerinse.* Prerinse uses water to separate loosely adhered particles (dirt, residue, etc.), considering two basic factors: (1) performance of the cleaning after the production cycle is completed to get ready for the next workday and (2) predetermined cleaning criteria: (a) the method to be used for specific surfaces (vessels, components, pipelines), (b) the period of rinsing, and (c) temperature. For both factors, most food processing plants have established appropriate policies for the prerinse.
2. *Clean.* Under most circumstances, soaking, scrubbing, and more soaking characterize any cleaning process. The goal is to remove sticky residues or particles from the surface. Of course, cleaning detergents or solutions are used in the soaking and scrubbing. The chemical reactions are the standard: saponification, hydrolysis, emulsification, dispersion, and so on. As usual, all chemical reactions are time and temperature dependent.
3. *Postrinse.* This is no different from rinsing cooking utensils after they have been scrubbed and soaked. This stage removes all detergents/sanitizers used and any particles left behind.

For all three stages, the water used must comply with rigid standards to avoid damage to equipment, corrosion, and status of microbiological presence.

Apart from manual cleaning, we have the clean-in-place (CIP) procedure, which uses a circulation system of chemical solutions pumped through the equipment "in place." Much food processing equipment is designed to have this built-in feature. Any automatic process has inherent problems that must be dealt with in a manner dictated by circumstances.

The use of a circulatory method in cleaning is dependent on two groups of factors: (1) substances used in the detergent or cleaning solutions and (2) the variables. Substances used can include an array of chemicals: caustic soda, acid, and so no. Obviously, the concentration of such chemicals is a critical factor. The variables include contact temperature, contact time, flow rate between surfaces, and substances in cleaning solution.

HOUSEKEEPING

Again, we can use our home as an example. We keep the inside clean by dusting, vacuuming, sweeping, and so on. We keep the outside of our house clean by removing garbage, leaves, droppings, peeling paints, and so on. It is of paramount importance that a food processing plant is clean both inside and out.

For internal housekeeping, part of the information was discussed in the cleaning process we presented earlier. However, we still have to worry about cleaning windows, debris under a counter or in the corner of a room, garbage cans, and so on. Most food companies hire regular maintenance crews to do the job. Unfortunately, the plant manager still has to develop policies to implement and evaluate procedures.

The environment of a food processing plant has always been a problem, including, as it does, garbage, birds, insects, rodents, and so on. Housekeeping for the immediate vicinity outside a food plant requires close monitoring.

Many professionals consider housekeeping as "non-glamorous" and "menial." However, it is so important that it requires complete attention from the management. The reason is simple. Regulatory officials from local, county, and state levels are serious about this aspect of food processing. If the company ships products across state lines, the U.S. Food and Drug Administration has the authority to issue warning letters about any unacceptable conditions, including sloppy housekeeping.

We will briefly analyze the two components of housekeeping.

The Master Schedule

Every manufacturing company, food or otherwise, has a master schedule of cleaning. Obviously, food particles attract rodents and other undesirable creatures, and their cleaning or removal is of utmost importance to the plant operation. A cleaning schedule essentially has the following components:

- *Coverage:* Rooms, storage areas, toilets, offices, freezers, walls, ceilings, and so on.
- *Frequency:* Each area requires a different frequency of cleaning, days (1, 2, 3, 4, etc.), weekly, and so on.

A cleaning schedule is meaningful only if the methods of cleaning are appropriate and the schedule is enforced or implemented. Also, the frequency of cleaning must be carefully evaluated in conjunction with the methods of cleaning. This is because a process of cleaning may increase the dust load in the air, which may in turn contaminate other surface areas.

Some areas require frequent cleaning and others do not. A storage room with infrequent traffic may be cleaned once a week, while a storage room with frequent traffic may need to be cleaned once a day.

Dust

- Most dry cleaning methods (e.g., wiping with a rag, vacuum cleaners, brooms, brushes, pressurized air) increase dust in the air.
- Since dust particles are charged electrically, they will adhere to any surfaces that are electrically or electrostatically charged. This results in contamination.
- Dust contamination is heightened when the environment, including surfaces, is moist, resulting in molds. When molds occur on piping; the backs of tanks, ducts and cables; the corners of ceilings; and other places that are obscured from vision, the problem increases.
- Dust moves from room to room by normal airflow from temperature differences or window and door drafts, resulting in further contamination.
- Dust dispersion is a risk that replaces the risk that has just been removed by cleaning.

The areas to be cleaned should be evaluated with great care:

- Although most objects (e.g., vats, holding tanks) are raised from the floor with a space for cleaning, it is still difficult to clean this part of the floor because the space is too narrow and hidden from view.
- Corners always pose a problem for cleaning. Special devices such as suction hoses are needed to keep them clean. These are places where insects, rodents, and other undesirable creatures will thrive.
- Bottoms of most equipment pose a problem in cleaning. Crawling on one's knees does not always solve the problem. Customized devices may be needed.

Wet cleaning by hand or machine is acceptable. Modern technology has made available gel, foam, aerosols, and special equipment. However, the water hose is still the method of choice in most food companies. Wet cleaning must take the following into consideration: (1) All material that can absorb moisture, such as cardboard boxes, pallets, and so on, must be removed. (2) After wet cleaning, the surfaces must be dried carefully. (3) A proper draining system should be in place and be maintained clean and free of debris around the openings.

U.S. GOVERNMENT ENFORCEMENT TOOLS

The FDA is charged with protecting American consumers by enforcing the federal Food, Drug, and Cosmetic Act and several related public health laws. What does it do when there is a health risk associated with a food product?

When a problem arises with a product regulated by the FDA, the agency can take a number of actions to protect the public health. Initially, the agency works with the manufacturer to correct the problem voluntarily. If that fails, legal remedies include asking the manufacturer to recall a product, having federal marshals seize products if a voluntary recall is not done, and detaining imports at the port of entry until problems are corrected. If warranted, the FDA can ask the courts to issue injunctions or prosecute those that deliberately violate the law. When warranted, criminal penalties—including prison sentences—are sought.

However, the FDA is aware that it has legal responsibility to keep the public informed of its regulatory activities. To do so, the FDA uses press releases and fact sheets. The FDA uses this tool before, during and after a health hazard event related to a food product. Some of these are briefly described below, emphasizing the sanitation deficiencies of affected food products.

PRESS RELEASES AND FACT SHEETS

FDA Talk Papers are prepared by their press office to guide FDA personnel in responding with consistency and accuracy to questions from the public on subjects of current interest. Talk Papers are subject to change as more information becomes available.

The regulatory tools used by the FDA are discussed below.

DATA ON UNSANITARY PRACTICES

For the FDA to enforce its laws and regulations, it must have specific data regarding the sanitary practices of a food processing plant. The FDA has a number of ways to determine if a food product is associated with unsanitary conditions in a food processing plant or if a food processing plant has sanitary deficiencies. They include (1) product monitoring, (2) activities based on reports from the public, (3) activities based on reports from other government agencies, and (4) establishment inspection reports.

Product Monitoring

Product monitoring is as old as the beginnings of modern food processing. At present, local, county, state, and federal health authorities conduct market food product sampling and analyses to determine the wholesomeness of food. Such monitoring is restricted by the availability of allocated budget and resources. However, the FDA has the most resources, and its monitoring efforts produce the most results.

When products are found to be unsanitary (pathogens, rats, insects, glass, metal, etc.) by the FDA, it will implement standard procedures to warn the public, remove such products from the market, and take a variety of other actions, which will be discussed later in this chapter.

Activities Based on Reports from the Public

The FDA has a website and an 800 number for the public to report health hazards including those related to the sanitation of food products. Since the es-

tablishment of such convenient means of communication, there has been an increasing number of consumers reporting products that pose health risks, such as glass in baby food, dead insects in frozen dinners, and so on. Occasionally, so-called whistle blowers, that is, employees of food companies, inform the FDA of products with contaminants from unsanitary practices. Based on the data provided by the public, the FDA implements standard procedures to handle any potential health hazards related to the products reported.

Activities Based on Reports from Other Government Agencies

Health care providers frequently are the source of information that eventually reveals the unsanitary practices of food companies. These people include physicians, pharmacists, nurses, dentists, public health personnel, and others. Most of these reports involve injury (e.g., food poisoning) and product abnormality (e.g., decomposed or spoiled contents). Their reports become a vital source of leads for the FDA to enforce its laws and regulations.

Establishment Inspection Reports

Inspection of a food processing plant by a government authority is the basis on which the government can decide if the food manufactured in the plant is wholesome and poses no economic fraud. The frequency and intensity of the inspection process will depend on resources and budgets, especially for nonfederal agencies. The FDA, as a federal agency, has more authority and resources and a larger budget.

The framework for inspecting a plant covers the following: (1) the basics (preparation and references, inspectional authority), (2) personnel, (3) plants and grounds, (4) raw materials, (5) equipment and utensils, and (6) the manufacturing process (ingredient handling, formulas, food additives, color additives, quality control, and packaging and labeling).

After an inspection is completed, the inspector gives the plant management a copy of the report. If there are sanitation deficiencies, the management will be expected to correct them.

The data collected from this inspection procedure and other sources discussed earlier become the central operation base on which the FDA will fulfill its legal responsibility to make sure that all deficiencies are corrected to reduce any hazard to the health of the consuming public.

The interesting part of this process is the enforcement of compliance. We have seen the manner in which the FDA compiles data on the sanitation of a food product and a food processing plant. We will now proceed to the regulatory activities the FDA uses to assure compliance.

RECALLS

FDA Consumer magazine has published several articles on the recall of food products in this country. The following information has been compiled from these public documents.

Misunderstanding

Recalls are actions taken by a firm to remove a product from the market. Recalls may be conducted on a firm's own initiative, by FDA request, or by FDA order under statutory authority.

The recall of a defective or possibly harmful consumer product often is highly publicized in newspapers and on news broadcasts. This is especially true when a recall involves foods, drugs, cosmetics, medical devices, or other products regulated by FDA.

Despite this publicity, FDA's role in conducting a recall is often misunderstood, not only by consumers, but also by the news media, and occasionally even by the regulated industry. The following headlines, which appeared in two major daily newspapers, are good examples of that misunderstanding: "FDA Orders Peanut Butter Recall," and "FDA Orders 6,500 Cases of Red-Dyed Mints Recalled."

The headlines are wrong in indicating that the agency can "order" a recall. FDA has no authority under the federal Food, Drug, and Cosmetic Act to order a recall, although it can *request* a firm to recall a product.

Most product recalls regulated by the FDA are carried out voluntarily by the manufacturers or distributors of the product. In some instances, a company discovers that one of its products is defective and recalls it entirely on its own. In others, the FDA informs a company of findings that one of its products is defective and suggests or requests a recall. Usually, the company will comply; if it does not, then the FDA can seek a court order authorizing the federal government to seize the product.

This cooperation between the FDA and its regulated industries has proven over the years to be the quickest and most reliable method for removing potentially dangerous products from the market. This

method has been successful because it is in the interest of the FDA, as well as industry, to get unsafe and defective products out of consumer hands as soon as possible.

The FDA has guidelines for companies to follow in recalling defective products that fall under the agency's jurisdiction. These guidelines make clear that the FDA expects these firms to take full responsibility for product recalls, including follow-up checks to assure that recalls are successful.

Under the guidelines, companies are expected to notify the FDA when recalls are started, to make progress reports to the FDA on recalls, and to undertake recalls when asked to do so by the agency.

The guidelines also call on manufacturers and distributors to develop contingency plans for product recalls that can be put into effect if and when needed. FDA's role under the guidelines is to monitor company recalls and assess the adequacy of a firm's action. After a recall is completed, the FDA makes sure that the product is destroyed or suitably reconditioned and investigates why the product was defective.

The FDA has stated the following guidelines several times in its magazine *FDA Consumer.*

Categories

The guidelines categorize all recalls into one of three classes, according to the level of hazard involved. *Class I* recalls are for dangerous or defective products that predictably could cause serious health problems or death. *Class II* recalls are for products that might cause a temporary health problem or that pose only a slight threat of a serious nature. *Class III* recalls are for products that are unlikely to cause any adverse health reaction but that violate FDA regulations.

The FDA develops a strategy for each individual recall that sets forth how extensively it will check on a company's performance in recalling the product in question. For a Class I recall, for example, the FDA would check to make sure that each defective product has been recalled or reconditioned. In contrast, for a Class III recall the agency may decide that it only needs to spot-check to make sure the product is off the market. Detailed regulations have been promulgated on FDA recalls in the U.S. Code of Federal Regulations.

Even though the firm recalling the product may issue a press release, the FDA seeks publicity about a recall only when it believes the public needs to be alerted about a serious hazard. For example, if a canned food product, purchased by a consumer at a retail store, is found by the FDA to contain botulinum toxin, an effort would be made to retrieve all the cans in circulation, including those in the hands of consumers. As part of this effort, the agency could issue a public warning via the news media to alert as many consumers as possible to the potential hazard.

The FDA also issues general information about all new recalls it is monitoring through a weekly publication, "FDA Enforcement Report."

Before taking a company to court, the FDA usually notifies the responsible person of the violation and provides an opportunity to correct the problem. In most situations, a violation results from a mistake by the company rather than from an intentional disregard for the law.

There are several incentives for a company to recall a product, including the moral duty to protect its customers from harm and the desire to avoid private lawsuits if injuries occur. In addition, the alternatives to recall are seizures, injunctions, or criminal actions. These are often accompanied by adverse publicity, which can damage a firm's reputation.

A company recall does not guarantee that the FDA will not take a company to court. If a recall is ineffective and the public remains at risk, the FDA may seize the defective products or obtain an injunction against the manufacturer or distributor.

The recalling firm is always responsible for conducting the actual recall by contacting its purchasers by telegram, mailgram, or first-class letters with information, including (1) the product being recalled, (2) identifying information such as lot numbers and serial numbers, (3) the reason for the recall and any hazard involved, and (4) instructions to stop distributing the product and what to do with it.

The FDA monitors the recall, assessing the firm's efforts.

Initiating a Recall

A firm can recall a product at any time. Firms usually are under no legal obligation to even notify the FDA that they are recalling a defective product, but they are encouraged to notify the agency, and most firms seek the FDA's guidance. The FDA may request a recall of a defective product, but it does so only when agency action is essential to protect the public health.

When a firm undertakes a recall, the FDA district office in the area immediately sends a "24 Hour Alert to Recall Situation" notifying the relevant FDA center (responsible for foods and cosmetics, drugs, devices, biologics, or veterinary medicine) and the FDA's Division of Emergency and Epidemiological Operations (DEEO) of the product, recalling firm, and reason for the recall. The FDA also informs state officials of the product problem, but for routine recalls, the state does not become actively involved.

After inspecting the firm and determining whether there have been reports of injuries, illness, or other complaints to either the company or to the FDA, the district documents its findings in a recall recommendation (RR) and sends it to the appropriate center's recall coordinator. The RR contains the results of FDA's investigation, including copies of the product labeling, FDA laboratory worksheets, the firm's relevant quality control records, and when possible, a product sample to demonstrate the defect and the potential hazard. The RR also contains the firm's proposed recall strategy.

The Strategy

The FDA reviews the firm's recall strategy (or, in the rare cases of FDA-requested recalls, drafts the strategy), which includes three things: the depth of recall, the extent of public warnings, and effectiveness check levels.

The depth of recall is the distribution chain level at which the recall will be aimed. If a product is not hazardous, a recall aimed only at wholesale purchasers may suffice. For more serious defects, a firm will conduct a recall to the retail level. And if public health is seriously jeopardized, the recall may be designed to reach the individual consumer, often through a press release.

But most defects don't present a grave danger. Most recalls are not publicized beyond their listing in the weekly Enforcement Report (mentioned earlier). This report lists the product being recalled, the degree of hazard (called "classification"), whether the recall was requested by the FDA or initiated by the firm, and the specific action taken by the recalling firm.

A firm is responsible for conducting "effectiveness checks" to verify—by personal visits, by telephone, or with letters—that everyone at the chosen recall depth has been notified and has taken the necessary action. An effectiveness check of level "A" (check of 100% of people that should have been notified) through "E" (no effectiveness check) is specified in the recall strategy, based on the seriousness of the product defect.

The Health Hazard Evaluation

When the center receives the RR from the district office, it evaluates the health hazard presented by the product and categorizes it as Class I, Class II, or Class III. An ad hoc health hazard evaluation committee of FDA scientists, chosen for their expertise, determines the classification . Classification is done on a case-by-case basis, after considering the potential consequences of a violation.

A Class I recall involves a strong likelihood that a product will cause serious adverse health consequences or death. A very small percentage of recalls are Class I.

A Class II recall is one in which use of the product may cause temporary or medically reversible adverse health consequences or in which the probability of serious adverse health consequences is remote.

A Class III recall involves a product not likely to cause adverse health consequences.

For Class I and Class II, and infrequently for Class III, the FDA conducts audit checks to ensure that all customers have been notified and are taking appropriate action. The agency does this by personal visits or telephone calls.

A recall is classified as "completed" when all reasonable efforts have been made to remove or correct the product. The district notifies a firm when the FDA considers its recall completed.

Planning Ahead

The FDA recommends that firms maintain plans for emergency situations requiring recalls. Companies can minimize the disruption caused by the discovery of a faulty product if they imprint the date and place of manufacture on their products and keep accurate and complete distribution records.

A "market withdrawal" is a firm's removal or correction of a distributed product that involves no violation of the law by the manufacturer. A product removed from the market due to tampering, without evidence of manufacturing or distribution problems, is one example of a market withdrawal.

A "stock recovery" is another action that may be confused with a recall. A stock recovery is a firm's removal or correction of a product that has not yet been distributed.

Even though the firm recalling the product may issue a press release, the FDA seeks publicity about a recall only when it believes the public needs to be alerted about a serious hazard. For example, if a canned food product, purchased by a consumer at a retail store, is found by the FDA to contain botulinum toxin, an effort would be made to retrieve all the cans in circulation, including those in the hands of consumers. As part of this effort the agency also could issue a public warning via the news media to alert as many consumers as possible to the potential hazard.

WARNING LETTERS

Under FDA regulations, a prior notice is a letter sent from the FDA to regulated companies about regulatory issues. One such notice is the warning letter. If the establishment inspection report includes a list of sanitary deficiencies, the FDA may send a warning letter to the food company to ask for proper correction of such deficiencies.

QUALITY ASSURANCE

Many college graduates in food science, food technology, and food engineering work for food processing plants. Eventually, many of them become operational managers in the company. At this stage, they realize the significance of quality assurance. They are responsible not only for the quality of the finished product, but also for its wholesomeness and safety for public consumption.

The principles and procedures for quality assurance are as applicable and beneficial to small plants as to larger plants. In many cases, quality control systems can be more efficiently administered in small plants because of a simpler organizational structure and more direct communication among employees. Although quality control is not the same as quality assurance (in general, "control" refers to one aspect of "assurance"), some professionals equate quality control systems with quality assurance. To avoid this issue and for ease of discussion, we use these terms (quality assurance, quality control, and quality control systems) interchangeably.

COST VERSUS BENEFIT

Quality assurance or control is a good management tool. A quality control system specifically tailored to the volume and complexity of a plant operation can be cost effective.

A properly designed and operated total quality control system will minimize the likelihood of mistakes during processing, give an indication of problems immediately, and provide the necessary information quickly so that the problems can be located and corrected in a timely manner. As a result, production delays are reduced, the need for reprocessing or relabeling is lessened, and the possibility of product recall and condemnation is reduced.

PRODUCT CONSISTENCY IMPROVED

Quality control systems provide the information necessary to consistently produce a uniform quality product at a predicted cost. Some processors have questioned whether the cost of implementing a total quality control system would be recovered unless the quality of the plant's product had been so poor that the plant suffered reduced sales and a high return of product.

It is true that a plant with a poor product would benefit most. In even the best plants, however, the lack of a quality control system results in a product that is more variable and not as well defined.

With organized controls and objective sampling, the plant has more extensive and precise information about its operation. As a result, management has better control, and product quality is stabilized. Records from a quality control system define product quality at the time of shipment and are helpful in dealing with claims of damage or mishandling during shipment.

EQUIPMENT COSTS

Contrary to the impression or idea that quality control systems require highly trained technicians and expensive equipment, a plant quality control system can be fairly simple and inexpensive and still be effective.

The expense of equipment is related to the type and complexity of products and operations and the volume of production. In most cases, a total quality control system in a small plant would require only inexpensive thermometers, calculators, knives, grinders, and existing testing equipment used for traditional inspection and quality assurance. If necessary, samples may be submitted to commercial laboratories.

The technical skills in food science, mathematics, and statistics necessary to establish a quality control system are available from trade associations and professional societies at a reasonable one-time charge. This assistance can be utilized to define defects, defective units, and critical control points and to establish corrective actions for the system. Once the technical details of the system are established, it can be operated by plant personnel familiar with the processing operation. It is not necessary to hire a quality control technician.

A critical control point is a point in the food processing cycle where loss of control would result in an unacceptable product. Such points may include the receipt of raw meat just before use, processing and storage operations, and delivery of the product to the customer.

The FDA and USDA have special programs designed to assist in identifying critical control points and setting up quality control systems in small food processing plants. They will also provide on-site assistance in the start-up of the system.

ELEMENTS OF A TOTAL QUALITY CONTROL SYSTEM

The first step in developing a plant quality control system is to outline the processes that occur in the plant. An easy way to do this is to visualize the physical layout of the plant operation. The building may be small and consist of only one or two rooms, or it may be large and contain many rooms. Make a list of the rooms or areas, and draw a flow diagram of the production process, starting with the incoming or receiving area and ending with the shipping area for finished goods.

For each room or area, list the activities that occur there, making special note of those that are unusual or are important relative to the process or product. For each, spell out the controls that are imposed—or should be imposed—whether precise or flexible, written or not written. Examples would include raw materials examined, ingredients weighed, meters used, scales calibrated, equipment cleaned, bills of lading examined, or trucks checked.

Identify the FDA or USDA inspection regulations that apply to each area of the processing plant and list them. The GMP regulations promulgated by the FDA are the most appropriate.

For each processing area, designate the person responsible for the controls or inspection—the name of a plant employee or an outside contractor. How often is the control or inspection check to be done? What records are to be kept? This information can be compiled by a clerical or administrative employee, and the FDA or USDA inspector can assist.

When this exercise has been completed, a rough outline of a total quality control system has been developed. It can be compared, area by area, to descriptions of the elements in the sample system in the appendix of the manual that records the system. If a company does not have such a manual, it is recommended that it start one.

As each element is reviewed, note where controls may be missing. The outline that remains, with missing controls added, is another step closer to a total quality control system.

The final step is to convert this outline into a written format, as though it were a set of instructions for plant employees. In reality, it can be the operating manual for the persons responsible for maintaining quality control in the plant.

GENERAL ELEMENTS OF TOTAL QUALITY CONTROL

We have just completed a discussion of the general outline of a plant quality control system. Within the system are various elements, determined by the type of operation in the plant.

In this section, the specific operations will be discussed, and the elements of a good quality control system will be outlined.

Receiving

Examples of controls:

- Examining (and possibly sample) incoming lots.
- Verifying identification marks.
- Checking carriers.
- Logging deliveries.

A plant's total quality control system will include written instructions for checking incoming raw materials such as raw fruits, flour, frozen fish, spices, salt, liquid ingredients, additives, and extenders and for recording the results. These materials must be verified for wholesomeness (free from indications of mishandling, decomposition, infestation), acceptability for intended use, and approval for use.

It may also be desirable at this point—although it is not mandatory—to test for composition (fat, moisture, etc.) to assure proper blending of formulated products. It is preferable to run the most frequent tests on products likely to have the most variation. For example, biological cultures and frozen

orange juice need more frequent analysis than frozen dough or dried beef. Sampling plans utilizing statistical quality control procedures are helpful in inspecting incoming lots. These plans are easy to use and may be obtained from several sources, including government booklets.

It is good practice to prepare a suppliers' or buyers' guide outlining the specifications for ingredients, additives, and other products bought outside the plant.

The air temperature and product temperature in the receiving area should be checked often enough to assure that the company's requirements are being met. This would include checks of freezers, doors, door seals, incoming railroad cars, and trucks. The quality control plan should include procedures for taking corrective action in the event a product is contaminated during shipment.

The receiving log should be checked to assure that entries are accurate and up to date and that all requirements regarding incoming products and materials are met. The log will be useful in indicating trends, so problems can be spotted early. The person who checks the log can keep a record of the dates and the results of the verifications.

Lots moved from the receiving area to other areas of the plant should be periodically checked to assure that their identity is properly maintained.

In preparing written instructions for the receiving area, identify the various checks to be made, who is to make them, when they are to be made, and how and where the information will be recorded.

Manufacturing

Examples of controls:

- Verifying wholesomeness.
- Verifying identification, weight, or volume of ingredients.
- Verifying ambient temperature.
- Handling of rejected ingredients or product.

Although ingredients may have been checked earlier for wholesomeness and acceptability, it is a good idea to make another check just prior to actual use in the manufacturing process. This recheck does not need to be painstaking. It should be ample to assure that unacceptable ingredients are not used and that ingredients are correctly identified and eligible for use in the product. The frequency of these rechecks can be reduced for small, low-volume plants.

A method for controlling the weight of each ingredient is also essential in order to assure a uniform and consistent finished product that complies with the company's quality requirements for the products and FDA's GMP regulations for the products.

Maintaining the correct temperature in an area is also important to good product quality. Occasional checks should be made during the shift, and a record should be kept of the findings. This will take only a small amount of time and effort on the part of a plant employee, but will identify any situations requiring correction. Inexpensive recording thermometers are useful for maintaining a record of room temperature.

Occasionally, unacceptable ingredients or materials will arrive in the manufacturing area, and procedures should be outlined for these situations. Remember, good management sets realistic and effective controls for dealing with these situations. The procedures that are outlined must be diligently followed.

In cases where a finished product must meet certain requirements, such as fat or moisture limits, consider sampling each lot. Sampling plans may be designed to fit each condition and type of analysis.

For the purpose of verifying formulation or checking wholesomeness, a lot can be each batch during each shift, several batches from the shift, or the shift's entire production. For the purpose of laboratory testing, a lot may consist of one day's production or several days' production of an item, depending on the volume and type of product.

Records of all inspections and tests must be made available to state and federal inspectors and maintained on file.

Packaging and Labeling

Examples of controls:

* Verifying label approval.
* Verifying accuracy of labeling.
* Checking temperatures.
* Finished product sampling.

Since this is one of the last steps prior to shipping, it is essential that no regulatory requirement be overlooked.

Checks must be made to assure that all labels have been approved by state and federal regulators and that proper labels are being used. Particular attention should be paid to the new nutrition labeling. It must be verified that illustrations represent the product, that net weight and count declarations are accurate, and that packaging meets the company's specifica-

tions. The temperatures of frozen products, as well as the condition of all containers and cases, should be checked and the findings recorded.

A net weight control program must assure that all lots leaving the plant meet with FDA's requirements for standardized foods as well as other applicable requirements. The sampling rate should be appropriate for the volume, type of product, size of package, and degree of accuracy desired. For instance, cartons of wholesale volumes need less frequent checks than retail packages.

Where applicable, routine systematic sampling, inspection, or analysis of a finished product must be part of the approved total quality control system, especially for a product going to retail outlets.

State and federal regulators in the plant, in regional offices, or in Washington can consult with processors on sampling, including rates, targets, and limits.

Shipping

Examples of controls:

* "First in, first out."
* Record of shipments.
* Checking order sizes and temperatures.
* Checking containers and carriers.

Records of the destination of products shipped from the plant are important to good quality control. In the event recall is necessary, the records will pinpoint the amount and exact location of the product.

The procedure for knowing the destination of each shipment should be explained in the quality control system. The plant may find it beneficial to have some type of container coding and dating system. This would identify the date of processing and packaging for returned goods. Occasional quality control checks should be made to verify the adequacy of the container codes and to verify order sizes; temperatures (where applicable); and the condition of containers, rail cars, or trucks used for shipping. These controls need not be complicated, but they must be adequate to assure effectiveness.

General Sanitation

Examples of controls:

* Rodents and pests.
* Product contamination.
* Employee hygiene.
* Facilities and environmental appearance.

A procedure to check the overall sanitation of plant facilities and operations, including outside adjacent areas and storage areas on plant property, should be included in a total quality control system.

In a total quality control system, a designated plant official will make the sanitation inspection and record the findings. If sanitation deficiencies are discovered, a plan for corrective action is necessary. Corrective action might include recleaning, tagging a piece of equipment, or closing off an area until a repair is completed.

A frequent systematic sanitation inspection procedure should be used where product contamination is possible, as from container failure, moisture dripping, or grease escaping from machinery onto product or onto surfaces that come into contact with product.

Good employee hygiene should be continuously emphasized through special instruction for new employees and properly maintained, adequate toilet facilities and , where applicable, appropriate facilities for such needs as breaks (e.g., regular, after minor accidents), smoking, breast feeding, change of clothing, lockers for personal items, vending machines, and so on. Clean work garments in good repair, good personal hygiene practices such as hand washing, periodic training, and the cleaning of floors and walls in nonproduction areas are signs of effective sanitation. Plant management will want to use a number of techniques to assure the continued effectiveness of this phase of the quality control system.

Employee Training

Examples of controls:

- New employee orientation.
- Refresher training.

When new employees begin work at a plant, it is useful to acquaint them with all aspects of the plant. The quality control system should provide for instruction of new employees on the plant's operations and products and on good hygiene practices.

A number of questions concerning hygiene should be addressed in this instruction. What basic things should any new employee know about food handling and cleanliness? Why is cleanliness essential? What are the standards—in other words, what does clean mean? Why are product temperatures important? What is a cooked product? What occurs if something is accidentally soiled? Which chemicals (cleaners, sanitizers, insecticides, food additives) are around? Does the new employee use or have any responsibility for any of these? How does the employee become acquainted with the operation and products? Whom does the employee consult if questions or problems arise?

Make a list of all the items that need to be covered in employee orientation and indicate generally how and when the orientation will be performed.

Employee training should not end with orientation; it should include an ongoing program to continually remind employees of the importance of good sanitation.

How are employees continually reminded of important functions, such as personal hygiene after a visit to the restroom? Will posting a sign or poster that fades over time communicate the appropriate level of importance? There are many ways of continuing employee training and maintaining sensitivity. Plant managers may find that occasionally changing methods will help emphasize management's commitment.

A brief description of the methods and time schedule for assuring that employees do not become unconcerned or indifferent is helpful.

COMPLETING THE TOTAL QUALITY CONTROL SYSTEM

When the details of the elements discussed in previous sections are compiled, the result is essentially the plant's "operating manual." It will also serve as the plant's total quality control system.

Upon completion, it should be reviewed. In some cases, a definition or description may be needed for such points as control limits, variability in weights, or number of defects per sample. Also, all critical control points should be covered.

In addition, those sections of the FDA's GMP regulations applicable to the operations of the plant must be listed. For each, identify the specific part of the quality control system that is designated to assure compliance.

If one or more full-time quality control personnel are employed at the plant, an organizational chart should be included showing how they fit into the plant's management structure. If there are no full-time quality personnel, identify who will assume specific responsibilities for quality control and list all other duties of that employee.

When the proposed total quality control system is completed, it is ready to be submitted to the company's management. Let us wish the best of luck to the officer who prepares the plan.

Part II
Applications

8
Bakery: Muffins

N. Cross

BACKGROUND INFORMATION

HISTORY OF MUFFINS

English muffins originating in London were made from yeast dough, in contrast to the quick bread muffins served in early America. Muffins are described as a quick bread since "quick-acting" chemical leavening agents are used instead of yeast, a "longer acting" biological leavening agent. Muffins have become increasingly popular as a hot bread served with meals or eaten as a snack. Freshly baked muffins are served in restaurants and bakeries, and consumers can buy packaged ready-to-eat muffins from grocery stores and vending machines. With the availability of dry mixes, frozen muffin batter, and predeposited frozen muffins available on the wholesale market, it is possible for restaurants and small bakeries to serve a muffin of a consistently high quality.

HEALTH CONCERNS

The economic burden of chronic disease is a worldwide problem. Chronic diseases contributed to 60% of the deaths worldwide in 2001 [World Health Organization (WHO) 2003a]. The increasing rate of obesity and the ageing of the population are expected to impact the burden of chronic disease.

The information in this chapter was derived from a chapter in *Food Chemistry Workbook,* edited by J. S. Smith and G. L. Christen, published and copyrighted by Science Technology System, West Sacramento, California, ©2002. Used with permission.

Those with obesity are at greater risk and have an earlier onset of the chronic diseases of diabetes, cardiovascular disease, cancer, and stroke. Ageing increases the risk for all chronic diseases. Nearly one-quarter of the population in developed countries is made up of those above 60 years of age, with expectations for the numbers to increase to one-third of the population by 2025 (WHO 2002a).

The problems of overweight and obesity are growing rapidly around the world and coexist with malnutrition in developing countries (WHO 2003a). Surveys of U.S. adults done in 1999–2000 showed that 64% of adults were overweight and 30% were obese (Flegal et al. 2002). The percentage of children and adolescents in the United States who are overweight has tripled in the past 30 years, with 15% of 6–19 year olds being overweight in 1999–2000 (Ogden et al. 2002).

Obesity rates have increased threefold or more in some parts of North America, Eastern Europe, the Middle East, the Pacific Islands, Australasia, and China since 1980 (WHO 2002b). The prevalence rates of overweight and obesity are growing rapidly in children and adults in such countries as Brazil and Mexico, where malnutrition and obesity coexist in the same household (Chopra 2002). Countries with the highest percentage (5–10%) of overweight preschool children are from the Middle East (Qatar), North Africa (Algeria, Egypt, and Morocco), and Latin America and the Caribbean (Argentina, Chile, Bolivia, Peru, Uruguay, Costa Rica, and Jamaica) (de Onis and Blossner 2000).

Globalization of food and the availability of energy-dense snack foods and fast foods have had a significant impact on dietary patterns and the incidence of chronic disease in both developing and developed countries (Hawkes 2002). For example, Coca-cola and Pepsi soft drinks and McDonald's, Pizza Hut, and Kentucky Fried Chicken fast foods are now available worldwide (Hawkes 2002). Changes in dietary patterns combined with a sedentary lifestyle have increased the rates of obesity and chronic disease. Dietary factors related to chronic disease are excessive intakes of calories, fat—especially saturated fat—, and sodium, and low intakes of fruits and vegetables and wholegrain breads and cereals (WHO 2001).

National dietary guidelines recommend limiting intakes of total fat, saturated fat, trans fat, cholesterol, free sugars, and sodium, and they promote dietary fiber from wholegrain breads and cereals and fruits and vegetables (WHO 2003b). In 2002, consumers in the United States reported making food choices in an effort to avoid fat, sugar, calories, and sodium and to increase fiber intake [National Marketing Institute (NMI) 2003]. Consumers chose fat free foods or foods low in fat 74–80% of the time and selected low calorie foods and low sodium foods 76% and 67% of the time, respectively. High fiber foods were chosen 75% of the time, and 40% of respondents reported using organic foods (NMI 2003).

The food industry has responded to concerns of consumers and public health officials by developing "healthy" food products, lower in saturated fat, trans fat, cholesterol, sodium, sugar, and calories. New ingredients have been developed by food scientists in the government and industry to use as fat replacers and sugar replacers in preparing baked products that are lower in calories and in saturated and trans fats (Table 8.1, Table 8.2). The newest category of ingredients is concentrated bioactive compounds with specific health benefits (Table 8.3) (Pszczola 2002a). These ingredients are added to formulations during food processing to enhance the health benefits of specific food products or to develop "functional foods." Individual foods such as apples, blueberries, oats, tomatoes, and soybeans are being marketed as functional foods because of the health benefits of components of these foods. For example, diets that include oat fiber and soy protein lower serum cholesterol, and lycopene in tomatoes reduces the risk of prostate cancer. Apples and blueberries contain unique antioxidants shown to reduce the risk for cancer (Pszczola 2001). Examples of bioactive ingredients available to the baking industry are OatVantage™ (Nature Inc., Devon, Pennsylvania), a concentrated source of soluble fiber, and FenuPure™ (Schouter USA, Minneapolis, Minnesota), a concentrated source of antioxidants from fruits and vegetables.

FOOD LABELING AND HEALTH CLAIMS

The Nutrition Labeling and Education Act (NLEA) issued by the Food and Drug Administration (FDA) in the United States in 1990 required food labels to include nutritional content on all packaged foods to be effective in 1994 (FDA 2003). Information required on the nutrition facts portion of the food label are the serving size and the amount per serving of calories, protein, fat, saturated fat, cholesterol, carbohydrates, fiber, sodium, calcium, vitamins A and C, and iron. A 1993 amendment to the NLEA author-

Table 8.1. Ingredients Used as Fat Replacers in Baked Products

Brand Name	Composition	Supplier
Carbohydrate based		
Beta-Trim™	Beta-glucan and oat amylodextrin	Rhodia USA, Cranbury, NJ
Fruitrim®	Dried plum and apple puree	Advanced Ingredients, Capitola, CA
Just Like Shorten™	Prune and apple puree	PlumLife division of TreeTop, Selah, WA
Lighter Bake™	Fruit juice, dextrins	Sunsweet, Yuma City, CA
Oatrim®	Oat maltodextrin	Quaker Oats, Chicago, IL
Paselli FP	Potato maltodextrin	AVEBE America, Inc., Princeton, NJ
Z-Trim	Multiple grain fibers	U.S. Department of Agriculture
Low and noncaloric, lipid-based		
Enova™	Triglycerides modified by substituting short- or medium-chain fatty acids	Archer Daniels Midland/Kao LLC, Decatur, IL
Benefat®	Triglycerides modified by substituting short- or medium-chain fatty acids	Danisco Culter, New Century, KS
Salatrim/Caprenin		Proctor and Gamble, Cincinnati, OH
Olestra/Olean®	Sucrose polyester	Proctor and Gamble, Cincinnati, OH

Table 8.2. Ingredients Used as Sugar Replacers in Baked Products

Sweetener	Brand Name	Sweetness Compared to Sucrose	Supplier
Acesulfame-K	Sunett®	200% sweeter	Nutrnova, Somerset, NJ
Sucralose	Splenda®	600% sweeter	Splenda, Inc., Ft. Washington, PA

Table 8.3. Ingredients Marketed for Specific Health Benefits

Brand Name	Composition	Health Benefit	Supplier
Caromax™ Carob Fiber	Carob fruit fiber; soluble fiber, tannins, polyphenols, lignan	Lower serum cholesterol	National Starch & Chemical, Bridgewater, NJ
FenuPure™	Fenugreek seed concentrate; galactomannan	Regulate blood glucose; lower serum cholesterol	Schouten USA, Inc., Minneapolis, MN
Fibrex®	Sugar beet fiber; soluble fiber, lignan	Lower serum cholesterol; regulate blood glucose	Danisco Sugar, Malmo, Sweden
MultOil	Diglycerides + phytosterols	Lower serum cholesterol	Enzymotec, Migdal HaEmeq, Israel
Nextra™	Decholesterolized tallow and corn oil; free of trans fat	Reduce the risk for coronary heart disease	Source Food Technology, Durham, NC
Novelose 240	Corn fiber; high amylose, resistant fiber	Reduce risk for colon cancer	National Starch & Chemical, Bridgewater, NJ
Nutrifood®	Fruit and vegetable liquid concentrates; source of antioxidants—carotenoids, anthocyanins, polyphenols	Reduce risk for chronic diseases—cancer, diabetes, and cardiovascular disease	GNT USA, Inc., Tarrytown, NY GNT Germany, Aachen, Germany
OatVantage™	Beta-glucans, a soluble fiber	Lower serum cholesterol	Nurture, Inc., Devon, PA

Table 8.4. Health Claims Approved for Food Labeling in the United States

Food Component	Health Claim
Calcium	Osteoporosis
Dietary fat	Cancer
Dietary saturated fat and cholesterol	Coronary heart disease
Fiber-containing grain products, fruits, and vegetables	Cancer
Sodium	Hypertension
Folate	Neural tube defects
Dietary sugar alcohol	Dental caries
Fruits, vegetables, and grain products that contain fiber, particularly soluble fiber	Coronary heart disease
Soy protein	Coronary heart disease
Whole grain foods	Heart disease and certain cancers
Plant sterols/stanol esters	Coronary heart disease
Potassium	High blood pressure and stroke

Source: FDA/CFSAN 2002b.

ized food manufacturers to add health claims related to specific food components (FDA 2003) (Table 8.4). However, for many "functional foods," the scientific evidence to meet FDA criteria to make health claims is lacking (Wahlqvist and Wattanapenpaiboon 2002). A 2003 amendment to the NLEA requires that trans fatty acids be listed under saturated fat on the food facts label by January 1, 2006 (FDA 2003).

The Codex Alimentarius Commission of the Food and Agriculture Organization of the United Nations World Health Organization (FAO/WHO) Codex Guidelines on Nutrition Labeling adopted in 1985 are similar to the NLEA implemented by the FDA in 1994 (FAO/WHO 2001a). The Codex Alimentarius Commission adopted the Codex Guidelines for the use of Nutrition Claims on food labels in 1997 (FAO/WHO 2001b). Codex standards are voluntary, and each country within the United Nations is free to adopt food-labeling standards. The European Union, which includes 15 member states in Europe, also sets guidelines for nutrition labeling and nutrition claims, subject to requirements of the individual member states.

The Food Standards Agency of the United Kingdom (FSA) was established in 2000 as the regulatory agency to set policy for food labeling in Great Britain and Northern Ireland (FSA 2003a). The Food Standards Australia New Zealand (FSANZ) specifies the requirements for food labeling in these countries (FSANZ 2003). Health Canada published new food labeling regulations January 1, 2003, making nutrition labeling mandatory for most foods and allowing diet-related health claims on food labels for the first time (Health Canada 2003).

FOOD LABELING STANDARDS FOR ORGANICALLY GROWN FOODS

The Organic Foods Production Act of 1990 passed by the U.S. Congress required the U.S. Department of Agriculture (USDA) to develop certification standards for organically produced agricultural products [Agricultural Marketing Service (AMS)/USDA 2003]. Producers who meet the standards may specify the percentage of the product that is organic on the food label if 70% or more of the ingredients in the product are organically grown (AMS/USDA 2003). The Codex Alimentarius Commission has also published standards for labeling organically grown foods (FAO/WHO 2001c). Organic fruits and vegetables are produced without using conventional pesticides, petroleum-based fertilizers, or sewage sludge–based fertilizers. Animal products identified as organic come from animals given organic feed but are not given antibiotics or growth hormones. Food products that have been developed through genetic modification cannot be labeled as organically grown foods (AMS/USDA 2003, FAO/WHO 2001c).

INGREDIENT LABELING FOR POSSIBLE ALLERGENS

The Codex Alimentarius Commission of the FAO/WHO and the Food and Drug Administration's Center for Food Safety and Applied Nutrition (FDA/CFSAN) require that food labels list all ingredients known to cause adverse responses in those with food allergens or sensitivities (FAO/WHO 2001a, FDA/

ORA 2001). The FDA requires that ingredients from the eight foods that account for ~90% of all food allergies be listed. These foods are peanuts, soybeans, milk, eggs, fish, shellfish, tree nuts, and wheat (FDA/ORA 2001). Codex standards require listing ingredients from these same eight foods plus all cereals that contain gluten (rye, barley, oats, and spelt), lactose, and sulphite in concentrations of 10 mg/kg or more (FAO/WHO 2001a). Gluten, lactose, and sulphite are listed on food labels because these substances cause distress for some, even though these substances are not considered allergens. Individuals with celiac disease or gluten intolerance eliminate all sources of gluten from the diet. A small percentage of individuals lack lactase, the enzyme needed to digest lactose, and avoid dairy products and all other foods with lactose additives.

Food processing plants are required to follow good manufacturing practices (GMP) to avoid possible cross-contamination with trace amounts of allergens during processing. An example of possible cross-contamination is using the same plant equipment to prepare "nut free" muffins after the equipment has been used to prepare muffins with nuts (Taylor and Hefle 2001). An example of GMP is dedicating food-processing plants to the production of allergen free foods (Taylor and Hefle 2001).

Small bakeries, defined by the number of employees or annual gross sales, and restaurants are exempt from FDA food labeling requirements. Food labeling to identify foods that have been genetically modified through bioengineering (GM) is voluntary (FDA/CFSAN 2001). However, because of consumer concerns about GM foods, managers of bakeries may choose to include a statement on the ingredient label such as "we do not use ingredients produced by biotechnology" (FDA/CFSAN 2001, 2002a). Consumers with food allergies have learned to read the list of ingredients on the food label to identify any possible sources of allergens. Managers of small bakeries that use nuts or soy flour in their operation but are unable to follow GMP because of the added cost may choose to alert consumers with a statement on the ingredient label, such as "this product was made on equipment that also makes products containing tree nuts." Making a decision to sell bakery products made with organic ingredients requires assessing the market for these products, the availability of organic ingredients, and the expected income from the operation.

RAW MATERIALS PREPARATION: SELECTION AND SCALING OF INGREDIENTS

Muffins made by large commercial bakeries are cake-type muffins, while those made in the home or small institutions are bread muffins. The differences between cake and bread muffins are that cake muffins are higher in fat and sugar and use soft wheat flours. A common problem encountered in bread-type muffins is tunnel formation resulting from overdevelopment of gluten. However, this problem is avoided in cake muffins since sugar, fat, and soft wheat flours interfere with gluten development and prevent tunnel formation. Bread muffins contain 12% of both fat and sugar, while cake muffins contain 18–40% fat and 50–70% sugar (Benson 1988).

Formulas for a standard cake muffin and a bran muffin are shown in Table 8.5. Ingredient formulas used by commercial bakeries are based on the weight of flour at 100% (Gisslen 2000). The amounts of other ingredients are a percentage of flour weight (baker's percent).

For example,

(total weight of muffin ingredient ÷ total weight of flour) × 100 = % of the ingredient

If the weight of another ingredient is the same weight as flour, the percent for that ingredient is also 100%. The advantage of using baker's percent is that batch sizes can be easily increased or decreased by multiplying the percent for each ingredient by the same factor. Weighing all ingredients, including liquids is faster and more accurate than using measurements, especially in large commercial bakeries.

FLOUR

Flour is the primary ingredient in baked products. Flour represents 30–40% of the total batter weight in most cake muffins (Benson 1988). Most muffin formulas contain a blend of cake or pastry flour and a high-protein flour such as bread flour, or all bread flour (Willyard 2000). The protein in flour is needed to provide structure in quick breads made with limited amounts of sugar. Flour contains starch and the proteins glutenin and gliadin, which hold other ingredients together to provide structure to the final baked product. Hydration and heat promote gela-

Table 8.5. Muffin Formulas Listed by Baker's Percent and Weight

Ingredient	Basic Cake Muffin, (Baker's %)	Weight (g)	Bran Muffin (Baker's %)	Weight (g)
Flour	100.00	990	—	—
Bread flour	—	—	50.00	4,545
Cake flour	—	—	18.75	1,704
Bran	—	—	31.25	2,842
Sugar	60.00	5,455	31.25	2,842
Baking powder	5.00	455	1.50	136
Baking soda	—	—	2.20	220
Salt	1.25	114	1.50	136
Milk powder	7.50	682	12.50	1,136
Molasses	—	—	37.50	3,409
Shortening	40.00	3,636	18.75	1,704
Whole eggs (liquid)	30.00	2,727	12.50	1,136
Honey	—	—	19.00	1,727
Water	60.00	5,455	100.00	990
Raisins	—	—	25.00	2,273
Total	303.75	27,616	316.70	32,790

Mixer: Hobart N-50 with 5 quart bowl and paddle agitator.

Directions for basic cake muffin formula:
   Blend dry ingredients together by mixing for 1 minute at low speed.
   Add shortening and eggs and mix for 1 minute at low speed.
   Add water and mix for 1 minute at low speed.
Scaling weight: 2.5 ounces batter
Yield: 2 1/2 dozen muffins
Bake: at 205°C for 19–21 minutes in a gas-fired reel oven.

Directions for bran muffin formula:
   Blend dry ingredients and mix for 1 minute at low speed.
   Add shortening, eggs, honey, molasses and 50% (4.5 kg) of the water and mix for 1 minute at medium
      low speed.
   Add the remaining water and mix for 1 minute at low speed.
   Add raisins and mix at low speed for 3 minute or until raisins are dispersed.
Scaling weight: 3 ounces batter
Yield: 3 dozen muffins
Bake: at 193°C for 20–25 minutes in a gas-fired reel oven.

Sources: Benson 1988, Doerry 1995b.

tinization of starch, a process that breaks hydrogen bonds, resulting in swelling of the starch granule, which gives the batter a more rigid structure (McWilliams 2001e).

Substituting whole wheat flour, wheat germ, rolled oats, or bran for part of the flour is an excellent way to increase fiber. Other flours used in muffins include cornmeal, soy, oat, potato, and peanut. An acceptable product is possible when cowpea or peanut flours are substituted for 25% or when whole-wheat flour or corn meal is substituted for 50% of all-purpose flour (Holt et al. 1992). Acceptable muffins have been prepared when soy protein flour was substituted for 10–20% (Sim and Tam 2001) or 100% of all-purpose flour (Bordi and others 2001). None of these flours contain glutenin or gliadin except whole wheat, and large pieces of bran in whole wheat flour cut and weaken gluten strands. Thus, there is minimal gluten development when these flours are used; however, the muffins tend to be crumbly and compact unless other modifications are made in the formula.

SUGAR

Amounts of sugar in muffins range from 50 to 70%, based on flour at 100% (Benson 1988). Sugar contributes tenderness, crust color, and moisture retention in addition to a sweet taste. Sucrose promotes tenderness by inhibiting hydration of flour proteins and starch gelatinization. Sugar is hygroscopic (attracts water) and maintains freshness. Corn syrup, molasses, maple sugar, fruit juice concentrates, and honey are used as sweeteners for flavor variety. Honey or molasses is often used as a sweetener in whole wheat or bran muffins to cover the bitter flavor of the bran (Willyard 2000). The quantity of liquid will need to be decreased if these sweeteners are used instead of sucrose because of the high water content in these syrups.

Chemical changes in sugars during baking contribute characteristic flavors and browning. Caramelization of sugar is responsible for the brown crust of muffins. Caramelization involves dehydration and polymerization (condensation) of sucrose (McWilliams 2001c). Reducing sugars such as dextrose, corn syrup, or high fructose corn syrup are often added to muffins at levels of 1–3% to increase crust color (Willyard 2000). Reducing sugars react with amino acids in flour, milk, and eggs to form a complex responsible for the flavor and brown crust of muffins. The reaction between the aldehyde or ketone group in reducing sugars and the amino acids in protein is described as the Maillard reaction (McWilliams 2001e). This Maillard reaction, together with caramelization, contributes to the characteristic flavor and color of the crust of a baked muffin. Crust temperatures reach 100°C and above, which lowers water activity. Both the high temperature and low water activity are necessary for the Maillard reaction to occur (McWilliams 2001f).

Sugar replacers such as acesulfame-K and sucralose (see Table 8.2) can be substituted for all or part of the sugar. Sugar replacers, however, do not contribute to tenderness, browning, or moisture retention; thus, other formula modifications are necessary for an acceptable product. For example, a small amount of molasses or cocoa may be added to substitute for color from the caramelization of sucrose. The shelf life of muffins prepared without sugar would be very limited.

FAT

Muffins contain 18–40% fat based on flour at 100% (Benson 1988). Fat contributes to the eating quali- ties of tenderness, flavor, texture, and a characteristic mouthfeel. Fat keeps the crumb and crust soft and helps retain moisture, and thus contributes to keeping qualities or shelf life (McWilliams 2001d). Fat enhances the flavor of baked products since flavor components dissolve in fat. Both shortening and vegetable oils are used in muffins.

To meet the demands of the consumer, muffin formulas are being modified to reduce total fat, saturated fat, trans fat, and calories, and to increase the amount of monounsaturated and polyunsaturated fat. Canola oil and flaxseed meal are being added to muffins to increase the proportion of monounsaturated fat. Muffins made with reduced fat and polyunsaturated fatty acids (13% safflower oil) were comparable in sensory and physical characteristics to the standard muffin made with shortening at 20% (Berglund and Hertsgaard 1986). Low fat and fat free muffins are available ready-to-eat and as frozen batters or dry mixes.

Various fat replacers have been classified by their macronutrient bases (see Table 8.1). Carbohydrate- and lipid-based fat replacers can be used to prepare muffins acceptable to the consumer. Lipid-based fat replacers that have the same chemical and physical characteristics as triglycerides are described as fat substitutes (Akoh 1998). These products provide the same characteristics as fat but with fewer calories. Monoglycerides, diglycerides, and modified triglycerides are examples of fat substitutes that replicate the mouthfeel and sensory qualities of baked products made with shortening.

Enova™ (Archer Daniels Midland KAO LLC, Decatur, Illinois) is an example of a diglyceride that is lower in calories than other oils and is being marketed as beneficial in weight management (Pszczola 2003). Benefat® (Danisco Culter, New Century, Kansas) and Caprenin (Proctor and Gamble, Cincinnati, Ohio) are examples of triglycerides modified by substituting shorter-chain fatty acids (Akoh 1998). Sucrose polyesters of six to eight fatty acids are marketed as Olean® (Procter and Gamble, Cincinnati, Ohio), a fat substitute with the same physical qualities as shortening without the calories since sucrose polyesters are not digested or absorbed in the human intestinal tract.

A commercial shortening product (Nextra™) (Source Food Technology, Durham, North Carolina) made from decholesterolized tallow and corn oil is being marketed to the baking industry as a trans-free fat to replace shortening (Pszczola 2002b). Other methods used by the food industry to decrease the amount of trans fat are (1) blending hydrogenated fat high in stearic acid with unhydrogenated oils and (2) interesterfying (rearranging) unhydrogenated oils with saturated fat–based oils (Hunter 2002).

Carbohydrate-based fat replacers are described as fat mimetics. For example, cellulose, corn syrup, dextrins, fiber, gum, maltodextrins, polydextrose, starches, and fruit-based purees. Z-trim, developed by a U.S. Department of Agriculture scientist, is a mixture of plant fibers (Inglett 1997). Fat mimetics replicate the mouthfeel and texture of fat in baked products and extend shelf life by binding water and trapping air (American Dietetic Association 1998). Acceptable low fat cake muffins (5% fat) used 2% pregelatinized dull waxy starch and corn syrup (3.6%) to replace fat (Hippleheuser et al. 1995).

Fruit purees or pastes of one or more fruits—apples, dates, figs, grapes, plums, prunes, and raisins are being promoted as fat replacers. Just Like Shorten™ is a mixture of dried prunes and apples. The fruit purees have humectant properties, promote tenderness and moistness, increase shelf life, and can replace some of the sugar and/or fat in muffins and cakes.

Formulas will need to be developed based on adjustments in ingredients when fat replacers are substituted for all or part of the fat in the formula. New formulas need to be prepared, the muffins evaluated using the muffin scorecard (Table 8.6), and the shelf life evaluated. Several formula adjustments may be necessary before an acceptable muffin is developed.

LEAVENING AGENTS

The amount of baking powder used in muffins varies between 2 and 6% based on flour at 100%, with lower amounts in muffins with ingredients that increase acid (Benson 1988). Gases released by a leavening agent influence volume and cell structure. During baking, heat increases gas volume and pressure to expand cell size until proteins are coagulated (McWilliams 2001c). Stretching of the cell walls during baking improves texture and promotes tenderness (McWilliams 2001c).

The quantity of leavening used in a baked product depends on the choice of leavening agent as well as other ingredients. Formulation of baking powders considers the amount of leavening acids needed to neutralize baking soda or sodium bicarbonate, an alkaline salt. Double-acting baking powder (most commonly used in muffins) contains both slow- and fast-acting acids (McWilliams 2001e). Fast-acting acids are readily soluble at room temperature, while slow-acting acids are less soluble and require heat over extended time to release carbon dioxide. Formulations of slow- and fast-acting acid leavening agents control the reaction time and optimize volume (Borowski 2000). An example of a formulation to neutralize sodium bicarbonate is a mixture of slow- and fast-acting acids—monocalcium phosphate monohydrate (a fast-acting acid) combined with sodium aluminum sulfate (a slow-acting acid). Development of baking powder requires consideration of the unique neutralizing value (NV) and the rate of reaction (ROR) (the percent of carbon dioxide released during the reaction of sodium bicarbonate with a leavening acid during the first eight minutes of baking) (Anonymous 2003b, Borowski 2000).

Baking soda is used in addition to double-acting baking powder when muffins contain acidic ingredients such as sour cream, yogurt, buttermilk, light sour cream, molasses, and some fruits and fruit juices (McWilliams 2001e). Baking soda in the amount of 2–3% in addition to baking powder is added to acidic batters (Benson 1988).

Sodium carbonate is a product of an incomplete reaction in formulas with excess sodium bicarbonate. Excess sodium carbonate results in a muffin with a soapy, bitter flavor and a yellow color because of the effect of an alkaline medium on the anthoxanthin pigments of flour (McWilliams 2001f). Also, formulas with too much baking powder or soda result in a muffin with a coarse texture and low volume because of an overexpansion of gas, which causes the cell structure to weaken and collapse during baking. Inadequate amounts of baking powder will result in a compact muffin with low volume. Figures 8.1 and 8.2 show different chemical reactions for fast-acting and slow-acting baking powders (McWilliams 2001e).

WHOLE EGGS

Liquid eggs contribute 10–30% of muffin batter based on flour at 100%, and dried eggs contribute 5–10% (Benson 1988). Eggs provide flavor, color, and a source of liquid. Upon baking, the protein in egg white coagulates to provide structure. Adding egg whites to muffin batter provides structure to the finished product and a muffin that is easily broken without excessive crumbling (Stauffer 2002). Substituting egg whites for whole eggs, however, will result in a dry, tough muffin unless the formula is adjusted to increase the amount of fat (Stauffer

Table 8.6. Scorecard for Muffins

Evaluator:	Product:	Date:

External Qualities

	Score

1a. Volume
    Specific Volume: $\pi r^2 \times$ height = weight in grams (cm)
    1 = low volume, compact cells; 5 = light with moderate cells;
    7 = large volume, large cells and/or tunnels
	Score

1b. Contour of the surface
    1 = absolutely flat; 3 = somewhat rounded; 5 = pleasingly rounded;
    7 = somewhat pointed; 9 = very pointed	

1c. Crust color
    1 = much too pale; 3 = somewhat pale; 5 = pleasingly golden brown;
    7 = somewhat too brown; 9 = much too brown	

Internal Qualities	

1d. Interior color
    1 = much too white; 3 = somewhat white; 5 = pleasingly creamy;
    7 = somewhat too yellow; 9 = much too yellow	

1e. Cell uniformity and size
    1 = much too small; 3 = somewhat thick; 5 = moderate;
    7 = somewhat too large; 9 = numerous large tunnels	

1f. Thickness of cell walls
    1 = extremely thick; 3 = somewhat thick; 5 = normal thickness;
    7 = somewhat too thin; 9 = much too thin	

1g. Texture
    1 = extremely crumbly; 3 = somewhat crumbly; 5 = easily broken;
    7 = slightly crumbly; 9 = tough, little tendency to crumble	

1h. Flavor
    1 = absolutely not sweet enough; 3 = not nearly sweet enough;
    5 = pleasingly sweet; 7 = somewhat too sweet; 9 = much too sweet	

1i. Aftertaste 1 = extremely distinct; 3 = somewhat distinct; 5 = none	

1j. Aroma
    1 = lack of aroma; 5 = sweet and fresh aroma;
    9 = sharp, bitter or foreign aroma	

1k. Mouthfeel
    1 = gummy, cohesive; 3 = somewhat gummy; 5 = tender, light and moist;
    7 = somewhat dry and tough; 9 = tough and hard to chew	

Overall Acceptability 1 = very unacceptable; 3 = somewhat acceptable; 5 = very acceptable	

Source: Adapted from McWilliams 2001a.

$$3CaH_4(PO_4)_2 + 8NaHCO_3 \longrightarrow Ca_3(PO_4)_2 + 4Na_2HPO_4 + 8CO_2 + 8H_2O$$

| Monocalcium Phosphate | Sodium bicarbonate | Tricalcium phosphate | Disodium phosphate | Carbon dioxide | Water |

Figure 8.1. Formation of bicarbonate of soda from a fast-acting acid salt.

Step 1.

$$Na_2(Al)_2(SO_4)_4 + 6H_2O \longrightarrow 2Al(OH)_3 + Na_2SO_4 + 3H_2SO_4$$

| Sodium aluminum sulfate | Water | Aluminum Hydroxide | Sodium sulfate | Sulfuric Acid |

Step 2.

$$3H_2SO_4 + 6NaHCO_3 \longrightarrow 3Na_2SO_4 + 6H_2CO_3$$
$$\downarrow \longrightarrow 6CO_2 + 6H_2O$$

| Sulfuric Acid | Sodium bicarbonate | Sodium Sulfate | Carbonic acid | Carbon dioxide | Water |

Figure 8.2. Formation of bicarbonate of soda and carbon dioxide from a slow-acting acid salt.

2002). Fat in the yolk acts as an emulsifier and contributes to mouthfeel and keeping qualities.

NONFAT DRY MILK POWDER

Milk powder represents 5–12% of the muffin batter based on flour at 100% (Benson 1988). Milk powder is added to dry ingredients, and water or fruit juice is used for liquid in muffin formulas. Milk powder binds flour protein to provide strength, body, and resilience—qualities helpful in reducing damage during packing and shipping (Willyard 2000). In addition, milk powder adds flavor and retains moisture. The aldehyde group from lactose in milk combines with the amino group from protein upon heating, contributing to Maillard browning.

SODIUM CHLORIDE

The amount of salt in muffins is 1.5–2% based on flour at 100% (Benson 1988). The function of so-

dium chloride is to enhance the flavor of other ingredients. Sodium chloride may be omitted from the formula without compromising flavor if other ingredients such as dried fruit or spices are added for flavor.

LIQUIDS

Liquids perform several functions in baked products (Benson 1988). These include dissolving dry ingredients, gelatinization of starch, and providing moistness in the final baked product. Insufficient liquid results in incomplete gelatinization of the starch and a muffin with insufficient structure to support expansion of air volume. The muffins will have nonuniform cell structure, overly crumbly texture, low volume, and a dip in the top.

ADDITIONAL INGREDIENTS

Other ingredients are often added to muffins for variety in flavor, texture, and color, and to increase specific nutrients or health components such as fiber, vitamins and minerals, or antioxidants from fruit and vegetable extracts. Part of the flour may be replaced with cornmeal, bran, whole wheat, oat, or other flours to increase the fiber content. Adjustments in the amount of water in the formula are necessary when whole wheat flour, bran, or other concentrated sources of fiber are added because fiber absorbs a great deal of water (Willyard 2000). An example of a concentrated source of fiber is Caromax™ (National Starch and Chemical, Bridgewater, New Jersey) (Pszczola 2001). Nutrifood® (GNT USA, Tarrytown, New York), a liquid concentrate marketed as a blend of the antioxidants—carotenoids, anthocyanins, and polyphenols—is an example of a bioactive ingredient (Pszczola 2002a).

Other ingredients can be substituted for part of the liquid. For example, applesauce, bananas, shredded carrots, or zucchini. Variations in texture are achieved by adding fresh fruit such as apples or blueberries or dried fruit such as dates, raisins, or apricots. Nuts and poppy seeds complement the flavor of sweet muffins, while grated cheese, whole-kernel corn, green peppers, chopped ham, and bacon add interest to corn muffins. Added flavorings include cinnamon, nutmeg, allspice, cloves, and orange or lemon zest. Topping mixtures such as chopped nuts, cinnamon, and sugar are added to the batter after depositing.

PROCESSING

STAGE 1: MIXING

There are two primary methods for mixing muffins—the cake method and the muffin method. The cake method involves creaming sugar and shortening together, then adding liquid ingredients, and finally adding dry ingredients. The muffin method of mixing involves two to three steps. First, dry ingredients are mixed together; second, shortening or oil and other liquids are mixed together; and third, the liquids are added to the dry ingredients and mixed until the dry ingredients are moistened. Additional ingredients are added at the end of the mixing cycle or after depositing the muffin batter. Institutional or commercial bakeries use a mixer on slow speed for three to five minutes. Inadequate mixing results in a muffin with a low volume since some of the baking powder will be too dry to react completely.

STAGE 2: DEPOSITING

The traditional size of muffins is two ounces, although today muffins are marketed in a wide range of sizes from one-half ounce mini-muffins to muffins five ounces or larger in size (Willyard 2000). For institutions or bakeries, small batter depositors are available that will deposit four muffins at a time. Also available are large piston-type depositors that maintain accurate flow of the batter (Benson 1988).

STAGE 3: BAKING

Many physical and chemical changes occur in the presence of heat to transform a liquid batter into a final baked muffin. Solubilization and activation of the leavening agent generates carbon dioxide that expands to increase the volume of the muffin. Gelatinization of starch and coagulation of proteins provide permanent cell structure and crumb development. Caramelization of sugars and Maillard browning of proteins and reducing sugars promote browning of the crust. Reduced water activity facilitates Maillard browning as well as crust hardening (McWilliams 2001f).

The choice of oven, baking pans, and baking temperature influences the final baked product (Benson 1988). A good flow of heat onto the bottom of the pan is necessary to produce a good product. Muffin tins are usually placed directly on the shelf or baking surface. The appropriate oven temperature is related to scaling and the type of oven. Standard two-ounce muffins are baked at 204°C or slightly higher in a deck oven. Deck ovens may be stacked and are often used in small retail bakeries since these are less expensive and easier to maintain than reel or rotary ovens. Reel ovens consist of an insulated cubic compartment six or seven feet high. A Ferris wheel–type mechanism inside the chamber moves four to eight shelves in a circle, allowing each shelf to be brought to the door for adding or removing muffin tins from the shelves (Matz 1988). Retail bakers often prefer the reel oven since several hundred to several thousand pounds of batter can be baked each day. Rack ovens may be stationary, or the racks may be rotated during baking.

STAGE 4: COOLING

Products should be cooled prior to wrapping. This allows the structure to "set" and reduces the formation of moisture condensation within the package.

Condensed moisture creates an undesirable medium that promotes yeast, mold, and bacterial growth and spoilage.

STAGE 5: PACKAGING

Muffins may be wrapped individually, in the tray in which they are baked, or transferred into plastic form trays for merchandizing (Benson 1988). The shelf life of muffins is three to five days for individually wrapped muffins and four to seven days for six or more muffins packaged in trays and wrapped in foil or plastic wrap. The storage life of muffins is significantly influenced by exposure to oxygen and moisture (Rice 2002). Cake muffins have a longer shelf life than bread muffins because of their high sugar content and lower water activity (Willyard 2000). Added ingredients, such as cheese, ham, and dried fruits that are high in sodium or sugar content, reduce water activity and increase shelf life.

FINISHED PRODUCT

A muffin fresh out of the oven will vary in appearance based on the formula (whether the formula is for a cake or bread muffin), the size of the muffin (mini-muffin or mega-muffin), and the desired shape, flat or mushroom-shaped tops to the traditional bell-shaped muffin (Willyard 2000). In general, a desirable muffin product has a symmetrical shape, a rounded top that is golden brown in color, cells that are uniform and moderate in size, and a sweet flavor and pleasant aroma; it is also tender and moist, is easily broken apart, is easy to chew, and has a pleasant aftertaste.

MUFFIN EVALUATION

Bakers can use Table 8.6, Scorecard for Muffins, to evaluate muffins during the process of developing or modifying muffin formulas. Large commercial bakeries may use more sophisticated methods to evaluate bakery products, such as gas chromatography to evaluate flavor components.

Volume

Compact muffins with small cells or large muffins with peaked tops and tunnels are undesirable in all types of muffins. Diameter is a more important criteria than volume for evaluating mushroom and flat-topped muffins. For bell-shaped muffins, volume is a quality that can be evaluated objectively by measuring the height and diameter ($\pi r^2 \times$ height). The volume can be determined indirectly by measuring the circumference of a cross section of the muffin in cubic centimeters and dividing by the weight in grams. This can be done by measuring the height of the muffin at the highest point, then slicing off the top of the muffin and measuring the diameter of the muffin.

Contour of the Surface

The muffin should be rounded and golden brown in color with a pebbled surface.

Color of Crust

Crust color should be a pleasing golden brown, not pale or burnt.

Interior Color

Crumb color should be a pleasant creamy color, not white and not too yellow. Crumb color will be darker with wholegrain flour or added ingredients such as nuts or dried fruits, or spices.

Cell Uniformity and Size

Cell structure can be evaluated by making a vertical cut in the muffin to form two equal halves and then making an ink print or photo copy (McWilliams 2001b). A desirable muffin should have a uniform cell structure without tunnels.

Thickness of Cell Walls

Uniform thick-walled cells are desirable. Coarseness, thin cell walls, uneven cell size, and tunnels indicate poor grain.

Texture

Texture depends on the physical condition of the crumb and is influenced by the grain. A desirable muffin should be easily broken and slightly crumbly. Extreme crumbling and toughness with lack of crumbling are undesirable characteristics.

Flavor

An acceptable muffin should have a pleasingly sweet flavor. Flat, foreign, salty, soda, sour, or bitter tastes are undesirable.

Aftertaste

An acceptable muffin should have a pleasant, sweet aftertaste, not bitter or foreign.

Aroma

Aroma is recognized by the sense of smell. The aroma may be sweet, rich, musty, or flat. The ideal aroma should be pleasant, fresh, sweet, and natural. Sharp, bitter, or foreign aromas are undesirable.

Mouthfeel

Mouthfeel refers to the textural qualities perceived in the mouth. Characteristics can be described as gritty, hard, tough, tender, light, and moist. A desirable muffin is tender, light, and moist and requires minimal chewing.

APPLICATIONS OF PROCESSING AND PRINCIPLES

Processing Stage	Processing Principles	References for More Information on the Principles Used
Selection and scaling of ingredients		Anonymous 2003a
Flour	Starch gelatinization, cell structure and volume, Maillard browning	Willyard 2000
Sugar	Flavor, tenderizer, crust quality,moisture retaining, reduction of water activity	Willyard 2000
Fat	Flavor, tenderizer, moisturizing	Willyard 2000
Milk powder	Binding effect on flour protein, flavor, crust color, Maillard browning, moisture retention	Willyard 2000
Whole eggs	Protein coagulation, emulsification, flavor, color	Willyard 2000, Stauffer 2002
Liquid	Hydration of flour proteins and starch, solvent for salt, sugar, leavening agent, cell structure and volume, moisture in final baked product	Doerry 1995a
Chemical leavening	Generation of carbon dioxide, volume and cellular structure	Borowski 2000
Salt	Flavor enhancer	Willyard 2000
Additional ingredients	Variety in flavor, texture and nutritive value	Willyard 2000
Mixing	Dispersion of ingredients, hydration of flour proteins and starch	Doerry 1995a
Depositing	Scaling of muffin	Benson 1988
Baking	Solubilization and activation of leavening agent, gelatinization of starch, coagulation of protein, caramelization of sugar, reduction of water activity, crumb development, color development, flavor development, crust formation, Maillard browning	Gisslen 2000
Cooling	"Setting" of structure, water evaporation	Doerry 1995a
Packaging	Retention of moisture, retention of flavor	Rice 2002

GLOSSARY

Allergen—a substance that causes an abnormal immune response in individuals with an allergy to that substance. The most common food allergens are peanuts, milk, eggs, wheat, soy, tree nuts, fish and shellfish.

AMS/USDA—Agricultural Marketing Service/U.S. Department of Agriculture.

Anthoxanthin—a naturally occurring color pigment in plants and wheat flour; the pigment turns yellow in the presence of an alkaline medium for example, the crumb is yellow when excessive amounts of baking soda have been added to the muffin batter.

Antioxidant—natural occurring compounds found in plant foods that have possible health benefits by quenching free radicals and thus preventing cancer, cardiovascular disease, and other chronic diseases.

Baker's percent—term used by the baking industry to describe the amount of each ingredient by weight for a "recipe" or formula compared to the weight of flour at 100%; also described as flour weight basis.

Caramelization—chemical changes in sucrose (dehydration and polymerization) in response to heat during baking; caramelization gives the characteristic color and flavor in baked products.

Carbon dioxide—gas produced by chemical leavening agents that expands muffin batter during baking.

Cell structure—an internal characteristic of baked products; a desirable cell structure is uniform with moderately sized cells. Factors that influence cell structure are the muffin formula, the mixing process, and baking temperature.

Chemical leavening agent—agents made of a mixture of alkaline bicarbonates and a leavening acid phosphate that is activated by water and baking temperatures to generate carbon dioxide, which expands the muffin batter during baking.

Coagulation—changes in the structure of protein in milk and eggs during baking that binds together muffin ingredients; denaturation of protein breaks weak chemical bonds and allows formation of stronger bonds among strands of protein, causing a "clumping" of protein.

Color pigments in fruits and vegetables—important sources of antioxidants; for example, pranthocyanidin in blueberries, lycopene in tomatoes, and lutein in spinach.

Crumb—an internal characteristic of baked products that describes the texture related to tenderness or ease in breaking into pieces from very crumbly to tough with little tendency to crumble.

Deck oven—a type of oven used in commercial bakeries, small bakeries, or restaurants. A deck oven may consist of single or multiple ovens stacked vertically; each oven has individual temperature controls.

DHHS/FDA—Department of Health and Human Services, Food and Drug Administration.

Emulsifying agent—an ingredient having both polar and nonpolar groups allowing for attraction of both polar (water) and nonpolar (oils) ingredients. Emulsifying agents improve keeping qualities of muffins by dispersing water throughout the batter.

FAO/WHO—Food and Agricultural Organization of the World Health Organization.

Fat replacers—ingredients used to replace fat in baked products to meet consumer demand for "healthier" foods lower in calories and saturated fat. Fat replacers replicate the mouthfeel and keeping qualities of fat by attracting water.

FDA—U.S. Food and Drug Administration.

FDA/CFSAN—Food and Drug Adminstration/Center for Food Safety and Applied Nutrition.

FDA/ORA—Food and Drug Administration/Office of Regulatory Affairs

Formula—term used instead of "recipe," by the baking industry; the weight of each ingredient is determined based on the weight of flour at 100%.

Formula percent—term used by the baking industry to describe the amount of each ingredient by weight for a "recipe" or formula compared to the weight of all ingredients.

FSA—Food Standard Agency of the United Kingdom.

FSANZ—Food Standards Australia and New Zealand.

Functional foods—foods marketed to have specific health benefits; for example, a health benefit of including oats in the diet is lowering blood cholesterol.

Gelatinization—changes in the starch granules of flour (breaking of hydrogen bonds and swelling) in the presence of water and heat; starch gelatinization gives structure to quick breads.

GMO (genetically modified organism)—refers to new plant varieties developed using genetic engineering or biotechnology, and ingredients made from GMO plants, for example corn meal made from genetically modified corn.

GMP—good manufacturing practices.

Hydration—the addition of liquids to dry ingredients in the preparation of quick breads; hydration promotes starch gelatinization which gives structure to the final baked product.

Hygroscopic—a quality of attracting water molecules; sugar in muffin batter attracts water and contributes to the moistness and keeping qualities of baked products.

Lactose—the disaccharide made of glucose and galactose and found in milk. Both lactose and protein in milk contribute to Maillard browning in baked products.

Maillard browning—a change in color that occurs during the baking process as a result of the reaction between an aldehyde or ketone group from sugar and the amino acids from protein sources in the batter such as milk, soy, and eggs.

Mouthfeel—refers to the textural qualities perceived in the mouth. Characteristics can be described as gritty, hard, tough, tender, light and moist.

NLEA—Nutrition Labeling and Education Act.

Nutraceuticals—naturally derived compounds from food, botanicals and dietary supplements marketed to prevent disease or to treat specific medical conditions. For example, plant sterol esters are added to vegetable oil spreads; the health benefit of these spreads is lowering of serum cholesterol.

Neutralizing value—the parts of sodium bicarbonate that will be neutralized by 100 parts of a leavening acid such as monocalcium phosphate.

NV—neutralizing value.

Organic—term used on food labels to identify agricultural products produced under specific guidelines as defined by regulatory agencies such as the U.S. Department of Agriculture's National Organic Program and the Joint FAO/WHO Food Standards Programme Codex Alimentarius Commission on Organically Produced Foods. Organic fruits and vegetables are raised without using conventional pesticides, petroleum-based fertilizers, or sewage sludge-based fertilizers. Animal products identified as organic come from animals that have access to the outdoors and are given organic feed but not antiobiotics or growth hormones.

Rate of reaction (ROR)—the percent of carbon dioxide released during the reaction between sodium bicarbonate and a leavening acid phosphate under standard conditions of temperature and pressure.

Reel oven—ovens with a Ferris wheel–type mechanism to move four to eight shelves in a circle allowing muffin tins to be moved to the front of the oven for removal.

Scaling—a term used by the baking industry to describe the weighing of ingredients.

Shelf life—the "keeping" qualities of baked products such as moistness and tenderness; sugar and fat extend the shelf life of quick breads.

Sodium aluminum sulfate—a slow-acting acid used in combination with a fast-acting acid such as monocalcium phosphate in double-acting baking powder that acts as a leavening agent in quick breads.

Sodium bicarbonate—commonly called baking soda, a leavening agent used in combination with acid ingredients such as sour cream, yogurt, buttermilk or fruit juice in muffin batter. Baking powder includes both an acid salt (monocalcium phosphate) and an alkaline salt (sodium bicarbonate).

Sodium chloride—commonly called salt and added to baked products to enhance other flavors.

Sugar replacers—calorie free or reduced calorie ingredients used in baked products to give sweetness with less calories than sugar; sugar replacers are used to meet the demands of consumers for "healthier" foods.

Trans fat—the form of fat in partially hydrogenated vegetable oil or "shortening," used in commercial bakery products. Diets high in trans fat raise LDL cholesterol and increase the risk for cardiovascular disease.

Water activity—the ratio of vapor pressure in food compared to the vapor pressure of water. Meats and fresh fruits and vegetables have high water activity; the addition of salt or sugar to foods lowers water activity because salt and sugar attract and hold water.

WHO—World Health Organization.

ACKNOWLEDGMENT

Ron Wirtz, Ph.D., former library director, American Institute of Baking, and currently head, Education and Information Services, Greenblatt Library, Medical College of Georgia for assistance with locating references and publications for this chapter.

REFERENCES

Akoh CC. 1998. Scientific status summary. Fat replacers. Food Technol 52(3): 47–53.

Agricultural Marketing Service, U.S. Department of Agriculture (AMS/USDA). 2003. The National Organic Program. Background information. http://www.ams.usda.gov/nop/FactSheets/Backgrounder.html. Accessed on July 8, 2003.

American Dietetic Association. 1998. Position of The American Dietetic Association: Fat replacers. J Am Diet Assoc. 98(4): 463–468.

Anonymous. 2003a. Flour composition, formula methods. The Encyclopedia of Baking. Kansas City, Mo.: Sosland Publishing Co.

Anonymous. 2003b. I. Leavening acids. http://www.gallard.com/baking.htm . Accessed on June 1, 2003.

Benson RC. 1988. Technical Bulletin. Muffins. American Institute of Baking. 10(6): 1–4.

Berglund PT, DM Hertsgaard. 1986. Use of vegetable oils at reduced levels in cake, pie crust, cookies, and muffins. J. Food Sci 51(3): 640–644.

Bordi PL, CU Lambert, J Smith, R Hollender, ME Borja. 2001. Acceptability of soy protein in oatmeal muffins. Foodserv Res Int. 13(2): 101–110.

Borowski R. 2000. Leavening basics. Baking and Snack. November 1. http://bakingbusiness.com/archives/archive_article.asp?ArticleID=36622. Accessed on June 19, 2003.

Chopra M. 2002. Globalization and food: Implications for the promotion of "healthy" diets. In: Globalization, Diets and Noncommunicable Diseases, 1–16. Geneva: World Health Organization.

de Onis M, M Blossner. 2000. Prevalence and trends of overweight among preschool children in developing countries. Am J Clin Nutr. 72(4): 1032–1039.

Department of Health and Human Services, U.S. Food and Drug Adminstration (DHHS/FDA). 2003. Food labeling: Trans fatty acids in nutrition labeling: Consumer research to consider nutrient content, and health claims. Federal Register. 68(133): 41434–41438.

Doerry WT. 1995a. Chapter 3. Cake baking. In: WT Doerry. Baking Technology. Vol. 2, Controlled Baking, 138–190. Manhattan, Kans.: American Institute of Baking.

___. 1995b. Chapter 6. Cake muffins. In: Baking Technology. Vol. 2, Controlled Baking, 208–213. Manhattan, Kans.: The American Institute of Baking.

Flegal KM, MD Carroll, CL Ogden, CL Johnson. 2002. Prevalence and trends in obesity among U.S. adults, 1999–2000. JAMA. 288(14): 1723–1727.

Food and Agriculture Organization, World Health Organization (FAO/WHO). 2001a. Codex general standards for the labeling of prepackaged foods. Codex Alimentarious—Food Labelling—Complete Texts—Revised 2001. Rome: Joint FAO/WHO Food Standards Programme Codex Alimentarius Commission. http://www.fao.org/DOCREP/005/Y2770E. Accessed on July 3, 2003.

___. 2001b. Codex guidelines on nutrition labeling. Codex Alimentarious—Food Labelling —Complete Texts—Revised 2001. Rome: Joint FAO/WHO Food Standards Programme Codex Alimentarius Commission. 2001. http://www.fao.org/DOCREP/005/Y2770E. Accessed on July 3, 2003.

___. 2001c. Section 2. Description and definitions. Codex Alimentarious—Organically Produced Foods. Rome: Joint FAO/WHO Food Standards Programme Codex Alimentarious Commission. http://www.fao.org/DOCREP/005/Y2772E/Y2772E00.htm. Accessed on July 8, 2003.

Food and Drug Administration (FDA). 2003. Food labeling: Trans fatty acids in nutrition labeling, nutrient content claims, and health claims. Federal Register 68(133): 41434–41505.

Food and Drug Administration, Center for Food Safety and Applied Nutrition (FDA/CFSAN). 2001. Guidance for industry. Voluntary labeling indicating whether foods have or have not been developed using bioengineering. http://www.cfsan.fda.gov/~dms/bio/abgu.html. Accessed on July 7, 2003.

___. 2002a. Food labeling and nutrition. Information for industry. http://www.cfsan.fda.gov/~dms/lab-ind.htm. Accessed on July 3, 2003.

___. 2002b. Food labeling and nutrition. Information for industry. A Food labeling guide. Appendix C. Health Claims. http://www.cfsan.fda.gov/~dms/flg-6C.html. Accessed on July 3, 2003.

___. 2002c. Food labeling and nutrition. Small business food labeling exemption. http://www.cfsan.

fda.gov/~dms/lab-ind.htm. Accessed on July 3, 2003.

Food and Drug Adminstration, Office of Regulatory Affairs (FDA/ORA). 2001. Compliance Policy Guide: Compliance policy guidance for FDA staff. Sec. 555.250. Statement of policy for labeling and preventing cross-contact of common food allergens. Apr 19. http://www.fda.gov/ora/compliance_ref/cpg/cpgfod/cpg555-250.htm. Accessed on July 5, 2003.

Food Standards Australia New Zealand (FSANZ). 2003. Food Labelling. http://www.foodstandards.gov/au/whatsinfood/foodlabelling.cfm. Accessed on June 25, 2003.

Food Standards Agency of the United Kingdom (FSA). 2003a. About us. http://www.food.gov.uk/aboutus/. Accessed on July 11, 2003.

___. 2003b. Claims on labels. http://www.food.gov.uk/foodlabelling/claimson-lables/ Accessed on July 11, 2003.

Gisslen W. 2000. Chapter 1. Basic principles. In: W Gisslen. Professional Baking, 3rd edition, 3–16. New York: John Wiley and Sons, Inc.

Health Canada. 2003. Nutrition Labelling website. http://www.hc-sc.gc.ca/hpfb-dgpsa/onpp-bppn/labelling-etiquetage/index-e.html. Accessed on June 25, 2003.

Hawkes C. 2002. Marketing activities of global soft drink and fast food companies in emerging markets: A review. In: Globalization, Diets and Noncommunicable Diseases, 1–78. Geneva: World Health Organization.

Hippleheuser AL, LA Landberg, FL Turnak. 1995. A system approach to formulating a low-fat muffin. Food Technol 49(3): 92–95.

Holt SD, KH McWatters, AVA Resurreccion. 1992. Validation of predicted baking performance of muffins containing mixtures of wheat, cowpea, peanut, sorghum, and cassava flours. J Food Sci 57(2): 470–474.

Hunter JE. 2002. Trans fatty acids: Effects and alternatives. Food Technol. 56(12): 140.

Inglett GE. 1997. Development of a dietary fiber gel for calorie-reduced foods. Cereal Foods World. 42(3): 82–83, 85.

Matz SA. 1988. Chapter 9. Oven and baking. In: Equipment for Bakers, 319–362. McAllen, Texas: Pan Tech International.

McWilliams, M. 2001a. Chapter 3. Sensory evaluation. In: Foods: Experimental Perspectives, 4th edition, 33–57. Upper Saddle River, N.J.: Prentice Hall.

___. 2001b. Chapter 4. Objective evaluation. In: Foods: Experimental perspectives, 4th edition, 59–81. Upper Saddle River, N.J.: Prentice Hall.

___. 2001c. Chapter 6. Physical aspects of food preparation. In: Foods: Experimental Perspectives, 4th edition, 97–119. Upper Saddle River, N.J.: Prentice Hall.

___. 2001d. Chapter 12. Fats and oils in food products. In: Foods: Experimental Perspectives, 4th edition, 245–265. Upper Saddle River, N.J.: Prentice Hall.

___. 2001e. Chapter 17. Dimensions in baking. In: Foods: Experimental Perspectives, 4th edition, 381–413. Upper Saddle River, N.J.: Prentice Hall.

___. 2001f. Chapter 18. Baking applications. In: Foods: Experimental Perspectives, 4th edition, 415–499. Upper Saddle River, N.J.: Prentice Hall.

National Marketing Institute (NMI). 2003. Health and wellness trends report 2003. Harleyville, Pa.: NMI

O'Brien Nabors L. 2002. Sweet choices: Sugar replacements for foods and beverages. Food Technol 56(7): 28–30, 34, 45. 41.

Ogden CL, KM Flegal, MD Carroll, CL Johnson. 2002. Prevalence and trends in overweight among U.S. children and adolescents, 1999-2000. JAMA 288(14): 1728–1732.

Pszczola DE. 2001. Antioxidants: From preserving food quality to quality of life. Food Technol. 55(6): 51–59.

___. 2002a. Evolving ingredient components offer specific health value. Food Technol 56(12): 50–71.

___. 2002b. Bakery ingredients: Past, present, and future directions. Food Technol 56(1): 56–72.

___. 2003. Putting weight-management ingredients on the scale. Food Technol 57(3): 42–57.

Rice J. 2002. Packed for life. Bakery and Snack. February 1. http://www.bakingbusiness.com/archives_article.asp?ArticleID=48958. Accessed on June 6, 2003.

Sim J, N Tam. 2001. Eating qualities of muffins prepared with 10% and 20% soy flour. J Nutr Recipe Menu Dev. 3(2): 25–34.

Stauffer CE. 2002. Eggs: Extra benefits. Baking and Snack. February 1. http://www.bakingbusiness.com/tech/channel.asp?ArticleID=48984. Accessed on January 20, 2003.

Taylor SL, SL Hefle. 2001. Scientific status summary. Food allergies and other food sensitivities. Food Technol 44(9): 68–93.

Wahlqvist ML, N Wattanapenpaiboon. 2002. Can functional foods make a difference to disease prevention and control? In: Globalization, Diets and Noncommunicable Diseases, 1–18. Geneva: World Health Organization.

Willyard M. 2000. Technical Bulletin. Muffins (Update). American Institute of Baking 22(10): 1–6.

World Health Organization (WHO). 2001. Nutrition and NCD Prevention. Department of Noncommunicable Disease Prevention and Health Promotion, Geneva. http://www.who.int/hpr/nutrition/index.shtml. Accessed on June 5, 2003.

___. 2002a. 1. Global ageing: A triumph and a challenge. In: Active Ageing. A policy framework, 6–18. Geneva: WHO

___. 2002b. Overview. In: Reducing Risks, Promoting Healthy Life. The World Health Report 2002, 7–14. Geneva: WHO

___. 2003a. 2. Background. In: Diet, Nutrition and the Prevention of Chronic Disease: Report of a Joint WHO/FAO Expert Consultation. WHO Technical Report Series 916, 4–12. Geneva: WHO

___. 2003b. 5. Population nutrient intake goals for preventing diet-related chronic diseases. In: Diet, Nutrition and the Prevention of Chronic Disease: Report of a Joint WHO/FAO Expert Consultation. WHO Technical Report Series 916, 54–70. Geneva: WHO

9

Bakery: Yeast-leavened Breads

R. B. Swanson

BACKGROUND INFORMATION

The essential ingredients in yeast-leavened bread are wheat flour, water, yeast, and salt. However, most bread produced in the United States and elsewhere incorporates small amounts of additional ingredients. These nonessential ingredients allow the baker to compensate for flour deficiencies and the production procedures chosen and to extend shelf life. They may also add color or desirable flavor attributes that improve consumer acceptability. Sugar, shortening (fat), and milk or milk products are frequently added. Use of yeast foods, dough improvers including surfactants and enzymes, and mold inhibitors is common in commercially produced breads. White-pan bread, the most commonly produced bread in the United States, is the focus of this chapter.

WHITE-PAN BREAD VERSUS VARIETY BREADS

White-pan breads are identified as any bread, other than a variety bread. Variety bread formulations often include meals or grits other than wheat flour in varying proportions. Whole wheat, rye, oats, barley, and millet are typical grain choices. Other breads classified as variety breads include those leavened with a starter, such as sourdough and salt-rising breads. These types of variety breads rely on both

The information in this chapter has been derived from a chapter in *Food Chemistry Workbook,* edited by J. S. Smith and G. L. Christen, published and copyrighted by Science Technology System, West Sacramento, California, ©2002. Used with permission.

yeast and bacterial fermentation. White hearth breads, including the French, Italian, and Vienna types, also fall in the variety bread category. The flavor and the crust and crumb characteristics of variety breads differ to varying degrees from white-pan breads. Production procedures also vary, with the extent to which variety bread production is similar to that of white-pan bread depending on the specific product being made (Pyler 1988).

WHITE-PAN BREAD QUALITY CRITERIA

Loaf volume, expressed as cubic centimeters (cc) per unit of weight, is the major criterion used to assess bread quality. Loaf shape, height, and length and the relative proportions of loaf height and length are part of this quality assessment. In white-pan breads, the loaf should have a rounded top without sharp corners or protruding sides and ends. Both the thickness of the crust and the break and shred should be uniform. Break and shred refers to the rupture along the side of the loaf where the upper crust meets the sidewalls and the vertical streaking associated with this rupture. Desirable crust color ranges from the deep golden brown of the top crust to the light golden brown of the sides and bottom. A thin, tender crust is preferable in white-pan bread. A desirable crumb structure has small thin-walled, oval cells that are readily compressed. These crumb characteristics are associated with a large volume increase. Acceptable grain, which is defined as crumb cell size, can be either open or close; open grain is characterized by large individual cells, whereas close grain exhibits small cells. Grain that is uniformly open or close, or that exhibits a continuous range of sizes is acceptable. Crumb color should be a creamy white without streaks or spots. Flavor, which includes both taste and aroma, should be pleasing and characteristic of the grain in the formulation; it is assessed subjectively (Anonymous 1987, Pyler 1988)

RAW MATERIALS PREPARATION

Wheat flour comprises 55–60% of white-pan bread. Wheat flour characteristics are determined by the wheat(s) selected, the milling process, and the treatments applied postmilling.

WHEAT SELECTION

Selection among available wheats is based on the intended end use. Three commercially significant

wheat species are important in North America: *Triticum compactum* (club wheat), which is used in cake and pastry flours, *T. durum*, which is used in pasta production, and *T. aestivum*, the most common varieties, which are used in a wide range of wheat-based products in North America and elsewhere. It is the preferred wheat species wherever yeast-leavened breads and related dough-based products are produced.

T. aestivum classes include hard red winter (HRW), hard red spring (HRS), soft red winter (SRW), hard white (HW) and soft white (SW). Hard versus soft refers to kernel characteristics and is related to how tightly the starch granules are packed in the protein matrix as well as to the extent of adherence between the protein and the starch. Relative hardness of the wheat kernels influences milling characteristics. Hard wheats exhibit greater resistance to grinding than soft wheats during the milling process. Hard wheats are often, although not always, higher than soft wheats in protein. Red and white refers to kernel color, which is determined by whether or not there is a red pigment in the outer layers of the wheat kernel. Spring and winter refers to growth habit. Winter wheat, which requires below-freezing temperatures to form the grain heads, is planted in the fall and harvested in the spring. Spring wheats are planted in the spring and harvested in late summer or early fall. Spring wheats do not require below-freezing temperatures to form grain heads. Millers blend wheats for uniformity in protein content and baking quality (Atwell 2001). Typical tests conducted on wheat prior to milling include moisture, bulk density, protein, and sprout damage. The results of these tests help the miller determine the blend characteristics.

WHEAT KERNEL STRUCTURE

The wheat kernel is composed of three distinct parts: the bran, the germ, and the endosperm. The milling process is designed to separate the endosperm from the germ and the surrounding bran. The starchy endosperm comprises about 85% of the wheat kernel and is the major constituent of flour; it is moderately high in protein content and is the location of about 80% of the protein in the wheat kernel. The bran, which is high in fiber and mineral content (ash), comprises about 14% of the kernel by weight and includes the outer layers of the kernel. Millers consider the aleurone to be part of the bran. This specialized layer of enzymatically active cells separates the bran from the endosperm and is considered botanically to be part of the endosperm. The germ (embryo) contains most of the lipids and is high in nutrients (Atwell 2001). The wheat kernel is depicted in Figure 9.1.

MILLING

The wheat selected is milled with a dry process. When wheat arrives at the mill, it contains foreign material that will affect the appearance, functionality, and mill operation (Fig. 9.2). Cleaning occurs either before or after the blending process. Cleaning is usually a dry process involving several steps. Magnets are used to remove ferrous materials; a stoner removes foreign materials such as small stones and mud balls that differ in specific gravity from wheat. A milling separator screens impurities that are larger and smaller than the wheat kernels, such as corn, mustard seeds, or soybeans. Wheat kernels also undergo a dry scour. In this step, the wheat kernels are impelled against a screen to abrade the surface. This removes impurities in the crease, which are otherwise very difficult to eliminate.

The controlled addition of moisture for up to 36 hours, called tempering or conditioning, accentuates the differences in grain components (Fig. 9.2). At a moisture content of about 15–16%, maximum milling efficiency and optimum performance of the resulting flour in the final product is achieved. The bran becomes tougher and does not powder during grinding, which facilitates its removal. Tempering makes the endosperm more friable. The germ becomes more pliable so that it can be flattened, and therefore more easily removed in the subsequent sifting process. If not done earlier, blending of wheats may occur after tempering. Unsound wheat kernels are also removed at this point.

After tempering, the grain kernels are broken open by shearing as they pass through a series of five to six break rollers (Fig. 9.2). These corrugated break rollers rotate at different speeds in opposite di-

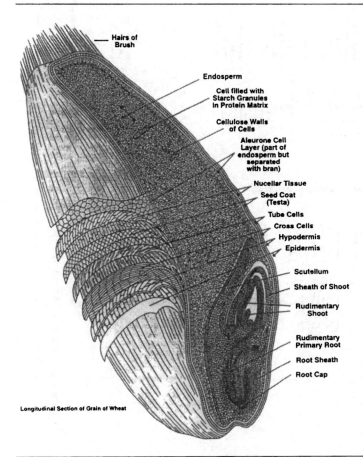

Hairs of Brush

Endosperm

Cell filled with Starch Granules in Protein Matrix

Cellulose Walls of Cells

Aleurone Cell Layer (part of endosperm but separated with bran)

Nucellar Tissue

Seed Coat (Testa)

Tube Cells

Cross Cells

Hypodermis

Epidermis

Scutellum

Sheath of Shoot

Rudimentary Shoot

Rudimentary Primary Root

Root Sheath

Root Cap

Longitudinal Section of Grain of Wheat

Figure 9.1. The wheat kernel. (Reprinted with permission from the North American Millers' Association, Washington, D.C.)

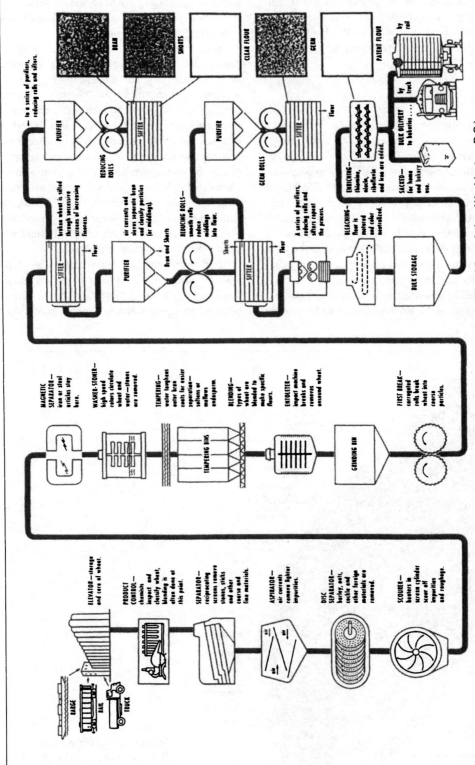

Figure 9.2. The wheat flour milling process. (Reprinted with permission from the North American Millers' Association, Washington, D.C.)

186

rections, breaking the wheat kernel into coarse particles and thereby exposing the endosperm. Following each break, the sheared and crushed kernels pass through a series of sieves that separates the material into three general size categories— the coarsest fragments are sent through the next break, the medium-sized particles are primarily endosperm and are known as middlings, and the finest particles are break flour. Because particle size is based in part on composition, sieving separates the components as well as the particles. With repeated passage through the break rolls, the amount of endosperm present in the coarse particles decreases. After the fifth or sixth break, the remaining coarse particles are primarily bran. Most of the germ is removed by the third break (Bass 1988).

Finally, the medium-sized particles (middlings) from all the breaks are passed through a series of smooth reduction rolls (Fig. 9.2). After each passage, in which the middlings are reduced in size and the adhering bran is loosened, the particles are sieved (or sifted). The flour-sized particles removed after each passage through the reduction rolls are known as a millstream. Each millstream has different characteristics, and millstreams can be blended to produce the grades of flour desired (Villanueva et al. 2001). The earlier millstreams, which tend to be higher in starch, may be combined to produce patent flour. Depending on the characteristics desired, 40–95% of the millstreams may be combined to produce patent flours. The remaining millstreams comprise clear flour. High quality clear flour, which contains a greater number of millstreams, is a creamy to grayish color, and may be used in variety breads. When all of the millstreams are combined, the resultant product is known as straight flour. One hundred pounds of cleaned wheat will yield about 72 pounds of straight flour and 28 pounds of by-products; thus, there is a 72% extraction rate. The by-products, also known as shorts, are composed of bran, germ, and some endosperm (Bass 1988).

The flour produced by milling is composed of endosperm agglomerates and fragmented starch and protein. The starch and protein content of any flour reflects the wheat or blend of wheats from which it was milled.

Because the sizes of the protein and starch particles are very similar, sieving does not separate these fractions. Further reducing the particle size of flour milled in the conventional process allows the protein and starch to be separated by a stream of turbulent air due to differences in particle size, shape, and density. Centrifugal force applied to the suspended particles yields two flour fractions that differ in starch and protein content. This latter milling process, known as air classification, allows production of flours that differ in relative proportions of starch and protein for specialized applications (Bass 1988)

POSTMILLING TREATMENTS

Postmilling treatment includes the incorporation of maturing and bleaching agents and enrichment. Oxidants such as benzoyl peroxide are added to bleach (whiten) the yellow pigments in the flour. Xanthophylls dominate the yellow pigments present. The bleaching effect is limited to the pigments present in the endosperm; any bran present, which is an indication of an inferior flour and is reflected in a higher ash content, resists the effects of the bleaching agents. Therefore, use of bleaching agents does not mask inferior flours. In the United States, maturing agents, including potassium bromate (at levels less than 50 ppm), azodicarbonamide (at levels less than 45 ppm), and ascorbic acid (at levels less than 200 ppm), which accelerate the natural aging process of the flour and improve baking quality, may also be incorporated. Some agents, for example, acetone peroxide, function as both bleaching and maturing agents. Legal limits as well as specific maturing agents allowed vary with country. Treatment levels vary with the wheat variety, the conditions of growth, and length of storage prior to milling. The degree of extraction during milling and the specific millstreams combined, as well as the intended end use and processing method chosen, also influence treatment levels selected. Although both maturing and bleaching can be accomplished naturally by storing flours for several weeks to months, the natural process is inconsistent. In addition, time and space requirements and the increased potential for insect infestation limit the practical use of storing flours rather than adding maturing and/or bleaching agents.

Flours may also be supplemented with enzymes, such as amylases and lipooxygenases, that improve their bread making performance. Lipooxgenases, added as soy flour, function as bleaching agents and dough improvers. Addition of α-amylase, in the form of diastatic malt or a fungal supplement, corrects a flour deficiency.

Enrichment, when added at the flour mill, is usually in the form of a premix containing the required nutrients. In the United States, the five required nutrients include thiamine, riboflavin, niacin, iron, and folic acid. Calcium, an optional enrichment nutrient, is sometimes added (21 CFR 137.165). Alternatively, bakers may choose to add enrichment, allowing the levels incorporated to be adjusted to specific formulations. This facilitates greater ease in meeting the mandated enrichment levels (Pyler 1988) The specific nutrients and their required enrichment levels vary with country.

FLOUR SELECTION AND FUNCTIONALITY

In general, the proximate composition of the flour depends primarily on the type of wheat. Hard wheat flours, such as hard red winter (HRW) and hard red spring (HRS) wheat, are about 82% starch, 12.5% protein, 3.5% fiber, 1.5% lipids, and 0.5% ash (Mathewson 2000). These flours are preferred for bread making in the United States. Lower protein flours are commonly used in Europe.

PROTEINS

The superiority of HRW and HRS wheat flours for yeast-leavened breads is attributed to the large amounts of high quality cohesive proteins present. Differences in protein content and quality among wheats affect loaf volume and the fineness, uniformity, and extensibility of the crumb grain (Zghal et al. 2001). For bread production in the United States, good quality protein at about the 12% level is desirable. Within wheat type, variety and environmental factors including nitrogen and sulfur availability, heat stress, water stress, and insect damage can influence protein quality. In addition, storage conditions can alter protein quality postharvest (Wrigley and Bekes 1999).

Wheat flour proteins have traditionally been sequentially extracted with salt solutions, 70% alcohol, 1% acetic acid, and reducing agents or alkali. Four fractions—albumin, globulin, gliadin, and glutenin are found. Although similar in solubility, none of the fractions is a single chemical entity. The albumin and globulin fractions each account for about 10% of the total flour protein. Gliadin and glutenin are known as the gluten proteins; these storage proteins account for about 80% of the protein present in flour. Levels increase as total flour protein increases.

The gluten proteins are responsible for dough properties. This viscoelastic protein complex, which is formed after hydration and mechanical manipulation, is responsible for structural support in yeast-leavened products. Factors that influence bread making quality are total amount of gluten proteins, relative proportion of gliadin to glutenin present, and the molecular weight distribution within each gluten protein fraction (Kolster and Vereijken 1993, Menkovska et al. 2002).

The hydrated gluten complex, which has been subjected to mechanical manipulation, is a cohesive, elastic, extensible fibrillar matrix covered with a protein membrane. Because gluten proteins form the continuous phase in dough, they govern dough properties. Gliadin, which is extensible and tacky, contributes extensibility and plasticity, whereas glutenin contributes elasticity and cohesiveness as well as extensibility (Wrigley 1994). It is the balance of these rheological properties that determines bread making quality of a particular wheat.

Glutenin exhibits a wide range of molecular sizes, up to tens of millions of Daltons (Da). It is one of the largest protein molecules in nature, and its large surface area appears to foster aggregate formation via intermolecular disulfide bonding. These polymers are composed of two main groups of polypeptide chains—high- and low-molecular-weight glutenin subunits. The ratio of these glutenin subunits affects dough rheology. Wheats with higher quantities of the high-molecular-weight glutenin subunits produce doughs with greater strength and stability, whereas increased levels of low-molecular-weight glutenins increase dough extensibility (MacRitchie 1999). Both mixing time and loaf volume increase with an increase in the ratio of high- to low-molecular-weight glutenin (Uthayakumaran et al. 2000).

Gliadin molecules interact with each other and glutenin (Fig. 9.3), limiting excessive interactions among glutenin polymers. Gliadin ranges in molecular size from 30,000 to 70,000 Da and consists mainly of monomeric proteins that interact via hydrogen bonding and hydrophobic interactions. These molecules are smaller, more globular, and more symmetrical than glutenin, and they exhibit a reduced surface area; therefore, they are less likely to interact with other proteins. Some gliadin fractions are relatively high in sulfur-containing amino acids, and these fractions also participate in intramolecular disulfide bonding. Some polymeric gliadin molecules, similar to glutenin subunits of lower molecular weight, are also present. These polymeric gliadin proteins are incorporated through

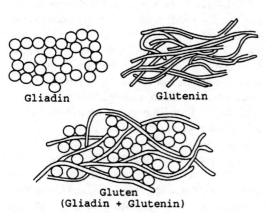

Gliadin Glutenin

Gluten
(Gliadin + Glutenin)

Figure 9.3. Schematic representation of gliadin, glutenin, and gluten (Bietz et al. 1973).

intermolecular disulfide bonding into glutenin (Huebner et al. 1997). Increased gliadin levels tend to decrease mixing time and resistance to dough breakdown while increasing dough extensibility. Loaf volume is decreased (Uthayakumaran et al. 2001).

Development of the gluten structure involves disulfide (-S-S-) and sulfhydryl (-SH) groups. Existing disulfide bonds are broken, and new ones are formed. With the formation of new disulfide bonds, glutenin polymers align, forming the basic structure of gluten network. The sulfhydryl groups reduce the disulfide bonds, facilitating molecular rearrangement. The covalent bonds formed primarily contribute cohesiveness to the gluten complex (Buskuk et al. 1997). In the presence of oxidants or maturing agents, -SH groups are removed, stabilizing the dough structure. Although the mechanism of action appears to be the same for all oxidants, their effects vary because they act for different time periods and at different stages of dough development. Some, like azodicarbonamide, have their effect during mixing; others, such as potassium bromate, react during fermentation and baking. Oxidants may be added during bread production, as well as during flour milling (Pyler 1988, Ranum 1992). Labeling requirements vary with the agent selected.

Although covalent bonding is responsible for continuity of the gluten network (Bloksma 1990), other types of bonding also play a structural role. Inter- and intramolecular hydrogen bonding, due primarily to the high number of amide groups provided by the amino acid glutamine in both glutenin and gliadin, is also an important contributor to the rheological

properties of gluten. The relatively high levels of amino acids with aliphatic and aromatic groups result in hydrophobic interactions as well. Hydrogen and hydrophobic bonding contribute elasticity and plasticity. Ionic bonding, due to the presence of charged amino acid residues, also contributes cohesiveness by increasing dough rigidity and reducing extensibility. In commercial bread production, sodium chloride and other mineral salts incorporated as yeast food make additional ions available (Pomeranz 1988). Gluten content and strength may be further enhanced by incorporation of a dry form of gluten, vital gluten, during bread production (Ranhotra et al. 1992, Weegels and Hamer 1992).

Water-soluble proteins, which include albumins and globulins, comprise only 10–15% of the flour proteins. They are important sources of flour enzymes (Pyler 1988). Enzymes impact flour and dough properties, in particular dough elasticity and stickiness, gassing, and the final crumb structure in breads (Mathewson 2000, Obel 2001) In addition to native enzymes, nonwheat sources of several enzymes are typically added during either milling or bread production to enhance flour functionality. Lipooxygenases and amylases are often incorporated, and proteases may be added. Denaturation temperatures vary with the enzyme source. When enzymes denature at oven temperatures, they are not considered part of the final product and therefore are not listed on the ingredient label.

In sound wheat, protease activity is low and probably has little impact on bread making. Therefore, supplementation with proteases, which act on gluten proteins, is necessary if their effects on dough rheology are desired. These compounds are added during bread production. Proteases reduce mixing time by reducing the resistance to mixing. They improve flow characteristics of the dough by decreasing the elasticity introduced by machine mixers. Finally, they improve gas retention by increasing the extensibility and pliability of the gluten complex. They may also affect product color and flavor by providing new amino acid groups to participate in the Maillard reaction (Mathewson 2000).

Although naturally occurring in flour, additional lipooxygenase is usually added as a soy flour supplement. The substrate on which lipooxgenase acts varies with the enzyme source (Pyler 1988). Lipooxegenase also influences color by oxidizing the yellow flour pigments. In addition, it strengthens the gluten and increases dough mixing tolerance. The oxidative effect of lipooxygenase on polyunsaturated fatty acids, tocopherols, carotenoids, and ascorbic acid may have a minor negative impact on nutritive quality. Although the resultant compounds also typically play a role in oxidative rancidity, levels present in bread are low and actually contribute to the wheaty flavor of bread (Mathewson 2000, Pyler 1988). Amylases, the most abundant enzymes in flour, are discussed in the carbohydrate section, below.

CARBOHYDRATES

Starch forms the bulk of the bread dough and has several important roles in its structure. The surface of the starch granule interacts to form a strong union with gluten. Starch also dilutes the gluten to the desired consistency. Further, it is a source of sugar (maltose) through the action of amylase on the starch granules damaged during the milling process (Sandstedt 1961).

Gelatinization is the process in which starch granules absorb water, swell, and break down, releasing amylose from the granule (Atwell et al. 1988). Gelatinization of the starch, which occurs at 60–70°C (140–158°F), allows the gas-cell film to stretch. Thus, the starch competes with gluten for water, resulting in setting and rigidity of the gluten film (Sandstedt 1961). In bread dough, the starch granules are embedded in the fibrillar protein network (Bechtel et al. 1978, Bloksma 1990).

Amylases can hydrolyze α-1,4-glycosidic linkages in carbohydrates, including starch. In flour, this activity is influenced by the degree of starch damage during the milling process. Therefore, starch damage is carefully controlled in the milling process because excessive levels of damage and the resulting amylase activity are detrimental to bread quality. When damaged starch is hydrolyzed by amylase, absorbed water is released, making the dough softer (Mathewson 2000, Obel 2001).

Amylases, which produce maltose subsequently used in fermentation, may be present naturally or added during flour milling and/or bread production. Any residual sugar remaining postfermentation can participate in the Maillard reaction during baking. Although wheat flour contains significant amounts of β-amylase, it is usually deficient in α-amylase. Beta-amylase, an exoenzyme, systematically splits the starch amylose chains into maltose units; starch amylopectin chains are split into maltose units and dextrins because β-amylase does not act within the branch point of the amylopectin molecule. Alpha-amylase, an endoenzyme, which acts at random on the α-1,4-glycosidic linkages of both amylose and amylopectin, yields products that range in size from maltose to oligosaccharides. In the United States, malted barley flour and diastatic malt syrup, the traditional sources of amylases, are customarily added at the flour mill and during bread production, respectively. In addition, microbial amylases, which are added by the baker or miller, have become available in recent years (Mathewson 2000).

Both water-soluble and water-insoluble hemicelluloses are present in wheat flour. This flour component is often referred to as pentosans because polymers of the pentose sugars D-xylose and L-arabinose dominate. Wheat flour contains 2–3% pentosans, 75–80% of which are insoluble in water. Water-insoluble pentosans improve crumb uniformity and elasticity, although deleterious effects on crumb grain and texture have also been reported. Water-soluble pentosans help regulate hydration, dough development characteristics, and dough consistency (Shelton and D'Appolonia 1985)

LIPIDS

Both free and bound lipids are present in flour. Each lipid fraction is composed of polar and nonpolar lipids, although the ratio differs among lipid fractions. The polar lipids include glycolipids and phospholipids. The nonpolar lipids are mainly triglycerides. Although constituting a small percentage of the flour weight, the polar lipids, specifically glycolipids, play an important role in dough development. The glycolipids are bound to gliadin through hydrogen bonding and to glutenin through hydrophobic interactions in the dough (Fig. 9.4). During baking, the polar lipids are translocated and bound to starch. Presence of these polar lipids facilitates proper dough expansion during fermentation and baking (Chung 1986, Chung et al. 1978).

OTHER ESSENTIAL BREAD INGREDIENTS

WATER

Water is the most common liquid used in commercial baking. It comprises approximately 33–40% of the dough by weight. Water is responsible for hydration of the dry ingredients in the bread formula and for forming the gluten complex during mixing. Starch granules absorb water, facilitating gelatinization during baking. Water also serves as a dispersing

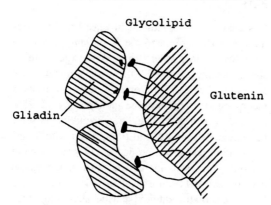

Figure 9.4. Model of the glutenin-glycolipid-gliadin complex (Chung et al. 1978)

medium for other ingredients, including yeast, and as a solvent for solutes (salt, sugar). Mineral salts naturally occurring in water may affect bread dough properties. Their importance is determined by the mineral concentration and the strain of yeast used. Hard waters containing high levels of calcium and magnesium ions may toughen the gluten, resulting in a tightening effect on dough.

However, fermentation may be retarded if soft water is added. Soft waters, which lack these minerals and may be slightly acidic, may produce soft, sticky dough with impaired gas retention. Fermentation rate may be increased if excessively alkaline water is used, which retards yeast enzyme activity (Anonymous 1995, Pyler 1988). The pH of natural water is between 6 and 8; most municipal water supplies are adjusted to a pH between 7.1 and 8.5. Periodic fluctuations in pH and mineral composition often occur even if the same source is used (Pyler 1988).

Chilled water is often used in commercial operations to avoid excessive heat from mixing and dough development process. Temperatures above 27°C (81°F) during dough development may over stimulate the yeast and adversely affect the gluten and starch. Alternatively, a portion of the water may be added as ice or a mechanically refrigerated mixer bowl may be used. Upon removal from the mixer, desirable dough temperatures are 25.5–27°C (78–81°F) (Pyler 1988).

YEAST

The primary baker's yeast is *Saccharomyces cerevisiae*. Several different strains are available. Different available forms include compressed, active dry, and instant. These forms differ primarily in moisture content and the need for refrigerated storage. Handling requirements also differ. During dough fermentation, yeast has three major functions: leavening, dough maturation, and flavor development. Leavening involves the enzymatic conversion of fermentable sugars into ethanol and carbon dioxide. Yeast, a living organism, is capable of both anaerobic and aerobic fermentation. However, anaerobic fermentation dominates because the oxygen present is rapidly consumed.

Fermentable sugars include glucose, fructose, and mannose. However, sucrose and maltose are the sugars typically present in yeast doughs. In addition to a small amount of sugar present in the flour, sucrose is added to yeast doughs, and maltose is formed when amylases act on the flour starch. During bread production, sucrose is rapidly converted to glucose and fructose by invertase (sucrase), and maltose is hydrolyzed to glucose by maltase. Both invertase and maltase are yeast cell enzymes. Because maltose is not hydrolyzed in the presence of glucose, it is available for use only when other sources of sugar have been greatly reduced. Thus, maltose is an important source of fermentable sugars only in the later stages of dough fermentation (DuBois 1984, van Dam and Hille 1992).

CO_2 is produced by yeast in the aqueous phase, and it is initially dissolved in this phase. However, once the water is saturated, additional CO_2 produced moves into preexisting gas bubbles. Because the aqueous phase remains saturated with CO_2, the CO_2 in the gas bubble cannot diffuse out. The viscoelastic properties of the dough allow the gas bubbles to stretch (Hoseney 1994). Therefore, gas production and retention during dough fermentation have significant effects on the loaf volume (Collado and De Leyn 2000).

Gas production rate is a function of temperature and pH. It increases as temperature increases up to 35°C (95°F). Gassing rate is also maximized when pH is between 4.5 and 5.5 (DuBois 1984). Excess sugar (> 5% flour-weight basis) and salt (> 2% flour-weight basis) decrease gassing power through an osmotic effect on yeast cells. Balancing relative levels of salt and sugar in the bread formula allows the fermentation rate to be controlled (Pyler 1988, van Dam and Hille 1992).

Yeast influences on dough maturation include lowering pH, altering interfacial tension in the dough phases due to ethanol production, and physically weakening dough as CO_2 expands (Pyler 1988). The distinctive flavor of yeast-leavened breads is partially due to the metabolic products of yeast. (Labuda et al. 1997, van Dam and Hille 1992). Yeast cells are uniformly distributed in the spaces between sheets of gluten (Bechtel et al. 1978).

SALT

Salt (sodium chloride) has three major functions in yeast-leavened breads: flavor, inhibition or control of yeast activity, and strengthening of gluten. An additional effect is its inhibitory action on spoilage microorganisms. A typical level of usage is 2% on a flour-weight basis. Flavor effects include imparting a salty taste or eliminating a flat, insipid taste, increasing the perception of sweetness, masking off flavors, and improving flavor balance (Gillette 1985). Yeast activity is retarded by salt through an osmotic effect on yeast cells, promoting more even fermentation with temperature fluctuations. Because salt reduces the rate of gas production, the proof time necessary to achieve the desired loaf volume is increased. Under ideal temperature conditions, this reduction in gas production rate may be negative (Pyler 1988). However, when salt incorporation is inadequate, the yeast ferments excessively. The result is dough that is gassy and difficult to process. In the absence of salt, the crumb of the baked loaves has an open grain and poor texture (Matz 1992). Salt's strengthening effect on gluten is particularly beneficial when soft water is used. The strengthening effect may be through direct interaction with the flour proteins (Galal et al. 1978). Salt's effects on dough simulate those of oxidizing agents.

OPTIONAL INGREDIENTS

SUGAR

Sucrose or corn syrup is usually incorporated in yeast-leavened breads to increase the rate of initial fermentation; otherwise, amylolytic activity is required to produce a substrate for the yeast. Flavor and color effects are also found when sugar is incorporated at higher levels. Reducing sugars not used by yeast are available for Maillard browning of the crust during baking. (DuBois 1981, Shelton and D'Appolonia 1985). In commercial yeast-bread production, liquid sugars are often used. Liquid handling systems afford easier and more sanitary handling.

FATS

Fats are optional ingredients in yeast-bread production. When incorporated, fat is present as discrete particles uniformly distributed between gluten sheets (Bechtel et al. 1978). Desirable qualities are achieved when 2–5% fat on a flour-weight basis is incorporated. Bread volume increases by 15–25% because fat allows the dough to expand longer prior to setting. Palatability is also improved with fat incorporation into bread products. The grain is more uniform, fineness is increased, moisture perception is increased, and texture is softened. Flavor may also be enhanced. In addition, plastic shortening may strengthen the sidewalls of bread, minimizing misshapen final loaves. An effect on shelf life is also found, as the resulting softer texture decreases the perception of staling (Stauffer 1998). Shortenings used in bread products may also be carriers for dough conditioners (surfactants).

YEAST FOODS

Yeast foods, a mixture of inorganic salts, are added for two major purposes: (1) to adjust the mineral composition of water and (2) to provide nitrogen and minerals for yeast. Typical active ingredients are yeast nutrients (ammonium salts), oxidants (usually potassium bromate or iodate), and pH regulators. Selection is based on flour, water supplies, and mixing and fermentation practices. The effect on physical dough characteristics results in increased loaf volume and symmetry, a finer grain, and a softer texture (Pyler 1988).

SURFACTANTS

Surfactants (or surface-active agents) act primarily as dough conditioners and staling inhibitors. Use of surfactants in yeast-leavened breads results in an increase in bread volume, a crust and crumb that are more tender, a finer and more uniform crumb cell structure, and staling inhibition (Pyler 1988). The most commonly used surfactants are mono- and diglycerides, which primarily contribute softness. Both crumb softening and anti-staling effects are due to a complex with amylose. Sodium stearoyl lactylate (SSL), another widely used surfactant, complexes with gluten proteins during gluten development, strengthening the dough. The increased dough strength improves dough-handling properties and increases tolerance to processing variables.

Other commonly used dough strengtheners are lecithins, ethoxylated monoglycerides, sorbitan monosterate, and diacetyl tartaric esters of mono- and diglycerides (Anonymous 1995).

Surfactants are amphophilic, that is, they have both hydrophilic and hydrophobic groups. Therefore, surfactants serve as a bridge between immiscible phases. The relative proportion of the hydrophilic and hydrophobic groups differs among surfactants. Monoglycerides and SSL are examples of surfactants that are predominantly hydrophobic and hydrophilic, respectively (Anonymous 1995).

MOLD INHIBITORS

Commercial bakery products, including breads, usually contain mold inhibitors. Calcium propionate, a naturally present metabolite in Swiss cheese, is most commonly used in yeast-leavened products. Typical levels are between 0.25 and 0.38% flour-weight basis in breads and rolls. Levels used are self-controlling, as high levels are associated with increased proof time and a cheese-like flavor (Pyler 1988, Ranum 1999). In addition to inhibiting mold growth, this compound also increases the calcium content of the bread. Sorbates may also be used as mold inhibitors. Unlike calcium propionate, sorbates are typically sprayed on the surface of baked products because of their inhibitory effect on yeast growth (Ranum 1999).

MILK PRODUCTS

Incorporation of milk products in yeast-leavened bread can improve both its nutritional and its eating quality. Nonfat dried milk, dairy blends, or dairy substitutes, which are incorporated in their dry form, are most often used commercially. Dairy blends are a combination of various dairy products including whey, caseinates, and nonfat dried milk. Dairy substitutes may include these dairy-based components as well as additional ingredients such as soy and/or corn flours, and soy protein. Both have been formulated to equal nonfat dried milk in functionality at a lower cost. Typical levels of up to 6% on a flour-weight basis are used. Nonfat dried milk incorporation at the 6% level reportedly increased loaf volume and improved grain and texture, crust color, and break and shred. Depending on the level of use and the bread production system employed, modification of the bread formula may be necessary to optimize resultant bread quality (Pyler 1988).

Available nonfat dried milks are classified as low-heat, medium-heat, and high-heat types. These milks differ in the extent to which the fluid milk is preheated prior to drying. High-heat nonfat dried milks are necessary to obtain good baking quality in yeast-leavened breads. Dry milks that have not been subjected to the preheating process produce slack doughs and result in decreased loaf volumes. However, incorporation of high-heat nonfat dried milk strengthens the gluten structure because the casein proteins present interact with the flour gluten proteins. Fermentation tolerance is also improved, which ensures more consistent bread quality on a day-to-day basis, and crumb firming is retarded postbaking, enhancing shelf life.

Nonfat dried milk incorporation also increases total production time because mixing, fermentation, recovery, and proofing times are increased. Longer fermentation and proofing times can be overcome by slightly increasing yeast levels. Additional moisture incorporation is also necessary to allow for hydration of the dry milk product present. In addition, baking times and temperatures should be adjusted to avoid excessive browning of the crust due to the increased level of residual sugars (lactose) present. When used at levels higher than 4%, adjustment of dough pH eliminates deleterious effects on quality (Pyler 1988).

BREAD PRODUCTION PROCEDURES

The sponge and dough method is the most popular bread production method in the United States. Other bread production techniques include the straight-dough process and the continuous process (Do-Maker and Am-Flow). The straight-dough method is used primarily in smaller commercial operations and for variety bread production. At present, the continuous process has limited use for yeast-bread production in the United States, and the equipment required is no longer being manufactured. Variations on this process continue to be used elsewhere.

Accelerated dough making procedures have also been developed. The Chorleywood procedure, a short-time procedure in which the dough is mixed under a partial vacuum, is popular in the United Kingdom. A no-time procedure that uses increased oxidant levels to speed development is common in Australia (Hoseney 1994). These accelerated dough making procedures eliminate the time-consuming bulk fermentation step in the sponge and dough or straight-dough processes. While the sponge and dough process takes about 6.5 hours of production time, and the straight-dough procedure is completed in approximately 4.5 hours, the short-time and no-time procedures require as little as 2 hours. Other adjustments required when accelerated dough production procedures are used include an increase in the level of yeast, a reduction in the sweetener level, and an increase in dough absorption. Further information on these expedited production systems can be found in Pyler (1988, 699–706).

SPONGE AND DOUGH PROCEDURES

The process outlined below is for the sponge and dough method, which yields soft bread with a fine cell structure and a well-developed flavor. It is the standard used for comparing the quality of various breads in the United States.

SPONGE FORMATION AND FERMENTATION

A portion of the flour (50–70% of the total), part of the water, the yeast, the yeast food, and any enzyme supplement used (Fig. 9.5A) are mixed to form a smooth, homogenous mass; gluten development is limited to that necessary to retain the gas produced by the fermenting yeast. This undeveloped dough is the sponge (Fig. 9.5B). Consistency ranges from stiff to soft, depending on the proportion of ingredients incorporated. Subsequently, the sponge is combined with the remaining dough ingredients, and the bread dough is developed. The major fermentative activity of the yeast occurs in the sponge.

The sponge is typically allowed to ferment for three to five hours at 23–26°C (74–78°F) and 75–80% relative humidity (Fig. 9.5C). Fermentation time increases as the percentage of the flour incorporated decreases. During fermentation, the volume of the sponge increases four to five times and the sponge ultimately collapses. At this point, 66–70% of the time required for sponge fermentation has elapsed. Additional fermentation time is required for optimal sponge development. pH is reduced with fermentation, and gas retention properties of the flour, vigorous yeast action, and flavor are developed. The desirable temperature for the fermented sponge is about 30°C (86°F) (DuBois 1981, Pyler 1988).

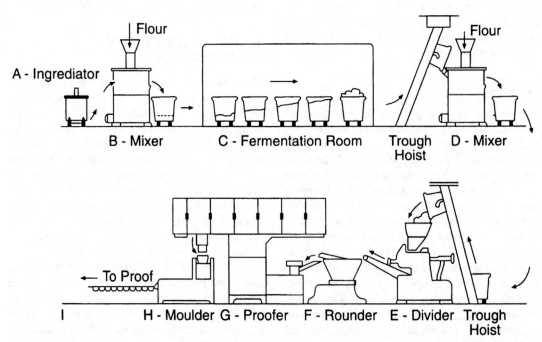

Figure 9.5. The sponge and dough bread production line: **A.** Ingrediator, **B.** Sponge mixer, **C.** Sponge fermentation room, **D.** Dough mixer, **E.** Dough divider, **F.** Dough rounder, **G.** Intermediate proofer, **H.** Dough moulder, **I.** Conveyor to the final proofer and oven (Seiling 1969).

ADDING AND MIXING THE NONSPONGE INGREDIENTS

Next, the fermented sponge is returned to the mixer and combined with the remaining ingredients, except salt (Fig.9.5D). Salt is typically incorporated during the last two to three minutes of mixing (DuBois 1981).

DOUGH DEVELOPMENT

Mixing continues until optimum development or optimum hydration has occurred (Hoseney 1985). Initially, the objective of mixing is to uniformly blend all the dough ingredients. This produces wet and sticky dough. As mixing continues and the gluten structure begins to form, the dough becomes drier and more elastic, and the dough mass becomes cohesive (Pyler 1988). Air incorporated decreases dough density and forms cells into which CO_2 later diffuses (Hoseney 1985, 1994). In the final stage of mixing, the dull dough acquires a satiny sheen and can be stretched into a smooth, uniformly thick sheet of dough. The dough has a dry appearance. Overmixing results in dough that is increasingly less elastic and more soft and extensible. Overmixed dough will pull into long, cohesive strands. Continued mixing results in dough disintegration (Pyler 1988)

Neither overmixed nor undermixed doughs hold up well in subsequent bread production operations. Factors that influence the time required for optimum development include flour strength, use of oxidizing and reducing agents, time of salt addition, enzyme supplementation, temperature, absorption level, sponge consistency, and pH (Galal et al. 1978, Hoseney 1985, van Dam and Hille 1992). The dough is allowed to rest prior to dough makeup (DuBois 1981).

DOUGH MAKEUP

DOUGH DIVISION AND ROUNDING

The first step in dough makeup is dividing the bulk dough into individual units of predetermined size (Fig. 9.5E). Because dough is divided on a volumetric basis rather than by weight, the entire process must occur within 20 minutes to ensure individual units of equal size. The individual dough pieces are irregular in shape. The cut surfaces are sticky, because aeration has been reduced due to the compression and shearing that occurs as the bulk dough is di-

vided (Pyler 1988). Compression also serves to subdivide the gas cells, resulting in a finer grained product (Hoseney 1985).

The rounding step (Fig. 9.5F), which follows, imparts a continuous nonsticky skin that facilitates the retention of CO_2 within the dough unit (while CO_2 continues to be produced by the yeast). The rounding operation also realigns the glutenin fibrils that were disrupted during the dividing step. In addition, it redistributes the gas cells, which results in bread that has a finer crumb structure and is more symmetrical in shape (Pyler 1988).

INTERMEDIATE PROOF

The rounded dough pieces are allowed to undergo a brief rest period. This recovery period usually lasts from 4 to 12 minutes, often under ambient temperatures and humidity conditions. It is commonly referred to as the intermediate proof (Fig. 9.5G). Dough that has undergone an intermediate proof, exhibits increased pliability and elasticity, and the surface is dry. These dough characteristics are essential to successful molding (Pyler 1988).

SHEETING, MOLDING, AND PANNING

The resulting dough units are shaped and molded prior to panning (Fig. 9.5H). First, the dough is sheeted or passed between closely spaced rollers to yield a thin and uniform dough layer. This step expels gas and redistributes gas cells, influencing final crumb grain. Next, the sheeted dough is curled into a relatively tight cylinder. Finally, the dough cylinder is subjected to pressure to lengthen the dough unit, seal the seams, and expel any trapped air. The molded and shaped dough units are deposited into the baking pans with the seam side down to prevent opening of the loaf during final proofing and baking (Pyler 1988).

FINAL PROOFING

Final proofing conditions include temperatures in the range of 32–54°C (90–130°F) and a relative humidity of 60–90%. Proof times typically range from 55 to 65 minutes (Fig. 9.5I). Generally, dough units are proofed to height or volume rather than for a fixed time. The dough has limited flow properties; thus, the volume increase is due to expansion. Flour strength, oxidant and dough conditioner selected, melting point of shortening selected, and conditions during dough development and makeup (including the degree of fermentation) all influence the final proofing conditions chosen (Pyler 1988).

FINISHED PRODUCT

BAKING

The overall objective of the baking process is to transform the dough into a light, porous, flavorful product. Acceptable results require baking temperatures of 191–235°C (375–455°F) (Marston and Wannan 1976, Pyler 1988). Baking time and temperature are influenced by dough formulation. Lean doughs are baked at higher temperatures for shorter periods of time. Rich doughs, high in sugar and dairy ingredients, will brown excessively if baked under conditions used for lean doughs (Pyler 1988).

The first change in the panned dough unit is the formation of an expandable surface skin. Steam injection is essential during the initial stage of baking. It prevents premature formation of a dry, inelastic skin that inhibits loaf expansion and results in tears in the crust. Steam also provides the moisture necessary for surface starch to undergo gelatinization, resulting in the desirable glossy crust. In addition, steam facilitates more rapid movement of heat into the loaf (Marston and Wannan 1976).

The initial increase in dough temperature accelerates enzymatic activity and growth of yeast. The dough becomes increasingly fluid as enzymes degrade starch granules in the crumb, although swelling of starch granules in the crumb is limited until temperatures approach 70°C (158°F). When the dough temperature has reached 50–60°C (122–140°F), yeast and bacteria have been killed, and most enzymes are inactivated. The surface skin also thickens and becomes less elastic. Loaf volume increases by about one-third the volume of the panned, unbaked dough unit. This rapid increase in volume in the initial phase of baking is commonly called oven spring (Marston and Wannan 1976, van Dam and Hille 1992).

Oven spring is caused by movement of CO_2 from the aqueous dough phase into the preexisting gas cells. Most of the CO_2 is produced by yeast during fermentation. This movement begins at about 49°C (120°F). Pressure exerted by the gas within the cells increases with temperature increase. The gluten film surrounding the gas vacuoles stretches as the gas cells expand. Bread made with stronger flours exhibits greater resistance to gas-cell coalescence, re-

sulting in a final loaf with fewer crumb defects (Zghal et al. 2001). Cell expansion is also influenced when alcohols, principally ethanol, are converted to gases; vaporization of water also plays a role. Volume increases continue until temperatures of about 79°C (175°F) are reached (DuBois 1984, Pyler 1988). In doughs that contain shortening, the fat interacts with starch and gluten, further increasing oven spring (Stauffer 1998).

During the second stage of baking, the crumb approaches 100°C (212°F) (Martson and Wannan 1976). During this stage, the primary changes are moisture evaporation, starch gelatinization, and protein coagulation. Starch granules gelatinize when the dough temperatures reach 60–70°C (140–158°F), whereas gluten and other proteins denature at 80–90°C (176–194°F) (Stauffer 1998). Because the denatured protein loses its water-binding capacity, water is transferred from the protein to the starch. Absorption of water by the starch facilitates gelatinization; gluten and starch interact to produce semirigid films that surround the gas cells. As the gas cells expand, the starch granules elongate, and the gluten film stretches until it ruptures (Hoseney 1994, Pyler 1988).

The third phase of baking is characterized by firming of the crumb cell walls at about 95°C (203°F) and the development of the desired crust color as temperatures reach 160°C (320°F) (Marston and Wannan 1976). Crust browning is attributable to both caramelization and Maillard reaction. Both reactions also contribute flavor and aroma compounds (Mathewson 2000). However, the contribution of caramelization is a minor one. The flavor of yeast-leavened bread is complex (Chang et al. 1995). Ingredients, fermentation products, mechanical degradations, chemical degradations, and thermal reaction products all play a role (Labuda et al. 1997, Shelton and D'Appolonia 1985).

STALING

Bread quality rapidly deteriorates after baking, resulting in decreased consumer acceptance and economic losses. This overall deterioration in quality is typically referred to as staling. It includes loss of flavor, toughening of the crust, firming of the crumb, an increase in crumb opacity, and a decrease in soluble starch (Hoseney 1994). The crust of freshly baked bread is crisp and dry. During staling, the crust toughens due to the migration of water from the crumb to the crust. The result is a soft and leathery crust and a firm crumb. Resistance of the bread crumb to deformation (firmness) is the attribute most commonly used to assess staling. Firmness is assessed by sensory evaluation as well as with instruments.

Instrumental tests involve the compression of bread samples one or more times between parallel plates. In addition to textural changes, there is a decrease in the pleasant aromas and flavors associated with fresh bread. Stale bread also develops a flat, papery trait and a bitter taste; alteration in flavor is assessed by sensory panelists (Caul and Vaden 1972).

Crumb staling has long been attributed to starch retrogradation. Retrogradation, which involves both amylose and amylopectin, is the recrystallization of the starch. Starch granules swell during baking, and amylose partially escapes from the granule, while the amylopectin becomes distended. The softness of the fresh product is attributed to extensible starch granules interlaced in a gel network of amylose. During storage, amylopectin molecules associate within the swollen starch granules, which results in more rigid granules and a firmer crumb. Crystallite growth also causes increased opacity of the crumb because of a change in the refractive index (Hoseney 1994).

Protein appears to play a role in bread staling as well, with protein quality influencing the rate of crumb firming. As the starch granules swell, hydrogen bonds form between the partially solubilized starch molecules and gluten, resulting in firming.

Because flour with lower quality protein tends to be more hydrophilic, it can form stronger bonds, and firming rate is increased. In addition, lower quality protein is usually associated with lower loaf volumes, which allows for increased starch concentration that promotes association within the amylopectin molecules and between starch and gluten (Martin et al. 1991). Incorporation of shortening and mono- and diglycerides decreases the rate of bread firming by decreasing the swelling of starch granules (Stauffer 1998, Martin et al. 1991). This, in turn, decreases the starch surface area exposed and results in fewer starch-gluten cross-links (Martin et al. 1991). Incorporation of bacterial or fungal α-amylase, which results in the formation of low-molecular-weight dextrins also retards bread firming. These intermediate-sized dextrins appear to interfere with the starch and protein association. Similar effects are not found with α-amylase supplementation with malted barley flour. This traditional source of α-amylase, which is incorporated during milling to supplement the low levels naturally present in wheat, produces larger dextrins. Indeed, it has been proposed that these larger dextrins actually act to cross-link protein fibrils. Therefore, malted barley flour may actually enhance the rate of bread firming (Martin and Hoseney 1991). Moisture content also affects crumb firming; the higher the moisture content, the slower the firming rate and the lower the firmness of the final product (He and Hoseney 1990).

APPLICATION OF PROCESSING PRINCIPLES

Processing Stage	Processing Principle(s)	References for More Information on the Principles Used
Ingredient selection	Milling, flour composition, protein quantity and quality, postmilling flour treatments	Atwell 2001, Bass 1988, 21 CFR 137.105, 21 CFR 137.165, Obel 2001, MacRitchie 1999, Wheat Protein symposium (papers) 1999, Wrigley 1994
Essential		
Flour	Leavening, reducing sugars, amylases, pH, flavor, aerobic/anaerobic conditions	van Dam and Hille 1992, Labuda et al. 1997
Yeast	Osmotic effect, flavor, antimicrobial	Galal et al. 1978
Salt	Hydration, pH, mineral salts, solvent dispersion	Pyler 1988
Water	Palatability, yeast fermentation	DuBois 1981, Shelton and D'Appolonia 1985
Nonessential		
Sugar	Palatability, yeast fermentation	DuBois 1981, Shelton and D'Appolonia, 1985
Fat	Palatability, crumb structure, loaf volume, shelf life	Stauffer 1998
Yeast food	Water mineral composition, dough rheology	Pyler 1988
Surfactants	Dough conditioning, staling inhibition, bread quality	Anonymous 1995, Pyler 1988
Mold inhibitors	Mold growth, shelf life, nutritional quality	Ranum 1999
Milk products	Palatability, nutritional quality, loaf volume	Pyler 1988
Sponge formation/ fermentation	Hydration, fermentation, pH	DuBois 1981; Hoseney 1985, 1994; Pyler 1988
Adding and mixing of nonsponge ingredients	Hydration, temperature, osmotic effect	DuBois 1981; Hoseney 1985, 1994; Pyler 1988
Dough development	Gluten development, air incorporation, oxidizing/ reducing agents, surfactants, enzymes, water absorption, temperature	Hoseney 1985, 1994; Pyler 1988
Dough division/rounding	Crumb grain	Hoseney 1985, Pyler 1988
Intermediate proof	Dough rheology	Pyler 1988
Sheeting, molding, panning	Crumb grain, loaf symmetry	Pyler 1988
Proofing	Temperature, relative humidity, fermentation	Pyler 1988
Baking	Yeast/enzyme activity, moisture content/vaporization, gas diffusion/solubility, starch gelatinization, protein denaturation, Maillard browning, flavor	Chang et al. 1995, Hoseney 1994, Labuda et al. 1997, Marston and Wannan 1976, Mathewson 2000
Shelf life	Staling—flavor and texture (starch and protein fractions, amylases), mold inhibition	He and Hoseney 1990, Hoseney 1994, Martin and Hoseney 1991, Martin et al. 1991, Ranum 1999

ACKNOWLEDGMENTS

Thanks are expressed to Marcy L. McEleveen for her assistance in preparing this chapter.

GLOSSARY

Accelerated dough making procedures—bread production procedures that exhibit a reduction in the time required for bulk dough fermentation. In general, involves intensive high-speed mixing and/or chemical dough development. Common in Great Britain and Australia.

Aleurone—specialized layer of enzymatically active cells that separates the starchy endosperm in wheat from the bran; botanically part of the endosperm, considered to be part of the bran by millers. Removed in the milling process.

Amylose—the essentially linear starch molecule composed of glucose units joined via α-1,4-glycosidic linkages.

Amylopectin—the branched starch molecule composed of glucose units joined via α-1,4-glycosidic linkages with α-1,6-glycosidic linkages at the branch points.

Bran—one of three anatomical parts of the wheat kernel; the outermost kernel layers made up of the outer pericarp and seed coat that are high in cellulose, hemicellulose, and minerals. Removed in the milling process, a major component of the shorts or mill feed.

Carbonyl-amine browning—Maillard browning; a key contributor to the flavor, aroma, and brown crust color of baked yeast breads. Nonenzymatic browning caused by the heat-induced reaction of an amine from a protein with a reducing sugar.

Continuous dough process—bread production procedure that produces bread with a fine, uniform cell structure that lacks characteristic bread aroma and flavor. The procedure uses a liquid preferment to maximize yeast fermentation, the preparation of a preliminary dough containing high levels of oxidizing agents that is combined with the preferment and remaining dough ingredients to form the final dough, and high-speed dough development under pressure.

Clear flour—millstreams remaining after those combined to produce patent flour are removed; 5–60% of the millstreams.

Endosperm—one of three anatomical parts of the wheat kernel; the largest part of the wheat kernel, composed primarily of starch embedded in a protein matrix. The major constituent of wheat flour.

Extraction rate—the percentage of flour recovered from ground and sieved wheat during the milling process, typically about 72%; 100 pounds of wheat that yields 72 pounds of flour and 28 pounds of mill feed has a 72% extraction rate.

Fermentation—the increase in bulk dough mass associated with use of carbohydrates by yeast to produce alcohol and CO_2.

Final Proofing—subjecting the sheeted, molded, degassed dough unit to appropriate temperature and humidity conditions for the appropriate period of time to allow the dough to regain its extensibility and aeration, immediately prior to baking.

Gelatinization—the disruption of the molecular order within the starch granule characterized by granular swelling, crystallite melting, loss of birefringence, and increased starch solubility in the presence of water and heat.

Germ—one of three anatomical parts of the wheat kernel; the embryo that is located at the base of the grain kernel. High in protein, lipid, ash, and thiamine. Removed in the milling process.

Gliadin—the cohesive protein fraction in wheat that is soluble in 70% alcohol and ranges in molecular size from 30,000 to 70,000 Daltons. In combination with glutenin is a major constituent of gluten. Primarily contributes extensibility and plasticity to the gluten complex.

Glutenin—the cohesive protein fraction in wheat that can be dispersed in alkali or dilute acid; molecular size ranges up to tens of millions of Daltons. In combination with gliadin is a major constituent of gluten. Primarily contributes elasticity and cohesiveness to the gluten complex.

Gluten—the hydrated visocoelastic wheat protein complex (gliadin + glutenin) formed with mechanical manipulation that is responsible for the bread making properties of wheat.

Hard wheat—wheat in which the starch granules are tightly packed in the protein matrix, and the protein and starch are closely associated; typically high in good quality cohesive proteins, making it suitable for yeast-product production.

HRS—hard red spring (wheat)

HRW—hard red winter (wheat)

HW—hard white (wheat)

Loaf volume—the overall indicator of bread quality, expressed as cubic centimeters per unit weight.

Maillard browning—carbonyl-amine browning; a key contributor to the flavor, aroma, and brown crust color of baked yeast breads. Nonenzymatic browning caused by the heat-induced reaction of an amine from a protein with a reducing sugar.

Middlings—the medium-sized particles that result from passage of the broken grain kernels through the break rolls during milling. Middlings are subsequently passed through the reduction rolls to produce straight flour.

LEEDS TRINITY UNIVERSITY

Millfeed—shorts; the by-products of flour milling consisting of bran, some endosperm, and germ particles.

Millstream—the material produced after each grinding and sieving step in the milling process that meets the particle size for classification as flour. Different millstreams differ in chemical composition due to variation in germ and bran content as well as the protein gradient in the wheat endosperm.

Monomeric proteins—nonaggregated with respect to covalent bonding—gliadin.

Oven spring—the sudden increase in dough volume in the initial phase of baking. Caused by rapid expansion of existing gases and the increase in yeast fermentation activity when subjected to heat.

Patent flour—the combined millstreams removed at the beginning of the reduction system in flour milling; these millstreams are more refined and higher in starch than straight flour. 40–95% of the total millstreams obtained.

Polymeric proteins—aggregated with respect to covalent bonding—glutenin

Shorts—millfeeds; the by-products of flour milling consisting of bran, some endosperm, and germ particles.

Sponge and dough procedure—a two-step yeast-leavened bread procedure in which the major fermentative action occurs in a preferment or sponge. The sponge typically contains most of the flour and water and all of the yeast, yeast food, and any enzyme supplement. Its consistency varies from stiff to soft. The fermented sponge is then mixed with the remaining ingredients to produce a yeast-leavened dough. The most popular yeast-leavened bread procedure in the United States. Serves as the standard for bread quality.

SRW—soft red winter (wheat).

SSL—sodium stearoyl lactylate.

Starch retrogradation—recrystallization of starch after gelatinization; a major factor in staling of the bread crumb.

Staling—overall deterioration in quality postbaking.

Straight-dough procedure—a single-step, yeast-leavened bread procedure in which all dough ingredients are mixed in a single batch. In the United States, used primarily in small commercial operations and for variety breads.

Straight flour—all millstreams generated by the milling process; straight flour = patent flour + clear flour.

Surfactant—surface-active agents that act primarily as dough conditioners and staling inhibitors.

SW—soft white (wheat).

REFERENCES

Anonymous. 1987. 1987 Reference Source. Statistical Reference Manual and Specifications Guide for Commercial Baking. Sosland Publishing Co., Merriam, Kans.

___. 1995. Reference Source 95–96. Statistical Reference Manual and Specifications Guide for Commercial Baking. Sosland Publishing Co., Merriam, Kans.

Atwell EA, LF Hood, DR Lineback, E Varriano-Marson, HF Zobel. 1988. The terminology and methodology associated with basic starch phenomena. Cereal Foods World 33:308–311.

Atwell WA. 2001. An overview of wheat development, cultivation and production. Cereal Foods World 46:59–62.

Bass EJ. 1988. Chapter 1. Wheat flour milling. In: Y Pomeranz, editor. Wheat Chemistry and Technology, vol. 2, 3rd edition, 1–68. AACC, St Paul, Minn.

Bechtel DB, Y Pomeranz, A deFrancisco. 1978. Breadmaking studied by light and transmission electron microscopy. Cereal Chem. 55:392–401.

Bietz JA, Huebner, F.R. and Wall, J.S. 1973. Glutenin. The strength protein of wheat flour. The Bakers Digest 47(1): 26–34, 67.

Bloksma AH. 1990. Dough structure, dough rheology and baking quality. Cereal Foods World 35:237–244.

Buskuk W, RL Hay, NG Larsen, RG Sara, LD Simmons, KH Sutton. 1997. Effect of mechanical dough development on the extractability of wheat storage proteins from bread wheat. Cereal Chem. 74:389–395.

Caul JF, AG Vaden. 1972. Flavor of white bread as it ages. Baker's Digest 46(1): 39–43.

Chang C-Y, LM Seita, E Chambers IV. 1995. Volatile flavor components of breads made from hard red winter wheat and hard white winter wheat. Cereal Chem. 72:237–242.

Chung OK. 1986. Lipid-protein interactions in wheat flour, dough, gluten and protein fractions. Cereal Foods World 31:242–256.

Chung OK, Y Pomeranz, KF Finney. 1978. Wheat flour lipids in breadmaking. Cereal Chem. 55:598–618.

Collado M, I De Leyn. 2000. Relationship between loaf volume and gas retention of dough during fermentation. Cereal Foods World. 45(5): 214–218.

DuBois DK. 1981. Fermented doughs. Cereal Foods World 26:617–622.

___. 1984. What is fermentation? It's essential to bread quality. The Bakers Digest 58(1): 11–14.

Galal AM, E Varriano-Marston, JA Johnson. 1978. Rheological dough properties as affected by organic acids and salt. Cereal Chem. 55:683–691.

Gillette M. 1985. Flavor effects of sodium chloride. Food Technol. 39(6):47–52, 56.

He H, RC Hoseney. 1990. Changes in bread firmness and moisture during long-term storage. Cereal Chem. 67:603–605.

Hoseney RC. 1985. The mixing phenomenon. Cereal Foods World 30:453–457.

___. 1994. Bread baking. Cereal Foods World 39:180–183.

Huebner FR, TC Nelson, OK Chung, JA Bietz. 1997. Protein distribution among hard red winter wheat varieties as related to environment and baking quality. Cereal Chem. 74:123–128.

Kolster P, JM Vereijken. 1993. Evaluating HMW glutenin subunits to improve bread-making quality of wheat. Cereal Foods World 38:76–82.

Labuda I, C Stegmann, R Huang. 1997. Yeasts and their role in flavor formation. Cereal Foods World 42:797–799.

MacRitchie F. 1999. Wheat proteins: Characterization and role in flour functionality. Cereal Foods World 44(4): 188–193.

Marston PE, TL Wannan. 1976. Bread baking: The transformation from dough to bread. The Bakers Digest 50(4): 24–28,49.

Martin ML, RC Hoseney. 1991. A mechanism of bread firming. II. Role of starch hydrolyzing enzymes. Cereal Chem. 68:503–507.

Martin ML, KJ Zeleznak, RC Hoseney. 1991. A mechanism of bread firming. I. Role of starch swelling. Cereal Chem. 68:498–503.

Mathewson PR. 2000. Enzymatic activity during bread baking. Cereal Foods World. 45(3): 98–101.

Matz SA. 1992. Bakery Technology and Engineering, 3rd edition. AVI/Van Nostrand Reinhold, New York.

Menkovska M, D Knezevic, M Ivanoski. 2002. Protein allelic composition, dough rheology and baking characteristics of flour mill streams from wheat cultivars with known and varied baking qualities. Cereal Chem. 79:720–725.

Obel, L. 2001. Putting enzymes to work in bakery applications. Cereal Foods World 46(9): 396–398.

Pomeranz Y. 1988. Chapter 5. Composition and functionality of wheat flour components. In: Y Pomeranz, editor. Wheat Chemistry and Technology, vol. 2, 219–343. American Association of Cereal Chemist, Inc., St Paul, Minn.

Pyler EJ. 1988. Baking Science and Technology, 3rd edition. Sosland Publishing Co. Merriam, Kans.

Ranhotra GS, JA Gilroth, GJ Eisenbraun. 1992. Gluten index and breadmaking quality of commercial dry glutens. Cereal Foods World 37:261–263.

Ranum P. 1992. Potassium bromate in bread baking. Cereal Foods World 37:253–258.

___. 1999. Encapsulated mold inhibitors —The greatest thing since sliced bread? Cereal Foods World 44(5): 370–371.

Sandstedt RM. 1961. The function of starch in the baking of bread. The Bakers Digest 35(3): 36–44.

Seiling S. 1969. Equipment demands of changing production requirements. The Bakers Digest 43(5): 54–56, 58–59.

Shelton DR, BL D'Appolonia. 1985. Carbohydrate functionality in the baking process. Cereal Foods World 30:437–442.

Stauffer CE. 1998. Fats and oils in bakery products. Cereal Foods World 43(3): 120–126.

Uthayakumaran S, FL Stoddard, PW Gras, F Bekes. 2000. Effects of incorporated glutenins on functional properties of wheat dough. Cereal Chem. 77:737–743.

Uthayakumaran S, S Tomoskozi, AS Tatham, AWJ Savage, MC Gianibelli, FL Stoddard, F Bekes. 2001. Effects of gliadin fractions on functional properties of wheat dough depending on molecular size and hydrophobicity. Cereal Chem. 78:138–141.

van Dam HW, JDR Hille. 1992. Yeast and enzymes in breadmaking. Cereal Foods World 37:245–251.

Villanueva RM, MH Leong, ES Posner, JG Ponte, Jr. 2001. Split milling of wheat for diverse end-use products. Cereal Foods World 46:363–369.

Weegels PL, RJ Hamer. 1992. Improving the bread-making quality of gluten. Cereal Foods World 37:379–385.

Wheat Protein symposium (papers). 1999. Cereal Foods World 44(8): 562–589.

Wrigley CW. 1994. Wheat proteins. Cereal Foods World 39:109–110.

Wrigley CW, F Bekes. 1999. Glutenin—Protein formation during the continuum from anthesis to processing. Cereal Foods World 44:562–565.

Zghal MC, MG Scanlon, HD Sapirstein. 2001. Effects of flour strength, baking absorption and processing conditions on the structure and mechanical properties of bread crumb. Cereal Chem. 78:1–7.

10
Beverages: Nonalcoholic, Carbonated Beverages

D. W. Bena

BACKGROUND INFORMATION

HISTORY OF SOFT DRINKS

The first carbonated beverage, of sorts, was provided by nature and dates back to antiquity, when the first carbonated natural mineral waters were discovered—although they weren't usually used for drinking. Instead, owing to their purported therapeutic properties, the ancient Greeks and Romans used them for bathing. It wasn't until thousands of years later, in 1767, that the British chemist Joseph Priestley was credited with noticing that the carbon dioxide (CO_2) he introduced into water gave a "pleasant and acidulated taste to the water in which it was dissolved" (Jacobs 1951). The history of carbonated soft drinks (CSDs) is somewhat sparse during its early evolution, but most agree that the development of CSDs is due, in large part, to pharmacists.

Today, carbonated beverages are primarily recognized for their refreshing and thirst-quenching properties. In the early to middle 1800s, however, it was these pharmacists that experimented with adding "gas carbonium," (CO_2) to water and supplementing its palatability with everything from birch bark to dandelions in the hopes of enhancing the curative properties of these carbonated beverages [National Soft Drink Association (NSDA) 2003]. "Soft drinks," a more colloquial yet very common name for carbonated beverages, distinguish themselves from "hard drinks," since they do not contain alcohol in their ingredient listing (NSDA 1999). This is in clear contrast to other beverages, such as distilled spirits, beer, or wine. These nonalcoholic, carbonated beverages are also called "pop" in some areas of the world, due to the characteristic noise made when the gaseous pressure within the bottle is released upon opening of the package (Riley 1972). Figure 10.1 provides a brief illustration of the major milestones in the history of American soft drinks.

CSDs, pop, soda—whatever the moniker given to these beverages—one thing is clear: they have been an important part of our popular culture for decades, and will continue to be for many years to come.

SOFT DRINK FACTS AND FIGURES

Few people consciously consider how something as ostensibly simple as soda pop can markedly affect the economy on several fronts. The National Soft

---

The information in this chapter has been modified from the *Beverage Education Handbook,* copyrighted by Daniel W. Bena, ©2003. Used with permission.

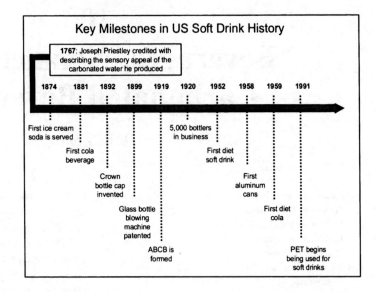

Figure 10.1. Key milestones in the
U.S. beverage industry.

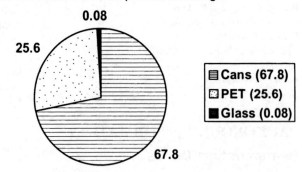

Figure 10.2. Distribution of cans, PET, and
glass CSD packages.

Drink Association (NSDA), founded in 1919 as the American Bottlers of Carbonated Beverages (ABCB), today represents hundreds of beverage manufacturers, distributors, franchise companies, and support industries in the United States. According to NSDA, Americans consumed nearly 53 gallons of carbonated soft drinks per person in 2002, and this translated into retail sales in excess of $61 billion. Nearly 500 bottlers operate across the United States, and they provide more than 450 different soft drink varieties, at a production speed of up to 2000 cans per minute on each operating line! Figure 10.2 summarizes the apportionment of total soft drink production in the year 2000.

Finally, as an industry, soft drink companies employ more than 183,000 people nationwide, pay more than $18 billion in state and local taxes annually, and contribute more than $230 million to charities each year. Few could argue that the soft drink industry has earned its place in the history of the American (and global) economy!

CARBONATION SCIENCE

Before discussing the process of manufacturing carbonated soft drinks, it is important to establish some fundamental chemical/physical concepts with regard to the carbonation process itself. Simply put, in the

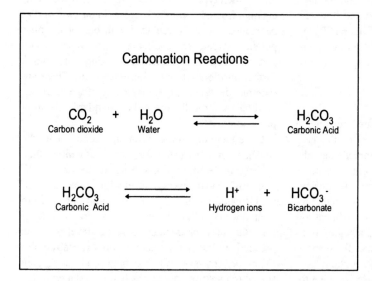

Figure 10.3. Carbonation reactions.

beverage industry, carbonation is the introduction of CO_2 gas into water, as depicted in Figure 10.3.

The favorable results of this simple combination are many: (1) the carbonation provides the characteristic refreshing quality for which carbonated beverages are most popular, (2) the dissolved CO_2 acts as both a bacteriostat and a bactericide, and (3) the CO_2 dissociates in aqueous medium to form carbonic acid, which depresses the pH of the solution, thereby making the product even more protected from microbial harm (Granata 1946). All in all, from a microbiologic perspective, carbonated soft drinks are innately very safe beverages.

Once the CO_2 is introduced into the water, which will ultimately join with flavors and sweeteners to form the complete beverage, the beverage technologist must understand how to measure and express the level of carbonation. The accepted convention in the beverage industry is not to measure CO_2 as a true concentration, expressed in parts per million (ppm), or milligrams per liter (mg/l). Instead, carbonation is expressed in volumes. The volume concept is ultimately based on the physical gas laws of Henry, Boyle, and Charles, wherein pressure, temperature, and volume are closely interdependent. The colder the liquid, the more gas can be dissolved within it. Even within the industry, however, there is some confusion over what the exact definition of a volume is (Medina 1993), usually arising from the temperature included in the definition. For our purposes, we will define one volume based on the Bunsen coeffi-

cient, described by Loomis as, "The volume of gas (reduced to 0°C and 760 mm) which, at the temperature of the experiment, is dissolved in one volume of the solvent when the partial pressure of the gas is 760 mm" (Loomis 1928). More informally, and to put this concept in perspective, consider a 10-ounce bottle of carbonated beverage, representing roughly 300 ml of liquid. If this carbonated beverage were prepared at one gas volume, the package would contain approximately 300 cc of CO_2. We would consider this very low carbonation from a sensory perspective and would have a barely noticeable "fizz" upon removal of the closure. Imagine, however, that for the same 300 ml of liquid, we carbonate to four gas volumes (a level typical of many products on the market today). This means that roughly 1200 cc of CO_2 have been introduced into the same 300 ml volume of liquid. More gas into the same amount of liquid and the same vessel size—imagine the increase in pressure contained within the bottle. This example explains why the characteristic "pop" of soda pop is heard when a bottle is uncapped or a can is opened!

For the purposes of this text, the discussion of carbonation has been somewhat oversimplified in order to make the concept more easily understood. As with any industry, the more one investigates any given topic, the more complicated and scientifically intense the subject usually becomes. Carbonation, for example, can be affected by a variety of factors, including other solids present in the liquid being carbonated, the temperatures of the gas and the liquid, the atmospheric pressure/altitude, and how far CO_2 varies from ideal gas behavior (Glidden 2001). These factors are cited merely for consideration, but are outside the scope of this chapter.

PROCESS OVERVIEW

The process of manufacturing carbonated beverages has remained fundamentally the same for the last several decades. Certainly, new equipment has allowed faster filling speeds, more accurate and consistent fill heights, more efficient gas transfer during carbonation, and other improvements, but the process remains one of cooling water, carbonating it, adding flavor and sweeteners, and packaging it in a sealed container. Figure 10.4 illustrates the overall process that we will be discussing throughout this chapter, in somewhat more detail, as we continue to build upon the basic foundation. As we proceed, the figures depicted will become more complete, as each critical process to carbonated beverage manufacture is explained.

Carbonated beverage production begins with careful measurement of the formula quantities of each component into the syrup blending tank. Critical components include the *concentrate*, which contains the bulk of the flavor system; the sweet-

ener, which typically includes the nutritive sweeteners high fructose syrup or sucrose (in the case of diet beverages, these are replaced with one of the high potency sweeteners available); and water, which generally begins as municipal drinking water and is further purified within the beverage plant. These are then blended to assure homogeneity of the batch according to carefully prescribed standard operating procedures.

Once blending in the syrup tank is complete, the finished syrup is tested for correct assembly, then pumped to the mix processor, where the syrup is diluted to finished beverage level with chilled, carbonated, treated water (often a 1:6 dilution of syrup to treated water, although this varies by product). After this, the now carbonated beverage-level solution proceeds to the filler, where it is fed (usually volumetrically, by gravity) into bottles or cans, then sealed (capped in the case of bottles, seamed in the case of cans). Then, the finished product is either passed through a warmer, in order to avoid excessive condensation from forming (depending on the type of secondary packaging used), or sent directly to secondary packaging. This can include plastic or cardboard cases, shrink wrap, stretch wrap, or even more innovative devices. After packaging, the product is palletized and stored in the warehouse until it is ready for distribution.

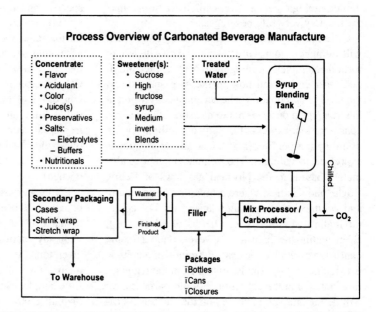

Figure 10.4. Process overview of carbonated beverage manufacture.

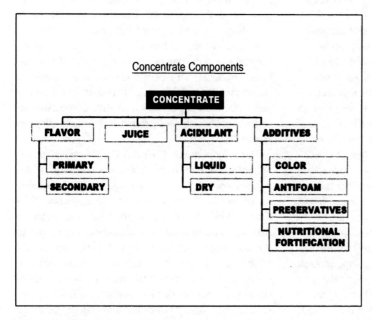

Concentrate Components

Figure 10.5. Concentrate components.

RAW MATERIALS PREPARATION

CONCENTRATE

In the carbonated soft drink industry, "concentrate" refers to a mixture of many different categories of ingredients, illustrated in Figure 10.5.

The most notable of these, and indeed, the topic of many urban legends surrounding its utter secrecy, is the flavor component. This is where the proprietary formulations of essential oils, which combine to form the characteristic flavor of the trademark beverage, are found. Flavor components can include a single primary component or may be distributed in various ways among multiple components—for example, a high potency sweetener supplied as a dry salt as part of a secondary flavor component. In general, the majority of flavor systems include primary flavor components, and these fall into three broad categories:

1. *Simple mixtures.* These are perhaps the simplest of the flavor categories to understand, but they also represent the minority of those in existence. Here, a combination of miscible liquids or easily soluble solids are blended together to form a homogenous aqueous mixture. Because so many essential flavor oils are not readily water soluble, the beverage technologist must abandon the idea of the simple mix-

ture for one of the other, more flexible categories of flavors.

2. *Extracts.* As the name implies, this category of flavors involves extracting the desired flavor constituents from essential oils. Simply put, the extraction solvent—usually ethanol (although sometimes propylene glycol is used)—is used to partition those flavor constituents that are soluble in the solvent, but not freely soluble in the water directly. In this way, these flavor compounds become fully dissolved in the ethanol first. Then, this ethanolic extract (which is, in effect, an ethanolic solution of the flavor compounds) is added to water. Since ethanol is freely miscible with water, it acts as a carrier vehicle to help dissolve or disperse the otherwise water-insoluble flavor constituents (Woodruff 1974). Today, equipment for both batch and continuous liquid extraction of flavor oils is available, and more novel approaches have also been developed (e.g., gas extraction, supercritical fluid extraction, and other patented processes).

3. *Emulsions.* This third category is likely the largest, encompassing the bulk of the flavor systems available today. In the carbonated beverage industry, oil-in-water (or o/w) emulsions are the standard. This model involves an oil (lipophilic) internal phase and an aqueous

(hydrophilic) external phase being made compatible by the use of a surfactant (or emulsifier). Surfactants are compounds that are amphiphilic; that is, there are both hydrophilic and lipophilic portions of the same molecule! This facilitates a decrease in the surface tension when oil and water are mixed together, and allows the lipophilic portion to align with the oil while allowing the hydrophilic portion to align with the water (Banker 1996). In so doing, the emulsifier forms a bridge, of sorts, between the two phases and allows them to be dispersed, without gross separation, for the desired length of time (generally at least as long as the technical shelf life of the beverage).

Since carbonated beverages are of low pH, owing in part to the carbonic acid from the dissolved CO_2, but also to the acid components of the formulas, acid hydrolysis is one of the major concerns of the beverage flavor developer. By positioning itself between the oil and water phases, the emulsifier protects the sensitive flavor oils from chemical degradation in this acidic environment. In addition, the emulsifier protects the flavor oils from oxidation by the naturally dissolved oxygen in the water, which constitutes the aqueous phase. So, a well-designed and prepared emulsion can dramatically extend the sensory shelf life of the flavor system and the overall physical stability of the beverage.

In addition to the flavors, Figure 10.5 also depicts a variety of other components that may be part of the concentrate. These include juices, which must be handled and stored carefully in order to preserve their quality, acidulants (both liquid and dry), and a host of other additives, depending on their desired function (e.g., antifoam, preservatives, nutrients, etc.).

WATER

Water is the major component in carbonated beverages and represents anywhere from 85 to near 100% of the finished product. Interestingly, it is unlike any other ingredient, since we rarely have the number of options for water supply that we have with other raw materials! Obviously, then, particular diligence must be employed when selecting a water supply. Beverage plants use water from ground supplies, surface supplies, or both. Ground supplies include springs, deep and shallow wells, and artesian aquifers. Surface supplies include rivers, lakes, streams, and reservoirs. Within these sources, there is wide variation in type and content of inorganic (e.g., metals, minerals, sulfate, chloride, nitrate), organic (e.g., volatile organics, natural organic matter), microbiologic (bacteria, viruses, protozoa), and radiologic (radionuclides, alpha- and beta-activity) components. Table 10.1 provides a relative comparison of some characteristics of ground and surface supplies (Bena 2003).

One critical point of which to be aware is that municipal treatment plants should not normally be depended upon to consistently supply water suitable for the needs of most carbonated beverage manufacturers. While the municipality treats the water so that it is safe to drink and aesthetically pleasing to the consumer (potable and palatable), they cannot afford to consider the needs of all industrial end users, so they may not consistently supply a water of the high quality needed for producing a finished carbonated beverage product and assuring the beverage a long shelf life. There is also the possibility of contamination of the city water as it passes through the distribution system from the municipal treating plant to the beverage plant. This is particularly true with respect to organic matter and metal content, such as iron. The quality of the water used for carbonated soft drinks must be considered from several perspectives:

- *Regulatory compliance.* The water used must be in compliance with all presiding local and national laws and guidelines. This jurisdiction is generally clear in the United States, between the Environmental Protection Agency and the Food and Drug Administration. However, as you consider international beverage locations, the regulatory picture sometimes becomes cloudy.
- *Beverage stability.* Intuitively, as the major ingredient in carbonated soft drinks, the constituents in water can have a profound impact on the overall quality and shelf life of beverage products. For example, if alkalinity is not controlled, the acidic profile of the beverage formulas will be compromised, making the beverage more susceptible to microbial growth and spoilage.
- *Sensory.* Many contaminants, even at levels within drinking water standards, may adversely affect the finished beverage. For example, some algae produce compounds (geosmin and methyl isoborneol) that are sensory active at levels as

Table 10.1. Comparison of Ground and Surface Water Supplies

Parameter	Groundwater	Surface Water
Total dissolved solids	Higher	Lower
Suspended solids	Lower	Higher
Turbidity and color	Lower	Higher
Alkalinity	Higher	Lower
Total organic carbon	Lower	Higher
Microbiology:		
Protection from bacteria and viruses	Highly protected	Highly susceptible
Protection from protozoa	Almost completely protected	Highly susceptible
Presence of iron and/or manganese bacteria	Common	Rare
Hydrogen sulfide gas	Common	Uncommon
Aeration/dissolved oxygen	Lower	Higher
Temperature	More consistent	More variable
Flow rate	Very slow (1 m/day)	Very fast (1 m/sec)
Flow pattern	Laminar	Turbulent
Susceptibility to pollution through surface run-off	Low	High
Time for a contaminant plume to resolve	Very Long—often decades, potentially centuries!	Usually short—days/ months; sometimes years

Sources: Bena 2003.

low as nanograms per liter (Suffet 1995). These can result in "dirty, musty" flavor and aroma in finished products.

- *Plant operations.* Water for nonproduct (auxiliary) uses must also meet the performance standards of the carbonated soft drink producer. These standards and guidelines are usually enacted to prevent corrosion (e.g., from high chloride content in heat exchangers) and scaling (e.g., from hardness salts in boilers), which may result in premature equipment failure and/or loss of operational efficiency.

Whether the beverage plant has its own well, or the water supply comes from a modern municipal treatment plant, each individual water supply presents its own particular problems. In most, if not all, cases, the incoming raw water that supplies a beverage plant already meets the applicable standards for potability of drinking water. The beverage producer then further purifies the water to meet the quality necessary for its products. This treatment can take many forms, but the three largest categories of in-plant beverage water treatment are (1) conventional lime treatment systems (CLTS), (2) membrane systems (including reverse osmosis, nanofiltration, and ultrafiltration), and (3) ion exchange. Volumes have been written about each treatment modality, and a detailed discussion is beyond the focus of this chapter. However, a brief summary of each treatment category is provided below (Bena 2003).

- *Conventional lime treatment systems (CLTS).* This treatment chain represents the majority of most beverage treatment armadas worldwide, although the balance is quickly shifting in favor of membrane technologies. CLTS involves the addition of a coagulant (as an iron or aluminum salt), hydrated lime (for pH control), and chlorine (for oxidation and disinfection) to a reaction tank. The agitation is gently controlled over the course of a two-hour retention time, during which a floc begins to form, grow, and settle, bringing contaminants with it to the bottom of the tank, where they await discharge. Figure 10.6 illustrates what happens in this reaction vessel.

  Historically, and as little as 25 years ago, conventional lime treatment was regarded as the ideal treatment for raw water of virtually any quality. Indeed, this system, coupled with the required support technology—fine sand filtration, granular activated carbon, polishing filtration, and ultraviolet irradiation—does address a broad range of water contaminants. The advantages and disadvantages of conventional lime treatment are summarized in Table 10.2.

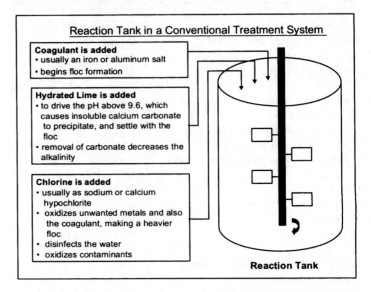

Reaction Tank in a Conventional Treatment System

Coagulant is added
• usually an iron or aluminum salt
• begins floc formation

Hydrated Lime is added
• to drive the pH above 9.6, which causes insoluble calcium carbonate to precipitate, and settle with the floc
• removal of carbonate decreases the alkalinity

Chlorine is added
• usually as sodium or calcium hypochlorite
• oxidizes unwanted metals and also the coagulant, making a heavier floc
• disinfects the water
• oxidizes contaminants

Reaction Tank

Figure 10.6. Reaction tank in a conventional lime treatment system.

Table 10.2. Advantages and Disadvantages of CLTS

Advantages	Disadvantages
Removes alkalinity and hardness	Does not effectively reduce nitrate, sulfate, or chloride concentration
Removes organic debris, particulates, and natural organic matter (NOM)	Sludge formation and disposal requirements
Reduces metal concentrations (iron, manganese, arsenic, others) and some radionuclides	May promote the formation of disinfection by-products (trihalomethanes) under certain conditions
Reduces some color compounds (tannins), off tastes, and off odors	Often difficult to operate consistently in waters with very low dissolved solids
Reduces bacteria, virus, and protozoan populations	Relatively large space requirements on plant floor ("footprint")

Source: Bena 2003.

• *Membrane technology.* Clearly, this technology has seen the most growth in recent years with the advent of more resistant membrane construction materials and more flexible rejection characteristics. Included in this category is the prototype of the cross flow, polymeric membrane filtration systems—reverse osmosis—in addition to nanofiltration and ultrafiltration (both polymeric and ceramic). By carefully controlling the membrane pore size during manufacture, and the applied pressure during operation, reverse osmosis membranes can effectively remove in excess of 99% of many dissolved species—down to the ionic level (for example, dissolved calcium or

sulfate). Table 10.3 illustrates the relative capabilities of the three major membrane processes with regard to a variety of possible constituents in the incoming water (Brittan 1997).

Since reverse osmosis is often the cited membrane standard against which the performance of other membrane filtration systems are judged, the advantages and disadvantages of reverse osmosis are listed in Table 10.4.

Also worth mentioning, though not discussed among this group, are the hybrid technologies, which include novel membrane and ion exchange utilization. Examples are electrodialysis technology for re-

Table 10.3. Relative Comparison of Reverse Osmosis, Nanofiltration, and Ultrafiltration

Component	Reverse Osmosis	Nanofiltration	Ultrafiltration
Alkalinity	95–98%	50–70%	None
TDS	95–98%	50–70%	None
Particulates	Nearly 100%	Nearly 100%	Nearly 100%
Organic matter	Most >100 MW	Most > 200 MW	Some > 2000 MW
THM precursors	90+%	90+%	30–60%
Sodium	90–99%	35–75%	None
Chloride	90–99%	35–60%	None
Hardness	90–99%	50–95+%	None
Sulfate	90–99%	70–95+%	None
Nitrate	90–95%	20–35%	None
Protozoa	Near 100%	Near 100%	Near 100%
Bacteria	Near 100%	Near 100%	Near 100%
Viruses	Near 100%	Near 100%	Near 100%
Operating pressure	200–450 psi	100–200 psi	80–150 psi

Source: Adapted from Brittan 1997.
Note: Removal percentages are approximate. Actual performance is system specific.

Table 10.4. Advantages and Disadvantages of Reverse Osmosis

Advantages	Disadvantages
Removes nearly all suspended material and greater than 99% of dissolved salts in full-flow operation,	Pretreatment must be carefully considered and typically involves operating costs for chemicals (acid, antiscalant, chlorine removal).
Significantly reduces microbial load (viruses, bacteria, and protozoans).	Does not produce commercially sterile water.
Removes nearly all natural organic matter (NOM)	Membranes still represent a substantial portion of the capital cost and may typically last 3–5 years.
May be designed as a fully automated system with little maintenance.	Low solids water may be aggressive toward piping and equipment, so this must be considered for downstream operations.
Requires relatively small space on the plant floor ("footprint")	High-pressure inlet pump is required.

Source: Bena 2003.

moval of ionic species in water, and continuous electrodeionization.

- *Ion exchange.* This technology is routinely utilized for partial or complete demineralization of the water supply, softening, and dealkalization, or it can be customized for selective removal of a specific contaminant (for example, denitratization). In simplest terms, ion exchange involves using a selective resin to exchange a less desirable ion with a more desirable ion. Of course, a great deal of chemical research goes into the development of these selective resin materials, but the functional outcome remains straightforward. For

example, softening resins are often employed to remove hardness (calcium and magnesium) from the water entering boilers and heat exchangers. In this application, the hardness ions are not wanted. The softening resin (for example, a sodium zeolite clay) is charged with active and replaceable sodium ions. When the hard water passes across the softening bed, the resin has a selectivity for calcium and magnesium, so it replaces them for sodium. The result is that the water exiting the softener is virtually free of calcium and magnesium (since they were replaced by sodium) and is safe to use in boilers and other equipment, since it will no longer have the tendency to form scale.

To supplement the major treatment systems mentioned above, the carbonated beverage producer often utilizes a host of other support technologies, including activated carbon filtration (to remove organic contaminants and chlorine), sand filtration (to remove particulates), and primary and secondary disinfection (using chlorine, ozone, ultraviolet, heat, or a combination). By the time the treated water is finished, it is microbially and chemically safe, clear, colorless, and ready to be used for syrup and beverage production.

SWEETENERS

The two major categories of sweetener types are nutritive (i.e., they provide some caloric value) and high potency (i.e., the type used in diet beverages, since they are many times sweeter than sucrose and are generally noncaloric). There are several high potency sweeteners available to the worldwide beverage developer (aspartame, acesulfame potassium, and others), and they are almost exclusively, if not always, included as part of the concentrate flavor system as a dry substance package. As such, their quality can be more easily controlled by the vendor, as with any of the other concentrate ingredients, and minimal intervention is needed at the carbonated soft drink manufacturing facility. These high potency sweeteners, therefore, will not be addressed in this chapter. However, a concise treatise on the topic is provided by the International Society of Beverage Technologists (Koch 2000).

Next to water, however, the nutritive sweeteners represent the second most prevalent ingredient in the finished beverage. The most common nutritive sweeteners used in the carbonated soft drink industry are sucrose and high fructose syrups, with sucrose (from cane or beet) being the most common internationally. Within the United States, nearly all the nutritive sweetener used in carbonated beverages is high fructose corn syrup (HFCS, either 42 or 55%). In 1996, the U.S. corn refining industry produced over 21 billion pounds of high fructose corn syrups, representing only about 12% of the total corn crop (Hobbs 1997).

Although high fructose syrups may be obtained from other starting materials, like wheat or tapioca starch, corn remains the most prevalent starting material. A starch slurry is first digested by the addition of alpha-amylase enzyme, resulting in gelatinization and ultimate dextrinization of the starting starch. Then, glucoamylase enzyme is added to obtain an enriched glucose syrup (95% glucose). The glucose syrup is then purified via particle filtration, activated carbon adsorption, and cation and anion exchange. Then, evaporation brings the solids content within range for effective passage through an isomerization column containing the glucose isomerase enzyme. This enzyme converts much of the 95% glucose syrup to fructose, which is again purified, as before, and evaporated. The result is HFCS-55 of high quality. In some formulas and/or markets, HFCS-42 is used, which is simply a blend of the HFCS-55 with the 95% glucose stream to result in a product that is 42% fructose. The generic process by which cornstarch is transformed to high fructose corn syrup is illustrated in Figure 10.7 (Boyce 1986).

In general, HFCS-55 (55% fructose) is a highly pure ingredient, due in large part to the activated carbon, cation, and anion exchange steps required of the process. However, the most recent research highlights the occurrence of potent sensory-active compounds that could form via chemical or microbial pathways in HFCS, including isovaleraldehyde, 2-amino acetophenone, and maltol (Finnerty 2002). When properly produced and stored, no additional treatment is necessary at the beverage plant.

Sucrose, though the clear exception in the North American beverage industry, continues to be the mainstay for international beverage markets. It may be obtained from sugar cane or sugar beet, following two distinct separation and purification schemes, as depicted in Figure 10.8 (Galluzzo 2000).

The three indicators of sucrose quality generally recognized by the sugar industry are color, ash, and turbidity. Internationally, depending on the quality of the available sucrose, it is not uncommon to subject the incoming granular or liquid sucrose to additional treatment at the beverage plant. Ash, or residual inorganic minerals, remains difficult to adequately treat at the carbonated soft drink plant, so great effort is made to source sucrose that has an acceptable ash content (as defined by the individual company specifications). Turbidity is easily remedied at the beverage plant via an in-line filtration step, often incorporating diatomaceous earth as a filter aid. Color, considered by some as the primary indicator of sucrose quality, can also be treated at the beverage plant but typically requires hot treatment through activated carbon. This treatment removes color and many sensory-active compounds, and also serves to render the sucrose free of most viable microorganisms. Figure 10.9 (Galluzzo 2000) briefly

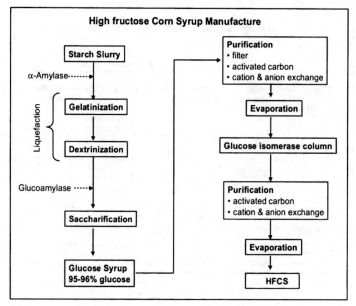

Figure 10.7. High fructose corn syrup manufacture.

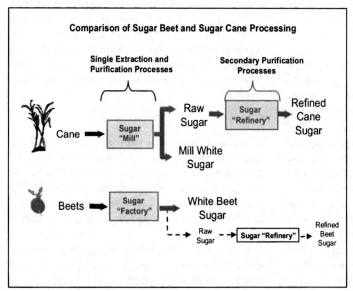

Figure 10.8. Cane versus beet sugar process flow.

summarizes the handling and treatment of sucrose at a carbonated soft drink facility.

Liquid sucrose, usually commercially available at a concentration of 67 Brix (equivalent to 67% sucrose, by weight), is sometimes used for the production of carbonated soft drinks. Two distinct disadvantages of using liquid sucrose instead of granulated sucrose are that (1) the end user ultimately pays for shipping 33% water, since the ingredient is only 67% sucrose solids, as compared to granulated sucrose, which is 100% sucrose solids, and (2) this water also means that the liquid has a higher water activity than

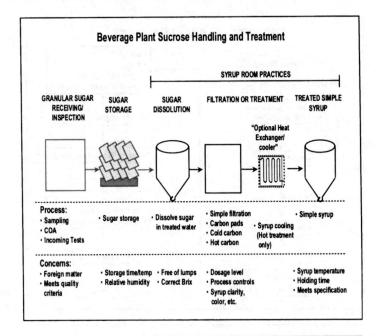

Figure 10.9. Sucrose handling and treatment at the beverage plant.

granulated sucrose, making it much more susceptible to microbial spoilage. With liquid sucrose operations, absolutely diligent transport and handling procedures are imperative.

The last, less common type of nutritive sweetener used in this industry is medium invert sugar (MIS). Chemically, this product has similarities to both sucrose and high fructose corn syrup. With MIS, the starting material is liquid sucrose, which is then treated with one of three processes: (1) heat and acid, (2) ion exchange, or (3) invertase enzyme. The end result of any of these processes is that roughly 50% of the starting sucrose is transformed into invert sugar, an equimolar mixture of glucose and fructose. At this point, the inversion process is stopped, and the final commercial product contains 50% sucrose, 25% glucose, and 25% fructose. This gained favor over liquid sucrose in the beverage industry for two main reasons: (1) the finished material is 76 Brix, versus 67 Brix for liquid sucrose, so less water is shipped, and (2) MIS has a much lower water activity and is therefore much more microbiologically stable.

In summary, the producers of carbonated soft drinks have several options at their disposal for providing the sweetness, which is so characteristic of these products, to the consumer. Internationally, sucrose is the major sweetener used, while in the

United States, high fructose corn syrup is preferred. Irrespective of the type of sweetener, the beverage industry has treatment methods at its disposal to assure that this ingredient consistently meets the high standards of chemical and microbial quality necessary for use in the production of syrup and beverage.

CO_2

At normal temperatures and pressures, CO_2 is a colorless gas, with a slightly pungent odor at high concentrations. When compressed and cooled to the proper temperature, the gas turns into a liquid. The liquid in turn can be converted into solid dry ice. The dry ice, on absorbing heat, returns to its natural gaseous state.

We learned a little of the history of carbonation earlier in this chapter, since the concept is so critical to the production of carbonated soft drinks. Just as critical is the quality of the CO_2 used in this application. For many years, the quality of CO_2, as an ingredient, was minimized, largely because there were no uniformly available methods with which to test the gas. Those procedures that were available required special expertise to properly sample and handle this cryogenic gas. The standards of quality of the CO_2 used in beverages were traditionally relegated to the U.S. Compressed Gas Association (CGA), whose quality verification levels were incorporated into a beverage company's specification system. Then, in 1999, the International Society of Beverage Technologists (ISBT) developed the Quality Guidelines and Analytical Procedure Bibliography for Bottler's Carbon Dioxide (McLeod 2001). This was a cooperative effort by CO_2 suppliers, end users, testing labs, and allied businesses to completely update the obsolescent guidelines that had been used for decades. The Guidelines are only available for purchase through ISBT (www.bevtech.org); they include parameters related to health/safety, sensory quality of the CO_2 or the finished beverage, and good manufacturing practices at the supplier.

CO_2 may be obtained and purified from a number of different feed gas sources, the majority of which are listed in Table 10.5.

There are other more exotic sources, which are often the result of CO_2 being generated as a side product during an organic chemical synthesis. In addition to commercial supplies, some carbonated beverage plants produce and purify their own CO_2. The most common feed gas sources for these applications are combustion (where the flue gas is recovered, concentrated, then purified) and breweries (where the CO_2 generated from microbial metabolism is recovered and purified). Whether supplied commercially or in-house, the CO_2 used in carbonated soft drinks is of high quality (> 99.9% CO_2); in most cases, it even exceeds that of medical grade gas.

The liquid CO_2 that is delivered to beverage plants is generally stored in large bulk receivers, which are vertically or horizontally oriented steel tanks with urethane foam or vacuum insulation. In the most common arrangement, CO_2 is withdrawn from the liquid phase at the bottom of the tank and vaporized by one of several methods. Due to this withdrawal, the equilibrium between vapor and liquid in the tank remains dynamic. The air gases (oxygen, nitrogen) partition into the vapor phase of the vessel and are routinely purged to maintain the purity of the CO_2 within the bulk receiver. Similarly, some components preferentially partition, in trace amounts, into the liquid phase of the CO_2 (liquid CO_2 is an excellent solvent). Many beverage plants choose to subject the freshly vaporized CO_2 to one final step of purification just prior to the point of use. This is usually a simple filtration through activated carbon alone, or through a mixed adsorbent bed of carbon (to remove organic contaminants), a silica-based desiccant (to remove moisture), and a molecular sieve (to remove sulfur compounds and some oxygenates).

In addition to the quality considerations already discussed, CO_2 safety is a key consideration for beverage industry technologists. Carbon dioxide is not usually considered to be a toxic gas in the generally accepted sense of the term (i.e., poisonous) and is normally present in the atmosphere at a concentration of approximately 0.03% (300 ppm). Under normal circumstances, CO_2 acts upon vital functions in a number of ways, including respiratory stimulation, regulation of blood circulation, and acidity of body fluids. The concentration of CO_2 in the air affects all of these. High concentrations are dangerous upon extended exposure, due to increased breathing and heart rates and a change in the body acidity. OSHA (U.S. Occupational Safety and Health Administration) establishes regulations governing the maximum concentration of CO_2 and the time-weighted average for exposure to CO_2. These regulations should be reviewed before installation of any CO_2 equipment, and the requirements should be fully met during operation and maintenance.

Since CO_2 is heavier than air, it may accumulate in low or confined areas. Adequate ventilation must be provided when CO_2 is discharged into the air. At lower levels where CO_2 may be concentrated, self-contained breathing apparatus or supplied-air respirators must be used. Filter-type masks should not be used. Appropriate warning signs should be affixed outside those areas where high concentrations of CO_2 gas may accumulate, and lock-out/tag-out procedures should be followed, as appropriate (Selz 1999).

SYRUP PREPARATION

Most carbonated beverage formulas begin with a simple syrup, which is usually a simple combination

Table 10.5. Feed Gas Sources for Carbon Dioxide

Combustion
Wells/geothermal (natural CO_2 wells)
Fermentation (breweries, ethanol plants, etc.)
Hydrogen or ammonia plants
Phosphate rock
Coal gasification
Ethylene oxide production
Acid neutralization

Source: Adapted from CGA-6.2. 2000, Table 3, page 5.

of the nutritive sweetener (sucrose, HFCS, MIS) and treated water. In some cases, it may also contain some of the salts outlined in the specific beverage document, depending on the order of addition that is required. Once the sweetener is completely dissolved, and the simple syrup is a homogenous batch, then the flavor and remaining components are added to form the finished syrup. All simple syrup should be filtered before being pumped to the finished syrup blending/storage tanks.

- *Using granulated sucrose.* Accurate weighing of granulated sugar is important. Granulated sugar is normally received in bulk form or in bags. Internationally, receipt in 50- or 100-pound jute or paper bags is not uncommon. It is extremely important that the sugar received by either means should be dry and free of lumps. Moist sugar creates two immediate and serious problems: (1) moist sugar can have high microbial counts, much of which will be yeast. Yeast is a serious problem to carbonated beverages, since it can lead to fermentation and eventual spoilage of the finished product. (2) Moist sugar makes accurate measuring difficult, since the moisture content is being weighed with the sucrose solids. This makes final control of the batch difficult and inconsistent.

   Sugar in lumps will create difficulties in making simple syrup and will take longer to dissolve. Lump sugar is usually an indication that the sugar was not fully dried during refinery production or was stored improperly (Delonge 1994a). Never use bulk sugar systems when faced with wet or even slightly moist sugar. It will cause bridging (flow restriction) in silo storage and make effective handling impossible. It is critical that any bulk sugar supply be consistently dry and that the storage environment be controlled to assure constant low humidity. Even the most

modern silo can bridge when faced with a moisture problem.

   Granulated sugar should always be added slowly to the treated water already measured into the tank. While sugar is being added, the tank agitator should be in constant operation. The agitation should continue until the sugar is completely dissolved. After the sugar has been completely dissolved and the simple syrup has been filtered into the blending/storage tank, the syrup is checked for sugar content (Brix). Table 10.6 outlines intuitive, but useful, reasons for off-target Brix readings.

- *Using liquid sugars.* There are three main types of liquid sugars that are used for syrup production, as discussed earlier: liquid sucrose, medium invert sugar, and high fructose syrups. Making simple syrup from liquid sucrose is similar to the procedure employed when using granulated sugar. The first step is to check the Brix of the liquid sucrose to find out how much water must be added to the batch to bring the Brix of the simple syrup to the level required by the formula. Most companys' beverage documents include a table that specifies how much liquid sucrose and additional treated water should be added to the batch, based on Brix. When liquid sucrose supplies are received at the plant, they should be accompanied by an analysis sheet comparing the tank load against the company standards.

- *Medium invert sugar* is resistant to microbial spoilage when being transported from supplier to plant, and while in storage. However, good sanitation procedures, as well as special precautions to prohibit secondary infection, are still required. When liquid invert shipments are received at the plant, they should be accompanied by an analysis sheet comparing the tank load against company standards. The formula document should

Table 10.6. Possible Brix Errors During Simple Syrup Production

High Brix	Low Brix
Weighing error—excess sugar	Weighing error—short sugar
Faulty scale	Faulty scale
Instrument error	Not weighing sugar bags
Too little water	Too much water
	Instrument error
	Moist sugar

Source: Delonge 1994a.

include a table that specifies how much of the sweetener and additional treated water should be added to the batch, based on Brix and the percent inversion. When testing for Brix in MIS samples, a correction factor must be used on refractometer readings to compensate for the nonsucrose solids that are a result of the inversion process.

- *Using high fructose syrup (HFS).* For liquid sugars in general, a sample should be taken before the sugar is accepted, and the analysis should confirm that the material is within standards. The installation, including receiving station, pumps, air blower/ultraviolet lamp, tanks, and piping/fittings, should be of approved materials (stainless steel) and in accordance with the individual beverage company's design guidelines. High fructose syrup is subject to crystallization, so storage temperatures should be controlled (generally maintained between 75°F/24°C and 85°F/29°C) by the use of indirect heating. The receiving station is a critical point and should be fully cleaned and hot sanitized before every delivery. As with MIS, when testing for Brix in HFS samples, a correction factor must be used on refractometer readings to correct to true Brix and compensate for the nonsucrose solids.

No matter what type of nutritive sweetener is used, once the simple syrup has been correctly prepared in the mixing tank, it should be pumped through the syrup filter into the storage tank so that the other concentrate components may be added. Most simple syrups will be in a range between 60 and 65 Brix, which makes them extremely susceptible to microbial spoilage, with yeast as the most likely culprit. Be sure to recognize and respect any time constraints included in the syrup preparation instructions. For example, a general rule of thumb is that simple syrup should not be kept longer than four hours before converting it to finished syrup. If hot sugar processing is used, remember to allow the simple syrup to reach ambient temperature prior to the addition of concentrate. This will help minimize thermal degradation of the flavor oils. Also, it is very important to add the individual components in the specific order detailed in the syrup preparation instructions. Incorrect order of addition can lead to a variety of problems, including changes in viscosity, flavor degradation, nutrient breakdown, and precipitation of insoluble materials in the syrup tank.

CARBONATION

Earlier in this chapter, we discussed the history, theory, and principle of introducing CO_2 gas into water to produce a carbonated beverage. We also addressed the importance of the quality of this CO_2 and of the treated water used to dissolve it. In this section, we will discuss the practical aspects of carbonation control.

Mix processing refers to the process of combining the finished syrup, treated water, and CO_2 in the correct proportions to meet beverage specifications. In addition to the proportioning function, mix processing will usually incorporate deaeration, mixing, carbonating, and cooling, depending on the manufacturer's design and the type of products being handled. The design of mix processing systems will vary from one manufacturer to another, incorporating the features that the manufacturer feels are advantageous to controlling production.

The primary function of the carbonating unit (carbonator) is to add CO_2 to the product. It must be carbonated to a level that, after filling and closing, results in a product within the standards for beverage carbonation. Some carbonating units incorporate cooling in the same tank or unit. The product can be slightly precarbonated with CO_2 injection and then exposed to a CO_2 atmosphere directly where cooling is in progress. Other systems separate the carbonating and cooling steps. The three most common forms of carbonating technology incorporate one or a combination of the following: (1) conventional (atmospheric exposure) introduction, (2) CO_2 injection, or (3) CO_2 eduction.

The ability of water, or beverage, to absorb CO_2 gas is largely dependent on the efficiency of the carbonating unit (Jacobs 1959). Other factors that influence CO_2 absorption include (1) product type, (2) product temperature, (3) CO_2 pressure, (4) time and contact surface area, and (5) air content. If the water temperature rises, the gas pressure must be increased if the same absorption of CO_2 is to be maintained. Conversely, if the temperature of the water or beverage entering the carbonating unit drops, the CO_2 becomes more soluble, and the pressure must be decreased to keep the volumes of carbonation within standards. Automatic CO_2 controls compensate for fluctuations in temperature, pressure, and flow. This allows the carbonating unit to produce a constant CO_2 gas absorption. Such controls are standard in modern processing units, which are available either as basic units or with computer interfaces to track the variation in product temperature, pressure, flow, and final CO_2 gas volumes absorbed during operating hours.

In many ways, this is a gross oversimplification of a process that, to this day, sometimes eludes strict control. Certainly, equipment has dramatically improved over the years, but loss of CO_2 remains a significant issue in terms of overall plant productivity. New membrane carbonation systems hold great promise for continuing this evolution, by helping to carbonate, at least in theory, more precisely and accurately than ever before. It has yet to be seen if these systems will endure the economic challenges, industry acceptance, and rigors of time.

FILLING, SEALING, AND PACKING

In the most fundamental terms, this section will address introducing the now freshly prepared and carbonated finished beverage into the package and sealing it in a manner that will preserve its integrity. This is simple in theory, but sometimes challenging in application. The bottle-filling unit includes bottle handling/transfer components, a filling machine, and a capper/crowner.

The purpose of the filler is to fill returnable and nonreturnable bottles to a predetermined level. It should do this efficiently, while minimizing foaming, and deliver the bottle to a crowner or closure machine to be sealed, or in the case of cans, to the lid seamer. A discussion of the design and engineering of filling machines is beyond the scope of this text and is normally relegated to the specific operating manuals supplied by the respective equipment vendor.

Carbonated beverage fillers, to prevent the loss of CO_2 from the freshly carbonated beverage, must be counterpressured. The advantage in using CO_2 gas for counterpressure purposes at the filler bowl is to reduce product air content. With can fillers, this is possible because the counterpressure gas is nor-

mally purged from the can to the atmosphere as part of the filling process. Most bottle fillers presently in use vacate the counterpressure gas back into the filler bowl as the bottle is being filled. The empty bottle moving into the sealing position (at the filling valve) already contains air. Even if the counterpressure gas is CO_2, vacating this mixture (air and CO_2) back into the filler bowl assures that the bowl will contain (predominantly) air. This can negate the advantage of CO_2 as a counterpressure gas and can actually waste CO_2 to the point of economic disadvantage. In place of CO_2, air or nitrogen is sometimes used as the counterpressure gas.

Imagine what happens when a carbonated beverage is agitated and then quickly uncapped. Sometimes, this same type of foaming can occur during filling. Foaming at the filler, even in small amounts, can cause a number of problems. Some of these deal with product quality, others with economics or plant operation. They are summarized in Table 10.7.

The cause(s) of foaming in a filling operation can range from a simple problem that can be corrected quickly to one requiring extensive trial and error testing. Many times, the troubleshooting exercise requires a combination of technical skill, creativity, and experience. Some causes of foaming at the filler are summarized in Figure 10.10 (Bena 2001).

When the problem is a single valve or occurs for a short period of time, it is usually easy to troubleshoot and correct. On-going foaming problems can be extremely difficult to correct. Manuals supplied by the manufacturer of the filler/mix processor usually address troubleshooting of foaming problems in detail and should be consulted. If the problem persists, contact the filler manufacturer.

One of the problems that can result from excessive foaming at the filler, aside from the poor aesthetics of sticky packages, is the formation of mold colonies on the external walls of the package. This might also be evident in the thread areas of bottles when the cap is removed. Proper sealing of the newly filled package is a critical step in the process-

Table 10.7. Problems Resulting from Foaming at the Filler

Quality	Economics / Operations
Underfilled package	Impact on filling speed
Product residue on bottle	Loss of CO_2 and product
Incorrect CO_2 level	Increased BOD (biochemical oxygen demand) to the drain (sewer surcharge)
	Increased cost of clean-up

Source: Delonge 1994b.

Some Causes of Foaming at the Filler

- ✓ Syrup overagitation
- ✓ Dirty bottles
- ✓ Dirty filler bowl
- ✓ Exceesively carbonated product
- ✓ Warm product (inadequate refrigeration)
- ✓ Line leaks (air introduction)
- ✓ Valve failure
- ✓ Vent tube spreading rubber--wrong position or missing
- ✓ Vent tubes--scored, missing, loose, incorrect size
- ✓ Inadequate drainage after washing or rinsing
- ✓ Particulates in water
- ✓ Improper carbo-cooler operation
- ✓ Inadequate carbo-cooler capacity, or operating beyond capacity
- ✓ Incorrect centering cup insert
- ✓ Product characteristics (more common in diets)
- ✓ Incorrect setting of valve operating/snift cams
- ✓ Bent valve operating levers

- ✓ Glass quality and configuration
- ✓ Excess air or dissolved oxygen in water
- ✓ High syrup temperatures; Warm bottles
- ✓ Too high a liquid level in bowl
- ✓ Frequent start/stop operation of filler (overagitation of product)
- ✓ Carbo-cooler outlet valve not opening fully
- ✓ Incorrect bowl pressure setting
- ✓ Hot or contaminated CO_2
- ✓ Damaged snift ferrule
- ✓ Rough transfer on A-frame
- ✓ Silicate or carbonate scale/deposits from water
- ✓ Worn valve liquid seal (skirt)
- ✓ Dirty valve screens
- ✓ Poor/leaking counterpressure seal
- ✓ Worn pump seals on water, syrup, or beverage transfer pumps (air eduction)

Figure 10.10. Some causes of foaming at the filler.

---

ing of carbonated soft drinks. The closures can be a variety of different types, including crimp-on metal crowns on glass bottles, screw-on metal or plastic caps on plastic bottles, or a lid seamed onto a can body. Each of these applications requires different equipment, but the overriding objectives are the same: (1) withstand the pressure from the CO_2 in this closed system, (2) provide the consumer with a safely sealed product, and one with tamper evidence, (3) prevent leakage of product out of the package, and (4) help contribute to the visual appeal of the overall package.

After proper application of the closure or lid, some beverage manufacturing plants pass the bottles and cans through a warmer, which is a tunnel of water sprays of carefully controlled temperature. The purpose is to bring the temperature of the filled packages (still cold from the chilled carbonated water introduced at the mix processor) up to close to ambient temperature. The main reason for this is to prevent excessive condensation, which can lead to problems, depending on the secondary and tertiary packaging used.

For example, in the United States and in many countries internationally, it is common to place bottles of carbonated beverage into rigid plastic crates for transport to a retail outlet. In these instances, warming is not usually needed, since the plastic

crates are essentially inert and allow for adequate airflow and ventilation of the product. Some producers, however, perhaps because of a particular marketing promotion, will shrink wrap multiple bottles together, then place them in a cardboard case box, and then stack them on a pallet that is stretch wrapped for structural stacking integrity. In the second example, if the bottles were not warmed after filling, there is a high probability that the excess condensation would be trapped (by the shrink wrap), absorbed by the cardboard (presenting a mold risk), and then subjected to a "greenhouse" effect from the poor ventilation of the stretch wrap. It is evident that a beverage producer's job is not complete simply because the product makes it safely to a sealed container!

QUALITY CONTROL AND ASSURANCE

In this section, we will distinguish quality *control* from quality *assurance:* Control will refer to testing typically performed by the beverage plant either immediately, on-site, or at a local contract lab; assurance will refer to the subject of a broader, usually centrally managed, program (e.g., frequent testing of the product from the trade by a central corporate laboratory). Typically, the bulk of testing performed in a carbonated beverage facility falls under the cat-

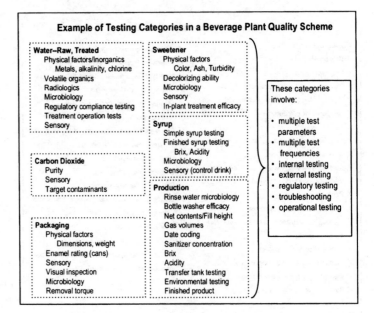

Figure 10.11. Example of testing categories in a beverage plant quality scheme.

egory of quality control. Each company prescribes its own specific testing protocol, including the parameters to test, analytic test methods to apply, and frequency. In addition, a rigorous quality program would clearly outline the actions to be taken (and by whom) in the event that this testing demonstrates an out-of-specification situation.

Since there is no single protocol for all plants to follow, Figure 10.11 summarizes the major categories of testing to consider when evaluating a beverage plant's quality monitoring scheme. This list is by no means exhaustive, but it does provide an idea of how rigorous the monitoring and control in a beverage plant should be.

In addition to this quality control scheme, most larger beverage companies have developed formalized quality assurance schemes, which are usually under centralized corporate management. The programs generally include some auditing function for compliance to standards and guidelines that includes visits to the production plants and sampling of finished products from the trade. These programs vary in terms of their focus and rigor, but trade sampling provides perhaps the best representation of what the consumers in a particular market are receiving. From this perspective, the data obtained are of extreme value and must be reviewed in concert with in-plant and external data, in order to provide the best overall picture of quality performance.

FINISHED PRODUCT

Low pH, high acidity, carbonation, and often ingredients that provide some natural antimicrobial activity (e.g., d-limonene in citrus oils) . . . all of these combine to make carbonated soft drinks a robust category of beverages. Of course, robust is a relative term and does not imply that carbonated beverages are completely immune to problems in the finished product. The formulas, however, go a long way toward providing a margin of designed product safety.

In fact, for non-fruit-juice-containing carbonated beverages, the types of problems that are typically encountered in the trade are relatively few, and rarely, if ever, present a health or safety threat to the consumer. Microbiologically, we have already mentioned the possibility of having mold form where the overall moisture in the environment is not controlled. For example, remember the scenario of freshly filled bottles, moist with condensation, that are shrink-wrapped, palletized, and stretch wrapped. The resulting greenhouse effect could easily provide the necessary conditions in which mold could grow. In finished product, although these beverages might contain a variety of organisms, these organisms will not remain viable under the conditions of the beverage. Only aciduric organisms can multiply, and these include some molds, yeasts, lactic acid bacteria, and acetic acid bacteria (Ray 2001). Of these, the clear majority of microbial problems are caused by spoilage yeast. This spoilage normally refers to any condition that affects the design appearance, flavor, or aroma of the product and is usually a problem of aesthetics where carbonated soft drinks are concerned.

In addition, as with any packaged product, the packaging materials can be a source of finished product problems. For example, misapplication of closures may occur, where removal torque is so high that consumers have difficulty opening the bottles. In areas of the world where returnable bottles are used, depending on their handling, they can become badly scuffed, presenting an unappealing look to the consumer.

Many problems with finished product can be—and are—averted before the product ever leaves the beverage facility. This is due, in large part, to diligent monitoring of the soft drink manufacturing process from beginning to end. We have already learned that the raw materials are held to high standards of quality upon receipt, and some—like water, sucrose, and CO_2—are often further purified within the beverage plant itself. Then, these raw materials are combined into a finished syrup, which is checked against standards of assembly and quality. This finished syrup is then diluted and carbonated, filled, and sealed to form the final beverage. The final product is tested chemically, microbially, and sensorially to assure that it meets the highest standards of its trademarked brand.

This said, the summary above represents only a small portion of the quality systems that overarch most finished products, and are clearly beyond the scope of this text. Suffice to say that many beverage companies begin to control quality as far back in the supply chain as possible . . . so far that some companies own their own citrus groves in order to strictly control the quality of the orange juice used in their orange-juice-containing carbonated beverages! In addition, as the principles of hazard analysis and critical control points (HACCP) become more commonplace in the beverage industry, many bottlers and canners are voluntarily formulating their own HACCP plans to formalize the monitoring and control of their processes. All of this is done with a single, predominant goal in mind . . . to provide the consumer with consistently high quality, great tasting, refreshing beverages.

APPLICATION OF PROCESSING PRINCIPLES

Table 10.8 provides recent references for more details on specific processing principles.

Table 10.8. References for Principles Used in Processing

Processing Principle	References for More Information on the Principles Used
Raw materials preparation	Journal of the American Water Works Association, 95(6), June 2003. Sugar Knowledge International (SKIL) websites (http://www.sucrose.com) Corn Refiners Association 2002.
Carbonation	Proceedings ISBT, 2000. Effects of Air on Carbonation. Proceedings ISBT, 1993. Carbonation in Aqueous Systems.
Filling, sealing and packing	Giles 2002.
Quality control and assurance	Food Safety 9(3), June/July 2003. Foster 2003. American Society for Quality website (http://www.asq.org) Clemson University website for quality control (http://deming.eng.clemson.edu/pub/tutorials/qctools/homepg.htm)

GLOSSARY

Absorption—ability of a porous solid to take up or hold a liquid or gas within itself, like a sponge takes up water.

ABCB—see *NSDA*.

Acid beverage floc—a type of floc that appears in the form of fluffy balls or granular strands at acid pHs. Disintegrates on shaking, and reforms on standing. The result of the agglomeration of any negatively charged polysaccharide with a positively charged protein molecule. As the process progresses, other particles, such as dextrans, colloidal species, and silicates are trapped and become part of the growing floc.

Activated alumina—one type of filter medium used to remove water vapor, alcohols, and trace levels of some odor-active volatile oxygenate impurities, usually from CO_2.

Activated carbon—a highly porous filter medium of vegetable origin (coconut shell, peat, wood, etc.) treated to develop a large surface area in order to create strong adsorptive forces. Used for decolorizing liquids, deodorizing, and removing contaminants. Activated carbon is well recognized for its effective removal of a wide range of organic impurities from CO_2. It is conventionally employed for purification of potable water.

Aldehydes—a broad class of organic compounds whose members are often highly flavor active (e.g., benzaldehyde, synthetic cherry/almond flavor) and that may be present as contaminants in CO_2 (especially from fermentation sources). One class of compounds in the even broader category known as volatile oxygenates.

Amylase—an enzyme that cleaves the amylose portion of starch into smaller units. Used by refineries to improve filtration characteristics of sugar solutions.

Amylose—linear, helical form of starch; forms a blue color with iodine.

Ash—inorganic constituents of sugar. May be measured gravimetrically by weighing the residue after combustion of a sample or by conductance.

Beets—a biennial root crop cultivated in temperate climates for its sugar content, which is extracted and purified to yield beet sugar.

Boyle's law—when applied to gases, if the temperature of a given quantity of gas is held constant, the volume of the gas varies inversely with the absolute pressure.

Bridging— formation of a bridge inside granulated sugar silos, due to high moisture of the stored sucrose. Problematic because it results in a resistance to free flow.

Brix—for solutions containing only sugar and water, 1 Brix = 1% sugar.

Carcinogenic— known or suspected of causing cancer in animals or humans.

Catalyst—any substance of which a fractional percentage notably affects the rate of a chemical reaction without itself being consumed or undergoing a chemical change.

CGA—U.S. Compressed Gas Association.

Charles' law—if the pressure on a given quantity of gas is held constant, the volume will vary directly as the absolute temperature; similarly, if the volume is held constant, the pressure will vary directly as the absolute temperature.

Clarification—a unit process used to remove suspended solids and colloidal substances from sugar.

CLTS—conventional line treatment systems.

CO_2—carbon dioxide; may exist as vapor, liquid, or solid (dry ice), depending on the conditions of temperature and pressure.

COC/COA—certificate of conformance/certificate of analysis; a combined document that attests that every load of CO_2 or sugar delivered to an end user conforms to their specifications (COC) and, in addition, the actual delivery of CO_2 or sugar has been tested for a variety of required parameters (COA).

Color (of sucrose)—one of the triad of quality parameters for sucrose, along with turbidity and ash.

Combustion—burning, or rapid oxidation.

Contact time—the length of time an adsorbent is in contact with a liquid prior to being removed by the filter.

COS—carbonyl sulfide; one of the high risk contaminants possible in CO_2, since it is essentially odorless by itself, but it can hydrolyze (chemically convert) to hydrogen sulfide in beverage and result in the unpleasant odor of rotten eggs.

CSD—carbonated soft drink.

Desiccant—a material capable of removing water vapor from CO_2 or other gas; common example is silica gel.

Dew point—the temperature at which water vapor begins to condense to water liquid out of a vapor-gas mixture. For CO_2, the dew point may be directly related to the moisture content of the CO_2 using a simple comparison table available from the Compressed Gas Association.

Enzyme—organic molecules of proteinaceous origin that catalyze chemical reactions, such as the cleavage of glycosidic linkages in starch (amylases).

Ethanol—C_2H_5OH; ethyl alcohol; grain alcohol; the alcohol used in alcoholic beverages and spirits, and one of the main products of fermentation.

Feed gas—usually the raw, impure CO_2 gas stream that enters a CO_2 refinery for further processing and purification.

Fermentation—a chemical change induced by a living microorganism (usually yeast, mold, or fungi) or an enzyme, which usually involves the decomposition of sugars or starches to yield ethanol and CO_2.

Flue gas—the mixture of gases that results from combustion and leaves a furnace by way of the chimney flue; usually contains oxygen, nitrogen, carbon dioxide, water vapor, and other gases.

Fructose—a monosacccharide with a molecular weight of 180.2. Like glucose, it is a reducing sugar. It is sweeter than glucose and very soluble in water.

Geosmin—a colorless, neutral oil with a pungent, earthy aroma at very low concentrations (< 1 ppb); metabolite of some algae (usually from soil); sometimes present in water supplies, sugar, etc.

Glucose—a monosaccharide that forms the backbone of starch.

HACCP—hazard analysis and critical control points.

HFCS—high fructose corn syrup.

ICUMSA—The International Commission for Uniform Methods of Sugar Analysis.

Inversion—the process whereby the sucrose molecule is cleaved into one molecule of glucose and one molecule of fructose.

Invert—a mixture of glucose and fructose.

Ion exchange—a process that uses synthetic, polymer-based materials, known as resins (in the form of small beads), to decolor and/or demineralize sugar solutions. Widely used for demineralization in the water and wine industries.

ISBT—International Society of Beverage Technologists.

Micron—a unit of length (μm) equal to 10^{-6} meters or 39/1,000,000 inches.

Mill—a cane sugar factory that processes raw cane juice into intermediate quality cane sugars such as milled white sugars.

MIS—medium invert sugar.

Molecular sieve—refers to a broad class of macroporous molecular adsorbents that can adsorb water and a variety of other constituents in both liquid and vapor phase. Many categories are available, but they frequently contain aluminosilicate or aluminophosphate components.

NSDA—National Soft Drink Association (formerly American Bottlers of Carbonated Beverages).

OSHA—U.S. Occupational Safety and Health Administation.

PET—polyethylene terephthalate.

Polysaccharides—long-chain polymers (such as starch) made up of repeating units of monosaccharides like glucose and fructose. May be branched or linear.

Raw sugar—an intermediate product used as feedstock in sugar refineries. Color varies from light brown to dark brown.

Refinery—(1) a manufacturing facility that reprocesses raw sugar into refined sugar, using a variety of unit operations, such as affination, clarification, decolorization, evaporation, crystallization, and finishing. (2) A commercial-scale CO_2 purification facility.

Refined sugar—the granular sugar obtained from a refinery process. Generally the highest purity sugar; may normally be used without further purification in bottling plants if it meets the end user specifications.

Relative humidity—the partial pressure of water vapor in air divided by the vapor pressure of water at the given temperature. Thus RH = $100p/p_s$.

Scale—the mineral deposit (calcium carbonate and other lime salts) that may form on the inside of evaporators.

Self-manufacture—the process that refers to the production of purified CO_2 by a noncommercial-scale producer, usually on the grounds of the beverage facility. Self-manufacturers fall into two general categories: (1) production plants, where a fuel is burned in a processing system with the specific intent of producing, capturing, and purifying the evolved CO_2, and (2) extraction plants, where the flue gases off an existing boiler are captured, and the CO_2 is removed and purified.

Silica gel—a desiccant, which is a filter medium that removes water vapor.

Starch—a polymer of glucose units with a backbone of α-1,4 linkages.

Sucrose—a disaccharide composed of one unit of glucose and one unit of fructose. It has the empirical formula $C_{12}H_{22}O_{11}$ and a molecular weight of 342.3.

Turbidity—(1) Any insoluble particle that imparts opacity to a liquid. (2) One of the key measures of sucrose quality, along with color and ash.

Volatile oxygenates—potential oxygenated impurities in CO_2, which may be intensely flavor active. Examples are aldehydes and ketones.

ACKNOWLEDGMENTS

The author wishes to thank the management team at PepsiCo International for the unparalleled support and encouragement they provided during the development of this manuscript. In addition, thanks go to the following people for providing their time and valuable input during the review of this manuscript:

Harry C. DeLonge
Past President, International Society of Beverage
   Technologists
President, Trilake Group
382 Leedsville Road
Amenia, NY 12501
(845) 373-8863

Lynda A. Costa
Manager, Ingredient Quality
PepsiCo International Concentrate Operations
350 Columbus Avenue
Valhalla, NY 10595
(914) 742-4893

REFERENCES

Banker G. 1996. Disperse systems. Modern Pharmaceutics, 3rd edition. Marcel Dekker, New York.

Bena D. 2001. Management and Control of Thread Mold. Pepsi-Cola Technical Bulletin.

___. 2003. Water Use in the Beverage Industry. In: YH Hui, editor. Food Plant Sanitation. Marcel Dekker, New York.

Boyce C, editor. 1986. Novo's Handbook of Practical Biotechnology. Novo Industri, Denmark.

Brittan PJ. 1997. Integrating conventional and membrane water treating systems. International Society of Beverage Technologists Short Course for Beverage Production, Florida, 1997.

CGA-6.2. 2000. Commodity Specification for CO_2. Table 3, p. 5. U.S. Compressed Gas Association.

Corn Refiners Association. 2002. Nutritive sweeteners from corn. Free for download at http://corn.org/web/pubslist.htm

Delonge H. 1994a. Sugar and Sugar Handling. Pepsi-Cola Production Manual, vol.2.

___. 1994b. Carbonation. Pepsi-Cola Production Manual, vol. 2.

Finnerty M. 2002. Sensory testing of high fructose corn syrup. Proceedings of the International Society of Beverage Technologists.

Foster T, editor. 2003. Beverage Quality and Safety. Institute of Food Technologists and CRC Press.

Galluzzo S. 2000. Sugar Quality Tool. PepsiCo Beverages International.

Giles G, editor. 2002. PET Packaging Technology. Sheffield Academic Press.

Glidden J. 2001. White Paper: Net Contents Determination by Weight. Pepsi-Cola internal document.

Granata AJ. 1946. Carbonic Gas and Carbonation in Bottled Carbonated Beverage Manufacture, Beverage Production and Plant Operation. NSDA, Washington, D.C.

Hobbs L. 1997. Sweeteners and sweetener handling systems. International Society of Beverage Technologists Short Course for Beverage Production, Florida International University, 1997.

Jacobs M. 1959. Manufacture and Analysis of Carbonated Beverages. Chemical Publishing Company, New York.

Jacobs MB. 1951. Chemistry and Technology of Food and Food Products, 2nd edition. Interscience Publishers, New York

Koch R. 2000. Worldwide high intensity sweetener marketplace. Proceedings of the International Society of Beverage Technologists.

Loomis AG. 1928. Int. Crit. Tables, vol. 3, no. 260.

McLeod E, executive director. 2001. Quality Guidelines and Analytical Procedure Bibliography for Bottler's CO_2. International Society of Beverage Technologists.

Medina, AS. 1993.Carbonation Volumes in Aqueous Solution. Proc. ISBT, 237.

National Soft Drink Association (NSDA). 1999. Frequently Asked Questions About Soft Drinks. NSDA, 1101 16th Street NW, Washington, D.C. 20036.

National Soft Drink Association (NSDA), 2003. The History of America and Soft Drinks Go Hand in Hand. NSDA, 1101 16th Street NW, Washington, D.C. 20036.

Ray B. 2001. Fundamental Food Microbiology, 2nd edition. CRC Press.

Riley John J. 1972. A History of the American Soft Drink Industry: Bottled Carbonated Beverages, 1807–1957. Arno Press.

Selz P. 1999. CO_2 Product Literature. Toromont Process Systems, Inc.

Suffet IH, editor. 1995. Advances in Taste and Odor Treatment and Control. Cooperative Research Report of the American Water Works Association Research Foundation and the Lyonnaise des Eaux. AWWA-RF, Denver, Colo.

Woodruff JG. 1974. Beverage acids, flavors, and acidulants. In: Beverages: Carbonated and Noncarbonated. AVI Publishing, Westport, Conn.

11
Beverages: Alcoholic, Beer Making

S. F. O'Keefe

HISTORY OF BEER MAKING

The origins of beer making are lost in human prehistory, before the advent of writing. The first fermentations were most likely fruits or honey, which required no preparation before sugars could "accidentally" be fermented into ethanol. Early man appreciated the effects of ethanol and the fact that fermentation and ethanol production resulted in a longer shelf life. Beer making came later, because the technology to sprout grains, dry them, and mix them with water is more complex than simply letting fruit spontaneously ferment.

The polar ice cap during the last ice age was at a maximum around 18,000–20,000 years ago. As the ice fields receded, the mild wet climate was ideal for wild grains and animals that fed on them. By 8000 B.C. the ice fields had disappeared. Many of the large animals that prehistoric humans hunted for survival became extinct due to climatic changes or perhaps overhunting. The fertile plains of the Nile, Tigres, and Euphrates rivers likely saw the beginnings of agriculture and domestication of animals. By 5000 B.C., civilizations were flourishing in these areas. Spelt, millet, wheat, and barley were grown and exported. Beer and bread production were linked and, by 3000 B.C., were a major export from Egypt. Interestingly, The Book of the Dead depicts beer and barley cakes, which were used for either beer or bread production.

It has been suggested that the transformation of prehistoric society from subsistance on hunting and gathering to agriculture was a result of the need for stable supplies of barley for use in bread and beer making (Katz and Voigt 1986). Clay tablets with recipes for beer making in the form of poems to brewing deities have been dated to 7000 B.C. (Hardwick 1995a). Other scholars suggest that firm archaeological evidence for fermentation including brewing only pushes as far back as 3500–4000 B.C. (Cantrell 2000). The fertile plains of lower Mesopotamia (currently between Iraq and Iran) or the Nile are the likely birthplace of barley cultivation, which would provide an ample supply of grain for brewing.

Mesopotamians are often given credit for the first brewing, although others argue that brewing may have originated first in eastern Africa. Beer not only

The information in this chapter has been derived from a chapter in *Food Chemistry Workbook*, edited by J. S. Smith and G. L. Christen, published and copyrighted by Science Technology System, West Sacramento, California, ©2002. Used with permission.

provided a pleasant beverage, but also calories and a source of water that was likely safer than contaminated waters. Beer drinkers may have had a "selective advantage" over others due to the increased nutritional value that the beer provided to them and their offspring (Katz and Voigt 1986). The intoxicating effect of alcohol in beer was undoubtedly also a significant factor in spread and influence of beer during early human history (Arnold 1911). Sumerians and Egyptians placed religious significance on drinking.

Brewing probably spread throughout Europe and Africa early in its history. In 1268, before he achieved sainthood, Saint Louis IX of France in 1268 enacted laws to ensure the quality of beer. Later, in Germany of the fifteenth and sixteenth centuries, laws were put into place to punish makers of bad beer with beatings, banishment, or death. The special technology and yeast used for brewing lager beer were discovered in Bavaria, and Munich became famous for it's dark, sweet, full-flavored lager. The yeast was smuggled to Czechoslovakia in 1842 by a Bavarian monk (Miller 1990). The original Pilsner was first brewed in the Burgerlisches brewhouse in Plzen, then in Bohemia. The pale color and high hopping of this new beer style rapidly became famous and was quite different from the darker, more malty, less hoppy beer brewed in Bavaria at that time. With the introduction of this new brewing technology, the cities of Plzen and Budweis became reknowned for the quality of their beer, and lager brewing in this pale style quickly spread to other parts of the world. The names Pilsner and Budweiser originally referred to beers produced in these cities, although today Pilsner has become a generic term for a beer brewed using lager yeast, and the name Budweiser has been adopted by a large American brewer. Lager yeast and fermentation technology were taken from Bavaria to America in the 1840s, and by 1844 Frederick Lauer was brewing lager in Pennsylvania (Arnold and Penman 1933, Salem 1880). Today, beer making is practiced in most countries, and beer is enjoyed worldwide.

OVERVIEW OF THE BREWING PROCESS

Beer is made using several distinct steps. There are many different procedures for the actual manufacture of different styles of beer, and an overview of the basic process can be seen in Figure 11.1.

Malting is the process whereby barley is allowed to germinate and is heated to stop further metabolism of complex carbohydrates. Many of the flavors associated with malted barley are developed during this process, through enzymatic and nonenzymatic reactions including caramelization and Maillaird reactions. Malted barley is milled to crush the endosperm and allow rapid hydration and enhanced enzymatic action.

Mashing occurs when the malted barley is mixed with water and starch is converted by α- and β-amylases to simpler sugars, which can be metabolized by yeast. By taking into account α- and β-amylase activities at different temperatures and the ability of various yeast strains to metabolize oligosaccharides, it is possible to produce beer with no, little, or high residual sweetness.

After starch conversion in the mash, the wort (pronounced wert), containing sugars, proteins, and other soluble components, is filtered from the husks in a process termed lautering. The sweet wort is boiled for 90+ minutes with the addition of hops. Hops impart the bitterness and much of the aroma to a beer; they also provide some protection against microbial spoilage. The boiled hopped wort is cooled rapidly, and after it reaches an appropriate temperature, yeast is added. The fermentable carbohydrate is metabolized to ethanol and carbon dioxide. The beer may be stored for weeks to months to develop flavors for lagers or for a much shorter time for ales.

PROCESSING OF BEER

SELECTION OF BARLEY AND OTHER GRAINS

Barley is the fourth most important cereal crop in the world after wheat, rice, and corn, and the major food use of barley is in beer making. Two main types of barley used in beer making are two-row and six-row, which indicate the number of rows of barley kernels arranged in the plant head. The two-row malt *Hordeum distichon* is commonly grown in Europe and the western United States and Canada, whereas the six-row barley *Hordeum vulgare* is more common in the upper midwest of the United States. Wheat, rye, and sorghum may also be used for beer making. Some brewers routinely use adjuncts, such as rice, corn, or sugars, which results in beer with less malt flavor and a very pale color. Regulations developed in Bavaria (the reinheitzgebot) in 1516 prohibited use of anything except

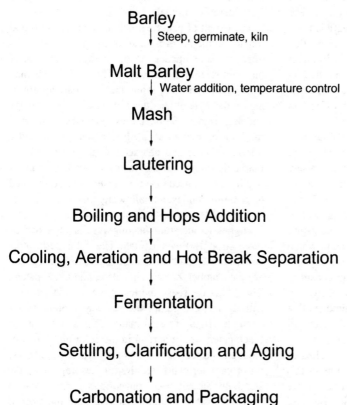

Figure 11.1. Flow chart of manufacturing process for beer.

water, hops, malted wheat, and malted barley in beer; at that time, the role of yeast in fermentation was not known. Additives that have been proposed or used over the years include enzymes, antioxidants, clarifying agents, water treatments (salts), hop extracts, adjuncts, malt extracts (liquid, dried), fruit, various flavors, and so on.

Selection of grain will, to a large extent, define the style or type of beer produced. Styles of beer vary considerably, ranging from dry, light-bodied, yellow, highly carbonated light beer with little hop aroma to jet black, creamy, sweet, very malty stouts, or even acidic and very complex lambics. The style of beer is dependent on ingredient selection, mashing, yeast and/or other microflora, and conditioning.

MALTING (STEEPING, GERMINATION, KILNING)

The malting process results in development of amylase enzymes required for conversion of starch to fermentable sugars. Nonmalted grains are used in some styles of beer, but levels are limited because they do not contain sufficient enzyme activity to convert starch to sugars. The malting process also develops flavors, which are carried to the finished beer. Consistent germination is facilitated by initial equilibration of the barley and careful size selection. The extent of germination (also known as modification) and degree of roasting or special processing can lead to malts that have different flavor, color, and starch:sugar ratios, all from the same barley. The extent of modification will also affect the amount of enzyme activity in the malt. However, it is important to arrest germination before the barley embryo has sufficient time to utilize too much of the starch present in the grain. The size of the rootlet (cull) or the acrospire length relative to the total grain length are used to determine the extent of modification of the grain.

To prepare the barley for germination, the barley is steeped (soaked) in water for two to three days at a cool temperature (12–15°C). The water is replaced several times to aid in microbial control and to replace oxygen. Barley kernels germinate to form acrospires (the embryonic plant) and rootlets. The starch-rich endosperm begins to be degraded via protease and amylase enzymes in a process called modification. The extent of modification will affect the starch and fermentable sugar levels and the levels of enzymes present. After about one week, the germination is stopped by drying below 50°C to less than 5% moisture.

Barley that is malted just enough to pass the reinheitzegebot might be germinated to a point where the acrospire is 25% of the length of the grain. Many continental lager malts have 50% acrospire development and are considered poorly modified. Many well-modified malts have had acrospire growth to 75% of the grain length. The more the acrospire growth, the greater the enzyme development, which is good for brewing. Conversely, the starchy endosperm is consumed by respiration, and considerable losses can be expected if the grain grows too much. European lager brewers had techniques that could be applied to poorly modified malts and still provide adequate hydrolysis to make beer.

Malting occurs when the modified grain is heated at 80°C or higher. The temperature and time of this heating affect the final color of the malt and, more importantly, the resulting enzyme activity. Because of the destructive effect of moist heat on enzymes, malt is usually first dried at a relatively low temperature to 2–3% moisture before high temperature exposure. The temperatures of kilning are typically 80–105°C for pale malts on a traditional 24-hour malting cycle. Temperatures may reach 225°C for several hours for dark-roasted specialty malts, and enzymes are completely denatured by this treatment. The final step in malting is the screening out of the rootlets (culls), which are undesirable.

Kilning results in Maillard and other chemical reactions that lead to color development and formation of flavor and aroma compounds. In a typical Maillard reaction, a reducing sugar in an open-chain aldose or ketose form reacts with a free amine. The Schiff base rearranges to a glucosylamine and undergoes Amadori rearrangement to a ketosamine. Since there are many different possible combinations of reducing sugars and amino reactants, Maillard reactions are very complex.

Further reactions form reductones that can polymerize to darkly colored melanoidins, or via an α-dicarbonyl and an amino group (Strecker degradation), may form any of a complex series of heterocyclic compounds. Many different Strecker degradation products with importance to beer flavor have been identified in malted barley.

The extent of heat treatment can have a profound impact on the resulting malt from barley. The malts differ in color impact, flavor, extract yield, impact on foam retention, and so on. The brewmaster may select a mixture of different malts that are appropriate for the style of beer desired, based on knowledge and experience. Some barley and wheat material may not be malted at all. Nonmalted grains add a different flavor and character to the beer and can markedly improve head (foam) stability. Chit malts that have been used in Germany are only germinated to a minimum extent, allowing their use in beer making under reinheitzgebot regulations.

Higher temperature heating results in a darker, more strongly flavored malt. The color descriptors commonly applied to malt, in increasing darkness, are pale, amber, brown, chocolate, and black patent. The darker the roast, the lower the residual enzyme activity but the higher the color and bitter flavor strength. Highly roasted malts such as chocolate or black patent are important in the color and flavor of stout and porter beers. Other malt types, called crystal or caramel malts, require no mashing, since the starch has already been converted in a process called stewing. Instead of immediately drying the steeped barley, it is heated moist to a temperature appropriate for amylase activity (65–70°C) for several hours. The endosperm is partially degraded to sugars, then the barley is kilned at high temperature. The dry heat causes the sugars to caramelize, resulting in caramel flavor and deeper color. Barley stewed for shorter times and kilned at lower temperatures avoids caramelization and color development. This type of malt is called dextrin or Cara-pils™ malt and has low fermentibility. Dextrin malts increase sweetness and body in a beer without increasing color or producing a caramel flavor. Vienna or Munich malts are kilned at a slightly higher temperature than pale malt to develop a richer malt flavor and deeper color in beer. Malt extracts (syrups or powders) can be produced but are not widely used commercially because of cost and less satisfactory flavor. Candy sugar is added to some Belgian high-gravity beers and adds to the complex flavor profiles.

The color of the malt is often described in Lovibond units (20, 60, 120, etc.), where increasing value refers to a darker, more heavily roasted malt. Generally, the heavier the roasting, the higher the color that will be imparted to the beer, but other variables such as water hardness also affect the color of finished beer. Typical pale malts have Lovibond values (expressed as °L) below 1–3 °L, whereas black patent and roast barley may have Lovibond values > 500 °L.

INGREDIENT SELECTION

The brewer must choose appropriate amounts and types of malt, hops, yeast, and water to make a beer of a particular style. Selection of ingredients is based on experience or recipes developed for a particular beer style. Since the raw materials are variable (malt flavor, enzyme activity, yeast activity, hop aroma, and α-acids, etc.), only with a great deal of experience and careful ingredient testing will beer be brewed with consistent flavor, color, and alcohol content. For these reasons, brewing is considered an art as much as a science.

MASHING

Water can vary greatly in its mineral content (hardness), color, and presence of disinfectants (ozone, chlorine, etc.). The minerals commonly present in water include sodium, calcium, magnesium, nitrate, chloride, sulfate, and bicarbonate. Temporary hardness is due to bicarbonates that can be removed during boiling. Permanent hardness results from sulfates, chlorides, or other minerals that cannot be removed by boiling. Boiling drives carbon dioxide out of solution and forces the following chemical equilibrium to the left:

$$CO_3^{-2} + CO_2 + H_2O \leftrightarrow 2\,HCO_3^{-}$$

After boiling, insoluble calcium carbonate precipitates and can be removed. High levels of magnesium ions counteract this reaction.

Several brewing regions were successful in large part because the mineral content of their water was ideal for a particular style of beer. Plzen, in the Czech Republic, has a very low water hardness with about 25 mg/l as calcium carbonate ($CaCO_3$), and the majority of hardness is temporary. Burton-on-Trent, England, has an extremely hard water, with over two-thirds as permanent hardness and a total hardness of over 900 mg/l. To achieve this water hardness, brewers may add minerals to the brewing water in a process often referred to as Burtonizing. Before adjustment of water minerals became widely employed, brewers simply had to build breweries in areas that had the water available that allowed pro-

duction of high quality beer of the desired style. Water hardness influences yeast growth and metabolism as well as other factors such as extraction of color compounds, hop flavors, rates of enzyme activity, and so on. The water of Burton-on-Trent is used so successfully for pale ales because it allows beer to be brewed with lighter color and better flavor balance than is possible in areas with softer water. The hardness was appropriate and necessary for pale ales of the highest quality, and once brewers realized this, many breweries were built in the area to take advantage of the water.

Calcium is very important since it is a major contributor to total hardness and can stabilize α-amylase. High levels of calcium limit the extraction of colored compounds and improve precipitation of protein and yeast. Addition of calcium will promote precipitation of calcium phosphate, reducing the pH. Magnesium is generally present at one-tenth the concentration of calcium, and its salts are more soluble. High levels of magnesium may contribute to undesirable astringent flavors. Sodium provides a palate fullness at low levels but can make a beer appear too salty. Bicarbonate may cause a high pH if present in excess and has been used to increase pH if needed. Sulfate promotes a dry, bitter flavor in beer. Excess sulfate may cause flavor defects and is a source of hydrogen sulfide during fermentation. Due to the complexity of salts relating to enzyme activity, pH, and solubility of flavor and color compounds, it is not surprising that particular attention must be paid to water quality to reproduce a desired style of beer.

The malt must be milled (ground or crushed) to facilitate starch conversion prior to placement in a vat for the mashing operation. This crushing is very important for two reasons: (1) inadequate grinding prevents complete starch hydrolysis, and (2) overgrinding results in slow or stuck (set) lautering. The barley husk is useful in lautering since it forms a natural filter bed, allowing rapid separation of the sugar-containing wort from the insoluble husks, hops, and other precipitates.

The barley is heated with water and is called a mash at this stage. The time-temperature conditions of mashing differ depending on the style of beer brewed. The mashing procedure may involve an acid rest, protein rest, saccharification rest, and lauter rest (mash off). Each rest refers to a time-temperature combination that is ideal for various enzymatic reactions. Mashing is basically the controlled enzymatic hydrolysis of phytin, protein, starches, and simpler

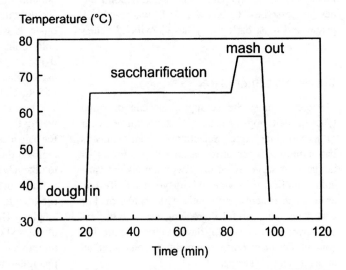

Figure 11.2. Time-temperature of an infusion mashing process.

sugars. Acid and protein rests are not always used in mashing. Phytin hydrolysis in the acid rest occurs at 35°C and results in release of phytic acid from phytin and precipitation of calcium and magnesium phosphates. An acid rest is used in cases when the water pH and grain selection result in a pH that is higher than the normal range 5.0–5.5. A protein rest is used when the nitrogen content of the barley is high enough to result in protein-derived chillhaze in the finished beer. The six-row barley varieties have higher nitrogen and may require a protein rest for wort clarification. All mashing techniques require a saccharification step to release fermentable sugars from starch.

The simplest mashing procedure is infusion mashing, a technique that is commonly used in British breweries (Fig. 11.2). This requires well-modified malts and no adjunct addition. Malted barley is added with water in a mash tun (tank) in ratios to produce a thick mash, which is heated to saccharification temperatures (62–65°C). The mash is held at this temperature until saccharification is complete. This method of mashing has the advantage that equipment is relatively simple.

Decoction mashing was developed for beer making with poorly modified malts, that is, those with low levels of diastatic enzymes (Fig. 11.3). Portions of the mash are removed (typically the heaviest one-third), boiled separately from the remaining mash, and then returned, raising the temperature of the entire mash. This sequence may be repeated two or three times. The purpose of the boiling step is to ge-

latinize or solubilize the starch, which is more easily hydrolyzed in this form by the limited enzymes in the mash. Although the boiling destroys enzyme activity in the boiled portion, the remaining enzymes act much more quickly on the gelatinized starch. Also, the stepped increase in the mash temperature allows the activity of enzymes that may be inactivated rapidly at higher temperatures. For example, β-glucanases will be active at 35–40°C and can help degrade β-glucans. Decoction mashing is more complex than infusion mashing because it requires several mash tuns and the transfer of mash back and forth between the tanks. Formerly, when temperature control was not as easy, it provided a simple mechanism to step the temperature of the mash. Modern malts are usually well modified, and the decoction mashing in many breweries has been simplified from three decoctions to two, or even one. Samuel Adams Boston Lager is an example of an American beer that uses a single decoction step.

Step mashing (temperature program) involves heating the mash stepwise at several different temperatures. Double mashing is used when adjuncts are employed in the beer. The barley is heated with water to a relatively low temperature. Cereal adjuncts (corn, rice) are boiled separately, to increase starch solubility, which results in more efficient amylitic activity. The two mashes are then combined, whereupon the temperature increases to a level appropriate for amylitic enzyme activity. Similar in some respects to the decoction procedure, an intermediate temperature step at 55°C may be held for some time

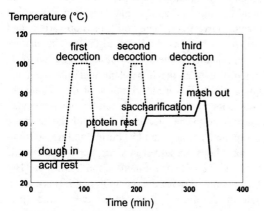

Figure 11.3. Time-temperature of a decoction mashing process.

to promote β-amylase activity. This method is useful for making dry or low calorie beer.

ENZYMES IN MASHING

The β-amylase attacks starch at α-1-4 bonds at the nonreducing end, hydrolyzing sequential maltose units from the chain (Fig. 11.4). This enzyme requires soluble starch; however α-amylase, which attacks α-1-4 bonds randomly, has some activity against ungelatinized starch. After α-amylase hydrolysis, a new site for β-amylase action is formed. Hydrolysis is limited near the α-1-6 branch points in amylopectin. Enzymes that can cleave the α-1-6 bonds are present in the unheated barley but are

largely destroyed during kilning. The dextrins that remain after hydrolysis (limit dextrins) contribute to the sweetness, body, and mouthfeel of the finished beer, but also contribute calories. There has been a lot of interest in β-glucans in beer because their presence negatively affects lautering rates and enzyme activities (Speers et al. 2003).

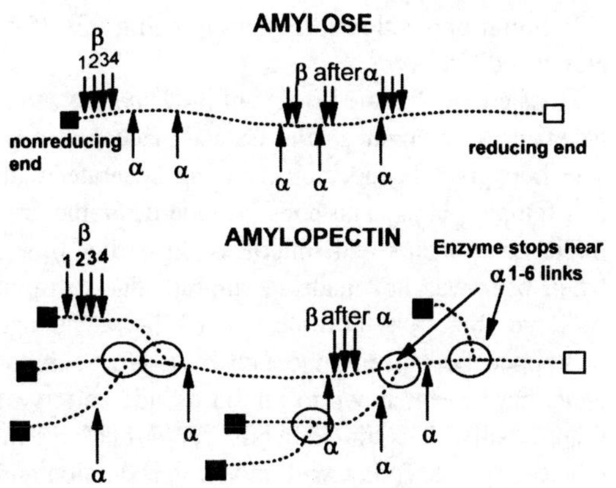

The thermal stabilities of α- and β-amylases in a mash are shown in Figure 11.5. Enzymatic activity of β-amylase is promoted at lower temperature, resulting in a great deal of starch being converted to fermentable sugar. Thus, a low temperature results in a higher ratio of fermentable to nonfermentable carbohydrate and results in a dryer, higher alcohol beer. A higher temperature mash results in fewer fermentable sugars and more complex, unfermentable sugars. This results in a product with (1) lower alcohol, because less carbohydrate is converted to ethanol, and (2) higher sweetness, because the oligosaccharides that are not fermented still provide a sweet sensory response. The trend for light, low calorie beers means that brewers tend to optimize the mash to provide a highly fermentable mixture of sugars to the yeast. Low calorie beers are produced using conditions of mashing and or yeast strains that almost completely attenuate (ferment) the carbohydrates in the beer. This removes the sweetness resulting from dextrins and unfermentable carbohydrates. Brewing low calorie beer can be aided by the addition of glucoamylase during mashing or by using green malt that contains α-1-6-glucosidase.

After the mashing is complete, it is heated to 75–77°C for up to 30 minutes (this step is often referred to as mash off). Enzymes are inactivated by the high temperatures, and the mash becomes easier to lauter (filter) because of decreased viscosity. It is

Figure 11.4. Action of α- and β-amylases on amylose and amylopectin.

Survival (% activity) of α- and β- amylase in mash

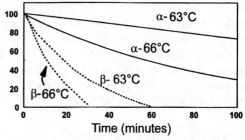

Figure 11.5. Survival of amylase activity in mash.

important that the temperature not exceed 75–77°C because this may result in extraction of unconverted starch or proteins, which can cause haze problems in the finished beer.

LAUTERING

The purpose of lautering is to remove insoluble malt husk and other materials from the sweet wort. The mash is transferred to a lauter tun, which has a porous bottom. In some brewing operations, mashing and lautering are conducted in the same vat. Lautering is basically a two-step process: filtration and extraction. The rate of filtration will depend on a number of factors, including the permeability of the filter bed, the depth of the filter bed, the viscosity of the liquid (related to temperature and sugar concentration), and the pressure applied across the bed.

The bed permeability is strongly affected by the size and shape of the husk particles that form the filter bed. Ideally, milling will leave the husks largely intact, forming a relatively permeable bed. Malt that is too finely ground can result in low permeability and slow lautering. A high percentage of wheat in the grainbill may also result in low lautering rates.

The mash is allowed to run off the bottom of the lauter tun and is pumped to the wort kettle. The first runnings of wort may be recirculated to ensure efficient filtration and removal of less soluble components and fines. Hot water (72–75°C) is sprayed on top of the mash to extract the soluble sugars. It is important that the temperature of the water in the grain bed be below 75°C and that the pH not rise above 5.7–6.0 to prevent extraction of unconverted starch, protein, astringent tannins, or other unwanted materials from the grain.

Because lautering can be time-consuming, modern techniques for wort separation have been developed, including the use of circulating rakes that cut into shallow (20–50 cm) grain beds. Effective lautering generally requires 90–120 minutes. Mash filters are mechanical filter presses that use a press cloth and may force water (referred to as sparge water) through the grain bed and compress the grain to remove sugar solutions. The presence of β-glucans and arabinoxyloses can slow filtration during lautering, and brewers try to control their levels during malting and by malt selection.

BOILING AND HOP ADDITION

After lautering, the clarified, sugar-rich mash is now called wort or sweet wort. The wort is boiled in a process that denatures and precipitates proteins and tannins. Hops are added, and the mixture is boiled for up to two hours. The boil time, hop variety, and amount of hops used will greatly affect the character and quality of the beer. Boiling is required to efficiently extract the hop resins and oils from the hops; it also aids in denaturation of microbial contaminants and removal of compounds that may cause haze in the final beer. Some of the aromatic components of hops are lost during boiling, so traditionally hops are added in steps, early in the kettle boil and again near the end of the kettle boil. Hot break is the term for the precipitate that forms as a result of boiling. This precipitate, called trub (pronounced troob), is usually removed by centrifugation, decanting, or filtration. Trub removal is usually associated with a better flavored product and is essential for quality lagers. Cold break is another precipitate that forms after the wort is cooled to fermentation temperature. Cold break is not always removed prior to fermentation because it is thought to aid yeast growth and promote a rapid fermentation.

Hops

Originally, beer was made without hops. Although hops were known in Roman times, the earliest documented use of hops for beer making was in Bavaria in the eighth century; however, other accounts suggest that monks in Gaul were the first to use hops. Many other plants were used for bittering or flavoring beer in various parts of the world. The use of hops spread throughout Europe and was common by the 1600s, and their use today is almost universal. In England, early terminology separated unhopped "ale" from hopped "beer," and there was considerable initial opposition in Britain to using this "foreign weed" in beer.

Hops are the female flowers of the *Humulus lupulis* plant, a perennial in the Cannabinaceae family. The hop plant is dioecious, having separate male and female plants. The hops provide desirable aromatic components and bitterness. In earlier times, when beer was not made in summer due to rapid spoilage, there may have been a stability advantage in hopped beer over unhopped beer. Hop components have been shown to inhibit a wide variety of gram positive bacteria and some fungi, but have no effect on yeast. Today, with modern production and pasteurization techniques, the use of hops is primarily for aroma and taste.

Hop utilization is defined as the amount of hop bittering compounds extracted into the wort. Hops are composed of flavor compounds (aromatic essential oil, bitter resins) as well as amino acids, and so on. The resins can be separated into hard resins (insoluble in hexane) and soft resins (soluble in hexane). The soft resins are primarily composed of α- and β-acids. The α-acids are mainly humulone, cohumulone, and adhumulone, while β-acids are composed of lupulone, colupulone, and adlupulone (Fig. 11.6). The α-acids are more important than β-acids for beer bittering, but neither class survives unchanged in the finished beer. Generally, the ratio of α- to β-acids in soft resin is 1:1, but ratios from 0.5:1 to 3:1 have been reported. Hops that have high levels of α-acids are described as high α hops. Both α- and β-acids are oxidized during storage, which may lead to cheesy odors. The bittering value and aroma of hops deteriorate during storage, and much emphasis is placed on proper storage and processing of hops to maintain high quality.

During boiling, the α-acids isomerize to iso-α-acids. This reaction is of great importance since the iso-α-acids are more soluble in wort than the corresponding α-acids. Without isomerization, bitterness would be very low due to limited solubility of α-acids. Solubility is affected by pH and temperature, and at pH ~5 and ~100°C, humulone is much more soluble than the corresponding β-acid, lupulone (~250mg/l versus 9mg/l). Solubility of humulone at 25°C at pH ~5 is only 50 mg/l, so much of the solubilized humulone would be precipitated on cooling.

Isomerization of humulone to *cis* and *trans* isohumulone is shown in Figure 11.7. The *cis* and *trans* iso acids are equally bitter, but hydrolysis may lead to the more soluble but weakly bitter humulinic acid. Oxidation of β-acids during storage results in

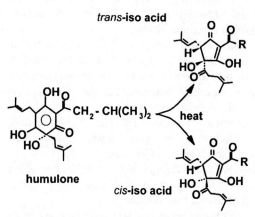

Figure 11.7. Isomerization of humulone (α-acid) during heating.

production of bitter compounds that may be important in beer bittering if old hops are used.

About 50% of total α-acids are solubilized as iso acids in the boiling wort. However, after fermentation the residual amount is usually between 10 and 40%. Utilization depends on the time of boiling (extent of isomerization of α-acid to iso-α-acids), the rate of hop usage and the gravity of the wort. Utilization is increased by longer boiling times and lower gravities (Fig. 11.8). However, the effect of time appears to be more important than gravity or hop usage rates. A 75-minute boil is fairly typical,

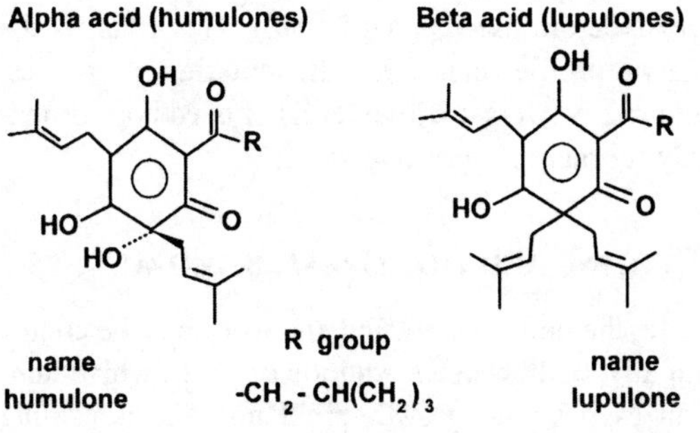

Alpha acid (humulones)

Beta acid (lupulones)

name	R group	name
humulone	$-CH_2-CH(CH_2)_3$	lupulone
cohumulone	$-CH(CH_2)_3$	colupulone
adhumulone	$-CH(CH_3)-CH_2-CH_3$	adlupulone

Figure 11.6. Alpha- and beta-acids in hops.

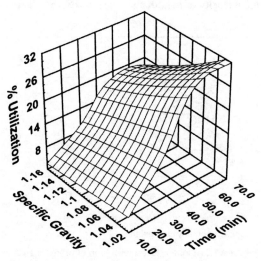

Figure 11.8. Utilization (%) of α-acids as a function of boiling time and wort specific gravity.

but boil time may range from one to two hours. Extended boiling results in darkening color in beer, which may be undesirable.

There are two general styles of hops, one for bitterness and one for aroma, but all hop varieties contribute to both. Some varieties are very aromatic but impart little bitterness (having low levels of α-acids), whereas others are highly bitter, but have poor or unbalanced aroma. It is less expensive to bitter with high α hops because lower amounts are required to provide a given bitterness. For these reasons, most brewmasters select a combination of hop varieties to attain the desired balance of bitterness and aroma for their finished beer.

The differences among hop varieties relate to the levels of essential oils that contribute to aroma and α-acids, which contribute bitterness. Some of the compounds that have been identified in hop essential oils and are associated with the pleasant hoppy flavor in beer include terpenoids, sesquiterpenoids (humulene epoxides), and cyclic ethers.

Oxidation of the hop isohumulones via light-induced reaction can cause the development of an offensive, skunky odor (sunstruck odor). This is due to an oxidative scission of the isopentenyl group on the isohumulone molecule, with free radical interaction and formation of 3-methyl-2-butene-1-thiol (isopentenyl or prenyl mercaptan; Fig. 11.9). The aroma threshold for prenyl mercaptan in beer is only 50 parts per trillion. Off flavors of this sort may also be formed by light-induced formation of hydrogen sulfide or methyl mercaptan. The flavor is more problematic in clear or green bottles, where the intensity of light that reaches the beer is greater.

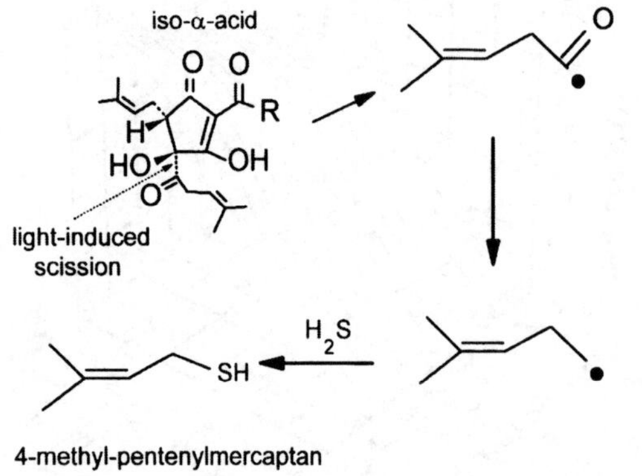

Figure 11.9. Light-induced formation of prenyl mercaptan (skunky off flavor).

Some protection is seen if the ketone is reduced to an alcohol on the side chain where scission takes place. This approach has been patented and is used by a large brewer that markets its beer in clear bottles.

Cascades, an American variety of hops, is known for the citrus-like aroma characteristics that they provide a beer. Likely limonene, α-terpineol, geraniol, or other compounds present in the hops are responsible for this characteristic. Because of the high volatility of many of the aroma compounds, hops are added late in the boiling, or even after fermentation (dry hopping). These late additions do not add to the bitterness since there is no opportunity for isomerization of the α-acids. However, they provide a strong hop aroma to the beer. Although the chemistry of hop aroma is still poorly understood, brewers have over the years developed and chosen hop varieties that either provide a high α-acid level (used for inexpensive bittering) or that are known for the fine floral hop odors they can provide a beer. High α-acid varieties include Cluster, Chinook, Eroica, Galena, Nugget, Yeoman, and Brewers Gold, while lower α-acid, high aroma/flavor varieties include Fuggles, Hallertauer, Saaz, Tettnanger, and Goldings.

Because hops deteriorate during storage, they may be processed to pellets or extracted using supercritical CO_2. Hop plugs or pellets are made by compressing milled hops; they have been widely accepted because they are more convenient to transport and store, are more easily dispersed in the kettle, maintain better quality during storage, and can be prepared with consistent α-acid levels. Hop pellets may also be processed to isomerize the α-acids, which prevents loss of these acids and minimizes off-odor formation during storage.

Lambic brewers do what would be unthinkable for most brewers by aging their hops for several years before use (Guinard 1990). This changes the flavor profile and alters the bittering properties. Lambic brewers use high levels of aged hops primarily for their preservative value.

COOLING AND HOT BREAK REMOVAL

After the boil is completed, the wort must be cooled rapidly to allow quick addition of yeast, which minimizes bacterial growth problems. Rapid cooling also promotes an efficient break (precipitation) of undesirable components. Some chemical reactions that occur in cooling wort can be detrimental to the flavor. For example, S-methylmethionine (SMM) is

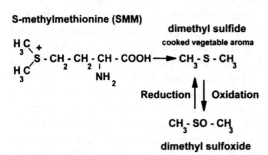

Figure 11.10. Formation of dimethyl sulfide (cooked vegetable off flavor).

converted to dimethyl sulfide (DMS), which has a cooked asparagus/vegetable aroma, by heating (Fig. 11.10). During vigorous boiling, this compound is volatilized and removed. However, slow cooling promotes formation of DMS but does not allow for loss via volatilization and can promote undesirable vegetable flavors in a beer. The SMM level is affected by the degree of roasting of the barley.

The potential alcohol formation depends on the amount of sugar present in the wort, the composition of the sugar (fermentable vs. nonfermentable), and the characteristics of the yeast selected. Sugar levels are estimated by measuring the specific gravity or the refractive index of worts. The specific gravity of the wort is tested and adjusted before yeast pitching to ensure appropriate alcohol levels and flavor characteristics in the finished beer. Since the specific gravity changes as the sugar is used up and alcohol is produced during fermentation, the original and terminal specific gravities are measured precisely. Original gravities range from 1.030 to 1.090, depending on the type of beer being produced. The final gravities range from 1.005 to 1.020. Traditionally, Scottish ales had very high terminal gravities of around 1.055 (Noonan 1993).

YEAST PITCHING AND FERMENTATION

The role of yeast in fermentation was largely unknown and poorly understood until experiments were conducted in the mid- and late 1800s by Mitcherlich, Pasteur, Buchner, and others. The first serious work to select and develop yeast strains for brewing was done at the Old Carlsberg brewery in Denmark. Until then, brewing was often inconsistent due to diseases called yeast infection or yeast turbidity. Cultures of yeast used at that time were mixed, having several strains or species of yeast. Emil Hanson, at Carlsberg, was the first to isolate pure strains of yeast and to use them in brewing in the late 1880s. Many strains of brewing yeast have been selected, and they are often closely guarded by major breweries.

Two major species of yeast are used in brewing: *Saccharomyces cerevisiae* and *S. uvarum (carlsbergensis)*. Although the species of yeast used in ale making is the same as that used in bread making, there are many differences between the two. Yeast strains for bread have been selected for fast growth and gas generation. Beer made with bread making strains would impart yeasty flavors, appropriate for bread, but not for beer. Beer yeasts must withstand ethanol at concentrations of 3–12% and be able to completely ferment the sugars provided by mashing.

The ale yeast, *S. cerevisiae*, tends to form a skin at the top of a fermentation and is thus called a top-cropping yeast. Its optimum temperature for activity is 13–21°C. *S. uvarum* settles to the bottom during fermentation and has an optimal activity below 10°C. Yeast strains within the different species have been developed over the years based on their brewing performance. It is only recently, with serological and genetic analysis, that precise relationships between various strains can be determined. Powdery strains flocculate (clump and precipitate) poorly, remaining in suspension and attenuating the wort more completely. Break strains flocculate rapidly and may precipitate before completely fermenting the wort. These strains must be roused or mixed up in the wort periodically to achieve complete attenuation.

The chemical reactions catalyzed by yeast are complex and include converting glucose to carbon dioxide and ethanol. The yeast produces ethanol as shown in the following chemical equation:

$$C_6H_{12}O_6 \leftrightarrow 2CO_2 + 2C_2H_5OH$$

Fusel alcohols are higher molecular weight alcohols that are derived principally from amino acid deamination and reduction of resulting oxo acids. Isoamyl alcohol, isobutanol, and other fusel alcohols may result in solvent, rose, or other floral off flavors. Factors that may lead to elevated levels of fusel alcohols are yeast strain, elevated amino acid levels in wort, high-temperature fermentation, continuous agitation, low yeast pitching rate (innoculation), and high ethanol concentration.

SETTLING AND RACKING

Fining is sometimes used to promote yeast flocculation and precipitation of proteins that may result in chill haze (haze that develops in chilled beer and disappears as beer warms up). Several different types of polysaccharides have been used, including isinglass (sturgeon swimbladder collagen), animal gelatin, and Irish moss (a seaweed that contains carageenans). Other polysaccharides such as alginates (propylene glycol alginate) can be used for head retention and are commercially available in high purity. Fining agents act as solid particles to which yeast adhere and flocculate. Powdery yeast strains tend to remain in suspension and more completely attenuate the wort, but their use may require aggressive fining or extended storage for acceptably clear beer.

CONDITIONING AND CARBONATION

After clarification of the fermented wort by settling and racking, beer is stored for a period to mature and develop flavors. Many of the maturation processes require yeast contact. Flavor mellowing is in part due to the formation of esters from alcohols and acids during storage. Secondary fermentation in beer is a slow fermentation that is controlled by low yeast numbers and/or low temperature. Bottle and cask conditioning, krausening, and lagering are three such techniques.

The traditional British practice of cask aging involves transfer of settled and racked beer to small oak casks, the addition of priming sugar and isinglass for clarification, and tightly closing the cask. The oak is chosen so as not to impart flavor to the beer. Hops may be added at this time in a procedure called dry hopping. Dry hopping adds a rich volatile aroma that cannot be achieved by hopping during kettle boil or fermentation because the volatiles are rapidly lost. The cask is stored for a period of time so the yeast can ferment the priming sugar and carbonate the beer before settling out. Because tapping the cask requires entry of air into the cask, the shelf life of the cask is quite short after opening. Traditional cask conditioning is becoming more popular in Britain as the interest in traditional ale increases. However, because of the inherent instability of cask beer, metal kegs are often used, with artificial carbonation and sterile filtration or pasteurization before filling.

Bottle conditioning is rarely practiced today, although some Belgian ales and other specialty beers (Sierra Nevada Pale Ale) are still bottle conditioned. The residual yeast are revived by adding sugar to the beer just prior to bottling, producing carbonation in the bottle. The amount of CO_2 needed to carbonate the beer can be exceeded by injudicious addition of sugar, and may result in exploding glass bottles, a rather common problem with amature brewers. Since the yeast is active in the bottle, a slight sediment of yeast is produced by bottle carbonation. This sediment is avoided by decanting, especially when cold. Many consumers prefer bottled beer with no sediment, and large breweries now practice artificial carbonation of fully clarified beer. This may be combined with trapping the CO_2 produced during fermentation and reusing it for carbonation.

Another traditional way to develop carbonation and mature flavors is via Krausening. In this procedure, a portion of unfermented or freshly fermenting wort is added back to the fermented beer. This results in yeast activation and action on the sugars in the wort. The beer is held at a fairly low temperature (10°C), and a slow secondary fermentation ensues. This process results in clarification, carbonation, and flavor maturation. Lagering is a process where incompletely fermented beer is transferred to cold storage (1–8°C) for a period of several weeks to months. The beer slowly finishes fermenting and develops characteristic flavors and clarity.

Diacetyl gives a butterscotch, or buttery, flavor to many lagers. At a low level, this may be desirable, but at high levels, diacetyl is a flavor defect. Diacetyl can be reduced to acetoin and 2,3-butanediol, which do not have important flavor impacts. This is traditionally accomplished by prolonged cold storage (lagering) or a short storage period of warm conditioning (14–16°C), called a diacetyl rest or ruh storage.

Filtration is often practiced for final clarification in large-scale breweries. Filtration may be preceded by treatment with adsorbents (silica gel or PVPP—polyvinylpolypyrrolidone) that remove the proteins that may cause chill haze. After filtration, beer may be force carbonated, bottle filled, and pasteurized. Pasteurization increases the shelf life substantially but alters the flavor. Sterile filtration techniques have been developed that eliminate the pasteurization step and allow bottled draft beer production.

Foam (head) retention is considered an important characteristic of a fine beer. The head of a beer is controlled by many diverse factors, ranging from the presence of detergent in the glass to the method of pouring. Many cask or keg beers are served using special pumps that generate a heavy layer of foam, some using nitrogen-pressurized dispensing through tiny holes in the dispensing tap. The nitrogen dispensing system (3:1 ratio of nitrogen:carbon dioxide) used by Guinness and some other brewers provides finer bubbles, resulting in a creamier head and smoother flavor. Recently, devices have been developed to improve the foam in canned beer. A small plastic device with pinholes is inserted into the can, which is dosed with nitrogen. On opening the can, the release of pressure causes the beer to stream out of the pinhole, and nitrogen promotes fine bubbles in the beer, resulting in a greatly improved head of foam.

The formation and stability of the head on a beer is an important quality attribute. Stability of foam appears to be primarily related to beer polypeptides, especially larger ones, levels and types of hop acids, and other factors. Addition of unmalted cereal to the grist leads to improved head retention, apparently due to high glycoprotein levels. Viscosity effects of dextrins tend to increase foam stability. Propylene glycol alginate is sometimes added for foam stabilization. This charged polysaccharide acts to protect the foam from the negative effects of partial glycerides or free fatty acids.

The art of brewing is becoming more scientific as knowledge and understanding increase. Part of the enjoyment of a fine beer lies in an understanding of the chemistry of beer making, the analysis of the flavor, color, and other characteristics that result from the brewer's art.

APPLICATION OF PROCESSING PRINCIPLES

Table 11.1 provides recent references for more details on specific processing principles.

Table 11.1. More References on Specific Processing Stages and the Principles Involved in the Manufacture of Beer

Processing Stage	Processing Principle	References for More Information on the Principles Used
Malting	Germination, heat-derived flavors	Goldammer 1999, Briggs et al. 1981
Mashing	Directed enzyme action	Houge et al. 1982, Briggs et al. 1981
Boiling and hops	Stabilize by removing microflora and proteins	Miller 1990, Houge et al. 1982, Lewis and Young 1995
Fermentation	Develop flavors and ethanol from sugars	Lewis and Young 1995, Houge et al. 1982, Helbert 1982
Conditioning	Develop flavors	Tressl et al. 1980, Houge et al. 1982

GLOSSARY

Adjunct—nonbarley grain or sugar source used in beer making.

Ale—style of beer made using top-cropping yeast *Saccharomyces cerevisiae.*

Barley—world's fourth most important cereal crop; primary use is in beer making.

Hops—female flowers of the hop plant used to provide bitterness and aroma to beer.

Lager—style of beer made using bottom-cropping yeast *Saccharomyces uvarum.*

Lambic—style of beer made with complex mixed culture of microorganisms.

Malt—sprouted and dried barley used as enzyme and sugar source in beer.

Malting—process of sprouting and drying barley that produces enzymes required for mashing.

Mash—mixture of grains and water with temperature controlled to affect enzyme activities.

Pilsner—straw yellow, bitter, highly carbonated lager beer style originating in Plzen, Czech Republic.

REFERENCES

Arnold JP. 1911. Origin and History of Beer and Brewing. Wahl-Henius Institute of Fermentology, Chicago, Ill.

Arnold JP, F Penman. 1933. History of the Brewing Industry and Brewing Science in America. U.S. Brewers Association, Chicago Ill.

Angelino SAGF. 1991. Chapter 16. Beer. In: H Maarse, editor. Volatile Compounds in Foods and Beverages, 581–616. Marcel Dekker, Inc., New York.

Briggs DE, JS Hough, R Stevens, TW Young. 1981.
    Malting and Brewing Science. Vol.1, Malt and
    Sweet Wort, 2nd edition. Chapman and Hall, New
    York.
Cantrell PA, II. 2000. Chapter 3.1. Beer and ale. In:
    KK Kiple, K Conee Ornelas, editors. The
    Cambridge World History of Food. Cambridge
    University Press, New York.
Ensminger AH, ME Ensminger, JE Koonlande, JRK
    Robson. 1994. Foods and Nutrition Encyclopedia,
    vol. 1, 2nd edition. CRC Press, Boca Raton, Fla.
Goldammer T. 1999. The Brewers Handbook. KVP
    Publishers, Clifton, Va.
Guinard JX 1990. Lambic. Brewers Publications,
    Boulder, Colo.
Haggblade S, WH Holzapfel. 1989. Chapter 5.
    Industrialization of Africa's indigenous beer brew-
    ing. In: KH Steinkraus, editor. Industrialization of
    Indigenous Food Fermentations, 191–283. Marcel
    Dekker, Inc., New York.
Hardwick WA. 1995a. Chapter 2. History and an-
    tecedents of brewing. In: WA Hardwick, editor.
    Handbook of Brewing, 37–51. Marcel Dekker, Inc.,
    New York.
___. 1995b. Chapter 4. An overview of beer making.
    In: W.A. Hardwick, editor. Handbook of Brewing,
    87–95. Marcel Dekker, Inc., New York.
Helbert JR. 1982. Chapter 10. Beer. In: G. Reid, edi-
    tor. Industrial Microbiology, 403–467. AVI
    Publishing Inc., Westport, Conn.
Hockett EA. 1991. Barley. Chapter 3. In: KJ Lorenz,
    K Kulp, editors. Handbook of Cereal Science and
    Technology, 133–136. Marcel Dekker, Inc., New
    York.
Houge JS, DE Briggs, R Stevens, TW Young. 1982.
    Malting and Brewing Science. Vol. 2, Hopped Wort
    and Beer, 2nd edition. Chapman and Hall, New
    York.

Katz SH, MM Voigt. 1986. Bread and beer: The early
    use of cereals in the human diet. Expedition
    28:23–34.
Lewis MJ, TW Young. 1995. Brewing. Chapman and
    Hall, New York.
Mares W. 1984. Making Beer. A.A. Knopf, New York.
Mecredy JM, JC Sonnemann, SJ Lehmann. 1974.
    Sensory profiling of beer by a modified QDA
    method. Food Technology 28:36–37, 40–41.
Miller D. 1990. Continental Pilsner. Brewers
    Publications, Boulder, Colo.
Noonan G J. 1993. Scotch Ale. Brewers Publications,
    Association of Brewers, Boulder, Colo.
Pasteur L. 1879. Studies on Fermentation. The
    Diseases of Beer. Translated from French by F
    Faulkner and D Constable Robb. MacMillan and
    Co., London.
Salem FW. 1880. Beer, Its History and Economic
    Value as a National Beverage. F.W. Salem and Co.,
    Hardford, Conn.
Smith G. 1994. The Beer Enthusiasts Guide. Storey
    Publishing, Pownal, Vt.
Speers, RA, Y-L Jin, AT Paulson, RJ Stewart. 2003.
    Effects of ß-glucan, shearing and environmental
    factors on the turbidity of wort and beer. Journal of
    the Institute of Brewing 109:236–244.
Strating J, BW Drost. 1988. Limits of beer flavor
    analysis. In: G Charalambous, editor. Frontiers of
    Flavor, 109–121. Elsevier Science Publishing, Inc.,
    Amsterdam.
Tressl R, D Bahri, M Kossa. 1980. Formation of off-
    flavor components in beer. In: G Charalambous, ed-
    itor. The Analysis and Control of Less Desirable
    Flavors in Foods and Beverages, 293–318.
    Academic Press, New York.

12
Grain, Cereal: Ready-to-Eat Breakfast Cereals

J. D. Culbertson

BACKGROUND INFORMATION

Many of the principles of ready-to-eat (RTE) cereal manufacture are similar for all products. The origin of many RTE cereal products can be traced to Battle Creek, Michigan, where the Kellogg brothers, C.W. Post, and others discovered and developed many novel processing technologies for turning raw cereal grains into breakfast cereals. In most processing schemes, raw grains are first cooked in some manner to gelatinize the starches present. The cooked grains are then flattened (flaked), formed (extrusion), shredded, or expanded (puffed). The moisture added during gelatinization must then be removed, usually through high-temperature drying, which is referred to as toasting.

The initial cereal grains are quite bland, which requires the development or addition of flavors. Many cereal products use caramelization or Maillard reactions, which occur in toasting, to generate desirable flavors in the finished food. Since fortification is important for product marketing, vitamins are usually a part of a cereal product formulation. For many reasons, including product protection, moisture barrier properties, flavor retention, and tamper evidence, RTE cereal packaging is a very important part of processing.

This chapter will concentrate on the making of a flaked corn grit product. Many grains can be flaked using the technologies discussed in the chapter. The most important discovery in regard to the ability to make flakes out of a cooked cereal grain relates to the chemistry of the starch contained within the endosperm. Early attempts to flake grains resulted in dismal failure. It was only after cereal pioneers discovered that cooked grains need to cool and "temper" to allow for the starch to reassociate (retrograde) that successful flaking was discovered. This simple discovery revolutionized the use and consumption of grains by humans.

Other processing techniques include puffing, shredding, and extrusion. These use processing steps

---

The information in this chapter has been derived from a chapter in *Food Chemistry Workbook,* edited by J. S. Smith and G. L. Christen, published and copyrighted by Science Technology System, West Sacramento, California, ©2002. Used with permission.

similar to those used in flaking, with several notable differences. They will be discussed after the corn flake process. Like flaking, successful puffing, shredding, and processing of extruded pellets rely on starch retrogradation for proper product processing.

RAW MATERIALS PREPARATION: DRY MILLING FIELD CORN

Field corn is an entirely different product than the sweet corn with which consumers are familiar. Field corn is allowed to mature and partially dry in the field prior to harvest in the fall. During harvesting, the kernels are removed from the cob by shelling. Field corn is typically dried on the farm prior to delivery to grain terminals or mills to prevent the growth of mold during storage. Corn kernels are dry milled to separate the germ and bran layers from the endosperm. The majority of the oil in a corn kernel is located in the germ. Removal of the oil assists in protecting the finished food from oxidative rancidity (see the section Finished Product: Packaging, below, for further information on rancidity). The bran layer (hull) is removed because it contains a variety of fibers (e.g., cellulose, hemicelluloses) that interfere with many processing procedures including cooking and flaking. Inclusion of the hull would also result in flakes that were very dissimilar in appearance and texture. The milled endosperm is referred to as a flaking grit. The bran and germ factions are further processed into other products including corn bran and corn oil. Typically, U.S. No. 1 or 2 yellow dent corn is used for the production of flaking grits (Caldwell and Fast 1990).

The typical flaking grit is approximately one-third the size of the original kernel. Each finished cornflake is essentially one processed grit, although two or three grits may occasionally stick together and result in a large flake. In the cereal industry these large flakes are called overs.

PROCESSING

STAGE 1: ADDITION OF COOKING LIQUOR

The grits must be cooked prior to flaking. Much of the flavor of cornflakes is due to the addition of sugars, proteins (or amino acids), and salt to the cooking water (cooking liquor). In a typical formulation six pounds of sucrose, two pounds of malt syrup, and two pounds of salt are added to 100 pounds of grits with enough water to yield cooked grits containing about 28–34% moisture (Caldwell et al. 1990). Salt is added to improve flavor. Malt syrup contains reducing sugars (maltose and glucose) and proteins or free amino acids that are critical to the creation of desired flavors and colors due to nonenzymatic browning, as discussed in the following sections. Malt syrup is made from barley using a controlled germination step. The barley is equilibrated to about 18% moisture, which enhances the synthesis of starch-degrading enzymes in the barley kernel. Sprouting of the seed is common. The barley is held for approximately four to six days, during which time much of the starch present in the barley is converted to maltose and other reducing sugars. Malt is the dried and ground germinated barley kernels produced in this process. Malt syrup is the concentrated water extract of the dried malt. Malt syrup used in cereal manufacture does not contain residual active starch-degrading enzymes since they would soften the grit and destroy desirable milling properties. The sugars that are not reducing (e.g., sucrose) do not react during the cooking process and may contribute to the residual sweetness of the product (Kujawski 1990). The sugar, malt syrup, and salt are dissolved in water to make the cooking liquor that is added to the grits. This is usually accomplished using a batch kettle. A batch kettle is simply a vessel that uses some form of agitator to stir the water as the sugar, malt syrup, and salt are added. The addition of dry ingredients can be automated in a variety of ways.

If the finished cornflakes are fortified, heat stable vitamins and minerals may also be added in the cooking liquor (Borenstein et al. 1990). Addition at this step in processing improves distribution in the finished product. An example would be a source of dietary iron.

STAGE 2: COOKING

Weighed amounts of raw corn grits and cooking liquor are loaded into batch cookers. Batch cookers are cylindrical stainless steel steam pressure cookers that are typically four to eight feet long and rotate at one to four revolutions per minute (rpm) during processing (Caldwell and Fast 1990). The tumbling action of the cookers provides sufficient agitation to keep the grits separated while cooking. The grits are cooked for approximately two hours at 15–18 psig of steam pressure. Cooking is complete when the original hard, white grits have turned a golden brown color and are soft and translucent. Incomplete cooking results in grits with white centers that will carry through processing and result in cornflakes with white spots. The moisture content of the cooked grits should be 28–34%.

Reactions of Interest during Cooking

The reactions of interest during the cooking process are (1) starch gelatinization, (2) Maillard browning, (3) Strecker degradation, and (4) lipid oxidation.

Starch gelatinization is the process in which starch granules absorb water that changes the appearance and texture of the grit. Thermal energy disrupts the bonds within the amylose and amylopectin fractions in starch granules and allows water and other molecules to hydrogen bond with the exposed hydroxyl groups of the glucose polymers. The result is a change in the appearance and texture of the grit. The original hard, white grits become translucent, pliable, and somewhat rubbery.

During cooking desirable changes in flavor and color occur. The grits turn golden brown and obtain cooked or toasted flavors. The reactions in Figure 12.1 illustrate the major pathways for the formation of flavors and colors in cornflake processing. Note that Maillard browning is the predominant browning pathway during cooking of grits.

Named for the French chemist who first studied it, the Maillard reaction is the primary source of color and flavor changes during the cooking process (Daniel and Weaver 2000). The reaction proceeds readily at typical grit cooking temperatures (ca.

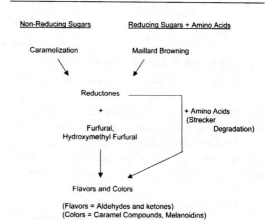

Figure 12.1. Major pathways of flavor and color formation in cornflake processing.

120°C or 248°F). The Maillard reaction is also called the carbonyl-amine reaction and is due to the condensation of an aldehyde or ketone with an amine to form aldosylamines or ketosylamines (Bean and Setser 1992). This reaction is illustrated in Figure 12.2.

Reductones may form reactive intermediates such as 5-hydroxymethyl-2-furaldehyde (hydroxymethyl-furfural or HMF) or important cooked flavors such as maltol and isomaltol. Reactants such as HMF may self-polymerize or react and/or polymerize with amino acids (discussed below) or proteins to form tan to golden brown polymers. If the polymers

---

$$C_6H_{12}O_6 + R\!-\!NH_2 \;\rightarrow\; RN\!=\!CH\!-\!(CHOH)_4\!-\!CH_2OH + H_2O$$

Glucose   Amino Acid   Aldosylamine (Schiff's Base)

Rearrangement yields highly reactive intermediates such as 1,2-eneaminol or 2,3-eneaminol shown below:

$$RNH\!-\!CH\!=\!COH\!-\!(CHOH)_3\!-\!CH_2OH \qquad RNH\!-\!CH_2\!-\!COH\!=\!COH\!-\!(CHOH)_2\!-\!CH_2OH$$

1,2 -eneaminol   2,3 -eneaminol

These further rearrange to form reactive intermediates called "reductones" such as 1- or 3-deoxyosulose shown below:

$$HCO\!-\!CO\!-\!CH_2\!-\!(CHOH)_2\!-\!CH_2OH \quad or \quad CH_3\!-\!CO\!-\!CO\!-\!(CHOH)_2\!-\!CH_2OH$$

3- deoxyosulose   1- deoxyosulose

Figure 12.2. The Maillard reaction: A condensed view.

$$HOOC-CNH_2-R_1 \; + \; R_2-CO-CO-R_3 \; \rightarrow \; R_2-CO-CNH_2-R_3 \; + \; CHO-R_1$$

| Amino Acid | Reductone | Amino ketone | Strecker |
| | (Dicarbonyls) | | Aldehyde |

Figure 12.3. The Strecker degradation.

are nitrogen free, they are known as caramel colors. If they contain nitrogen, they are called melanoidins.

Reductones generated in Maillard browning may further react with free amino acids in a sequence known as Strecker degradation, which is illustrated in Figure 12.3.

Aminoketones formed in the reaction may condense with other intermediates to form melanoidin pigments, or they may condense with another molecule of an aminoketone to form alkypyrazines (Namiki and Hayashi 1983). Pyrazine formation is of interest due to their carcinogenicity in rodent studies (Shibamoto and Bjeldanes 1993). Strecker aldehydes, formed from the residual carbon skeleton of the participating amino acid, as shown above, also contribute to the flavor of the cooked product.

The products of nonenzymatic browning have also been shown by various researchers to have antioxidative effects (Bean and Setser 1992), and products in which caramelization, Maillard reactions, and Strecker degradations occur are thought to have improved shelf life. It is not known which of the intermediates or products of these reactions are re-sponsible. Lipid oxidation proceeds via a free radical mechanism (Dziezak 1986). Peroxide decomposition is a key step in oxidative rancidity: it not only forms two new radicals that further promote oxidation, but is also the mechanism by which volatile fragments are formed from the original fatty acids. These reactions are outlined in Figure 12.4.

From 18-carbon fatty acids, carbonyls (aldehydes and ketones) of 3, 5, 6, 9, and 12 carbons are common. The mixture depends on the food system (catalysts and pathways of oxidation) and the number of double bonds in the original fatty acid. Examples of common carbonyls released during the oxidation of lipids in cereals include malondialdehyde, pentane, hexanal, hexenal, nonanal, and dodecanal. These volatiles are responsible for the undesirable aromas from rancid products often described as the aroma of wet cardboard or freshly mown grass.

Based on their structure, it is feasible that glycosylamines, reductones, and caramel/melanoidin pigments may act to scavenge free radicals from a food system and therefore delay the onset of oxidative rancidity. Theorized reactions are shown in Figure 12.5.

Initiation (L = Lipid)	$LH \rightarrow L^\bullet + H^\bullet$	Electron removal by oxidant; Formation of free radicals
Propagation	$L^\bullet + O_2 \rightarrow LOO^\bullet$	Formation of peroxy radical
	$LOO^\bullet + LH \rightarrow LOOH + L^\bullet$	Hydroperoxide; new lipid radical
	$LOOH \rightarrow LO^\bullet + {}^\bullet OH$	Hydroperoxide decomposition
	$LO^\bullet + {}^\bullet OH + 2\,LH \rightarrow LOH + 2\,L^\bullet + H_2O$	Formation of two new radicals

Figure 12.4. Pathways of lipid oxidation.

Possible antioxidant activities of browning intermediates (BRN-OH):

Preferential oxidation:	$BRN\text{-}OH \rightarrow BRN\text{-}O^\bullet$	(Formation of a stable radical)
Removal of oxygen:	$4\,BRN\text{-}OH + O_2 \rightarrow 4\,BRN{=}O + 2\,H_2O$	
Radical scavenging:	$L^\bullet, LO^\bullet, {}^\bullet OH + BRN\text{-}OH \rightarrow LH, LOH, H_2O + BRN\text{-}O^\bullet$	

Figure 12.5. Mechanisms of inhibiting lipid oxidation.

We have concentrated on Maillard reactions because caramelization of nonreducing sugars such as sucrose requires much higher temperatures (generally greater than 150°C or 302°F) and is not thought to occur during the cooking process in cereal manufacture.

STAGE 3: DELUMPING

Even with the constant rotation of the cooker, cooked grits are often agglomerated into large masses that must be separated into individual pieces that can be dried, tempered, and flaked. A common feature of most lump-breaking machines is the incorporation of large volumes of air, which is drawn through the equipment. The air helps cool the product and leads to the formation of a skin on the surface of the grit that reduces its stickiness. Lump-breaking machines generally consist of rotating drums that have finger-like projections on their surfaces. A common design is to have two rotating drums whose projections intermesh. As larger agglomerates pass between the projections, they are broken up into individual particles or grits.

STAGE 4: INITIAL GRIT DRYING

Cooked grits with a moisture content of 28–34% are too soft to be flaked. The grits are dried to about 14–17% moisture using forced-air dryers operating at 250°F (121°C) or less (Miller 1990). A typical dryer configuration consists of a wide slotted conveyor belt that passes through a chamber in which the temperature, humidity, and airflow can be controlled. Drying must be carefully controlled to prevent case hardening (the formation of a thick, tough skin on the grit surface), which would greatly reduce the rate at which moisture could be removed from the grit. The last section of the dryer is designed to cool the grit back to ambient temperature by passing ambient or cooled air over the grits.

STAGE 5: TEMPERING

In cereal manufacture, tempering usually follows a drying or cooling step and is the period during which the product is held in bins to allow for the equilibration of moisture within and among the particles. Grits must be cooled to less than 100°F (38°C) prior to tempering to prevent the darkening of the product due to Maillard browning (Miller 1990). Originally, tempering required 18–24 hours;

however, the use of modern grit driers with controlled humidity has reduced this to approximately 2–3 hours. Tempering bins are simple in design and usually consist of a large, enclosed, slow-moving conveyor belt or screw. The speed of the belt or screw revolutions is adjusted to allow for proper grit tempering prior to flaking.

During tempering, the gelatinized starch in the grit begins to retrograde. This reassociation of amylose and amylopectin chains results in an increasingly crystalline structure in the gel and an increase in the firmness of the cooked grit. Retrogradation is accelerated by the use of temperatures below 80°F (27°C) during tempering. Properly tempered grits flake readily, while grits that have not undergone the proper degree of retrogradation tend to be gummy and impossible to flake.

STAGE 6: FLAKING

Flaking of tempered grits is accomplished on a flaking mill. The mill consists of two massive steel cylinders called rolls. The position of one roll is adjustable so that the distance between the two rolls (the roll gap) can be set to produce a flake of the desired thickness (Fast et al. 1990a). In processing, the roll gap is commonly referred to as the nip. Flake thickness dictates the texture of the finished product and must be monitored carefully. In general, thin flakes are crisp while thick flakes are tough. If the flakes are too thin, excessive breakage will occur in the later stages of processing, such as packaging. The rolls rotate inwardly towards themselves and pull grits through the gap. The most popular sizes of rolls are 20 and 26 inches in diameter and 30 and 40 inches in length. Rolls are made of chilled iron or alloy-iron casting. The flaking pressures generate large amounts of heat, which could eventually cook flakes to the rolls and stop production. Flaking rolls are hollow so that cooling water can pass through their interior and cool them, preventing temperature increases on the roll surface.

STAGE 7: TOASTING

Flakes are difficult to toast uniformly on a flat surface because the edges toast or brown more rapidly than the center. Therefore, flakes with perfect color in the center would have overtoasted, dark brown edges. For this reason, rotary toasting or fluidized bed ovens, which toss or suspend flakes in heated air, are the standard for corn flake manufacture.

Most toasting ovens operate at 450–600°F (232–315°C) (Fast et al. 1990b). A fluidized bed, toasting oven consists of an insulated chamber into which heated air is delivered at high velocities, which suspends the product over a conveyor belt. Typical ovens are between 30 and 60 feet (10–20 m) long. Suspending the product improves the efficiency of heat transfer and moisture removal from all regions of the flake and produces a uniformly toasted finished product. Flakes travel through the oven in about three minutes, during which time their moisture content is reduced from 14–17% to 2–3%.

During toasting, product temperatures are elevated to the degree that caramelization of non reducing sugars such as sucrose occurs. The reaction results in the dehydration of the sugar molecules and the formation of reductones (such as 1 or 3-deoxyosulose), HMF, and caramel or melanoidin pigments. Amino acids may react (Strecker degradation) with caramelization intermediates in a fashion identical to those produced by Maillard browning. Compounds that inhibit lipid oxidation may also be formed during toasting operations. It is perhaps more critical that they be formed in this step, since the lipids are under severe oxidative stresses during the toasting operation.

STAGE 8: VITAMIN APPLICATION

After toasting, flakes may be sprayed topically with the vitamins that would not have endured cooking, drying, flaking, or toasting (Kujawski 1990). Examples would include vitamins A, B_1 (thiamin), C (ascorbate), and E (tocopherol). Flakes are conveyed under a fine mist of an oil-based or emulsified vitamin spray. Sprays are highly concentrated to limit the amount of moisture or lipid that is added to the toasted flake. Added moisture would be detrimental to flake texture because it would reduce crispness.

STAGE 9: COATING

Many cereal products are given a final topical application of sugars or flavors. These are applied in a variety of ways including sprays and coating drums. Sprays are typically applied onto a moving conveyor belt of product. Sufficient mixing must occur to ensure proper distribution of the flavor or coating. For many applications, the most efficient means of coating the product is to apply a spray inside a rotating drum where the product is gently tumbled. These coating drums are usually inclined planes that allow the product to tumble down via gravity, or inclined screws that transport the product up an incline while

gently mixing it. Flakes, puffed, shredded, and extruded products can all be coated in this manner. In some cases, the products are put through a short drying process to remove the moisture associated with sugar application. Sugars that crystallize readily, such as sucrose, are preferred for coatings, because they will form opaque glazes that give the product a "frosted" appearance. Since a frosted appearance is a desirable characteristic, many coated products contain 50% or more of the sugared coating. Crystalline sucrose is not highly hygroscopic and will not cause moisture to deposit on the flake surface. Subsequently, the flakes remain separate and pourable. If the coating contains substantial amounts of glucose and fructose it can be very hygroscopic and not only pull moisture from the ambient air, but also cause separate cereal pieces to glue together. Some processors use small amounts of oil to prevent clump formation.

Coatings can also be used to extend the length of time a cereal product can be soaked in milk before becoming soggy. This is commonly called "bowl life" in the cereal industry. Besides sugars, dextrins and maltodextrins and other carbohydrate-based polymers can be sprayed onto the cereal pieces to extend bowl life. Polymeric materials are used in systems where bowl life needs to be extended, but added sweetness is not desired.

ALTERNATIVES TO THE FLAKING PROCEDURE

Puffing

There are two traditional methods used for puffing. One uses high-temperature ovens or towers. The other uses puffing "guns." The concept in all three is that quick exposure to very high temperatures will cause the moisture in the grain to covert rapidly to steam, which will expand the endosperm as it tries to escape. The puffing of popcorn is a similar process, where the moisture inside the kernel is converted to steam, which eventually causes the hull to rupture. When the steam is released, the endosperm of the popcorn kernel is puffed and expanded. Rice and corn can be puffed using all three techniques, while wheat and oats only work well using puffing guns.

For corn or rice puffing, endosperms are prepared in the same manner as flaking grits up through the tempering stage. Rice is then passed through a flaking mill in which the roll gap is set so that the kernels are just slightly compressed (or bumped). Bumping is required for proper expansion and is believed to facilitate expansion by partially rupturing the kernel's native structure. Exposing the kernels to very hot (550–650°F, 290–345°C) oven chambers will result in puffing. Some processes use puffing towers, in which cooked, tempered kernels fall by gravity through a zone of very hot air or even a natural gas flame.

Puffing guns are so called because of the audible nature of their operation. In some types, water and grain are added to a heated puffing gun, which is closed, locked, and rotated for approximately 20 minutes. When the pressure reaches approximately 200–250 psi, the gun is opened rapidly. The sudden drop in pressure causes the hot grains to rapidly expand. The puffed mixture is then screened to remove unpuffed kernels and may be dried to achieve the desired final moisture, usually between 2 and 4%.

Shredding and Extrusion

Besides flaking and puffing, which we have discussed, there are two other basic methods for manufacturing RTE breakfast cereals that deserve mention: shredding and extrusion. The chemistry of what occurs in their processing is essentially similar to what we have already learned from the flaking process, but the process itself is quite different.

Shredding. The manufacture of shredded wheat products involves the shredding of wheat kernels into long strands that can be laid out in large mats (webs), which are pressed and cut to the appropriate size. Wholegrain wheat is traditionally used for making shredded products. It is cooked in rotary cookers in water only. After cooking, it is tempered to allow for moisture equilibration and retrogradation of the starch. The kernels are then passed through shredding rolls that contain V-shaped grooves to cut the kernel into long strands. Interestingly, a single wheat kernel can provide a shred that is over two feet (30 cm) long. The shreds are assembled into webs that are up to one to two inches (3–4 cm) thick and three to nine feet (1–3 m) wide. They usually contain 10–20 layers of shreds. The mats are pressed into the appropriate sizes for the finished food, and the edges of the pieces are sealed by this pressing operation. Cutting of the web produces the proper size sealed pieces. The cereal pieces are then toasted using several different types of ovens, including fluidized bed driers and band ovens. Band ovens are simple stainless steel conveyors that carry product through a hot oven environment. The final moisture content is generally below 5%. Besides vitamin application, concentrated sucrose and other sugar solutions can then be applied using sprays to provide a frosted appearance.

Extrusion. Basic extrusion involves the use of various types of grain flours and can include particulate material such as bran. Extrusion has been widely adopted by the industry because it allows for the mixing of various cereal grain flours and other ingredients to produce unique products. It is also more efficient because it combines many of the typical processing stages into one. Extruded materials, called doughs, can be toasted, puffed, flaked, and even shredded to make a number of different types of cereal products. Cereal flours and other ingredients are metered into one end of a long screw inside a steel barrel. Modern extruders may have multiple screws. The screws serve several functions, including mixing, shearing, and pressurization of the extruder barrel. On the end of the barrel is a forming die, which allows the pressure inside the barrel to build. As the pressure builds, the temperature of the mixture increases. Extrusion is a very complex process. Modern extruders have multiple screws and a number of zones where heating, steam injection, and cooling can occur. In addition, vitamins, flavors, and colors may be metered into various zones in the extruder. As the dough is extruded, it can be sliced off the face of the die to form the basic shape of the cereal. Depending on the temperature and pressure at the die, products can be flakelike, rings, or any of a number of other shapes and textures. If the last stage of the barrel is high temperature, significant puffing can be achieved. Extrusion can also be used to produce dough pellets for further processing. The pellets may be tempered and pressed into flakes, puffed in puffing ovens or guns, or shredded into web-based products.

Since many extruded products do not see temperatures high enough to cause browning or caramelization inside the extruder barrel, they are many times toasted in another operation. Many extruded products would have poor color if not toasted in this manner. Vitamin sprays and coatings can be applied using methods similar to that used for flaked products.

FINISHED PRODUCT: PACKAGING

Product packaging has many functions including product protection, identification, tamper resistance, and consumer attraction and appeal. From a food chemistry standpoint, packaging serves to protect the product from undesirable moisture gain and often serves as a delivery system for antioxidants (Coulter 1988). Moisture gain leads to loss of flake crispness and acceptability.

The instability of lipids is a particular problem in dry cereal products such as cornflakes, even though they contain relatively low levels of unsaturated fats (0.5–2%)(Coulter 1988). Lipid oxidation leads to the formation of peroxides in the food product. Peroxides decompose to form a variety of volatile aldehydes and ketones that are associated with the aroma of "rancid" food products (Schmidt 2000). In most cases, antioxidants are required to stabilize the product. Commonly used antioxidants in cereal products are butylated hydroxyanisole (BHA) and butylated hydroxytoluene (BHT). Their use is limited to 50 parts per million, either singly or as a total concentration. Antioxidants may be added directly prior to cooking, but BHA and BHT are nonpolar and disperse rather poorly in this situation. Additionally, BHA and BHT are volatile and may not carry through the manufacturing process. Several decades ago it was discovered that due to their volatility, these antioxidants could be incorporated into the packaging material and would quickly migrate into the product after packaging (Coulter 1988). Since BHA, BHT, and the lipid they are trying to protect are all nonpolar, the antioxidants, entering the product in this fashion, will tend to concentrate where they are needed most.

Common packaging materials for cornflakes are waxed paper or various polymer films such as polyethylene (Monahan 1988). Because of their wide application for most types of cereal packaging equip-ment, polymeric films are predominantly used. Important considerations in packaging materials are water vapor transmission and flavor barrier properties. Depending on the formulation and type of processing, cereal products vary in their need for moisture protection. Cereals with many hygroscopic components, such as sugars and other simple carbohydrates, may require more substantial packaging materials to protect them from rapid loss of texture due to moisture uptake during storage prior to sale. Manufacturers might also consider the region of the United States or the world where the product may be marketed. For example, products sold in very humid regions of the United States might be packaged in films that provide higher levels of moisture protection than those sold in more arid regions.

Flavor barrier properties are also very important. Many flavors used in cereal products are volatile. Loss of flavor during storage may be due to absorption of the volatiles by the packaging materials or by their migration through the liner itself. Manufacturers of packaging materials have developed a number of products that have the ability to trap or otherwise slow down the process by which volatiles are lost. This type of packaging material is a flavor barrier.

Once the proper film is selected, the packaging equipment is adjusted for its use. In a typical bag-in-box operation, pouches or bags are formed from the film, filled with product, and placed in the carton.

APPLICATION OF PROCESSING PRINCIPLES

Table 12.1 provides recent references for more details on specific processing principles.

Table 12.1. References for Principles Used in Processing

Processing Stage	Processing Principle(s)	References for More Information on the Principles Used
Milling	Remove hull	Potter and Hotchkiss 1995
Cooking	To gelatinize starch; flavor/color development	Daniel and Weaver 2000
Grit cooling/tempering	Separate cooked grits; allows for starch retrogradation	Caldwell et al. 1990.
Flaking	Produce flakes of desired thickness	Caldwell et al. 1990
Toasting	Develop flavor/color/crispness	Daniel and Weaver 2000
	Develop antioxidants	Fast et al. 1990b, Antony et al. 2002
Vitamin application	Add nutrients	Borenstein et al. 1990
Packaging	Moisture protection; deliver antioxidants	Monahan 1988

GLOSSARY

Amino acids—building blocks of protein; contain a carboxylic acid and amino functional group; amino group participates in Maillard browning.

Antioxidants—molecules that scavenge free radicals and oxygen in foods; slow down oxidative rancidity.

Antioxidants, carry through—the ability of an antioxidant to survive processing and be present in finished food.

Antioxidants, from nonenzymic browning—intermediates of caramelization or the Maillard reaction have demonstrated antioxidant activities in food processing.

BHA (butylated hydroxyanisole)—common synthetic phenolic antioxidant .

BHT (butylated hydroxytoluene)—common synthetic phenolic antioxidant.

Bowl life—length of time a cereal remains crisp when exposed to milk.

Bumping—processing of tempered rice kernels by passing them through a flaking mill with the roll gap set wide. Disrupts native kernel structure to allow for desired expansion during puffing.

Caramelization—series of dehydration reactions in food sugars that leads to the formation of brown polymers and a variety of caramel flavors.

Carbonyl-amine—reaction between amino group on amino acid or protein and an intermediate of caramelization or Maillard reaction.

Case hardening—the formation of a dense, hard layer on the surface of a material being dried; due to drying at too fast a rate; prevents further efficient drying in the product.

Cooking liquor—mixture of sugars, malt syrup, salt, and flavors in which corn grits are cooked.

Cooking—heating of corn grits in cooking liquor at elevated temperature and pressure to rapidly cause the gelatinization of starch contained in the grit; cookers rotate to assist in grit dispersal and heat distribution.

Cornflakes—product made from cooked, flavored, toasted, fortified corn grits.

Delumping—breaking up lumps of corn grits coming from cooking operation.

Dry milling—removal of germ and bran layer of dry corn kernels by abrasion.

Dryers, rotary—used to remove excess moisture from cooked corn grits to allow for tempering.

Dryers, fluidized bed—high velocities of heated air used to suspend and toast flaked corn grits to achieve desired color and flavor.

Flaking—flattening of cooked, tempered corn grit by passing through a small gap between two massive steel rolls.

Flavor barrier—properties of certain polymers that block the migration of flavor molecules through a food package.

Gelatinization—use of heat and moisture to disrupt the hydrogen bonding between strands of amylase and amylopectin in starch granules.

Grits—corn endosperm obtained by dry milling.

HMF (hydroxymethylfurfural)—an intermediate of browning pathways that can form food flavors and colors.

Hygroscopic—the property of a material that relates to its ability to remove moisture from the surrounding environment.

Maillard browning—reaction of carbonyl and amino groups to cause the dehydration of sugar molecules; can occur at ambient temperature; requires reducing sugars as reactants.

Malt syrup—mixture of dried malted barley and water. Malted barley is prepared by allowing barley to germinate and produce amylases that degrade the barley starch into maltose and other sugars.

Nip—another term for roll gap; the distance between the rolls of a flaking mill.

Nonreducing sugars—sugars with a free aldehyde group on carbon 1 or a free ketone group on carbon 2; sucrose is an example.

Overs—large cereal flakes that result from flaking two or more cooked corn grits at once. Due to grits sticking together after cooking and cooling.

Oxidative rancidity—see Rancidity, oxidative

Packaging materials—materials used to form pouches or bags for placement of finished food. Older forms included waxed paper. More modern materials are polymers such as polyethylene.

psi—pounds per square inch.

psig—pounds per square inch gauge.

Puffing guns—devices that use heat and pressure to cause cooked, tempered cereal grains to expand rapidly creating a porous structure.

Puffing tower—device through which cooked, tempered rice or corn can be passed by gravity. Hot air or natural gas flames heat the product extremely fast, causing puffing of the kernel.

Pyrazines—heterocyclic ring structures that contain carbon and nitrogen; mutagenic and carcinogenic in mice and rats.

Rancidity, oxidative—reactions of molecular oxygen and unsaturated fatty acids that lead to formation of off flavors and odors.

Rancidity, volatiles from oxidative—fragments of original fatty acids contained in the triglyceride; usually six to nine carbon aldehydes and ketones that have an aroma similar to drying paint, cut grass, or wet cardboard.

Reducing sugars—sugars with a free aldehyde on carbon 1 or a free ketone on carbon 2; capable of reacting in the Maillard reaction; can react with metal ions and reduce them to a valence of zero.

Reductones—intermediates of browning pathways that can donate electrons to other molecules.

Retrogradation—the reassociation of amylase and amylopectin strands in a gelatinized starch mixture.

Roll gap—distance between the two rolls of a flaking mill. See Nip.

RTE—Ready-to-eat.

Shred—a thin strand made by passing wheat kernels through special rolls that peel the wheat kernel into a long shred.

Strecker degradation—the reaction of amino acids and products of browning reactions that leads to the release of carbon dioxide and the formation of new aldehydes; can ultimately lead to the formation of pyrazines.

Tempering—storage of cooked, dried corn grits to allow for moisture equilibration and starch retrogradation.

Toasting—use of high temperatures to accelerate caramelization and the Maillard reaction in flaked corn grits; usually done using fluidized bed toasting ovens.

Toasting oven—use of high temperature, high velocity air to toast corn flakes or other cereal products.

Vitamin stability—the ability of a vitamin to withstand processing parameters including high temperatures and oxygen exposure. Heat stable vitamins can be added in the initial cooking stage; less stable vitamins are sprayed on the finished, toasted cereal flakes.

Web—an ordered accumulation of shreds that can be pressed and cut into individual cereal pieces.

REFERENCES

Antony SM, Han IY, Rieck JR, Dawson PL. 2002. Antioxidative effect of Maillard reaction products added to turkey meat during heating by addition of honey. J Food Sci 67(5):1719–1724.

Bean MM, CS Setser. 1992. Chapter 3. Polysaccharides, sugars, and sweeteners. In: JA Bowers, editor. Food Theory and Applications, 69–198. Macmillan Publishing Co., New York.

Borenstein B, E Caldwell, HT Gordon, L Johnson, TP Labuza. 1990. Chapter 10. Fortification and preservation of cereals. In: RB Fast, EF Caldwell, editors. Breakfast Cereals and How They Are Made, 188–199. American Association of Cereal Chemists, Inc., St. Paul, MN.

Caldwell EF, RB Fast. 1990. Chapter 1. The cereal grains. In: RB Fast, CF Caldwell, editors. Breakfast Cereals and How They Are Made, 1–34. American Association of Cereal Chemists, Inc., St. Paul, MN.

Caldwell EF, RB Fast, C Lauhoff, RC Miller. 1990. Chapter 3. Unit operations and equipment. I. Blending and cooking. In: RB Fast, EF Caldwell, editors. Breakfast Cereals and How They Are Made, 56–78. American Association of Cereal Chemists, Inc., St. Paul, Minn.

Coulter RB. 1988. Extending shelf life by using traditional antioxidants. Cereal Foods World 33:207–210.

Daniel JR, CW Weaver. 2000. Chapter 5. Carbohydrates: Functional properties. In: GL Christen, JS Smith, editors. Food Chemistry: Principles and Applications, 55–78. Science Technology System, West Sacramento, Calif.

Dziezak JD. 1986. Antioxidants—the ultimate answer to oxidation. Food Technol. 40:94–97.

Fast RB, GH Lauhoff, DD Taylor, SJ Getgood. 1990a. Flaking ready-to-eat breakfast cereals. Cereal Foods World 35:295–298.

Fast RB, FJ Shouldice, WJ Thompson, DD Taylor, SJ Getgood. 1990b. Toasting and toasting ovens for breakfast cereals. Cereal Foods World 35:299–310.

Kujawski DM. 1990. Designing flavors for breakfast cereals. Cereal Foods World 35:312–314.

Midden TM. 1989. Twin screw extrusion of corn flakes. Cereal Foods World 34:941–943.

Miller BD. 1990. Chapter 4. Unit operations and equipment. II. Drying and dryers. In: RB Fast and EF Caldwell, editors. Breakfast Cereals and How They Are Made, 79–122. American Association of Cereal Chemists, Inc., St. Paul Minn.

Monahan EJ. 1988. Packaging of ready-to-eat breakfast cereals. Cereal Foods World. 33:215–221.

Namiki M, T Hayashi. 1983. Chapter 2. A new mechanism of the Maillard reaction involving sugar fragmentation and free radical formation. In: The Maillard Reaction in Foods and Nutrition, 21–46. ACS Symposium Series 215. American Chemical Society, Washington, D.C.

Potter N, J Hotchkiss. 1995. Chapter 17. Cereal grains, legumes and oilseeds. In: N Potter, J Hotchkiss, editors. Food Science, 5th edition, 381–408. Chapman Hall, New York.

Schmidt K. 2000. Chapter 7. Lipids: Functional properties. In: GL Christen, JS Smith, editors. Food Chemistry: Principles and Applications, 97–123. Science Technology System, West Sacramento, CA.

Shibamoto T, LF Bjeldanes. 1993. Chapter 10. Toxicants formed during food processing. In: T Shibamoto, LF Bjeldanes, editors. Introduction to Food Toxicology. Academic Press, San Diego, CA.

13
Grain, Paste Products: Pasta and Asian Noodles

J. E. Dexter

INTRODUCTION

Asian noodles and pasta are the two general categories of paste products. Asian noodles and pasta appear similar, but they are differentiated by a fundamental difference in manufacturing procedures. Asian noodles usually are formed by passing the dough through sheeting rolls. Most dried pasta products are formed by extruding the dough through a die, but some pasta products, predominately marketed as fresh pasta, are formed by passing the dough through sheeting rolls.

Paste products are found in one form or another all over the world. Pasta and Asian noodles are popular for many reasons. They are versatile, natural, and wholesome foods. The manufacturing processes are relatively simple, and products are available in a variety of shapes. Dried paste products can be stored conveniently for long periods and can be easily handled and transported. Modified atmosphere packaging has greatly increased the shelf life and popularity of fresh pasta and noodles.

Asian noodles are ancient foods that probably originated in northern China as much as 6000 years ago (Chen 1993, Hatcher 2001, Hou and Kruk 1998, Miskelly 1993). The art of noodle making was well developed during the Han dynasty (206 B.C. to 220 A.D.). By the Sung dynasty (960–1279 A.D.), noodles had taken on diverse forms by variations in cooking and preparation procedures (Huang 1996). Noodles spread from China to other parts of Southeast Asia and became firmly established in Japan by the sixteenth century (Nagao 1981).

Pasta has been known to Mediterranean civilizations for many centuries, but its origin is obscure. A popular story credits Marco Polo with introducing noodles to Italy after his travels to the Far East in 1292. This story can be discounted on the basis of

The information in this chapter has been derived from two chapters in *Food Chemistry Workbook,* edited by J. S. Smith and G. L. Christen, published and copyrighted by Science Technology System, West Sacramento, California, ©2002. Used with permission.

other historical evidence (Agnesi 1996, Martini 1977, Matsuo 1993). Etruscan art found in the tombs of Relievi, northwest of Rome, suggest that a type of pasta product was known in 600 B.C. Also, there are references to lasagna in essays of ancient Greek and Roman authors. In Genoa, Italy, a will dated 1279, before the return of Marco Polo, bequeaths, among other things, a basket full of macaroni. Whatever the origin, Italy is regarded as the home of pasta. Names of various types of pasta such as macaroni, spaghetti, lasagna, and so on are all Italian.

Noodles and pasta products are sufficiently different in their raw material requirements and processing regimes that this chapter will deal with each separately, beginning with extruded pasta.

Figure 13.1. Pasta products come in a myriad of sizes and shapes as illustrated by this display of selected products. (Courtesy of Canadian Grain Commission.)

BACKGROUND INFORMATION ON PASTA

Pasta processing initially was a simple procedure performed by artisans or *pastaio* (Kempf 1998). Flour and water were mixed and kneaded into dumplings. Eventually dough was sheeted and cut into strips. Pasta was marketed exclusively in fresh form until it was discovered that the coastal climate of Italy had an ideal climate for drying (Martini 1977).

Mechanization of pasta manufacturing began during the industrial revolution. The first mechanical devices for pasta processing were developed in the early 1700s. Extrusion presses were used by then, hydraulic presses were designed in the mid-1800s, and late in the nineteenth century kneaders came into use (Agnesi 1996, Marchylo and Dexter 2001, Matsuo 1993). It was not until the early twentieth century that drying cabinets became available. Pasta processing remained a batch manufacturing process until the 1930s, when continuous extrusion using an extrusion auger revolutionized the process. Continuous pasta dryers soon followed (Davis 1998). The first continuous automatic production line that processed semolina into pasta ready for packaging was designed in 1946 (Marchylo and Dexter 2001). Other important advancements have followed. Application of vacuum during mixing and extrusion minimizes oxidation of yellow pigments and improves pasta color. Teflon inserts in bronze dies improve product uniformity and impart a smoother surface and an enhanced appearance. Higher drying temperatures improve pasta texture and hygiene and, because drying times are shorter, enable more compact lines for a given capacity. Computerization increases production efficiency and product consistency.

Pasta comes in a myriad of shapes and sizes, some of which are illustrated in Figure 13.1. Most pasta shapes can be classified into two groups: long goods and short goods. The most familiar type of long goods is spaghetti, which may have a diameter ranging from 1.5 to 2.5 mm. Hollow extruded goods, commonly referred to as macaroni, come in various lengths and forms. The most popular form of short goods in North America is probably elbow macaroni. Short goods include not only various tube-shaped products, but also more exotic shapes like shells, letters, stars, bow ties, and wagon wheels.

RAW MATERIAL FOR PASTA PRODUCTS

BASIC INGREDIENTS

The main raw ingredient for premium quality pasta is semolina milled from high quality durum wheat. Semolina is a granular product comprised of evenly sized endosperm particles. Durum wheat is very hard and therefore lends itself to producing a high yield of semolina. Some durum wheat flour is inadvertently produced as a by-product of semolina milling. It is of lower value than semolina and is used in lower grade pasta. Lower grade pasta may also contain a granular product, referred to as farina, milled from common wheat (a different wheat species that is predominately intended for baked goods and Asian noodles) and/or common wheat flour. Processed maize and various other cereals are used occasionally in pasta but are much less satisfactory raw materials (Baroni 1988). Other ingredients include water and optional ingredients such as egg, spinach, tomato, herbs, and so on, and vitamins and minerals for nutritional enrichment (Marchylo and Dexter 2001).

DURUM WHEAT REQUIREMENTS

Many pasta manufacturing plants are integrated durum wheat milling and pasta processing units. A discussion of durum wheat milling is beyond the scope of this chapter. The durum wheat milling process has been described by many authors (Bass 1988; Bizzarri and Morelli 1988; Dexter and Marchylo 2001; Feillet and Dexter 1996; Posner and Hibbs 1997; Sarkar 1993, 2003).

Millers purchase durum wheat on the basis of specifications that ensure a good yield of semolina and others that satisfy demands by pasta manufacturing customers. A universally accepted specification is minimum test weight (weight per unit volume), an index of kernel plumpness that is strongly related to semolina yield (Dexter et al. 1987). Most millers will specify a minimum percentage of hard vitreous kernels because nonvitreous kernels are soft and result in lower semolina yield (Dexter et al. 1988). There will also be limits on foreign material and physical defects such as broken kernels, shrunken kernels, and damaged kernels. Kernels with surface discoloration are tolerated in very low amounts in premium quality durum wheat to avoid excessive specks in semolina, thereby protecting pasta appearance (Dexter and Edwards 1998a). Millers may also demand a minimum amount of yellow pigment because yellowness is an important aesthetic aspect of pasta.

Other important durum wheat specifications are minimum protein content and an estimation of gluten properties (elasticity and extensibility) based on internationally recognized strength tests (Feillet and Dexter 1996). As will be discussed in more detail later, protein content is the primary factor in pasta cooking quality, but gluten properties also contribute to pasta cooking quality.

SEMOLINA PARTICLE SIZE, REFINEMENT, AND COLOR

Pasta manufacturers require millers to meet numerous semolina specifications to achieve the desired pasta quality (Feillet and Dexter 1996, Marchylo

and Dexter 2001). Semolina should be uniform in particle size. For most modern pasta presses, best results are obtained when about 90% of the particles have a diameter between 200 and 300μm. Uniform particle size ensures that semolina will flow freely. Coarse particles, particularly those over 500μm, will not absorb water adequately during mixing and kneading, thereby imparting unsightly white blotches in the finished product (Antognelli 1980).

Aesthetics play a major role in pasta marketing. In most countries consumers prefer pasta that is bright amber yellow. The natural yellow pigmentation in durum wheat semolina is imparted mainly by xanthophyll, a carotenoid pigment (Irvine 1971). Consumers also generally prefer pasta that exhibits no evidence of brownness and a minimum of bran specks.

Variations in semolina color are due to intrinsic differences between durum wheat varieties, physical damage to wheat due to adverse growing conditions, and the efficiency of semolina milling. Less refined semolina will have more bran specks. Most pasta manufacturers establish a maximum permitted speck count to protect pasta appearance. It is also common practice to specify minimum levels of yellowness and brightness in semolina, as measured by a spectrophotometer.

Activities of enzymes associated with poor color increase as semolina becomes less refined. Lipoxygenase, which is concentrated in the germ (Bhirud and Solsulski 1993), catalyzes the oxidation of yellow pigments, resulting in less intense pasta color (Irvine 1971). Peroxidase and polyphenol oxidase, which are concentrated in the outer layers of wheat kernels (Fraignier et al. 2000, Hatcher and Kruger 1993), cause browning of pasta during processing.

A widely used specification to guarantee semolina refinement is a maximum semolina ash content, a measure of mineral content. Outer layers of wheat kernels have ash contents ≥ 6%, compared to < 1% for starchy endosperm (Dexter and Marchylo 2001). Ash content also increases from the center to the outer regions of the endosperm. Accordingly, ash content is strongly correlated to semolina extraction rate (amount of semolina derived from a given amount of wheat) and milling efficiency.

SEMOLINA PROTEIN QUANTITY AND QUALITY

As metioned earlier, exclusive of processing conditions, the texture of cooked pasta depends primarily on the quantity of protein and the properties of the gluten protein (Feillet and Dexter 1996, Marchylo and Dexter 2001). The viscoelastic nature of wheat gluten is unique among cereals and makes wheat ideally suited for making a wide range of products from various types of bread to pasta and noodles. Gluten proteins form a viscoelastic mass during hydration, mixing, and kneading that imparts plasticity and elasticity to pasta dough. The gluten proteins form a network that envelopes the hydrated starch granules. As protein content increases, the gluten network becomes more extensive, less starch is leached during cooking, and pasta becomes firmer and less sticky and maintains its structure and texture better when overcooked.

Differences in intrinsic cooking quality among durum wheat varieties of comparable protein content are due mainly to qualitative differences in gluten proteins (Feillet and Dexter 1996, Marchylo and Dexter 2001). Gluten is comprised of gliadin proteins, which are single polypeptide chains, and glutenins, which are made up of multiple chains referred to as subunits. Glutenin chains are bound together by disulphide bonds into large polymers (MacRitchie 1992). Gliadin proteins impart the viscous component to dough, and glutenin polymers impart the elastic component. Accordingly, a high proportion of gliadin protein is associated with a weak extensible dough. Glutenin polymers are among the largest proteins known, having molecular weights estimated into the millions. Durum wheat dough elasticity is determined by the proportion of total glutenin and the molecular size of the glutenin polymers.

French researchers made the important discovery that durum wheat varieties that contain a specific gliadin, known as gamma-45, have stronger gluten and intrinsically better pasta cooking quality than varieties containing another specific gliadin, known as gamma-42 (Damidaux et al. 1978). This discovery has greatly aided durum wheat breeders by providing a robust marker for gluten strength that can be used in early generation selection. While gamma gliadin-45 is a useful marker, it is now generally agreed that its presence is not the actual reason for superior pasta cooking quality. The actual cause is believed to be the presence of specific low molecular weight glutenin proteins that are linked genetically to either the gamma-45 or the gamma-42 gliadins (Payne et al. 1984).

Pasta manufacturers almost always establish a minimum acceptable protein content to ensure ac-ceptable pasta cooking quality. They also often establish gluten strength specifications because of the secondary importance of gluten strength in imparting superior pasta cooking quality.

A minimum gluten index is widely used as a strength specification. To determine gluten index, wet gluten is washed from semolina or ground wheat, the gluten ball is centrifuged over a screen, and the proportion of wet gluten that does not pass through the screen is determined (Cubadda et al. 1992). The higher the percentage retained on the screen, the stronger the gluten.

Gluten strength is also associated with dough properties. The two most popular dough tests used by the pasta industry are the alveograph and the mixograph (Marchylo and Dexter 2001). In the alveograph test, sheeted dough is inflated into a bubble by air pressure. Parameters derived from a recorded curve include P (pressure), which is related to the height of the curve; a measure of elasticity, L (length of the curve), which is related to extensibility; and W (work), the area under the curve, which is related to the energy required to inflate the bubble. The mixograph is a recording dough mixer widely used in the United States. Mixing time to peak dough consistency and the shape of the curve are good measures of gluten strength.

PASTA MANUFACTURING

Pasta manufacturing is a relatively simple process (Antognelli 1980, Baroni 1988, Dalbon et al. 1996, Dick and Matsuo 1988, Feillet and Dexter 1996, Hummel 1966, Marchylo and Dexter 2001, Mar-

Figure 13.2. A modern long goods pasta processing line. (Courtesy of Bühler.)

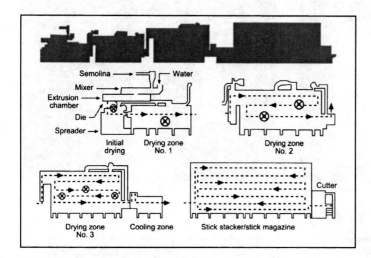

Figure 13.3. Schematic representation of a modern continuous long goods line. Arrows indicate the path taken along the production line. Symbols in dryers represent fans.

chylo et al. 2004; Matsuo 1993, Milatovic and Mondelli 1991). The basic elements of pasta processing common to all forms of pasta are hydration, mixing, kneading, extrusion to give the pasta the desired shape, predrying, drying, and cooling prior to packaging. A photo of a modern long goods pasta line is shown in Figure 13.2, and a schematic representation of a long goods pasta line is shown in Figure 13.3.

HYDRATION, MIXING, AND KNEADING

Modern pasta plants feature computer control to optimize processing conditions (Fig. 13.4a). Semolina is combined with water and optional ingredients above the press (Fig. 13.4b). Proportions of ingredients are controlled automatically by dosers, which measure by weight or by volume. The amount of water added to achieve optimum dough properties for extrusion varies from about 25 to 30 kg water per 100 kg semolina (Dalbon et al. 1996). The initial moisture content of semolina is about 14%, making the final water content of pasta dough approximately 30–35%.

In traditional presses there is a mixing stage during which water is uniformly incorporated into semolina within a paddle mixer (Fig. 13.4c). Depending on the length and intensity of the mixing process and the characteristics of the raw material, lumps of various sizes up to 2 or 3 cm in diameter are formed. Paddles are angled so that dough is advanced continuously through the mixer at a specified rate. Kneading of the mixed agglomerated particles is achieved by driving the mixture through a cylinder by beveled

helical plates under a vacuum of 50–70 cm Hg. Dough passes into the vacuum chamber through a rotary seal. The mixing and kneading process takes about 15 minutes.

Vacuum performs two important functions. First, oxygen is removed, reducing oxidation of yellow pigments and giving pasta a more intense color. Second, vacuum prevents the formation of air bubbles. If air bubbles are present, the dried pasta is less bright, less translucent, and has less mechanical strength.

EXTRUSION

The vacuum section of the mixer empties directly into an extrusion chamber. Pasta dough is extruded under vacuum by means of an auger. Modern high capacity presses have two or three augers and have production capacities of 1000–8000 kg/h, and occasionally more. The configuration of the auger (pitch to diameter ratio) is designed to impart a pressure of about 100 atmospheres. Heat is developed during extrusion. The combination of heat, pressure, and shear during the extrusion process makes the gluten network within the dough continuous, and the dough becomes plastic and translucent (Matsuo et al. 1978). Cold water is circulated around the extrusion chamber to control the temperature at 45–50°C to prevent denaturation of the gluten protein. If gluten is denatured at this stage, the physical properties of the pasta dough and the texture of the cooked pasta will be adversely affected (Abecassis et al. 1994).

The dough is received from the extruder by the

Figure 13.4. Stages in pasta processing. **A.** Modern lines have computer control. **B.** A view above a press showing the doser, which dispenses semolina and water, and the mixing chamber. **C.** A close-up of a paddle mixer showing the lumps of dough. **D.** Dies for long goods extrusion. **E.** Trimming spaghetti and spreading strands onto sticks for conveyance through the dryers. **F.** Removing spaghetti from sticks at the end of the line. (Photos courtesy of Canadian Grain Commission.)

head, which uniformly distributes the dough to a die. Dies are carefully designed to distribute pressure uniformly throughout the dough to make extrusion as uniform as possible (Fig. 13.4d). For long goods, the dough is extruded from a long rectangular die, which forms the product into parallel curtains. The strands are looped over metal rods by a process known as spreading (Fig. 13.4e). No matter how well the die is designed, strands will not be equal in length, and are trimmed by a cutter bar. The trimmed material is sent to a chopper and is recycled to the press.

Pasta nests, which are particularly popular in South America in soups, are delicate shapes that are difficult to manufacture. The extrusion and initial shaping process is similar to long goods. The product then goes to nesting machines that determine the final shape. Special shaping tubes in the nesting machine receive the curtain of strands, which are cut and wound into nests and conveyed to the dryer on trays.

For short pasta, the dough passes from the extrusion auger to a head designed to take circular dies (Figs 13.5a,b). Short goods dies of varying configuration are used to produce products with a range of forms and size. Beneath the short goods die is a rotary cutter with one or more blades that cuts the product to the desired length. Some short goods, commonly known as Bologna pasta, are cut out from a sheet (Fig. 13.5c). The sheet is continuously extruded and passes through rollers that stamp out the desired shape. Waste is recycled back to the extrusion press.

An innovation in press design is the Polymatik press introduced by Bühler, a pasta equipment manufacturer, in 1995 (Dexter and Marchylo 2001). A small Polymatic press is shown in Figure 13.5d. The Polymatik press mixes and develops pasta dough in 20 seconds. A twin-screw extruder forms the dough, which is sent directly to the extrusion auger. The entire system is under vacuum, which assures excellent pasta color. Other advantages of this system include rapid changeover of dies, an advantage when multiple short goods forms are being manufactured, and a clean-in-place (CIP) system that allows excellent sanitation.

DRYING AND COOLING

Drying is a critical part of the pasta manufacturing process. The temperature and humidity of the drying chambers must be carefully controlled. If strands are dried too quickly, the surface will harden, and the strands will fracture (check) due to stresses set up as the moisture trapped within the interior attempts to migrate through the surface.

During the 1970s pasta equipment manufacturers began to promote high-temperature (HT) (> 60°C) drying. Within a few years, HT drying became firmly established. Advantages of HT drying include better hygiene, especially for egg products, shorter drying time, which permits more compact drying lines for a given capacity, improved pasta color, and better cooked pasta texture (Dexter et al. 1981). Initially, temperatures recommended during HT drying were 70–80°C. Recently, ultra high-temperature (UHT) drying programs have been developed with temperatures in excess of 100°C (Pollini 1996). Before the advent of HT drying, drying times for long goods varied from 20 to 30 hours. HT drying has reduced drying times for long goods to 8–14 hours. Typical low-temperature and HT drying diagrams are illustrated in Figure 13.6.

Immediately after extrusion and trimming, the long goods strands pass under fans, which dry the surface to prevent stretching, and the strands are conveyed to the first drying chamber. Typically, a long goods dryer consists of three zones. In the first zone, known as the predrier, readily removable water is removed quickly while the dough is still plastic. The rapid moisture removal during the predrying phase minimizes microbiological activity, strengthens the pasta shape for subsequent handling, inhibits enzymatic activity, and reduces the total drying period. By the end of predrying the moisture content is 17–18%.

High temperature is not applied until late in the predrier stage or following, because its beneficial effect on cooked pasta texture is negated if the moisture content is above 20% (Dexter et al. 1981). The actual mechanism for the enhancement of cooked pasta texture by HT drying is not completely understood, but it has been attributed to stabilization of the gluten network and/or to modifications to starch pasting properties (Antognelli 1980, Dalbon et al. 1985, Feillet and Dexter 1996, Marchylo et al. 2004, Vansteelandt and Delcour 1998, Zeifel et al. 2003).

Application of high temperature when the pasta is at too high a moisture content promotes gluten network denaturation, reducing the ability of the gluten network to stabilize the pasta structure during cooking. In addition, if application of high temperature occurs above a threshold moisture content, which varies depending on the temperature applied, starch

Figure 13.5. Stages in pasta processing. **A.** Dies for short goods. **B.** View of press extruding hollow shorts goods. A revolving knife cuts the product, which is then conveyed to a predrier. **C.** Stamping out Bologna pasta (bow-ties). The extruded sheet is visible in upper background. **D.** A pilot-scale version of the Bühler Polymatik press, which features a new design with a twin-screw extruder (on the right) feeding directly into the extrusion chamber. (Photos B and C courtesy of the Canadian Grain Commission, Photos A and D courtesy of the Canadian International Grains Institute.)

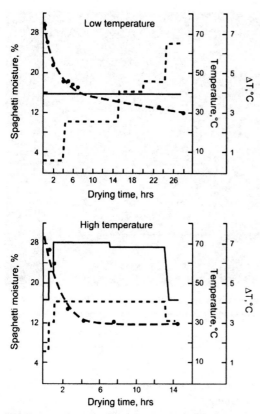

Figure 13.6. Typical low temperature and high temperature drying cycles. Dashed line indicates pasta moisture content. Solid line indicates dry bulb temperature. Dotted line indicates wet bulb temperature depression (ΔT). The greater (ΔT), the lower the relative humidity in the drying chamber.

reaction occurs when carbonyl groups, usually reducing sugars, condense with free amino groups from amino acids, peptides, and proteins (Ames 1990, Sensidoni et al. 1999). The initial phase of the Maillard reaction occurs without influencing pasta color, but significant loss of the essential amino acid lysine may occur to the detriment of protein nutritional quality (Dexter et al. 1984). Advanced Maillard reaction results in greater redness in the pasta, which is aesthetically undesirable. In addition, nutritional problems are exacerbated due to further loss of lysine and the formation of unnatural compounds whose safety is dubious (Resmini et al. 1996).

Maillard reaction is favored during HT drying when pasta moisture falls below 16% (Pagani et al. 1992, Resmini et al. 1996). Therefore, to mitigate the Maillard reaction, modern HT drying programs limit application of high temperature late in the drying process. The beneficial effect of high temperature on pasta texture is still evident, with less nutritional loss (Dexter et al. 1984). It has been suggested that the loss of nutritional value under properly controlled HT and UHT drying conditions should not be considered a serious defect because, in general, pasta products are not consumed as a source of essential amino acids (Pollini 1996).

During the remaining drying stages, the moisture content of the pasta is reduced to approximately 12.5% by a series of alternating ventilation and rest periods. After the final drying zone, there is a cooling zone. Cooling conditions must be carefully controlled to avoid the checking that can arise if there is a moisture imbalance within the pasta. The dried product now enters the stacker, which stores sufficient production so that packaging need not be carried out continuously.

Short goods predriers and driers vary in design, depending on the configuration, thickness, and specific weight of the pasta. In the initial stage, known as the shaking predrier, the product is conveyed on a vibrating tray. Dryers are either rotating drum units or continuous belt units. The rotary units are best for small shapes that must be continually stirred during drying, whereas belt units are most common for larger shapes. In belt drying units the product moves from the upper level of the drier to the next level by falling onto a belt moving in the opposite direction. Product is conveyed from the bottom of one drying unit by a moving belt to a bucket conveyor, which elevates the product to the top of the next drying unit.

Specially designed dryers are required for nests. The product moves along in trays and passes from

gelatinization will occur. If starch gelatinization occurs during drying, the strands disintegrate more during cooking, more solids are lost to the cooking water, and there is increased surface stickiness. Heat treatment of starch at low moisture is beneficial to cooking quality because starch pasting properties are modified, and the onset of gelatinization is delayed during cooking. The result is better surface characteristics (less stickiness) and reduced loss of solids to the cooking water due to less starch exudation (Vansteelardt and Delcour 1998).

Negative aspects of HT drying, if conditions are not carefully controlled, are deterioration of color, which affects aesthetic appeal, and nutritional issues (Dexter and Marchylo 2001). HT drying can induce nonezymatic browning due to Maillard reaction (Feillet et al. 2000, Pagani et al. 1992). The Maillard

Figure 13.7. A. Machine packaging of spaghetti. **B.** Pasta products display. Packaging in plastic film is popular for pasta because it attractively presents the product to the consumer, and it provides a good vapor barrier. (Photos courtesy of the Canadian Grain Commission.)

the upper level to the one below on a slide in order to cushion the jolt of landing.

PACKAGING

The final stage of processing is packaging (Fig. 13.7a). In the case of long goods, the strands are first removed from sticks and cut to length. Trimmings are ground into "regrinds" and recycled to the mixer. Regrinds, particularly under HT and UHT drying conditions, adversely affect the color and texture of pasta, so care must be taken not to introduce an inordinate amount, particularly in long goods (Fang and Khan 1996, Milatovic and Mondelli 1991).

In modern plants, packaging is done automatically, or semiautomatically (Varriano-Marston and Stoner 1996) (Fig. 13.7a). Many kinds and sizes of packaging are used (Fig. 13.7b). The most popular packaging materials are plastic film and cardboard. Plastic film offers the best product presentation and the best protection from dampness and insects. The vapor barrier associated with plastic film is an important consideration: dried pasta will readily adsorb moisture, and rapid moisture change can induce checking. Cardboard packaging offers the best protection during handling. Many countries have stringent labeling laws that make it mandatory to

provide ingredient and nutritional information on packaging.

Fresh high-moisture pasta that is packaged immediately following preparation without drying is a small but rapidly growing sector of the pasta industry. These products are at risk from microbial contamination, so they are generally modified-atmosphere packaged to prolong shelf life (Giese 1992). A modified atmosphere is established in a barrier (plastic) package by either drawing a vacuum on the package and back flushing with the desired gas composition (usually carbon dioxide and/or nitrogen) or by a continuous gas flush with the desired mixture.

BACKGROUND INFORMATION ON ASIAN NOODLES

Asian noodle products are very diverse (Hatcher 2001, Hou and Kruk 1998, Wu et al. 1998). Most noodles are prepared from common wheat flour. It has been estimated that 30–40% of wheat flour consumption in Southeast Asia is as noodles (Miskelly 1993). Starch noodles, rice noodles, and buckwheat noodles are also popular (Miskelly 1993, Hatcher 2001).

Ingredients, flour specifications, manufacturing procedures, and final preparation of wheat-based noodles vary from region to region (Miskelly 1998). The two main wheat noodle classifications are white salted noodles and yellow alkaline noodles. Popular noodle preparations include fresh (raw uncooked), partially cooked (boiled Hokkien style), dried, steamed and dried (traditional instant), and steamed and fried (instant, ramen style) (Chen 1993, Wu et al. 1998).

The evolution of noodles has continued into modern times. A significant recent event was the development of the fully automated production of deep-fried packaged instant noodles (ramen style) in Japan in 1957 (Chen 1993, Miskelly 1993, Nagao 1981). Instant noodles were quickly accepted in Asia as a convenience food. In the 1970s instant noodles were successfully introduced into the United States, and they have also become established in Australia and Europe.

The diversity of Asian noodles makes it impossible to consider all forms in this chapter. We will focus on some popular types of wheat noodles, which encompass most of the range of quality requirements and processing options.

RAW MATERIALS FOR WHEAT NOODLE PRODUCTS

GENERAL FLOUR MILLING CONSIDERATIONS

To achieve the right flour quality, it is common for millers to blend various wheat types (hard and soft, and of varying protein content and gluten properties) prior to milling, or to blend flours from various wheat types to optimize noodle-making potential. A bright flour that is free from bran specks is a universal requirement of all manufacturers making premium noodles (Hatcher and Symons 2000a,b). Noodle manufacturers commonly require millers to provide flour that meets specified brightness and ash content values to assure satisfactory flour refinement. Accordingly, wheat intended for noodle production must be reasonably free of physical defects that may adversely affect flour quality (Dexter and Edwards 1998b). Wheat with a white seed coat is preferred when milling for noodles because bran specks in flour are less conspicuous than for wheat with a red seed coat (Ambalamaatil et al. 2002).

The textural attributes of cooked white salted noodles improve as flour particle size becomes finer and mechanical starch damage becomes lower (Hatcher et al. 2002). Regardless of noodle type, a fine flour (100% of particles passing through a 130 μm sieve) with uniform particle size distribution is desirable to ensure uniform water distribution within the specified mixing time (Chen 1993, Hatcher 2001, Miskelly 1998). Coarse flour particles absorb water slowly, whereas very fine flour particles will absorb water quickly. Very fine particles also are often accompanied by high levels of mechanically damaged starch, which can cause cooked noodles to have a sticky surface and poor eating quality (Moss et al. 1987).

Further discussion of common wheat milling is beyond the scope of this chapter. However, common wheat flour milling has been documented by many authors (Bass 1988; Izydorczyk et al. 2002; Owens 2001; Posner and Hibbs 1997; Sarkar 1993, 2003).

RAW MATERIAL REQUIREMENTS FOR WHITE SALTED NOODLES

White salted noodles have a simple formula comprising flour, water, and 1–3% common salt (Crosbie 1994). The general consumer preferences for white salted noodles are a bright clean creamy color, a soft but elastic texture, and a smooth surface (Miskelly 1993, Nagao 1981, Oh et al. 1985).

The texture of cooked white salted noodles is strongly affected by starch properties (Azudin 1998, Baik et al. 1994, Guo et al. 2003, Konik et al. 1992, Morris 1998, Moss et al. 1986, Oda et al. 1980, Toyokawa et al. 1989). Starch consists of two carbohydrate polymers, amylose and amylopectin. Amylose is a linear polymer, whereas amylopectin is a branched polymer with one of the highest molecular weights known among naturally occurring polymers (Lineback and Rasper 1988). Starch is deposited in wheat endosperm in granules. Starch granules are insoluble in cold water, but when suspended in water at room temperature they swell. When wheat starch is heated in water, as when noodles are being cooked, it undergoes a series of changes known as gelatinization and pasting (Morris 1990). The viscosity (thickness) and stability (rate of breakdown of viscosity) of the starch paste are related to starch composition and are an intrinsic characteristic of wheat varieties. Desirable attributes of the starch component for Asian noodles include low amylose content, high swelling power, low initial gelatinization temperature, and high paste viscosity (Noda et al. 2001, Seib 2000). Some millers stipulate specific wheat varieties because of minimum paste viscosity specifications demanded by white salted noodle manufacturers. The rapid visco-analyzer is widely used to determine noodle flour starch pasting properties (Batey et al. 1997).

Protein content is an important secondary factor affecting white salted noodle texture. To achieve the desired soft elastic texture, flour protein content ideally should be between 8 and 11% (Miskelly 1998, Nagao 1981). Higher protein content not only results in white salted noodles that are too firm, but also reduces noodle brightness (Miskelly 1984, Oh et al. 1985).

RAW MATERIAL REQUIREMENTS FOR ALKALINE NOODLES

Alkaline noodles contain alkaline salts (often referred to as *kansui*) or sodium hydroxide, which raise dough pH to 9–11.5. Numerous combinations of alkaline salts, including sodium and potassium carbonates, bicarbonates and phosphates, and sodium hydroxide are used, depending on regional preferences (Miskelly 1996). Inclusion of alkaline salts in noodle formulas imparts a characteristic aroma and flavor, a yellow color, and a firm, elastic texture. The yellow color is associated with naturally occurring flavones in flour, which are colorless at acidic pH but turn yellow at the high pH of alkaline noodles (Miskelly 1993, 1996). The hue and the intensity of the yellow color is affected by the alkaline salt used, the length of time after sheeting, the protein content, and the degree of refinement of the flour (Kruger et al. 1994, Moss et al. 1986). A highly refined flour is a primary quality requirement for alkaline noodles to maximize noodle brightness and minimize visible bran specks (Chen 1993, Hatcher 2001, Hatcher and Symons 2000a,b). Polyphenol oxidase enzymes, which reside in the outer region of the wheat kernel, have been implicated in the deleterious brown color that develops over time in raw alkaline noodles made from poorly refined flour (Kruger et al. 1992).

Alkaline salts also toughen noodle dough (Edwards et al. 1996) and affect the pasting characteristics of starch (Miskelly 1996). Alkaline noodles prepared with NaOH are stickier and less firm than alkaline noodles made with blends of sodium and potassium carbonate, possibly because NaOH adversely affects gluten development, resulting in more disruption of the protein network during cooking (Moss et al. 1986). NaOH also solubilizes starch. Regardless of the alkaline salts used, the preference for a firm texture means that a flour protein content of over 11% is generally preferred because the extent of the protein matrix in the noodle is directly related to cooked noodle texture (Crosbie et al. 1999, Miskelly 1998, Moss et al. 1987).

Starch properties affect the elastic texture of cooked alkaline noodles (Akashi et al. 1999). There is not clear consensus on the importance of peak paste viscosity as an alkaline noodle textural determinant (Batey et al. 1997, Crosbie et al. 1999, Konik et al. 1994). However, paste viscosity stability (resistance to breakdown) is positively related to noodle smoothness, but negatively related to noodle firmness (Konik et al. 1994, Miskelly and Moss 1985, Wu et al. 1998).

HANDMADE-NOODLE MANUFACTURING

In most countries, the manufacture of noodles is fully or partially mechanized, but in China an estimated 80% of noodles are still handmade (Miskelly 1993). Noodles can be formed by hand by either the hand-stretched method or the hand-cut method (Chen 1993, Huang 1996).

Hand-stretched noodles are often made in restaurants in China by hand swinging (Miskelly 1993). The noodle dough is skillfully pulled and stretched into a long rope by repeated folding in half between each drawing stage (Figs 13.8a–d). Raw noodles are cooked immediately after preparation.

An alternative method for preparing hand-stretched noodles is to lightly stretch noodle dough strips with intermediate rest stages (Chen 1993). In the final stretching stages the dough is coiled around bamboo rods, which are pulled away from each other until the dough has doubled its original length (Fig. 13.9a). The stretching is repeated until the desired thickness is reached. The complete process can take from 12 to 30 hours. The noodles may be marketed as dried (Fig. 13.9b) or steamed noodles. The superior mouth feel and firmness of hand-stretched noodles has maintained their popularity.

The hand-cut noodle process is the forerunner of the modern mechanized process. Hand-cut noodles are made by pressing wet flour into dough balls, which are progressively combined and kneaded until an elastic dough is obtained (Chen 1993, Hatcher 2001, Huang 1996, Miskelly 1993). After a rest period the dough is rolled out into a thin sheet, folded over, and cut into strings with a knife. Hand-cut noodles are primarily marketed as fresh noodles.

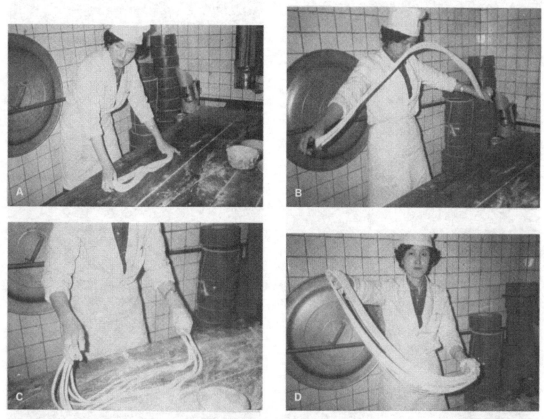

Figure 13.8. Handmade stretched noodles. **A.** Rolling noodle dough in preparation for making stretched noodles by hand swinging. **B.** Stretching noodles by hand swinging. **C.** Dough is doubled over and stretched repeatedly until desired thickness is achieved. **D.** Stretched noodles upon completion of hand swinging. (Photos courtesy of the Canadian Grain Commission.)

MECHANIZED NOODLE MANUFACTURING

PROCESSING TO THE RAW NOODLE STAGE

With the exception of Japan and Korea, the noodle processing industry in Asia is still dominated by small-scale cottage industries (Figs 13.9c,d). The exception is steamed and fried instant noodles, which are produced almost exclusively in modern automated plants. The initial stages of noodle manufacturing up to the cutting stage are common to all noodles regardless of formulation and final form.

The initial stage of noodle manufacturing is hydration and mixing. Either horizontal mixers or pin mixers can be used. Water addition ranges from about 30 to 35% of flour by weight, depending on flour properties and noodle type. Optimal water addition is critical to achieve the best surface charac-

teristics, color, and cooked noodle texture (Hatcher et al. 1999). Mixing time varies depending on the mixer and generally ranges from 5 to 20 minutes (Kim 1996, Miskelly 1996, Wu et al. 1998). At this stage, a crumbly dough is formed (Fig. 13.10a).

After mixing, the crumbly dough is transferred to a hopper and compressed into a dough sheet by passing through sheeting rolls with a roll gap of about 3 cm (Hatcher 2001, Miskelly 1996, Nagao 1981). In automated plants, two dough sheets are formed during the first compression pass, and they are combined and laminated into a single sheet during a second compression pass (Fig. 13.10b).

Following compression, the dough should be rested for 15–30 minutes prior to further sheeting to achieve higher quality noodles. Resting makes the dough more elastic, imparting superior sheeting properties and better firmness following cooking

Figure 13.9. The noodle industry in much of Asia is dominated by small-scale cottage industry. **A.** Preparing hand-stretched noodles by stretching on bamboo poles. **B.** Drying hand-stretched noodles. **C.** Reducing yellow alkaline noodles. The hopper and compression rolls can be seen in the background. **D.** Preparing raw noodles for market. (Photos A and B courtesy of the Canadian International Grains Institute, Photos C and D courtesy of the Canadian Grain Commission.)

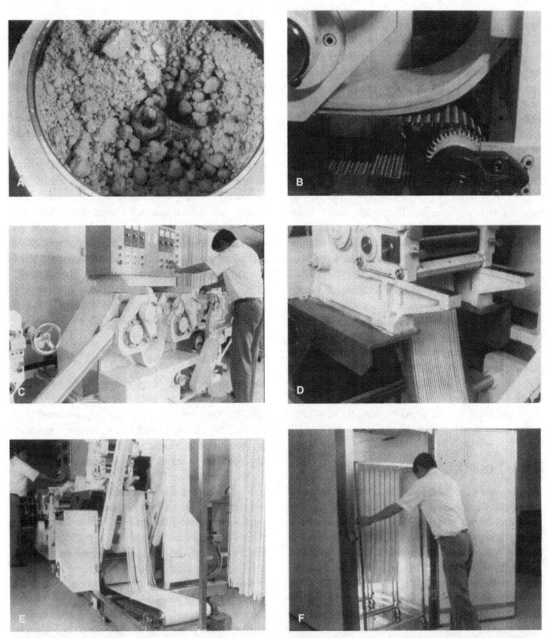

Figure 13.10. Manufacturing of Asian noodles on the Canadian International Grains Institute pilot-scale noodle line. **A.** Crumbly dough following mixing. **B.** Combining and laminating two dough sheets following the first compression pass. **C.** Reduction of noodle dough thickness by successive sheeting passages. Note the conveyor belt, which rests the dough following the compression stage by slow passage of the dough to the first reduction sheeting. **D.** Cutting the noodles. **E.** Cutting the noodles to length and spreading onto sticks in preparation for drying. **F.** Transferring noodles into dryer. (Photos courtesy of the Canadian International Grains Institute.)

(Hatcher 2001, Moss et al. 1987). In automated noodle plants, the resting stage is often eliminated, although it can be retained by slow passage of the dough on a conveyor belt prior to the first reduction sheeting (Fig. 13.10c). In some automated plants there is a rest stage after one or more intermediate reduction passes.

The compressed dough sheet is now reduced in thickness by passage through a series of sheeting rolls of gradually reducing roll gap (Fig. 13.10c). The number of reduction passes varies, but is usually between three and seven (Miskelly 1996). The final thickness of the noodle dough varies from about 1 to 2.5 mm, depending on the noodle type. By the final reduction passage the dough should achieve full gluten development with the formation of a uniform gluten matrix (Moss et al. 1987). However, due to the low moisture content of noodle dough, gluten development is still not achieved to the same extent as in bread dough (Dexter et al. 1979).

The finished dough sheet is now passed through a pair of slotted cutting rolls to produce noodle strands (Fig. 13.10d). Noodle width and shape (square or rectangular) varies, depending on the dimensions of the cutting roll slots. Cutting the noodle strands to length is the final stage in raw noodle processing.

DRIED NOODLES

Noodles that have been dried have the advantage of long shelf life, although this is offset by the requirement for longer cooking time, which leads to a softer stickier cooked product (Chen 1993). In some parts of Asia, where the climate is suitable, noodles can be dried outdoors. In modern automated plants, wet noodles are hung and spread onto sticks (Fig. 13.10e) and dried in chambers that closely control the rate of drying by regulating temperature and relative humidity (Fig. 13.10f) (Chen 1993, Nagao 1981). The drying process is critical to the quality of dried noodles. Incorrect drying conditions can cause strand stretching, cracking, warping, and splitting.

BOILED NOODLES

In Japan, white salted noodles are often precooked prior to marketing (Nagao 1981). The noodles are boiled in water (pH 5–6) for 10–25 minutes, depending on noodle thickness and the desired texture. The boiled noodles are washed and cooled in running water. Cooked noodles, packed in plastic pouches with a packet of soup or sauce, are a convenience food. The packed precooked noodles need to be cooked only two to three minutes prior to eating.

Partially cooked alkaline noodles, known as Hokkien noodles, are popular in Southeast Asia (Chen 1993, Miskelly 1996, Moss et al. 1987). The noodles are boiled for about one minute and immediately sprayed with cold water. In automated plants, the water in the boiling baths is continually replenished. If the water is not replenished, starch, dextrins, and alkaline salts accumulate, resulting in increased cooking loss and surface stickiness. The noodles are cooled and then coated with oil to keep the strands from sticking together. The noodles are marketed in bamboo baskets, trays, or polyethylene bags. The noodles must be reboiled or fried prior to consumption. The precooking stage in Hokkien noodles arrests the darkening seen in raw alkaline noodles by inactivating polyphenol oxidase enzymes (Chen 1993, Shelke et al. 1990).

STEAMED AND FRIED (INSTANT) NOODLES

Instant noodles can be produced either by boiling and drying or by steaming and frying. Fried instant noodles have literally exploded in popularity since the mid 1970s. Fried instant noodles have taken 90% of the market in Korea and have become the most popular form of noodle in Japan (Azudin 1998, Kim 1996). The product is popular because it is convenient and can be stored for several months. There are two common forms—(1) square or round noodles blocks, which are sold in bags, and (2) cup noodles, which are sold in Styrofoam cups (Chen 1993, Hatcher 2001, Kim 1996). The noodles are accompanied by a soup base, which is separately packaged.

The term *instant* is a little misleading because the product must be cooked or reheated prior to consumption. The noodle blocks sold in bags usually are cooked in boiling water for three to four minutes prior to serving. The cup noodles are marketed as a convenient snack food and are ready for eating one to three minutes after pouring hot water into the cup. The cup noodle strands usually are thinner and less densely packed than noodle blocks to facilitate rapid hydration.

The formulation of fried instant noodles includes salt (1.5–2%) and usually a small amount of alkaline salts (about 10% of salt), although some high quality products are produced without alkaline (Kim 1996). After passing through the cutter, the noodle dough is continually fed into a traveling net con-

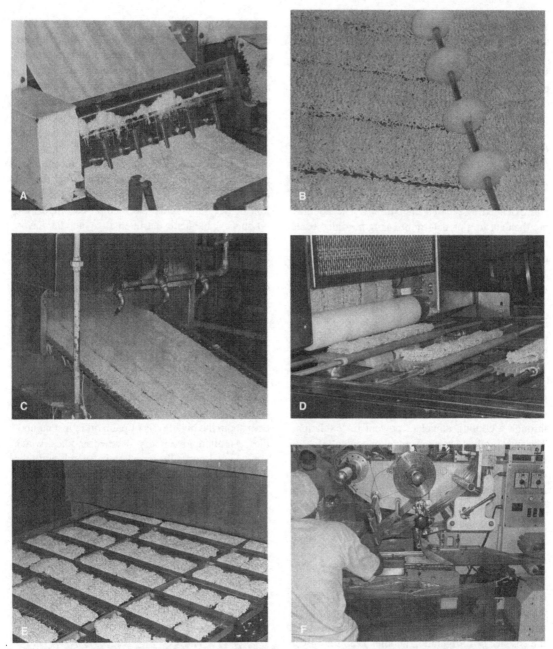

Figure 13.11. Automated manufacturing of steamed and fried instant noodles. **A.** Cutting and waving. The wave is imparted by a speed differential between the strands and the conveyor. **B.** Separating strands before conveyance to steamer. **C.** Noodle strands immerging from steamer. **D.** Cutting and folding steamed noodles back into a double layer. The double layer is formed by simultaneously cutting to length and pushing a double length of strands from the middle through the rollers. **E.** Steamed noodle blocks in baskets prior to frying. **F.** Packaging. (Photos courtesy of the Canadian Grain Commission.)

veyor that moves more slowly than the cutting rolls above it. The speed differential between the noodle strands and the net conveyor imparts a wave to the strands (Figs 13.11a,b). The waving procedure improves strand separation and makes steaming more efficient (Miskelly 1996). The waved noodle strands proceed through a steam tunnel, where they are steamed at 150–250 kPa for varying lengths of time (100–240 seconds), depending on noodle quality (Fig. 13.11c). Following steaming, starch gelatinization is about 80% complete (Kim 1996). The steamed noodles are immediately cooled by fans and extended to separate the strands. The noodles are then cut to a predetermined length, folded back to form a double layer (Fig. 13.11d), and placed individually in square baskets for noodle blocks (Fig. 13.11e) or round baskets for cup noodles. The noodle baskets travel to a tunnel fryer where they are immersed in hot oil and deep fried at 140–150°C for about 1–1.5 minutes.

Palm oil is the most popular oil for instant noodles. In the frying process, excess water is removed, oil is incorporated into the noodles, and more starch gelatinization occurs. The noodles come out of the fryer at over 140°C. The noodles are drained and cooled immediately to room temperature by passing through a cooling tunnel to prevent oil oxidation. The final product has an average oil content of 20% (range about 18–26%) and a moisture content of less than 10% (Chen 1993). In the final stage of processing, the cooled noodles and the accompanying soup base packet are automatically packaged into a bag or cup (Fig. 13.11f).

EXTENDING NOODLE SHELF LIFE

The high ambient temperatures and high humidity of Southeast Asia pose severe limitations on the keeping quality of fresh and boiled noodles, even where careful sanitary practices are maintained. In the absence of refrigeration noodles keep for only a few days.

In Japan, the rapid growth in supermarkets has lead to an interest in preservative technology (Nagao 1981, Wu et al. 1998). The shelf life of fresh boiled noodles can be extended to several months by using modified-atmosphere packaging and retort pouch packaging (Miskelly 1996). Noodles with extended shelf life can be prepared by soaking in acid prior to low-temperature thermal processing and cooling (Wu et al. 1998). However, many people do not like the acid taste of acidified noodles, which varies according to the type and concentration of acid used (citric, malic, lactic, gluconic, or acetic) in the soaking liquor.

For instant noodles the quality of the oil is an important consideration. During processing the oil must be turned over frequently to prevent accumulation of free fatty acids and thermal decomposition products, which are detrimental to noodle flavor. The addition of antioxidant preservatives to the oil has been shown to extend shelf life (Rho et al. 1986). However, even under ideal processing and storage conditions, the shelf life of instant noodles is limited to five or six months.

RECENTLY DEVELOPED NOODLE PRODUCTS

Improvements in noodle processing technology and development of new products continue. Health conscious consumers are increasingly expressing concern about the high levels of palm oil in instant noodles. An alternative is dry steamed noodles, which are steamed normally to gelatinize starch and then dried using hot air (Hatcher 2001). The main drawback to dry steamed noodles is a longer cooking time than instant noodles.

Precooked chilled or frozen white salted and alkaline noodles are becoming more common in Japan. The noodles are optimally cooked, requiring only a rapid defrosting or heating to return to optimal texture (Hatcher 2001). They are mainly sold to restaurants but are becoming more available in supermarkets and convenience stores.

APPLICATION OF PROCESSING PRINCIPLES

Table 13.1 provides recent references for more details on specific processing principles.

Table 13.1. Specific Processing Stages and the Principle(s) Involved in the Manufacturing of Pasta and Asian Noodles

Processing Stage	Processing Principle(s)	References for More Information on the Principles Used
Extrusion of pasta	Forming dough into shapes by forcing through a die	Abecassis et al. 1994, Dalbon, et al. 1996
HT drying of pasta	Moisture reduction, water activity, enhanced cooking quality, improved storage stability	Baroni 1988, Pollini 1996
HT drying of pasta	Loss of nutritional value by nonenzymatic browning (Maillard reaction)	Feillet et al. 2000, Ames 1990.
Precooking of instant noodles	Starch gelatinization	Morris 1990, Lineback and Rasper 1988
Packaging	Preservation during storage and handling	Giese 1992, Varriano-Marston and Stoney 1996

GLOSSARY

Alkaline salts—Various combinations of sodium and potassium carbonates, bicarbonates, and phosphates used as an ingredient in yellow alkaline noodles often referred to as *kansui*.

Amylopectin—A branched glucose polymer that is a component of starch and has one of the highest molecular weights known among naturally occurring polymers.

Amylose—A linear polymer of glucose that is a component of starch.

Ash content—A measure of mineral matter in semolina and flour commonly used as a specification for estimating semolina and flour refinement.

Checking—Fracturing of pasta due to improper drying conditions.

CIP—Clean in place.

Flavones—Naturally occurring compounds, colorless in flour, that turn yellow at the high pH of alkaline noodles.

Gliadin—Single polypeptide chain gluten proteins that impart extensibility to gluten.

Gluten—Proteins found in wheat endosperm that form a viscoelastic mass when hydrated and mixed.

Glutenin—Multiple chain proteins bound together by disulfide bonds into large polymers that impart elasticity (resistance to extension) to gluten.

High-temperature (HT) drying—Drying of pasta above 60°C.

Hokkein noodles—Partially cooked alkaline noodles popular in Southeast Asia.

Instant noodles—Noodles precooked by steam and/or frying and marketed, accompanied with a soup base, as dried rapid-preparation noodles in blocks or cups.

Lipoxygenase—An enzyme that catalyzes the oxidation of yellow pigments in semolina, resulting in less intense pasta color.

Maillard reaction—Condensation of reducing sugars and free amino groups during HT and UHT drying that induces browning of pasta and loss of nutritional quality.

Polyphenol oxidase—An enzyme associated with undesirable browning of pasta and noodles.

Regrinds—Reground dried pasta trimmings that are recycled back to the extrusion press.

Semolina—A granular flour produced from durum wheat that is the primary raw material for the manufacture of premium quality pasta products.

Spreading—The looping of extruded pasta long goods strands over metal rods and trimming to length prior to conveyance to dryers.

Ultra high-temperature (UHT) drying—Drying of pasta above 100°C.

REFERENCES

Abecassis J, R Abbou, M Chaurand, M-H Morel, P Vernoux. 1994. Influence of extrusion conditions on extrusion speed, temperature, and pressure in the extruder and on pasta quality. Cereal Chem. 71:247–253.

Agnesi E. 1996. The history of pasta. In: J Kruger, R Matsuo J Dick, editors. Pasta and Noodle Technology, 1–12. American Association of Cereal Chemists, St. Paul, Minn.

Akashi H, M Takahashi, S Endo. 1999. Evaluation of starch properties of wheats used for Chinese yellow-alkaline noodles in Japan. Cereal Chem. 76:50–55.

Ambalamaatil S, OM Lukow, DW Hatcher, JE Dexter, LJ Malcolmson, BM Watts. 2002. Milling and quality evaluation of Canadian hard white spring wheats. Cereal Foods World 47:319–327.

Ames JF. 1990. Control of the Maillard reaction in food systems. Trends Food Sci. Tech. 1:150–154.

Antognelli C. 1980. The manufacture and applications of pasta as a food and as a food ingredient: A review. J. Food Technol. 15:125–145.

Azudin NM. 1998. Screening of Australian wheat for the production of instant noodles. In: AB Blakeney and L O'Brien, editors. Pacific People and Their Food, 101–121. American Association of Cereal Chemists, St. Paul, Minn.

Baik B-K, Z Czuchajowska, Y Pomeranz. 1994. Role and contribution of starch and protein contents and quality to texture profile analysis of Oriental noodles. Cereal Chem. 71:315–320.

Baroni D. 1988. Manufacture of pasta products. In: G Fabriani, C Lintas, editors. Durum: Chemistry and Technology, 191–216. American Association of Cereal Chemists, St. Paul, Minn.

Bass EJ. 1988. Wheat flour milling. In: Y Pomeranz, editor. Wheat: Chemistry and Technology, 3rd edition, vol. 2, 1–68. American Association of Cereal Chemists, St. Paul, Minn.

Batey IL, BM Curtin, SA Moore. 1997. Optimization of rapid-visco analyzer test conditions for predicting Asian noodle quality. Cereal Chem. 74:497–501.

Bhirud PR, FW Sosulski. 1993. Thermal inactivation kinetics of wheat germ lipoxygenase. J. Cereal Sci. 58:1095–1098.

Bizzarri O, A Morelli. 1988. Milling of durum wheat. In: G Fabrianni, C Lintas, editors. Durum: Chemistry and Technology, 161–189. American Association of Cereal Chemists, St. Paul, Minn.

Chen PMT. 1993. Noodle manufacturing technology. In: Grains and Oilseeds: Handling Marketing, Processing, 4th edition, vol. 2, 809–829. Canadian International Grains Institute, Winnipeg, Canada.

Crosbie GB. 1994. The relationship between starch swelling properties, paste viscosity and boiled noodle quality in wheat flours. J. Cereal Sci. 13:145–150.

Crosbie GB, WJ Lambe, H Tsutsui, RF Gilmore. 1992. Further evaluation of the flour swelling volume test for identifying wheats potentially suitable for Japanese noodles. J. Cereal Sci. 15:271–280.

Crosbie, GB, AS Ross, T Moro, PC Chiu. 1999. Starch and protein quality requirements of Japanese alkaline noodles (Ramen). Cereal Chem. 76:328–334.

Cubadda R, M Carcea, L Pasqui. 1992. Suitability of the gluten index test for assessing gluten strength in durum wheat and semolina. Cereal Foods World 37:866–869.

Dalbon C, M Pagani, P Resmini, M Lucisano. 1985. Effects of heat treatment applied to wheat starch during the drying process. Getreide Mehl Brot 39:183–189 [in German].

Dalbon, C, D Grivon, M Pagani. 1996. Continuous manufacturing process. In: J Kruger, R Matsuo J Dick, editors. Pasta and Noodle Technology, 13–58. American Association of Cereal Chemists, St. Paul, Minn.

Damidaux R, J-C Autran, P Grignac, P Feillet. 1978. Mise en évidence de relations applicable en sélection entre l'électrophorégrammes des gliadins et les propriétés viscoélastiques du gluten de Triticum durum Desf. C.R. Acad. Sci. Paris, Sér. D, 287:701–704.

Davis R. 1998. The evolution of pasta equipment in Europe. Pasta Journal (December): 18–23.

Dexter JE, NM Edwards. 1998a. The implications of frequently encountered grading factors on the processing quality of durum wheat. Association of Operative Millers Bulletin (October): 7165–7171.

___. 1998b. The implications of frequently encountered grading factors on the processing quality of common wheat. Association of Operative Millers—Bulletin (June): 7115–7122.

Dexter JE, BA Marchylo. 2001. Recent trends in durum wheat milling and pasta processing: Impact on durum wheat quality requirements. In: J Abecassis, J-C Autran, P Feillet, editors. Proc. International Workshop on Durum Wheat, Semolina and Pasta Quality: Recent Achievements and New Trends, November 27, 2000, Montpellier, France, 139–164. Institute National de la Recherche, Montpellier, France.

Dexter JE, RR Matsuo, BL Dronzek. 1979. A scanning electron microscopy study of Japanese noodles. Cereal Chem. 56:202–208.

Dexter JE, RR Matsuo, BC Morgan. 1981. High temperature drying: Effect on spaghetti properties. J. Food Sci. 46:1741–1746.

Dexter JE, R Tkachuk, RR Matsuo. 1984. Amino acid composition of spaghetti: Efffect of drying conditions on total and available lysine. J. Food Sci. 49:225–228.

Dexter JE, RR Matsuo, DG Martin. 1987. The relationship of durum wheat test weight to milling performance and spaghetti quality. Cereal Foods World 32:772–777.

Dexter JE, PC Williams, NM Edwards, DG Martin. 1988. The relationships between durum wheat vitreousness, kernel hardness and processing quality. J. Cereal Sci. 7:169–181.

Dick JW, RR Matsuo. 1988. Durum wheat and pasta products. In: Y Pomeranz, editor. Wheat: Chemistry and Technology, 3rd edition, vol. 2, 507–547. American Association of Cereal Chemists, St. Paul, Minn.

Edwards NM, MG Scanlon, JE Kruger, JE Dexter. 1996. Oriental noodle dough rheology: Relationship to water absorption, formulation, and work input during dough sheeting. Cereal Chem.73:708–711.

Fang K, K Khan. 1996. Pasta containing regrinds: Effect of high temperature drying on product quality. Cereal Chem. 73:317–332.

Feillet P, JE Dexter. 1996. Quality requirements of durum wheat for semolina milling and pasta production. In: J Kruger, R Matsuo J Dick, editors. Pasta and Noodle Technology, 95–131. American Association of Cereal Chemists, St. Paul, Minn.

Feillet P, J-C Autran, C Icard-Vernière. 2000. Pasta brownness: An assessment. J. Cereal Sci. 32:215–233.

Fraignier, M-P, N. Michelle-Ferrière, K. Kobrehel. 2000. Distribution of peroxidases in durum wheat (Triticum durum). Cereal Chem. 77:11–17.

Giese, J. 1992. Pasta: New twists on an old product. Food Tech. 51(2): 118–126.

Guo G, DS Jackson, RA Graybosch, AM Parkhurst. 2003. Asian salted noodle quality: Impact of amylose content adjustments using waxy wheat. Cereal Chem. 80:437–445.

Hatcher DW. 2001. Asian noodle processing. In: G. Owens, editor. Cereals Processing, 131–157. Woodhead Publishing Ltd., Cambridge, UK.

Hatcher DW, JE Kruger. 1993. Distribution of polyphenol oxidase in flour millstreams of Canadian common wheat classes milled to three extraction rates. Cereal Chem. 70:51–55.

Hatcher DW, SJ Symons. 2000a. Assessment of Oriental noodle appearance as a function of flour refinement and noodle type by image analysis. Cereal Chem. 77:181–186.

___. 2000b. Influence of sprout damage on oriental noodle appearance as assessed by image analysis. Cereal Chem. 77:380–387.

Hatcher DW, JE Kruger, MJ Anderson. 1999. Influence of water absorption on the processing and quality of oriental noodles. Cereal Chem. 76:566–572.

Hatcher DW, MJ Anderson, RG Desjardins, NM Edwards, JE Dexter. 2002. Particle size and starch damage effect on the processing and quality of white salted noodles. Cereal Chem. 79:64–71.

Hou G, M Kruk. 1998. Asian noodle technology. In: G Ranhotra, editor. AIB Technical Bulletin, vol. 20, no. 12, 1–10. American Institute of Baking, Manhattan, Kans.

Huang S. 1996. China —The world's largest consumer of pasta products. In: J Kruger, R Matsuo, J Dick, editors. Paste Products: Chemistry and Technology, 301–329. American Association of Cereal Chemists, St. Paul, Minn.

Hummel Ch. 1966. Macaroni Products, 2nd edition. Food Trade Press, Ltd., London, U.K.

Irvine GN. 1971. Durum wheat and paste products. In: Y Pomeranz, editor. Wheat: Chemistry and Technology, 2nd edition, 777–798. American Association of Cereal Chemists, St. Paul, Minn.

Izydorczyk M, SJ Symons, JE Dexter. 2002. Fractionation of wheat and barley. In: L Marquart, JL Slavin, RG Fulcher, editors. Whole Grain Foods in Health and Disease, 47–82. American Association of Cereal Chemists, St. Paul, Minn.

Kempf C. 1998. From buttons to bowties: The evolution of pasta machinery in the United States. Pasta Journal (October): 14–26.

Kim S-K. 1996. Instant noodles. In: J Kruger, R Matsuo, J Dick, editors. Paste Products: Chemistry and Technology, 195–225. American Association of Cereal Chemists, St. Paul, Minn.

Konik CM, DM Miskelly, PW Gras. 1992. Contribution of starch and non-starch parameters to eating quality of Japanese white salted noodles. J. Sci. Food Agric. 58:403–406.

Konik CM, LM Mikkelson, R Moss, PJ Gore. 1994. Relationship between physical starch properties and yellow alkaline noodle quality. Starch/Starke 46:292–299.

Kruger JE, BC Morgan, RR Matsuo, KR Preston. 1992. A comparison of methods for the prediction of Cantonese noodle color. Can. J. Plant Sci. 72:1021–1029.

Kruger JE, MH Anderson, JE Dexter. 1994. Effect of flour refinement on raw Cantonese noodle color and texture. Cereal Chem. 71:177–182.

Lineback DR, VF Rasper. 1988. Wheat carbohydrates. In: Y Pomeranz, editor. Wheat: Chemistry and Technology, 3rd edition, vol. 1, 277–372. American Association of Cereal Chemists, St. Paul, Minn.

MacRitchie F. 1992. Physiochemical properties of wheat proteins in relation to functionality. In: J.E. Kinsella, editor. Advances in Food and Nutrition Research, vol. 36, 1–87. Academic Press, London, U.K.

Marchylo BA, JE Dexter. 2001. Pasta production. In: G. Owens, editor. Cereals Processing, 109–130. Woodhead Publishing Ltd., Cambridge, U.K.

Marchylo BA, JE Dexter, LM Malcolmson. 2004. Improving the texture of pasta. In: D Kilcast, editor. Texture in Food. Vol. 2: Solid Foods, 475–500. Woodhead Publishing Ltd., Cambridge, United Kingdom.

Martini A. 1977. Pasta and Pizza, 14–25. St. Martin's Press, New York.

Matsuo RR. 1993. Durum wheat: Production and processing. In: Grains and Oilseeds: Handling, Marketing, Processing, 4th edition, vol. 2, 779–807. Canadian International Grains Institute, Winnipeg, Canada.

Matsuo RR, JE Dexter, BL Dronzek. 1978. Scanning electron microscopy study of spaghetti processing. Cereal Chem. 55:744–753.

Milatovic L, G Mondelli. 1991. Pasta Technology Today. Chirotti Editori, Pinerolo, Italy.

Miskelly DM. 1984. Flour components affecting paste and noodle colour. J. Sci. Food Agric. 35:463–471.

___. 1993. Noodles—A new look at an old food. Food Australia 45:496–500.

___. 1996. The use of alkali for noodle processing. In: J Kruger, R Matsuo, J Dick, editors. Paste Products: Chemistry and Technology, 227–273. American Association of Cereal Chemists, St. Paul, Minn.

___. 1998. Modern noodle-based food—Raw material needs. In: AB Blakeney, L O'Brien, editors. Pacific People and Their Food, 123–142. American Association of Cereal Chemists, St. Paul, Minn.

Miskelly DM, HJ Moss. 1985. Flour quality requirements for Chinese noodle manufacture. J. Cereal Sci. 15:271–280.

Morris C. 1998. Evaluating the end-use quality of wheat breeding lines for suitability in Asian noodles. In: A.B. Blakeney and L. O'Brien, editors. Pacific People and Their Food. American Association of Cereal Chemists, St. Paul, Minn.

Morris VG. 1990. Starch gelation and retrogradation. Trends Food Sci. Tech. 1:2–6.

Moss HJ, DM Miskelly, R Moss. 1986. The effect of alkaline conditions on the properties of wheat flour dough and Cantonese-style noodles. J. Cereal Sci. 4:261–268.

Moss R, PJ Gore, IC Murray. 1987. The influence of ingredients and processing variables on the quality and microstructure of Hokkien, Cantonese and instant noodles. Food Microsturcture 6:63–74.

Nagao S. 1981. Soft wheat uses in the Orient. In: Y Yamazaki, C Greenwood, editors. Soft Wheat: Production, Breeding, Milling and Uses, 267–304. American Association of Cereal Chemists, St. Paul, Minn.

Noda T, T Tohnooka, S Taya, I Suda. 2001. Relationship between physicochemical properties of starches and white salted noodle quality in Japanese wheat flours. Cereal Chem. 78:395–399.

Oda M, Y Yasuda, S Okazaki, Y Yamauchi, Y Yokoyama. 1980. A method of quality assessment for Japanese noodles. Cereal Chem. 62:441–446.

Oh NH, PA Seib, AB Ward, CW Deyoe. 1985. Noodles. IV. Influence of flour protein, extraction rate, particle size, and starch damage on the quality characteristics of dry noodles. Cereal Chem. 62:441–446.

Owens, G. 2001. Wheat, corn and coarse grains milling. In: G. Owens, editor. Cereals Processing, 27–52. Woodhead Publishing Ltd., Cambridge, U.K.

Pagani MA, P Resmini, L Pellegrino. 1992. Technological parameters that influence Maillard reaction during pasta processing. Tecnica Molitoria 43:577–592 [in Italian].

Payne PI, EA Jackson, LM Holt. 1984. The association between gamma gliadin 45 and gluten strength in durum wheat varieties: A direct causal link or the result of genetic linkage? J. Cereal Sci. 2:73–81.

Pollini CM. 1996. THT technology in the modern industrial pasta drying process. In: J Kruger, R Matsuo, J Dick, editors. Pasta and Noodle Technology, 59–74. American Association of Cereal Chemists, St. Paul, Minn.

Posner ES, AN Hibbs. 1997. Wheat Flour Milling. American Association of Cereal Chemists, St. Paul, Minn.

Resmini P, MA Pagani, L Pelligrino. 1996. Effect of semolina quality and processing conditions on nonenzymatic browning in dried pasta. Food Australia 48:362–367.

Rho KL, PA Seib, OK Chung, DS Chung. 1986. Retardation of rancidity in deep-fried instant noodles (ramyon). JAOCS 63:251–257.

Sarkar AK. 1993. Flour milling. In: E. Bass, editor. Grains and Oilseeds: Handling, Marketing, Processing, 4th edition, vol. 2, 604–653. Canadian International Grains Institute, Winnipeg, Manitoba.

___. 2003. Grain milling operations. In: A Chakraverty, AS Mujumdan, GS Raghavan, HS Ramaswamy, editors. Handbook of Postharvest Technology: Cereals, Fruits, Vegetables, Tea and Spices, 253–325. Marcel Dekker, New York.

Seib PA. 2000. Reduced-amylose wheats and Asian noodles. Cereal Foods World 45:504–512.

Sensidoni A, D Peressini, CM Pollini. 1999. Study of the Maillard reaction in model systems under conditions related to the industrial process of pasta thermal VHT treatment. J. Sci. Food Agric. 50:317–322.

Shelke K, JW Dick, YF Holm, KS Loo. 1990. Chinese wet noodle formulation: A response surface methodology study. Cereal Chem. 67:338–342.

Toyokawa H, GL Rubenthaler, JR Power, EG Schanus. 1989. Japanese noodle qualities. II. Starch components. Cereal Chem. 66:387–391.

Varriano-Marston E, F Stoner. 1996. Pasta packaging. In: J Kruger, R Matsuo, J Dick, editors. Pasta and Noodle Technology, 75–94. American Association of Cereal Chemists, St. Paul, Minn.

Vansteelandt J, JA Delcour. 1998. Physical behavior of durum wheat starch *(Triticum durum)* during industrial pasta processing. J. Agric. Food Chem. 46:2499–2503.

Wu TP, WY Kuo, MC Cheng. 1998. Modern noodle-based foods—Products range and production methods. In: AB Blakeney, L O'Brien, editors. Pacific People and Their Food, 37–90. American Association of Cereal Chemists, St. Paul, Minn.

Zeifel C, S Handschin, F Escher, B Conde-Petit. 2003. Influence of high temperature drying on structural and textural properties of durum wheat pasta. Cereal Chem. 80:159–167.

14
Dairy: Cheese

S. E. Beattie

BACKGROUND

The processing of milk into cheese is a relatively simple task that involves basic aspects of food chemistry. In the conversion of milk into cheese, we see the importance of water activity, oxidation/reduction potential, and pH; lipid, carbohydrate, and protein chemistry; and mineral-protein interactions. What is amazing is that there are several hundred different cheese varieties made throughout the world, and all have similarities in manufacture and in the chemistry that accompanies the making of cheese.

CATEGORIES OF CHEESE

It is unknown how many different cheeses are found in the world. A publication by the U.S. Department of Agriculture (USDA, Handbook 54, 1953, revised 1978) describes over 400 cheeses and lists the names of 800 more. It would seem that each country develops a unique style with which to convert perishable milk into a more stable product such as cheese. So categorizing cheese becomes a bit of a problem. The type of aging that is used, the type of microorganisms involved, how the milk is handled, the amount of moisture in the cheese, and how the curd is handled are all likely parameters for categorizing cheeses.

Different ways to categorize cheese might include (1) coagulation type, (2) ripening method, and (3) texture.

Coagulation Type

Acid only. Cottage cheese, cream cheese, and Neufchatel are examples of cheeses that utilize an acid only for coagulating the protein in the milk. These cheeses typically have higher moisture (50–80%) and contain significant quantities of residual lactose.

Heat and acid. Ricotta and queso blanco are examples of cheeses that have a heat step included in the precipitation of the casein

The information in this chapter has been derived from a chapter in *Food Chemistry Workbook,* edited by J. S. Smith and G. L. Christen, published and copyrighted by Science Technology System, West Sacramento, California, ©2002. Used with permission.

protein of the milk. Because of the heat step these cheeses are typically a bit lower in moisture (50–70%). Since no microbial growth has occurred, the cheese retains a significant amount of lactose. Additionally, the whey proteins have been retained with the cheese.

Acid and enzymes. This describes how many cheeses are made. Acid is produced by added or naturally occurring lactic acid bacteria (LAB), and a coagulating enzyme (rennet or chymosin) is added to form the curd. Cheddar, Swiss style, brick, and many other cheeses are made in this manner. The amount of residual lactose in the curd is usually very low.

Ripening Method

Fresh unripened cheeses. These cheeses are not aged and are consumed shortly after manufacture. Mozzarella, cottage, ricotta, and cream cheeses are examples.

Soft, surface mold ripened cheeses. Camembert and Brie are the best-known examples of these cheeses that use LAB to produce acid in the milk and an enzyme (rennet or chymosin) to cause coagulation of the milk protein. The surface growth of white *Penicillium caseicolum* mold gives the cheese its characteristic flavor and texture. These cheeses have little residual lactose.

Internally mold ripened. Gorgonzola, Roquefort, and blue are examples of cheeses that are ripened throughout by the growth of the blue-green mold *Penicillium roquefortii*. The curd is formed by acid produced from bacteria, and coagulation occurs with the addition of rennet. The microflora of the cheese is responsible for the typical flavor.

Surface bacteria ripened. Limburger, brick, Port du Salut and Tilsiter are examples of cheeses that rely upon a surface smear of bacteria and yeasts to form the flavor and texture of these cheeses. The curd is formed by LAB and rennet.

Internally bacteria ripened stirred curd. Colby, Gouda, and Edam are examples of cheeses that have acid produced by LAB and rennet curd formation, followed by washing of the curd to reduce acid development. The flavor of these cheeses is relatively mild.

Internally ripened curd. Cheddar, provolone, and Romano cheeses rely upon LAB and rennet to form the curd. The LAB produce the majority of flavor as these cheeses ripen.

Internally ripened secondary culture. Emmentaler, Jarlsburg, and Swiss cheeses are produced much like hard, internally ripened cheeses, except that a secondary culture of a bacterium that utilizes lactic acid to produce carbon dioxide is added to the milk. The carbon dioxide forces open the eyes of these holed cheeses.

Texture

Very hard. Parmesan and Romano are examples of this type of cheese in which the curd is cooked to a relatively high temperature (50°C), causing it to dry out. The aging process of these cheeses is typically over one year after the curd has been formed. Moisture content is usually less than 32%.

Hard. Cheddar, Colby, Swiss style, Gouda, and many other cheeses fall into this category. Moisture ranges from 37 to 45%.

Semisoft. This is a very diverse group of cheeses and includes Gorgonzola, Limburger, brick, and Muenster. The texture is relatively soft and nearly spreadable. Moisture content is in the 43–50% range. Part of the texture is derived from proteolysis.

Soft. These cheeses are characterized as being relatively easy to spread. Brie, cream, Neufchatel, and ricotta are examples of these cheeses, which have a moisture content up to 55%. In some of these cheeses the texture is a result of proteolysis.

MILK QUALITY AND COMPOSITION

Milk for cheese manufacture must be of high quality with regard to microbiological and chemical parameters. High numbers of somatic cells from the animal are undesirable. The milk must be absolutely free from antibiotics. Raw, heat-treated, or pasteurized milk may be used in cheeses. The use of pasteurized or heat-treated milk reduces flavor variations in many cheeses by killing adventitious microorganisms. If milk is not pasteurized, then there is a risk of pathogenic bacteria surviving in the cheese and causing illness in the consumer. For hard ripened cheeses (e.g., Cheddar, brick, Asiago) made from raw milk, the curing process may eliminate pathogens. By U.S. law, raw milk cheese must be cured for at least 60 days at a temperature above 4°C.

For most cheeses, raw milk must be maintained at 4°C or less to maintain microbiological quality. Because of the potential for undesirable bacterial growth, temperature-abused milk may be the source of off flavors in the finished cheese.

The most important constituents of milk for cheese manufacture are casein protein, fat, lactose, and the calcium phosphate complex. Casein and mineral content are directly related to the firmness of the eventual curd. The breed of animal will determine the protein and fat content of the milk. Milk from Jersey cows has one of the highest casein protein contents of all breeds of cow. This will increase the yield of cheese; however, these animals do not produce as much milk as the more common Holstein cow. On the other hand, Holstein cows have milk with lower protein content but produce more milk. Mineral levels in milk can be affected by stage of lactation, season, and mastitis.

Milk from mastitic cows will have a higher number of somatic cells and possibly bacteria. When stored at a cold temperature, proteolytic and lipolytic enzymes from the bacteria and somatic cells may degrade casein, which results in decreased yields of cheese and potential bitterness. Lipolytic enzymes from bacteria and/or animal may cause flavor rancidity in the final cheese. Somatic cell numbers as low as 300,000 may be enough to reduce yields of Cheddar cheese.

MILK PROTEINS

Milk contains two general categories of proteins: serum (whey) proteins and caseins. The caseins are the proteins responsible for the structure of cheeses. The four casein proteins associate in a quaternary structure known as a micelle. These micelles are relatively large and are present in the milk as particles that are in colloidal suspension in the serum portion of the milk. Micelles are aggregations of smaller subunits (submicelles) that are complexes of many molecules of the four casein proteins.

The four casein proteins are called α_{s1}, α_{s2}, β, and κ. The genes and primary structures of each of these proteins have been sequenced. α_{s1}, α_{s2}, and β appear to be distantly related to each other because of structural similarities in various regions of the proteins (Swaisgood 1985, 1992). These three proteins are known as the calcium sensitive caseins since at low calcium concentrations they are precipitated if κ-casein is not present. α_{s1}, α_{s2}, and β caseins all have regions where the serine residues are phosphorylated, resulting in areas of net negative charge. α_{s2}-casein has several of these groups of charge and is therefore the most hydrophilic of the calcium sensitive caseins. Calcium ions are attracted to the phosphorylated serine residues. This interaction may form calcium phosphate bridges, which help stabilize the casein interactions. β-casein is the most hydrophobic of the caseins and contains a small, very polar domain, with the bulk of the protein being hydrophobic and interacting with the hydrophobic core of the submicelle (Swaisgood 1992). In proteins, the hydrophobic effect is important in stabilizing secondary, tertiary, and quaternary structure. At low temperatures, water molecules become more ordered in structure, which relieves some of the hydrophobic-effect forces and may allow unfolding or disassociation of hydrophobic areas of proteins. Thus, at low temperature (< 4°C), the hydrophobic association of β-casein with the casein submicelle is weakened, and β-casein may dissociate from the submicelle.

The exact nature of the interactions of these caseins is unclear; however, it is known that they form spherical submicelles that aggregate into the larger micelles. The submicelles have variable amounts of each of the individual caseins but are stabilized by hydrophobic interactions, calcium phosphate, and κ-casein. Submicelles containing κ-casein are found on the exterior of the micelle. κ-casein is divided into hydrophobic and hydrophilic sections, with the hydrophilic end having a net negative charge (Swaisgood 1992). The hydrophilic section has several glycosylated threonine residues as well as phosphorylated serine residues. Thus the sugars and charged molecules are hydrophilic and are associated with the surrounding water, while the hydrophobic end is buried in the rest of the submicelle and associates with the hydrophobic core. On submicelles containing κ-casein, the hydrophilic end forms a kind of hairy layer that helps prevent aggregation of the micelles because of steric hindrance and electrostatic repulsion (Brule and Lenoir 1987). The larger micelles are stabilized by hydrophobic interactions, calcium phosphate bridges, and κ-casein. A portion of the micelle calcium phosphate is in equilibrium with soluble serum calcium phosphate. This equilibrium is shifted with temperature, pH, and other factors. As pH decreases, the amount of soluble calcium in the serum portion increases with a concomitant decrease in calcium associated with the micelles (van Hooydonk et al. 1986). Thus, the ionic attraction of micelles for each other is lost, and the caseins dissociate from each other, resulting in a decrease in the number of caseins or micelles that participate in curd formation and a reduced yield of cheese. This is the reason for the addition of calcium chloride to the cheese milk.

MILK LIPIDS

The lipids in milk are found mainly as triacylglycerols. These neutral lipids are found within milk fat globules secreted by the dairy animal directly into the milk. The milk fat globule is composed of a phospholipid membrane surrounding neutral triacylglycerols. Milk fat globules are in colloidal suspension in the milk, but they are relatively large (0.1–10 μm) and may aggregate and rise to the surface of the milk. Homogenization reduces the milk fat globule size and prevents aggregation of the droplets. Additionally, homogenization makes the neutral lipids and some phospholipids available as substrate for endogenous and exogenous lipases. Milk is generally heat treated to inactivate lipases before it is homogenized. Most cheeses are made from milk that has not been homogenized. An exception is some blue-veined cheeses that rely on lipolysis for flavor, which makes having small fat droplets an advantage to flavor development.

The fatty acid composition of ruminant lipids is very complex and includes a great diversity of fatty acids. This diversity of fatty acids is a result of the rumen bacteria altering consumed lipids. It is the fatty acid profile of milk lipids that gives milk its smooth melting characteristic in the mouth and provides the unique flavor of some cheeses. The fatty acid profiles of milk from several ruminants are given in Table 14.1.

Fatty acids can be categorized based upon branching (relatively rare in most animal and plant fats), degree of unsaturation, and chain length. Chain length plays a significant role in the functionality of the fatty acid. The short-chain and branched fatty acids (4–10 carbons long) are volatile and have a strong aroma. As can be seen in Table 14.1, the amounts of these shorter chain fatty acids are relatively high in milk. When freed from the triacylglyceride, they contribute to the overall aroma of cheese. The longer chain unsaturated fatty acids are sources of off flavors caused by autoxidation of the double bonds in the fatty acid. While excess free fatty acids cause a rancid flavor in some cheeses (Cheddar for example), these fatty acids are important to the flavor of blue-veined cheeses and some Italian hard cheeses.

MILK CARBOHYDRATES AND OTHER ORGANIC COMPOUNDS

Perhaps the most important component of milk with respect to cheese making is the disaccharide lactose. A variety of LAB ferments the sugar into lactic acid,

Table 14.1. Total Fat and Fatty Acid Profiles of Ruminant Milk

| Fatty Acid (Common Name) | Amount of Fatty Acid/100 g Milk | | | |
	Goat	Cow	Sheep	Indian Buffalo
Butyric (4:0)	0.128	0.119	0.204	0.276
Caproic (6:0)	0.094	0.070	0.145	0.153
Caprylic (8:0)	0.096	0.041	0.138	0.071
Capric (10:0)	0.260	0.092	0.400	0.141
Lauric (12:0)	0.124	0.103	0.239	0.167
Myristic (14:0)	0.325	0.368	0.660	0.703
Palmitic (16:0)	0.911	0.963	1.622	1.999
16:1 isomers	0.082	0.082	0.128	0.142
Stearic (18:0)	0.441	0.444	0.899	0.682
18:1 isomers	0.977	0.921	1.558	1.566
18:2 isomers	0.109	0.083	0.181	0.070
18:3 isomers	0.040	0.053	0.127	0.076
Total lipid (%)	4.14	3.66	7.00	6.89

Source: U.S. Department of Agriculture, Agricultural Research Service 2003.

which subsequently reduces the pH of the milk sufficiently to precipitate the milk proteins. Additionally, the reduced pH aids in the activity of the milk-clotting enzyme, chymosin. Two general types of LAB are found in cheeses—homofermentative and heterofermentative. Homofermentative LAB ferment lactose into lactic acid only. Strains of *Lactococcus* are the most important homofermentative LAB. The heterofementative LAB, on the other hand, produce lactic acid, ethanol, and carbon dioxide from lactose. Heterofermentative organisms are not usually used as the starter cultures in most cheeses, but they may be found as adventitious contaminates that contribute to flavor.

Citric acid is an organic acid that occurs naturally in milk at low levels. It is metabolized by certain LAB to diacetyl, the characteristic flavor of butter. Adventitious microorganisms may also use the compound to form a variety of flavor compounds.

RAW MATERIALS PREPARATION

While there are literally hundreds of different varieties of cheese, most cheeses are made using similar procedures and materials. The interaction of enzymes, pH, milk components, microorganisms, and environment all impact the final cheese flavor, texture, and appearance. Figure 14.1 shows a generalized flowchart for the manufacture of ripened cheeses.

PRETREATMENT OF MILK

Milk for cheese manufacture may be raw, heat treated, or pasteurized. Heat treatment is considered any heating that would not be considered pasteurization. Since heat treatment and pasteurization kill many of the microorganisms in raw milk, vat-to-vat variation in cheese flavor and yield is reduced. Heat

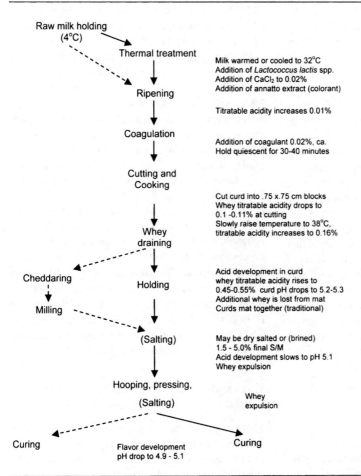

Figure 14.1. Generalized flowchart for cheese making (Wilster 1980).

treatment of milk may also help to reestablish the colloidal calcium phosphate balance in the casein micelles, and return β-casein to the micelle (Guinee 2003). Heat treatment in excess of normal pasteurization temperatures will cause denaturation of whey proteins. The whey protein, β-lactoglobulin, is especially important because it will essentially coat the casein micelle upon denaturation (Green and Grandison 1987). This will physically prevent chymosin from hydrolyzing the κ-casein.

Homogenization of milk to reduce the fat globule size is practiced for some cheese varieties but not for most cheeses. For Cheddar cheese standardization of milk to a casein:milk fat ratio of 0.67–0.72 is considered to be optimum; this ratio is higher for other cheeses such as Roquefort or other blues. At this level, the amount of milk fat that is left in the curd will be within legal limits and will not interfere with draining of whey or moisture loss from the curd.

While many cheeses are made from full fat milk, certain cheeses commonly are made from reduced or nonfat (skim) milks or milk that has been supplemented with fat. The separation of the fat from the milk does not change the protein content, but it does change the final cheese flavor profile.

UNRIPENED, ACID-COAGULATED MILK

Some fresh cheeses rely upon heat and the addition to the milk of a food grade acid to cause precipitation of the caseins. Lactic, citric, and acetic acids are the acids commonly used for this type of cheese. The curd is formed by bringing the pH of the milk near the isoelectric point of the casein micelle. The isoelectric point of a protein is the pH at which the overall net ionic charge on the protein is zero. Most proteins have a net positive or negative charge; this net charge causes the protein molecules to repel each other by electrostatic repulsion. Casein micelles aggregate as the pH is lowered toward a pH of 4.7. As the pH of the milk approaches this value, the repulsive charges on the casein micelles are reduced. This allows the micelles to aggregate as large curd particles. These curd particles can then be filtered from the serum (whey) portion of the milk.

CURD FORMATION

INITIAL RIPENING OF MILK

Initial ripening of cheese milk entails adding lactic acid starter culture or allowing naturally occurring

LAB to grow. This initial period allows the starter to become acclimated to the milk and allows acid production to begin. Typically, cheeses that utilize naturally occurring LAB are allowed to ripen over a longer period of time since the numbers of the bacteria are low. In contrast, LAB starter cultures added to milk result in high numbers of bacteria immediately and a shorter ripening time.

If there was a heat treatment, the milk is cooled to approximately 30–32°C. Calcium chloride may be added to enhance micelle structure and thereby maintain yield. The plant extract, annatto, may be added at this time for color. Cheeses without annatto extract are cream colored. Lactic acid starter cultures in a variety of forms may be added to the milk at this stage. Cultures may be frozen, freeze dried, actively growing, or from a mother culture. The LAB used commonly in cheese production are strains of *Lactococcus lactis* ssp. *cremoris* or *lactis*. These homofermentative bacteria ferment lactose into lactic acid, which reduces the pH of the milk and subsequent curd. Fermentation of the lactose to lactic acid is fundamentally important to the quality of many cheeses (Fig. 14.2.)

Depending upon the source of the LAB, added or naturally occurring, the milk is allowed to ripen for an hour to overnight. This ripening period allows the LAB to begin metabolizing lactose to lactic acid, which increases the titratable acidity (the actual amount of acid in the milk) from 0.10% to 0.16–0.2% and slightly decreases the pH. Milk contains a variety of buffering agents, including milk proteins and citrate, that resist dramatic changes in pH as the amount of lactic acid increases. The gradual decrease in pH is important for optimum activity of the proteolytic enzyme—chymosin (rennet). While a coagulated mass could be obtained using just LAB or chymosin, the combination results in a much firmer coagulum. The production of lactic acid is important because it enhances chymosin activity, yields a firmer curd, increases the rate of whey loss from curd, and serves as a preservative for the finished cheese. The milk is left undisturbed once the chymosin is added to allow a firm gel to form.

ENZYMATIC COAGULATION OF THE MILK PROTEINS

Milk coagulation or clotting is the formation of a gel from the caseins. The enzymatic process is generally in two stages: (1) enzymatic action and (2) the resulting aggregation of micelles.

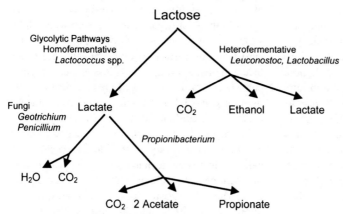

Figure 14.2. Fate of lactose during cheese making.

The addition of a milk-clotting enzyme causes the slightly acidic milk to coagulate into a gel. Milk-clotting protease enzymes are derived from several sources, including calves' stomachs (rennet or chymosin), the mold *Mucor miehei,* and recombinant forms of chymosin from *E. coli* and various yeasts that contain the bovine gene for chymosin. Chymosin is an acid protease that hydrolyzes proteins at specific sites in the primary structure of the protein. Chymosin cleaves κ-casein between positions 105 (phenylalanine) and 106 (methionine), which divides the chain into two discrete regions, one hydrophobic and one hydrophilic. The products of the enzyme reaction are the water-soluble component, glycomacropeptide, and the hydrophobic portion, para-κ-casein. Upon cleavage, the glycomacropeptide diffuses into the serum portion of the milk, and the para-κ-casein stays with the casein micelle. When κ-casein is cleaved, the casein micelle is destabilized, with loss of a net negative charge. Because repulsive micelle surface charges and steric interference are reduced, hydrophobic interactions between micelles can occur. This results in aggregations of micelles. As observed by electron microscopy, the micelles form chains that gradually aggregate into a network (Brule and Lenoir 1987).

The pH of the cheese milk is continually dropping as lactose is metabolized by the starter culture. The added calcium chloride acts to maintain the equilibrium of the colloidal calcium with the serum calcium, and thus the integrity of the micelles is maintained. Calcium bridges stabilize the casein network. This prevents yield loss caused by loss of unassociated submicelles into the serum. The coag-

ulated milk is a gel with milk-fat globules, starter bacteria, and whey all entrapped in a casein network. The moisture content is about the same as that of the original milk, 87%.

CUTTING, COOKING, SALTING, AND FORMING OF THE CURD

The handling of the cut curd is a major determinant of the final cheese type. Cooking and handling differences lead to different end products. For example, high cooking temperatures (up to 50°C) lead to a firm cheese such as Parmesan, while lower cooking temperatures result in softer curd and softer cheeses.

Once the milk has been coagulated, the gel is cut with knives into small pieces. Cutting aids in the loss of moisture and causes some loss of fat from the curd. Syneresis occurs as the curd is lightly heated and stirred. As the curd is heated, casein continues to aggregate into a tighter network. This causes the curd to contract, which results in expulsion of whey and some fat. Syneresis does not cause a change in the hydration level of the casein proteins. During curd shrinkage, fat globules are disrupted and become subject to loss or more available for lipolytic enzymes. The loss of moisture from the curd is important to the keeping quality of the final product. Loss of free moisture and subsequent salting helps reduce the water activity of the curd and prevents flavor defects associated with whey. The moisture content in the final cheese will vary depending upon the type of cheese and can range from 18–27% for Romano and Parmesan, respectively, to upwards of 55% for some soft cheeses such as cream cheese. After cooking, the resultant curd is much firmer and smaller in size. The pH of the curd at this stage ranges from 5.9 to 6.2.

The whey portion of the milk contains a dilute solution of serum proteins, lactose, and minerals. Whey is a major waste problem for cheese manufacturers, as it is very dilute and has a high pollution coefficient. The lactose in the whey is in equilibrium with the lactose in the curd. This residual lactose is important to the starter culture as a nutrient. Once the whey has been drained from the curd, the only lactose available for metabolism is in the curd. Reduction in the pH of the curd keeps the growth of nonstarter bacteria to a minimum, helps with further syneresis, protects flavor, and helps form the cohesive mat of the curd.

Depending upon the type of cheese, separation of the whey from the curd may be accomplished by one of several methods. Dipping, straining, draining, and possibly pressing are the major ways that the whey is removed from the curd. In Cheddar and similar cheeses, the cooked curds settle in the bottom of the cheese vat and form a mat. The free whey is drained from the vat, and the curds fuse together to form a fibrous mat. The term "Cheddaring" comes from the treatment of the matted curd. Originally, the mat was cut into blocks that were stacked on each other, and the position of the individual blocks was rotated so that each was exposed to the force of the other blocks. This process forced more whey from the curd. During this time, further acid was developed by fermentation of the curd lactose. A pH drop to less than 5.8 is necessary for the formation of a fibrous mat that will lend itself to Cheddaring. Deformation and hydrophobic association of the casein micelle strands are believed to be the cause of the fibrous network. Some modern Cheddar cheese is made in automated systems where traditional Cheddaring is not practiced. The curd is held at 38–39°C for a period of time sufficient to allow the pH to develop.

In blue-veined cheeses, curd is dipped from the vat and placed into hoops to drain. This practice keeps the individual curd pieces from matting together to form a cohesive structure. Because the ripening mold is aerobic, the curds need to be somewhat open to allow some air circulation for mold growth to occur. Blue-veined cheeses may also be "needled": the curd block is pierced to form air holes throughout the block. These holes allow oxygen to get to the mold.

Salting is useful for several reasons, including taste and texture, preservation by lowering water ac-

tivity and inhibiting of bacterial growth, and control of the final pH of the cheese. Salt-in-moisture is the controlling factor in each case (Lawrence and Gilles 1987). Because sodium chloride diffuses into the curd from the surface, several gradients are established in the curd fragment. A loss of calcium phosphate occurs with salting, and a pH gradient is established because LAB continue to produce acid at the center of the curd but not at the surface.

In Cheddar-type cheeses, after the mat has formed and been cut into blocks, the blocks are mechanically cut into small rectangles (fingers) prior to salting. Milling of the fingers makes smaller curd particles, which helps with further whey removal and allows salt to diffuse into the curd more uniformly. Dry sodium chloride is added to the milled curd, and the mixture is allowed to "mellow." Mellowing allows the salt to dissolve and be absorbed into the moist curd. In extreme cases, these gradients can lead to a defect in Cheddar cheese known as seaminess. This defect is characterized by the pressed cheese showing discrete curd fragments and a potentially crumbly texture because the curd did not fuse properly upon pressing (Bodyfelt et al. 1988). Because of the desiccating effect of salt, calcium orthophosphate dihydrate crystals may appear on the surface of the cheese curd (Lawerence and Gilles 1987).

In other types of cheeses, the salting is done using a brine solution. The cheese block is soaked in the brine for a period of time (depending upon block size), and the salt diffuses into the cheese. This type of salting is used for brick and Swiss types of cheeses. Dry salting may also be used. In this case, the salt is rubbed onto the surface of the cheese and the sodium chloride diffuses into the cheese.

Hooping gives a characteristic shape to the final cheese. In some cases, pressing may be used with hooping to remove more whey and cause the curd to consolidate into one mass of defined shape. Pressing is done by mechanical and/or vacuum methods. Vacuum methods remove air pockets between the curd particles, thereby reducing the incidence of the "open" defect. Open cheese is characterized by irregular openings in the cheese. The blocks of cheese may be vacuum packaged also. If the cheese was made from raw milk it must undergo curing for at least 60 days at temperatures above 5°C to kill any pathogenic microorganisms. Most cheese must undergo some period of ripening or aging to develop the characteristic flavor.

CHEMISTRY OF CHEESE RIPENING/AGING

The chemistry of cheese aging is one of the more complex biochemical transformations that occur in foods. The relatively flavorless curd is transformed into a product with hundreds of compounds that contribute to the overall flavor of the final cheese. These transformations of the proteins, carbohydrates, lactic acid, and lipids are caused by several things, including (1) enzymes present in the milk, (2) added enzymes, (3) enzymes released from various microorganisms, (4) metabolism of added and adventitious microorganisms, and (5) spontaneous reactions caused by the low oxidation/reduction (Eh) potential of the final block of cheese.

The ripening process will be described in terms of changes to each of the major components of the milk and the causes of those changes.

METABOLISM OF CARBOHYDRATE AND LACTIC ACID

As described above, the lactose portion of the milk is metabolized into lactic acid by glycolytic pathways in the starter cultures (Fig. 14.2). The resultant change in pH impacts the type of microflora that develops on or in the cheese. In semi-hard cheeses and hard cheeses, such as Cheddar and Parmesan types, lactic acid coupled with low Eh helps prevent the outgrowth of many microorganisms, including some pathogens.

Metabolism of lactose to lactic acid occurs by two different pathways (Choisy et al. 1987). Lactose is first transported across the membrane while being phosphorylated at the C-1 position. The phosphorylated galactose moiety is first metabolized to tagatose, then to dihydroxyacetone-P, which is converted to glyceraldhyde-3-P. The glucose moiety is converted to glyceraldhyde-3-P through the Embden-Meyerhoff pathway. Glyceraldehyde-3-P is first metabolized to pyruvate and subsequently to lactic acid. For every molecule of lactose, not quite four molecules of lactic acid are produced (Reaction I).

In surface-ripened and blue-veined cheeses, the fungal microflora completes the oxidation of lactose by metabolizing it into carbon dioxide and water. This is an important step in the microbial ecology of these cheeses. The loss of acid causes an increase in the pH of the curd to levels that favor the growth of bacteria that are important flavor producers but are unable to grow at lower pH values. These bacteria grow and metabolize other components of the curd into a variety of flavor compounds.

Important to the eye and flavor development in Swiss-type cheeses is the metabolism of lactic acid by *Propionibacterium freundreichii* (Fig. 14.2). This anaerobic bacterium is added to the cheese milk with the starter culture. The low Eh of the cheese block and the presence of a carbon/energy substrate, lactic acid, favors growth of the bacterium. In this reaction (Reaction II), lactic acid is metabolized to acetate and propionate.

$$\text{I} \quad \underset{\text{Lactose}}{C_{12}H_{24}O_{12}} \rightarrow \underset{\text{Pyruvate}}{CH_3COCOO^-} \rightarrow \underset{\text{Lactate}}{CH_3CHOHCOO^-}$$
$$\rightarrow CO_2 + H_2O$$

$$\text{II} \quad \underset{\text{Lactic acid}}{3\ CH_3CHOHCOO^-} \rightarrow \underset{\text{Proprionate}}{2\ CH_3CH_2COO^-}$$
$$+ \underset{\text{Acetate}}{CH_3COO^-} + CO_2 + H_2O$$

The production of carbon dioxide in the cheese block causes the characteristic eye formation in Swiss-style cheeses. Additionally, propionibacteria produce flavor compounds that give these cheeses flavor.

CHANGES IN PROTEIN

The protein of the cheese curd is primarily the caseins from the milk. Relatively little whey protein is left in the curd after the block is formed. The casein provides structure to the block and eventually is broken down by native milk enzymes, enzymes from the starter culture, and enzymes produced by secondary cultures. The overall reaction for casein breakdown is shown in Figure 14.3. The importance of proteolytic enzymes in the flavor and texture of cheeses cannot be overemphasized. The resulting peptides and amino acids are important flavors and flavor precursors and cause pH changes in the cheese.

There are several different sources of the proteases and peptidases found in cheese. These sources are (1) endogenous (from the animal): plasmin and cathepsin D; (2) exogenous refined: coagulating enzymes (mainly chymosin); and (3) exogenous microbial based: starter lactic acid bacteria, nonstarter lactic acid bacteria, adventitious bacteria, and secondary cultures.

Proteases and peptidases have specific amino acid recognition sequences that fit into the binding and active sites of the enzyme. The secondary and terti-

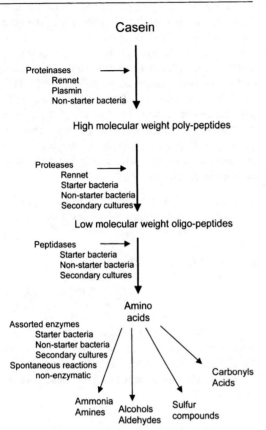

Casein

Proteinases
Rennet
Plasmin
Non-starter bacteria

High molecular weight poly-peptides

Proteases
Rennet
Starter bacteria
Non-starter bacteria
Secondary cultures

Low molecular weight oligo-peptides

Peptidases
Starter bacteria
Non-starter bacteria
Secondary cultures

Amino acids

Assorted enzymes
Starter bacteria
Non-starter bacteria
Secondary cultures
Spontaneous reactions
non-enzymatic

Carbonyls
Acids

Ammonia
Amines

Alcohols
Aldehydes

Sulfur
compounds

Figure 14.3. Fate of casein proteins in cheese ripening (after Fox et al. 1995).

ary structures of the protein reduce the availability of some likely sequences because they may be buried within the globular mass of the protein. An important consideration is the fact that hidden bonds may become available when a protease acts upon the protein distant from the hidden bond. Thus, a peptide that is formed may be hydrolyzed more rapidly than if it were still a part of the original protein. An example of this appears to occur in α_{s1}-casein when chymosin cleaves a large peptide from the protein. This large peptide then is further degraded by chymosin and plasmin (Guinee 2003).

The two proteins that are most degraded in cheese curds are α_{s1}-casein and β-casein. In Cheddar cheese, these two proteins are broken into a variety of peptides by plasmin, residual chymosin, and proteases associated with the lactic starter culture. One consequence of plasmin hydrolysis of the hydrophobic β-casein is the production of hydrophobic peptides. There is some indication that these peptides

are a component of the bitterness found in Cheddar cheeses (Farkye 1995). But in blue-veined cheeses, the proteases produced by the mold *Penicillium roqueforti* are responsible for most of the proteolysis that results in a softening of the curd. The fungi associated with soft ripened cheeses such as Camembert and Brie also produce proteases and peptidases. Indirectly, it is the activity of these enzymes that lead to the soft texture that characterizes these cheeses. Fungal metabolism of the liberated amino acids includes a deamination step that increases the free ammonia at the cheese surface. This basic compound diffuses into the cheese, elevating the pH and allowing the activity of endogenous enzymes (plasmin and chymosin) to continue proteolysis. The pH change also affects the calcium phosphate interactions in the micelles, causing softening (van Hooydonk et al. 1986). Thus, endogenous and exogenous enzymes cause softening in the soft ripened cheeses.

An extremely proteolytic organism, *Brevibacterium linens,* is found in the surface smear of a variety of cheeses including Limburger and Brie. This organism does not start growing until the pH of the cheese has increased as a result of the metabolism of several yeasts. As described above, the fungi oxidize lactic acid to carbon dioxide and water.

Breakdown of the proteins changes the texture of the curd, leading to a softer curd. This may be evidenced by comparison of a very sharp Cheddar cheese compared with a mild Cheddar—the very sharp Cheddar has been aged for upwards of 18 months while the mild Cheddar has been aged for 60 days. Typically, the textures of the cheeses are dramatically different: the very sharp Cheddar is very soft, and the mild is relatively firm and rubbery.

Proteolysis results in peptides that are further degraded by peptidases to individual amino acids. Peptidases may be from several sources including actively metabolizing bacteria and fungi or lysed cells. The amino acids serve as building blocks and energy sources for a variety of microbes associated with cheese. Typically, amino acids may be used as such, without alteration, or they may undergo several reactions, including deamination or transamination, demethiolation, decarboxylation, and others (Sousa et al. 2001). Through these reactions, additional flavor compounds are formed, including ammonia, sulfur-containing compounds, aldehydes, ketones, and alcohols (Fig. 14.4). The nitrogen liberated from some of these reactions may be apparent in very aged Brie or Camembert, where a pronounced ammonia smell indicates a potentially overripened cheese. As could be expected in these cheeses, the interior is very fluid because of the amount of proteolysis that has occurred from both the endogenous and microbial proteases/peptidases. Another example of amino acid metabolism occurs in surface-ripened cheeses such as Limburger or Tilsiter. In these cheeses, the microbial flora, especially *Brevibacterium linens,* on the surface metabolize the protein into amino acids. During the metabolism of methionine, the terminal sulfur group is cleaved from the amino acid. The resulting methanethiol (CH_3SH) has a very low order threshold (0.06–2 ppb; Sable and Cottenceau 1999). This is about 1000–10,000-fold less than acetic acid. Importantly, methanethiol and compounds that are formed from it are found in a variety of cheeses, including Cheddar, and are important contributors to the overall flavor of these cheeses. In cheeses that are not surface ripened, the source of the sulfur compounds may be from compounds other than methionine.

Other carbonyl-containing compounds formed during the metabolism of amino acids may be further reduced into alcohol. The low oxidation/reduction potential of the cheese matrix provides a good reducing environment for producing these flavor compounds.

CHANGES IN LIPIDS

The fatty acid profile of bovine triacylglycerols includes a great proportion of volatile short-chain fatty acids that are important to cheese flavor when they are hydrolyzed from the triacylglycerol. If free fatty acids are present in milk or some cheeses, the perception is of a rancid flavor. However, in some cheeses, the fatty acids are extremely important to the typical flavor of the cheese. The amount of free fatty acids found in Mozzarella is approximately 14-fold and 90-fold less than that found in Romano and blue-veined cheeses, respectively (Woo and Lindsay 1984, Woo et al. 1984,).

Fatty acids are a source of carbon and energy for microorganisms. The chemistry of the lipids of cheese begins with hydrolysis of the fatty acid from the glycerol backbone either by lipases associated with the microbial flora of the cheese (Fig. 14.5) or by endogenous milk lipase. Butyric (C4) and other short-chain fatty acids are highest in the cheeses such as Roquefort, Parmesan, and Romano; however, Cheddar and similar cheeses owe their "cheesiness" to these short-chain fatty acids. The "cheesiness" flavor is reduced in reduced-fat cheeses because of the reduction in these short-chain fatty acids (Lindsay 1982).

The production of other flavor compounds from fatty acids occurs during the metabolism of the microflora of the cheese. *Penicillium roquefortii* is used in blue-veined cheeses to produce the characteristic color and flavor. Spores and mycelia of this organism produce methyl ketones from fatty acids during metabolism by β-oxidation of fatty acids. Incomplete oxidation of even numbered fatty acids appears to result in the production of methyl ketones with 2-heptanone and 2-nonanone present in the

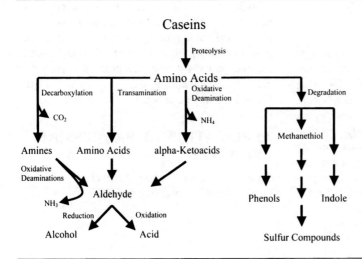

Figure 14.4. Fate of amino acids in curd.

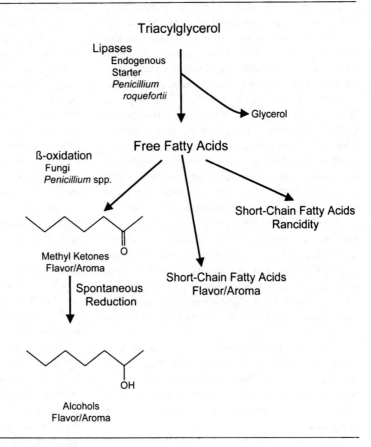

Figure 14.5. Fate of triacylglycerols during cheese ripening.

greatest quantities. These compounds are volatile and contribute greatly to the aroma of blue veined cheeses. In the reducing environment of the cheese, these methyl ketones may be reduced into corresponding alcohols that are also flavor compounds.

FINISHED PRODUCT

The manufacture of cheese is a complex interaction between naturally occurring changes in milk or curd and technology. Some cheese manufacturing processes have been modified to allow nearly continuous processing and automation. In some cases, this has resulted in a very uniform product without much of the traditional flavor associated with the cheese. Some would argue that current automated methods of Cheddar cheese manufacture have produced a very bland product without much character.

Other cheese manufacturing processes have not lent themselves to automation, and the traditional low technology handmade methodology is used.

The complexity of cheese chemistry has provided a wealth of research opportunities for food scientists. With this research effort, many new discoveries have been made about the role of microorganisms, enzymes, and environment in cheese chemistry. The advent of new demands on our food supply, such as reduced-fat but full flavored products, opens new doors of opportunities to study and research the flavor and texture of cheeses.

APPLICATION OF PROCESSING PRINCIPLES

Table 14.2 provides recent references for more details on specific processing principles.

Table 14.2. References for Principles Used in Processing

Processing stage	Processing Principle(s)	References for More Information on the Principles Used
Pretreatment of milk	Heat treatment or pasteurization	Lawrence and Gilles 1987
	Homogenization	Swaisgood 1985
Unripened acid-coagulated milk	Acid concentration of proteins	Guinee 2003, Swaisgood 1985
Curd formation	Concentration/precipitation of protein	Guinee 2003
	Syneresis or drying of curd	
	Water activity reduction	
	pH reduction	Lawrence and Gilles 1987
Chemistry of cheese ripening/aging	Enzymatic modification of components	Fox et al. 1995
	Competitive microflora	Sousa et al. 2001

ACKNOWLEDGMENTS

Dr. Earl Hammond, Department of Food Science and Human Nutrition, Iowa State University, and Dr. Anand Rao, Burnsville, Minnesota, are thanked for their critical review of the manuscript.

GLOSSARY

Casein—proteins that are found in milk synthesized in the mammary gland; the major protein component of cheese; there are four: α_{s1}, α_{s2}, β, and κ.

Chymosin—(rennet) a protease enzyme that is used for cheese coagulation.

Coagulation—formation of a gel of casein micelles; caused by reduction in pH or chymosin activity.

Hooping—practice of putting curd particles into a shape (hoop).

Isoelectric point—the pH at which the casein micelle has no net charge; pH 4.6.

LAB (lactic acid bacteria)—a group of bacteria used in cheese manufacture that metabolize lactose into lactic acid.

Lactose—the major disaccharide found in milk composed of glucose and galactose.

Micelle—the quaternary structure of casein proteins.

Pressing—application of force to curd to remove whey.

Rennet—a protease enzyme isolated from calves' stomachs used for cheese coagulation.

Syneresis—the loss of water (whey) from curd particles.

Whey—a by-product of cheese manufacture that is a dilute solution containing serum proteins, residual lactose, and lactic acid.

REFERENCES

Bodyfelt FW, J Tobias, GM Trout. 1988. Sensory evaluation of cheese. In: Sensory Evaluation of Dairy Products, 300–375. New York: AVI/Van Nostrand Reinhold.

Brule G, J Lenoir. 1987. The coagulation of milk. In: Cheesemaking: A Eck, editor. Science and Technology, 1–20. Paris: Lavoisier Publishing, Inc.

Choisy C, M Desmazeaud, JC Gripon, G Lamberet, J Lenoir, C Tourneur. 1987. Microbiological and biochemical aspects of ripening. In: A Eck, editor. Cheesemaking: Science and Technology, 62–100. Paris: Lavoisier Publishing, Inc.

Dalgleish DG. 1987. The enzymatic coagulation of milk. In: PF Fox, editor. Cheese: Chemistry, Physics and Microbiology, vol. 1, 63–96. London: Elsevier Applied Science.

Farkye N. 1995. Contribution of milk-clotting enzymes and plasmin to cheese ripening. Advan. Exp. Med. Biol. 367:195–207.

Fox PF, TK Singh, PLH McSweeney. 1995. Biogenesis of flavour compounds in cheese. Advan Exp. Med. Biol. 367:59–98.

Green ML, AL Grandison. 1987. Secondary (non-ezymatic) phase of rennet coagulation and post-coagulation phenomena. In: PF Fox, editor. Cheese: Chemistry, Physics and Microbiology, vol. 1, 97–134. London: Elsevier Applied Science.

Guinee TP. 2003. Role of protein in cheese and cheese products. In: PF Fox, PLH McSweeney, editors. Advanced Dairy Chemistry: Proteins, 3rd edition, 1083–1174. New York: Kluwer Academic/Plenum Publishers.

Lawrence RC, J Gilles. 1987. Cheddar cheese and related dry-salted cheese varieties. In: PF Fox, editor. Cheese: Chemistry, Physics and Microbiology, vol. 2, 1–44. London: Elsevier Applied Science.

Lindsay RC 1982. Quantitative analysis of free fatty acids in Italian cheeses and their effects on flavor. Proceedings of the Annual Marschall Cheese Symposium. Visalia, Calif. Available at http://www.marschall.com/marschall/proceed/index.htm.

Sable S, G Cottenceau. 1999. Current knowledge of soft cheeses flavor and related compounds. J. Agric. Food Chem. 47:4825–4836

Sousa MJ, Y Ard, PLH McSweeney. 2001. Advances in the study of proteolysis during cheese ripening. Int. Dairy J. 11:327–345.

Swaisgood HE. 1985. Characteristics of fluids of animal origin: Milk. In: O Fennema, editor. Food Chemistry, 791–828. New York: Marcel Dekker, Inc.

___. 1992. Chemistry of the caseins. In: PF Fox, editor. Advanced Dairy Chemistry, vol. 1, 63–110. New York: Elsevier Applied Science.

U.S. Department of Agriculture, Agricultural Research Service. 2003. USDA National Nutrient Database for Standard Reference, Release 16. Nutrient Data Laboratory Home Page, http://www.nal.usda.gov/fnic/foodcomp

Van Hooydonk ACM, HG Hagedoorn, IJ Boerrigten. 1986. pH induced pysico-chemical changes of casein micelles in milk and on their renneting. Neth. Milk Dairy J. 40:281–296.

Wilster GH. 1980. Practical Cheese Making. Corvallis: Oregon State University Press.

Woo AH, RC Lindsay 1984. Concentrations of major free fatty acids and flavor development in Italian cheese varieties. J. Dairy Sci. 67: 960–968.

Woo AH, S Kollodge, RC Lindsay. 1984. Quantification of major free fatty acids in several cheese varieties. J. Dairy Sci. 67:874–878.

15
Dairy: Ice Cream

K. A. Schmidt

BACKGROUND INFORMATION

Ice cream is a frozen food made from milk fat, milk solids-not-fat, sweeteners, and flavorings; a variety of fruits, nuts, and other items also may be added. Ice cream in the United States has a legal definition, which can be found in the Code of Federal Regulations (CFR 2003b), which specifies solids, fat, and air contents. These specifications state that vanilla ice cream must contain a minimum of 10% milk fat by weight, a minimum of 20% milk solids and at least 192g of total food solids per liter of ice cream, with each liter of ice cream weighing a minimum of 540 g. Other ice cream categories exist, such as re- duced calorie ice creams, which in the United States must meet the nutrient claims that comply with "re- duced fat." (CFR 2003a) These legal requirements often dictate the types and ratios of ingredients used in frozen desserts as well as some of the processing conditions. Because minimum contents (except air content) normally are stated in the federal require- ments, commercial ice creams vary considerably in body, flavor, melt, and texture characteristics. Re- cent statistics have shown that 61% of all frozen dessert products manufactured in the United States fall into the ice cream category and 26% into the nonfat and low fat ice cream category. The remain- ing portions of frozen dessert products consist of frozen yogurt (5%), water ices (4%), sherbets (3%), and other (1%) categories [International Dairy Foods Association (IDFA) 2002].

In 2001 approximately 6,116,560,000 liters of frozen desserts were made in the United States, with an annual per capita consumption of 21.5 liters, re- flecting both the size of the industry and the popu- larity of the final products. The most popular frozen dessert flavor sold in U.S. supermarkets in 2001 was vanilla; thus, vanilla ice cream will be used as the model product throughout this chapter (IDFA 2002).

Ice cream processing is basically a two-step process—the mix making and the mix freezing. Mix is the liquid product consisting of milk ingredients— fat and milk solids-not-fat—sugar, flavor (perhaps), and water. Optional mix ingredients such as corn

The information in this chapter has been derived from a chapter in *Food Chemistry Workbook,* edited by J. S. Smith and G. L. Christen, published and copyrighted by Science Technology System, West Sacramento, California, ©2002. Used with permission.

Contribution 03-385-B from the Kansas Agricultural Experiment Station. This material is based upon work supported by the Cooperative State Research Services, U.S. Department of Agriculture. The author acknowledges the cooperation of the Dairy Processing Plant, Kansas State University for Figures 15.2, 15.3, and 15.4.

syrup solids, whey, whey protein powders, caseinates, colors, egg solids, and stabilizers and emulsifiers may be used, depending upon the desired end product. In most countries, the mix must be pasteurized to assure a pathogen-free product; however, the minimum times and temperatures may vary with the country and process choice. Mixes may be homogenized, but all are cooled prior to the freezing process. Additional steps after pasteurization and cooling may include aging, flavoring, and coloring. The second major process step is freezing and hardening of the final product. During this step, mix is frozen in equipment referred to as a "freezer," cooled during the hardening stage, and subsequently distributed to markets. Many other frozen desserts, such as sherbet and sorbets, are made using a process similar to that for ice cream, but formulations differ, as do some of the ingredient choices and final product requirements. The U.S. Code of Federal Regulations provides guidance in formulating and manufacturing other frozen dessert products that meet legal specifications.

RAW MATERIALS PREPARATION

Typical ingredients received into an ice cream making operation would include milk fat sources, milk solids sources, sweeteners, stabilizers and emulsifiers, colors, flavors, and particulate materials—nuts, fruits, and candy pieces. All ingredients should be analyzed for quality and composition to ensure that the preparation of the final product complies with legal requirements, company specifications, and consumer expectations. Ingredients can arrive as liquids that may require refrigerated storage, powders that may only require ambient storage, or frozen products that may require frozen storage. Each ingredient should be maintained in appropriate storage facilities (dry storage, refrigerated silos, frozen ingredient coolers) to maintain the quality and integrity of the ingredients, and they should be checked periodically to assure usability in the final product.

Ice cream mixes are made to specifications of milk fat, milk solids-not-fat, and total solids contents. Therefore, once individual ingredient composition is known, formulations will be calculated to balance the milk solids, milk fat, and the total solids in the mix. A short list of potential ingredients and their functionality in the mix is presented in Table 15.1. The table only shows a partial list of ingredients—other ingredients that may be used include whey products and other milk protein sources; a wide range of hydrocolloids such as alginates, and carboxymethyl cellulose; and an even wider selection of particulate pieces, flavors, and colors, as the number of frozen dessert flavors has continued to increase to meet market demands. Dependent on the source of "flavor and color" materials, the recommended usage level will vary considerably, as will

Table 15.1. Typical Ingredients, Usage Levels, and Sources for Vanilla Ice Cream Mixes

Ingredient Category	Usage Level	Sources	Function
Colors	0.001–?	Caramel	Characteristic color for flavor
Emulsifiers	0.05–0.25%	Monoacylglycerides, diacylglycerides, spans, tweens	Mix emulsification, mix deemulsification
Flavors	0.001–?	Vanilla extract, vanillin	Flavor base
Milkfat	10% minimum	Butter/butter oil, condensed milks, creams, milks, plastic cream	Air incorporation, foam stabilization, flavor mouth feel, texture
Milk solids, nonfat		Butter, condensed milks, creams, milks, nonfat dry milk, plastic cream	Emulsification, flavor, melt quality, solids content, texture, water binding properties
Stabilizers	0.1–0.5%	Guar gum, locust bean gum, microcrystalline cellulose, xanthan gum	Air incorporation, body and texture, melt quality, viscosity increase, water binding properties
Sweeteners	12–16%	Corn syrup, corn syrup solids, dextrose, sucrose	Body and texture, depress freezing point, enhance flavor, sweetness, viscosity

Sources: Adapted from the Code of Federal Regulations and Marshall and Arbuckle 1996.

the company's specifications for color and flavor for a particular product. Generally, ingredient selection is based on price and availability. Thus, exact formulations may vary throughout the year, dependent on market demands, ingredient availabilities, and production capacities. For instance, liquid sugar sources may be easier to handle in one ice cream facility, and thus, syrups may be an important part of a formulation at that facility, but not at a facility that predominately uses dried products. Prices of nonfat dry milk or condensed milk may fluctuate throughout the year, affecting the exact composition of the least-cost formulation, so that ingredients and their usage percentages may vary, but the final mix formulation will balance for required amounts of specific ingredients. Most plants formulate and manufacture "white" and "chocolate" mixes that will serve as the base for many different flavors of ice cream. For example, a white mix could be the base for cookies and cream ice cream as well as a strawberry ice cream. The difference is the flavoring, coloring, and particulate added to the mix to make the final product.

PROCESSING STAGE 1

The main steps of ice cream processing are depicted in the flowchart in Figure 15.1. Each of these steps is discussed in further detail below.

BLENDING

Blending disperses the dry ingredients into the liquid components for creation of a product that is as uniform as possible (Fellows 2000). Different types of equipment can be used, based on the blending objective. For instance, dry ingredients may be preblended prior to addition into the liquid ingredients. In this case, mixing is done to create a more homogeneous powder, despite differences in particle size and density. Liquid ingredients are normally blended in a mix vat, with gentle agitation to prevent degradation of the milk fat. Eventually the dry ingredients are added to the liquid ingredients via a pump and possibly recirculation to aid dispersion. Advantages of blending include the production of a homogeneous product, enhanced powder hydration, and minimization of product losses by preventing dry ingredients from sinking to the bottom and not being fully incorporated into the final product. Disadvantages of blending are the need for equipment, additional process time, and additional energy input (Fellows 2000, Spreer 1998).

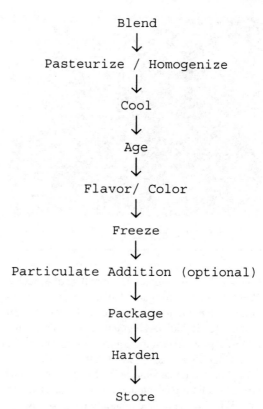

Figure 15.1. Flowchart for ice cream processing.

In smaller ice cream plants, mix ingredients will be weighed or metered and then blended together. At large, commercial ice cream production facilities, dry ingredients may be preblended separately and then added to warmed (30–40°C) liquid ingredients. Because dried ingredients often are used in the ice cream mix, adequate and thorough blending through agitation or recirculation is necessary to initiate protein and polysaccharide rehydration, to suspend colloidal materials, and to solubilize sugars and salts (Marshall and Arbuckle 1996). If raw cream is incorporated, excessive agitation can lead to destruction of milk fat globule membranes, initiating undesirable enzymatic activity or fat coalescence, which can contribute to undesirable flavor and mouthfeel characteristics in the final products (Keeney n.d.).

PASTEURIZATION

The purpose of pasteurization is to inactivate any and all pathogens that are in the mix. Most countries have minimum time-temperature pasteuriza-

Figure 15.2. An HTST pasteurizer for ice cream mix.

tion standards for mixes (Marshall and Arbuckle 1996). In the United States, batch pasteurization requires a minimum temperature of 68.3°C for 30 minutes (CFR 2003b). High-temperature short-time (HTST) pasteurization requires a minimum of 79.4°C for 25 seconds (CFR 2003b). The relationship of time and temperature is a function of the microbial load, fat content, and microbial inactivation rate. HTST is the most commonly used pasteurization choice in the United States because of its energy efficiency and speed. A small HTST pasteurizer is shown in Figure 15.2.

In the HTST design, three heat exchange sections exist: regenerator, heating, and cooling. All three sections consist of stainless steel plates that channel the flow of product to prevent cross-contamination. In the HTST pasteurizer, further cooling and heating need to occur to allow the ice cream mix to achieve the target temperatures (Spreer 1998). The HTST pasteurizer design allows for rapid heat transfer between unpasteurized and pasteurized products. Separate streams are maintained in the regenerator section, where pasteurized product is cooled while unpasteurized product is heated. As the cold product flows over the plate, heat transfers from the pasteurized side to warm the unpasteurized product, and the temperature of the unpasteurized mix increases from 4 to 60°C during its flow through the regenerator. Warm, unpasteurized mix then flows through the heater section of the HTST pasteurizer over a similar series of plates, but instead of pasteurized mix, a hot medium (usually water) flows on the opposite side (Alfa Laval n.d.). Product emerges from this section at the desired pasteurization temperature.

The mix is then held at that temperature as it flows through a holding tube, which is designed to ensure that transit time is a minimum of 25 seconds, before passing through a flow diversion valve. This valve has temperature sensors that constantly monitor the mix temperature to assure the pasteurization temperature has been maintained. When the temperature has been maintained for the proper time, the pasteurized mix flows into the regenerator section to warm the unpasteurized mix while the pasteurized mix is being cooled to ~15 to 23°C. The partially cooled pasteurized mix then flows into the cooling section. In the cooling section, pasteurized mix passes on one side and a coolant (cold water that may contain glycol) passes on the other side. Generally, temperature change is from 20 to 4°C (Alfa Laval n.d.)

Timing pumps as well as other design criteria will maintain the proper flow rate and pressure differential to prevent cross-contamination. In well-designed systems, mix dwell times can range from 60 to 120 seconds from incoming cold (7°C) unpasteurized mix to cold (7.3°C) pasteurized mix (Alfa Laval n.d.)). Quick cooling protects against microbial growth and initiates milk fat crystallization and water binding by polysaccharides and proteins (Marshall and Arbuckle 1996). The high apparent viscosity is noticeable as the cooled mix is collected in a refrigerated vat.

Two other pasteurization processes can be used for ice cream mixes: ultra high-temperature (UHT) and batch pasteurization. UHT pasteurization can be used to produce ice cream mixes that are frozen at later dates. In a UHT system, higher temperatures (> 140°C) are used, perhaps in conjunction with longer times (2–12 seconds) (Spreer 1998), and the UHT mixes may be considered commercially sterile. In this case, UHT mixes are cooled to room temperature and are aseptically filled into containers. A batch pasteurizer relies on longer times than HTST but lower temperatures to inactivate the pathogens. In a batch pasteurizer, mix ingredients are blended together in a jacketed vat that is equipped with an agitator. When the ingredients reach the necessary pasteurization temperature, holding time begins. Regulations in the United States dictate the necessary attached equipment, such as a recording temperature chart, airspace heater, and indicating thermometer, to ensure adequate heat transfer throughout the product.

Although the main objective of pasteurization is to inactivate pathogenic microorganisms, other reactions occur. The heat melts the emulsifiers and fat, denatures some of the whey proteins, and aids hydrocolloid hydration and stabilization (Keeney n.d., Marshall and Arbuckle 1996). Melting of the milk fat allows for interaction among the lipid fractions, effective homogenization of the fat globules, and better control of fat crystallization later in the process. Protein denaturation increases emulsifying capacity, protein-stabilizer interactions, and water-binding capacity. The high pasteurization temperatures allow for complete hydrocolloid hydration and dispersion stabilization. Usually, batch-pasteurized mixes are more viscous than HTST-pasteurized mixes because batch pasteurization may induce additional protein denaturation and also enhance hydrocolloid hydration. If the mix is too viscous, heat transfer rates will decrease during the freezing process. Disadvantages of pasteurization predominately focus on the financial output to maintain and operate the equipment. Properly designed, operated, and maintained pasteurization equipment can be expensive. In addition, both batch and HTST operations need refrigerated pasteurized mix vats, which contribute to the operating cost for the company.

HOMOGENIZATION

Homogenization is the size reduction of particles into a more uniform distribution in the liquid phase of the system; the final result is a more consistent product (Spreer 1998). Homogenization uses the principles of restricted flow, high pressure, and diverted flow to reduce particle size. In the case of ice cream mixes, the homogenization creates smaller diameter milk fat globules (< 2 μm) that are more evenly dispersed, and thus aids in mix emulsion stabilization (Goff 1999, Spreer 1998).

The principle of homogenization is to force liquid to flow under high pressure through a narrow orifice, usually just slightly larger than the diameter of the particle to be homogenized. As the liquid flows through the narrow orifice, velocity increases, and turbulence and cavitation may result. These forces cause the fat globule to disintegrate and form more, smaller particles (Spreer 1998). Factors that affect the overall homogenization effect include pressure, temperature, orifice size, and design.

The advantages of homogenization include the greater surface area of the fat globules, viscosity enhancement, and greater stability. Disadvantages are mostly chemical in nature. Homogenized milk fat is more sensitive to light-induced oxidation, and protein stability can also be affected (Alfa Laval n.d., Bodyfelt et al. 1988, Goff 1999, Spreer 1998).

Because of their relatively high fat content, mixes are homogenized in two stages (two passes through the homogenizer); the first stage is set at a pressure of 13–15 MPa, the second at 3–5 MPa. Mix temperatures should be in the range of 50–66°C to assure efficient homogenization (Fellows 2000, Spreer 1998). Because homogenization reduces fat globule size, the number of fat globules increases, and new membranes—consisting of some of the added proteins, emulsifiers, and original materials—form on these new, smaller fat globules (Marshall and Arbuckle 1996). The incorporated proteins and emulsifiers promote the desirable whipping characteristics that aid fat destabilization and foam formation during freezing (Marshall and Arbuckle 1996). In most HTST and UHT systems, the homogenizer is part of the system itself, in which mix is piped from/to the homogenizer and then back to the pasteurizer, and sometimes serves as a flow control mechanism during the pasteurization process. Because flow is quick through both of these pasteurization systems, homogenization may occur before (HTST and some UHT) or after (batch or perhaps UHT) pasteurization (Fellows 2000, Goff 1999, Varnam and Sutherland 2001). In a batch pasteurization operation, mix is homogenized after pasteurization to prevent rancidity problems (Varnum and Sutherland 2001).

COOLING

After the mix is homogenized and pasteurized, it is cooled quickly to ≤ 7.3°C. The objective is to cool the product as quickly as possible and then maintain the cold temperature to prevent microbial growth or proliferation (Fellows 2000). Cooling is usually done with a heat exchanger, which is a part of the HTST system, as described above in the pasteurization section (Alfa Laval n.d.) If the mix is commercially sterilized in a UHT system, the mix may not be required to be cooled to below 7.3°C: it may only be cooled to ambient temperatures (20–25°C) and then aseptically filled into containers, sealed, and then placed in storage. For a batch pasteurizer, various cooling equipment that is separate from the vat can be used; one of the most commonly used would be a plate heat exchanger (Fellows 2000, Spreer 1998). This equipment relies on a cold medium (cold water or refrigerated water containing glycol or a similar component to lower the freezing point) passing on one side of a stainless steel plate, while the warmer pasteurized mix passes over the other side. As products pass, the mix is cooled and the coolant warmed. Cooling rate is controlled by the number of plates, coolant temperature, and product flow rate (pump speed) (Fellows 2000, Spreer 1998). The design of this equipment is such as to protect the pasteurized mix from air and "coolant" contamination.

Advantages of cooling focus on food safety, but cooling also initiates milk fat crystallization and water binding by polysaccharides and proteins (Marshall and Arbuckle 1996). The high apparent viscosity is noticeable as the cooled mix is collected in a refrigerated vat. Disadvantages include energy, labor, and time inputs, as well as the capital investment in equipment and its maintenance and operation. The high viscosity of the mix may hamper high product recovery and heat transfer rates, which could add to the cost of the process. If the equipment is not clean and sanitized, it can be a point of potential contamination with unwanted components.

PROCESSING STAGE 2

FLAVORING AND COLORING

Prior to freezing, the pasteurized mix is metered into another tank where color and flavor are added (if indicated in the specifications); this is often referred to as the flavor tank. As discussed earlier, a large quantity of "white mix" can be made and then used as a base for assorted flavors of the frozen product. This subdivision occurs to meet daily production quotas, for example 60% for vanilla, 25% for chocolate chip, 10% for strawberry, and 5% for mint chocolate chip. Thus, a designated amount of mix will be pumped into a flavor tank and adequately flavored and colored, as indicated in the product specifications and formulation sheets.

A flavor tank is generally a smaller version of a mix tank. The flavor tank may or may not be jacketed with a coolant to maintain low temperatures (the larger the tank, the greater the likelihood of a refrigerated tank). Its most important attribute is the ability to agitate well and to be closed (i.e., not open to environmental contamination). The main advantage of the flavor tank is the production of a more homogenous product, assuring that the product contains consistent flavor/color. A tank design in which maximum product recovery is achieved would be an advantage to a processor. Disadvantages are similar to those shared above: additional equipment to maintain and clean and a potential source of contamination.

In an ice cream operation, the process of flavoring and coloring is relatively simple. The appropriate amounts of liquid flavor (usually suspended in an alcohol base) and color are measured/metered and slowly incorporated into the flavor tank. Sufficient agitation ensures a homogenous product prior to freezing. Most flavors are suspended in an alcohol base, which could denature the casein proteins; thus, flavor addition should be done slowly. Because the mix ingredients have been pasteurized and homogenized, agitation control is not as critical as in the initial blending step, where excessive agitation may initiate undesirable chemical reactions, especially with milk fat. However, because the mix will not be heat treated prior to consumption, all ingredients (colorings, flavorings, particulates, variegates, etc.) added postpasteurization to the mix or ice cream must be of high quality to ensure a safe, wholesome ice cream product.

FREEZING

The purpose of freezing ice cream is two-fold—formation of the foam structure and initiation of the freezing process (Goff 1999, Marshall and Arbuckle 1996). Freezing in the ice cream industry refers to the process in which the temperature decreases from 4°C to approximately −6°C with simultaneous air incorporation, in a piece of equipment known as the freezer (Fig. 15.3).

In the freezer, liquid mix is subjected to vigorous agitation and cold temperatures. The freezer consists of a jacketed barrel that contains refrigerant, which

Figure 15.3. A continuous freezer used for ice cream processing.

maintains a very cold temperature at the interior barrel wall. A freezer has an interior dasher that contains horizontal blades (Alfa Laval n.d.) (Figs 15.3 and 15.4 contain more detail). When the freezer is turned on, the dasher moves in a circular manner, agitating or "whipping" the mix and forcing the mix into contact with the cold barrel wall. The mix contacts the barrel wall and freezes along the cold surface, and the dasher blades scrape the thin layer of frozen mix from the surface of the barrel and resuspend the frozen mix in the unfrozen mix (Alfa Laval n.d., Marshall and Arbuckle 1996). As more mix cools and freezes, air will be incorporated, as the vigorous agitation forces air into the partially frozen mix that contains these small ice crystals, producing a frozen foam known as ice cream. The foam is formed as the fat globules destabilize and aggregate around air bubbles, providing stability to the foam (Goff 1999). The formation of numerous, tiny air cells is desirable in a high quality ice cream, since air cells help control the size of the ice crystals (Marshall and Arbuckle 1996). The viscosity of the mix that flows around the air cells may help control water movement during hardening and storage (Keeney n.d.). The amount of air whipped into the product (overrun) can be measured by volume (Keeney n.d.):

$$\% \text{ overrun}$$
$$= \frac{(\text{weight of mix} / \text{volume}) - (\text{weight of ice cream} / \text{volume})}{\text{weight of ice cream} / \text{volume}} \times 100$$

In the United States, ice cream legally cannot contain more than 100% overrun (equal volumes of air and mix) (CFR 2003b). As air incorporation increases, the perception of warmness increases, whereas flavor impact generally decreases. Most ice creams on the market are manufactured with 90–95% overrun, except for speciality ice creams, which may have a lower air content (Marshall and Arbuckle 1996). Without air, ice cream would be very dense, crumbly, and unpalatable (Bodyfelt et al. 1988, Marshall and Arbuckle 1996). Other frozen desserts may have different targeted overrun values. Generally the overrun can be set at the freezer; it normally only requires slight adjustment throughout the production run.

Most manufacturers freeze ice cream in a continuous freezer, which continuously pumps mix into the freezer barrel and extrudes ice cream from the barrel. Dwell times are very short, generally 45–180 seconds (Alfa Laval n.d.). At such a rate, the ice cream is frozen quickly, producing tiny ice crystals. Ice crystals are smallest at the time of discharge from the freezer and generally increase during storage and distribution. As temperature fluctuations occur in a frozen product, smaller ice crystals melt, and the liquid water migrates to larger ice crystals to refreeze and form larger ice crystals when the temperatures decrease again (Goff 1999, Marshall and Arbuckle 1996). Batch freezers also may be used to freeze ice cream (Marshall and Arbuckle 1996). In this process, a specific quantity of mix is added to the freezer and then the refrigerant and agitator are started. Once the desired temperature and air incorporation are reached, all the frozen product is discharged; then the process is repeated with a new batch of mix (Marshall and Arbuckle 1996).

Advantages of quick freezing are the ability to incorporate air and initiate many small-sized ice crystals. Generally these two events coincide with high quality ice cream. The disadvantage of quick freezing is greater reliance on the equipment and auxiliary equipment (compressors, etc.) to maintain low temperatures.

In commercial ice cream plants, most ice cream is frozen in the continuous freezer style. Mix is continually pumped in, and product is discharged at a constant speed. The next step, packaging, needs to be totally in sync with freezer production because freezer stoppage is difficult. Maximizing product output with minimized dwell time is desirable.

PACKAGING

The purposes of packaging are product identity, product integrity, and product safety; conveying nu-

Figure 15.4. The dasher being placed inside the ice cream freezer.

tritional data in accordance to federal specifications; and enhancing consumer appeal. Vanilla ice cream can be packaged into containers of many different sizes (single service 113 ml to 12.5 liter containers) and materials including paperboard, plastic, and foil laminates (Marshall and Arbuckle 1996). The equipment used to package ice cream varies, but most are based on a weight fill control mechanism that is automated, with an automatic closure/lid machine either incorporated or detached. The objective is to fill the ice cream container as quickly as possible, with minimum disruption of the air cells and ice crystals in the product.

Packaging ensures that an adequate amount of product is provided to the consumer in an attractive shape and that unwanted contaminants in the food product are prevented. Disadvantages are the added cost to the product, and introduction of another "barrier" that will need to be further frozen (see next step). For container sizes and shapes that don't pack tightly, shipping costs may be increased, as "less product" and "more air" may be involved in a shipment.

For ice cream, the packaging process needs to be rapid. The ice cream may be at $-6°C$ as it is discharged into the containers, and the exposure to warmer environmental temperatures in the manufacturing facility may result in melting; therefore, one production consideration is the need to package the product and move it into the hardening room quickly (Marshall and Arbuckle 1996). For the most part, packaging materials for ice cream have not been highly researched. Desirable properties for ice cream packaging materials include light weight and the ability to prevent light penetration and moisture loss.

HARDENING

As soon as possible after packaging, ice cream is placed in a hardening facility, which is normally kept at -30 to $-35°C$ or lower with forced air movement. The purpose of hardening is to continue freezing the ice cream as quickly as possible to minimize ice crystal size and stabilize the foam (Marshall and Arbuckle 1996). The smaller the ice crystal, the smoother and more acceptable the ice cream will be (Bodyfelt et al. 1988). Because of the solids content (especially those compounds that affect the colligative properties) of ice cream, part of the ice cream will always remain unfrozen during commercial frozen storage conditions. This, along with physical changes such as temperature fluctuation, allows deleterious reactions such as ice crystal growth and lactose crystallization to occur during storage, which affects the shelf life of the product (Alexander 1999).

Hardening equipment varies considerably. Some manufacturing plants use hardening tunnels or spiral freezers, wherein packaged products are conveyed via belts (usually interlock or chained) through a cold (-30 to $-55°C$) space with considerable air movement (300–600 cfm) (Fellows 2000). The timing of the conveyor belts can be adjusted to account for product mass and load on the freezer or individual package size (Fellows 2000). Other plants use hardening rooms, where product is placed into a cold space with high air movement, but no product is moved. Generally, given equivalent loads and product, transported product will achieve final temperature at a faster rate than stationary product due to cold air flowing completely around the individual products. When ice cream is filled into rectangular boxes, contact-plate freezers may be used. In this case, the boxes of ice cream are placed directly onto refrigerant-filled shelves that may reach -60 to $-70°C$. These shelves are then contained within a space designed to maintain the cold temperatures. The shelves rotate, and at the point of discharge, the ice cream boxes emerge at -15 to $-20°C$ or the desired temperature. Dependent upon the refrigerant used, dwell times can be as short as 90 minutes.

The advantages of hardening are preservation of ice cream and improved quality. Rarely is ambient temperature such that the quality of ice cream is maintained without melting. Usually, the quicker the hardening process the smaller the ice crystal size and the smoother the product texture (Bodyfelt et al. 1988, Goff 1999, Marshall and Arbuckle 1996). Disadvantages of hardening are predominately associated with the cost of cold air space(s). Refrigeration and air movement are expensive and can be hazardous to the personal safety of workers. Larger ice cream production loads require larger hardening spaces or a greater number of hardening spaces.

FROZEN STORAGE

After leaving the hardening room, ice cream is placed in frozen storage. The desired storage temperature is $-15°C$ or less. The lower temperatures ensure that as much water as possible is maintained as ice in the product, and thus less water is available to form larger sized crystals. From the frozen storage facility, the ice cream is ready for distribution and consumption. However, ice cream remains a dynamic system; thus, a part of the ice cream remains unfrozen (Marshall and Arbuckle 1996). As temperatures increase or decrease, ice melts or water freezes. The nature of this process is that the smallest of ice crystals melt first and refreeze into larger sized crystals (Goff 1999, Marshall and Arbuckle 1996). Thus, the storage stability and quality of ice cream are highly dependent upon stabilizing the air cells and ice crystals in the frozen form and then maintaining that structure with cold temperatures.

Most frozen storage rooms are insulated cold spaces with some air movement, where ice cream is maintained at the coldest temperatures possible. For obvious reasons, the ice cream is stored quiescently and is often "sleeved" or "banded" into larger grouping sizes for ease of shipping and handling. The colder the temperature, the better conditions are for maintaining ice cream quality.

FINISHED PRODUCT

Ice cream as a finished product can take many forms not only from the container size, but also in appearance, body, texture, and flavor. Product appearance can vary tremendously as consumer preferences for specific brand selections indicate their desire for products that contain specific ingredients. For all products, the product must conform to the specifications set by the governing bodies—federal and state. But specifications for remaining aspects such as color, particulate size, shape, and amount are set by the individual company to meet their consumers' preferences and needs. In almost all cases, ice cream is eaten from the freezer and enjoyed in its frozen state. From a safety perspective, strict sanitary guidelines and temperature controls are necessary to maintain the quality of the original product.

APPLICATION OF PROCESSING PRINCIPLES

Three major processing principles are used as the basis for ice cream production: pasteurization, homogenization, and temperature reduction. All three of these processing conditions affect the final quality of the end product. Pasteurization not only ensures adequate inactivation of pathogens, but also decreases the overall microbial population, inactivates some enzymes, melts fats, and enhances the

hydration properties of powdered ingredients. Homogenization predominately affects the milk fat globules by reducing their size, with a secondary effect of altering the composition of the milk fat globule membrane. Freezing initiates lowering of product temperature and forms the foam structure that is necessary for the unique eating quality of the product. Hardening further lowers the temperature, which affects the ice, fat, and sugar crystal size and shapes. Frozen storage is necessary until consumption to maintain product integrity.

GLOSSARY

Blending—the act of dispersing at least two different types of ingredients (dry and dry; dry and liquid; liquid and liquid) for creation of a homogenous product.

CFR—Code of Federal Regulations.

cfm—cubic feet per minute.

FDA—U.S. Food and Drug Administration.

Hardening—a term used in the ice cream industry to refer to the process in which the frozen, packaged ice cream product is further subjected to colder temperatures to continue freezing the ice cream as quickly as possible to minimize ice crystal size and stabilize the foam.

HTST (high temperature short time)—a pasteurization methodology that uses higher temperatures for shorter times to achieve destruction of all pathogenic bacteria in the liquid product.

Homogenization—a process combining flow direction and pressure to reduce the particle size, achieving a more uniform distribution in a liquid system.

Ice cream—a frozen food made from milk fat, milk solids-not-fat, sweeteners, flavorings, water, and possibly a variety of other compounds such as fruits, nuts, candies, and so on.

Ice cream freezing—the process wherein mix temperature is decreased from 4°C to approximately −6°C with simultaneous air incorporation.

Ice cream mix—the liquid product consisting of milk ingredients, milk fat and milk solids-not-fat, as well as the sweetener, flavor (perhaps), and water.

IDFA—International Dairy Foods Association.

Overrun—the amount of air/volume incorporated into ice cream.

Pasteurization—the heat treatment used to inactivate any and all pathogenic microorganisms contained within the product.

UHT (ultra high temperature)—a continuous heat treatment that uses very high temperatures and longer times to ensure destruction of almost all bacteria and inactivation of most enzymes.

REFERENCES

Alexander RJ. 1999. Sweeteners Nutritive. Eagen Press, Eagen, Minn.

Alfa-Laval. n.d. Dairy Handbook. Alfa-Laval, Food Engineering, AB, P.O. Box 65, S-221 00 Lund, Sweden.

Bodyfelt FW, J Tobias, GM Trout. 1988. Sensory evaluation of ice cream and related products. In: The Sensory Evaluation of Dairy Products. AVI, Westport, Conn.

Chandran R. 1998. Dairy Based Ingredients. Eagen Press, Eagen, Minn.

Code of Federal Regulations (CFR). 2003a. Title 21. Part 101 Food Labeling. Food and Drug Administration, Department of Health and Human Services, Washington, D.C.

Code of Federal Regulations (CFR). 2003b. Title 21. Part 135. Frozen Desserts. Food and Drug Administration, Department of Health and Human Services, Washington, D.C.

International Dairy Foods Association (IDFA). 2002. The Latest Scoop. IDFA, Washington, D.C.

Fellows PJ. 2000. Food Processing Technology Principles and Practice, 2nd edition. CRC Press, New York.

Goff HD. 1999. Http://www.uoguelph.ca/foodsci/dairyedu/html.

Keeney PG. n.d. Commercial Ice Cream and Other Frozen Desserts. The Pennsylvania State University Extension Circular, Rv10M1172. U. Ed. 3-102.

Marshall RT, WS Arbuckle. 1996. Ice Cream, 5th edition. Chapman and Hall, New York.

Spreer E. 1998. Milk and Dairy Product Technology. Marcel Dekker, Inc., New York.

Varnam AH, JP Sutherland. 2001. Ice cream and related products. In: Milk and Milk Products Technology, Chemistry and Microbiology. Aspen Publishers, Inc., Gaithersburg, Md.

16
Dairy: Yogurt

R. C. Chandan

BACKGROUND INFORMATION

Fermented dairy foods have constituted a vital part of human diet in many regions of the world since times immemorial. They have been consumed ever since the domestication of animals. Evidence for the use of fermented milks comes from archeological findings associated with the Sumerians and Babylonians of Mesopotamia, the Pharoahs of northeast Africa and Indo-Aryans of the Indian subcontinent (Chandan 1982, 2002, Tamime and Robinson 1999). Ancient Indian scriptures, the Vedas, dating back some 5000 years mention *dadhi* and buttermilk. Also, the ancient Ayurvedic system of medicine cites fermented milk *(dadhi)* for its health giving and disease fighting properties (Aneja et al. 2002).

Historically, products derived from fermentation of milk of various domesticated animals resulted in conservation of valuable nutrients that otherwise would deteriorate rapidly under the high ambient temperatures prevailing in South Asia and the Middle East. Thus, the process permitted consumption of milk constituents over a period significantly longer than was possible for milk itself. Concomitantly, conversion of milk to fermented milks resulted in the generation of a distinctive viscous consistency, smooth texture, and unmistakable flavor. Furthermore, fermentation provided food safety, portability, and novelty for the consumer. Accordingly, fermented dairy foods evolved into the cultural and dietary ethos of the people residing in the regions of the world to which they owe their origin.

Milk is a normal habitat of a number of lactic acid bacteria, which cause spontaneous souring of milk held at bacterial growth temperatures for an appropriate length of time. Depending on the type of lactic acid bacteria gaining entry from the environmental sources (air, utensils, milking equipment, milkers, cows, feed, etc.), the sour milk attains a characteristic flavor and texture.

The diversity of fermented milks may be ascribed to (a) use of milk obtained from various domesticated animals, (b) application of diverse microflora, (c) addition of sugar, condiments, grains, fruits, and so on to create a variety of flavors and textures, and (d) application of additional preservation methods, for example, freezing, concentrating, and drying. The fermented foods and their derivatives constitute a staple meal or an accompaniment to a meal and may be used as a snack, drink, dessert, and condiment, either spread or used as an ingredient in cooked dishes.

Major fermented milk foods consumed in different regions are listed in Table 16.1.

Milk of various domesticated animals differs in composition and produces fermented milk with a characteristic texture and flavor (Table 16.2).

The milk of various mammals exhibits significant differences in total solid, fat, mineral, and protein content. The viscosity and texture characteristics of yogurt are primarily related to its moisture content and protein level. Apart from quantitative levels, protein fractions and their ratios play a significant role in gel formation and strength. Milk proteins further consist of caseins and whey proteins that have distinct functional properties. In turn, caseins are comprised of α_{s1}-, β-, and κ-caseins. The ratio of casein fractions and the ratio of casein to whey protein differ widely in milks of various milch animals. Furthermore, pretreatment of milk of different species prior to fermentation produces varying magnitudes of protein denaturation. These factors have a profound effect on the rheological characteristics of

Table 16.1. Major Fermented Dairy Foods Consumed in Different Regions of the World

Product Name	Major Country/Region
Acidophilus Milk	United States, Russia
Ayran/eyran/jugurt	Turkey
Busa	Turkestan
Chal	Turkmenistan
Cieddu	Italy
Cultured buttermilk	United States
Dahi/dudhee/dahee	Indian subcontinent
Donskaya/varenetes/kurugna/ryzhenka/guslyanka	Russia
Dough/abdoogh/mast	Afghanistan, Iran
Ergo	Ethiopia
Filmjolk/fillbunke/fillbunk/surmelk/taettemjolk/tettemelk	Sweden, Norway, Scandinavia
Gioddu	Sardinia
Gruzovina	Yugoslavia
Iogurte	Brazil, Portugal
Jugurt/eyran/ayran	Turkey
Katyk	Transcaucasia
Kefir, Koumiss/Kumys	Russia, Central Asia
Kissel maleka/naja/yaourt/urgotnic	Balkans
Kurunga	Western Asia
Leben /laban/laban rayeb	Lebanon, Syria, Jordan
Mazun/matzoon/matsun/matsoni/madzoon	Armenia
Mezzoradu	Sicily
Pitkapiima	Finland
Roba/rob	Iraq
Shosim/sho/thara	Nepal
Shrikhand	India
Skyr	Iceland
Tarag	Mongolia
Tarho/taho	Hungary
Viili	Finland
Yakult	Japan
Yiaourti	Greece
Ymer	Denmark
Zabady/zabade	Egypt, Sudan

Sources: Adapted from Chandan 2002, Tamime and Robinson 1999.

Table 16.2. Proximate Composition of Milk of Mammals Used for Fermented Milks

	% Total Solids	% Fat	% Total Protein	% Casein	% Whey Protein	% Lactose	% Ash
Cow	12.2	3.4	3.4	2.8	0.6	4.7	0.7
Cow, Zebu	13.8	4.6	3.3	2.6	0.7	4.4	0.7
Buffalo	16.3	6.7	4.5	3.6	0.9	4.5	0.8
Goat	13.2	4.5	2.9	2.5	0.4	4.1	0.8
Sheep	19.3	7.3	5.5	4.6	0.9	4.8	1.0
Camel	13.6	4.5	3.6	2.7	0.9	5.0	0.7
Mare	11.2	1.9	2.5	1.3	1.2	6.2	0.5
Donkey	8.5	0.6	1.4	0.7	0.7	6.1	0.4
Yak	17.3	6.5	5.8	–	–	4.6	0.9

Sources: Adapted from Chandan and Shahani 1993, Chandan 2002.

fermented milks, leading to bodies and textures ranging from drinkable fluid to firm curd. Fermentation of the milk of buffalo, sheep, and yak produces a well-defined custard-like body and firm curd, while milk of other animals tends to generate a soft gel consistency.

Cow's milk is used for the production of fermented milks, including yogurt, in a majority of the countries around the world. In the Indian subcontinent, buffalo milk is used widely for dahi making, using mixed mesophilic cultures (Aneja et al. 2002). In certain countries, buffalo milk is the base for making yogurt using thermophilic cultures. Sheep, goat, or camel milk is the starting material of choice for fermented milks in several Middle Eastern countries.

Yogurt is a semisolid fermented product made from a standardized milk mix by the activity of a symbiotic blend of *Streptococcus thermophilus* (ST) and *Lactobacillus delbrueckii* subsp. *bulgaricus* (LB) cultures.

In the United States, the past two decades have witnessed a dramatic rise in per capita yogurt consumption from nearly 2.5 pounds to 7.4 pounds. The increase in yogurt consumption may be attributed to yogurt's perceived natural and healthy image, providing to the consumer convenience, taste, and wholesomeness attributes. Table 16.3 summarizes recent trends in the consumption of refrigerated yogurt in the United States.

Figure 16.1 illustrates various forms of yogurt in the U.S. market.

In the year 2002, yogurt sales in the United States topped $2.6 billion. From 1995 to 2002, as a snack and lunchtime meal, yogurt consumption grew by 60%. As breakfast food, yogurt consumption increased 75% during the same period.

RAW MATERIALS PREPARATION

DAIRY INGREDIENTS

Yogurt is a Grade A product. Grade A implies that the milk used must come from Food and Drug Administration (FDA) supervised Grade A dairy farms and Grade A manufacturing plants as per regulations enunciated in Pasteurized Milk Ordinance (U.S. Department of Health and Human Services 1999). Yogurt is made from a mix standardized from whole, partially defatted milk, condensed skim milk, cream, and nonfat dry milk. Supplementation of milk solids-not-fat (SNF) of the mix with nonfat dry milk is frequently practiced in the industry. The FDA specification calls for a minimum of 8.25% nonfat milk solids. However, the industry uses up to 12% SNF or nonfat milk solids in the yogurt mix to generate a thick, custard-like consistency in the product. The milk fat levels are standardized to 3.25% for full fat

Table 16.3. Annual Total and Per Capita Sales of Refrigerated Yogurt in the United States

Year	Total sales (million pounds)	Per capita sales (pounds)
1982	600	2.6
1987	1074	4.4
1992	1154	4.5
1997	1574	5.8
1998	1639	5.9
1999	1717	6.2
2000	1837	6.5
2001	2003	7.0
2002	2135	7.4

Source: International Dairy Foods Association 2003.

LEEDS TRINITY UNIVERSITY

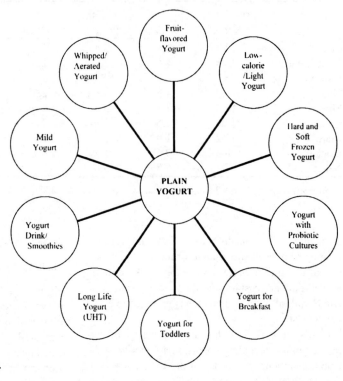

Figure. 16.1. Types of commercial yogurt.

yogurt. Reduced fat yogurt is made from mix containing 2.08% milk fat. Low fat yogurt is manufactured from mix containing 1.11% milk fat. Nonfat yogurt mix has milk fat level not exceeding 0.5%. These fat levels correspond to the Food and Drug Administration requirement for nutritional labeling of nonfat, reduced fat, and low fat yogurt (Chandan 1997). All dairy raw materials should be selected for high bacteriological quality. Ingredients containing mastitis milk and rancid milk should be avoided. Also, milk partially fermented by contaminating organisms and milk containing antibiotic and sanitizing chemical residues cannot be used for yogurt production. The procurement of all ingredients should be based upon specifications and standards that are checked and maintained with a systematic sampling and testing program by the plant quality control laboratory. Since yogurt is a manufactured product, it is likely to have variations according to the quality standards established by marketing considerations. Nonetheless, it is extremely important to standardize and control the day-to-day product in order to meet consumer expectations and regulatory obligations associated with a certain brand or label.

YOGURT STARTERS

Spontaneous souring of milk yields uncontrollable flavor and texture characteristics with food safety concerns. Modern industrial processes utilize defined lactic acid bacteria as a starter for yogurt production. A starter consists of food grade microorganism(s) that on culturing in milk predictably produce the attributes that characterize yogurt. The composition of yogurt starter is shown in Table 16.4. Also, shown are some additional organisms found in yogurt or yogurt-like products marketed in various parts of the world. Most of the yogurt in the United States is fermented with *Streptococcus thermophilus* (ST) and *Lactobacillus delbrueckii* subsp. *bulgaricus* (LB). In addition, optional bacteria, especially those of intestinal origin, are incorporated in the starter or the product. *Lactobacillus acidophilus* is commonly added as additional culture to commercial yogurt. Other cultures added belong to various *Lactobacillus* and *Bifidobacterium* species. Both ST and LB are fairly compatible and grow symbiotically in milk medium. However, the optional organisms do not necessarily exhibit compatibility with

Table 16.4. Required and Optional Composition of Yogurt Bacteria

Required by FDA Standard of Identity for Yogurt	Optional Additional Bacteria Used or Suggested
Streptococcus thermophilus (ST)	
	Lactobacillus acidophilus
	Lactobacillus casei
	Lactobacillus casei subsp. *rhamnosus*
Lactobacillus delbrueckii	*Lactobacillus reuteri*
subsp. *bulgaricus (LB)*	*Lactobacillus helveticus*
	Lactobacillus gasseri ADH
	Lactobacillus plantarum
	Lactobacillus lactis
	Lactobacillus johnsoni LA1
	Lactobacillus fermentum
	Lactobacillus brevis
	Bifidobacterium longum
	Bifidobacterium breve
	Bifidobacterium bifidum
	Bifidobacterium adolescentis
	Bifidobacterium animalis
	Bifidobacterium infantis

Source: Adapted from Chandan 1999.

LB and ST. Judicious selection of strains of LB, ST, and the optional organisms is necessary to ensure the survival and growth of all the component organisms of the starter. Nevertheless, product characteristics, especially flavor, may be slightly altered when yogurt culture is supplemented with optional bacteria. In some countries of Europe, *Lacobacillus bulgaricus* is replaced with *Lactobacillus lactis* to market "mild" yogurt.

Commercial production of yogurt relies heavily on the fermentation ability of, and the characteristics imparted by, the starter. Satisfactory starter performance requires rapid acid development; development of typical yogurt flavor, body, and texture; exopolysaccharide secreting strains to enhance the viscosity of the yogurt; scale-up possibilities in various production conditions, including compatibility with the variety and levels of ingredients used and with fermentation times and temperatures; survival of culture viability during the shelf life of the yogurt; probiotic properties and survival in the human gastrointestinal tract for certain health attributes; and minimum acid production during distribution and storage at 4–10°C until yogurt is consumed.

The activity of a starter culture is determined by direct microscopic counts of culture slides stained with methylene blue. This exercise also indicates physiological state of the culture cells. Cells of ST grown fresh in milk or broth display pairs or long chains of spherical, coccal shape. Under stress conditions of nutrition and age (old cells, cells exposed to excessive acid, colonies on solid media, milk containing inhibitor), the cells appear oblong in straight chains that resemble rods. Acid-producing ability is measured by pH drop and titratable acidity rise in 12% reconstituted nonfat dry milk medium (sterilized at 116°C for 18 minutes) incubated at 40°C for 8 hours. A ratio of ST to LB of 3:1 gives a pH of 4.20 and titratable acidity of 1.05% under the above conditions.

The influence of temperatures of incubation on the growth of yogurt bacteria is shown in Table 16.5. Acid production is normally used as a measure of growth of a yogurt culture.

However, growth of the organisms is not necessarily synonymous with their acid-producing ability. Differences in acid liberated per unit cell mass, which are related to both environmental effects and genetic origin, have been recorded.

Yogurt fermentation constitutes the most important step in its manufacture. To optimize parameters for yogurt production and to maintain both a uniformity of product quality and cost effectiveness in the manufacturing operation, an understanding of the factors involved in the growth of yogurt bacteria is important.

Table 16.5. Growth Temperature Profile of Yogurt Bacteria

Growth Temperature	*Strptococcus thermophilus* °C	*Lactobacillus delbrueckii* subsp. *bulgaricus* °C
Minimum	20	> 15
Maximum	50	50–52
Optimum	39–46	40–47

Source: Adapted from Chandan and Shahani 1993.

Collaborative growth of ST and LB is a unique phenomenon. Yogurt starter organisms display an obligate symbiotic relationship during their growth in milk medium. Although they can grow independently, they utilize each other's metabolites to effect remarkable efficiency in acid production. In general, LB has significantly more cell-bound proteolytic enzyme activity, producing stimulatory peptides and amino acids for ST. The relatively high aminopeptidase and cell-free and cell-bound dipeptidase activity of ST is complementary to the strong proteinase and low peptidase activity of LB. Urease activity of ST produces CO_2, which stimulates LB growth. Concomitant with CO_2 production, urease liberates ammonia, which acts as a weak buffer; consequently, milk cultured by ST alone exhibits a considerably lower titratable acidity or high pH of coagulated mass. Formic acid formed by ST as well as by heat treatment of milk accelerates LB growth. The rate of acid production by yogurt starter containing both ST and LB is considerably higher than that of either of the two organisms grown separately.

Yogurt organisms are microaerophilic in nature. Heat treatment of milk drives out oxygen. It also wipes out competitive flora. Furthermore, heat produces sulfhydryl compounds, which tend to generate reducing conditions in the medium. Accordingly, rate of acid production in high-heat–treated milk is considerably higher than in raw or pasteurized milk.

However, there are inhibiting factors for yogurt culture growth. Proper selection of ST and LB strains is necessary to avoid possible antagonism between the two organisms. Also, certain abnormal milks (mastitic cows, hydrolytic rancidity in milk) are inhibitory to their growth. Seasonal variations in milk composition resulting in lower micronutrients (trace elements, nonprotein nitrogenous compounds) may affect starter performance. Natural inhibitors secreted in milk (lactoperoxidase thiocyanate system, agglutinins, lysozyme) are generally destroyed by proper heat treatment. Antibiotic residues in milk

and entry of sanitation chemicals (quaternary compounds, iodophors, hypochlorites, hydrogen peroxide) have a profound inhibitory impact on the growth of yogurt starter. Yogurt mixes designed for manufacture of refrigerated or frozen yogurt may contain appreciable quantities of sucrose, high fructose corn syrup, dextrose, and corn syrups obtained from various degrees of starch hydrolysis (dextrose equivalent). The sweeteners exert osmotic pressure in the system, leading to progressive inhibition and decline in the rate of acid production by the culture. Being a colligative property, the osmotic based inhibitory effect would be directly proportional to concentration of the sweetener and inversely related to the molecular weight of the solute. In this regard, solutes inherently present in the milk solids-not-fat part of the yogurt mix, accruing from starting milk and added milk solids and whey products, would also contribute toward the total potential inhibitory effect on yogurt culture growth.

The acid-producing ability of yogurt culture in mixes containing 8.0% sucrose is fairly good. Commercial strains that are relatively osmotolerant may allow use of higher levels of sugars without interruption in acid production during yogurt manufacture.

Bactereophages, virus-like microbes, kill bacteria by their lytic action. Phage infections and the accompanying loss in rate of acid production by lactic cultures result in flavor and texture defects as well as major product losses in fermented dairy products. Occasionally, serious economic losses in the yogurt industry have been attributed to phage attack. In general, thermophilic starters have not been threatened by phage attack as much as mesophilic starters, which are largely used in cheese production. In view of the dramatic increase in volumes of products that utilize thermophilic cultures (e.g., Mozzarella cheese, Swiss cheese, yogurt), phage inhibition of LB and ST is now encountered in yogurt plants. It is known that there are specific phages affecting ST and LB and that ST is relatively more susceptible than LB.

The yogurt fermentation process is relatively fast (2.5–4 hours). It is improbable that both ST and LB would be simultaneously attacked by phages specific for the two organisms. In the likelihood of a phage attack on ST, acid production may be carried on by LB, causing little or no interruption in production schedule. In fact, a lytic phage may lyse ST cells, spilling cellular contents in the medium, which could conceivably supply stimulants for LB growth. This rationale may explain partially why the yogurt industry has experienced a low incidence of phage problems. Nonetheless, most commercial strains of yogurt cultures have been phage typed. Specific phage sensitivity has been determined to facilitate starter rotation procedures as a practical way to avoid phage threats in yogurt plants. The ST phage is normally destroyed by heat treatment of 74°C for 23 seconds. This phage proliferates much faster at pH 6.0 than at 6.5 or 7.0. Methods used for phage detection include plaque assay, detection of inhibited acid production (litmus color change), enzyme immunoassay, ATP (adenotriphosphate) assay by bioluminescence, and measurement of changes in impedance and conductance

Phage problems in yogurt plants cannot be ignored. Accordingly, adherence to strict sanitation procedures would ensure prevention of phage attack.

SWEETENERS

Nutritive carbohydrates (mainly sucrose) are used in yogurt manufacture. Sucrose is the major sweetener used in yogurt production. High intensity sweeteners (e.g., aspartame, sucralose, neotame, acesulfame K, etc.) are used to produce light yogurt containing about 60% of the calories contained in normal, sugar-sweetened yogurt. Low levels of crystalline fructose may be used in conjunction with aspartame and other high intensity sweeteners to round up and improve overall flavor of light yogurt. The level of sucrose in yogurt mix appears to affect the production of lactic acid and flavor by yogurt culture. A decrease in the characteristic flavor compound (acetaldehyde) production has been reported at 8% or higher concentration of sucrose (Chandan 1982, Chandan and Shahani,1993). Sucrose may be added in a dry, granulated, free-flowing, crystalline form or as a liquid sugar containing 67% sucrose. Liquid sugar is preferred for its handling convenience in large operations. However, storage capacity (in sugar tanks), heaters, pumps, strainers, and meters are required. The corn sweeteners, primarily glucose, usually enter yogurt via the processed fruit flavor in which they are extensively used for their flavor enhancing characteristics. Up to 8% corn syrup solids are used in frozen yogurt.

Commercial yogurts have an average of 4.06% lactose, 1.85% galactose, 0.05% glucose, and 4.40 pH.

STABILIZERS

The primary purpose of using a stabilizer in yogurt is to produce smoothness in body and texture, impart gel structure, and reduce syneresis. The stabilizer increases shelf life and provides a reasonable degree of uniformity of product. Stabilizers function through their ability for form gel structures in water, thereby leaving less free water for syneresis. In addition, some stabilizers complex with casein. A good yogurt stabilizer should not impart any flavor, should be effective at low pH values, and should be easily dispersed in the normal working temperatures in a food plant. The stabilizers generally used in yogurt are gelatin; vegetable gums like carboxymethyl cellulose, locust bean, and guar; seaweed gums like alginates and carrageenans; whey protein concentrates; and pectin.

Gelatin is derived by irreversible hydrolysis of the proteins collagen and ossein. It is used at a level of 0.3–0.5% to get a smooth shiny appearance in refrigerated yogurt. Gelatin is a good stabilizer for frozen yogurt. The term "Bloom" refers to the gel strength as determined by a Bloom gelometer under standard conditions. Gelatin of a Bloom strength of 225 or 250 is commonly used. The gelatin level should be geared to the consistency standards for yogurt. Amounts above 0.35% tend to give yogurt of relatively high milk solids a curdy appearance upon stirring. At temperatures below 10°C, the yogurt acquires a pudding-like consistency. Gelatin tends to degrade during processing at ultrahigh temperatures, and its activity is temperature dependent. Consequently, the yogurt gel is considerable weakened by a rise in temperature.

The seaweed gums impart a desirable viscosity as well as gel structure to yogurt. Algin and sodium alginate are derived from giant sea kelp. Carrageenan is made from Irish moss and compares with 250 Bloom gelatin in stabilizing value. These stabilizers are heat stable and promote stabilization of the yogurt gel by complex formation with Ca^{+2} and casein.

Among the seed gums, locust bean gum or carob gum is derived from the seeds of a leguminous tree.

Carob gum is quite effective at low pH levels. Guar gum is also obtained from seeds and is a good stabilizer for yogurt. Guar gum is readily soluble in cold water and is not affected by the high temperatures used in the pasteurization of yogurt mix. Carboxymethyl cellulose is a cellulose product and is effective at high processing temperatures. Whey protein concentrate is commonly used as a stabilizer, exploiting the water-binding property of denatured whey proteins. Pectins are obtained from fruit and are a good choice for "natural" or organic types of yogurt.

The stabilizer system used in yogurt mix preparations is generally a combination of various vegetable stabilizers to which gelatin may be added. Their ratios as well as the final concentration (generally 0.5–0.7%) in the product are carefully controlled to get the desired effects.

FRUIT PREPARATIONS FOR FLAVORING YOGURT

The fruit preparations for blending in yogurt are specially designed to meet the marketing requirements for different types of yogurt. They are generally present at levels of 10–20% in the final product. A majority of the fruit preparations contain natural flavors to boost the fruit aroma and flavor.

Flavors and certified colors are usually added to the fruit-for-yogurt preparations for improved eye appeal and better flavor profile. The fruit base should meet the following requirements. It should (a) exhibit the true color and flavor of the fruit when blended with yogurt, and (b) be easily dispersible in yogurt without causing texture defects, phase separation, or syneresis. The pH of the fruit base should be compatible with yogurt pH. The fruit should have zero yeast and mold populations in order to prevent spoilage and to extend shelf life. Fruit preserves do not necessarily meet all these requirements, especially of flavor, sugar level, consistency, and pH. Accordingly, special fruit bases are designed for use in stirred yogurt. They generally contain 0.1% artificial flavor or 1.25% natural flavor, 0.1% potassium sorbate and an appropriate level of coloring. The pH is adjusted to 3.8–4.2, depending on the particular fruit.

Calcium chloride and certain food-grade phosphates are also used in several fruit preparations. The soluble solids range from 60 to 65% and viscosity is standardized. Standard plate counts on regular fruit bases are generally less than 500 CFU/g.

Coliform count and yeast and mold counts of non-aseptic fruit preparations are less than 10 CFU/g. The fruit flavors vary in popularity in different parts of the country and during different times of the year. In general, the more popular fruits are strawberry, raspberry, blueberry, peach, cherry, orange, lemon, purple plum, boysenberry, spiced apple, apricot, and pineapple. Blends of these fruits are also popular. Fruits used in yogurt-base manufacture may be frozen, canned, dried, or combinations thereof. Among the frozen fruits are strawberry, raspberry, blueberry, apple, peach, orange, lemon, cherry, blackberry, and cranberry. Canned fruits are pineapple, peach, mandarin orange, lemon, and cherry. The dried fruit category includes apricot, apple, and prune. Fruit juices and syrups are also incorporated in the bases. Sugar in the fruit base protects fruit flavor against loss by volatilization and oxidation. It also balances the fruit and the yogurt flavor. The pH control of the base is important for fruit color retention. The color of the yogurt should represent the fruit color in intensity, hue, and shade. The fruit bases are obtained by cooking fruit with sugars, fruit juices, flavor, color, and stabilizer, followed by quick cooling and packaging in pails or totes. The base should be stored under refrigeration to obtain optimum flavor and extend shelf life. To avoid unnecessary contamination of yogurt, aseptically packaged, sterilized fruit preparations are now preferred by yogurt manufacturers.

PROCESSING

The sequence of processing stages in a yogurt plant is given in Table 16.6.

PRODUCTION OF YOGURT STARTERS

Frozen culture concentrates available from commercial culture suppliers have received wide acceptance in the industry. Reasons for their use include convenience and ease of handling, and reliable quality and activity. The concentrates are shipped frozen in dry ice and stored at the plant in special freezers at $-40°C$ or below for a limited period of time specified by the culture supplier.

The starter is the most crucial component in the production of yogurt of high quality and uniformity. An effective sanitation program including filtered air and positive pressure in the fermentation area should significantly control airborne contamination. The result would be a cleaner plant environment,

Table 16.6. Sequence of Processing Stages in the Manufacturing of Yogurt

Step	Salient Feature
Milk procurement	Sanitary production of Grade A milk from healthy cows is necessary. For microbiological control, refrigerated bulk milk tanks should cool to 10°C in 1 hour and <5°C in 2 hours. Avoid unnecessary agitation to prevent lipolytic deterioration of milk flavor. Milk pickup from dairy farm to processing plant is in insulated tanks at 48-hour intervals, as appropriate.
Milk reception and storage in manufacturing plant	Temperature of raw milk at this stage should not exceed 10°C. Insulated or refrigerated storage up to 72 hours helps in raw material and process flow management. Quality of milk is checked and controlled.
Centrifugal clarification and separation	Leucocytes and sediment are removed. Milk is separated into cream and skim milk or standardized to desired fat level at 5°C.
Mix preparation	Various ingredients to secure desired formulation are blended together at 50°C in a mix tank equipped with a powder funnel and an agitation system.
Heat treatment	Using plate heat exchangers with regeneration systems, milk is heated to temperatures of 95–97°C for 7–10 minutes, well above pasteurization treatment. Heating of milk kills contaminating and competitive microorganism, produces growth factors by breakdown of milk proteins, generates microaerophilic conditions for growth of lactic organisms, and creates desirable body and texture in the cultured dairy products.
Homogenization	Mix is passed through extremely small orifice at pressure of approximately 1700 MPa (2000–2500 psi), causing extensive physicochemical changes in the colloidal characteristics of milk. Consequently, creaming during incubation and storage of yogurt is prevented. The stabilizers and other components of a mix are thoroughly dispersed for optimum textural effects.
Inoculation and incubation	The homogenized mix is cooled to an optimum growth temperature (42 °C). Inoculation is generally at the rate of 0.5–5%, and the optimum temperature is maintained throughout incubation period to achieve the desired titratable acidity. A pH of 4.5 is commonly used as an endpoint of fermentation. Quiescent incubation is necessary for product texture and body development.
Cooling, fruit incorporation and packaging	The coagulated product is cooled to 5–22°C, depending upon the style of yogurt. Using fruit feeder or flavor tank, the desired level of fruit and flavor is incorporated. The blended product is then packaged.
Storage and distribution	Storage at 5°C for 24–48 hr imparts in several yogurt products desirable body and texture. Low temperatures ensure desirable shelf life by slowing down physical, chemical, and microbiological degradation.

Sources: Adapted from Chandan and Shahani 1995, Mistry 2001, Robinson et al. 2002.

which in turn would promote optimum fermentation conditions for yogurt bacteria. Accordingly, fermentation time would be predictable, and production schedules would be maintained. Also, clean environment should enhance the quality and shelf life of the product.

Many plants use frozen direct-to-vat or freeze-dried direct-to-vat cultures for yogurt production. However, for cost savings, large yogurt manufacturers prefer to make bulk starters in their own plant from frozen bulk cultures. The medium for bulk starter production is antibiotic-free, nonfat dry milk reconstituted in water at 10–12% solids level. Following reconstitution of nonfat dry milk in water, the medium is heated to 90–95°C and held for 30–60 minutes. Then the medium is cooled to 43°C in the vat. The next step is inoculation of frozen bulk culture. The frozen culture is thawed in the can in cold or lukewarm water that contains a low level of sanitizer until the contents are partially thawed. The culture cans are emptied into the starter vat in an aseptic manner, and bulk starter medium is pumped over the partially thawed culture to facilitate mixing and uniform dispersion.

The incubation period for yogurt bulk starter ranges from four to six hours; the incubation temperature (43°C) is maintained by holding hot water in the jacket of the tank. The fermentation must be quiescent (lack of agitation and vibrations) to avoid phase separation in the starter following incubation. The progress of fermentation is monitored by titratable acidity measurements at regular intervals. When the titratable acidity is 0.85–0.90%, the fermentation is terminated by turning the agitators on and replacing the warm water in the tank jacket with iced water. Circulating iced water drops the temperature of starter to 4–5°C. The starter is now ready to use, following a satisfactory microscopic examination of a methylene blue stained slide of the starter. A morphological view helps to ensure healthy cells in the starter and maintenance of desirable ST/LB ratio. A ratio of 3:1 in favor of ST produces a mild-flavored yogurt.

MIX PREPARATION

A yogurt plant requires a special design to minimize contamination of the products with phage and spoilage organisms. Filtered air is useful in this regard. The plant is generally equipped with a receiving room to receive, meter or weigh, and store milk and other raw materials. In addition, facilities include a process and production control laboratory, a dry storage area, a refrigerated storage area, a mix processing room, a fermentation room, and a packaging room.

The mix processing room contains equipment for standardizing and separating milk, pasteurizing and heating, and homogenizing along with the necessary pipelines, fittings, pumps, valves, and controls. The fermentation room housing fermentation tanks is isolated from the rest of the plant. Filtered air under positive pressure is supplied to the room to generate clean room conditions. A control laboratory is generally set aside where culture handling, process control, product composition, and shelf life tests are carried out to ensure adherence to regulatory and company standards. There is also a quality control program, established by laboratory personnel. A utility room is required for maintenance and engineering services needed by the plant. The refrigerated storage area is used for holding fruit, finished products, and other heat-labile materials. A dry storage area at ambient temperature is primarily utilized for temperature-stable raw materials and packaging supplies.

Standardization of milk for fat and milk solids-not-fat content results in fat reduction and in an increase of 30–35% in lactose, protein, mineral, and vitamin content. The nutrient density of yogurt mix is thus increased over that of milk. Specific gravity changes from 1.03 to 1.04 g/ml at 20°C. Addition of stabilizers (gelatin, starch, pectin, agar, alginates, gums, and carrageenans) and sweeteners further impacts physical properties.

HEAT TREATMENT

The common pasteurization equipment consists of a vat, plate, triple-tube, scraped, or swept surface heat exchanger. In yogurt processing, a plate heat exchanger and high-temperature short-time (HTST) pasteurization system is commonly used. The mix is subjected to much more severe heat treatment than in conventional pasteurization temperature-time combinations. Heat treatment at 85°C for 30 minutes or 95–99°C for 7–10 minutes is an important step in manufacture. The heat treatment (1) produces a relatively sterile medium for the exclusive growth of the starter; (2) removes air from the medium to produce a more conducive medium for microaerophilic lactic cultures to grow; (3) effects thermal breakdown of milk constituents, especially proteins, releasing peptones and sulfhydryl groups, which provide nutrition and anaerobic conditions for yogurt culture; and (4) denatures and coagulates whey proteins of milk, thereby enhancing the viscosity, leading to a custard-like consistency in the product. The intense heat treatment during yogurt processing destroys all the pathogenic flora and most vegetative cells of all microorganisms contained therein. In addition, milk enzymes inherently present are inactivated. Consequently, the shelf life of yogurt is assured. From a microbiological standpoint, destruction of competitive organisms produces conditions conducive to the growth of desirable yogurt bacteria. Furthermore, expulsion of oxygen, creation of reducing conditions (sulfhydryl generation), and production of protein-cleaved nitrogenous compounds as a result of heat processing enhance the nutritional status of the medium for growth of the yogurt culture.

Physical changes in the proteins as a result of heat treatment have a profound effect on the viscosity of yogurt. Whey protein denaturation, of the order of

70–95%, enhances water absorption capacity, thereby creating smooth consistency, high viscosity, and stability from whey separation in yogurt.

HOMOGENIZATION

The homogenizer is a high-pressure pump that forces the mix through extremely small orifices. It includes a bypass for safety of operation. The process is usually conducted by applying pressure in two stages. In the first stage, pressure of the order of approximately 14 MPa (2000 psi) reduces the average diameter of the average milk fat globule from approximately 4 micrometers (μm) (range 0.1 to 16 μm) to less than 1 micrometer. The second stage uses a pressure of 3.5 MPa (500 psi) and is designed to break the clusters of fat globules apart, with the objective of inhibiting creaming in the milk. Homogenization aids in texture development and alleviates surface creaming and syneresis problems. Since homogenization reduces the fat globules to an average of less than 1 μm in diameter, no distinct creamy layer (crust) is observed on the surface of yogurt produced from homogenized mix. In general, homogenized milk produces soft coagulum in the stomach, which may enhance digestibility.

The homogenized mix is brought to 43°C by pumping it through an appropriate heat exchanger. It is then collected in fermentation tanks.

FERMENTATION

Fermentation tanks for the production of cultured dairy products are generally designed with a cone bottom to facilitate draining of relatively viscous fluids after incubation.

For temperature maintenance during the incubation period, the fermentation vat is usually insulated and covered with an outer surface of stainless steel. The vat is equipped with a heavy-duty, multispeed agitation system, a manhole containing a sight glass, and appropriate spray balls for CIP (clean-in-place) cleaning. The agitator is often of the swept surface type for optimum agitation of relatively viscous cultured dairy products. For efficient cooling after culturing, plate or triple-tube heat exchangers are used.

Contribution of the Culture to Yogurt Texture and Flavor

The starter is a critical ingredient in yogurt manufacture. The rate of acid production by yogurt cul-

ture should be synchronized with plant production schedules. When frozen culture concentrates are used, an incubation period of five hours at 43°C is required for yogurt acid development. With bulk starters at 4% inoculum level, the incubation period is 2.5–3.0 hours at 43°C.

Proper fermentation with yogurt culture leads to the formation of typical flavor compounds. Lactic acid, acetaldehyde, acetone, diacetyl, and other carbonyl compounds constitute key flavor compounds of yogurt. The production of flavor by yogurt cultures is a function of time and the sugar content of yogurt mix. Acetaldehyde production in yogurt takes place predominantly in the first one to two hours of incubation. Eventually, an acetaldehyde level of 23–55 ppm develops in yogurt. The acetaldehyde level declines in later stages of incubation. Diacetyl varies from 0.1 to 0.3 ppm and acetic acid varies from 50 to 200 ppm. These key compounds are produced by yogurt bacteria. Diacetyl and acetoin are metabolic products of carbohydrate metabolism in ST. Acetone and butane-2-one may develop in milk during preformentation processing.

The milk coagulum during yogurt production results from the drop in pH due to the activity of the yogurt culture. The streptococci are responsible for lowering the pH of a yogurt mix to 5.0–5.5, and the lactobacilli are primarily responsible for further lowering of the pH to 3.8–4.4. Several mucopolysaccharide-producing strains of yogurt culture are utilized in the yogurt industry. The texture of yogurt tends to be coarse or grainy if it is allowed to overferment prior to stirring, or if it is disturbed at pH values higher than 4.6. Incomplete blending of mix ingredients is an additional cause of coarse texture. Homogenization and high fat content tend to favor smooth texture. Gassiness in yogurt is a result of excessive CO_2 and hydrogen production that may be attributed to defects in starters or contamination with spore-forming *Bacillus* species, coliform bacteria, or yeast. In comparison with plate heat exchangers, cooling with tube-type heat exchangers causes less damage to yogurt structure. Further, loss of viscosity of yogurt may be minimized by use of well-designed booster pumps, metering units, and valves in yogurt packaging.

The pH of yogurt during refrigerated storage continues to drop. Higher storage temperatures accelerate the drop in pH. As a result of fermentation by yogurt bacteria, several changes take place in yogurt mix.

Changes in Milk Constituents

Among the carbohydrate constituents, the lactose content of yogurt mix is generally around 6%. During fermentation, lactose is the primary carbon source, which leads to an approximately 30% reduction in lactose during the fermentation process. However, a significant level of lactose (4.2%) survives in yogurt after fermentation. One mole lactose gives rise to one mole of galactose, two moles of lactic acid, and energy for bacterial growth. Although there is a large excess of lactose in the fermentation medium, lactic acid buildup beyond 1.5% acts as a growth inhibitor, progressively inhibiting further growth of yogurt bacteria. Normally, the fermentation period is terminated by a temperature drop to 4°C. To achieve this, yogurt mass is pumped through a heat exchanger. To smoothen the texture, a texturizing cone is inserted in the pipe leading to the heat exchanger. At 4°C, the culture is live, but its activity is drastically limited to allow fairly controlled flavor in marketing channels.

Lactic acid production results in coagulation of milk, beginning at a pH below 5.0 and ending at a pH of 4.6. The texture, body, and acid flavor of yogurt owe their origin to lactic acid produced during fermentation.

Small quantities of organoleptic moieties are generated through carbohydrate catabolism, via volatile fatty acids, ethanol, acetoin, acetic acid, butanone, diacetyl, and acetaldehyde. Homolactic fermentation in yogurt yields lactic acid as 95% of the fermentation output. Lactic acid acts as a preservative.

Hydrolysis of milk proteins is easily measured by liberation of $-NH_2$ groups during fermentation. After 24 hours, free amino groups in yogurt double. Proteolysis continues during the shelf life of yogurt, doubling free amino groups again in 21 days storage at 7°C. The major amino acids liberated are proline and glycine. The essential amino acids liberated increase 3.8- to 3.9-fold during yogurt storage, indicating that various proteolytic enzymes and peptidases remain active throughout the shelf life of yogurt. The proteolytic activity of the two yogurt bacteria is moderate, but it is quite significant in relation to symbiotic growth of the culture and production of flavor compounds (Chandan 1989, Chandan and Shahani 1993).

A weak lipase activity results in the liberation of minor amounts of free fatty acids, particularly stearic and oleic acids. Individual esterases and lipases of yogurt bacteria appear to be more active towards short-chain fatty acid glycerides than towards long-chain substrates. Since nonfat and low fat yogurts comprise the majority of yogurt marketed in the United States, lipid hydrolysis contributes little to the product attributes.

Both ST and LB are documented in the literature to elaborate different oligosaccharides in yogurt mix medium. As much as 0.2% (by weight) mucopolysaccharides have been observed in yogurt in 10 days of storage. In stirred yogurt, drinking yogurt, and reduced-fat yogurt, exopolysaccharides can contribute to smooth texture, higher viscosity, lower synerisis, and better mechanical handling. However, excessive shear during pumping destroys much of this textural advantage because viscosity generated by the mucopolysaccharides is susceptible to shear. Most of the polysaccharides elaborated in yogurt contain glucose, galactose, and minor quantities of fructose, mannose, arabinose, rhamnose, xylose, or N-acetylgalactosamine, individually or in combination. Molecular weight is of the order of 0.5–1 million Daltons (Da). An intrinsic viscosity range of 1.5–4.7 dl/g^{-1} has been reported for exopolysaccharides of ST and LB. The polysaccharides form a network of filaments visible under a scanning electron microscope. The bacterial cells are covered by part of the polysaccharide, and the filaments bind the cells and milk proteins. Upon shear treatment, the filaments rupture from the cells, but maintain links with casein micelles. Ropy strains of ST and LB are commercially available. They are especially appropriate for stirred yogurt production.

Other interesting metabolites are elaborated in yogurt. Yogurt organisms generate bacteriocins and several antimicrobial compounds. Benzoic acid (15–30 ppm) in yogurt has been detected and associated with the metabolic activity of the culture. These metabolites tend to exert a preservative effect by controlling the growth of contaminating spoilage and pathogenic organisms that gain entry postfermentation. As a result, the product attains extension of shelf life and a reasonable degree of safety from food-borne illness. As a consequence of fermentation, yogurt organisms multiply to a count of 10^8–10^{10} CFU/g^{-1}. Yogurt bacteria occupy some 1% of the volume or mass of yogurt. These cells contain cell walls, enzymes, nucleic acids, cellular proteins, lipids, and carbohydrates. Lactase or β-galactosidase activity of yogurt has been shown to contribute a major health-related property. Clinical studies have concluded that yogurt containing live and active culture can be consumed by several millions of lactose-deficient individuals without developing gastrointestinal distress or diarrhea (Chandan 1989, Chandan and Shahani 1993, Fernandes et al. 1992).

Yogurt is an excellent dietary source of calcium, phosphorus, magnesium, and zinc in human nutrition. Research has shown that bioavailability of the minerals from yogurt is essentially equal to that from milk. Since yogurt is a lower pH product than milk, most of the calcium and magnesium occurs in ionic form. The complete conversion from colloidal form in milk to ionic form in yogurt may have some bearing on the physiological efficiency of utilization of the minerals.

Yogurt bacteria during and after fermentation affect the B-vitamin content of yogurt. The processing parameters and subsequent storage conditions influence the vitamin content at the time of consumption of the products. Incubation temperature and fermentation time have a significant effect on the balance between vitamin synthesis and vitamin utilization of the culture. In general, there is a decrease of vitamin B_{12}, biotin, and pantothenic acid and an increase of folic acid during yogurt production. Nevertheless, yogurt is still an excellent source of the vitamins inherent to milk.

APPLICATION OF PROCESSING PRINCIPLES

MANUFACTURING PROCEDURES

Plain Yogurt

Plain yogurt is the basic style and forms an integral component of the fruit-flavored yogurt. The steps involved in the manufacturing of set-type and stirred-type plain yogurts are shown in Figure 16.2. Plain yogurt normally contains no added sugar or flavors in order to offer the consumer natural yogurt flavor for consumption as such, or the option of flavoring the yogurt with other food materials of the consumer's choice. In addition, it may be used for cooking or for salad preparation with fresh fruits or grated vegetables. In most recipes, plain yogurt is a substitute for sour cream that provides a lower calorie and lower fat alternative. The fat content may be standardized to the levels preferred by the market. Also, the size of package may be geared to market demand. Polystyrene and polypropylene cups and lids are the chief packaging materials used in the industry.

Fruit-flavored Yogurt

Figure 16.3 illustrates manufacturing outline for fruit-on-the-bottom style yogurt.

In this yogurt type, two-stage fillers are used. Typically, 59 ml (2 ounces) of fruit preserves or special fruit preparations are layered at the bottom, followed by 177 ml (6 ounces) of inoculated yogurt mix on the top. The top layer may consist of yogurt mix containing stabilizers, sweeteners, and the flavor and color indicative of the fruit on the bottom. After placing lids on the cups, incubation and setting of the yogurt takes place in the cups. When a desirable pH of 4.4–4.5 is attained, the cups are placed in refrigerated rooms with high-velocity forced air for rapid cooling. At the time of consumption, the fruit and yogurt layers are mixed by the consumer. Fruit preserves have an FDA standard of identity. A fruit preserve consists of 55% sugar and a minimum of 45% fruit that is cooked until the final soluble solids content is 68% or higher (65% in the case of certain fruits). Frozen fruits and juices are the usual raw materials. Commercial pectin, 150 grade, is normally utilized at a level of 0.5% in preserves, and the pH is adjusted to 3.0–3.5 with a food-grade acid, namely, citric, during manufacturing of the preserves.

Stirred style yogurt or blended yogurt is produced by blending the fruit preparation thoroughly in the fermented yogurt base obtained after bulk culturing in fermentation tanks. Figure 16.4 shows a process flow outline.

Stabilizers, especially gelatin, are commonly used in this form of yogurt unless milk solids-not-fat levels are relatively high (12–14%). In this style, cups are filled with an in-line blended mixture of yogurt and fruit. Pumping of the yogurt-fruit blend results in considerable loss of viscosity. Upon refrigerated storage for 48 hours, the clot is reformed and viscosity is recovered, leading to a fine body and texture. Overstabilized yogurt possesses a solid-like consistency and lacks a refreshing character. Spoonable yogurt should not have the consistency of a drink. It should melt in the mouth without chewing.

Several variations of this procedure exist in the industry. Fruit incorporation is conveniently effected by the use of a fruit feeder at a 10–20% level. Prior to packaging, the stirred-yogurt texture can be made smoother by pumping it through a valve or a cone made of stainless steel mesh.

The incubation times and temperatures are coordinated with plant schedules. Incubation tempera-

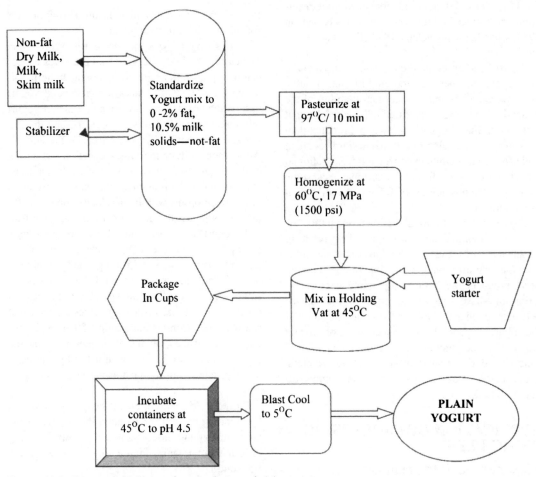

Figure. 16.2. Process flow diagram for manufacture of plain yogurt.

tures lower than 40°C in general tend to impart a slimy or sticky appearance to yogurt.

Aerated Yogurt

This category of yogurt resembles mousse in that the product acquires a novel, foam-like texture. The aeration process is similar to an ice cream process, but the degree of overrun (extent of air content) is kept around 35–50%. Foam formation is facilitated by use of appropriate emulsifiers, and the stability of foam is achieved by using gelatin in the formulation.

Heat-treated Yogurt

The shelf life of yogurt may be extended by heating the yogurt after culturing to inactivate the culture and

the constituent enzymes. Heating to 60–65°C extends the shelf life to about 12 weeks at 12°C. Ultra high-temperature (UHT) treatment and aseptic packaging ensures shelf life even more, even with room temperature storage. However, these treatments destroy the "live and active" nature of yogurt, which is a desirable consumer attribute to maintain. Current Federal Standards of Identity for refrigerated yogurt permit the thermal destruction of viable organisms with the objective of shelf life extension, but the parenthetical phrase "heat treated after culturing" must show on the package following the yogurt labeling. The postripening heat treatment may be designed to ensure destruction of starter bacteria, contaminating organisms, and enzymes. Concomitantly, it is critical to redevelop the texture and body of the yogurt by appropriate stabilizer and homogenization processes.

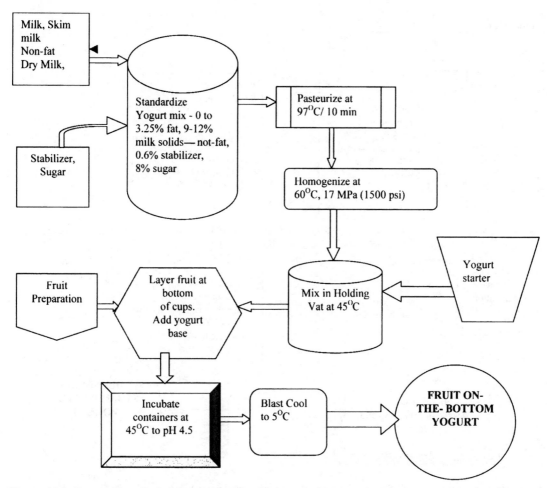

Figure. 16.3. Process flow diagram for manufacture of fruit-on-the-bottom style yogurt.

Frozen Yogurt

Both soft serve and hard-frozen yogurts have gained popularity in recent years. Consumer popularity of frozen yogurt has been propelled by its low fat and nonfat attributes. The recently developed frozen yogurt is a very low acid product that resembles ice cream or ice milk in flavor and texture. In some instances, the blend is pasteurized to ensure destruction of newly emerging pathogens, including *Listeria* and *Campylobacter* in the resulting low-acid food. To provide live and active yogurt culture in the finished product, frozen culture concentrate may be blended with the pasteurized product. Figure 16.5 shows a typical manufacturing outline for frozen yogurt.

Yogurt Beverages

Also called drinking yogurt, yogurt smoothie, and yogurt drink, this product is made in a procedure similar to that used for stirred style or blended yogurt (see Fruit-flavored Yogurt, above). However, fruit preparations generally consist of juices and purees. The stabilizers used are a nonthickening type (e.g., pectins, gums, modified starch) used to control whey separation during the product shelf life. The recent trend is to include fructo-oligosaccharide prebiotics such as inulin and to fortify with a significant daily requirement of most vitamins and minerals. All the beverages marketed in the United States contain live and active cultures to qualify them as a functional food.

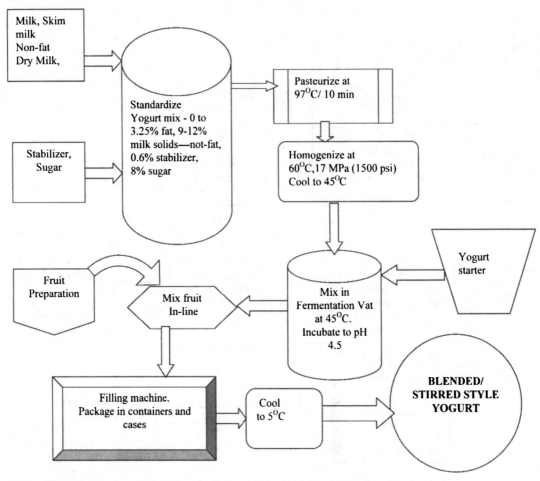

Figure. 16.4. Process flow diagram for manufacture of blended (stirred style) yogurt.

PACKAGING

Most plants attempt to synchronize the packaging lines with the termination of the incubation period. Generally, textural defects in yogurt products are caused by excessive shear during pumping or agitation. Therefore, positive drive pumps are preferred over centrifugal pumps for moving the product after culturing or ripening. For incorporation of fruit, it is advantageous to use a fruit feeder system. Various packaging machines of suitable speeds (up to 400 cups per minute) are available to package various kinds and sizes of yogurt products.

Yogurt is generally packaged in plastic containers varying in size from 4 to 32 ounces. The machines involve volumetric piston filling. The product is sold by weight, and the machines delivering volumetric measure are standardized accordingly. The pumping of fermented and flavored yogurt base exerts some shear on the body of the yogurt. Cups of various shapes characterize certain brands. Some plants use preformed cups. The cup may be formed by injection molding, where beads of plastic are injected into a mold at high temperature and pressure. In this type of packaging, a die-cut foil lid is heat sealed on to the cups. Foil lids are cut into circles and are procured from a supplier along with the preformed cups. A plastic overcap may also be used. In some cases, partially formed cups are procured and assembled at the plant. Other plants use roll stock, which is used in the form-fill-seal system of packaging. In this case, cups are fabricated in the plant by thermoforming, where plug is rammed into a sheet

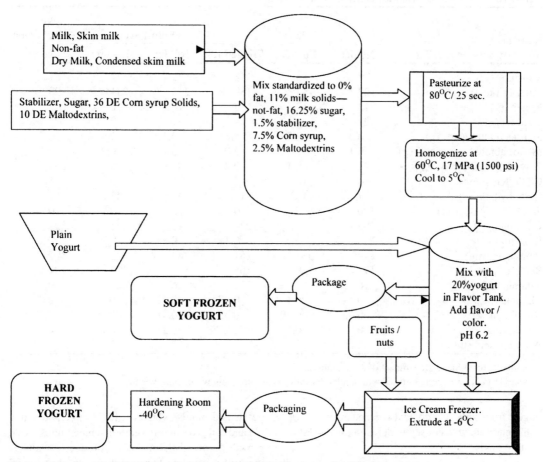

Figure. 16.5. Process flow diagram for manufacture of hard-frozen and soft-frozen yogurt.

of heated plastic. Multipacks of yogurt are produced by this process. Following the formation of the cups, they are filled with the appropriate volume of yogurt and are heat sealed with a foil lid. They are then placed in cases and transferred to a refrigerated room for cooling and distribution. In breakfast yogurt, a mixture of granola, nuts, chocolate bits, dried fruit, and cereal is packaged in a small cup and sealed with a foil. Subsequently, the cereal cup is inverted and sealed on the top of the yogurt cup. This package is designed to keep the ingredients isolated from the yogurt until the time of consumption. This system helps to maintain crispness in cereals and nuts, which otherwise would become soggy or interact adversely if mixed with yogurt at the plant level.

Some interesting innovations in yogurt packaging include the spoon-in-the-cup lid and squeezable tubes. The former adds convenience in eating yo-gurt, while the squeezable tubes add play value for children. In addition, yogurt in tubes is freeze-thaw stable, which adds another dimension of convenience and versatility to its use.

FINISHED PRODUCT

NUTRIENT PROFILE OF YOGURT

Typical composition and nutrient profiles for yogurt are shown in Table 16.7. In general, yogurt contains more protein, calcium, and other nutrients than milk, reflecting the extra solids-not-fat content. However, the data shown in the table for whole milk reflects no milk solids added to the yogurt mix. Therefore, its nutritional profile is similar to that of milk. Nevertheless, bacterial mass content and the products of the lactic fermentation further distinguish

Table 16.7. Typical Nutritional Composition of Yogurt

| Nutrient (per 8 oz. serving = 227 g) | Plain | | | Fruit-flavored | | Light, Vanilla/Lemon |
	Nonfat	Low Fat	Whole Milk	Nonfat	Low fat	Nonfat
Moisture	85	85	88	75	74	87
Calories (kcal)	127	144	139	213	231	98
Protein (g)	13	12	8	10	10	9
Total fat (g)	Tr	4	7	Tr	2	Tr
Saturated fatty acids (g)	0.3	2.3	4.8	0.3	1.6	0.3
Monosaturated fatty acids (g)	0.1	1.0	2.0	0.1	0.7	0.1
Polyunsaturated fatty acids (g)	Tr	0.1	0.2	Tr	0.1	Tr
Cholesterol (mg)	4	14	29	5	10	5
Carbohydrate (g)	17	16	11	43	43	17
Total dietary fiber (g)	0	0	0	0	0	0
Calcium (mg)	452	415	274	345	345	325
Iron (mg)	0.2	0.2	0.1	0.2	0.2	0.3
Potassium (mg)	579	531	351	440	442	402
Sodium (mg)	174	159	105	132	133	134
Vitamin A (IU)	16	150	279	16	104	0
Thiamin (mg)	0.11	0.1	0.07	0.09	0.08	0.08
Riboflavin (g)	0.53	0.49	0.32	0.41	0.40	0.37
Niacin (mg)	0.3	0.3	0.2	0.2	0.2	0.2
Ascorbic acid (mg)	2	2	1	2	2	2

Source: United States Department of Agriculture 2002.
Note: Data is for yogurts fortified with nonfat dry milk, except for plain whole milk yogurt.

yogurt from milk. Fat content is standardized to be commensurate with consumer demand for low fat to fat-free foods.

QUALITY CONTROL

A well-planned quality control program must be executed in the plant to maximize the keeping quality of the product. To deliver yogurt with the most desirable attributes of flavor and texture to the consumer, it is imperative to enforce a strict sanitation program and good manufacturing practices.

Refrigerated Yogurt and Yogurt Beverages

Shelf life expectations from commercial yogurt vary, but generally approximate six weeks from the date of manufacture is normal, provided temperature during distribution and retail marketing channels does not exceed 7°C. Lactic acid and some other metabolites produced by the fermentation process protect yogurt from most gram-negative psychrotrophic organisms. In general, most quality issues in a yogurt plant are not related to proliferation of spoilage bacteria. Most spoilage flora in yogurt are yeasts and molds, which are highly tolerant to low pH and can grow at refrigeration temperatures. Yeast growth during shelf life of the product constitutes more of a problem than mold growth. The fungal growth manifests within two weeks of manufacture, if yeast contamination is not controlled. To ensure maximum shelf life, several manufacturers use potassium sorbate to control the growth of yeasts and molds in the product.

The control of yeast contamination is managed by aggressive sanitation procedures related to equipment, ingredients, and the plant environment. Clean-in-place chemical solutions should be used with special attention to their strength and proper temperature. Hypochlorites and iodophors are effective sanitizing compounds for fungal control on food-contact surfaces and in combating environmental contamination. Hypochlorites at high concentrations are corrosive. Iodophors are preferred for their noncorrosive property because they are effective at relatively low concentrations.

Yeast and mold contamination may also arise from starter, packaging materials, fruit preparations, and packaging equipment. Organoleptic examination of yogurt starter may be helpful in eliminating the fungal contamination therefrom. If warranted, direct microscopic view of the starter may reveal the presence of budding yeast cells or mold mycelium filaments. Plating of the starter on acidified potato dextrose agar would confirm the results. Avoiding contaminated starter for yogurt production is essential.

Efficiency of equipment and environmental sanitation can be verified by enumeration techniques involving exposure of poured plates to the atmosphere in the plant or making a smear of the contact surfaces of the equipment, followed by plating. Filters on the air circulation system should be changed frequently. Walls and floors should be cleaned and sanitized frequently and regularly.

The packaging materials should be stored in dust-free and humidity-free conditions. The filling room should be fogged with chlorine or iodine regularly.

Quality control checks on fruit preparations and flavorings should be performed (spot checking) to ascertain sterility and to eliminate yeast and mold entry via fruit preparation. Refrigerated storage of the fruit flavorings is recommended.

Quality control programs for yogurt include control of product viscosity, pH, flavor, body and texture, color, fermentation process, and composition. Daily chemical, physical, microbiological, and organoleptic tests constitute the core of quality assurance. The flavor defects are generally described as too intense (acid), too weak (fruit flavor), or unnatural. The sweetness level may be excessive or weak, or may exhibit corn syrup flavor. The ingredients used may impart undesirable qualities such as staleness; old ingredients; uncleanness; and metallic, oxidized, or rancid flavors. Lack of control in processing procedures may cause overcooked, caramelized, or excessively sour flavor notes in the product. Proper control of processing parameters and ingredient quality assures good flavor. Product standards for fats, solids, viscosity, pH (or, titratable acidity), and organoleptic characteristics should be strictly adhered to. Syneresis or appearance of a watery layer on the surface of yogurt is undesirable and can be controlled by judicious selection of effective stabilizers and by following proper processing conditions.

In aerated yogurt, it is desirable to measure overrun to ensure uniformity of texture from day to day. Also, the weight of yogurt in the cup will be related to the degree of overrun. Accordingly, overrun control will ensure the weight of the product in the cup.

Frozen Yogurt

In hard-pack frozen yogurt, a coarse and icy texture may be caused by the formation of ice crystals due to fluctuations in storage temperatures. Sandiness may be due to lactose crystals that result from too high levels of milk solids. Soggy or gummy defects originate from levels of milk solids-not-fat or sugar content that are too high. A weak body results from an overrun that is too high and insufficient total solids.

Color defects may be caused by the lack of intensity or authenticity of hue and shade. Proper blending of fruit purees and yogurt mix is necessary for uniformity of color. The compositional control tests include fat, moisture, pH, overrun, and microscopic examination of the yogurt culture to ensure a desirable ratio of LB to ST. Good microbiological quality of all ingredients is necessary.

LIVE AND ACTIVE STATUS OF YOGURT CULTURES

Yogurt products enjoy the image of a health-promoting food. The type of cultures, their viability, and their active status are important attributes from the consumer's standpoint. Quality control tests are necessary to ensure the "live and active" status of the culture. As per National Yogurt Association standards, yogurt must pass an activity test. The cultures must be active at the end of shelf life. Samples of yogurt stored at a temperature of between 1 and 10°C for the duration of the stated shelf life are subjected to the activity test. In this test, 12% (w/v) solids nonfat dry milk is pasteurized at 92°C for 7 minutes and cooled to 43.3°C; then 3% inoculum of the material under test is added, and the milk is fermented at 43.3°C for four hours. The total yogurt organisms are enumerated in the test material both before and after fermentation by International Dairy Federation procedure (IDF 1988). The activity test is met if there is an increase of one log or more of yogurt culture cells during fermentation.

Generally, at the time of manufacture, yogurt should contain not less than 100 million CFU/g. Assuming that the storage temperature of yogurt through distribution channels and the grocery store is 4–7°C, a loss of one log cycle in culture viability is expected during the period between manufacture and consumption. Therefore, at the time of consumption, yogurt should deliver at least 10 million CFU of live yogurt organisms per gram of the product. In case yogurt receives temperature abuse, it is desirable to manufacture yogurt with even higher counts of viable culture to assure that at the consumption stage, the product contains at least 10 million CFU/g.

APPLICATION OF PROCESSING PRINCIPLES

See Table 16.8 for more details on references for the processing principles in the manufacture of yogurt.

Table 16.8. References for Details on Processing Principles

Processing stage	Processing Principle(s)	References for More Information on the Principles Used
Production of yogurt starters	Preparation of growth medium, inoculation of culture concentrate, and incubation to achieve strong and viable culture for yogurt production	Hassan and Frank 2001; Tamime 2002
Mix Preparation	To secure desired formulation, various ingredients are blended together at 50°C in a mixing vessel equipped with agitation and a powder funnel attached to a circulating pump to assist in liquefying dry powders.	Chandan 1997, Chandan and Shahani 1993
Heat treatment	Using plate heat exchangers with regeneration systems, the mix is quickly heated to 95–97°C and held at this temperature in a holding tube and quickly cooled to 43–45°C. The objective is to kill contaminating and competitive microorganisms, produce growth factors by breakdown of proteins, generate microaerophilic conditions for growth of yogurt culture, and create desirable body and texture in yogurt.	Chandan and Shahani 1993; Tamime and Robinson 1999
Homogenization	The yogurt mix is forced through an extremely small orifice at approximately 1700 MPa, causing extensive physico-chemical changes in the colloidal characteristics of milk. Consequently, creaming during incubation and storage of yogurt is prevented. The stabilizers and various other ingredients are thoroughly mixed for optimum texture and body generation in the product.	Tamime and Robinson 1999, Chandan and Shahani 1993
Inoculation and incubation	The homogenized mix is cooled to 43°C, the optimum growth temperature of yogurt culture. Inoculation rate of starter is 5%, and optimum temperature and quiescent conditions are maintained until the pH of the fermented base drops down to 4.4–4.5. To control further acid buildup, the fermented base is cooled down to 4°C. It is then mixed with fruit preparations and flavorings and packaged.	Tamime and Robinson 1999, Chandan and Shahani 1993

GLOSSARY

Acetaldehyde—a compound that characterizes the flavor profile of yogurt.

Amino acid—an organic acid containing both an amino ($-NH_2$) and an acidic ($-COOH$)group: the building blocks of proteins.

Ash—the residue left when a substance is incinerated at a very high temperature.

Casein micelles—large colloidal particles of milk; complexes of protein and salt ions, principally calcium and phosphorus.

CFU/g—colony forming units per gram.

Coliform count—a group of intestinal tract organisms; their presence in food indicates contamination with fecal matter.

Denaturation—the process that proteins undergo when subjected to heating, resulting in disruption of the noncovalent bonds that maintain their secondary and tertiary structure. Functional properties are influenced by denaturation.

Density—mass per unit volume.

Diacetyl—a chemical compound characterizing the aroma and flavor of butter, milk fat and fermented milks.

Fatty acids— a group of chemical compounds containing carbon and hydrogen atoms and a carboxylic group ($-COOH$) at the end of the molecule; formed by lipid hydrolysis; unattached fatty acids are free fatty acids.

FDA—U.S. Food and Drug Administration.

Functional Foods— foods shown by clinical trials to promote health, prevent disease, or help in the treatment of certain disorders.

Galactose— a monosaccharide or simple sugar formed as a result of hydrolysis of milk sugar lactose.

Gelation—the process of gel formation, in which whey proteins or certain stabilizers like gelatin create desirable texture by holding the free water of the food system.

HAACP (hazard analysis and critical control points)—a system of steps established for quality control and safety of food production through anticipation and prevention of problems.

Homogenization—a process for reducing the size of fat globules in milk products; upon storage at 7°C, no visible separation of cream layer is observed.

HTST—high temperature short time.

Hydrocolloids—gums and modified starch and other polysaccharides used for thickening and water binding in food systems.

Hydrolysis—as a result of enzyme action, proteins and glycerides of fats are broken down to their constituent amino acids and free fatty acids, respectively.

Hydrolytic rancidity—a flavor defect associated with the activity of enzyme lipase in which short-chain free fatty acids are liberated from milk fat, resulting in objectionable aroma and flavor.

IDF—International Dairy Federation.

IDFA—Internationl Dairy Foods Association.

Lactase—also called β-galactosidase; an enzyme that splits milk sugar into glucose and galactose.

Lactose—milk sugar, a disaccharide composed of glucose and galactose.

Lactose intolerance—maldigestion of lactose by certain individuals who experience abdominal pain, bloating, and diarrhea after consuming milk and dairy products containing lactose.

LB—*Lactobacillus delbrueckii* subsp. *bulgarius*.

Lipase—an enzyme that hydrolyzes fats, glycerides, or acylglycerols.

Low fat yogurt—product containing at least 8.25% solids-not-fat, with fat reduced to deliver not more than 3 g of fat per serving of 8 ounces.

Nonfat yogurt—product containing at least 8.25% solids-not-fat, with fat reduced to deliver not more than 0.5 g of fat per serving of 8 ounces.

Pasteurization—a process of heating fluid milk products to render them safe for human consumption by destroying 100% of the disease-producing organisms (pathogens). The process inactivates approximately 95% of all the microorganisms in milk.

Prebiotics—nondigestible food ingredients that improve the host's health by selectively stimulating the growth and/or activity of the beneficial bacteria of the colon.

Probiotics—live organisms introduced into the gastrointestinal system of humans to improve the balance or metabolic activity of beneficial microorganisms.

Protease—an enzyme that hydrolyzes proteins.

Proteolysis—the enzymatic breakdown of proteins to smaller fragments, peptides.

Psychrotrophic—refers to cold tolerant microorganisms capable of growth at 4–15°C.

SNF—solids-not-fat.

Somatic cell count—count of the mixture of dead epithelial cells and leucocytes that migrate into milk from the udder of the cow.

Specifications—a set of chemical, physical, and microbiological measurements required for acceptance of ingredients or products.

ST—*Streptococcus thermophilus.*

Standard of identity—a legal standard, maintained by the FDA, that defines a food's minimum quality, required and permitted ingredients, and processing requirements, if any. Applies to a limited number of staple foods.

Standardization—a step in processing in which milk fat and milk solids-not-fat are made to conform to certain specifications by removal, addition, or concentration of milk fat.

Syneresis—the separation of liquid from a gel.

Titratable acidity—test used for determining milk quality and monitoring the progress of fermentation; it measures the amount of alkali required to neutralize the compounds of a given quantity of milk/milk products and is expressed as percent lactic acid.

UHT (ultra high-temperature) treatment—heat treatment at a temperature of 135–150°C for a holding period of 4–15 seconds; sterilizes the product for aseptic packaging to permit storage at ambient temperature.

Ultra pasteurization—pasteurization of fluid milk and products at 125–138°C for a holding period of 2–5 seconds to kill all the pathogenic bacteria; permits storage at refrigeration temperature for an extended period.

Viscosity—resistance to flow; a measure of the friction between molecules as they slide past each other.

Whey—the watery fluid appearing on the surface after the curd is formed in the manufacture of fermented dairy products.

REFERENCES

Aneja RP, BN Mathur, RC Chandan, AK Banerjee. 2002. Technology of Indian Milk Productsk 158–182. Dairy India Yearbook, New Delhi, India.

Chandan RC, editor. 1989. Yogurt: Nutritional and Health Properties. Nat. Yogurt Assoc., McLean, Va.

___. 1982. Fermented dairy products. In: G. Reed, editor. Prescott and Dunn's Industrial Microbiology, 4th edition, 113–184. AVI Publishing Co., Westport, Conn.

___. 1997. Dairy-Based Ingredients. Eagan Press, St. Paul, Minn.

___. 1999. Enhancing market value of milk by adding cultures. J. Dairy Sci. 82:2245–2256.

___. 2002. Symposium: Benefits of Live Fermented milks: Present Diversity of Products. Proceedings of International Dairy Congress. CD-ROM, Paris, France.

Chandan RC, KM Shahani. 1993. Yogurt. In: YH Hui, editor. Dairy Science and Technology Handbook, vol. 2, 1–56. VCH Publicatons, New York.

___. 1995. Other fermented dairy products. In: G Reed and TW Nagodawithana, editors. Biotechnology, 2nd edition, vol. 9, 386–418. VCH Publications, Weinheim, Germany.

Fernandes CF, RC Chandan, KM Shahani. 1992. Fermented dairy products and health. In: BJB Wood, editor. The Lactic Acid Bacteria, vol. 1, 279–339. Elsevier Applied Science, New York.

Hassan A, JF Frank. 2001. Chapter 6. Starter cultures and their use. In: EH Marth, JL Steele, editors. Applied Microbiology, 2nd edition, 151–205. Marcel Dekker.

International Dairy Federation. 1988. Yogurt: Enumeration of Characteristic Organisms—Colony Count Technique at 37 C. IDF Standard No. 117A. Brussels, Belgium.

International Dairy Foods Association (IDFA). 2003. Dairy Facts. Washington, D.C.

Mistry VV. 2001. Chapter 9. Fermented milks and cream. In: EH Marth, JL Steele, editors. Applied Dairy Microbiology, 2nd edition, 301–325. Marcel Dekker.

Robinson RK, AY Tamime, M Wszolek. 2002. Chapter 8. Microbiology of fermented milks. In: RK Robinson, editor. Dairy Microbiology Handbook, 367–430. John Wiley and Sons, New York.

Tamime AY. 2002. Chapter 7. Microbiology of starter cultures. In: RK Robinson, editor. Dairy Microbiology Handbook, 261–366. Wiley-interscience.

Tamime AY, RK Robinson. 1999. Yogurt Science and Technology, 2nd edition. Woodhead Publishing Limited, Cambridge, England /CRC Press, Boca Raton, Fla.

U.S. Department of Agriculture (USDA). October 2002. Nutritive Value of Foods, 22–23. USDA, Washington, D.C.

U.S. Department of Health and Human Services (DHHS). Grade "A" Pasteurized Milk Ordinance. 1999 Revision. Publication no. 229. U.S. Department of Public Health, U.S. DHHS, Food and Drug Administration,Washington, D.C.

17
Dairy: Milk Powders

M. Carić

BACKGROUND INFORMATION

Drying was a common and very popular preservation method centuries ago. However, its real expansion began in the 1900s, with advance drying techniques suitable for industrial application for edible fluid products. Spray-drying (Masters 1985) allows a quick heat and mass transfer and produces a high quality product that, after reconstitution, closely resembles the original product and is economical to manufacture.

In addition to milk, various dairy and nondairy products are produced in dry form. These include whey powder, dry cream, dry milk–based beverages, infant formulas, casein, and other milk protein products.

The big step forward in spray-drying technology was the invention of instantization by Peebles (Peebles 1958, Peebles and Clary 1955) (Carić 1993). Further innovations were the development of membrane methods for concentrating and fractionating (1970s) prior to spray-drying (Singh and Newstead 1992) and the introduction of a three-stage dry-

ing procedure (1980s). All enhance the quality of dry products and reduce energy consumption.

More details on drying are given in Chapter 2, Food Dehydration.

RAW MATERIALS PREPARATION

Raw milk preparation consists of the following operations (Fig.17.1.): (1) receiving, (2) selection, (3) clarification, (4) cooling, and (5) storage.

RECEIVING AND SELECTION

Milk selected for preparing long-lasting products, like milk powders, must be of the highest possible quality (chemical, sensory, and bacteriological). The same strict quality criteria that apply to dairy products involving starter cultures also apply to raw milk for dry dairy products.

The acidity of raw milk must be of natural origin, that is, from CO_2, protein, phosphates, and citrates from milk, and fall below 0.15% lactic acid. Higher acidity and a high bacterial count have a negative effect on the solubility of the resulting milk powder.

High bacteria counts increase fat oxidation during storage of the powder. In the past, fat oxidation was accelerated by the presence of adventitious metal ions. At present, this is not the case, since in the dairy industry all equipment coming into contact with product is made of stainless steel. Antibiotics, detergents, and other chemicals, if present in raw milk, would result in a final product of lower quality and shorter shelf life.

CLARIFICATION

Clarfication of raw milk is carried out by filters positioned in pipelines and centrifugal separators, called centrifugal clarificators (Bylund 2003).

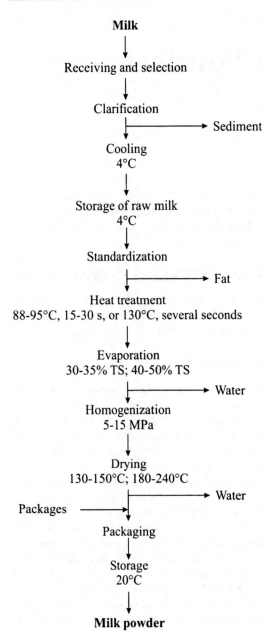

Figure 17.1. Flow chart of milk powder production

COOLING

Clarified milk is cooled in plate or tubular heat exchangers to 4°C.

STORAGE OF RAW MILK

Raw milk is stored at 4°C in huge isolated tanks, commonly located out of the plant buildings. These tanks are connected to the processing line by pipelines.

PROCESSING STAGE 1: STANDARDIZATION.

The ratio of milk fat to total solids is standardized by adjustment. Products include full fat milk powders, partially defatted milk powders, and skim milk powders. In skim milk powder production, fat in raw milk is reduced to 0.05–0.10%. Standardization or fat separation in all cases is carried out by centrifugal separators.

PROCESSING STAGE 2: HEAT TREATMENT

This operation is usually carried out at temperatures higher than pasteurization in order to destroy all pathogenic and most saprophytic microorganisms, inactivate enzymes, and activate SH groups of β-lactoglobulin, which act like antioxidants during powder storage. Most common are high-temperature short-time (HTST) regimes, for example, 88–95°C during 15–30 seconds. A temperature of 130°C has been used as well. Equipment used for heat treatment is usually an indirect type: plate or tubular heat exchangers. However, skim milk powder is produced by either a "high heat" or a "low heat" procedure. While the low-heat method includes HTST heat treatment, the high-heat method uses parameters of 85–88°C during 15–30 minutes. Such intensive heat treatment is used to produce skim milk powder intended for use in bakery products where a high degree of whey protein denaturation is desired. This high-heat treatment of milk is carried out in a "hot-well" vessel.

PROCESSING STAGE 3: EVAPORATION.

Evaporation in milk powder production is a cheaper way to eliminate water from raw milk prior to drying. Steam consumption is several times higher during spray-drying than during multiple-effect vacuum evaporation (about 10 times if steam recompression is also included in evaporation). Evaporation also increases the average diameter of powder particles in, and prolongs the shelf life of, the final product.

Milk that will be roller dried may be concentrated up to 33–35 % total solids, while for spray-drying, the concentration may be 40–50% total solids.

Higher concentrations than 35% total solids for roller drying will form a thick layer on the rollers, which can cause slower drying and induce irreversible changes in protein, lactose, and fat. Concentrations higher than 50% total solids for spray-drying will increase viscosity to such an extent as to jeopardize "atomization." In order to avoid negative changes in milk components due to high temperatures, evaporation is always carried out in a partial vaccum, thus decreasing the boiling temperatures (45–75°C). To reduce cost, multiple-effect evaporators of the tubular or plate type are used. This means that fresh steam is used only in the first effect, while vapors generated from milk are used in the subsequent effect. Vapors from this effect are used in the next effect, and so on. So, every subsequent effect has a lower boiling temperature, corresponding to a higher vacuum. The leading design in the dairy industry is the falling film tubular evaporator

(Knipschildt 1986). It is composed of vertical tubes 3–5 cm in diameter and 15 m long. They are arranged in a corpus, called calandria. The milk is introduced in the tubes at the top, and the interspace between the tubes is heated by steam (Fig. 17.2) .

Plate evaporators are more frequently used in plants of smaller capacity and in industries where a higher versatility of product assortment is necessary.

A further decrease of energy consumption by evaporation was achieved during the 1970s by introducing two systems for vapor recompression: thermal vapor recompression (TVR) and mechanical vapor recompression (MVR).

In TVR (Fig. 17.2), the product vapor from one calandria is compressed to a higher temperature by steam injection and used in the first effect. The vapor from one calandria is used as the heating medium in the next. In MVR, the heating medium in the first effect is also the vapor generated in the next effects, compressed by a turbocompressor or high-pressure fan to a higher pressure. As in other designs, the heating medium in each calandria is vapor from the previous calandria.

In addition to traditional evaporation, modern

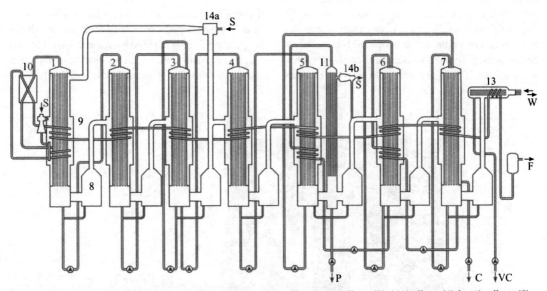

Figure 17.2. Falling film evaporator with TVR: (1) first effect, (2) second effect, (3) third effect, (4) fourth effect, (5) fifth effect, (6) sixth effect, (7) seventh effect, (8) vapor separator, (9) pasteurizing unit, (10) heat exchanger, (11) finisher, (12) preheater, (13) condenser, (14a,b) thermocompressor. F = feed. S = steam. C = condensate. VC = vacuum. W = water. P = product. (Courtesy APV Anhydro A/S)

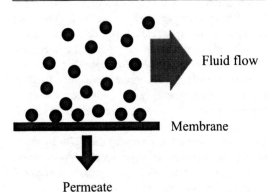

Fluid flow

Membrane

Permeate

Figure 17.3. Cross-flow filtration.

membrane methods such as ultrafiltration (UF), microfiltration (MF), reverse osmosis (RO), and demineralization are used for water removal and/or fractionation in dairy industry. The industrial application of UF, MF and RO (Fig. 17.3) was initiated by the introduction of cross-flow (instead of dead-end filtration) and by the invention of asymmetric membranes (Fig. 17.4). With the new patented asymmetric membranes, only the thin surface layer is an active part of the membrane. Being very thin, it permits much better water flux than earlier membranes. Permeate components that pass the active layer will pass the supporting porous layer easily. Deposit formation (fouling) is markedly decreased and forms only on the surface. UF and MF are membrane methods involving concentration and fractionation in which not only water but also other small molecules pass through the asymmetric semipermeable membrane; RO is a membrane method in which only water molecules pass through the membrane. However, concentrating by RO is cost effective only for up to 20–25% of total solids, so this method is usually combined with evaporation in milk powder production to achieve the lowest cost possible (Carić 1994).

PROCESSING STAGE 4: HOMOGENIZATION

Homogenization is not an obligatory operation in milk powder production, but if applied, it reduces free fat content, improving the final product quality. Free fat adversely affects milk powder solubility and increases susceptability to oxidative changes. By homogenization, free fat is transformed to fat globules: whey protein and casein micelles are adsorbed, and a fat globule membrane is formed (regenerated). In this way the free fat concentration is markedly reduced. Homogenization is carried out in continuous homogenizers common in the dairy industry, in partly concentrated milk or in the final concentrate (after evaporation), at pressures of 5 and 15 MPa.

PROCESSING STAGE 5: DRYING.

Drying is a basic operation in milk powder production. The prolonged shelf life of milk powder, that is, its preservation, is ensured by water removal to such an extent that neither microorganisms nor their spores can develop. This method is widely applied in

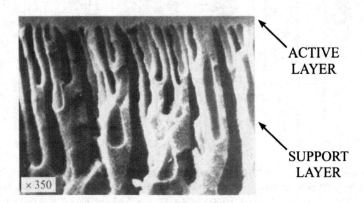

ACTIVE LAYER

SUPPORT LAYER

Figure 17.4. Cross section of asymmetric UF membrane.

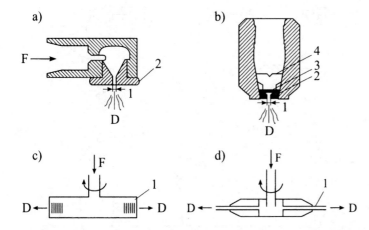

a)

b)

c)

d)

Figure 17.5. Atomizer design. (**a,b**) Pressure nozzle: (1) diameter of orifice, (2) orifice plug (groove), (4) core. (**c,d**) Centrifugal atomizer: (1) vertical slots or spokes. F = feed. D = droplets.

Some common industrial techniques for drying milk are (1) roller drying, at atmospheric pressure or under vacuum; (2) spray-drying, with either centrifugal or pressure atomization (Figs 17.5–17.8); (3) two-stage spray-drying; and (4) three-stage spray-drying (Figs 17.5–17.8).

Spray-drying is a dominant method for drying milk (and other fluid food systems); it results in product that is high in quality and cost effective compared to other methods such as roller drying. So far, all attempts to develop a better drying method have failed. It appears that spray-drying may remain dominant for the foreseeable future. Recent developments attempt to further improve the product quality and cost effectiveness of spray-drying techniques (Carić and Kalab 1987).

Roller drying is still used when milk is dried for specific purposes, such as confectionery (where lactose caramelization is preferred for particular products) or for feed blend production. Direct contact of concentrated milk with the hot roller surfaces during drying induces irreversible changes in milk components, for example, lactose caramelization, protein denaturation, and Maillard's reactions.

Powder characteristics in dried milk powder produced from roller drying, especially solubility, flavor, and appearance are inferior to those produced by spray-drying. The heating medium introduced into rollers is saturated steam at temperatures up to 150°C, which is almost the temperature reached by the milk during drying. A thin film of dry milk is removed by knives and brought by spiral conveyer to a hammer mill, to be pulverized.

In spray-drying, concentrated milk is dispersed

Figure 17.6. Spray drier of laboratory scale. Model Lab 1, APV Anhydro A/S at Faculty of Technology, Novi Sad University, Serbia and Montenegro.

("atomized") into small droplets in a spray-drying chamber, where it is exposed to a hot air flow (Figs 17.5–17.8). Inlet air temperature usually ranges from 180 to 240°C, while outlet air temperature is

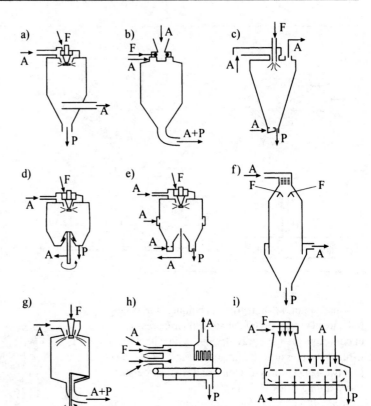

Figure 17.7. Drying chamber designs: (**a**) conical-based chamber with two-point product discharge, (**b**) conical-based chamber with single-point product discharge, (**c**) conical-based chamber with integrated static fluid bed, (**d**) inverted-base coned chamber, (**e**) inverted-base coned chamber with integrated static fluid bed, (**f**) tower-form (nozzle tower) chamber, (**g**) flat-based chamber with product sweeper, (**н**) box chamber with integrated screw conveyor, (**j**) box chamber with integrated conveying band. F = feed. A = air flow. P = product.

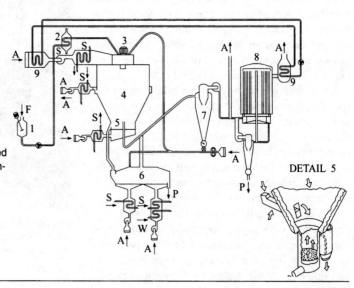

Figure 17.8. Three-stage drying: (1) feed tank, (2) concentrate preheater, (3) atomizer, (4) spray drying chamber, (5) integrated fluid bed, (6) external fluid bed (instantizer), (7) cyclone, (8) bag filter, (9) liquid coupled heat exchanger. F = feed. A = air. S = steam. W = water. P = product. Detail: (5) integrated fluid bed. (Courtesy of APV Anhydro)

DETAIL 5

under 100°C (70–90°C) which is the highest temperature of the milk as well (Carić and Milanović 2002).

Various drying chamber designs are shown in (Fig. 17.7.).

Air and milk flow are concurrent, countercurrent, or mixed (under an angle). There are two different types of atomizing devices: centrifugal (rotary) atomizers and pressure (nozzle) atomizers (Fig. 17.5). In order to achieve wide versatility in powder production, most dryers are now constructed to accommodate both atomizing devices. Milk is dispersed in centrifugal atomizers at 10,000–20,000 rpm or in pressure nozzles at 17–25 MPa. Dispersed milk forms droplets 20–150 μm in diameter. The resulting huge surface area permits quick heat and mass transfer (Carić 2002): (1) heat from hot air to milk (heat transfer) and (2) water from fluid milk droplets to air (mass transfer).

In both centrifugal and pressure atomization, milk droplets take a spherical shape. The final product has globular particles that contain vacuoles of occluded air (Figs 17.9 and 17.10.).

The most important advantages of spray-drying over other techniques are that (1) the drying process is performed at low temperatures and is very short (less than 30 seconds) and (2) the product is of excellent quality with no adverse effects, that is, heat-induced changes (Pisecky 1997).

Further enhancement of spray-dried reconstitution properties was achieved by the introduction of an instantization procedure based on agglomeration (Figs 17.9 and 17.10).

Milk powders are packaged in such a way as to protect the product from moisture, air, light, and insects. Container materials may include paper, multilayer boxes, bags, barrels with a polyethylene layer,

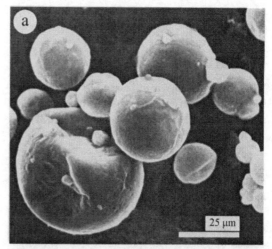
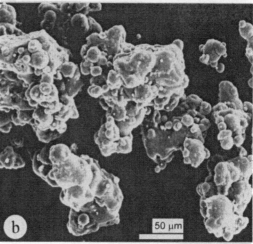

Figure 17.9. SEM (scanning electron microscopy) of spray dried milk powder. (**a**) One-stage dried. (**b**) Agglomerated.

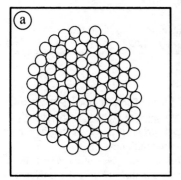
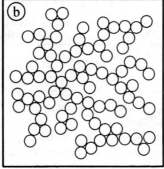

Figure 17.10. Shema of spray dried milk powder. (**a**) One-stage dried. (**b**) Agglomerated.

and cans covered with aluminum foil. It is especially important to protect powder from moisture because of its high hygroscopicity.

FINISHED PRODUCT

Milk powder quality is influenced to a great extent by various factors during processing and storage: (1) manufacturing parameters and procedures, (2) drying techniques, and (3) storage conditions.

The most important factor in this respect is drying technique. Spray-dried milk powder is of superior quality in powder structure, solubility, flowability, and flavor and color, compared with other industrially dried milk powders. There are, however, differences among various spray-drying systems. Properties of single-stage spray-dried powders differ markedly from those of two-stage or three-stage dried powders, that is, agglomerated (instantized) powders. Due to the small quantity of water (10–14 %) to be dried in the second drying stage, the powder gets a cluster-like structure. The void spaces among particles are easily filled up with water during reconstitution (great contact area), resulting in quick solubilization of the product (Figs 17. 9 and 17.10).

Milk is not the only product produced in powder form in the dairy industry. Other milk and dairy related products can be dried: for example, dry dairy beverages, dietetic dry products, dry cultured milk products, whey powder, whey protein products, casein and caseinates, coprecipitates, coffee whitener, dried ice cream mix, infant formulas, and various special blends (Carić 1994).

In sum, the essential processing principle during milk powder production is intense and quick heat and mass transfer. This was made possible by the development of spray-drying techniques in the last century. The main objectives of the process were to develop a product (1) that resembled, after reconstitution, the original as much as possible and (2) that had low production costs and good storing stability. This aim has been achieved by introduction of modern spray-drying techniques. In addition to the main features of spray-drying and instantization technology, the outstanding achievements in concentrating and drying milk or food are the development of multistage vacuum evaporation with thermal and mechanical vapor recompression resulting in better economy, and the introduction of membrane methods, which allowed numerous combinations of dairy-based powders with different compositions.

APPLICATION OF PROCESSING PRINCIPLES

The following table provides specific processing stages and the principle(s) involved in the manufacturing of dried milk.

Processing Stage	Processing Principles	References for More Information on the Principles Used
Standardization	Separation	Eyrich 1997, Bylund 2003
Heat treatment	HTST	Lopez-Fandino and Olano 1999
Evaporation	Heat evaporation	Holmstrom 1999, Carić 2002
Homogenization	High pressure homogenization	Lawrence, Clarke, and Augustin 2001, Bylund 2003
Drying	Spray drying	Straatsma et al. 1999, Carić and Milanović 2002.

GLOSSARY

Agglomeration—basic principle of instantizing milk (and other food) powders. By this method, 10–15 % water is left in the powder and removed later in a device called an instantizer or fluidized bed dryer, which incorporates air among powder particles. The product has a special "agglomerated" structure and dissolves easily. Incorporated air enables water to come into greater contact with powder by reconstitution. The contact surface is much greater than that of classic spray-dried milk powder.

Asymmetric membranes—new type of membrane developed and patented in the 1970s by the University of California at Los Angeles (Loeb and Sourirajan

1964]). Asymmetric membranes are composed of a very thin surface layer (0.1–1.0 μm) with pores of 2–20 μm in diameter and a relatively thick, porous supporting layer (20–100 μm).

Atomizing devices—device used to achieve a high surface-to-mass ratio in dispersed fluid; two atomizing device designs are used: centrifugal (rotary) and pressure (nozzle), enabling a fast heat transfer from hot air to milk droplets and mass (water) transfer from milk droplets to hot air.

Concentrating—operation that, if carried out prior to drying, results in a product higher in quality and lower in cost. For example, the steam consumption per kilogram of evaporated water is about three to six times higher during spray-drying, than that for a double-effect vacuum evaporator. In addition to evaporation, membrane methods are also used for concentrating.

Cross-flow filtration—filtration method developed simultaneously with the advance of asymmetric membranes. The flow of "dead end" filtration is normal to the membrane. The flow of cross-flow filtration is parallel to the membrane, with the permeate "cross flowing" through the membrane. Fouling from the formation of a deposit (sediment) on the membrane is thus reduced. From this technique, membrane methods such as UF, RO, and MF are made available for industrial application.

Drying—a preservation method in which long shelf life of the product is achieved by water removal so that microorganisms cannot develop. The measure of available water is a_w value.

Evaporation—evaporation by thermal energy: the first industrial technique developed for water removal (concentrating the fluid food system); achieved by evaporation in a partial vacuum with a continuous multiple-effect (3–7) vacuum evaporator. In the late 1980s, low cost and effective methods were developed: thermal vapor recompression and mechanical vapor recompression.

Heat and mass transfer—transfer of heat from hot air to milk and transfer of mass from milk droplets to air. Spray-drying disperses milk through small orifices by centrifugal or pressure atomizer to form a fine uniform particle with a diameter of 20–150 μm. The resulting large surface area enables intensive heat and mass transfer. Evaporated water from milk is removed simultaneously by hot air.

Instantizer (fluidized bed dryer)—a specially constructed dryer, where the last phase of drying during the instantization procedure takes place. It has a vibrating bottom where air passes through the thin layer of wet powder, removing water from agglomerated particles.

Instant characteristics—in corresponding equilibrium, wettability, penetrability, sinkability, dispersibility, and rate of dissolving. These properties of instant milk powder allow better reconstitution than for regular powder. Instantization improves the rate and completeness of powder reconstitution without changing its solubility.

Microfiltration (MF)—a membrane separation process where water and small/large molecules (proteins and fat) pass through the membrane, with bacteria concentrated in the retentate.

Milk powder structure—the shape of milk powder particles depends on the drying techniques. Milk powder produced by roller drying has particles of compact structure and irregular shape with no occluded air. The structure of spray-dried powder particles is spherical, containing one or more vacuoles of occluded air. Instant milk powder has agglomerated (clustered) structure, where more air is incorporated among (between) the powder particles.

MVR—mechanical vapor recompression.

Particle density—corresponds to a total volume of 1 cm^3.

Powder flowability—the ability of powder to flow; measured as the time in seconds necessary for a given volume of powder to leave a rotary device through slits.

Powder solubility—the ability of powder to dissolve in water.

Reverse osmosis (RO)—a membrane separation process where only water passes through the semipermeable membrane (permeate is pure water), while milk is concentrated (retentate).

Roller drying—one of two techniques for drying milk on an industrial scale (other is spray-drying). Roller drying is rarely used, except for particular purposes and/or by low capacity plants.

Specific dry milk products—spray-dried, milk-based products tailored to the needs of diverse consumer groups. There are various dried milk products for athletes, infants (formulas), reducing diets, convalescents, tailored dried-milk–based meals, and so on.

Spray-drying—one of two techniques for drying milk on an industrial scale (other is roller drying). At present, spray-drying is the dominant technique for drying milk, dairy products, and other edible fluid products. This technique includes dispersing ("atomizing") evaporated milk into fine droplets and exposing them to a flow of hot air in a spray-drying chamber, where rapid heat and mass transfer take place.

TVR—thermal vapor recompression.

Ultrafiltration (UF)—a membrane separation process where water and other small molecules (lactose and salt) pass through the semipermeable membrane (permeate), while macromolecules are concentrated (retentate).

Whey powder—whey, a by-product of cheese production, dried by the same drying techniques as are used for milk. Spray-drying of whey includes an additional operation: crystallization of lactose.

Whey protein powders—powders of different composition and properties obtained using various methods of fractionation prior to spray-drying for whey treatment.

REFERENCES

Bylund G. 2003. Centrifugal separators and milk fat standardisation systems. In: Dairy Processing Handbook, 91–115. Lund, Sweden: AlfaTetra Processing systems AB.

Carić M. 1993. Concentrated and dried dairy products. In: YH Hui, editor. Dairy Science and Technology Handbook. Vol. 2, Product Manufacturing, 257–300. New York: VCH Publishers.

Carić M. 1994. Concentrated and Dried Dairy Products, 249, New York: VCH Publishers.

Carić M. 2002. Milk powders: Types and manufacture. In: H Roginski, JW Fuquay, PF Fox, editors. Encyclopedia of Dairy Sciences, vol.1, 1869–1874, Academic Press.

Carić M, M Kalab. 1987. Effects of drying techniques on milk powders quality and microstructure: A review. Food Microstructure 6:171–180.

Carić M, S Milanović. 2002. Milk powders: Physical and functional properties of milk powders. In: H Roginski, JW Fuquay, PF Fox, editors. Encyclopedia of Dairy Sciences, vol.1, 1874–1880. Academic Press.

Eyrich L. 1997. Standardization for improved profitability. Scandinavian Dairy Information 11(2).

Holmstrom P. 1999. The component that revolutionised evaporation. Scandinavian Dairy Information (2): 18–19.

Knipschildt M. 1986. Drying of milk and milk products. In: RK Robinson, editor. Modern Dairy Technology , vol. 1, 131–234, London: Elsevier.

Lawrence A, PT Clarke, MA Augustin. 2001. Effects of heat treatment and homogenisation pressure during sweetened condensed milk manufacture on product quality. Australian Journal of Dairy Technology 56(3): 192–197.

Loeb S, S Sourirajan. 1964. U.S. Patent 3,133,132.

Lopez-Fandino R, A Olano. 1999. Selected indicators of the quality of thermal processed milk. Food Science and Technology International 5(2).

Masters K. 1985. Spray-drying. In: R Hansen, editor. Evaporation, Membrane Filtration and Spray-drying, 299–345. Vanlose: North European Dairy Journal.

Peebles DD, DD Clary, Jr. 1955. Milk treatment process. U.S. Patent 2,710,808.

Peebles DD. 1958. Dried milk product and method of making same. U.S. Patent 2,835,586.

Pisecky J. 1997. Handbook of Milk Powder Manufacture. Copenhagen: Niro A/S.

Singh H, DF Newstead. 1992. Aspects of proteins in milk powder manufacture. In: PF Fox, editor. Advanced Dairy Chemistry, 2nd edition. Vol. 1, Proteins, 735–765. London: Elsevier.

Straatsma J, G van Homvelingen, AE Steenbergen, P de Jong. 1999. Spray-drying of food products. II. Prediction of insolubility index. Journal of Food Engineering 42 (2): 73–77.

18
Fats: Mayonnaise

S. E. Duncan

BACKGROUND INFORMATION

INTRODUCTION

Mayonnaise is a creamy, pale yellow, mild-flavored food product frequently used in preparation of salads, sandwiches, and many other food products. Although consisting of relatively few ingredients and processing steps (Fig. 18.1), successful formulation and processing requires an understanding of the role of each ingredient and of the critical processing steps in creating the delicate structure.

Mayonnaise is a unique emulsion. The major component, oil, is dispersed throughout the lesser amount of the continuous aqueous phase. The structure of mayonnaise is easily disrupted because of this unusual relationship. Integration of processing and chemistry is essential to understanding the formation and stabilization of this product.

The Code of Federal Regulations (CFR 21.169.140) specifically describes mayonnaise and the ingredients permitted in the manufacture of the product (CFR 1993). Mayonnaise is a semisolid food in which vegetable oil(s) are emulsified with vinegar and/or lemon or lime juice, and egg-yolk containing ingredients, which may include egg yolks (liquid, frozen, dried), whole eggs (liquid, frozen, dried), or any of the yolk products in combination with liquid or frozen egg white. Mayonnaise must contain not less than 65% by weight of vegetable oil and 2.5% by weight of acetic or citric acid, as provided by vinegar or lemon/lime juices. Commercial mayonnaise generally contains 77–82% vegetable oil. Optional ingredients in the formulation include salt; a nutritive carbohydrate sweetener (sucrose); spice or natural flavoring; monosodium glutamate; sequestrants, such as ethylenediaminetetraacetic acid (EDTA), to preserve color and/or flavor; citric or malic acid; and crystallization inhibitors (i.e., oxystearin, lecithin, polyglycerol esters of fatty acids) (Table 18.1).

Saffron and turmeric, or any spice or flavoring that imparts a color simulating the color imparted by egg yolk, are not permitted in mayonnaise. Citric

The information in this chapter has been derived from a chapter in *Food Chemistry Workbook*, edited by J. S. Smith and G. L. Christen, published and copyrighted by Science Technology System, West Sacramento, California, ©2002. Used with permission.

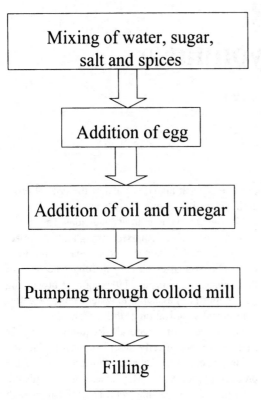

Mixing of water, sugar, salt and spices

⇓

Addition of egg

⇓

Addition of oil and vinegar

⇓

Pumping through colloid mill

⇓

Filling

Figure 18.1. Flowchart for the manufacture of mayonnaise.

starchy paste, about double the amount of acid, three times the amount of sugar, one-third the amount of salt, less than one-half the amount (not less than 30%) of vegetable oil, and a minimum egg yolk level (CFR 1993, Krishnamurthy and Witte 1996). The starch paste is prepared from native or chemically modified starches such as tapioca, wheat, rye, or cornstarch. Chemical modification of the starch can help improve physical stability against syneresis or acid breakdown. The starch paste, with vinegar and spices mixed in, is then mixed with a modified mayonnaise base to obtain the spoonable salad dressing. In addition to salt and sugar, optional ingredients include nonnutritive sweeteners, spices, monosodium glutamate, thickeners and stabilizers, sequestrants, and crystallization inhibitors. This chapter focuses strictly on the ingredients and processing of full fat mayonnaise.

The pH of mayonnaise ranges from 3.6 to 4.0 (Krishnamurthy and Witte 1996). Acetic acid, from vinegar, is the predominant acid, representing 0.29 to 0.5% of the total product. The sugar and salt in the formulation are dissolved in the aqueous phase. The aqueous phase contains 9–11% salt and 7–10% sugar, contributing to a relatively low water activity (a_w) of 0.929. Mayonnaise is studied in food processing to develop an understanding of the interaction of the primary ingredients (oil, vinegar and egg) in forming an emulsion.

or malic acid are limited to an amount not greater than 25% of the weight of the acids of the vinegar or diluted vinegar (calculated as acetic acid). All ingredients must be safe and suitable for food use.

Spoonable salad dressing differs from mayonnaise in that it contains a cooked or partially cooked

DESCRIPTION OF THE MAYONNAISE EMULSION

An Oil-in-Water Emulsion

Mayonnaise is a difficult emulsion to prepare (Tressler and Sultan 1982, Weiss 1983). Generally,

Table 18.1. Ingredients Formula for Mayonnaise

Ingredient[a]	Weight %	Emulsion phase
Vegetable oil	65–80	Oil
Egg yolk	7.0–9.0	Emulsifier
Vinegar (4.5% acetic acid)	9.4–10.8	Water
Sugar	1.0–2.5	Water
Salt	1.2–1.8	Water
Spices		Water
Mustard flour	0.2–2.8	
White pepper	0.1–0.2	
Oleoresin paprika, garlic, onion spices	0.1	
Water	To make 100%	Water

[a]Weight % of emulsifying ingredients is inversely related to weight % of oil in formula.

an emulsion forms with the major component of the formulation existing in the continuous phase. Minor components are dispersed throughout the continuous phase and compose the dispersed phase. However, in mayonnaise, the major component (oil) is forced, as fine droplets, to disperse throughout the lesser amount of the continuous aqueous phase. Commercially manufactured mayonnaise contains about 80% lipid, as does margarine (Krog et al. 1985). However, mayonnaise is an oil-in-water (o/w) emulsion, whereas margarine is a more stable water-in-oil (w/o) emulsion. When the mayonnaise emulsion breaks, the emulsion reverts back to a stable condition, where oil becomes the continuous phase and the aqueous portion becomes discontinuous (Tressler and Sultan 1982). Under these conditions, the aqueous phase does not disperse, and creaming (phase separation) readily occurs.

The high amount of oil in the product does not favor formation of an o/w emulsion. An emulsifier is needed to stabilize this unique dispersion. An emulsifier works at the surface of two otherwise immiscible liquids and functions to reduce the interfacial tension between the two phases by reinforcing the contact surface between them (Potter and Hotchkiss 1995). An adequate amount of an effective emulsifier is needed to coat the oil droplet during manufacture of the emulsion. The smaller the oil droplets in the aqueous phase, the larger the surface area of the oil droplets. Finer dispersions will require more emulsifier to surround the oil droplets and stabilize the emulsion.

In mayonnaise manufacturing, addition of synthetic emulsifiers is not permitted. The only source of emulsifiers for stabilization of mayonnaise is obtained from egg yolk (Potter and Hotchkiss 1995). Therefore, the egg yolk must be completely dissolved in the water phase before addition of the oil begins in order to achieve sufficient emulsification efficiency. Effective emulsifiers for oil-in-water emulsions, such as mayonnaise, are hydrophilic emulsifiers (Verlags 1994). Lecithin, a low-molecular-mass surfactant that occurs naturally in egg yolk, is an effective emulsifier. The oil droplets may also be stabilized by high-molecular-mass surfactants, such as proteins, found in egg white or egg yolk.

Physical Characteristics of the Emulsion

Functional properties such as spreadability, mouthfeel, emulsion stability, and salt release are affected by the dispersal of oil droplets in the aqueous phase (Fig. 18.2; Heertje 1993a). A maximum of 74% of the total volume of an ideal emulsion, in which all particles are of the same size, can be the dispersed phase when the droplets are spherical within the continuous phase (Depree and Savage 2001). Although the minimum of oil permitted in mayonnaise is 65%, most commercially manufactured mayonnaise has 77–82% oil (Weiss 1983). The more oil dispersed in the emulsion, the stiffer it will be (Weiss 1983). At 65% oil, the mayonnaise is thin whereas at 80–84%, the product is very thick and heavy bodied. At the high levels of oil usage, the product may become too rubbery and dry.

The change in mayonnaise texture and mouthfeel can be attributed to the dispersion of fine oil droplets within the aqueous phase and the interactions be-

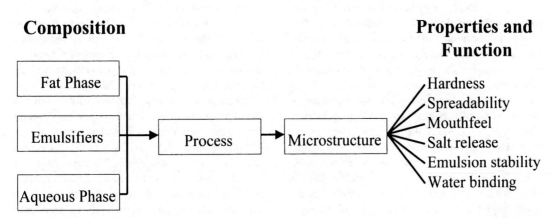

Composition

Properties and Function

Fat Phase

Emulsifiers → Process → Microstructure

Aqueous Phase

Hardness
Spreadability
Mouthfeel
Salt release
Emulsion stability
Water binding

Figure 18.2. Relation between composition, processing, structure, and function of fat spreads. (Heertje 1993a)

tween oil droplets. The emulsification system will become overloaded with greater than 84% oil. The droplets are too tightly packed, with a very thin film between them. A weak gel is formed by flocculation of oil droplets: interactions between droplets are dependent on van der Waals attractions, balanced by electrostatic and steric repulsion (Depree and Savage 2001). If the attractive forces are too strong, the droplets will coalesce; strong repulsive forces allow the droplets to slide by, and the product demonstrates a low viscosity and is prone to "creaming." Mechanical shock can easily cause oil droplets to coalesce and the emulsion to break. Mayonnaise is formulated to give maximum stability against coalescence.

The viscosity of emulsions depends on the volume fraction and the properties of the continuous phase (Krog et al. 1985). The viscosity of the dispersed oil phase is seemingly inconsequential if the droplets behave as rigid spheres. However, if the droplets become deformed because of tight packing (at greater than 80% oil), the viscosity of the droplets plays a significant role in the overall viscosity of the emulsion. The rheological behavior of mayonnaise is characterized as viscoelastic (Holcomb et al. 1990).

Microstructure of the Emulsion

The functional properties, especially the rheological and sensory properties, of the product are linked to the microstructure of the product (Fig. 18.2). Product composition and processing conditions are determinative factors in the formation of microstructure.

The mayonnaise emulsion is difficult to examine by ultrastructure techniques because of the high lipid content and fragility of the interfacial film surrounding the oil droplets (Holcomb et al. 1990). Successful techniques have demonstrated that mayonnaise contains lipid droplets that are tightly packed together. The high volume of oil causes the formation of a honeycomb structure of closely packed droplets (Heertje 1993b). An important aspect of oil droplets is their size and homogeneity of size distribution. Many droplets are spherical, but they vary considerably in droplet size, with smaller droplets (about 0.2 μm) packed in the interstices between larger droplets (Fig. 18.3). Droplets range in size from 2 to 25 μm (Langton et al. 1999). The average droplet size is approximately 2.2 μm (Krog et al. 1985).

Tight packing causes deformation of droplets in

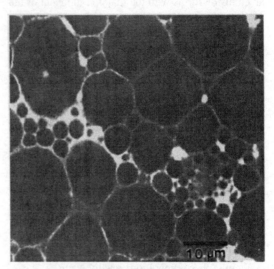

Figure 18.3. Distorted oil droplets in mayonnaise. CLSM. Fluorescent staining of the continuous water phase and the interface by Nile blue. (From Heertje 1993b)

hexagonal shape, contributing to a honeycomb appearance (Langton et al. 1999). Distorted droplets, polyhedral in shape, are found more frequently in products manufactured with > 80% oil (Krog et al. 1985). A high degree of distorted droplets may influence the stability and viscosity of the emulsion.

The aqueous phase surrounding the droplets is continuous, separating the oil droplets (Holcomb et al. 1990). The optional ingredients in the formulation (spices, sugar, salt, etc.) that enhance the flavor of the product are found in the continuous aqueous phase. Fragments of egg yolk granules are the predominant structure in the continuous phase. Observed as electron-dense particles, these fragments adhere to the interfacial film and to each other, forming a protein network. The protein network increases the viscosity of the mayonnaise and enhances the stability of the emulsion.

The surface film of the droplets is formed of coalesced low-density lipoproteins of egg yolk and microparticles from the yolk granules (Holcomb et al. 1990). The interfacial film is seen as a thin, electron-dense band when viewed under the high magnification of an electron microscope. The high degree of interfacial elasticity created by the interfacial film structures on the oil droplet surfaces contributes greatly to the stability of the droplets (Holcomb et al. 1990, Krog et al. 1985). Thus, when that film is thin or weak and a mechanical force is exerted, the droplets may easily coalesce.

RAW MATERIALS PREPARATION AND INGREDIENT FUNCTIONALITY

The ingredients used in the manufacture of a stable mayonnaise emulsion are very important. The proportions of egg and oil are balanced to obtain the desired body, viscosity, and texture.

OIL

Oil is the largest contributor to ingredient cost of mayonnaise manufacturing, based on use volume (Depree and Savage 2001). Economics must be balanced with the need for product stability and quality. Reducing the oil content reduces the investment, but reducing the proportion of oil also reduces the potential number of oil droplets and affects mayonnaise quality. Reducing the proximity of oil droplets weakens the interactions, and the emulsion is less stable. This may be overcome in a medium to low fat emulsion by reducing oil droplet size, which also contributes to a "creamier" appearance.

The quality of the oil is very important to the flavor of the product and the stability of the emulsion since such a high percentage of the product is composed of oil. Cottonseed, soybean, sunflower, safflower, corn, and olive oil are all used in the manufacture of mayonnaise (Krishnamurthy and Witte 1996). Unhydrogenated soybean oil is most commonly used because it is less costly. Highly saturated oils (e.g., palm oil) or peanut and similar oils that solidify at refrigerated temperatures are seldom used because they can cause the emulsion to break at cold temperatures (Depree and Savage 2001). Flavor quality is usually improved by using a deodorized oil. The other oils may be used if a nutritional claim is of interest or a unique flavor is desired. As little as 10% olive oil has a noticeable effect on flavor, contributing a unique gourmet flavor.

Unsaturated oils have a tendency to oxidize, which will affect the flavor and quality of the mayonnaise. Soy oil is typically rich in natural antioxidants, especially tocopherols, that help protect the oil from oxidation. Corn and sunflower oils, which have greater amounts of linolenic acid than soybean oil, are more susceptible to oxidation. Oils used for manufacture of mayonnaise should have no off flavor from oxidation. Iodine values may be used to indirectly predict the resistance of the oil to oxidation (Newton 1989). Alternative direct methods include the active oxygen method and the Schall oven test.

Soybean and cottonseed oils will form crystals at cold temperatures (Weiss 1983). These crystals will break the mayonnaise emulsion, causing the oil fraction to separate from the other ingredients (Jones and King 1993, Weiss 1983). Winterization of the oils will prevent this problem. Winterization involves chilling the oil and filtering out solid fat crystals (Jones and King 1993). A cold test is completed to determine the degree of winterization the oil has undergone (Newton 1989). The cold test value is reported as the length of time an oil sample can sit in an ice bath before cloudiness appears. Freezing of the aqueous phase will also cause the emulsion to break.

EGGS

Eggs are the most expensive ingredient on a per pound basis, and they contribute significantly to product performance and flavor quality. Eggs are also the most complex and least understood ingredient in the product. Egg quality can only be determined by performance testing of the egg product (Tressler and Sultan 1982). Low egg solids content can cause failure of the emulsion. This is correctable by increasing the amount of egg material proportionately. The amount and type of egg solids have an effect on emulsion viscosity and strength. Therefore, experience at manufacturing mayonnaise is very important when determining the amount of eggs to add to the formulation.

One critical element in the process that can only be determined by performance testing is the emulsifying capacity of the egg yolk (Holcomb et al. 1990). The egg yolk is a rich source of lecithin (a phospholipid) and proteins and lipoproteins, including lipovitellin, lipovitellinin, and liviten, which contribute to the emulsifying capacity of the egg yolk (Depree and Savage 2001). These emulsifying compounds in the egg yolk contribute only about 10% of the yolk weight. Commercial yolk contains about 43% solids and constitutes approximately 40% of the whole egg. Whole egg is 26% solids. Egg solids can be fortified by addition of more yolk than is normal. A common level of fortification is 33% solids. The protein in the egg white, which gels on addition of the acid component, assists in emulsification by forming a solid gel structure. The aqueous phase becomes more rigid as more emulsifier and colloidal solid matter is dispersed in this phase. If the egg content is too low, a stable emulsion cannot form.

Eggs used for manufacture of mayonnaise may be fresh, frozen, or dried yolks or whole eggs (Depree and Savage 2001). However, processing of egg yolks can disrupt egg yolk structure and reduce the emulsifying properties provided by the egg. Mayonnaise made with freshly broken eggs is usually weak in body. Frozen egg yolks will gel irreversibly on freezing at $-6°C$, becoming indispersible and useless for mayonnaise production. Mechanical processing (such as homogenization or colloid milling) or addition of enzymes (such as proteases and phopholipases) can inhibit yolk gelation. Addition of 10% salt, 10% sugar, or egg white will resolve the problem by permitting only partial gelation and is the generally accepted method. Extended storage of frozen salted or sugared yolk causes changes in the quality and function of the yolk. The thawed egg will be thick but dispersible, and the resulting mayonnaise will be thick and creamy. Dried eggs will disperse readily in the aqueous portion of the emulsion, resulting in a thicker product than one obtained from frozen egg at the same solids content. Pasteurization of yolks will not affect the emulsifying properties.

Pasteurization of eggs used for mayonnaise is recommended (Weiss 1983). This is to prevent contamination of the product with *Salmonella*. Pasteurization of salted yolk or salted fortified egg does not affect the emulsification properties of the egg. Egg yolk is resistant to heat below 65°C, but at temperatures above that point, denaturation of the egg yolk may start (Verlags 1994).

In addition to the effects on viscosity and emulsion stability, the egg yolk contributes color to the mayonnaise (Weiss 1983). The primary source of yellow color in the product comes from the egg yolk. No other coloring material is permitted. The color of the yolk is primarily dependent on the feed given the laying chickens. The oil does not contribute to the yellow color but may contribute a greenish cast, especially with safflower and olive oil (Krog et al. 1985).

ACIDS

Distilled vinegar is the most common source of acid used in the manufacture of mayonnaise because it is less costly (Weiss 1983), but citric and malic acids may be used also (CFR 1993). The vinegar must constitute not less than 2.5% by weight, calculated as acetic acid (CFR 1993). Vinegar strength is measured by "grain" (Weiss 1983). Vinegar of 100 grains strength is 10% acetic acid. Industrial vinegar is usually 100 or 120 grains. Vinegar flavor, and the subsequent mayonnaise flavor, varies with the levels of ethyl acetate and other flavor components produced as intermediates in the reaction that converts ethyl alcohol to acetic acid. Lemon or lime juice may be used as the acid source, in place of vinegar, at the same percentage by weight, and must be calculated as citric acid. Lemon or lime juice add flavor and are frequently used for gourmet products. Cider, malt, and wine vinegars, more costly than distilled vinegar, may also be used for gourmet products. They contribute unique flavors at small amounts. In excess, the flavor contribution is too extreme, resulting in a spoiled flavor. The dark color of the vinegar also is imparted to the mayonnaise. Charcoal filtration may be used to bleach the specialty vinegar, but this may also remove some flavor notes.

The addition of acids decreases the pH of the emulsion and affects the structure (Depree and Savage 2001). When the pH is near the isoelectric point of egg yolk proteins, the charge on the proteins is minimized, allowing the proteins on the droplet surface to be in close contact. The resulting flocculation increases the viscoelasticity and stability of the mayonnaise.

MUSTARD

Two types of mustard flour, white and brown, are commonly used in mayonnaise production for flavor contribution and to assist with emulsification (Depree and Savage 2001). Mustard flour possesses some emulsifying properties, which depend on the type of mustard used, the balance of the various ingredients in the formulation, and the process by which the product is prepared (Krishnamurthy and Witte 1996, Weiss 1983). It is most effective when added with egg yolk. Specks from the mustard flour may be visible in the mayonnaise. Mustard oil, obtained from mustard seed, does not contribute to color of the product to the same extent as mustard flour and may not function as an emulsifying adjunct.

The flavor contribution of mustard varies by ingredient selection as well. White mustard, which is odorless, is hot to the taste, whereas brown mustard has a sharp odor. Typically the two mustard varieties are proportionally blended to achieve the desired flavor and aroma levels. The "bite" imparted by mustard is attributed to a hydrolyzed glycoside, allyl isothiocyanate, in the mustard oil. Mustard oil retains its original flavor potency longer and does not contribute specks.

SALT, SUGAR, SPICES

The remaining ingredients, such as paprika, salt, and sugar, contribute to a balanced, smooth, rich flavor (Tressler and Sultan 1982). These ingredients also provide some physical stability and inhibition of microorganisms (Depree and Savage 2001). A tight emulsion results in mild flavor. A weak emulsion emphasizes sweetness, tartness, and saltiness, making a poorly balanced flavor especially apparent.

Viscosity increases with increasing salt concentration, up to 15%, in egg yolk (Depree and Savage 2001). Salt improves the characteristics of mayonnaise in three ways: (1) The surface-active materials of egg yolks become more available because salt assists in dispersing the egg yolk granules. (2) Salts neutralize the charges on proteins, which permits increased adsorption of proteins to the oil droplet surface and increases the strength of the droplet coating. (3) Adjacent oil droplets interact more strongly because of the neutralization of the charge on the droplet surface. These contributions help stabilize the emulsion, even if the pH of the mayonnaise is different from the isoelectric point of the egg yolk proteins. However, too much salt can adversely affect emulsion stability by causing egg yolk proteins to aggregate in the aqueous phase rather than at the surface of the lipid droplet. Sucrose does not function as salt does; it actually weakens the interaction between lipid droplets, possibly by shielding reactive groups, with a resulting decline in viscosity.

PROCESSING OF MAYONNAISE

EQUIPMENT

Most mayonnaise manufactured today is still completed as a batch or continuous batch operation. The intricacies of obtaining a consistent, stable emulsion still mean that mayonnaise production is more of an art than a science (Krishnamurthy and Witte 1996). Many factors are well understood independently; however, the interrelationships between these factors and the variability surrounding them is not as clear. Individual experience in manufacturing mayonnaise is helpful in yielding success, and many

major manufacturers of mayonnaise use proprietary techniques (Krishnamurthy and Witte 1996, Weiss 1983).

The equipment used for manufacturing mayonnaise must be stainless steel (Tressler and Sultan 1982, Weiss 1983). The vinegar will corrode ordinary steel and aluminum. Of primary importance in the process is some form of intensive mixer that will disperse the oil into fine droplets. Small batch operations may be completed in a small planetary mixer, such as a Hobart mixer, equipped with a paddle. This is frequently the case for manufacture of mayonnaise in gourmet restaurants. For commercial manufacture of mayonnaise, a colloid mill and other continuous-flow emulsifying mixers are used (Fig. 18.4).

The Dixie-Charlotte system, with capacities ranging from 15 to 200 gallons/batch, is most widely used for commercial mayonnaise (Krishnamurthy and Witte 1996, Tressler and Sultan 1982, Weiss 1983). Final volume is 60–1200 gallons/hour of finished product. The system is comprised of two Dixie mixers and a Charlotte colloid mill connected by appropriate piping, valves, and a rotary displacement

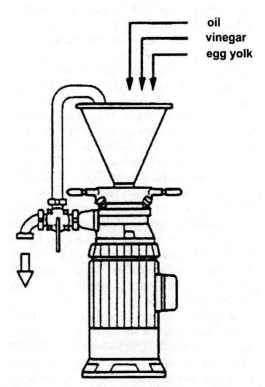

oil
vinegar
egg yolk

Figure 18.4. Toothed colloid mill. (Verlags 1994)

pump. The system is arranged so that one mixer is feeding the mixed mayonnaise formulation to the mill while another batch is being mixed in the other mixer. As one mixer is emptied, the mix from the other is beginning to be pumped, thus providing the continuous-flow/batch operation. The Dixie mixer is a deep circular tank fitted with three turbine mixers mounted side-by-side on a horizontal shaft near the bottom of the tank. The shaft is turned by a variable speed motor. The mayonnaise prepared in the mixer is complete but coarse, requiring no further processing, but the emulsion will be soft and similar to one made with a planetary mixer. The creamy texture of commercial mayonnaise as recognized today is obtained by pumping the soft, coarse emulsion through the Charlotte colloid mill.

MIXING

The mayonnaise mix preparation is started by adding mayonnaise from the previous run to a level in the Dixie mixer to reach the mixer shaft, giving the turbine blades a heavy material to work against while shearing the oil into fine droplets (Tressler and Sultan 1982, Weiss 1983). The fine droplets, in a loosely aggregated network like a foam, contribute to the special mouthfeel of mayonnaise (Depree and Savage 2001). Water (approximately one-third of the water phase), salt, flavors, sweeteners, seasonings, and optional acidulants are mixed to make a slurry (see Fig. 18.1; Krog et al. 1985, Verlags 1994). Egg is mixed in with low-speed agitation (Tressler and Sultan 1982, Weiss 1983). The fluid ingredients should be chilled to between 10 and 16°C when the high-speed colloid mill is used because the temperature will rise about 6°C during milling. Fluid ingredients for mayonnaise manufactured in small planetary mixers should be about 16–21°C. The addition of the egg to the water phase is important to allow the low hydrophilic properties of the egg yolk to function when the oil is added. This will prevent a phase inversion of the emulsion. The good mayonnaise operator, with careful observation and experience, can correct for egg performance variables by a slight modification of the process.

The sequential addition of vinegar gives a better product viscosity than is obtained with complete addition of vinegar at the process beginning. When oil and vinegar are added simultaneously, a water-in-oil emulsion with a viscosity similar to that of the oil from which it is made is formed (Krishnamurthy and Witte 1996, Weiss 1983). When vinegar is added sequentially, small oil droplets are formed, resulting in a more stable emulsion.

ADDITION OF VINEGAR AND OIL

Oil and vinegar are pumped or gravity fed from the supply tanks into the mixer (Tressler and Sultan 1982, Weiss 1983). Oil is slowly added in a thin stream initially (Verlags 1994). The mayonnaise is thin, and the first oil particles emulsified are quite large (Paul and Palmer 1972). The rate of oil addition is gradually increased as the mayonnaise starts to thicken (Verlags 1994). This prevents the mayonnaise from getting too thick for pumping by the colloid mill. As the oil level increases in the emulsion, the dispersed droplets become smaller and the mayonnaise becomes stiffer (Fig. 18.5; Paul and Palmer 1972). The addition of vinegar at any stage of emulsification causes coalescence of some oil droplets, and the mayonnaise temporarily becomes thinner. The addition of more oil is readily accomplished once the emulsion is started. The vinegar portion of the water phase is added in between additions of oil and, particularly near the end, one portion of vinegar is added between two portions of oil (Verlags 1994). However, addition of the egg yolk with addition of the oil phase leads to an unstable emulsion with a tendency for oil separation.

PUMPING AND MILLING

When the ingredients are mixed thoroughly, as observed by the experienced operator, the product is pumped to the colloid mill, where additional emulsification occurs (Holcomb et al. 1990). The colloid mill is a mechanical device with a high-speed rotor (3600 rpm) and a fixed stator (Krishnamurthy and Witte 1996, Weiss 1983). The mix cannot sit in the mixer because the mix may gel (Tressler and Sultan 1982, Weiss 1983). The longer the mix is held in the mixer prior to milling, the softer the final product will be. Timing is critical to balance the mix time and pumping time between the two Dixie mixers. As the mix is emptied from the mixer into the shaft line, adequate mayonnaise is retained in the bottom of the mixer to seed the next batch.

The colloid mill is operated at a rotational speed of approximately 3600 rpm (Tressler and Sultan 1982, Weiss 1983). The size of the mill opening influences the size of the oil droplets. The correct mill opening to yield the desired product characteristics is determined by trial and error. The most effective

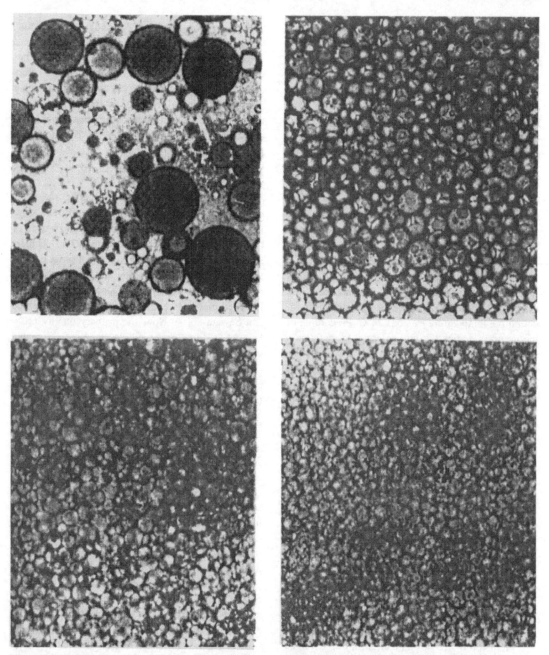

Figure 18.5. Mayonnaise. Showing the change in emulsion as oil is added. Magnification approximately 200x. Clockwise from upper left: (1) coarse emulsion formed after addition of 1 tablespoon oil (vinegar and seasonings were added to the egg yolk before the oil was added), (2) after addition of 1/4 cup oil, (3) after addition of 3/8 cup oil (oil spheres becoming smaller and mayonnaise stiffer), (4) after addition of 1/2 cup of oil. (Paul and Palmer 1972)

mill opening is the smallest that will not result in a broken emulsion. Making the oil droplets too small will increase the total surface area of the oil droplets, exceeding the limits of the emulsifying agents present (Krog et al. 1985). Product formulation and the emulsifying capacity of the egg yolk are factors in determining the optimal mill opening (Holcomb et al. 1990). The optimal mill opening is usually within the range of 25–40 mm. The high shear at low velocity results in a reduction in particle size and the development of the expected texture of commercial mayonnaise. The clearance between the rotor and the stator influences the amount of shear imposed, the viscosity of the final product, and the throughput of the mill.

Some mill heads are jacketed for circulation of a coolant (Tresller and Sultan 1982, Weiss 1983). This is to maintain the temperature of the product at less than 24°C. Emulsion failure will result if the product output temperature exceeds 24°C. Precooling of the liquid ingredients is still essential to maintain a sufficiently low temperature.

In a continuous production line, a dispersion of dry ingredients in water is initially prepared in a mixing tank at room temperature (Langton et al. 1999). The dispersion is mixed in-line with egg yolk and then dosed into the emulsification cylinder using a dosing pump. The emulsification cylinder includes a rotating shaft with pins and inlets for dosing the oil and vinegar placed at the beginning and end of the cylinder. Rotation speed can be adjusted. Oil is added initially into the aqueous mixture, followed with vinegar, resulting in a crude emulsion. The visco-rotor, which is a colloid mill, produces the final emulsion, with fine oil droplets. A scrape surface heat exchanger is used to cool the emulsion before filling.

FILLING

The emulsion is still flowable after milling but must be pumped into the appropriate containers quickly (Tressler and Sultan 1982, Weiss 1983). The emulsion will set up into a semisolid gel after a certain length of time. The time required for gelling is dependent on several interrelated factors associated with formulation, equipment, and the procedure. If the gel is disturbed, it will become soft even though it will gel again.

Retail packaging is usually glass or polyethylene plastics. Wholesale packaging may range from single serve (1 tablespoon) for food service distribution to one- to five-gallon poly packaging for institutional trade. Packaging choices consider the convenience to the user, the costs of material and distribution, and the shelf life of the product. Glass offers greater barrier protection from oxygen than many plastics, thus providing better protection against oxidation of the product. A minimum headspace in the container is recommended to reduce oxidation.

FINISHED PRODUCT

Historically, in homemade mayonnaise and early commercial mayonnaise, reduced quality and spoilage was attributed to creaming and coalescence of oil droplets (Depree and Savage 2001). Understanding of the physical and chemical processes involved in emulsion formation and stabilization has resulted in improved product stability and in shelf life measured in months instead of weeks (Depree and Savage 2001). Now, the primary quality problems related to storage and spoilage are associated with emulsion stability and oil quality. Commercially processed mayonnaise is stable for a reasonable time period (six months or more) but is classified as a semiperishable product (Krishnamurthy and Witte 1996).

EMULSION STABILITY AND TEXTURAL QUALITY

One sign of product aging is a thinning or less viscous product. Phase separation and thinning occur more readily with mechanical shock. Phase separation also occurs with exposure to low temperatures. Rapid addition of oil, unregulated agitation during emulsification, high storage temperatures, and excessive agitation during distribution also can contribute to emulsion destabilization.

Emulsion stability can be assessed in storage at elevated temperatures, by resistance to centrifugation, or by measuring the degree of creaming after 24 hours in mayonnaise diluted with an equal volume of water. (Depree and Savage 2001). These methods give an indication of the relative stability of different emulsions but are not correlated with practical storage times.

Measuring the texture or consistency of mayonnaise is another valuable quality measurement. A special viscometer with a weighted, perforated plunger is used to measure viscosity. The plunger falls through the sample, and consistency is reported in seconds (Krishnamurthy and Witte 1996). Penetration of a pointed rod into a sample from a defined height is a simple, but appropriate, method for quality control evaluation of viscosity. The distance the rod penetrates into the sample is inversely proportional to the viscosity of the sample. Alternatively, the Brookfield Helipath viscometer is a more elaborate tool for measuring viscosity. While these methods give an indication of product variation for limited parameters, they do not provide a complete description of the attributes of the body and texture of mayonnaise or salad dressing. Use of more complex analytical tools, such as a texture profile analysis, may be used to provide a more complete assessment of body and texture. Simulated shipping conditions may be needed to evaluate the stability of the mayonnaise emulsion when subjected to mechanical shock.

OIL QUALITY AND OXIDATIVE DETERIORATION

Poor quality oils will result in shortened shelf life of the product, primarily due to oxidative changes that impact flavor and odor. Only oil of the best quality should be used, and the flavor quality of the oil should be evaluated, by sensory evaluation or gas chromatography, prior to incorporation into the product. Other oil quality indicators for oxidation, such as peroxide value, also should be evaluated. Peroxide values for fresh deodorized oil should be zero, with an acceptable cutoff value of less than 1.0 mEq/kg fat at the time of use in mayonnaise processing (Krishnamurthy and Witte 1996). Peroxide values are best suited, however, for detecting the onset of autoxidation and related rancid flavor. Carbonyl values may be more useful in determining the degree of rancidity (Depree and Savage 2001). The egg components also are susceptible to oxidative deterioration.

Oxidative rancidity may occur on the surface of the mayonnaise as well as internally (Krishnamurthy and Witte 1996, Weiss 1983). A high proportion of oil is exposed to the aqueous phase because of the dispersion of small lipid droplets. Dissolved oxygen in the aqueous phase and air bubbles introduced and trapped within the emulsion during the mixing process contribute to the high probability of oxidative reactions in mayonnaise (Depree and Savage 2001). Energy (e.g., that emitted by light), in the presence of catalysts, reacts with unsaturated fats to form free radicals. These free radicals react with molecular oxygen to form peroxide radicals. Peroxide radicals

can propogate additional free radicals or decompose into aldehydes, ketones, and alcohols. These intermediate reaction products then interact to form stable compounds that contribute to the "rancid" flavor characteristic of spoiled mayonnaise. Metals, light, and plant pigments act as catalysts for oxidation.

Light energy at wavelengths of 365 nm as well as in the blue range of visible light promote oxidation and discoloration of mayonnaise, but wavelengths greater than 470 nm do not affect unsaturated fats (Lenneston and Lignert 2000 in Depree and Savage 2001). The light energy acts on photosensitizing agents, such as carotenoids, which then react with unsaturated fats. Cool fluorescent lights, such as those typically used in supermarkets, emit light in the wavelengths of concern. Packaging materials that block wavelengths in the UV and 410–450 nm range may help reduce oxidation problems.

MICROBIAL SPOILAGE AND PRODUCT SAFETY

Microbial contamination and spoilage is secondary to oxidative rancidity (Kirshnamurthy and Witte 1996). Commercially processed mayonnaise is rarely implicated in food-borne outbreaks, but homemade mayonnaise has been associated with illnesses from *Salmonella*. Raw eggs have been implicated as a primary source of infection (Radford and Board 1993). While use of pasteurized eggs is recommended, the U.S. Food and Drug Administration does allow the use of unpasteurized egg if the final product meets three criteria: (1) contains > 1.4% acidity (acetic acid) in the aqueous phase, (2) has a final pH of 4.1 or less, and (3) is held 72 hours before shipment to trade (Radford and Board 1993). The salt added to frozen egg yolk provides additional resistance to microbial spoilage during thawing (Tressler and Sultan 1982, Weiss 1983). Refrigerated temperatures protect *Salmonella* spp. against acidulants, so the higher holding temperature and time, prior to refrigeration, is recommended to allow the germicidal activity of the acidulant to be effective (Radford and Board 1993).

Microbiological safety can be assessed by titratable acid measurements and pH or challenge testing (Dodson et al. 1996 in Xiong et al. 2000). Survival of *Salmonella* as well as *Clostridium perfringens* and *Staphylococcus aureus* is influenced by product pH and the type of acidulant used in product preparation. The acid is the primary preservative against microbial spoilage in mayonnaise (Tressler and

Sultan 1982, Weiss 1983). At the pH range of mayonnaise (3.6–4.0), acetic acid exists (vinegar) primarily in the undissociated form, exerting maximum antimicrobial activity (Radford and Board 1993). The acetic acid in vinegar is more germicidal than citric acid, the primary acidulant in lemon juice (Radford and Board 1993). Egg ingredients such as egg yolk, egg white, or whole egg have similar effects on mayonnaise pH when the ratio of egg to vinegar is less than 2.5, and this relationship is the primary determinant of mayonnaise pH (Xiong et al. 2000). Oils containing low concentrations of phenolic compounds, such as sunflower and olive oil, contributed to a higher death rate from *Salmonella* Enteritidis than did virgin olive oil (Radford and Board 1993). The pH is decreased with the addition of salt and sugar but increased by oil, mustard, and pepper. Garlic and mustard, at concentrations of 0.3–1.5% (w/w), resulted in an increased rate of death from *Salmonella* Enteritidis, but salt at similar concentrations had a protective effect (Radford and Board 1993). The low water activity of mayonnaise

also contributes to a preservative effect. With o/w emulsions, the growth rate of microorganisms is not affected by the distribution of water, only by the chemical composition of the aqueous phase.

Separation of the emulsion is one of the first signs of microbial spoilage, although bubbles of gas and rancid aromas may precede the emulsion separation (Jay 1992). Yeasts, molds, and a limited number of bacteria, such as *Lactobacillus*, are the primary spoilage bacteria (Jay 1992). Sources of contamination include spices, raw eggs, and contaminated ingredients. In general, incorporation of mayonnaise in foods increases the safety of those foods. The commonly held belief that acid dressings, such as mayonnaise, are important vehicles in food poisoning outbreaks is without merit.

APPLICATION OF PROCESSING PRINCIPLES

Table 18.2 provides recent references for more details on specific processing principles.

Table 18.2. Manufacture and Application Principles for Mayonnaise Production

Processing Stage	Application Principles	References for More Information on this Principle
Mixing	Water activity, flavor, antimicrobial	Krishnamurthy and Witte 1996, Depree and Savage 2001
Addition of egg	Emulsifiers, protein as surfactant solids content, color, gelation, emulsifying capacity	Fennema 1996, Depree and Savage 200
Addition of oil and vinegar	Continuous phase, dispersed phase, pH, fat crystallization, autoxidation, emulsion stability grain, disassociated acids, coalescence, viscosity winterization, flavor, color, dispersion, reversion deodorization denaturation	Krishnamurthy and Witte 1996, Depree and Savage 2001, Radford and Board 1993, Heertje 1993b
Pumping through colloid mill	Gelation, surface area, emulsifying capacity	Krishnamurthy and Witte 1996, Heertje 1993
Filling	Viscosity, gelation, oxidation	Krishnamurthy and Witte 1996, Depree and Savage 2001

GLOSSARY

CFR—Code of Federal Regulations.

Colloid mill—mechanical device with a high-speed rotor and fixed stator; used to produce small droplets of the dispersed phase in an emulsion.

Continuous phase—liquid phase in which the dispersed phase exists.

Dispersed phase—liquid phase distributed in small droplets throughout the continuous phase of an emulsion.

EDTA—ethylenediaminetetraacetic acid.

Emulsifier—a surface-active compound with a hydrophilic head and a lipophilic tail; acts to reduce surface or interfacial tension for stabilizing emulsions and may contribute other functions.

Emulsion—dispersion of one immiscible liquid in another.

Lecithin—a phospholipid found in egg yolk that functions as a natural emulsifier.

o/w, w/o—oil-in-water and water-in-oil emulsions.

Sequestrant—compound that scavenges metal ions.

Water activity—a property of solutions; the ratio of vapor pressure of solution to the vapor pressure of pure water.

w/w—weight to weight.

REFERENCES

Code of Federal Regulations. 1993. Mayonnaise. Section 169.140. Federal Register, 533–534. Washington, D.C.

Depree JA, GP Savage. 2001. Physical and flavour stability of mayonnaise. Trends in Food Science and Technology 12:157–163.

Fennema O. 1996. Food Chemistry. Marcel Dekker, Inc., New York.

Heertje I. 1993a. Microstructural studies in fat research. Food Structure 12:77–94.

Heertje I. 1993b. Structure and function of food products: A review. Food Structure 12:343–364.

Holcomb DN, LD Ford, RW Martin, Jr. 1990. Chapter 8. Dressings and sauces. In: K Larsson, SE Friberg editors. Food Emulsions, 2nd edition. Marcel Dekker, Inc., New York.

Jay JM. 1992. Modern Food Microbiology, 4th edition, 243. Van Nostrand Reinhold, New York.

Jones LA, CC King, editors. 1993. Cottonseed Oil. National Cottonseed Products Assn., Inc., and Cotton Foundation, Memphis, Tenn.

Krishnamurthy RG, VC Witte. 1996. Chapter 5. Cooking oils, salad oils, and oil-based dressings. In: YH Hui, editor. Bailey's Industrial Oil and Fat Products, 5th edition, vol. 3, 193–223. John Wiley and Sons, Inc.

Krog NJ, TH Riisom, K Larsson. 1985. Chapter 5. Applications in the food industry: I. In: P Becher, editor. Encyclopedia of Emulsion Technology. Vol. 2, Applications, 321–384. Marcel Dekker, Inc., New York.

Langton M. E Jordansson, A Altskar, C Sorensen, A Hermansson. 1999. Microstructure and image analysis of mayonnaises. Food Hydrocolloids 13:113–125.

Newton S. 1989. Fats and oils: How do they perform? Prepared Foods 158(5): 178–185.

Paul PC, HH Palmer. 1972. Food Theory and Practice, 109–111. John Wiley and Sons, New York.

Potter NN, JH Hotchkiss. 1995. Food Science, 5th edition. Chapman and Hall, Inc., New York.

Radford SA, RG Board. 1993. Review: Fat of pathogens in home-made mayonnaise and related products. Food Microbiology 10:269–278.

Tressler DK, WJ Sultan. 1982. Mayonnaise and salad dressing. Food Products Formulary. Vol.2, Cereals, Baked Goods, Dairy and Egg Products, 377–382. AVI Publishing Co., Inc., Westport, Conn.

Verlags H. 1994. The structure and manufacture of mayonnaise and emulsified sauces. International Food Marketing and Technology.

Weiss TJ. 1983. Chapter 10. Mayonnaise and salad dressing. In: Food Oils and Their Uses, 2nd edition, 211–230. AVI Publishing Co., Westport, Conn.

Xiong R, G Xie, AS Edmondson. 2000. Modelling the pH of mayonnaise by the ratio of egg to vinegar. Food Control 11:49–56.

19
Fats: Vegetable Shortening

L. A. Carden and L. K. Basilio

BACKGROUND INFORMATION

SHORTENING DEFINED

The term "shortening" derives from a baking term used to describe a fat's ability to shorten gluten protein strands in batters and doughs; shortening the gluten strands tenderizes the product. The original shortening used by cooks was lard, which could be rendered from the fat of homegrown hogs. While many modern cooks find the aroma and texture of lard objectionable, it was widely used (before the production of vegetable shortening), and produced flaky, tender pastries and breads of good volume as well as other baked products. It also served as a cooking medium in the production of crispy fried foods. Vegetable shortening is an edible fat similar in consistency to lard—that is, it is a plastic fat containing no water—and is composed of partially hydrogenated vegetable oil. Most often, vegetable shortening is made from soybean oil blended with cottonseed oil, which increases the plasticity or spreadability of the fat, an important characteristic of shortening. Vegetable shortenings also are capable of incorporating air into flour mixtures, increasing the volume of the final baked good. This makes them invaluable both in baking and in preparation of icings and fillings. While vegetable shortening is used extensively in baking, it is also valuable as a frying fat. Shortening is widely used in deep-fat frying as well as pan frying, sautéing, and grill frying.

SOYBEAN FACTS AND FIGURES

The soybean is a legume planted in late spring and harvested in the fall. The soybean is not indigenous to the Americas, but was brought to the United States from China in the 1800s. Currently, soybeans are grown in 29 states. Use of the soybean was limited to animal forage until the early 1900s, at which time study of the legume by George Washington Carver led to discovery of uses for human consumption. Carver found that the soybean is an excellent source of high quality protein as well as oil. Crop production grew over the next 40 years, but oil processors relied on foreign sources of oil until

---

The information in this chapter has been derived from a chapter in *Food Chemistry Workbook,* edited by J. S. Smith and G. L. Christen, published and copyrighted by Science Technology System, West Sacramento, California, ©2002. Used with permission.

World War II, when those sources were eliminated. The domestic soybean then became their best source of vegetable oil. In 1956, American soybean growers began promoting American beans in Japan; today, soybeans grown in the United States provide the largest amount of beans worldwide [University of Illinois Urbana-Champaign (UIUC) n.d.].

Sixty pounds of beans (a bushel) will yield 11 pounds of oil, some of which will eventually become vegetable shortening. In 1996, 64.83 million metric tons of soybeans with a crop value of $16,317 million were produced. Eighty-two percent of the edible oil in the United States is produced by refining of soybeans. In 1996, 5.59 metric tons of soybean oil were consumed in the United States, about half of which was converted to vegetable shortening for baking and frying (UIUC n.d.). As a source of oil, the soybean offers a number of advantages that greatly outweigh its disadvantages. The fatty acids found in soybean oil are largely unsaturated, providing an oil that remains liquid over a range of temperatures; also, the fatty acids in soybean oil readily react to selective hydrogenation. In addition, soybean oil is easy to refine. Undesirable constituents are easily removed, and some of the natural antioxidants present survive the refining process, reducing the amount that must be added. The presence of linolenic acid, an 18-carbon polyunsaturated fatty acid, is detrimental to shelf stability. Linolenic acid is highly susceptible to oxidative rancidity, but it can be hydrogenated to decrease its sensitivity to oxygen and light (Pryde 1987).

RAW MATERIALS PREPARATION

SELECTION, HARVESTING, AND STORAGE OF SOYBEANS

The functionality and quality of vegetable shortening is dependent on several factors, beginning with the harvesting of soybeans. The highest quality oil is produced from fully mature soybeans that have not suffered damage from environmental conditions or from handling during harvest. Oil processed from soybeans harvested while still green will be off colored due to concentrations of chlorophyll. A season of heavy rains, hailstorms, and wind or extreme heat can produce field-damaged soybeans with high levels of phosphatides/gums, iron, and copper, which affects the functionality of the oil (Weiss 1983). Whole, clean beans should have moisture levels in the range of 13–14%; higher levels of moisture may

lead to problems during the extraction process. In addition, higher moisture content promotes microbial growth. Once harvested, soybeans should be stored in a controlled environment to retard respiration. Respiration during storage produces CO_2 and heat, which can affect the quality of the soybeans; good aeration and low temperatures in a controlled atmosphere prevent damage from the by-products of respiration (Woerful 1995)

PREPARATION OF BEANS

Before oil extraction, soybeans must be cleaned and dried to a moisture content of approximately 10%. During tempering, a rest period of 10 days, the hull of the bean loosens, releasing the cotyledon. After tempering, the beans are moved from storage silos into the refinery by way of belt conveyors and/or bucket elevators. They drop into storage units, pass over magnets to remove metal contaminants, and are sent through cracking rolls to break the hulls for easier removal. After the beans are cracked and dehulled, they are conditioned by steam softening in a steam-jacketed cooker. This softening process increases the pliability of the beans and denatures enzymes activated during cracking. After the conditioning process is completed, the beans are flaked—reduced to small particles—which increases the efficiency of the extraction process. Flaking rolls are designed to produce flakes that are about 0.254 mm thick (Mustakas 1987). Once flaking is completed, the beans are ready to enter the extractor.

PRODUCTION PROCESSES

EXTRACTION OF OIL

The three most common forms of extraction include hydraulic pressing, expeller pressing, and solvent extraction. Hydraulic pressing originated in Europe in the late 1700s and utilizes a machine-shop-type press that removes oil in batches. Because batch production is not economical, it is no longer used in the United States. Traditional expeller pressing also is no longer used (Mustakas 1987). However, an extrusion-expelling (E-E) process has been developed by Insta-Pro International, Triple "F," Inc., of Des Moines, Iowa. The E-E process involves use of a dry autogenous extruder that generates heat by friction, followed by screw pressing to remove the oil. Examination of the extracted oil reveals low levels of phosphatides and free fatty acids. The oil has a nutty roasted flavor (Wang and Johnson 2001).

Solvent extraction in a percolation extractor is currently the predominant method of removing oil from soybeans. The solvent used to pull oil from the flaked beans is n-hexane. A miscella consisting of n-hexane and oil is percolated through the flaked beans counter to their flow. The miscella absorbs oil from the beans, and contains 25–30% oil when it leaves the extractor (Mustakas 1987). The hexane in the miscella is recovered through a series of distillations that condense the solvent and concentrate the oil. The hexane passes directly into the solvent tank; the miscella increases in oil content from ~70% after the first distillation to 99% when the process is complete. The wet flakes, containing ~35% hexane and small amounts of water and oil, are processed to increase moisture level and remove hexane in a desolventizer/toaster, producing soybean meal, a major ingredient in animal feed (Mustakas 1987).

NATURAL REFINING OF EXTRUSION-EXPELLED OILS

A natural method of refining oil extracted by the E-E process has been reported by Wang and Johnson (2001). The crude oil is allowed to settle for two days at 5°C. This period of settling results in the removal of fines and some gums. Afterwards, the settled oil is water degummed at 60°C with agitation. As the degumming process progresses, vigorous agitation decreases to a more gentle action. The gums settle, and a clear oil remains, which is then refined by addition of an adsorbent. Wang and Johnson (2001) report use of Magnesol® as an effective adsorbent to remove free fatty acids from the degummed oil. The Magnesol® is removed by filtration after the free fatty acids are adsorbed. Deodorization is carried out at lower than usual temperatures. The oil produced by this natural process is not bleached; it is a golden-colored oil that is easily recognizable as "natural."

CHEMICAL REFINING

Extracted oil is refined to remove undesirable microconstituents, which occur naturally in the soy oil. Crude oil contains free fatty acids, phosphatides, coloring matter such as chlorophyll, and other insoluble substances such as meal fines, which can interfere with the quality of the oil. In the chemical refining process, weak alkali in the form of aqueous sodium hydroxide is added to the crude oil to saponify free fatty acids, affecting removal of phosphatides and color bodies. The percentage of alkali to be added must be calculated carefully to provide an amount of caustic that will remove the microconstituents without destroying triglycerides (Mounts and Khym 1987). Thorough combining of the oil and alkali is accomplished by rapidly mixing cool oil and alkali; the mixture is then heated to 75–82°C and fed into a centrifuge, which separates the mixture into oil (the light phase) and the heavy materials produced by alkali reaction. Free fatty acids react with the caustic to form soaps. Phosphatides hydrate and coagulate. Chlorophyll and other coloring matter are saponified. These impurities compose the heavy phase or "foots," also known as soapstock, which are separated from the oil during centrifuging. The oil phase is washed twice, centrifuged again to separate it from the water, and vacuum dried to reduce moisture to less than 0.1%. Sulfuric acid is added to the heavy phase to convert soaps back to free fatty acids, which can also be sold as an ingredient for animal feed (Mounts and Khym 1987).

BLEACHING

Bleaching of the oil is an important step in the production of oil for shortening. One of the sensory qualities of a good vegetable shortening is its snow-white color, but oil that has just completed the refining process is not colorless. Some pigments and odorants remain following the refining process. Xanthophylls, carotenes, and chlorophylls may remain in the refined oil, giving it a dark golden or greenish color. In oil destined for vegetable shortening, certain chlorophyll isomers not only may influence the color of the oil, but also can be extremely detrimental to oxidative stability. Their presence can result in a gray- or green-colored shortening, if not removed by bleaching, and can lead to oxidative rancidity in the oil. In addition, residual soaps and phosphatides as well as prooxidant metals, peroxides, and secondary oxidative products such as aldehydes and ketones may remain in the oil after refining. Bleaching is used to reduce these oxidation products and to further clean the oil, improving its flavor, color, and oxidative stability. The bleaching process involves usage of an adsorbent material that is mixed with the oil,; this combination of adsorbant and oil is heated in a vacuum vessel to remove any moisture from the adsorbent. Removal of the moisture activates the adsorbant (Weiss 1983). In some refineries, silica may be added to the oil first to absorb residual soaps and phospholipids; theoretically, less adsorbent will be needed to remove pigments and odorants when the silica is used (Carlson and Scott 1991). The adsorbent used most often in U.S. refineries is an acid-washed clay; either sulfuric acid or hydrochloric acid is added to the clay to catalyze the bleaching action. Currently, the bleaching process is continuous, proceeding under a vacuum at ~82°C (180–220°F) (Stauffer 1996). After bleaching is complete, the mixture is filtered to separate the bleaching clay from the oil. The "spent earth," the clay filled with impurities from the oil, is disposed of as mandated by environmental laws. Current strict environmental controls preclude the hauling of the spent earth to landfills; rather, the refinery must seek other disposal methods (Carlson and Scott 1991).

HYDROGENATION

Vegetable oils contain polyunsaturated fatty acids, which makes them highly susceptible to oxidative rancidity. Their shelf life can be greatly shortened when exposed to air and light. The purpose of hydrogenation is to modify the properties of an oil by altering the degree of saturation and the configuration of fatty acids in the lipid. Hydrogenation converts refined and bleached liquid vegetable oils into the solid or semisolid fats used to produce vegetable shortening. The degree of saturation of a fatty acid is determined by the number of double bonds: saturation increases as the number of double bonds decreases. A fully saturated fatty acid has no double bonds. Increasing the saturation of a fatty acid increases its melting point. A highly saturated fatty acid will be solid at room temperature, while a poly-unsaturated fat will maintain its liquid form at the same temperature. Stearic acid, for instance, is a common fatty acid in meat triglycerides and is solid at room temperature. Oleic acid occurs in a variety of vegetable oils including soybean oil; it is liquid at room temperature. Both of these fatty acids are composed of 18 carbons; the difference between the two is that stearic acid is fully saturated. Oleic acid, however, contains one double bond, a single point of unsaturation, resulting in a 51° difference in melting point. Stearic acid must be heated to 69.9°C to become liquid; oleic acid is liquid at 18.9°C. That single point of unsaturation in oleic acid also increases the instability of the fatty acid. Exposure to air can lead to the formation of free radicals due to the susceptibility of the double bond to reaction with oxy-

gen. To provide stability and increase functionality, oils are commonly hydrogenated as part of the production process. Hydrogenation is accomplished by exposing the oil to hydrogen gas in the presence of a catalyst (Wan 1991). Currently, the catalyst of choice is nickel. Research is ongoing with other catalysts that may be effective in conjugating fatty acids and may be used in the hydrogenation process in the future (Larock et al. 2001) Oil is hydrogenated in carbon steel vessels in which temperature, hydrogen gas pressure, agitation of the oil, and concentration of the nickel catalyst can be well controlled. After hydrogenation, the nickel catalyst is removed by filtration (Wan 1991). Hydrogenation results in the conversion of some of the double bonds to saturated bonds.

Hydrogenation also produces *trans* isomers of some unsaturated fatty acids; the *cis* configuration is the naturally occurring form. In a *trans* isomer, the position of the hydrogens attached to double-bond carbons is altered. They are relocated to positions opposite each other rather that parallel to each other. The *trans* isomer of an unsaturated fatty acid has a higher melting point than its corresponding *cis* form. Oleic acid present in an oil that is being hydrogenated may undergo conversion to its *trans* form, elaidic acid. The fatty acid still contains one point of unsaturation, but the change in position of the hydrogens attached to the double-bond carbons increases the melting point to 43.0°C. Thus, both the degree of saturation and the configuration of fatty acids affect their melting points.

The majority of food lipids are triglycerides consisting of three fatty acids attached to a glycerol backbone. While the degree of saturation of a fatty acid may be described simply in terms of the number of double bonds, the properties of a lipid cannot. One measurement used to characterize the properties of a fat is solid fat index (SFI). SFI is a measure of the ratio of solids to liquid present in a fat at a given temperature. This ratio is measured over a range of temperatures to derive the SFI profile of a fat (Wan 1991). The SFI profile of a shortening indicates its functionality.

Several factors influence the degree of saturation that occurs and the extent to which *trans* isomers are formed. These include hydrogen pressure, catalyst concentration, catalyst type, reaction temperature, and time of reaction (Wan 1991). When these parameters are controlled, fats of varying SFI profiles may be obtained. Because fatty acids with higher degrees of unsaturation are more reactive, they can be selected for hydrogenation by setting the parameters for the process. With high process temperature, low hydrogen pressure with low rate of agitation, and a high nickel concentration, the more highly unsaturated fatty acids will be hydrogenated first, prior to less unsaturated ones. Linolenic acid, an 18-carbon fatty acid with three double bonds (C18:3), will add hydrogen more readily under these conditions than linoleic acid, which has two double bonds (C18:2); oleic acid, with only one point of unsaturation, will be least reactive (Stauffer 1996). This selective process produces fats with differing SFI profiles, referred to as base stocks; base stocks may be blended to obtain vegetable shortenings with specific SFI profiles.

BLENDING

Blending of base stocks is done in order to produce a vegetable shortening with specific functional properties, as indicated by its SFI. An all-purpose shortening is considered to have acceptable plasticity when the SFI at room temperature is between 10 and 25 (Stauffer 1996). Plasticity is a term that describes a fat that is soft yet retains its structure to some degree. It is pliable when shear is applied. A spatula pulled across a handful of shortening will push the shortening in the same direction, flattening it in the hand. When the spatula is reversed, the flattened shortening will move with it on the return movement. Through all this application of force, the basic structure of the shortening is unchanged.

Depending on the application of the vegetable shortening in a food system, vegetable shortenings with certain SFI profiles are needed. The plastic range is important in applications such as cakes, where the shortening must maintain its structure during creaming in order to incorporate air into the batter. In other applications, such as confectionary coatings, shortening that retains a higher portion of solids at relatively higher temperatures buts melts around body temperature is desired. The confectionary shortening needs to dissolve away in the mouth to provide the mouthfeel associated with fine candies.

Another important factor to consider when blending base stocks is the crystalline structure of fats; although shortening appears to be solid at room temperature, in actuality it is a mixture of crystals in oil. The four crystalline forms found in fats are the alpha, beta prime, intermediate, and beta. The alpha crystalline form is very fine, needle shaped, and un-

stable, quickly converting into the more stable beta prime structure. In vegetable shortenings, the beta prime crystal is desired because it imparts a smooth, creamy texture, contributing to a fine texture in baked products. Think of the smooth, shiny surface of a just-opened can of vegetable shortening. That smoothness is evidence of the presence of beta prime crystals in the fat. Intermediate crystals are extremely unstable and slightly coarse in consistency; they form when a beta prime shortening is stored improperly at too warm a temperature. Intermediate crystals are so unstable that they recrystallize almost immediately into the much coarser, grainy beta crystals, which are considered undesirable for many applications (Wan 1991).

In order to produce a vegetable shortening with a smooth texture and an appropriate crystal form, different base stocks are blended. Vegetable oils, such as soybean, that contain a small percentage of palmitic acid (C16:0) or are comprised of only one or two types of triglycerides prefer the beta crystalline form, the most stable form of crystal (Stauffer 1996). Palm and cottonseed oils tend to form stable beta prime crystals; thus, in order to obtain a shortening that will maintain a smooth texture, the soybean oil typically is blended with approximately 5–10% palm or cottonseed oil (Weiss 1983). This produces a shortening that retains the beta prime crystalline form and thus has a smooth texture with ideal plasticity for incorporation of air.

DEODORIZATION

Fat is known to be a carrier of flavor in food preparation. If the fat has strong or objectionable flavors of its own, its usefulness as a flavor carrier is decreased. Shortening for use in frying or in baking should have a bland flavor with a free fatty acid content no greater than 0.05% by weight. Refined, bleached, and hydrogenated oil may still contain undesirable constituents, including free fatty acids that have escaped saponification during refining, odorants, and prooxidants, that can decrease the quality of the finished oil and contribute to formation of additional free fatty acids. Deodorization removes any remaining impurities that will volatilize under the conditions of use, leaving a bland, clear liquid. Oil is pumped into a deaerator, where oxygen is removed from the oil; the oil is pumped to the top of the deodorizer system and flows by gravity through a series of trays in which it is steam sparged and stripped of volatiles, stripped a second time, and cooled. During the cooling process, the oil is treated with 0.005–0.01% citric acid to chelate prooxidant trace metals and prevent metal-catalyzed oxidation. Antioxidants such as propyl gallate (PG) and tertiary butylhydroquinone (TBHQ) are added to increase the oxidative stability of the oil. Two other common antioxidants that may be added are butylated hydroxytoluene (BHT), and butylated hydroxyanisole (BHA). BHA and BHT lengthen the shelf life of the oil (Wan 1991). In oils that are not destined for use as baking shortenings, methyl silicone, an antifoaming agent, is added to increase the smoke point and reduce foam in frying oils. Oil that has been refined, bleached, and deodorized is referred to as an RBD oil.

PLASTICIZING

As mentioned previously, the creaming capabilities of shortening contribute greatly to its functionality. The shortening must be able to incorporate air bubbles into its structure to add volume to baked goods. The plasticity of the shortening influences its creaming ability. Shortening is plasticized by rapidly cooling the oil and injecting nitrogen gas into the shortening, during which time the triglycerides in the shortening form crystals. The goal is to produce a solid shortening with beta prime crystals, as discussed above. Careful cooling with sufficient agitation is necessary to plasticize the shortening and form the desired crystals. Without the proper controls, beta crystals will form, giving the shortening a coarse texture and reducing its creaming power.

The equipment used for plasticizing the RBHD (refined, bleached, hydrogenated, and deodorized) oil is called a Votator. It typically has two units, working units A and B. The A unit is a scraped-surface heat exchanger consisting of an internal cylinder that holds the shortening and an outer cylinder that contains the coolant. In the A unit, the oil is cooled to 15–20°C and some crystallization occurs; nitrogen gas is also added in the A unit (Wan 1991). While some crystallization occurs in the A unit, the shortening is still fluid when it is pumped into the B unit. The B unit whips nitrogen, at a level of 10–15%, into the shortening; as the shortening is agitated, further crystallization occurs and the mass begins to solidify. The nitrogen gas enhances the white color of the shortening and retards lipid oxidation. Lipid oxidation occurs when oxygen causes chemical changes in unsaturated fatty acids, resulting in the formation of off flavors (Wan 1991). From

the plasticizer, the now malleable mass moves through a homogenizing valve into package fillers.

PACKAGING

Shortening is packaged just after plasticizing. To avoid textural defects in the shortening, the fill temperature must be maintained between 27 and 29°C. Packaging guards against lipid oxidation by limiting exposure of the shortening to oxygen and light, extending the shelf life of the product. A variety of sizes of packages, ranging from the typical one-pound can that is sealed after filling to bulk drums containing 380 pounds of shortening, may be filled. The one-cubic-foot package is a popular fill for food service use as it is easily transferred to deep-fry cookers (Brekke 1987).

After packaging, many shortenings are tempered over a period of two to four days at a temperature between 27 and 29°C in order to extend the plastic range of shortening. As the shortening is tempered, the beta prime crystals are further stabilized, which improves the functionality of the shortening. Shortening that is not tempered becomes brittle when stored at cool temperatures (Brekke 1987, Stauffer 1996).

ANALYTICAL TESTING OF OILS AND FATS

Solid fat index has been discussed at one determinant of the quality and functionality of a fat. Several other analytical tests can be used to check the quality of a processed fat. As food fats are mixtures of triglycerides, they do not melt at sharp temperatures. The presence of varied fatty acids with differing saturation levels causes triglycerides to melt over a range of temperatures. Two methods are most commonly used to determine the melting point of a food fat.

The *complete or capillary melting point* is determined by chilling the fat in a capillary tube until it solidifies, then heating it in a water bath until the fat is completely liquefied. The temperature at which the fat becomes liquid is recorded as complete melting point and is equal to an SFI of 0. The method more often used to determine the melting point of a fat is the Wiley method. The fat is molded into a disk three-eighths inch in diameter and one-eighth inch thick. It is solidified and chilled for a minimum of two hours, after which it is suspended in an alcohol/water bath and slowly heated until the circular disk is altered to a spherical shape. The temperature at which this occurs is recorded as the Wiley melting point (Pomeranz and Meloan 1987, Stauffer 1996).

As a quality control, the *cloud point* of oils to be used for production of mayonnaise and/or salad dressings is often determined. The oil is held in an ice bath until it appears cloudy. A time to cloudiness of 20 hours is considered excellent. In quality control laboratories, a quick method, which gives results within one hour, is used. The oil is chilled at $-60°C$ for 15 minutes and then held at $10°C$ for 30 minutes (Stauffer 1996).

To classify the type of fat for marketing purposes, the degree of unsaturation can be determined by iodine value (IV). Iodine or another halogen is added to double bonds in fats and expressed as grams iodine absorbed by 100 g of fat. A small sample of the fat is reacted with reagent and then titrated with thiosulfate. The IV is calculated as the difference between the titration of a blank and the titration of the sample. Iodine value is not affected by the presence of *trans* fatty acids. Both *cis* and *trans* forms react with the iodine (Pomeranz and Meloan 1987, Stauffer 1996).

To meet food-labeling regulations, the fatty acid composition of oils and fats must be determined. High-performance liquid chromatography (HPLC) and gas-liquid chromatography (GLC) can both be used to separate and identify fatty acids. Gas-liquid chromatography is often preferred. The fatty acids are converted to methyl esters before separation by GLC. The total fatty acid content of a fat, as well as the distribution and position of fatty acids on the molecule, can be determined by GLC. GLC can also be used to separate *cis* and *trans* fatty acids; revisions of the labeling regulations may soon require inclusion of *trans* fats on food labels (Pomeranz and Meloan 1987, Stauffer 1996).

Polyunsaturated fats are susceptible to oxidative rancidity, an autoxidation process initiated by the presence of oxygen and leading to the formation of free radicals, resulting in hydroperoxides in the oil. The hydroperoxides impart very unpleasant aromas and flavors to oils and fats. Determination of oxidation is an important quality control in the oil industry. The peroxide value (PV) is the most frequently used test for oxidized fatty acids. The hydroperoxides formed in the fat will react with iodide ions, giving rise to iodine. Saturated potassium iodide is reacted with a sample of fat or oil dissolved in glacial acetic acid and chloroform. The iodine released by this reaction is titrated with sodium thiosulfate,

and the PV is expressed as milliequivelents of iodine per kilogram of fat (mEq/kg). PV indicates the degree of oxidation that has occurred but not the stability of the fat (Pomeranz and Meloan 1987, Stauffer 1996). To determine the stability of a fat, the active oxygen method (AOM) or the oil stabilty index (OSI) is used. In AOM, air is bubbled through the fat, a sample is withdrawn at intervals, and PV is determined. AOM stability is the time required to develop a peroxide concentration of 100 mEq/kg fat. The AOM has largely been replaced in the industry by the OSI, which automatically measures oil stability and gives results which coordinate well with AOM values (Stauffer 1996).

APPLICATIONS OF VEGETABLE SHORTENING IN FOOD PREPARATION

The processing of oil to hydrogenated vegetable shortening is designed to give the fat appropriate characteristics for use in food preparation. Shortening is most often used in baking. Selective hydrogenation of unsaturated fatty acids, blending of base stocks to provide proper crystal formation, and plasticizing of the fat all have an influence on the uses of shortening in baked products. Functions of hydrogenated vegetable shortening in baking include the tenderization of the product and incorporation of air to increase volume. Tenderization of the crumb in a baked product is related to the plasticity of the fat. In biscuits and pastry, the fat is worked into the flour mixture in large pieces that melt during baking, forming layers of fat within layers of flour mixture. An example is the cutting of shortening into flour for biscuits; instructions always direct the baker to work the fat into the flour until it is the size of peas. Small pieces of shortening are completely surrounded by flour. As the biscuits bake, the fat repels water, keeping it away from the flour proteins gliadin and glutenin and reducing the formation of the gluten complex. The result is a tender, flaky biscuit. It is the plasticity of the shortening, as described above, that allows it to spread within a dough to form layers within the flour mixture (Bowers 1992, McWilliams 2001). In products made with batters, flour mixtures with more water present, the degree of saturation of the shortening influences tenderness. Selective hydrogenation of oils intended for shortening provides the appropriate mix of mono- and polyunsaturated fatty acids with saturated ones. The unsaturated fatty acids are able to align themselves at the interface of the fat and flour mixture, where they block the passage of water to the gluten proteins. The fluidity of the shortening aids in this function (McWilliams 2001).

A second function of hydrogenated vegetable shortening in baked products is entrapment of air to add volume to a product. In cake batters and some other baked goods, fat is creamed with the sugar; eggs are added after the creamed mixture reaches the foamy stage. The plasticity of the fat enables it to disperse in the sugar, forming a foamy mixture that holds the air beaten into the mixture. The clumps or bubbles of fat are much finer in size and are more

equally dispersed throughout the mixture than in a dough. The sharp sugar crystals are able to cut into the fat, giving rise to tiny spaces where steam and CO_2 can collect during baking (McWilliams 2001). The combination of the three gases (air, CO_2, and steam) provides the desired volume to the cake.

APPLICATION OF PROCESSING PRINCIPLES

Table 19.1 provides recent references for more details on specific processing principles.

Table 19.1. References for Principles Used in Processing

Processing Stages	Processing Principles	References for More Information on the Principles Used
Selection, harvesting and storage of soybeans	Moisture content, pigmentation, respiration, oxidation, hydrolysis	Woerful 1995, Weiss 1983
Extraction of oil from soybeans	Solvent extraction, steam evaporation/distillation	Wang and Johnson 2001, Mustakas 1987
Refining of oil	Saponification, hydration, centrifugation	Wang and Johnson 2001, Mounts and Khym 1987
Bleaching of oil	Adsorption, acid-activated clay; peroxide and secondary oxidation product reduction; chelation, chlorophyll reduction	Carlson and Scott 1991, Stauffer 1996
Hydrogenation of oil	Saturation of fatty acids, melting point, lipid oxidation, double bond configuration, catalyst, formation of *trans* fatty acids, solid fat index	Wan 1991, Larock et al. 2001
Blending of base stocks	Solid fat index, crystalline structure, melting point	Stauffer 1996, Wan 1991
Deodorization of vegetable shortening with cool-down period	Reduction of free fatty acids, volatilization, anti-oxidants, chelating agents, emulsifiers, anti-foaming agents	Wan 1991, Stauffer 1996
Plasticizing of vegetable shortening	Crystal formation, heat exchange	Wan 1991, Stauffer, 1996
Packaging of vegetable shortening	Lipid oxidation, crystal formation, tempering	Brekke 1987, Stauffer 1996
Analytical testing of shortening quality		Pomeranz and Meloan 1987, Stauffer 1996

GLOSSARY: ACRONYMS

AOM—active oxygen method.
BHA—butylated hydroxyanisole.
BHT—butylated hydroxytoluene.
E-E process—extrusion-expelling process.
GLC—gas-liquid chromatography.
HPLC—high-performance liquid chromatography.
IV—iodine value.
OSI—oil stability index.
PG—propyl gallate.
PV—peroxide value.
RBD oil—refined, bleached, and deodorized oil.
RBHD oil—refined, bleached, hydrogenated, and deodorized oil.
SFI—solid fat index.
TBHQ—tertiary butylhydroquinone.

REFERENCES

Bowers J. 1992. Food Theory and Applications. Macmillan Publishing Company, New York.

Brekke OL. 1987. Chapter 19. Soybean oil food products—their preparation and uses. In: DR Erickson, EH Pryde, OL Brekke, TL Mounts, RA Falb, editors. Handbook of Soy Oil Processing and Utilization, 89–103. American Soybean Assoc. and AOCS, Champaign, Ill.

Carlson KF, JD Scott. 1991. Recent developments and trends: Processing of oilseeds, fats and oils. Inform 2(12): 1034–1060.

Larock RC, XS Dong, S Chung, CK Reddy, LE Ehlers. 2001. Preparation of conjugated soybean oil and other natural oils and fatty acids by homogeneous transition metal catalysts. Journal of American Oil Chemists Society 78(5): 447–453.

McWilliams M. 2001. Foods Experimental Perspectives, 4th edition. MacMillan Publishing Company, New York.

Mounts TL, FP Khym. 1987. Chapter 7. Refining. In: DR Erickson, EH Pryde, OL Brekke, TL Mounts, RA Falb, editors. Handbook of Soy Oil Processing and Utilization, 89–103. American Soybean Assoc. and AOCS, Champaign, Ill.

Mustakas GC. 1987. Chapter 4. Recovery of oil from soybeans. In: DR Erickson, EH Pryde, OL Brekke, TL Mounts, RA Falb, editors. Handbook of Soy Oil Processing and Utilization, 49–65. American Soybean Assoc. and AOCS, Champaign, Ill.

Pomeranz Y, CE Meloan. 1987. Food Analysis Theory and Practice, 2nd edition. An AVI book,Van Nostrtand Reinhold, New York.

Pryde EH. 1987. Chapter 2. Composition of soybean oil. In: DR Erickson, EH Pryde, OL Brekke, TL Mounts, RA Falb, editors. Handbook of Soy Oil Processing and Utilization, 49–65. American Soybean Assoc. and AOCS, Champaign, Ill.

Stauffer CE. 1996. Fats and Oils. Eagan Press, St. Paul, Minn.

University of Illinois Urbana-Champaign (UIUC). n.d. Web site developed by the College of Agricultural, Consumer, and Environmental Sciences: http://www.stratsoy.uiuc.edu/

Wan PJ. 1991. Introduction to Fats and Oils Technology. AOCS, Champaign, Ill.

Wang T, LA Johnson. 2001. Natural refining of extruded-expelled soybean oils having various fatty acid compositions. Journal of American Oil Chemists Society 78(5): 461–466.

Weiss TJ. 1983. Food Oils and Their Uses, 2nd edition. AVI Publishing Co., Westport, Conn.

Woerful JB. 1995. Chapter 4. Harvest, storage, handling, and trading of soybeans. In: DR Erickson, editor. Practical Handbook of Soybean Processing and Utilization. AOCS Press, Champaign, Ill.

20
Fats: Edible Fat and Oil Processing

I. U. Grün

BACKGROUND INFORMATION

Fats and oils are both mixtures of triacylglycerides. Thus, chemically they are essentially the same, and the differentiation into fats and oils is mostly arbitrarily based on the physical state of the mixtures at room temperature, that is, if they are solid or liquid. However, room temperature is not a well-defined term that would allow such differentiation easily. While it is obvious that room temperature in a trop-ical country might mean something very different than room temperature in a Scandinavian country, even within the United States, room temperature is lower in the winter than in the summer because humans consider a range of temperature of approximately 65–75°F (18–24°C) as comfortable. Thus, we will use the term fat throughout this chapter without consideration of whether the fat might be solid or liquid at room temperature.

Fats and oils are harvested from both the plant and the animal kingdoms. However, while we therefore ought to differentiate only between animal and plant fats, because of the unique composition of the fat of most fish, fats from fish are often categorized separately as marine oils. This separation also makes sense from a processing standpoint, as will be shown later. The processing of fats is easy to comprehend and to remember because it is a logical progression of steps, which ultimately yield a pure (≥ 99.9%) shelf-stable product.

RAW MATERIALS PREPARATION

EXTRACTION

The first step in fat production is, of course, the extraction or harvest of the fat. This is where the first major difference between animal, plant, and marine fats is encountered. While the rendering of animal fat is similar for all animal sources, fats from plant sources are extracted in numerous different ways. The goal of the extraction process is to get the highest yield of fat with the least amount of impurities.

The information in this chapter has been derived from a chapter in *Food Chemistry Workbook,* edited by J. S. Smith and G. L. Christen, published and copyrighted by Science Technology System, West Sacramento, California, ©2002. Used with permission.

Rendering of Animal Fat

The fat rendered from animals is located in the adipose tissue of animals. Intramuscular fat, which is known to be in part responsible for the tenderness and juiciness of steaks, is not rendered for fat collection. The adipose tissue, which contains between 70 and 95% fat, is trimmed, washed, and ground. The fat is then rendered from the ground adipose tissue by a wet or a dry rendering method. There are also two other rendering methods, slurry rendering and digestive rendering, which will not be discussed here because of their limited use. The most commonly used method is wet rendering with high heat (steam rendering), which is done in pressurized vessels. Steam is directly injected under high pressure into the trimmed fat, disintegrating the fat cells and releasing the fat. The layer of fat that rises to the top (tankage) is skimmed off and then centrifuged to rid it of water, yielding 99.5% pure fat. Because of the treatment of the fat with water at high temperatures, some hydrolysis of the triacylglycerides into free fatty acids occurs, and a low-temperature wet rendering method has been developed. However, because free fatty acids are easily removed, and different equipment is needed for the cold rendering method, it has not been widely adopted by industry. In dry rendering, the fat is extracted by drying the trimmed adipose tissue in steam-jacketed vessels. The fat is liquefied and drained off. The remaining tissue is pressed to extract the remaining fat.

Rendering of Marine Fats (Oils)

Marine fats have received considerable attention over the last two decades because they contain comparatively large amounts of long-chain omega-3 (also called n-3) fatty acids, which are indicated to have numerous health benefits. Although marine fats are rendered similarly to other animal fats, one important difference exists. While land animals have clearly identifiable fat storage areas (the adipose tissue), fish do not. Instead, fish are differentiated into lean fish and fatty fish. In the lean fish, such as cod, the fat is mostly stored in the liver, while in the fatty fish, such as herring, the fat is dispersed throughout the muscle tissue. Although some lean fish are used for marine fat production, hence the availability of, for example, cod liver oil, most marine fat is extracted from a small fatty fish, menhaden (*Brevoortia*), belonging to the herring family (Clupeidae). The oil is rendered by pressing the steam-cooked

fish and then separating the resulting liquid into aqueous and oil phases by centrifugation. The crude fish oil is highly susceptible to oxidation because of the omega-3 fatty acids and must be thoroughly refined before it can be used for human consumption. However, improvements in the fish oil processing industry, including proper deodorization and stabilization with antioxidants, have resulted in the availability of stable fish oils with a clean taste.

Extraction of Plant Fats

As mentioned previously, there are many methods for extracting fat from plants. Because of space limitations, several of the methods will receive only cursory attention. In almost all instances, the extraction of fat from plants requires extensive mechanical pretreatment of the plant tissue. Most plant fats are stored in the seeds, which can vary from soft-tissued fruits, such as avocadoes, to hard-shelled nuts. This variety of possible sources for plant fats clearly illustrates the need for a variety of approaches to extracting the fat. After a general cleaning step to remove foreign materials, such as sticks and stones, the pretreatment may include peeling, crushing, shelling, and/or dehulling, depending on the source of the fat. The extent of each of the pretreatments also depends certainly on the plant, as can be easily understood by comparing the dehulling of a peanut or a hazelnut with the dehulling of a sunflower seed or a palm kernel. As in the rendering of animal fat, some plants require a heat treatment (cooking) prior to the extraction; however, in extracting plant fats, the purpose is different than that in rendering animal fat. Cooking is usually done to coagulate proteins, rupture cell membranes, release fats out of protein-lipid interactions, and/or to break emulsions in the oilseeds. In addition, the seeds are often flaked prior to cooking in order to increase the surface area, especially when solvent extraction is subsequently used for the removal of the fats from the seeds. There are two major approaches for extracting the fat from the seeds: solvent extraction and mechanical extraction (pressing).

Pressing can be done in either a batch process or a continuous process. While batch processing is still used in some countries, continuous screw presses are used virtually exclusively in the United States. Continuous screw presses will extract the majority of the fat and leave a residual amount of fat in the seed below 5%.

In general, solvent extraction, which is most commonly done with hexane, is more efficient than mechanical extraction by pressing and can reduce the amount of fat that remains in the seed to below 3%. Because the relative amount of fat that partitions into the solvent over time decreases with each time increment, solvent extraction is more efficient than pressing, specifically for seeds with a low initial amount of fat. However, mechanical extraction works better for seeds with a high initial fat content. The mechanics of mingling the solvent with the flaked seeds are not as simple as one might think, because solvent cost, recovery, and cleaning are a large part of the operating costs. This operating cost is another reason why it is more efficient to use mechanical extraction for seeds with a high fat content. Usually, the seed flakes are successively washed with recovered solvent in order to increase yield and efficiency. Solvent extraction can be done using either a batch method or a continuous method. The continuous method is more common because of its higher efficiency. It involves a countercurrent extraction system: the flakes to be extracted are washed by solvent that already contains fat gained downstream in the extraction system. In other words, the fresh solvent essentially encounters flakes that have already lost most of their fat, because the fresh solvent enters the system at the end of the extraction system. The batch extraction system can also be set up as a countercurrent system by using solvent with an ever-increasing fat content to mix with flakes with lesser degrees of extraction. However, most batch systems use fresh solvent for each of the extraction steps, and depending on the extraction ratio of each step, the specific number of extraction steps needed to achieve a desired total extraction rate can be calculated.

Although the fat extraction may yield fat with a purity of up to 95%, the remaining impurities in this crude fat extract make the fat highly susceptible to oxidation. Thus, after being extracted, the fat must be cleaned to remove these impurities. This obligatory cleaning process is commonly called refining.

OBLIGATORY PROCESSING STEPS

DEGUMMING

The first step in cleaning the fat is usually degumming, which is the removal of phospholipids (sometimes incorrectly called "gums" because of functional properties that are similar to those of carbohydrate-based gums). Phospholipids have both lipophilic and hydrophilic moities that make them excellent emulsifiers, but they also allow faster spoilage of the fat because they are more susceptible to oxidation than triacylglycerides. Most phospholipids found in crude oils are diacylglycerides (in which the third alcohol group of the glycerol is esterified to phosphoric acid), which are called phosphatidic acids. The phosphatidic acids can have another moiety, such as choline, attached to the phosphoric acid, resulting in phosphatidyl choline, which is better known by the name of lecithin. Lecithin is an important by-product of the degumming process. In degumming, the fat is heated to about 165°F (74°C) and a small amount of water (1–3%) is added, which causes hydration of the phospholipids, making them soluble in the water and insoluble in the fat phase. Phospholipids that do not hydrate easily can be solubilized in the aqueous phase by a small amount of phosphoric acid. Other minor components, such as proteins and carbohydrates, are also removed as they enter the aqueous phase. Centrifugation is then used to separate the two phases.

NEUTRALIZATION

The neutralization step is also often referred to as caustic or alkali refining. However, because the term refining also refers to all of the combined steps of fat purification, the use of the term neutralization is recommended in order to avoid confusion.

Alkali

As mentioned above, some extraction procedures will generate free fatty acids. In addition, all fats contain a small amount of free fatty acids due to enzymatic actions in the plant or animal tissue prior to extraction. Because free fatty acids are more susceptible to oxidation than triacylglycerides, removal of free fatty acids is essential for the manufacture of a shelf-stable product. The easiest way to remove free fatty acids is by neutralization with alkali, such as sodium hydroxide, which essentially results in the formation of soaps. (Of course, the alkali must be in a low concentration to avoid saponification, that is, soap formation by breaking the ester bonds between the glycerol and the fatty acids. While saponification of the triacylglycerides, instead of just neutralization of the free fatty acids, might be desirable in soap production, it is considered a loss in the refining of edible fats.) The soaps are then removed by centrifugation, resulting in a fat with a free fatty acid content well below 0.05%. Recent research at the Texas Engineering Experiment Station improved this process by using sodium silicate. The advantage of this new process is that not only does the sodium silicate neutralize the free fatty acids, the excess sodium silicate then functions as an absorbent for the soap, allowing for removal of the soap by simple filtration instead of the more energy intensive centrifugation.

Distillation

A different approach to getting rid of the free fatty acids is vacuum distillation, also called steam refining or physical refining. Free fatty acids are considerably more volatile than triacylglycerides and can be removed by a simple distillation procedure (also see Deodorization, below). However, while the alkali neutralization step can be done without prior degumming, removal of free fatty acids by distillation is problematic when a crude fat has been insufficiently degummed because the heating step will cause foaming and darkening of the fat.

BLEACHING

With few exceptions, such as olive oil, consumers expect their fat to have little or no color. In addition, many pigments are prooxidants that will make a fat more susceptible to oxidation. Thus, most fats and oils are bleached in order to remove pigments. The bleaching process involves the use of an absorbent, such as Fuller's earth, which will also remove some minor residual impurities, such as soaps that may not have been removed in the neutralization step, chelated metals, and peroxides, that are the source of off flavors. The bleaching process is usually done under vacuum and is a continuous process. The amount of absorbent used is approximately 1%, but can vary between 0.2 and 2%, depending on the fat.

DEODORIZATION

Although deodorization is listed here before the optional processing steps because it is a mandatory step, it is usually the last step in the refining process. Thus, it is done after any optional processing step that may have been selected to modify fat functionality has been completed. Fats are excellent solvents for most flavor compounds, and many fats contain a variety of volatile chemicals that are odor active and

thus impart a smell and flavor to the fat. In addition, several of the flavor volatiles are secondary products of fat oxidation and impart a rancid quality to the oil. Because fats and oils, with some exceptions such as butter and margarines, are rarely consumed directly, but are ingredients in industrial and home food production, consumers expect most fats and oils to be bland in odor and flavor. The concentration of the various volatile components ranges widely, but rarely exceeds 1000 ppm or 0.1%. Nevertheless, due to the odor activity of many volatile compounds at far lower levels, their removal is a critical component of producing an acceptable fat. Because of the similarity in volatility of many odor compounds and the free fatty acids, this processing step can also be used in lieu of the neutralization step because it removes free fatty acids as well (also see Distillation, above). The deodorization process is essentially a steam distillation under vacuum and is based on the large difference in vapor pressure between the triacylglycerides and the volatile impurities. The vacuum lowers the boiling point and increases the vapor pressure of the various volatile components even further; the steam aids in the evaporation of the volatiles because it can come into intimate contact with the fat, allowing the volatiles to be "carried out" of the fat.

OPTIONAL PROCESSING STEPS

Fat manufacturers have the option of choosing one or several processing steps that are not required to yield a stable fat, but which, nevertheless, might be desirable because they can be used to change the characteristics and functionalities of a fat.

DEWAXING

Waxes are esters of long-chain free fatty acids and monohydroxyl alcohols. Most waxes have a high melting point, causing turbidity of a liquid fat over time. Waxes rarely influence the overall functionality of the fats and are removed by slightly chilling the fat. Because of their high melting point, waxes will crystallize out before the triacylglycerides do and can be filtered out (also see Winterizing/ Fractionation, below).

HYDROGENATION

Hydrogenation is probably the most commonly used optional processing step. The purpose of hydrogenation is the saturation of fats, which is the addition of hydrogens to the double bonds in the fats. Hydrogenation will increase the melting point of a triacylglyceride. Fats are hydrogenated for various reasons, including changing the plastic properties of the fat. However, most importantly, hydrogenation lowers the degree of unsaturation, which makes the fat more resistant to oxidation. A classical example is the partial hydrogenation of soybean oil. Soybean oil contains small amounts of linolenic acid (C18:3), which are responsible for flavor reversion (off flavor development). Partial hydrogenation of the oil eliminates the linolenic acid, resulting in a much more stable fat. Because of the differences in reaction rates, highly unsaturated fatty acids are hydrogenated quicker than fatty acids with fewer double bonds. In hydrogenation, the fat is mixed with hydrogen gas and a catalyst. The most commonly used catalyst is nickel. Hydrogenation is usually done in a closed vessel at high temperatures and pressures. A problem with hydrogenation that has recently gained considerable attention is the development of *trans* fatty acids. In nature, fatty acids contain almost exclusively *cis* double bonds. However, during hydrogenation the double bond is cleaved due to the addition of the catalyst at the double-bond site. Especially in partial hydrolysis situations, the catalyst may split off the fatty acid without hydrogenation taking place. In this case, the double bond may be reestablished in the *trans* configuration because, for the short time span that the catalyst is bound to the fatty acid, the original double bond becomes a freely rotatable single bond, allowing for *trans* fatty acid formation. Over the last decade, the hydrogenation process has undergone numerous modifications to reduce the formation of *trans* fatty acids because of their health implications.

INTERESTERIFICATION

Although interesterification can dramatically change the functionality of a fat, it is not commonly practiced because of a lack of control over the resulting fat. Unlike hydrogenation or winterizing, interesterification does not change the overall fatty acid profile of the fat; it rearranges the fatty acids within and among triacylglycerides by hydrolyzing and reesterifying ester bonds between the fatty acids and the glycerol molecules. The result is a fat with a narrower melting range due to more random distribution of fatty acids among the triacylglycerides. Because of the formation of *trans* fatty acids in hy-

drogenation, interesterification, in combination with blending, has recently received considerable attention as a possible replacement for hydrogenation.

WINTERIZING/FRACTIONATION

The term fractionation is almost self-explanatory: the fat is divided into fractions. Fractionation is based on the differences in melting point among the various triacylglycerides. Because the fractionation of the fat is based on temperatures well below room temperature, the process has also been termed winterizing. In winterizing, the temperature of the fat is lowered, which causes triacylglycerides with a high melting point to crystallize, that is, solidify. Triacylglycerides with high melting points contain a relatively larger share of either saturated fatty acids or *trans* fatty acids. Thus, winterization can be used to reduce the amount of *trans* fatty acids that were formed during hydrogenation, although there is no selectivity for *trans* fatty acids. Fats with a wide melting range, such as cottonseed oil and palm oil, require winterizing, but winterizing is also practiced with animal fats such as lard and tallow. The effect of winterizing can be easily modeled in a home setting by placing olive oil in a refrigerator. The commercial process often involves "seeding" the liquid fat (oil) with a solid fat, allowing for faster crystallization. The fat is then chilled via cooling coils at a specific rate that depends on the type of fat. The chill rate and agitation rate are crucial for proper crystallization. The solid, crystallized fat is then separated from the liquid fraction by means of filtration.

PLASTICIZING/TEMPERING

While the fractionation process is used to create a fat that remains a liquid (oil) at refrigeration temperatures, the plasticizing or tempering process is used to give a fat that is solid at room temperature a certain functionality. The process is essentially identical to that of winterizing, except that no fractions are separated. Because of the multitude of triacylglycerides it contains, a fat is never truly a complete solid. While winterizing allows the removal of high-melting-point triacylglycerides to obtain a fat that is liquid at refrigeration temperatures and above, even most solid fats (e.g., lard, tallow, cocoa butter) are not fully crystallized fats: they still contain liquid or noncrystallized triacylglycerides; hence, the SFI (solid fat index) measurement. In the plasticizing process a fat that is mostly solid at room temperature is heated well above its melting range. The rate of cooling it back down to room temperature and the degree of agitation will direct the crystallization, influencing not only the crystal size, but also, more importantly, the crystal type, which is of great importance in the manufacturing of foods containing solid fats such as chocolate products. For a thorough discussion of fat crystallization, the reader is referred to Schmidt (2000).

FINISHED PRODUCT

The product of fat refining is a 99.9% pure mixture of triacylglycerides that has a bland flavor and a free fatty acid content $\leq 0.05\%$. The peroxide value, a measure of the degree of oxidation, is ≤ 0.2. The color depends on the origin of the fat, but is usually a 2.0–3.0 on the red Lovibond color scale, which is equal to a light yellow.

APPLICATION OF PROCESSING PRINCIPLES

Table 20.1 provides recent references for more details on specific processing principles.

Table 20.1. References for Principles Involved in Fat Processing

Processing Stage	Processing Principles	References for More Information on the Principles Used
Rendering	To separate the lipid fraction (physical: difference in melting point; chemical: difference in solubility)	Williams and Hron 1996, Tufft 1996
Degumming	To remove phospholipids by difference in aqueous solubility	Carlson 1991, Haraldsson 1983
Neutralization	To remove free fatty acids (physical: distillation; chemical: salt formation due to the chemical reaction of the acids with a base followed by centrifugation)	Hodgson 1996, Carlson, 1991
Bleaching	To remove pigments by absorption	Hodgson 1996, Carlson 1991
Deodorization	To remove odors by evaporation/ distillation	Carlson 1991, 1996
Dewaxing	To remove waxes by melting point difference	Haraldsson 1983, Carlson 1991
Hydrogenation	To remove polyunsaturated fatty acids by chemically saturating the double bonds	Hastert 1996, Carlson 1991
Interesterification	To narrow the melting range by chemically breaking ester bonds and reforming ester bonds	Liu and Lampert 1999, Ainsworth et al. 1996, Ramamurthi and McCurdy 1996
Winterizing/fractionation	To narrow the melting range by separating triacylglycerides by differences in melting point to yield an oil	Krishnamurthy and Kellens 1996, Carlson 1991
Plasticizing/tempering	To narrow the melting range by separating triacylglycerides by differences in melting point to yield a fat	Lawson 1985

ACKNOWLEDGMENT

The author acknowledges the help of Drs. A. Clarke and Y.H. Hui in the proofreading of the manuscript, and the support of the University of Missouri Agricultural Experiment Station.

GLOSSARY

Adipose tissue—the tissue of animals that holds the fat cells.

Alkali refining—see *Caustic refining*.

Bleaching—removal of pigments.

Caustic refining— the removal of free fatty acids; also called neutralization.

Cis double bond—a double bond in which the hydrogen atoms on the carbon atoms that are connected by the double bond are on the same side of the molecule.

Cooking—a heating step in the extraction of fat from plants.

Crude fat—unprocessed fat extracted or rendered from plant or animal tissue.

Degumming—removal of phospholipids from a crude fat.

Deodorization—removal of compounds that have greater volatility than triacylglycerides from fats by distillation.

Dewaxing—removal of waxes from fats.

Diacylglyceride—a glycerol molecule with two of its three hydroxyl groups esterified to fatty acids.

Distillation—a method that separates chemicals based on their volatility using heat.

Fat—a mixture of triacylglycerides that is solid at room temperature.

Flavor reversion—oxidative spoilage of fats specifically found in fats containing linolenic acid, such as soybean oil.

Fractionation—dividing a mixture of compounds into fractions (also see Winterizing).

Fuller's earth—an absorbent used to remove pigments.

Hydrogenation—saturation (removal) of double bonds.

Hydrolysis—the breaking of the ester bond between the fatty acids and the glycerol molecule.

Interesterification—randomization of fatty acids within and among triacylglycerides by breaking and reestablishing ester bonds.

Lecithin—phosphatidyl choline—a phospholipid.

Linolenic acid—all-*cis*-9,12,15-octadecatrienoic acid

Marine oil—triacylglycerides extracted from fish, high in omega-3 fatty acids.

Monoacylglyceride—a glycerol molecule with one of its three hydroxyl groups esterified to fatty acids.

Neutralization—the removal of free fatty acids (also called caustic refining).

Oil—a mixture of triacylglycerides that is liquid at room temperature.

Omega-3 (n-3) fatty acid—a fatty acid that has the first double bond on the third carbon atom from the methyl end.

Oxidation—the degradation of fats by oxidation of the carbon atoms adjacent to the double bonds, resulting in low-molecular-weight compounds such as aldehydes and alcohols.

Peroxide—an early oxidation product in fat oxidation.

Phosphatidic acid—a diacylglyceride with the third hydroxyl group of the glycerol esterified to phosphoric acid.

Phospholipid—a lipid containing a phosphate group.

Physical refining—see *Steam refining*.

Pigments—organic chemicals that reflect light at specific visible wavelengths that impart color to the fat.

Plasticizing—changing the functionalities of fats by changing their crystal structure using tempering.

ppm—parts per million.

Rendering—the extraction of fat from animal tissue.

Refining—the combination of processing steps to purify a crude fat into an edible fat.

SFI (solid fat index)—percent of triacylglycerides in a fat that are solid.

Solvent extraction—use of an organic solvent to remove fats from plant tissues.

Steam refining—the removal of free fatty acids by steam distillation.

Tankage—layer of fat on top of the aqueous phase after steam rendering.

Tempering—heating and cooling of a fat with agitation that promotes the formation of a specific crystal structure.

Trans double bond—double bond in which the hydrogen atoms on the carbon atoms that are connected by the double bond are on opposite sides of the molecule.

Triacylglyceride—glycerol molecule with all three of its hydroxyl groups esterified to fatty acids.

Winterizing—the removal of triacylglycerides with high melting points by crystallizing them out at refrigeration temperatures.

REFERENCES

Ainsworth S, C Versteeg, M Palmer, MB Millikan. 1996. Enzymatic interesterifaction of fats. The Australian Journal of Dairy Technology 51(2): 105–107.

Belitz H-D, W Grosch. 1999. Lipids. Chapter 3. In: Food Chemistry, 152–236. Springer, Berlin, Germany.

Carlson KF. 1991. Fats and oils processing. INFORM 2(12): 1046–1060.

___. 1996. Chapter 6. Deodorization. In: YH Hui, editor. Bailey's Industrial Oil and Fat Products. Vol. 4, Edible Oil and Fat Products: Processing Technology, 5th edition. Wiley Interscience Publication, New York.

Duncan SE. 2000. Chapter 6. Lipids: Basic concepts. In: GL Christen, JS Smith, editors. Food Chemistry: Principles and Applications, 79–96. Science Technology System, West Sacramento, Calif.

Erickson DR. 1990. Edible fats and oils processing: Basic principles and modern practices. American Oil Chemists' Society, Champaign, Ill.

Gunstone FD. 1996. Fatty Acid and Lipid Chemistry. Blackie Academic and Professional, London, U.K.

Gunstone FD, FA Norris. 1983. Lipids in Foods: Chemistry, Biochemistry and Technology. Pergamon Press, Oxford, U.K.

Haraldsson G. 1983. Degumming, dewaxing and refining. Journal of the American Oil Chemists' Society 60(2): 251–256.

Hastert RC. 1996. Chapter 4. Hydrogenation. In: YH Hui, editor. Bailey's Industrial Oil and Fat Products, 5th edition. Vol. 4, Edible Oil and Fat Products: Processing Technology. Wiley Interscience Publication, New York.

Hernandez E, S Rathbone. 2002. TEES researchers develop cheaper, more efficient method of producing vegetable oil. Webpage accessed April 1, 2003. http://tees.tamu.edu/portal/page?_pageid=33,31327&_dad=portal&_schema=PORTAL&p_news_id=297

Hodgson AS. 1996. Chapter 3. Refining and bleaching. In: YH Hui, editor. Bailey's Industrial Oil and Fat Products, 5th edition. Vol. 4, Edible Oil and Fat Products: Processing Technology. Wiley Interscience Publication, New York.

Hui YH. 1996. Bailey's Industrial Oil and Fat Products, 5th edition. Vol. 4, Edible Oil and Fat Products: Processing Technology. Wiley Interscience Publication, New York.

Krishnamurthy R, M Kellens. 1996. Chapter 5. Fractionation and winterization. In: YH Hui, editor. Bailey's Industrial Oil and Fat Products, 5th edition. Vol. 4, Edible Oil and Fat Products: Processing Technology. Wiley Interscience Publication, New York.

Lawson HW. 1985. Chapter 6. Processing technology. In: Standards for Fats and Oils, 33–43. AVI Publishing Co., Westport, Conn.

___. 1985. Standards for Fats and Oils. AVI Publishing Co., Westport, Conn.

List GR. 2004. Decreasing *trans* and saturated fatty acid content in food oils. Food Technol. 58(1): 23-31.

Liu L, D Lampert. 1999. Monitoring chemical interesterification. Journal of the American Oil Chemists' Society 76(7): 783–787.

Ramamurthi S, AA McCurdy. 1996. Interesterification—Current status and future prospects. In: Development and Processing of Vegetable Oils for Human Nutrition, 62–86. AOCS Press, Champaign, Ill.

Schmidt K. 2000. Chapter 7. Lipids: Functional properties. In: GL Christen, JS Smith, editors. Food Chemistry: Principles and Applications, 97–113. Science Technology System, West Sacramento, Calif.

Tufft LS. 1996. Chapter 1. Rendering. In: YH Hui, editor. Bailey's Industrial Oil and Fat Products, 5th edition. Vol. 5, Industrial and Consumer Nonedible Products from Oils and Fats. Wiley Interscience Publication, New York.

U.S. Patent 6,448,423 (Hernandez, et al.). 2002. Refining of glyceride oils by treatment with silicate solutions and filtration. The Texas A&M University System, Sept. 10, 2002.

United Nations Industrial Development Organization. 1977. Guidelines for the Establishment and Operation of Vegetable Oil Factories. United Nations, New York.

Williams MA, RJ Hron. 1996. Chapter 2. Obtaining oils and fats from source material. In: YH Hui, editor. Bailey's Industrial Oil and Fat Products, 5th edition. Vol. 4, Edible Oil and Fat Products: Processing Technology. Wiley Interscience Publication, New York.

21
Fruits: Orange Juice Processing

Y. H. Hui

The information in this chapter, with minor modifications, has been derived from *The Orange Book,* ©1998, copyrighted and distributed by Tetra Pak Processing Systems AB, Lund, Sweden. Used with permission. The minor modifications affect only the headings of sections. Please consult the original document when it is necessary for the following reasons: (1) Since the original chapter has been processed (transcription, reproduction, and proofreading), there may be errors or deficiencies. If so, the author bears sole responsibility. (2) This chapter is only one of the 13 chapters in the book.

BACKGROUND INFORMATION ON AN ORANGE JUICE PROCESSING PLANT

Orange processing plants are located in the vicinity of the fruit growing area. Fruit should be processed as soon as possible after harvesting because fruit deteriorates quickly at the high temperatures found in citrus-growing areas. Orange products, on the other hand, are produced in a form that allows them to be stored for extended periods and shipped over long distances. In the orange industry, the basic unit of reporting crop and plant intake is commonly the fruit box.

A box of oranges is defined as containing 40.8 kg (90 pounds) of fruit. In Florida, the small to medium-size plants typically process 5–10 million boxes (200,000–400,000 tons) per season, the large plants up to 25 million boxes. Most Brazilian citrus plants have a much higher capacity. One of the world's largest orange juice plants, Citrosuco, at Mattão, Brazil, can take in 60 million boxes (2.4 million tons) of fruit during a season. In most other orange-growing regions, citrus processing plants are considerably smaller than those located in Florida and Brazil.

FRUIT RECEPTION

Fruit is delivered in trucks that discharge their loads at the fruit reception area. The fruit may be prewashed to eliminate immediate surface dirt and pesticide residue before any leaves and stems still attached to the fruit are removed. Pregrading by manual inspection to remove any unsuitable fruit follows. Sound fruit is conveyed to storage bins. Damaged fruit goes directly to the feed mill.

EXTRACTION

Extraction involves squeezing or reaming juice out of either whole or halved oranges by means of mechanical pressure. After final washing and inspection, the fruit is separated according to size into different streams or lanes. Individual oranges are directed to the most suitable extractor in order to achieve optimum juice yield. As the extraction operation determines juice yield and quality, the correct setting of extractor operating conditions is very important.

CLARIFICATION

After extraction, the pulpy juice (about 50% of the fruit) is clarified by primary finishers, which separate juice from pulp. The finishing process is a mechanical separation method based on sieving. The juice stream is further clarified by centrifugation. The pulp stream, containing pieces of ruptured juice sacs and segment walls, may then go to either pulp recovery or pulp washing.

FROZEN CONCENTRATED ORANGE JUICE (FCOJ) PRODUCTION

From the buffer/blending tanks and after clarification, juice goes to the evaporator. Within the evaporator circuit, the juice is first preheated and held at pasteurization temperature. It then passes through the evaporation stages of the process, where it is concentrated up to 66° Brix. During the evaporation process, volatile flavor components flash off, but they can be recovered in an essence recovery unit.

Juice concentrate is cooled and blended with other production batches as required to level out fluctuations in quality. It then goes to frozen storage in tanks or drums as FCOJ, sometimes for several years.

NOT-FROM-CONCENTRATE JUICE (NFC) PRODUCTION

Instead of being concentrated, juice may be processed at single strength as an NFC product. Clarified juice is pasteurized before storage. Deoiling may be required to reduce oil levels in the juice, and deaeration to remove oxygen is part of good practice. Since the product is consumed year-

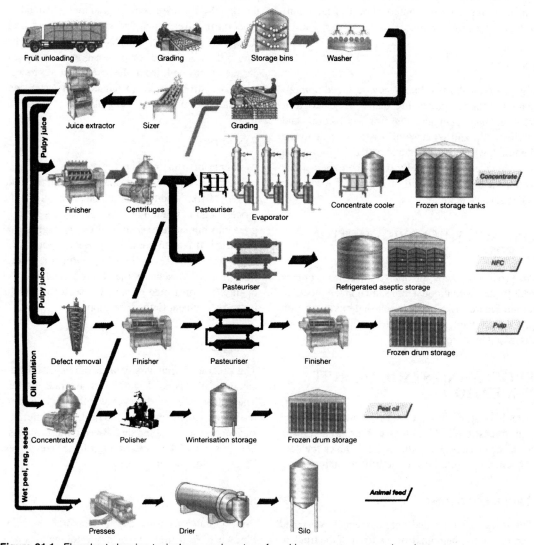

Figure 21.1. Flowchart showing typical processing steps found in an orange processing plant.

round but production is seasonal, NFC juice may be stored for up to one year. It is stored in bulk either frozen or under aseptic conditions.

PULP PRODUCTION

For pulp recovery, pulpy juice from the extractor is passed through a defect removal system, where undesirable pulp components, such as seed and rag, are removed. The clean pulp stream is then concentrated in a primary finisher. After heat treatment, the pulp slurry is typically concentrated further before being sent to frozen storage.

PULP WASH

If the pulp fraction is not recovered for commercial sale, pulp from the final juice finishers and clarifiers can be washed with water to recover juice solubles. This stream is called pulp wash and may, legislation permitting, be blended with juice concentrate.

PEEL OIL RECOVERY

Recovered peel oil represents some 0.3% of the fruit intake. The emulsion of oil and water coming from the extractor section is clarified by centrifugation in two steps. The purified oil contains dissolved waxes, which are removed by winterization (refrigeration) of the oil for a certain time.

FEED MILL

It is economically feasible to include a feed mill operation in larger processing plants. Rejected fruit from grading, peel and rag from extraction, and washed pulp and other solid waste are sent to the feed mill, where it is dried and pelletized for animal feed. Smaller plants usually truck their solid waste to a plant with a feed mill or dispose of it in other ways, such as landfill.

ORANGE JUICE PRODUCTION STEPS

All production steps for orange juice, from orange fruit to packaged product, are shown in the block diagram Figure 21.2. The steps carried out in the fruit processing plant, as highlighted in the figure, are discussed in more detail.

PROCESSING STAGE 1: FRUIT RECEPTION

After harvesting, fruit picked in the groves is loaded onto trucks (typically 20 tons in Florida) and taken to the processing plant. Figure 21.3 shows the subsequent processing flow at the fruit reception.

TRUCK UNLOADING

The trucks are unloaded onto a specially designed tipping ramp. The ramp lifts the front of the truck to allow the fruit to roll off the rear of the trailer directly onto a conveyor. The fruit is then conveyed to the prewash station. Alternatively, the truck may be reversed down a ramp so that the fruit is unloaded directly onto a conveyor.

PREWASHING, DESTEMMING, AND PREGRADING

The fruit may undergo initial washing to remove dust, dirt, and pesticide residues. Some processors have discontinued washing the fruit before bin storage because wet fruit in the bins can make downstream sanitation more difficult. The fruit then moves on to destemming and pregrading.

The roller conveyor of the destemming and pre-grading tables allows any leaves or twigs to fall through the conveyor bed. Pregrading by manual inspection removes rotten and visibly damaged fruit. Rejected fruit, known as culls, may be sent to the feed mill. Water used for prewashing is often condensate recovered from the evaporation process: there is a strong desire to reduce total water consumption in orange processing plants.

SAMPLING

A sample of fruit is taken for analysis from each truck. The main parameters analyzed are juice yield, degrees Brix, acidity, and color. This gives the processor an indication of fruit ripeness. As the fruit goes into bin storage, each load can be tagged and identified. It is then possible to select suitable fruit from various sources for blending during the extraction process to achieve the desired final product quality. The measured juice yield may also form the basis for payment to the fruit supplier.

FRUIT STORAGE

The pregraded fruit is stored in bins specially designed with inclined multilevel internal baffles. These distribute the fruit evenly in the bin to prevent too much weight pressing on fruit. The time fruit stays in the storage bins should be as short as possible, less than 24 hours. Storage for longer times, however, does occur.

SURGE BIN

Fruit is drawn from the storage bins into the surge bin, where fruit from one or more storage bins may be combined.

FINAL FRUIT WASHING

Thorough washing of the fruit is carried out immediately before the extraction process. The wash water may include a mild disinfectant to help reduce the microbial population on the fruit surface. Fresh water or condensate recovered from the evaporators is used for final washing.

FINAL GRADING

The fruit passes over a series of grading tables for final visual inspection, where damaged or unsuitable fruit is removed.

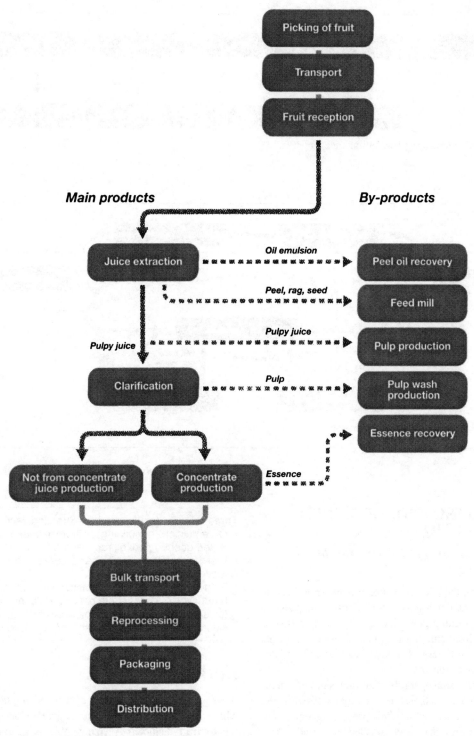

Figure 21.2. Production steps for orange juice.

Figure 21.3. Processing flow for fruit reception.

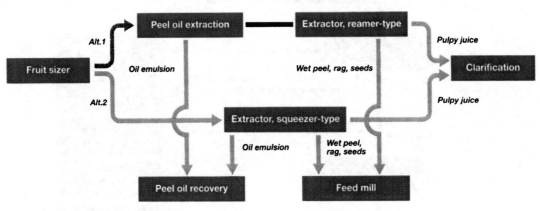

Figure 21.4. The juice extraction process.

PROCESSING STAGE 2: JUICE EXTRACTION

GENERAL CONSIDERATIONS AND FRUIT SIZING

The aim of the juice extraction process (Fig. 21.4) is to obtain as much juice out of the fruit as possible while preventing rag, oil, and other components of the fruit from entering the juice. These components may lead to bitterness in taste or other defects later during juice storage.

The extraction operation determines product quality and yield, and therefore has a major effect on the total economics of the fruit processing operation. Once the fruit has been washed and graded (inspected), it is ready for the extraction process. To optimize extractor performance, the raw fruit must be sorted according to size because individual extractors are set to handle fruit of only a certain size range.

Three streams result from the extraction section:
- Oil emulsion, containing oil from the peel and water, goes to peel oil recovery.
- Wet peel, with pulp, rag, and seeds, flows directly to the feed mill.
- Pulpy juice goes first to clarification and then to production of concentrate or NFC. Pulp intended for sale as pulp goes to pulp production. Residual pulp goes to pulp washing or the feed mill.

Fruit Sizing

After grading, the fruit passes over the sizing table, which divides the fruit into different streams according to fruit diameter. A sizing table is generally made up of a series of rotating rollers over which the fruit passes. The distance between the rollers is preset, and increases as the fruit travels over the table. Over the first set of rollers, the smallest fruit drops between the gaps onto a conveyor, which carries them to an extractor set for their particular size range. As the gap increases, larger fruit will pass between the rollers onto extractors set for their defined size range. In this way, all the fruit is selected to suit the individual settings of the extractors. There are normally two to three different size settings in an extractor line.

A well-functioning fruit sizer is essential to producing juice of high quality and/or yield. Depending on the extractor type, any fruit, small or big, can be oversqueezed or undersqueezed. If the fruit is oversqueezed, excessive rag and peel will get into the juice, with resulting bitterness. If the fruit is undersqueezed, insufficient yield will result. See next section.

Extractor Types

Two types of extractor dominate in orange processing plants, the squeezer type and the reamer type. For these two types there are two major brands, FMC (squeezer type) and Brown (reamer type). Both extraction systems are dedicated to citrus fruit. The reamer-type extraction system provides excellent separation of the orange components juice, oil, and peel. It works best—as regards both product quality and yield—with fruit that is round in shape and of uniform ripeness such as is found with Florida fruit. Squeezer-type extractors are less sensitive to the size and shape of the fruit but can lead to higher oil content in the juice and more damaged pulp than with reamer-type extractors. Adjustments to the standard squeezer-type extractor may be needed to keep oil levels low and/or improve pulp quality.

Globally, squeezer-type extractors are the most common. However, in Florida, the total installed extraction capacity is about equal for these two types of extractor. The major share of the NFC produced in Florida is extracted using reamer-type extractors.

Another type of extraction equipment is the rotary press extractor. These are more multipurpose machines that may also be used to process other types of fruit. With rotary press extractors, the fruit is cut in half, and the halves pass between rotating cylinders that press out the juice. Oil is extracted from the peel in a separate step prior to extraction. Although the extraction process is simple, both juice yield and juice quality are less optimal than those obtained with squeezer- and reamer-type extractors. Rotary press extractors, which have a high capacity per unit and require lower investment, are popular in the Mediterranean area. However, they are of minor im-

Figure 21.5. A squeezer-type orange juice extractor.

portance globally in comparison with squeezer- and reamer-type extractors.

Once installed in a plant, extraction systems are not easily interchangeable due to the different demands on the surrounding equipment.

THE SQUEEZER-TYPE EXTRACTOR

A squeezer-type extractor is shown in Figure 21.5. These are placed in lines in the extractor room with up to 15 extractors per line. Each extractor may be fitted with five heads, which are available in different sizes so that they can handle the type and quality of fruit available. Typical sizes are 2 3/8, 3, 4, and even 5 inches (used mainly for grapefruit). The head size for each extractor in a line is chosen to optimize the handling of sized fruit. The extractor separates the fruit into four parts—pulpy juice, peel, core (rag, seeds, and pulp), and oil emulsion.

The head of an extractor comprises an upper and a lower cup (see Fig. 21.6). The cups have metal fingers that mesh together as the upper cup is lowered onto the lower cup. A cutter comes up through the center of the lower cup to cut a hole through the skin in order to allow the inner parts of the orange to flow out. The cutter is part of the perforated strainer tube, sometimes referred to as the prefinisher.

Once the strainer tube has cut into the fruit, the upper cup squeezes down on the lower cup. This pressure initially forces the juice to burst out of the juice vesicles and pass through the perforations of the strainer tube. Some of the pieces of the ruptured juice sacs (i.e., pulp) will pass through with the

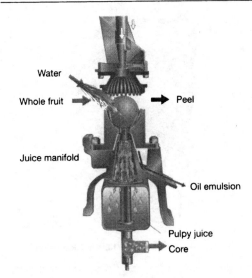

Water

Whole fruit ➡   ➡ Peel

Juice manifold

Oil emulsion

Pulpy juice
Core

Figure 21.6. Operation of the squeezer-type orange juice extractor.

juice. The upper cup continues to squeeze down on the lower cup to extract as much juice as possible.

Eventually, the downward pressure causes the peel to break up, disintegrate, and pass up through the fingers of each cup. Juice flows through the strainer tube into the juice manifold. The core material is discharged from the bottom of the strainer tube through the orifice tube.

As the peel is forced through the fingers of the cups during the last step of the extraction cycle, oil is released from the peel. The bits of peel are washed with recycled water to extract the oil from the oil sacs. The oil is discharged from the extractors as an emulsion with water.

With squeezer-type extractors, one item of equipment, the extractor, separates the fruit into four principal product streams in one basic step. It is claimed that contact is avoided between the juice and oil, and the juice and peel.

For successful operation of this equipment, the correct selection of cup size and adjustment of the cup and cutter operation are important. Too much pressure applied to fruit resulting from the use of undersize cups may result in peel entering the juice stream. If too little pressure is applied, the yield will drop.

The throughput of a five-head extractor will vary according to the quality and size of fruit. The standard operating speed is 100 rpm, or 500 oranges per minute. Fruit will not always flow to each cup: 90% utilization is high. A typical capacity for medium-sized fruit is 5 tons/hour of fruit per extractor, corresponding to about 2500 l/h of juice.

Modifications for Premium Pulp

As the pulpy juice passes through the holes in the strainer tube in the standard extractor, the pulp tends to be broken up into small pieces, typically $\leq$ 2–3 mm in length. This is acceptable if the pulp is intended for pulp wash and as commercial pulp for certain markets.

Market demands in the juice market are changing, and the need for more "natural" pulp that has been subjected to less shear is increasing. In a squeezer-type extractor of modified design, larger pulp pieces, up to 15–20 mm long, flow along with the juice stream. The main difference in design is the use of a modified strainer tube with larger openings that allow more pulp to remain in the juice stream. The pulp is subsequently separated from the juice and treated in a modified pulp recovery system. In 1995 there were a handful of such premium pulp lines in Florida.

Premium Juice Low-Oil Extractor

Certain fruit varieties (e.g., the Florida Valencia) will express more oil into the juice stream than other varieties. This can lead to oil content in the juice exceeding acceptable levels (e.g., 0.035%, the maximum level permitted in Florida for grade "A" juice).

This is a problem with NFC juice but is less so with juice intended for concentrate, because most of the oil will flash off in the evaporator. In the low-oil version of the squeezer-type extractor, the design of the strainer tube and orifice tube area is modified. This unit cuts a smaller core and puts less pressure on the fruit during extraction, thereby reducing the amount of peel oil that gets into the juice. These modifications may also lead to a reduction in juice yield. When the top spray of water is stopped, the amount of peel oil to be recovered is thereby reduced.

As an alternative, hermetic centrifuges or vacuum flashing can be used in conjunction with the standard extractors to deoil the single-strength juice. This allows a higher juice yield to be maintained during extraction; excess oil is then removed after the extraction process.

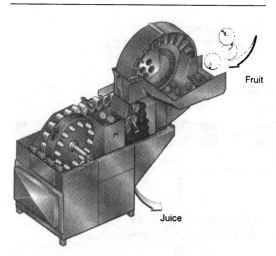

Figure 21.7. A reamer-type orange juice extractor.

THE REAMER-TYPE EXTRACTOR

The reamer-type extractor is based on the same principle as a typical manual kitchen squeezer used to prepare orange juice for breakfast. An extraction line comprises several extractors, and it is very important to set up each extractor to suit the size of fruit fed into it. A reamer-type extractor is illustrated in Figure 21.7.

Fruit is fed into the feed wheel and cut in half. The halves are oriented and picked up in synthetic rubber cups mounted on a continuous belt system. A series of nylon reamers (cone-shaped inserts that have ridges molded into the form from tip to base) are mounted on a rotating turntable.

The reamers, in the vertical plane for most models, enter each fruit half and rotate as they penetrate them. The speed of rotation varies as the reamer penetrates the fruit, being slower towards the end of the operation. Juice, pulp, rag, and seeds pass out through one outlet, and the peel segment passes out through the peel chute.

The juice and pulp are separated from the rag and seeds by a strainer, then passed on to the finishers. The size, pressure, and rotation speed of the reamer can be adjusted to suit the maturity, size, and quality of fruit.

The reamer-type system typically gives a better quality of pulp (longer and larger cell fragments) than standard squeezer-type extractors. Juice yields from the two systems are comparable.

The Oil Extraction System

Peel oil can be recovered from the peel using a separate oil extraction system, which is placed upstream of the juice extractors. It operates on the principle of puncturing oil sacs in the flavedo and washing the oil out to make an emulsion (see Fig. 21.8). In the first stage of the oil extraction system, whole fruit passes over a series of rollers with small but sharp needle-like projections. The oil glands are pricked rather than scraped open, so that little damage is done to the peel. Therefore, the amount of contaminating material washed away with the oil is minimal. This, in turn, makes the water stream separated from the emulsion cleaner and easier to recycle.

The rollers conveying fruit are placed in a water bath, and the oil from the pierced glands is washed out with water. After a finishing (straining) stage to remove any large particles of peel, the oil-water emulsion can be concentrated and polished in a series of centrifuges (see section on peel oil recovery). The water is recycled to a large degree.

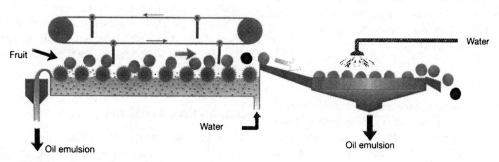

Figure 21.8. An oil extraction system.

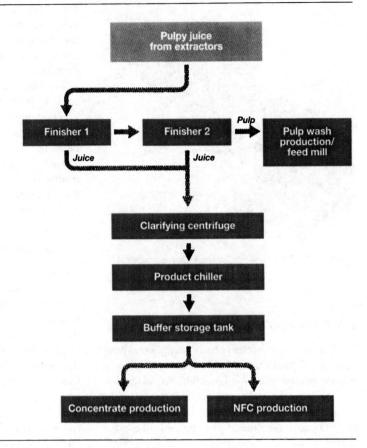

Figure 21.9. The clarification process.

Instead of the recently developed oil extraction system upstream of juice extraction, older installations incorporate peel shavers placed after the juice extraction stage. The outer layer of flavedo is literally shaved off from the peel mechanically. It is washed and pressed to remove the oil. The emulsion is then centrifuged in the conventional manner.

The reamer-type extraction system requires two separate steps to extract juice and oil from the fruit. Nevertheless, the oil emulsion is often considered cleaner and easier to centrifuge than emulsions from other types oil recovery systems, and the extracted juice has less contact with the oil.

Downstream of the Juice Extractors

The juice streams leaving either a squeezer-type extractor line or a reamer-type extractor system flow to clarification and then evaporation, or pasteurization if the end product is NFC juice. The oil emulsion flows to peel oil recovery for separation by centrifugation. Peel, rag, seeds, and other solid material are conveyed to the feed mill.

PROCESSING STAGE 3: CLARIFICATION

The juice leaving the extraction process is clarified because it contains too much pulp and membrane material to be processed in the evaporator or as NFC juice. Typical process steps in juice clarification are shown in Figure 21.9. Pulp levels in pulpy juice from the extractors are generally around 20–25% floating and sinking pulp. The juice is therefore finished, that is, pulp is removed from the juice. A finisher is basically a cylindrical sieving screen. There are two types of finisher: screw-type and paddle. Their operating principles are shown in Figures 21.10 and 21.11.

SCREW-TYPE FINISHERS

These include a stainless steel screw that conveys the pulpy juice through the unit and presses the pulp against the cylindrical screen. The juice flows through the screen holes.

The pulp is consequently "concentrated" inside

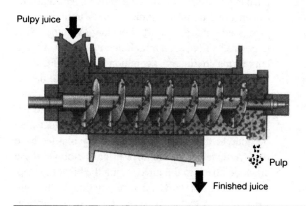

Pulpy juice

Pulp

Finished juice

Figure 21.10. Operation of screw-type finishers.

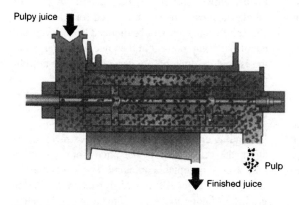

Pulpy juice

Pulp

Finished juice

Figure 21.11. Operation of paddle finishers.

the screen and is discharged at the end of the finisher. As pulp is discharged through a restricted opening, the resulting back pressure in the finisher helps to squeeze out more juice from the pulp mass.

PADDLE FINISHERS

These finishers incorporate a set of paddles rotating on a central shaft within the cylinder. The pulp is pushed against the screen by the paddles. Paddle finishers apply centrifugal force rather than pressure to separate the pulp from the juice. This usually provides gentler pulp treatment than screw finishers.

Two finishers are often placed in series at the end of the extraction line. The upstream primary finisher is not set as tight as the downstream secondary unit, and so will have a higher flow capacity.

The exact configuration of the clarification stage depends upon the manufacturer of the extractor system and the type of pulp that the processor wishes to recover. The pulpy juice stream from a reamer-type

system or premium pulp squeezer-type extractor may first pass through a classifying finisher (with larger holes) to remove peel and membrane pieces before pulp recovery. The standard squeezer-type extractor includes a prefinishing tube in the extractor, and the pulpy juice flows directly to the primary finisher.

CENTRIFUGAL CLARIFICATION

Typically, the pulp content in juice leaving the secondary finisher is about 12%. This pulp is predominantly sinking pulp. If the market requires a juice with lower sinking pulp content, the juice can be further clarified by centrifugation. A two-phase clarifier is normally used for this application. However, if the juice needs to be deoiled, a three-phase centrifuge can be used to lower the pulp content to some extent and at the same time deoil the juice.

Separation in the disc-stack centrifuge takes place in the spaces created between a number of conical discs stacked on top of each other to provide a large separation area. Most models rotate at between 4000 and 10,000 rpm. The accumulated solids can be discharged, without stopping the centrifuge, by rapidly opening an annular slot at the periphery of the rotating bowl. The clarified juice leaves the centrifuge under pressure. Clarification by centrifugation often leads to improved operation of the evaporator system by providing consistent pulp levels in the juice.

TURBOFILTERS

Turbofilters were introduced in Brazil during the mid-1990s as an alternative to screw and paddle finishers. Turbofilters are claimed to give a more stable level of sinking pulp in the finished juice than do conventional finishers. They incorporate a stainless steel conveyor, rotating faster than in screw finishers, which pushes the pulpy juice against a plastic screen. The pulp content of the juice can be adjusted by changing the inclination of the turbofilter.

BLENDING

After clarification, the juice often undergoes some degree of blending with juice from other batches in order to balance its flavor, color, acidity, and Brix levels before further processing. If intended for NFC juice production, the juice leaving the clarification section should be cooled to 4°C to minimize the potential for microbiological activity before it is passed into the buffer/blending tanks.

PROCESSING STAGE 4: NFC JUICE PRODUTION

The aim of NFC juice processing is to produce orange juice using a minimum of thermal processing. Nevertheless, the thermal treatment should be sufficient to ensure that the product is physically and microbiologically stable. Since fruit harvesting is seasonal and juice consumption is year-round, the product must be stable enough to be stored for several months up to one year so that seasons can be bridged.

In some instances, during the season, NFC juice is pasteurized and packaged for the retail market without long-term bulk storage. When this is the case, some blending may occur following the clarification step to minimize hourly variations in acidity and degrees Brix. Some pulp may also be added, depending upon market demands. More commonly, the juice is processed and stored in bulk under aseptic or frozen conditions for some months until it is reprocessed and packaged. For large-volume NFC juice production, such as is found in Florida, aseptic tank farms are the most common form of NFC juice storage. The reprocessing often involves the blending of juice from early- and late-season fruit in order to standardize degrees Brix, ratio, color, and so on. The addition of pulp to the consumer product may be done at this stage. Sometimes, if volatiles have been removed from the juice prior to storage, these are added back to the juice during the blending step.

The steps for NFC juice production up to bulk storage are shown in Figure 21.12. After clarification, but prior to buffer storage, the product should be cooled as soon as possible to prevent microbiological growth and enzymatic reactions. A plate or tubular heat exchanger can be used for cooling; the choice of exchanger will be dictated by the type and quantity of pulp present in the juice. However, cooling is seldom done before pasteurization in a traditional citrus processing facility.

OIL REDUCTION

Depending upon fruit variety and the extractor operation, the oil content in the juice from extraction may exceed acceptable amounts. The levels may be specified by a legal standard, for example Florida grade "A" juice should have a maximum oil content of 0.035%. Alternatively, the oil content may be decided on the basis of consumer preference. Acceptable levels of oil in juice ready for consumption range from 0.015 to 0.030%.

Reduction of oil can be achieved in different ways:

- *Adjusting the extractor.* Less pressure is applied to fruit during extraction, or a low-oil extractor (squeezer-type) is used. Both alternatives are likely to reduce juice yield.
- *Vacuum flashing of preheated juice.* This method can remove desirable volatiles from the juice along with the oil.
- *Centrifugal separation of the oil phase from the clarified juice.* With this method juice yield from the extractors can be maintained at a high level, and there is no heating of the juice.

Deoiling with Centrifuges

Removal of oil from single-strength juice with centrifuges has been practiced for years. It is a difficult

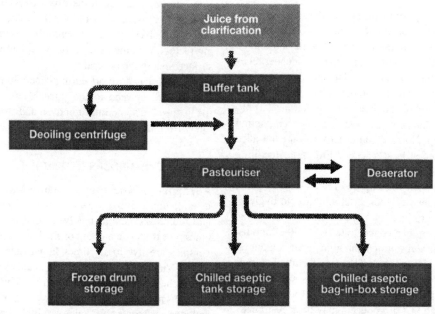

Figure 21.12. NPC production through to bulk storage.

separation task because the oil droplets are well emulsified, particularly in juice from squeezer-type extractors. Hermetic centrifuges give good results in separating oil even from juice coming from squeezer-type extractors.

In a hermetic centrifuge the rotating bowl is completely filled with liquid. This means that there are no air pockets and thus no free liquid surfaces in the bowl, which in turn avoids air entrainment and high shear forces. The feed enters the centrifuge bowl from underneath through a hollow spindle (Fig. 21.13). The smooth acceleration of the product as it enters the centrifuge prevents scattering of the oil globules, thereby enhancing separation. The hermetic (gastight) design also prevents loss of volatile components in the juice and ingress of oxygen.

In the deoiling of single-strength juice with hermetic centrifuges, oil concentrations can typically be reduced from 0.04–0.08% to 0.02–0.035%. In terms of juice yield, the use of a deoiling centrifuge in combination with standard extractors gives a yield increase of 2–4% over that of an extractor fitted with low-oil components.

The deoiled juice is buffer stored for a short period prior to pasteurization. Some blending to level quality variations may be carried out.

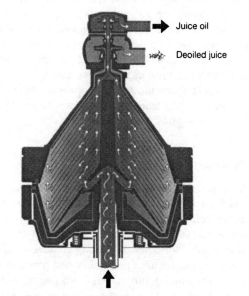

Figure 21.13. Operational principle of a hermetic centrifuge for deoiling juice.

PRIMARY PASTEURIZATION

Pasteurization prior to storage, the primary pasteurization, must achieve two goals: (1) inactivate the enzymes present in the juice and (2) make the juice microbiologically stable. It is carried out using tubular or plate heat exchangers. The choice of heat exchanger depends on the amount and type of pulp in the product and on the processor's preference. Tubular heat exchangers are best for juice containing floating pulp. Normally, after bulk storage the juice is pasteurized at least a second time prior to filling into retail packages.

The long shelf life required for NFC juice going to bulk storage demands strict attention to hygiene. Single-strength juice is more sensitive to microbial contamination than concentrate (where the high osmotic pressure resulting from high sugar content retards microbial growth). The use of chilled storage instead of frozen storage also imposes much stricter hygiene requirements for NFC juice production than that to which FCOJ producers may be accustomed.

Good manufacturing practice demands that the pasteurizer system be presterilized at 95°C or higher prior to production and that a clean-in-place (CIP) program be integrated with the control system. NFC juice volumes to be processed are normally large, so a high degree of energy recovery is advisable. Thermal treatment is a concern among many NFC juice producers. Excessive heat load on the juice should be avoided. Careful control of temperature and residence time using well-designed heat exchangers is important. A low temperature differential between the heating medium (hot water) and the product minimizes "shock" to the product.

Deaeration

Air tends to get mixed into the juice in the extractors and finishers. Some of the entrained air may escape during buffer storage, but juice going to pasteurization is normally saturated with dissolved oxygen. It also contains some free air. During product storage, oxygen present in juice in the dissolved state and as free bubbles may destroy a significant amount of the available vitamin C by oxidation. Air bubbles present in the product during pasteurization may also lead to insufficient heat treatment.

Deaeration as part of the pasteurization process is therefore recommended for the production of NFC juice. Deaeration is accomplished by passing the product through a vacuum chamber. Free air bubbles expand in a vacuum and tend to escape quite easily from the juice, although dissolved oxygen is more difficult to remove. Deaeration efficiency, or reduction of dissolved oxygen, depends on several operating factors, including the vacuum applied and juice surface area in the deaerator.

Volatiles that flash off during deaeration are condensed and returned to the juice stream. Alternatively, they are sometimes removed and stored separately from the bulk juice.

LONG-TERM FROZEN STORAGE

After primary pasteurization, orange juice is stored in bulk under either frozen or aseptic conditions. NFC juice production involves large product volumes. For the same amount of final juice, NFC juice volumes are five to six times larger than those for FCOJ. Vitamin degradation and changes in flavor during the storage period are minimized by freezing, but the energy and warehousing costs of freezing and storing frozen NFC juice are high. Freezing of NFC juice leads to handling problems because it freezes solid, whereas frozen orange concentrate is very viscous but still pumpable.

There are three major options for long-term storage of NFC juice: (1) frozen storage, (2) aseptic storage in tanks, and (3) aseptic storage in bag-in-box bulk containers.

Frozen storage of NFC juice is more appropriate to lower quantities of NFC product. Large-volume producers in Florida store NFC produce aseptically in very large tanks. NFC produce is mainly shipped in aseptic bag-in-box containers or frozen in drums.

Juice to be stored frozen is filled into mild steel 200-liter [55-gallon (U.S.)] drums lined with a polyethylene plastic bag. As the product is to be frozen, the net filling volume is about 170 liters (45 U.S. gallons). Alternatively, the juice may be poured into block formers and then frozen (on-site storage). The frozen product is usually kept −18°C or lower.

Thawing of NFC juice to make it ready for final processing also leads to some logistic and handling difficulties. It takes several days or weeks for bulk product in drums to thaw at ambient temperature. The outer layer of juice may be exposed to microbiological contamination during thawing, with a subsequent negative impact on product quality. Crushing systems enable more rapid handling but entail higher energy consumption and capital investment.

Systems for freezing larger blocks of juice that incorporate novel techniques for rapid freezing and thawing have been introduced, but none is in commercial use.

ASEPTIC STORAGE IN TANKS

As an alternative to frozen storage, NFC product may be chilled in aseptic tanks. Technology exists to build very large tanks, up to a capacity of four million liters, for the aseptic storage of juice. Unique fabrication techniques are used to coat the internal surface of the tanks. The tanks are sterilized prior to filling by flooding them with a sterilizing fluid (e.g., iodoform). As an alternative to fitting the tanks with cooling jackets, they are installed within a large refrigerated building. The preferred storage temperature is about −1°C, just above the freezing temperature of the juice.

The juice must be agitated periodically to avoid separation or sinking pulp and to maintain Brix uniformity. Pressurized nitrogen above the juice surface is often maintained to minimize the risk of vitamin C loss through oxidation. Normally, when product is required from these tanks, it is drawn off, blended with juice from another part of the season (and perhaps pulp), and repasteurized. In Florida, a large and growing share of NFC juice is stored in tank farms with very large aseptic tanks. However, this technology requires a substantial initial investment, and the value of product at risk when stored in such large tanks is considerable.

ASEPTIC STORAGE IN BAG-IN-BOX BULK CONTAINERS

As an alternative to aseptic tanks, the juice may be filled into 1000-liter (300-gallon) aseptic bag-in-box containers (Fig. 21.14). The bags, placed in bins, usually made from wood, are stored under refrigerated conditions. After storage, the product is accessed by opening the bag and pumping out the product. Alternatively, the bag can be emptied and the juice transferred aseptically to the filler.

"One ton" aseptic bag-in-box containers for NFC product storage require more labor for filling and emptying than large tanks do. However, bag-in-box containers allow added flexibility regarding storage capacity, as the investment required to store additional volumes of juice is moderate. A drawback of the aseptic tank approach is finite storage volume, unless a major investment is undertaken to have reserve capacity. Consequently, the bag-in-box solution is often preferred for start-up operations for

Figure 21.14. A filler for bag-in-box containers.

NFC juice production. NFC juice processors who already have aseptic tanks installed may also use bag-in-box containers to provide additional storage capacity and to ship NFC product.

For long-term storage of juice (six months or more), bag material with a very good oxygen barrier is recommended. Bags made with foil-based aluminium laminate offer higher protection against oxygen than bags made with metallized laminates, in which the aluminium layer is much thinner. Aseptic security during product filling and storage must be high. Any contamination may lead to blown bags during storage and shipment. Needless to say, a single blown bag during shipment can cause a lot of trouble. Several filling systems for aseptic bag-in-box containers evolved from conventional (nonaseptic) bag-in-box systems. A sterile chamber surrounds the filling head, and chemical sterilants are used for sterilization. Other systems were subsequently developed specifically for aseptic filling. An example of this type of system is shown in Figure 21.14. It incorporates a simple filling system (spout and filling valve), and steam is used as the sterilizing agent.

REPROCESSING OF NFC JUICE

In the United States, some NFC juice is moved in bulk by road and rail tankers to juice packers but most NFC product is filled into retail packages in Florida and distributed across the country. Shipping to Europe in bulk is done in frozen drums and aseptic bag-in-box containers. Overseas shipping of packaged product is at a cost disadvantage compared with bulk shipping. The additional time delay adds to difficulties with logistics and forecasting. Traditionally, NFC juice taken from storage at the fruit processor's site for reprocessing is blended with juice from a different part of the season and/or with pulp. The juice blend is then repasteurized prior to filling into consumer packages. The second pasteurization will add thermal impact to the product. Alternatively, specially designed equipment can be used to transfer juice from aseptic bulk bags to consumer packages via an aseptic tank, without the need for repasteurization. Bags containing juice from different production batches may be blended in the aseptic buffer tank. Such a transfer system is illustrated in Figure 21.14. It can be installed at the fruit processor's site for juice from on-site storage, or utilized for bags shipped to the juice packer.

PROCESSING STAGE 5: CONCENTRATE PRODUCTION

Globally, most orange juice is produced as concentrate. Juice from the clarification step is evaporated

to remove most of the water (Fig. 21.15). Currently, the most widely used citrus evaporators are of tubular design, which can handle very large flow rates. In addition to these, plate evaporators are installed in citrus plants for handling mainly small to medium volumes. During the 1970s and 1980s, there was a large expansion in concentrate capacity in the major citrus markets of Brazil and Florida. Today, little increase in evaporator capacity is needed in these regions, but new evaporators are being installed to satisfy the requirements of other orange-producing regions that are expanding.

TUBULAR EVAPORATOR SYSTEMS

The most common type of tubular evaporator system used for orange juice is the TASTE evaporator. It is generally described as a continuous, high-temperature short-time (HTST) evaporator of the long, vertical tube, falling film type. The name is an acronym for thermally accelerated short-time evaporator. It was designed and developed in Florida, and today this type of evaporator is manufactured in many different countries.

TASTE evaporators were designed for the large juice volumes commonly processed in large citrus plants, where evaporator capacities can exceed 100,000 kg/h of water evaporated. Versions that

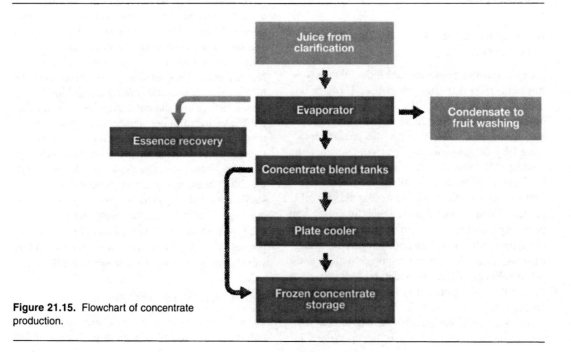

Figure 21.15. Flowchart of concentrate production.

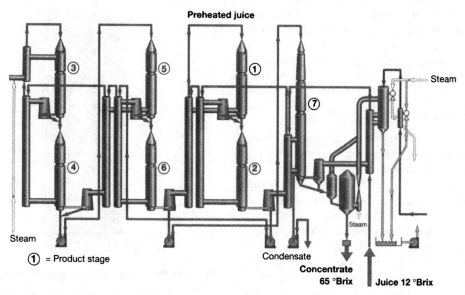

Figure 21.16. A simplified flow diagram of a tubular evaporator.

have as many as seven effects are installed (seven effects means basically that the steam is reused to evaporate water in seven steps). Such systems have very low specific steam consumption: only 1 kg of steam is used to evaporate 6 kg of water. However, additional effects increase the residence time for the product in the evaporator accordingly. These evaporator systems are dedicated to citrus fruit.

A flow diagram of an evaporator with seven product stages is presented in Figure 21.16. The juice is first preheated to 95–98°C. Holding at pasteurization temperature stabilizes the juice by means of microbial and enzyme inactivation. The product then passes through a number of stages under vacuum until a concentration of up to 66° Brix is achieved. By this time the product temperature has fallen to about 40°C. The residence time in the evaporator is typically five to seven minutes or longer.

Good distribution is of primary importance in the design of an evaporator. It ensures that all the product is uniformly treated and that the heat exchange surface is used to its maximum potential.

A special feature of the TASTE evaporator is the way in which the product is distributed across the tube bundle. The juice is fed into the distribution section at a temperature and pressure greater than in the entry zone of the tube bundle. The liquid is fed through a diverging expansion nozzle that converts all the product into a liquid/vapor mixture. The ex-

panding vapor accelerates the liquid/vapor mixture through a second nozzle and cone assembly. Further flash expansion of the vapor causes atomization of the liquid phase into a turbulent mist. The acceleration effect can cause mist velocities to exceed 50 m/s on leaving the tube bundle. The high degree of turbulence increases heat transfer rates and reduces burn-on, which helps to achieve long operating runs. The vapor and liquid are separated in a drywall separator at the outlet of each stage.

Homogenization

Sometimes, homogenization of concentrate is carried out within the evaporator system. Product then normally passes through a homogenizer prior to the last effect. At this stage, the concentration is approximately 40–42° Brix. Homogenization breaks down the pectin, thereby lowering the viscosity of the concentrate. This increases the efficiency of the final stage of the evaporator. It is also claimed that homogenization reduces the sinking pulp level in the product. This could permit juice with higher pulp levels to be fed to the evaporator.

Other Tubular Evaporation Systems

There are also other tubular evaporator systems of similar design for citrus plants, which include a conventional mechanical method for distributing product across the tube bundles. They incorporate thermal recompression to increase steam economy without increasing residence time. Relatively few of these evaporator systems have been installed for high product capacities.

PLATE EVAPORATOR SYSTEMS

For small to medium flow rates, plate evaporators can also be used. Plate systems can be designed for flexibility, and when installed in citrus plants, they are often used to process other types of juice outside the orange juice season. As the name implies, plate evaporators consist of plates clamped together in a frame with gaskets between them. Some advantages of plate evaporators over other types of evaporators are that (1) capacity increases are easily attained by adding more plates and (2) maintenance and inspection are easily carried out by opening the frame. However, the large number of gaskets is a drawback.

When cassettes (welded double plates) are used instead of single plates, the number of gaskets required is halved. In Tetra Alvap evaporators the heating medium (steam or vapor) passes through the space between the welded plates. Product channels are formed between individual cassettes separated by gaskets. This configuration allows ready inspection of product channels by opening the frame. A small temperature difference between the product and the heating medium is sufficient in cassette evaporators. This allows lower operating temperatures to be used than in traditional tubular evaporators.

There are two types of cassette evaporator: the falling film type and the rising film type. Both types are installed in small and medium-size citrus processing plants.

Falling Film Cassette Evaporator

In the falling film evaporator, the liquid product enters at the top and flows down over the plates. Evaporation takes place as the liquid travels down the plate, thereby reducing the quantity of liquid and increasing the vapor flow. The cassettes have a heating surface designed for evaporation rather than just liquid/liquid heat transfer (Fig. 21.17).

The pattern and corrugation of the plate take into account the change in liquid and vapor quantities across the plate so that the liquid film is maintained constant on the heating surface. As the corrugation becomes gradually less deep at the lower part of the

Figure 21.17. A cassette for a failing film evaporator.

plates, there is less surface area for the liquid to cover, and more space available for the vapor.

The product is distributed evenly over the heating surface by feed nozzles; no superheating is required. All metal-to-metal contact points are located on the steam side; there are none on the product side. This reduces product fouling and facilitates CIP.

The pressure drop over the unit is very small, which also allows for a lower operating temperature range than is possible in conventional evaporators. As the residence time of the product is well defined and short, the thermal impact on the product is minimized.

Evaporator capacities range from 1000 to 20,000 kg/h of evaporated water, and the units are normally configured in two to four effects combined with thermocompression to give a specific steam consumption down to 1 kg of steam per 5 kg evaporated water.

Rising Film Cassette Evaporator

In the rising film cassette evaporator, the product enters the bottom of the cassette and rises up over the heating surface as it boils (Fig. 21.18). No mechanical feed distribution device is needed, and even distribution is achieved through gravity. It is possible to evaporate products of higher viscosity and higher pulp content than in a falling film evaporator.

Compared with the falling film evaporator, the

Figure 21.18. A rising film cassette evaporator.

thermal impact on the product is somewhat higher due to the higher temperature difference required between the product and heating medium. However, the operating temperatures can still be kept well below those needed in a tubular evaporator.

The rising film evaporator handles capacities up to 50,000 kg/h of evaporated water and is commonly configured in two to four effects combined with thermocompression. The specific steam consumption is down to 1 kg steam per 5 kg evaporated water.

THE CENTRIFUGAL EVAPORATOR

Very gentle product treatment during evaporation is achieved in the centrifugal thin-film evaporator (Fig. 21.19). The heating surface consists of rotating cones. The combination of heating and centrifugal force allows a high degree of concentration to take place in one single pass, in a very short time, at a very low temperature. A typical residence time for concentrating orange juice from 12 to 65° Brix is about 10 seconds at a temperature of 50°C. These gentle conditions give the lowest possible thermal impact on the product.

The centrifugal evaporator (Centritherm) handles capacities from 50 to 5000 kg/h of evaporated water. It is configured in one effect and, consequently, has a specific steam consumption of approximately 1.1 kg steam/kg water evaporated. To increase the ca-

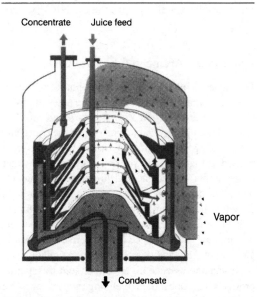

Figure 21.19. The operational principle of a centrifugal evaporator.

pacity and reduce the steam consumption, it can be combined with a cassette evaporator.

Although its capacity is too low and its steam consumption is too high for production of standard concentrates, the superior heat-transfer efficiency and gentle product treatment of the centrifugal thin-film evaporator are desirable features for producing premium concentrates that command a high market price.

ESSENCE RECOVERY

During evaporation, volatile juice components are stripped from the juice together with the water. These are often recovered in an essence recovery system connected to the evaporator. The essence process usually forms an integral part of the mass and thermal balance of the evaporator system. Dr. James Redd of Florida pioneered the development work in the design of essence recovery units, and the first commercial system was installed in 1963.

The vapors from the early product stages of the evaporator contain most of the volatiles from the juice. These are captured and sent to a still mounted on the evaporator. The important volatiles are separated from the water by distillation under vacuum and condensed by chilling. The product essence is a concentrated mixture of aqueous and oil soluble aroma compounds. This essence is separated into oil and aqueous phases by either decantation or centrifugation.

Water Phase Aroma and Essence Oil

The aqueous phase (called water phase aroma or essence aroma) contains the flavor top notes. It has an alcohol strength typically standardized at 12–15%. The oil phase (essence oil) holds the fruity and sweet-tasting flavors of fresh juice. It has different properties than those of peel oil. Add-back of water phase aroma and essence oil to concentrate has replaced the previous practice of adding single-strength juice (cut-back) to improve the flavor of concentrate.

In Florida, Valencia oranges are used to produce the best essence; little essence can be derived from early varieties of fruit, and it is often of poorer quality.

Aroma and essence oil are either sold as separate products to concentrate blending houses or juice packers, or purchased on contract by specialty flavor manufacturing companies.

CONCENTRATE STORAGE

After evaporation, the 65° Brix concentrate is chilled to −10°C. It is then routed to storage. Blending of different production lots and addition of peel oil and essences may be done on the way to concentrate storage. Storage takes place in bulk storage tanks or 200-liter drums with plastic liners. Drum storage is normally maintained at −20 to −25°C, while bulk storage in large tanks is often maintained at −10°C. Just prior to dispatch from the plant, concentrate drawn from different bulk storage tanks is often blended to meet product specifications. Concentrates are sometimes diluted with pulp, for example, to reduce the Brix level. Shipping of frozen concentrate involves drums, tank cars, or bulk tanker ships.

Concentrate is traded as FCOJ. The term "frozen" may be misleading: concentrate at 65° Brix does not freeze solid at −10°C due to its high sugar content. The most common concentration for FCOJ is 65–66° Brix, but bulk concentrates of lower Brix are also available. FCOJ of 55–58° Brix is typically supplied to dairies.

ALTERNATIVE CONCENTRATION METHODS

Alternatives to evaporation for concentrating orange juice have been developed and tested, but so far none are in commercial operation on a large scale. Lower Brix levels of the concentrate and often high operational costs in comparison with the evaporator systems in common use have prevented the commercialization of the new systems. Two methods that do not use heat for concentration are freeze concentration and membrane filtration.

Freeze Concentration

This method is based on the fact that during the freezing of sugar solutions, ice crystals are first formed, which can be separated out from solution, thereby increasing the sugar concentration. When freeze concentration is applied to juice, inactivation of enzymes is necessary. This may be accomplished by pasteurizing the juice before freezing or by pasteurizing the resulting concentrate.

Several studies have shown that, compared with conventional evaporation, freeze concentration yields superior flavor quality. However, the low temperatures involved lead to high viscosities in the concentrated products, which limit the degree of concentration that can be achieved and the amount of pulp and insoluble solids that may be present in the juice to be concentrated. Concentrates of up to 40° Brix can be obtained with this method.

Membrane Filtration

Membrane filtration is another method evaluated for concentrating orange juice without using heat, but the resulting high viscosity of of the concentrate reduces filtration efficiency and limits the degree of concentration that can be achieved. To minimize viscosity, the pulp is first separated from the juice, for example, by ultrafiltration (UF), to leave a clear liquid (serum), which is concentrated by reverse osmosis.

The pulpy stream, rich in enzymes, is pasteurized before being recombined with the serum concentrate. Mixing back of the insoluble solids stream, essentially at single-strength juice concentration, reduces the Brix value of the concentrate. Concentrations up to 42° Brix have been reported.

Concentration systems using other membrane processes have also been tested. However, the necessity to retain the sugars, acids, and aroma compounds in order to maintain a balanced citrus juice flavor puts tough demands on potential membrane systems.

PROCESSING STAGE 6: PEEL OIL RECOVERY

The oil-water emulsion, or oil frit, from the extraction process is sent to the peel oil recovery section.

Apart from the oil and water, other fruit substances are present in the emulsion. These include particles of peel and pulp, and soluble pectin and sugars. The aim of the peel oil recovery system is to recover pure oil by removing all other substances with as little oil loss as possible.

STRAINING AND CONCENTRATION STEP

The first step involves using a finisher as a straining method to remove large bits of peel and other parts of the orange that must not enter downstream centrifuges (Fig. 21.20). After straining, the oil emulsion, containing about 0.5–2.0% oil, enters the first stage centrifuge (also called a desludger or concentrator). The centrifuge concentrates the oil up to 70–90%.

The first centrifuge is a three-phase machine. The light phase is concentrated oil, the heavy phase is water, and the third phase is residual particulate matter. The control of solids discharge from the sludge space is critical to the overall performance of the oil recovery system. If the discharge frequency is set too high, product is lost; if the sludge space is allowed to fill up, separation efficiency is lost.

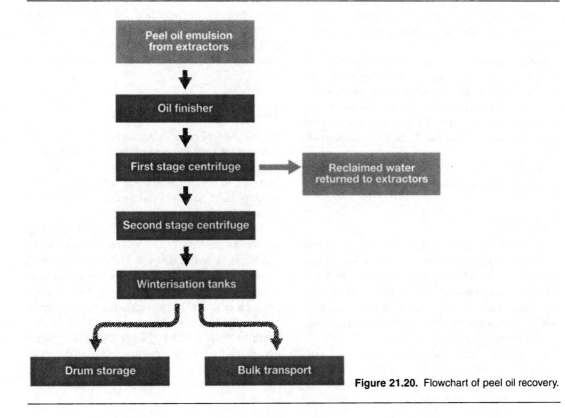

Figure 21.20. Flowchart of peel oil recovery.

The water stream is often recycled back to the oil extraction system as spray water, although it is important that some water be removed from the system to allow additional fresh water to enter it. Microbiological problems may occur if the same water is continuously recycled. Moreover, the centrifuged water contains desirable components such as soluble pectin. As the concentration of these components builds up in the emulsion, the oil separation efficiency decreases, thereby resulting in lower oil yields. Again, this limits the amount of water recycling possible.

The centrifuged water also contains microscopic particles of oil that are too small to be separated by the centrifuge. As this level of oil builds up with water recycling, the effectiveness of the water for extracting oil from the peel decreases. This will also lead to an overall drop in the efficiency of oil recovery.

The type of oil extraction used and the performance of the centrifuges will determine the amount of water that can be recycled. The cleaner the peel oil emulsion, the higher the oil yield from the peel oil recovery system and the larger the amount of water that can be recycled. The oil extraction system upstream of the reamer-type juice extractor is claimed to give a "less contaminated" oil emulsion than the one-step squeezer-type extraction system.

For oil recovery, the hermetic centrifuge has several advantages over the open-bowl-type design. The fully flooded bowl in the hermetic machine ensures that oil does not come in contact with air. The precise manner in which the interface between oil and water is controlled leads to higher separation efficiency.

A hermetic centrifuge for concentration of peel oil emulsion is shown in Figure 21.21.

POLISHING

The concentrated oil stream then passes to a second stage centrifugation process (polishing). Within this machine, the oil is further concentrated to > 99% purity. The flow rates are extremely small (1–2%) compared with the flow rates in the first stage or with flow rates used in juice clarification or deoiling of single-strength juice.

Since the product has already undergone one centrifugation process, virtually no solid particles remain in the product. For smaller capacities, a solid-bowl machine is used, and the water and oil are

Figure 21.21. Hermetic centrifuge for peel oil concentration.

continuously discharged. Periodic takedown removes any material that collects in the bowl periphery. For larger flow rates, a solids-ejecting polisher is used, in which water and oil leave the machine under pressure. Accumulated solids are discharged about once or twice per hour. One ton of fruit typically yields 200–300 liters of emulsion to the first centrifuge and 3–6 liters of concentrated oil to the polisher.

THE WINTERIZATION PROCESS

The polished oil contains trace amounts of dissolved wax derived from the peel of the fruit. At temperatures above 15 or 20°C, the wax is totally dissolved. However, at lower temperatures it may give a haze to the product. To avoid this problem, the polished oil is dewaxed, or winterized.

The winterization process involves precipitating the wax by causing it to crystallize and then settle. The oil is stored in tanks at 1°C or lower, which causes the waxes to come out of solution and sediment. The process typically takes 30 days or more, although at lower temperatures this period may be considerably shorter. The winterized oil is then decanted from the tank. Larger processors collect the sludge from different winterizing tanks so that once sufficient material has accumulated, the waxes can be removed by centrifugation to recover residual oil.

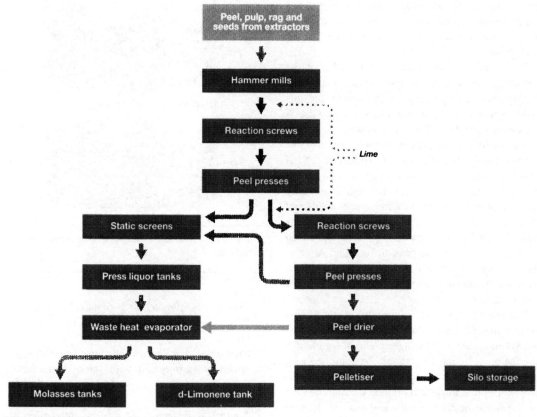

Figure 21.22. Flowchart of feed mill operations.

The winterized oil is packed in 200-liter (55-gallon) drums or road tankers. Normally the oil is stored under refrigeration ($-10°C$) and is traded as cold-pressed oil, or more accurately, cold-pressed peel oil, CPPO. It is used as a raw material in the flavor manufacturing industry and by concentrate blending houses and drink-base manufacturers.

PROCESSING STAGE 7: FEED MILL OPERATIONS

After juice extraction, about 50% of the fruit remains. Much of this residual fruit matter is seemingly low-grade material in the form of peel, rag, core, seeds, and pulp not used for commercial purposes. This waste is sent to a feed mill, installed in most larger processing plants.

Feed mill operations represent a significant part of the total plant running costs. The drying of solids and the evaporation of the liquid stream are energy intensive. Less waste and increased recycling of liquids in other parts of the plant are desirable for both economic and environmental reasons. Legislative pressure for environmental controls in citrus plants continues to increase.

The revenue from the sale of by-products from the feed mill makes a significant contribution to the overall profitability of orange processors. There are continuous developments in finding additional products that can be recovered from peel and other waste streams.

FEED MILL PROCESS STEPS

The feed mill receives rejected fruit from the grading tables in the reception area and waste material from juice processing. The overall moisture content of this combined material is 80%. Screw conveyors carry the material to the wet-peel bins of the feed mill. From here, it is broken down to small pieces by

hammer mills (see Fig. 21.22). Small amounts of
lime (0.15–0.25%) are added after this step to aid
the dewatering process. After a dwell time of 10–15
minutes, the mixture is conveyed or pumped to the
peel presses.

In the primary peel presses, some 10% of the
moisture is removed. Continuous screw presses have
largely replaced hydraulic batch presses for this
task. Further addition of lime and secondary pressing can remove 2 or 3% extra moisture

The liquid from the presses (press liquor) contains
approximately 9–15% soluble solids, much of
which is sugar solids. The oil content can be between 0.2 and 0.8%. The press liquor normally
flows over static screens to remove peel solids and
then on to the waste heat evaporator. The press
liquor is usually concentrated to 50° Brix and added
back to the peel residue prior to pressing. Alternatively, it may be concentrated to 72° Brix and used
as raw material for a fermentation process to make
citrus alcohol.

The press liquor contains a high amount of suspended materials and often includes sandlike matter. When decanter centrifuges are used for clarifying the press liquor, they should be equipped with
special internal tiles to minimize erosion. Clarification of the press liquor can prolong the running time
of the waste heat evaporator and reduce cleaning
time substantially, thereby contributing to greater
cost efficiency in running the feed mill. d-limonene
is stripped off in the waste heat evaporator and can
be recovered as a separate stream from the vapor
phase.

The pressed peel is dried in a rotary drier to a
moisture content of about 10% and then pelletized
to make animal feed. The vapor that comes from the
peel drier is used as heating medium in the waste
heat evaporator.

PROCESSING STAGE 8: PULP PRODUCTION

Floating pulp, that is, the larger solid particles in the
juice, mainly consist of small pieces of ruptured cell
sacs and segment walls. They are separated from the
juice in finishers. (The very small pulp particles
flow with the juice stream from the finisher. These
fine particles tend to sediment at the bottom of the
juice and are referred to as sinking pulp.) The pulp
stream from the finisher is handled in different
ways, depending on the end use of the pulp. The alternatives are

Standard squeezer-type extractor

Reamer-type extractor

Figure 21.23. Illustration of relative pulp sizes after
extraction.

- *Recovery for production of commercial pulp.*
  Pulp is used as add-back in juice and juice
  drinks.
- *Production of pulp wash,* the juice sugars obtained by washing pulp with water. The remaining material is sold as "washed pulp" or taken to
  the feed mill.
- *Routing to the feed mill* for drying into pellets
  for animal feed.

In the past, most pulp went to pulp washing and
the feed mill. However, now that the current market
trend is to add more pulp cells to the final juice, the
proportion of pulp from the extractors going to commercial pulp production is increasing. For most
processors, however, more pulp is obtained from the
fruit than is required by the juice industry for add-
back to juice.

The extractor type and operation will influence
the quality of the pulp produced. In some plants, the
extractors used for pulp production are adjusted to
optimize pulp quality rather than to maximize juice
yield. The visual difference between pulp from
reamer-type extractors (Brown) and standard
squeezer-type extractors (FMC) is illustrated in Figure 21.23.

PRODUCTION FACTORS THAT AFFECT COMMERCIAL PULP QUALITY

Some of the process conditions that have a significant
influence on pulp properties are given in Table 21.1.

Table 21.1. Influence of Process Conditions on Pulp Properties

Pulp Properties	Process Conditions
Cell length and fragmentation degree	Fruit variety and fruit maturity
	Size of the holes in the strainer tube (squeezer-type extractors)
	Extraction pressure
	Use of paddle or screw finisher
	Back pressure applied to the primary and final finishers (screw type)
	Equipment and operating conditions for the pulp stabilization step
Oil content	Extraction pressure. High pressure gives higher juice yield but also higher oil content in the pulpy juice stream.
Defects in final product	Depends on what type of equipment is used to separate defects from the pulpy juice stream
Pulp concentration (i.e., the concentration of pulp particles in pulp slurry)	Tightness applied in the finishers

PROCESS STEPS IN PULP PRODUCTION

The exact configuration of the pulp line will vary from plant to plant, and its design will depend on the type of extraction system and processor preference. The basic pulp production steps are shown in Figure 21.24.

Instead of pulp juice from the extractors, pulp from the primary finishers in juice clarification is sometimes taken as feed to the pulp production lines. Dilution with juice prior to the defect removal step may then be needed.

Extraction

During the juice extraction process, segment and cell sac walls are torn into pieces. Both the reamer-type extractor and the specially designed squeezer-type extractor used for premium pulp put less shear force on the pulp than the standard squeezer-type extractor. This results in larger and less fragmented pulp pieces. However, defects such as core and seeds also end up in the pulpy juice from the extractors. This imposes greater demands on the defect removal system. Sometimes, pulp from the primary

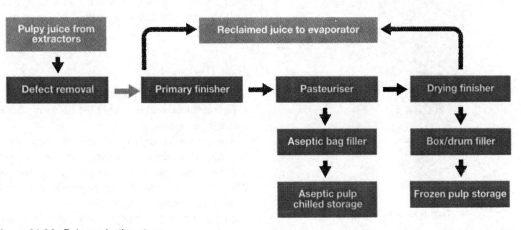

Figure 21.24. Pulp production steps.

finishers in juice clarification is conveyed to feed the pulp production line. Dilution with juice prior to defect removal may then be required.

Defect Removal

Defects are normally described as small fragments of peel, membrane, or seed. As the absence of defects in the final product is an important quality parameter, they have to be removed from the pulp/ juice slurry.

Defects are removed in a series of separation steps. The first step may be a classifying finisher. This is a paddle-type finisher that incorporates screens with large perforations that will allow juice and cells to pass through but will retain large seeds and pieces of membrane. The pulpy juice stream then goes to one or more hydrocyclones. If there are a lot of defects, two or more hydrocyclones are used in series. Hydrocyclones are based on gravity separation and remove defects that have a higher density than the pulp slurry.

Figure 21.25 shows the liquid and particle flow in a cyclone. The in-feed, which is tangentially introduced into the cone, starts moving in a downward spiral along the cyclone wall. As it nears the cone outlet, some of the product leaves through the underflow orifice, while most of it changes direction

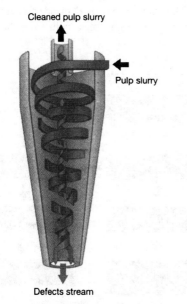

Cleaned pulp slurry

Pulp slurry

Defects stream

Figure 21.25. A hydrocyclone used for defect removal.

and flows upward to the cyclone overflow, taking an inner spiral path. If the density of the particles is higher than that of the liquid, the centrifugal force presses the particles against the cyclone wall from where they are pushed down and out through the bottom opening.

Separation in a cyclone is improved with lower solids concentration and lower liquid viscosity. As small, immature seeds are lighter than pulp slurry, they are difficult to remove. Thus, the quality of fruit delivered to the processor is important to the results of defect removal.

Concentration (Primary Finishers)

The "cleaned" stream from the defect removal system is normally concentrated prior to heat treatment. The reasons for this are twofold: (1) energy is saved by heating/chilling less product, and (2) less juice is subjected to additional heat treatment.

Concentration is done in a screw- or paddle-type primary finisher. Paddle finishers treat pulp particles more gently. The operation of the finisher can be adjusted so that the pulp concentration of the discharged pulp slurry is at the required strength for the downstream pasteurization step. In Florida, most processors operate so that the pulp slurry from the primary finisher has a typical pulp concentration of 400–500 g/l. In Brazil, there is a difference between plants—from 150–200 g/1 up to 500 g/1 pulp concentration. The lower range is due to using plate heat exchangers in the pasteurizer.

The pulp stream from the primary finisher to pasteurization cannot be kept constant; it will vary in both flow rate and pulp concentration (10–15%) during a production shift. Over a season, different fruit varieties and extractor settings will give wider variations.

Heat Treatment

The two objectives of pulp slurry pasteurization are (1) to inactivate enzymes and (2) to destroy relevant microorganisms.

The necessary degrees of enzyme deactivation and microbial reduction depend on how the pulp will be further processed and stored. The required deactivation determines the pasteurization conditions (temperature and time).

As the enzymes in oranges are located in the fruit cell walls, the enzyme concentration is significantly higher in pulp slurry than in clarified juice. To achieve complete inactivation of enzymes, more intensive heat treatment is needed for pulp slurry than for juice. However, complete enzyme inactivation is normally not required. Enzyme activity should be reduced to such an extent that the pulp (1) is stable during bulk storage and (2) will not lead to cloud separation in reconstituted juice.

If the downstream handling of heat-treated pulp is nonaseptic (e.g., the drying finisher), complete killing of microorganisms is not required. This is the case for pulp stored frozen, the most common storage method. In this case, the heat treatment is referred to as "stabilization." Typical heating conditions are 90–100°C for 30 seconds.

When pulp is to be stored chilled in aseptic bag-in-box containers, heat treatment may be referred to as "pulp stabilization/sterilization." Temperatures in excess of 100°C are normally used. A higher degree of enzyme inactivation is required for chilled storage than for frozen storage. Aseptic storage also requires that heat-treated pulp have no microbial activity. Furthermore, downstream equipment must not recontaminate the product.

Which Heat Exchanger?

The heat exchangers used for pasteurization of pulp slurry are typically of the tubular type. Any obstructions on the product side, such as contact points in a plate heat exchanger, should be avoided. Often heat exchangers incorporate a single product tube. With this type there is no risk of uneven product flow. However, throughput is limited due to the pressure drop.

A multitube heat exchanger (see Fig. 21.11) can process high pulp flow rates without the drawback of excessive pressure drops. The inlet to the parallel tubes requires careful design to ensure that pulp does not stick to tube entrances, causing blockage and uneven flow rates through the tubes.

Heat-treating pulp at concentrations much above 500 g/l is not really feasible in tubular heat exchangers because heat-transfer coefficients rapidly decline above this concentration. Efficient heat transfer is inhibited by the high cellulose content of the product. If tubular heat exchangers are used for higher pulp concentrations, they become very large, which entails slow heat-up and cool-down times, resulting in a loss of product quality. A pasteurization system for pulp using multitube heat exchangers can also have the dual function of pasteurizing NFC juice.

The nature of the pulp recovery process tends to entrain air into the product stream. This has to be considered in the design of heat treatment processes.

Concentration (Drying or Final Finisher)

Traditionally, the heat-treated pulp is further concentrated up to 950–1000 g/l using a final or drying finisher. Although still wet, it is called "dry" pulp because it will not release any free liquid when pressure is applied to it. The residual liquid is mainly adsorbed onto the cellulose membranes. The concentration of dry pulp is measured for product specification by a special method called Quick Fiber. The liquid in the pulp, essentially NFC juice, typically corresponds to 5–8% of pulp mass for standard pulp, and 9–13% for premium pulp. Thus, when pulp is added during reconstitution at the juice packer, the juice still present in the "dry" pulp will provide additional NFC juice.

Packing in Boxes/Drums for Frozen Storage

The concentrated pulp is normally packed in 20 kg corrugated cardboard boxes lined with a polyethylene bag and is then frozen. Freezing takes several days. Pulp may also be packed in drums (200 liter/55 gallon) for frozen storage. However, drums are not often supplied to juice packers as they are usually too large for the batches of reconstituted juice.

Packing in Aseptic Bag-in-Box Containers for Chilled Storage

If the stabilization process is modified to become a stabilization/sterilization process, it is possible to pack pulp aseptically and store it refrigerated. Packing is done directly after heat treatment. Hence, the aseptic pulp will be bulk stored at a much lower concentration than frozen pulp.

The disadvantage of packing pulp aseptically at a 500 g/l concentration is that a larger storage (and shipping) volume is needed than for the same amount of dry pulp. The advantage is that the pulp is much easier to handle because it is pumpable and needs no thawing or crushing. It also gives the possibility of enhancing the final product. When the aseptic pulp is added back to juice reconstituted from concentrate, juice present in the aseptic pulp (effectively NFC juice) may provide some of the desired flavor associated with NFC products. Aseptic pulp is produced by several processors in Florida.

PROCESSING STAGE 9: PULP WASH PRODUCTION

Pulp washing is carried out to recover juice solubles in pulp coming from the juice finishers and from the centrifuges in the clarification or deoiling process. Thorough pulp washing can increase the total yield of soluble solids by 4–7%, which contributes significantly to overall plant economics. The process steps are shown in Figure 21.26.

The juice sugars are reclaimed by a countercurrent washing system. The pulp/water slurry is strained through a finisher between the washing stages, and the separated "juice" is called pulp wash. Process development includes the use of static mixers to blend and allow equilibrium of soluble juice and pulp components during washing. The pulp stream is concentrated by evaporation. It is added back to concentrated orange juice (if the law permits) or used as a base for juice drinks.

The number of stages in a pulp washing system is chosen according to cost-effectiveness. A maximum of four stages can recover up to 50, 63, 75, and 80%, respectively, of the available juice sugars. The amount recovered depends on fruit variety and maturity.

DEBITTERING AND ENZYME TREATMENT

Pulp wash is high in limonin, which causes bitterness. Consequently, untreated pulp wash has limited use as add-back into high quality juice drinks. However, the bitter taste can be removed by a debittering process that uses ultrafiltration and adsorbtion of separated bitter components onto resin.

The high content of pectin in pulp wash leads to a greater increase in viscosity during evaporation than is seen with pure juice. This can lead to a limit of 40° Brix for pulp wash concentration. Therefore, breakdown of pectin by enzyme treatment is often included in the pulp washing process. Typical conditions are a retention time of up to one hour at 45°C in the reactor tank. After centrifugation, enzyme-treated pulp wash can be concentrated to the normal 65° Brix level and then blended with orange juice concentrate or packed in 200-liter drums and frozen.

WASHED CELLS

Washed cells can either be sent to the feed mill or be bulk packed in 25 kg cardboard boxes or 200-liter drums, which are stored frozen. The product is traded as washed pulp or washed cells and used in some drink applications

REGULATIONS FOR AND USE OF PULP WASH

Pulp wash is often used as a sugar source in formulated beverages and juice drinks, and as a clouding agent for providing body and mouthfeel. Adding back pulp wash to orange juice concentrate is allowed in Brazil. In Florida and other parts of the United States, up to 5% pulp wash may be added to concentrate, provided it is produced along with the juice extraction process. Nevertheless, the quality standards and marketing approach of some processors or organizations may still preclude the addition of pulp wash.

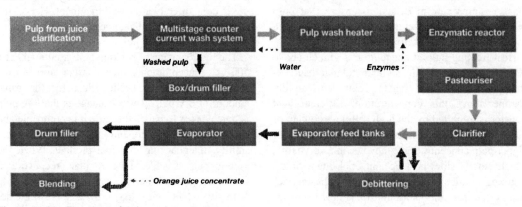

Figure 21.26. Flowchart of pulp wash production.

European Union regulations (Fruit Juice Directive of 1993) do not allow a product that contains pulp wash to be called orange juice. Pulp wash which has been debittered is, however, very difficult to detect in the juice product. Permission to use pulp wash in orange juice is under consideration in the European Union.

PROCESSING STAGE 10: ESSENCE RECOVERY

Essence recovery is an integral part of the evaporation process and is described under the section on concentrate production.

GLOSSARY

° Brix—concentration of all soluble solids in juice. It is not a measure of sugars only, although sugars make up the bulk of the solids in orange juice. Degrees Brix are determined by measurement of juice density or refractive index.

CIP—clean-in-place.

Cloud—source of opaque appearance of orange juice; formed by soluble and insoluble compounds released during juice extraction. The solid particles are kept in suspension by the presence of soluble pectin in the juice. Cloud is an important quality attribute of most citrus juices and contributes to their mouth feel.

Cold-pressed peel oil (CPPO)—oil derived from the peel of citrus fruits. Oil sacs are found in the surface of the peel and these are ruptured during oil extraction. The oil is recovered from the oil/water emulsion by mechanical means (as opposed to thermal processing). Also known simply as peel oil.

Deaeration—the process of removing air (oxygen) from juice. Dispersed air as free air bubbles is quite easily removed from juice but dissolved air requires an effective deaeration process.

Defects—factors that degrade citrus product quality; examples are small seeds or black specs in juice, poor color scores, out-of-range ratios.

Enzyme activity—measure of enzyme concentration in juice: the necessary inactivation of enzymes is achieved by heat treatment of juice.

Enzymes—proteins that catalyze biochemical reactions. As regards orange juice quality, pectin methyl esterase (PME) is the most important.

Essence—volatile components recovered from the evaporation process; separated into an aqueous phase (essence aroma) and an oil phase (essence oil).

Essence oil—source of specific flavor notes, mainly esters and carbonyls; contributes a floral fruity aroma and a juicy flavor to juice.

Evaporation—process of removing water from juice by heat.

Extraction—process of squeezing out juice from either whole or halved oranges by means of mechanical pressure; peel oil also obtained by a mechanical extraction process.

FCOJ (frozen concentrated orange juice)—the most common bulk orange juice product stored and shipped; produced commercially by concentrating juice up to 66° Brix by evaporation.

Finisher equipment—used to separate pulp from juice. This process is referred to as juice finishing.

Flash pasteurization—expression used for pasteurization carried out in a heat exchanger (during a very short period of time, a "flash") as opposed to tunnel pasteurization; there is no flash of product. Also referred to as high-temperature short-time (HTST) heat treatment.

Maillard reaction—nonenzymatic chemical reaction involving condensation of an amino group and a reducing group (sugars), resulting in the formation of intermediates that ultimately polymerize to form brown pigments (melanoidins).

NFC juice (not-from-concentrate juice) —natural, single-strength juice that has undergone neither concentration nor dilution during production.

POJ (pasteurized orange juice)— term used in Florida for NFC juice.

Press liquor—product stream in the feed mill area obtained by removing moisture from the citrus peel in a (screw) press. The press liquor is concentrated in a waste heat evaporator to form molasses.

Pulp—the solid particles in orange juice. Also the commercial name for the product, consisting of broken pieces of cell sacs and segment wall, added back to the final juice.

Pulp wash—process by which soluble solids (mainly sugars) are recovered from pulp. The soluble solids are leached from the pulp with water through a system of mixing screws and finishers. The liquid stream from a pulp wash system is referred to as pulp wash, secondary solids, or WESOS (water extracted soluble orange solids).

Shelf life—time period beyond which food product becomes unacceptable from a safety, sensorial, or nutritional perspective.

Single strength—the term assigned to juice at its natural strength, either directly from the extraction process or in a reconstituted form.

TASTE—thermally accelerated short-time evaporator.

Viscosity—measure of the "thickness" of a fluid. It affects the "body" of the juice and is created primarily by pectin-related stabilization of the cloud or colloids in the juice. The presence of insoluble material also contributes to increased juice body or viscosity.

Washed pulp—solid particles remaining from the pulp wash process. It is sold in frozen form for addition to fruit beverages, or recovered in the feed mill area for use as animal feed.

REFERENCES

Anonymous. 1985. Technical manual—Reconstituted Florida orange juice; Technical manual—Freshly squeezed Florida orange juice. Florida Department of Citrus, Scientific Research Department, University of Florida, IFAS-CREC.

Ashurst PR, editor. 1995. Production and packaging of non-carbonated fruit juices and fruit beverages, 2nd edition. Blackie Academic and Professional, U.K.

Kimball DA. 1991. Citrus processing: Quality control and technology. Van Nostrand Reinhold.

Nagy S, CS Chen, P Shaw, editors. 1993. Fruit juice processing technology. AgScience Inc., Fla.

J Redd, D Hendrix, C Hendrix Jr. Revised 1986. Quality control manual for citrus processing plants, vol. 1. AgScience Inc., Auburndale, Fla.

Redd J, D Hendrix, C Hendrix, Jr., editors. 1992. Quality control manual for citrus processing plants, revised edition, vol. 2. AgScience Inc., Auburndale, Fla.

Redd J, O Shaw, C Hendrix, Jr., D Hendrix, editors. 1996. Quality control manual for citrus processing plants, vol. 3. AgScience Inc., Auburndale, Fla.

Saunt, J. 1990. An illustrated guide to citrus varieties of the world. Sinclair International, Ltd., Norwich, England.

22
Meat: Hot Dogs and Bologna

T. Lawrence and R. Mancini

BACKGROUND INFORMATION

Hot dogs and bologna are defined as comminuted, cooked sausages that contain no more than 30% fat and no more than 40% combined fat and added water [Code of Federal Regulations (CFR) 2003b]. They are made from combinations of beef, pork, and poultry and may be smoked. Seventy-nine percent of American households purchased hot dogs in 1999, and hot dog sales for 2000 were $1.6 billion (Anonymous 2000). Bologna, as the name implies, originated in Bologna, Italy, and continues to be one of America's favorite luncheon meats.

FORMULATION AND INGREDIENTS

SELECTION OF MEAT SOURCES

Skeletal muscle trimmings with varying levels of fat, edible by-product meats, and mechanically separated tissue may be used as raw material. Raw material must be of high quality and have low microbial counts. Generally, hot dogs and bologna are a combination of beef, pork, and poultry: the specific combination is highly dependent upon market preferences and cost. However, species-specific (i.e., all beef) hot dogs and bologna are available. High-quality, mechanically separated, skeletal tissue can be used for up to 100% of the lean source. Likewise, good quality pork or beef by-product meats (i.e., hearts) are permitted, but they must be listed separately on the ingredient statement. Nevertheless, finished product color intensity is related to raw material pigment (myoglobin) concentration, rate of postmortem pH decline (pale, soft, and exudative vs. normal vs. dark, firm, and dry), and nonmeat ingredient selection and amount.

Skeletal muscle contains myosin, actin, and actomyosin (salt-soluble proteins). These proteins stabilize sausage batters by entrapping fat and binding

---

The information in this chapter has been derived from a chapter in *Food Chemistry Workbook,* edited by J. S. Smith and G. L. Christen, published and copyrighted by Science Technology System, West Sacramento, California, ©2002. Used with permission.

water in a matrix-like system. The ability of meat pieces to adhere to one another, retain moisture, and stabilize fat is often referred to as the "binding ability" of meat. However, not all meats have similar binding abilities. Thus, when formulating sausage batter, the binding ability of each raw material must be considered to meet a minimum requirement for bind.

The use of meat with a high content of connective tissue should be limited. Formulations with excessive levels of collagen will have decreased functional properties and poor product quality.

REWORK

A limited amount of broken pieces, ends, or misshapen cooked product, commonly known as rework, may be used in formulations. Due to the reduced water content, limited binding ability, and increased fat proportion of rework, excessive use in a formulation may result in an inferior product if the other raw materials do not compensate for its diminished functionality. The coagulated proteins in rework require that the percentage of rework in the formulation must not reduce the capability of the new batter to bind water and encapsulate fat. Thus, rework should be limited to 10% or less of the total batter formulation.

NONMEAT INGREDIENTS

Numerous nonmeat ingredients are incorporated in hot dog and bologna formulations because they enhance the finished product. All added ingredients must be food grade and be approved for the intended use.

Salts

Sodium and/or potassium chloride serve multiple functions. They facilitate extraction and solubilization of salt-soluble myofibrillar proteins, increase ionic strength and water-holding capacity, and enhance the flavor of the finished product. Salts also lower the water activity of the finished product, which decreases microbial growth. Salt changes the osmotic balance and electrostatic charges of proteins within muscle. Because the majority of the water is located between filaments, changing the electrostatic repulsion of filaments with salt also changes the amount of space between filaments that can be occupied by water.

Increasing the salt concentration influences the electrostatic balance and solubility of muscle proteins, a phenomenon also termed "salting out." In the manufacture of hot dogs and bologna, "salting out" is exploited in order to extract salt-soluble proteins and form a batter. In addition to affecting ionic strength, salt slightly disrupts protein structure, which exposes hydrophobic regions of the protein that are normally buried. This results in hydrophobic exclusion of the protein from the solution; thus, "salting out." Salt also weakens actomyosin linkages, resulting in dissociation of the two proteins and increasing their solubility.

Salt lowers water activity through its interaction with water via dipole-ion bonds. These water-ion bonds are strong, and they compete with microorganisms for water, minimizing the amount of water available for microorganism growth. However, salts may accelerate lipid oxidation by donating free electrons, which catalyze secondary autoxidation reactions. Hot dog and bologna formulations typically contain 2–3% salt.

Sweeteners

Sweeteners enhance flavor and offset the harshness of salt. They also improve the peelability of hot dogs. Commonly used sweeteners include corn syrup, corn syrup solids, and sucrose. Sweeteners such as dextrose (glucose) enhance the browning of grilled sausages, whereas other sweeteners such as sorbitol reduce undesirable browning during prolonged grilling (rotary grill). Like salts, sugars lower water activity, by forming hydrogen bonds with water, which disorganize the structure of water and lessen its availability for growth of microorganisms and chemical reactions. In addition, sugars enhance water binding by increasing ionic strength. Sweeteners normally are used at 0.5–2% of the formulation.

Spices

Various combinations of spices are used in small quantities to impart a desirable flavor profile to the finished product. To eliminate microbial contamination, spices commonly are sterilized (gamma irradiation). Most spices are finely ground to prevent visible specks of spice in the finished product, which consumers find unappealing. Essential oils, soluble extracts, and oleoresins (often on a dry carrier) are used as an alternative to dry ground spices because they further reduce the occurrence of specks and minimize bacterial problems. In addition, compared with ground spices, essential oils allow for more control of taste intensity. Aside from imparting desirable flavors, some spices serve as antioxidants (paprika; Aguirrezabal et al. 2000) or antimicrobials (clove, garlic, mustard; Prescott et al. 1999). Spices commonly found in hot dog or bologna formulations include allspice, cardamon, clove, coriander, garlic, ginger, mace, mustard, nutmeg, oregano, paprika, and pepper (black, white, or red).

Alkaline Phosphates

Sodium and/or potassium alkaline phosphates are included in the formulation to maximize the water-binding ability of meat (Molins 1991). The isoelectric point (point at which the net charge is zero) of muscle is approximately 5.1. At this point, a lack of charge repulsion minimizes interfilament spaces between actin and myosin, leaving little space for water and causing low water-binding ability. Alkaline phosphates increase pH and net charge, thereby increasing interfilament spaces and allowing more space for water. The addition of phosphates provides supplementary ion species, which increases ionic strength and enhances water-holding capacity. Phosphates improve yields by reducing moisture loss (shrink) during cooking and cooling. Phosphates also aid myofibrillar protein extraction by dissociating actomyosin (Claus et al. 1994). Additionally,

phosphates act as metal chelators and antioxidants. Phosphates readily bind free cations such as calcium and magnesium and thus are able to remove these ions from a solution. By removing free metal ions from solution, phosphates minimize prooxidant activity. When used at excessive levels (greater than 0.4% in the final product), phosphates may impart a soapy flavor to the final product. Phosphates are not highly soluble in water (especially cold water) and thus are difficult to disperse within a solution. Regulations permit phosphates at a maximum of 0.5% (5000 ppm) of meat block weight (CFR 2003c).

Sodium Nitrite

Sodium nitrite was originally used to control the outgrowth of *Clostridium botulinum* spores. This preservative inhibits *C. botulinum* reproduction as well as the germination of spores through various mechanisms. However, the exact mechanism is not completely understood. In addition, sodium nitrite fixes pigment color after it is chemically reduced to nitric oxide by cure accelerators such as sodium erythorbate. Following addition of sodium nitrite to the batter, nitric oxide is formed and combines with myoglobin to form nitric oxide myoglobin. Upon heating, nitric oxide myoglobin forms nitric oxide hemochrome, which produces the typical pink, cured meat color (Fig. 22.1). Sodium nitrite con-

CURED MEAT COLOR FORMATION

MYOGLOBIN
(Deoxymyoglobin, Oxymyoglobin, Metmyoglobin)
+
Sodium nitrite $\xrightarrow{\text{Cure accelerator}}$ Nitric oxide
↓
NITRIC OXIDE MYOGLOBIN
(dull red color)
↓
NITRIC OXIDE METMYOGLOBIN
(tannish brown color)
+
Heat
↓
NITRIC OXIDE HEMOCHROME
(typical cured meat color; pink, heat stable)
+
Light or oxygen
↓
Hemichromes
(Gray-brown faded cured meat color)

Figure 22.1. Cured meat color formation process.

tributes to the flavor associated with cured meats. It also acts as an antioxidant by decreasing heme iron oxidation. Because this ingredient is used in such small quantities, it is often added as a pink-colored "cure" that contains salt and 6.25% sodium nitrite. The pink color also helps to distinguish it from salt and minimizes formulation errors. Regulations limit its use to 0.0156% (156 ppm) of meat block weight (CFR 2003c). Nitrite should not be mixed with cure accelerators before it is added to the meat batter. Mixing the two ingredients prior to product formulation would prematurely accelerate the conversion of nitrite to nitric oxide. This would diminish the effectiveness and ability of nitrite to inhibit microbial growth and fix color.

Cure Accelerators

Cure accelerators such as sodium ascorbate and its isomer sodium erythorbate speed the curing process and enhance nitric oxide hemochrome formation by expediting the conversion of sodium nitrite to nitric oxide (Claus et al. 1994). These reducing agents promote chemical reduction of sodium nitrite to nitric oxide, which is the ligand necessary to produce cured color. Thus, by accelerating the conversion of nitrite to nitric oxide, cure accelerators maximize the amount of nitric oxide available to bind to myoglobin. This maximizes nitrosylmyoglobin formation and the development of cured color. Cure accelerators also help stabilize cured color in the retail product through their ability to serve as antioxidants, reducing agents, and chelators. Therefore, compounds such as ascorbic acid limit pigment and lipid oxidation, which in turn maximizes desirable color and appearance. By increasing the rate of sodium nitrite conversion, cure accelerators decrease residual sodium nitrite in the final product. Regulations limit their use to 0.0550% (550 ppm) of meat block weight (CFR 2003c).

Water/Ice

Water and ice provide a medium for the addition of water-soluble nonmeat ingredients (salt, sweeteners, phosphates, nitrite, cure accelerators). Water and/or ice aid in temperature control in the meat batter during chopping and emulsification. The addition of water lowers product cost, compensates for evaporative losses during cooking, and facilitates the production of low fat, low calorie products. Water should be filtered so that it does not contain high levels of dissolved calcium or iron salts. Failure to use filtered water may result in an unintended addition of prooxidant metals that are chelated by phosphates, thereby decreasing phosphate functionality.

Extenders and Binders

Extenders and binders reduce product cost, enhance water binding, and improve yield and slicing performance. Additionally, they can be used to increase protein content, modify color or flavor, improve texture, and assist in forming and stabilizing the batter matrix. Examples of extenders and binders include nonfat dry milk, soy proteins, sodium caseinate, gelatin, whey proteins, gums, and starches. They frequently are used in low fat, and fat free products. Regulations limit their use to 3.5% or less in the final product, and they must be indicated on the product label.

MANUFACTURING AND PROCESSING PROCEDURES

PREBLENDING OF MEATS

Initially, raw material particle size is reduced by grinding, chopping, or flaking. Grinders with a bone/cartilage separator are used to remove bone chips. Lean ground meats are mixed together with selected nonmeat ingredients (salt, sodium nitrite, and water) to form a preblend. Lean preblends are often stored for 6–24 hours to promote extraction of salt-soluble proteins (myosin, actin, and actomyosin). During this brief storage period, chemical analyses (fat, moisture, protein) are performed, and the results are used to determine the amount of additional raw material necessary to achieve the lean:fat target ratio. Addition of sodium nitrite to the lean preblend is necessary to initiate cured color development, offset the prooxidant effects of salt, and retard microbial spoilage.

FORMULATION

Cooked sausages generally contain added water, which is defined as [percentage moisture − (4 × percentage protein)] in the final product (Claus et al. 1994). Fat is restricted to a maximum of 30%, and the combined percentage of added water and fat must not exceed 40%. Two examples of percent fat and added water might include 10% fat and 30% added water, or 21% fat and 19% added water.

These requirements, plus restrictions on nonmeat ingredients (sodium nitrite, cure accelerators, phosphates, and extenders), mandate careful formulation for specific (targeted) product composition. In formulation, the weight loss during cooking (moisture evaporation) must also be taken into account.

Current industry practices use computer programs to determine the least-cost formulation. This formulation method utilizes information such as species, moisture, fat, and protein composition, binding ability, color contribution, collagen content, and raw material price. These variables determine the most economic formulation that will meet the finished product specifications.

MIXING AND BATTER FORMATION

Mixing further extracts salt-soluble proteins and evenly distributes fat and nonmeat ingredients throughout the ground raw material. During batter formation, fat, muscle, and connective tissue (discontinuous phase) particle size are reduced, and solubilized muscle proteins are released into the liquid phase (continuous phase). These released proteins encapsulate fat globules to form a stable meat batter and ensure a uniform texture and appearance in the final product. Simultaneous mixing and batter formation can be accomplished in a bowl chopper. The use of vacuum during chopping removes air from the product and increases product density. Batter formation can also be accomplished in an emulsion mill, which is a high-speed multiknife/plate grinder. Because bowl chopping or emulsion milling increases batter temperature, the temperature of the batter must be carefully monitored. If the batter temperature exceeds the melting temperature of the fat, the batter may destabilize, resulting in loss of functionality (Pearson and Gillett 1999).

CASINGS

The majority of hot dogs are stuffed into cellulose casings, which are manufactured from cotton linters or wood pulp. These inedible, small diameter (15–45 mm) hot dog casings often are purchased as shirred sticks. Through shirring, it is possible to pleat and compress a long casing (e.g., 25 m) into a stick 30 cm in length. Natural casings also can be used for hot dogs, adding a specialty appearance and texture.

Large diameter fibrous casings used for bologna have special reinforcement (regenerated cellulose) and are less elastic when wet than the smaller diameter casings used for skinless hot dogs. Fibrous casings have the strength necessary for automated slicing. Large fibrous casings vary in diameter from 50 to 250 mm and are commonly dyed red. Bologna casing may be up to six feet long. Small diameter bologna may be formed into rings for ring bologna. Moisture-permeable casings may be prestuck to allow for moisture evaporation and smoke permeation. In addition, this casing allows air to escape from the sausage batter, which minimizes undesirable air pockets, gel pockets, and fatting out.

STUFFING

During stuffing, the batter is forced into the casing. Stuffing pressure is critical and should be monitored in order to minimize variation in product weight, avoid product fatting out (low pressure/air pockets), and prevent casing rupture during stuffing and cooking due to excessive pressure. Stuffing shapes the product and provides a means of containment, suspension, handling, and separation during cooking.

A variety of stuffers are available for the manufacture of hot dogs and bologna. Some stuffers use hydraulic pistons to force the batter through a horn of selected size, whereas others use intermeshing augers (twin screws) or metal fingers (vanes) to push the meat batter through a stuffing horn. Vacuum stuffers remove air voids and help maintain consistent density. Low-density batters often contain air pockets that may fill with gelatin or fat during cooking, causing gelatin or fat pockets. Continuous stuffer/linkers are fed by a meat pump.

LINKING

Hot dogs are linked so that a specific number of links, depending on diameter, will make up precisely one pound or one retail package. Links are placed on smokesticks for cooking. Retail package size allows 4–10 hot dogs per pound and 4–12 inches in length.

COOKING AND SMOKING

Smoking/cooking involves a short drying period, followed by smoking and cooking. Upon heating, skeletal muscle proteins coagulate to form a stable gel matrix composed of protein, water, and fat. Within this matrix, proteins encapsulate fat and bind water, which distributes the immiscible phases (protein, water, and fat).

Cooking also pasteurizes the product, removes moisture, and forms the cooked meat flavor typically associated with hot dogs and bologna. Heat converts nitrosated myoglobin into a stable nitric oxide hemochrome pigment, which results in cured meat color. Heat denatures myoglobin, changing its conformation, causing it to unfold, and exposing residues that are normally buried. This allows for various intramolecular bonds between amino acids and the nonprotein moiety, resulting in the hemochrome responsible for the pinkish color of cured meat.

One common method of applying smoke to hot dogs and bologna is liquid smoke, which may be added to the batter before cooking, or misted (atomized) on the product during cooking. Smoking imparts a desired smoky flavor and color while darkening the exterior skin. During cooking and smoking, a surface skin or coagulated protein membrane is formed at the casing/meat interface. This membrane allows casing removal and, depending on its elasticity, imparts mouth-feel or "bite" to the hot dog. The phenolic compounds present in smoke also serve as antioxidants and antimicrobials.

Smoking and cooking cycles (time and temperature) depend upon product diameter and the type of smoking-heating system used. Smokehouses are generally one of two types: batch houses or continuous-flow houses. In either system, temperature and relative humidity are controlled for optimum smoke deposition, batter stability, pasteurization, and minimal moisture loss (yield control). Determination of product end-point temperature is dependent on product type, composition, and the microbial lethality necessary to pasteurize the product. Higher end-point cooked temperatures result in a longer product shelf life.

CHILLING

Hot dogs and bologna are normally chilled with a brine spray or shower (< 25°F). Compared to air chilling, brine chilling is faster, minimizes evaporative losses, inhibits microbial growth, facilitates casing removal, and increases product firmness, resulting in fewer broken hot dogs.

CASING REMOVAL OR SLICING

After chilling, inedible hot dog casings are stripped away by a peeling machine. Inedible casings are no longer needed because the surface skin or coagu-

lated protein membrane has already been formed at the casing/meat interface. This membrane now acts as a casing, holding the finished product together until it is consumed. The phenolic compounds present in smoke also serve as antioxidants and antimicrobials. Bologna casings may be stripped away or may be left on the product, depending upon processor and customer preference. Bologna commonly is sliced for retail sale, but may also be sold as chubs. Because natural casings provide a texture that is desired by many consumers, they are not removed.

PACKAGING AND LABELING

Packaging protects the product, enhances product appearance, minimizes weight loss, and maximizes shelf life. The package label also conveys important information to the consumer.

After peeling, hot dogs are collated, aligned into single packages (1 pound, 12 ounces, etc.), and vacuum packaged in barrier films.

Sliced bologna is commonly packaged in impermeable heat-shrinkable or formed plastic film and vacuum sealed. Bologna chubs are often vacuum packed in barrier films with the original casing intact.

Vacuum packaging is necessary to maintain a typical cured meat color under lighted display conditions by reducing photooxidation (ultraviolet/visible light or oxygen-induced fading of the nitric oxide hemochrome pigment). If vacuum packaging is not used and the product is exposed to atmospheric oxygen, the characteristic pink cured color quickly fades and turns gray.

Labels must include (1) product name, (2) ingredient list in descending order of predominance, (3) manufacturer name and address, (4) net weight of contents, (5) official inspection legend, (6) handling/storage instructions, and (7) nutritional information (CFR 2003a). Other label information could include a sell-by date, a Universal Product Code, a lot or batch code, cooking suggestions, and processor contact information. Special claims (i.e., low fat, fat free, low sodium) may also appear on the label, provided the product meets the requirements imposed by the particular claim.

QUALITY CONTROL

Quality control assures consistency of the finished product. Analytical and organoleptic measures used in quality control include (1) fat, moisture, and protein analysis, (2) shelf-life estimates, (3) sensory perception—flavor, color, odor, texture, (4) product net weight, (5) salt, nitrite, and erythorbate analyses, (6) collagen content, (7) product binding ability, and (8) vacuum package integrity. Although not covered in the scope of this chapter, the food safety program referred to as HACCP (hazard analysis and critical control points) is a vital component in the manufacture of hot dogs and bologna.

PROBLEMS/CAUSES

The following is a compilation of problems that may be encountered during the production of hot dogs or bologna and causes of those problems.

Problem	Causes
Rancidity—rancid odors and flavor	*Lipid oxidation.* Multiple causes include using impure salt contaminated with heavy metal ions, leaking packages, insufficient vacuum, excessive exposure to light, and temperature abuse.
Faded, undercured color	*Insufficient nitrite or reducing agent.* Insufficient time and/or temperature for color reaction to occur. Leaking packages. Excessive exposure to light. Oxidized myoglobin prior to curing.
Green spots	*Nitrite burn.* Excessive nitrite and/or insufficient reducing agent, or insufficient distribution of these ingredients. Also failure to get nitrite into meat pieces.
Fat caps	*Unintentional addition of air to the batter during emulsion process.* Insufficient air removal and/or poor stuffing pressure. Excessive collagen content in raw material. Air voids fill with gelatin during the cooking process. Too much fat relative to bind.
Green cores in bologna	*Insufficient thermal processing.* Bacteria growth causes greening. Incomplete cure reaction due to high pH phosphates and/or mechanically separated meat.
Poor peelability	*Lack of surface protein coagulation.* Excessive dehydration during chilling. Casings are beyond their shelf life or were stored improperly.
Fatting out	*Low density batters with air pockets.* Insufficient removal of air during grinding and stuffing. Result of the breakdown of the protein matrix, causing fat accumulation on the surface of the cooked hot dog.

GLOSSARY

Actin—salt-soluble protein known as the thin filament.

Actomyosin—complex of bound actin and myosin.

Antimicrobial—substance that retards the growth of microflora.

Antioxidant—substance that retards lipid oxidation.

Batter—a matrix of protein, fat, water, and nonmeat ingredients; also inaccurately referred to as an emulsion.

Binding ability—ability of proteins to bind fat, water, and other proteins and retain them during cooking.

Bologna—a fully cooked, mildly seasoned sausage.

Brine shower—a saturated saltwater solution applied to hot dogs and bologna after cooking in order to lower product temperature and minimize casing shrinkage.

Collagen—the predominant structural protein in connective tissue that forms gelatin upon heating.

Chelators—chemical compounds that bind metal ions, prohibiting their interference in chemical reactions.

Cure accelerators—chemical compounds that speed the conversion of sodium nitrite to nitric oxide.

DFD (dark, firm, dry)—a postmortem muscle phenomenon resulting in high ultimate pH (> 6.0) that occurs from antemortem depletion of muscle glycogen. Because of high pH, the muscle has a dark, firm, and dry appearance. Also known as "dark cutting."

Formulation—the sum of ingredients used to make a sausage product.

Frankfurter—a fully cooked, mildly seasoned smoked sausage, commonly known as a hot dog or wiener.

HACCP—hazard analysis and critical control points.

Ionic strength—a measure of the concentration of ions in a solution.

Least-cost formulation—formulation that meets a set of desired product specifications while also ensuring the lowest raw material expense.

Mechanically separated tissue—muscle that is mechanically separated from bone and connective tissue.

Myoglobin—water-soluble protein containing heme; responsible for meat color.

Myosin—predominant salt-soluble myofibrillar protein known as the thick filament; responsible for binding water and fat.

pH—a measure of hydrogen ion concentration. Normal meat pH is 5.6–5.9.

Preblend—ground meat and selected nonmeat ingredients (usually salt, sodium nitrite, water) that are blended and held for up to 24 hours to maximize extraction of salt-soluble proteins.

Prestuck casings—casings made with small holes to allow for moisture evaporation and smoke penetration. These casings also lessen the amount of air that gets trapped in a sausage, reducing undesirable air pockets and voids in the finished product.

PSE (pale, soft, exudative)—a postmortem muscle phenomenon resulting from accelerated postmortem metabolism. Because of low ultimate pH ($\leq$ 5.5) and greater than normal protein denaturation, the muscle has a pale, soft, and exudative (watery) appearance.

Rework—product that is aesthetically unacceptable for retail sale that is used in a future formulation. Use of rework must be carefully monitored in order to maintain protein functionality.

Salt-soluble protein—A protein that can be extracted from meat using salt.

Shirred casing—casing that is pleated so the storage length is decreased to approximately 1/75 of the use length. This process also simplifies the task of loading the casing onto the stuffing horn, which makes them highly efficient and very conducive to high speed manufacturing operations.

Sodium nitrite—a salt that is converted to nitric oxide, which reacts with myoglobin and, upon heating, forms nitric oxide hemochrome, the typical cured meat color; also responsible for inhibiting the outgrowth of *Clostridium botulinum* spores.

Surface skin—coagulated proteins that form the smooth, thin, skin at the perimeter of the product.

Water activity—a measure of the availability of water within food. The availability of water influences growth of bacteria, yeast, and fungi as well as the rates of enzymatic activity and lipid peroxidation.

Water-holding capacity—ability of meat to retain water during processing and storage.

Water-soluble protein—a protein that is easily dispersed or solubilized in an aqueous solution. The most abundant water-soluble protein in meat is myoglobin.

REFERENCES

Aguirrezabal MM, J Mateo, MC Dominguez, JM Zumalacarrequi. 2000. The effect of paprika, garlic and salt on rancidity in dry sausages. Meat Sci. 54:77–81.

Anonymous. 2000. 2000 state of the industry report. The National Provisioner 214(8): 50.

Code of Federal Regulations (CFR). 2003a. Title 9—Animal and Animal Products, Part 317—Labeling, marking devices, and containers. U.S. Government Printing Office, Washington, D.C. Available on the Web at http://www.access.gpo.gov/cgi-bin/cfrassemble.cgi.

___. 2003b. Title 9—Animal and Animal Products, Part 319—Definitions and standards of identity or composition. U.S. Government Printing Office, Washington, D.C. Available on the Web at http://www.access.gpo.gov/cgi-bin/cfrassemble.cgi.

___. 2003c. Title 9—Animal and Animal Products, Part 424—Preparation and Processing Operations. U.S. Government Printing Office, Washington, D.C. Available on the Web at: http://www.access.gpo.gov/cgi-bin/cfrassemble.cgi.

Claus JR, JW Colby, GJ Flick. 1994. Chapter 5. Processed meats/poultry/seafood. In: DM Kinsman, AW Kotula, BC Breidenstein, editors. Muscle Foods, 106–162. Chapman and Hall, New York.

Molins RA. 1991. Phosphates in Food. CRC Press, Boca Raton, Fla.

Pearson AM, TA Gillett. 1999. Processed meats, 3rd edition. Aspen Publishers, Inc., Gaithersburg, Md.

Prescott LM, JP Harley, DA Klein. 1999. Chapter 43. Microbiology of food. In: Microbiology, 4th edition. WCB/McGraw Hill, Boston, Mass.

23
Meat: Fermented Meats

F. Toldrá

BACKGROUND INFORMATION

The origin of fermented meats goes far back in time. Ancient Romans and Greeks manufactured fermented sausages, and in fact, the origin of words like sausage and salami may proceed from the Latin expressions *salsicia* and *salumen*, respectively (Toldrá 2002). The production and consumption of fermented meats expanded throughout Europe in the Middle Ages and were adapted to climatic con-

ditions (e.g., smoked in northern Europe and dried in Mediterranean countries). The experience in manufacturing these meats came to America with settlers (e.g., states like Wisconsin still have a good number of typical northern European sausages like Norwegian and German sausages).

Today, a wide variety of fermented sausages are produced; the variations depend on raw materials, microbial populations, and processing conditions. For instance, northern-type sausages contain beef and pork as raw meats, are ripened for short periods (up to three weeks), and are usually subjected to smoking. In these sausages, shelf life is mainly due to acid pH and smoking rather than drying. On the other hand, Mediterranean sausages mostly use only pork, are ripened for longer periods (several weeks or even months), and are not typically smoked (Flores and Toldrá 1993). Examples of different types of fermented sausages, according to the intensity of drying, are listed in Table 23.1. Undry and semidry sausages are fermented to reach low pH values, and are usually smoked and cooked before consumption. Shelf life and safety are mostly determined by pH drop and reduced water activity, as a consequence of fermentation and drying, respectively. The product may be considered stable at room temperature when pH < 5.0, and the moisture:protein ratio is below 3.1:1 (Sebranek 2004). Moisture:protein ratios are defined for the different dry and semidry fermented sausages in the United States, while water activity values are preferred in Europe.

RAW MATERIALS PREPARATION

There are several considerations listed in Table 23.2 that must be taken into account when producing fermented meats. The selection of the different op-

LEEDS TRINITY UNIVERSITY

Table 23.1. Examples of Fermented Meats with Different Dryness Degrees

Product	Type	Examples	Weight loss (%)	Drying/Ripening
Undry fermented sausages	Spreadable	German teewurst	< 10	No drying
		Frische mettwurst	< 10	No drying
Semidry fermented sausages	Sliceable	Summer sausage	< 20	Short
		Lebanon bologna	< 20	Short
		Saucisson d'Alsace		
Dry fermented sausages	Sliceable	Hungarian and Italian salami	> 30	Long
		Pepperoni	> 30	Long
		Spanish salchichón	> 30	Long
		French saucisson	> 30	Long

Sources: Lücke 1985, Roca and Incze 1990, Toldrá 2002

Table 23.2. Some Options in the Processing of Fermented Meats

Aspects	Options
Type of meat	Pork, beef, . . .
Quality of meat	Choose good quality. Reject defective meats (pork PSE and DFD), abnormal colors, exudation, . . .
Origin of fat	Choose either chilled or frozen (how long?) fats. Reject oxidized fats.
Type of fat	Control of fatty acids profile (excess of PUFA?)
Ratio	Choose desired meat:fat ratio
Particle size	Choose adequate plate (grinder) or speeds (cutter)
Additives:	
Salt	Decide concentration
Curing agent	Nitrite or nitrate depending on type and length of process
Carbohydrates	Type and concentration depending on type of process and required pH drop
Spices	Choose according to required specific flavor
Microflora	Natural or added as starter?
Starters	Choose microorganisms depending on type of process and product
Casing	Material and diameter depending on type of product
Fermentation	Conditions depending on type of starter used and product
Ripening/drying	Conditions depending on type of product
Smoking	Optional application. Conditions depending on type of product and specific flavor
Color	Depends on raw meat, nitrite, and processing conditions
Texture	Depends on meat:fat ratio, stuffing pressure, and extent of drying
Flavor	Choose adequate starter and process conditions
Water activity	Depends on drying conditions and length of process

tions, which will be discussed in the following sections, facilitates the choice of the most adequate conditions for the correct processing, safety, and optimal final quality.

INGREDIENTS

Lean meats from pork and beef, in equal amounts, or pork only are generally used. Quality characteristics such as color, pH (preferably below 5.8), and water-holding capacity are very important. Meats with pH higher than 6.0 are indicative of a type of pork meat known as DFD (dark, firm, and dry) that binds water tightly and is easily spoiled. Pork meat with another defect, known as PSE (pale, soft, and exudative), is not recommended for use because the color is pale, and the sausage would release water too fast, possibly causing the casing to wrinkle. Meat from older animals is preferred because of its more intense color, due to the accumulation of myoglobin, a sarcoplasmic protein, which is the natural pigment responsible for color in meat.

Pork back and belly fats constitute the main source for fats. Special care must be taken for the polyunsaturated fatty acid profile, which should be lower than 12%, and the level of oxidation (measured as peroxide value), which should be as low as possible (Demeyer 1992). Some rancidity may develop after long-term frozen storage since lipases present in adipose tissue are active even at temperatures as low as $-18°C$ and are responsible for the continuous release of free fatty acids that are susceptible to oxidation (Hernández et al. 1999). So, extreme caution must be taken with fats stored for several months as they may develop a rancid flavor.

ADDITIVES

Salt is the oldest additive, and it has been used in cured meat products since ancient times. Salt, at about 2–4%, exerts several functions, including (1) producing an initial reduction in water activity, (2) giving the meat a characteristic salty taste, and (3) contributing to an increased solubility of myofibrillar proteins.

Nitrites exert an important antimicrobial effect, especially against the growth and spore production by *Clostridium botulinum*. Nitrite also contributes to antioxidative stability as well as the typical cured meat color and flavor (Gray et al. 1981, Pegg and Shahidi 2000). The reduction of nitrite is favored by the presence of reducing substances such as ascorbic and eyrthorbic acids or their sodium salts. These substances contribute to the reduction of the formation of nitrosamines as the residual amounts of nitrite are very low.

Carbohydrates like glucose or lactose are used quite often as substrates for microbial growth and development. Disaccharides, and especially polysaccharides, may delay the growth and pH drop rate because they have to be hydrolyzed to monosaccharides by microorganisms.

Sometimes, additional substances may be used for specific purposes (Demeyer and Toldrá 2003). This is the case for glucono-delta-lactone, added at 0.5%, which may simulate bacterial acidulation. In the presence of water, glucono-delta-lactone is hydrolyzed to gluconic acid and produces a rapid decrease in pH. The quality is rather poor because the rapid pH drop drastically reduces the activity of flavor-related enzymes such as exopeptidases and lipases. Other substances that may be added include

(1) phosphates to improve resistance to oxidation, (2) vegetable proteins such as soy isolates to replace meat proteins, and (3) manganese sulphate as a cofactor for lactic acid bacteria (LAB).

Spices, either in natural form or as extracts, are added to give a characteristic aroma or color to the fermented sausage. There is a wide variety of spices like pepper, paprika, oregano, rosemary, garlic, onion, and so on; each one gives a particular aroma to the product. Some spices also contain powerful antioxidants. The most important aromatic volatile compounds may vary depending on the geographical and/or plant origin. For instance, garlic, which contributes a pungent and penetrating smell, is typically used in chorizo, and pepper is used in salchichón and salami. Paprika gives fermented meats a characteristic flavor and color due to its high content of carotenoids (Ordoñez et al., 1999). The presence of manganese in some spices, like red pepper and mustard, stimulates the activity of several enzymes involved in glycolysis and thus enhances the generation of lactic acid (Lücke 1985).

STARTERS

Typical fermented products were initially based on the development and growth of desirable indigenous flora, sometimes reinforced with back slopping, which is the addition of a previous ripened fermented sausage with adequate sensory properties. However, this practice usually yielded a high heterogeneity in product quality. The use of microbial starters, as a way to standardize processing as well as quality and safety, is relatively new. In fact, the first commercial use of microbial starters was in the United States in the 1950s, followed by Europe in the 1960s; since that time starter use has become widespread. Today, most fermented sausages are produced with a combination of lactic acid bacteria to achieve adequate acidulation, and two or more cultures to develop flavor and facilitate other reactions such as nitrate reduction.

In general, microorganisms used as starter cultures must satisfy several requirements, in accordance with the purposes of their use: nontoxicity for humans, good stability under the processing conditions (resistance to acid pH, low water activity, tolerance to salt, resistance to phage infections), intense growth at the fermentation temperature (i.e., 18–25°C in Europe or 30–35°C in the United States), generation of products of technological interest (e.g., lactic acid for pH drop, volatile compounds for aroma, nitrate reduction, secretion of bacteriocins, etc.), and lack of undesirable enzymes (e.g., decarboxylases responsible for amine generation). Thus, the most adequate strains have to be carefully selected and controlled because they have a very important role in the process and are decisive in determining final product quality. The most important microorganisms used as starters belong to one of the following groups: lactic acid bacteria (LAB), *Micrococcaceae*, yeasts, and molds (Leistner 1992). Main roles and functions for each group are shown in Table 23.3.

Lactic Acid Bacteria (LAB)

The most important function of LAB is the generation of lactic acid from glucose or other carbohydrates through either homo- or heterofermentative pathways. The accumulation of lactic acid produces a pH drop in the sausage. However, some undesirable secondary products such as acetic acid, hydrogen peroxide, acetoin, and so on may be generated in the case of certain species that use heterofermentative pathways. *Lactobacillus sakei* and *L. curvatus* grow at mild temperatures that are usual in the processing of European sausages, while *L. plantarum* and *Pediococcus acidilactici* grow well at higher temperatures (30–35°C), closer to the fermentation conditions in sausages produced in the United States. Lactic acid bacteria also have a proteolytic system, consisting in endo- and exopeptidases, that contributes to the generation of free amino acids during processing, and most LAB are also able to generate different types of bacteriocins with antimicrobial properties.

Micrococcaceae

This group consists of *Staphylococcus* and *Kocuria* (formerly *Micrococcus*), which are major contributors to flavor due to their proteolytic and lipolytic activity. Another important function is nitrate reductase activity, which is necessary to reduce nitrate to nitrite, contributing to color formation and safety. However, these microorganisms must be added in large amounts because they grow poorly or even die just at the onset of fermentation, when low pH conditions are prevalent. Preferably, low-pH–tolerant strains should be carefully selected. The species from this family also have an important catalytic function that contributes to color stability and, somehow, prevention of lipid oxidation.

Yeasts

Debaryomyces hansenii is the predominant yeast in fermented meats, mainly growing in the outer area of the sausage due to its aerobic metabolism. *D. hansenii* has good lipolytic activity and is able to degrade lactic acid. In addition, it has an important deaminase/deamidase activity, using free amino acids as substrates and producing ammonia as a subproduct that raises the pH in the sausage (Durá et al. 2002).

Molds

Some typical Mediterranean dry fermented sausages have molds on the surface. The most usual are *Penicillium nalgiovense* and *P. chrysogenum*. They contribute to sausage flavor through their proteolytic and lipolytic activity, and to sausage appearance in the form of a white coating on the surface. They also generate ammonia through their deaminase and deamidase activity, contributing to pH rise. Inoculation of sausages with natural molds present in the fermentation room is dangerous because toxigenic molds might grow. So, fungal starter cultures are mainly used as a preventive measure against the growth of other mycotoxin-producing molds. They also give a typical white color on the surface that is demanded in certain Mediterranean areas.

CASINGS

Casings may be natural, semisynthetic, or synthetic, but a common required characteristic is permeability to water and air. Natural casings are natural portions of the gastrointestinal track of swine, sheep, and cattle, and although irregular in shape, they have good elasticity, tensile strength, and permeability. Natural casings are typically used for traditional sausages because they present a homemade appearance. Semisynthetic casings are based on collagen that shrinks with the product and is permeable but cannot be overstuffed (Toldrá et al. 2004). Synthetic cellulose-based casings are nonedible, but they are preferred for industrial processes because of advantages such as controlled and regular pore size, uniformity for standard products, and hygiene. These casings are easily peeled off.

A wide range of sizes, between 2 and 15 cm, may be used, depending on the type of product. Of course, the diameter strongly affects fermentation and drying conditions. So, pH drop is more impor-

Table 23.3. Main Roles and Effects of Microorganisms in Fermented Meats

Group	Microorganisms	Primary Role	Secondary Role	Chemical Contribution	Effects
Lactic acid bacteria	*L. sakei, L. curvatus, L. plantarum, L. pentosus, P. acidilactici, P. pentosaceus*	Glycolysis	Proteolysis (endo and exo)	Lactic acid generation, generation of free amino acids	pH drop, safety, firmness, taste
Micrococcaceae	*K. varians, S. xylosus, S. carnosus*	Nitrate reductase, lipolysis, catalase	Proteolysis (exo)	Reduction of nitrate to nitrite; generation of free fatty acids, ready for oxidation; generation of free amino acids; degradation of hydrogen peroxide	Safety, aroma, taste
Yeasts	*Debaryomyces hansenii*	Lipolysis	Deamination/deamidation, transamination	Generation of free fatty acids, transformation of amino acids, lactic acid consumption and generation of ammonia	Aroma, taste, pH rise
Molds	*Penicillium nalgiovense, P. chrysogenum*	Lipolysis, proteolysis (exo)	Deamination/deamidation, transamination	Generation of free fatty acids, ready for oxidation, generation of free amino acids and their transformation, generation of ammonia	Aroma, taste, pH rise, external aspect

tant in large diameter sausages where drying is more difficult to achieve.

PROCESSING STAGE 1: COMMINUTION

A sample flow diagram for the processing of fermented sausages is shown in Figure 23.1. Chilled meats, pork alone or mixtures of pork and beef, and porcine fats are submitted to comminution in a grinder (Fig. 23.2). There are several plates with different hole sizes, depending on the desired particle size. Previous trimming for removal of connective tissue is recommended, especially when processing

undry or semidry fermented sausages where no further hydrolysis of collagen will occur. Salt, nitrate and/or nitrite, carbohydrates, microbial starters, spices, sodium ascorbate, and optionally, other nonmeat proteins are added to the ground mass, and the whole mix is homogenized under vacuum to avoid bubbles and undesired oxidations that affect color and flavor (Fig. 23.3). Grinding and mixing take several minutes, depending on the amount. Industrial processes may use a cutter, as an alternative to grinding and mixing, when the required particle sizes are small. The cutter consists of a slowly moving bowl, containing the meats, fat, and additives, that rotates against a set of knives operating with rapid rotation. The fat and meat must be prefrozen

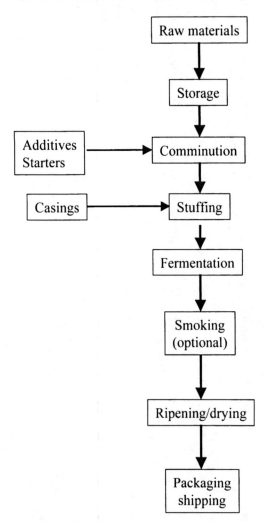

Figure 23.1. Flow diagram showing the most important stages in the processing of fermented sausages.

Figure 23.2. Grinding of meats and fats. There are many sizes of grinder plates to accord with the required particle size.

Figure 23.3. Detail of the batter after mixing in a vacuum mixer massager.

(−6 to −7°C) to avoid smearing of fat particles during chopping. This phenomenon (smearing) consists in a fine film of fat forming over the lean parts, which may reduce the release of water during drying (Roca and Incze 1990). The cutter operates under vacuum to avoid any damage by oxygen, taking only a few minutes, and the ratio of the rotation speed of the bowl to that of the knives determines the desired particle size.

PROCESSING STAGE 2: STUFFING

The mixture is stuffed under vacuum into casings, either natural collagen or synthetic, with both extremes clipped. The vacuum avoids the presence of bubbles within the sausage and disruptions in the casing. The stuffing must be adequate in order to avoid smearing of the batter, and temperature must be kept below 2°C to avoid this problem. Once stuffed (Fig. 23.4), the sausages are hung in racks and placed in natural or air-conditioned drying chambers.

PROCESSING STAGE 3: FERMENTATION

FERMENTATION TECHNOLOGY

Once sausages are stuffed, they are placed in computer-controlled, air-conditioned chambers and left to ferment for microbial growth and development. A typical chamber is shown in Figure 23.5. Temperature, relative humidity, and air speed must be carefully controlled to foster appropriate microbial growth and enzyme action. The whole process can be considered as a lactic acid, solid-state fermentation, where several simultaneous processes take place: (1) microbial growth and development, (2) biochemical changes, mainly enzymatic breakdown of carbohydrates, proteins, and lipids, and (3) physical changes, mainly acid gelation of meat proteins and drying.

Meat fermentation technology differs between the United States and Europe. High fermentation temperatures (30–35°C) are typical in U.S. sausages, followed by a mild heating process, as a kind of pasteurization, instead of drying, to kill any trichinellae. Thus, starters such as *L. plantarum* or *P. acidilactici*, which grow well at those high temperatures are typically used. In the case of Europe, different technologies may be found, depending on the location and climate. Historically, there is a trend towards short, processed-smoked sausages in cold and humid areas, like northern European countries; and long, processed-dried sausages in warmer and drier countries, as in the Mediterranean area. In the case of northern European (NES) countries, sausages are fermented for about three days at intermediate temperatures (25–30°C), followed by short ripening periods (up to three weeks). These sausages are subjected to rapid pH drop and are usually smoked for specific flavors (Demeyer and Stanhke 2002). On the other hand, Mediterranean sausages require longer processing times. Fermentation takes place at milder temperatures (18–24°C), for about four days, followed by mild drying conditions for a longer time, usually several weeks or months. *L. sakei* or *L. curvatus* are the LAB most often used as starter cultures (Toldrá et al. 2001). Time required for the

Figure 23.5. Example of a fermentation/drying chamber with computer control of temperature, relative humidity, and air rate.

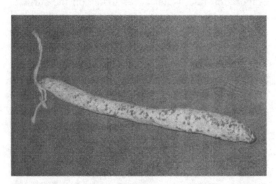

Figure 23.4. Sausage stuffed into a collagen casing, 80 mm diameter, and clipped on both extremes.

fermentation stage is a function of temperature and the type of microorganisms used as starters.

The technology is quite different in China and other Asian countries. Sausages are dried over charcoal at 48°C and 65% relative humidity for 36 hours and then at 20°C and 75% relative humidity for three days. Water activity rapidly drops below 0.80, although pH remains about 5.9, which is a relatively high value. Fermentation is relatively poor, and the sour taste, which is considered undesirable, is reduced. The Chinese raw sausage is consumed after heating (Leistner 1992).

MICROBIAL METABOLISM OF CARBOHYDRATES

The added carbohydrate is converted, during fermentation, into lactic acid of either the $D(-)$ or $L(+)$ configuration, or a mixture of both, depending on the species of LAB used as starter. The ratio between both L and D enantiomers depends on the action of L and D lactate dehydrogenase, respectively, and the presence of lactate racemase. The rate and final amount of lactic acid depend on the type of LAB species used as starter, type and content of carbohydrates, fermentation temperature, and other processing parameters. The accumulation of lactic acid produces a pH drop that is more or less intense, depending on its generation rate. Some secondary products, such as acetic acid, acetoin, and so on, may be formed through heterofermentative pathways (Demeyer and Stahnke 2002). Acid pH favors coagulation of protein, as it approaches its isoelectric point, and thus also favors water release. It also contributes to safety by inhibiting undesirable pathogenic or spoilage bacteria. The pH drop favors initial proteolysis and lipolysis by stimulating the activity of muscle cathepsin D and lysosomal acid lipase, both of which are active at acid pH, but an excessive pH drop does not favor later enzymatic reactions involved in the generation of flavor compounds (Toldrá and Verplaetse 1995).

PROCESSING STAGE 4: RIPENING AND DRYING

Temperature, relative humidity, and airflow have to be carefully controlled during fermentation and ripening to allow correct microbial growth and enzyme action while keeping adequate drying progress. The air velocity is kept at around 0.1 m/s, which is enough for a good homogenization of the environment. Ripening and drying are important for enzymatic reactions related to flavor development, and to get the required water loss necessary to produce reduction in water activity. The length of the ripening/drying period is 7–90 days; the length depends on many factors, including the kind of product, diameter, degree of dryness, fat content, desired flavor intensity, and so on. The reduction in a_w is slower in beef-containing sausages. The casing must stay attached to the sausage as it shrinks during drying. In general, products that are ripened longer tend to be drier and more flavorful.

PHYSICAL CHANGES

The most important physical changes during fermentation and ripening/drying are summarized in Figure 23.6. The acidulation produced during the fermentation stage induces protein coagulation and thus some water release. The acidulation also reduces the solubility of sarcoplasmic and myofibrillar proteins, and the sausage begins to develop consistency. The drying process is a delicate operation that must achieve equilibrium between two different mass transfer processes: diffusion and evaporation (Baldini et al. 2000). Water inside the sausage must diffuse to the outer surface and then evaporate to the environment. The two rates must be in equilibrium because a very fast reduction in the relative humidity of the chamber would cause an excessive evaporation of the sausage surface that would reduce the water content on the outer parts of the sausage and cause hardening. This is typical of sausages with a large diameter because of the slow water diffusion rate. The cross section of these sausages shows a darker, dry, hard outer ring. On the other hand, when the water diffusion rate is much higher than the evaporation rate, water accumulates on the surface of the sausage and causes wrinkled casings. This situation may happen in short-diameter sausages being ripened in a chamber with high relative humidity. The progress in drying reduces the water content, up to 20% weight loss in semidry sausages and 30% in dry sausages (Table 23.1). The water activity decreases according to the drying rate, reaching values below 0.90 for long-ripened sausages.

CHEMICAL CHANGES

There are different enzymes, from both muscle and microbial origin, involved in reactions related to color, texture, and flavor generation. These reac-

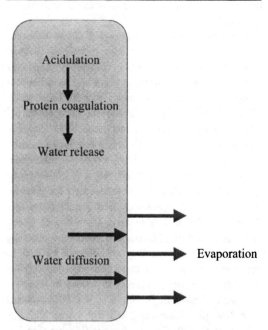

Figure 23.6. Scheme showing important physical changes during the processing of fermented meats.

peptides and free amino acids as final products, is proteolysis (Toldrá 1998). An intense proteolysis during fermentation and ripening is mainly carried out by endogenous cathepsin D, an acid muscle proteinase that is very active at acid pH. This enzyme hydrolyzes myosin and actin, producing an accumulation of polypeptides that are further hydrolyzed to small peptides by muscle and microbial peptidylpeptidases and to free amino acids by muscle and microbial aminopeptidases (Sanz et al. 2002). The generation of small peptides and free amino acids increases with the length of the process, although the generation rate is reduced at acid pH values because the enzyme activity is far from its optimal conditions. Free amino acids may be further transformed into other products such as volatile compounds, through Strecker degradations and Maillard reactions; ammonia, through deamination and/or deamidation reactions by deaminases and deamidases, respectively, which are present in yeasts and molds; and amines by microbial decarboxylases.

Another important group of enzymatic reactions, affecting muscle and adipose tissue lipids, is lipolysis (Toldrá 1998). Thus, a large amount of free fatty acids (between 0.5 and 7%) is generated through the enzymatic hydrolysis of triacylglycerols and phospholipids. Most of the observed lipolysis is attributed, after extensive studies on model sterile systems and sausages with added antibiotics, to endogenous

tions, summarized in Figure 23.7, are very important for the final sensory quality of the product. One of the most important groups of reactions, mainly affecting myofibrillar proteins, and producing small

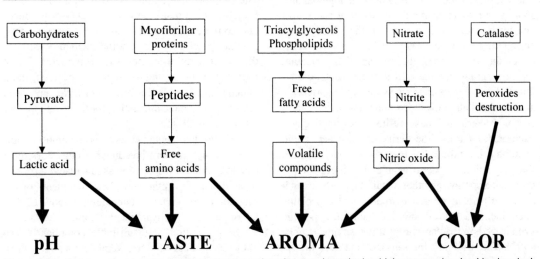

Figure 23.7. Scheme showing the most important reactions by muscle and microbial enzymes involved in chemical and biochemical changes affecting the sensory quality of fermented meats.

lipases that are present in muscle and adipose tissue (e.g., the lysosomal acid lipase, present in the lysosomes and very active at acid pH) (Toldrá 1992, Hierro et al. 1997, Molly et al. 1997).

Catalases are mainly present in microorganisms such as *Kocuria* and *Staphylococcus* and are responsible for peroxide reduction and, thus, contribute to color and flavor stabilization. Nitrate reductase, also present in those microorganisms, is important for reducing nitrate to nitrite in slow-ripened sausages with an initial addition of nitrate. This enzyme is inhibited at low pH and would not act in those sausages with very fast pH drop.

PROCESSING STAGE 5: SMOKING

Smoking is mostly applied in northern countries with cold and/or humid climates. Initially, it was used for preservation purposes, but today its contribution to flavor and color is more important (Ellis 2001). In some cases, smoking can be applied just after fermentation or even at the start of the fermentation. Smoking can be accompanied by heating at 60°C and has a strong impact on the final sensory quality properties. It has a strong antioxidative effect and gives a characteristic color and flavor to the product, which is now the primary role of smoking. The bacteriostatic effect of smoking compounds inhibits the growth of yeasts, molds, and certain bacteria.

SAFETY

The stability of the sausage against pathogen and/or spoilage microorganisms is the result of successive hurdles (Leistner 1992). Initially, the added nitrite curing salt is very important for the microbial stability of the mix. During the mixing under vacuum, oxygen is gradually removed and redox potential reduced. This effect is enhanced when adding ascorbic acid or ascorbate. Low redox potential values inhibit aerobic bacteria and make nitrite more effective as a bactericide. During the fermentation, lactic acid bacteria can inhibit other bacteria not only through the generation of lactic acid and subsequent pH drop, but also through other metabolic products such as acetic acid, hydrogen peroxide, and especially, bacteriocins, a kind of low-molecular-mass peptide synthesized in bacteriocin-positive strains (Lücke 1992). The drying of the sausage continues the reduction of the water activity to the low values (a_w below 0.92) that inhibit spoilage and/or the growth of pathogenic microorganisms . Thus, the correct interaction of all these factors assures the stability of the product.

Some food-borne pathogens that might be found in fermented meats are briefly described. *Salmonella* is more usual in fresh, spreadable sausages (Lücke 1985), but can be inhibited by acidification to pH 5.0 and/or drying to $a_w < 0.95$ (Talon et al. 2002). Lactic acid bacteria exert an antagonistic effect on *Salmonella* (Roca and Incze 1990). *Staphylococcus aureus* may grow under aerobic or anaerobic conditions and requires $a_w < 0.91$ for inhibition, but it is sensitive to acid pH. So, it is important to control the time elapsed before reaching the pH drop in order to avoid toxin production. Furthermore, this toxin is produced only in aerobic conditions (Roca and Incze 1990). *Clostridium botulinum* and its toxin production capability are affected by a rapid pH drop and low a_w even more than by the addition of LAB and nitrite (Lücke 1985). *Listeria monocytogenes* is limited in growth at $a_w < 0.90$ combined with low pH values and specific starter cultures (Hugas et al. 2002). *Escherichia coli* is rather resistant to low pH and a_w but is reduced when exposed to $a_w < 0.91$ (Nissen and Holck 1998). Adequate prevention measures consist in correct cooling and a hazards analysis and critical control point (HACCP) plan that includes application of good manufacturing practices, sanitation, and strict hygiene control of personnel and raw materials.

In recent years, most attention has been paid to biopreservation as a way to enhance protection against spoilage bacteria and food-borne pathogens. The bioprotective culture consists in a competitive bacterial strain that grows very quickly or produces antagonistic substances such as bacteriocins. Another precise biopreservation method consists in the direct addition of purified bacteriocins. Those bacteriocins belonging to group IIa (pediocin-like), which displays inhibition against *Listeria*, have been reported to be most interesting for the meat industry (Hugas et al. 2002).

Parasites like trichinae have been almost eliminated through modern breeding systems. Pork meat free of trichinae must be used as raw material for fermented sausages; otherwise, heat treatments of the sausage to reach internal temperatures above 62.2°C are required to inactivate them (Sebranek 2003).

The generation of undesirable compounds (see Table 23.4) depends on several factors. The most important is the hygienic quality of the raw materials. For instance, the presence of cadaverine and/or putrescine may indicate the presence of contaminat-

Table 23.4. Safety Aspects: Generation of Undesirable Compounds in Dry Fermented Meats

Compounds	Route of Formation	Origin	Concentrations (mg/100g)
Tyramine	Microbial decarboxylation	Tyrosine	< 16.0
Tryptamine	Microbial decarboxylation	Trytophane	< 6.0
Phenylethylamine	Microbial decarboxylation	Phenylalanine	< 3.5
Cadaverine	Microbial decarboxylation	Lysine	< 0.6
Histamine	Microbial decarboxylation	Histidine	< 3.6
Putrescine	Microbial decarboxylation	Ornithine	< 10.0
Spermine	Microbial decarboxylation	Methionine	< 3.0
Spermidine	Microbial decarboxylation	Methionine	< 0.5
Cholesterol oxides	Oxidation	Cholesterol	< 0.15

Sources: Adapted from Maijala et al. 1995, Shalaby 1996, Hernández-Jover et al. 1997, Demeyer et al. 2000.

ing meat flora. The processing conditions may favor the generation of biogenic amines, although the type of natural flora or microbial starters used for processing is the most important issue, because the presence of microorganisms with decarboxylase activity can induce the generation of biogenic amines. In general, tyramine is the amine generated in higher amounts; it is formed by certain LAB through enzymatic activity for the decarboxylation of tyrosine (Eerola et al. 1996). Tyramine releases noradrenaline from the sympathetic nervous system, and the peripheral vasoconstriction and increase in cardiac output results in higher blood pressure and risk for hypertensive crisis (Shalaby 1996). However, the estimated tolerance level for tyramine (100–800 mgkg^{-1}) is higher than for other amines (Nout 1994). The amines derived from foods are generally degraded in humans by the enzyme monoamine oxidase (MAO) through oxidative deamination reactions. Those consumers using MAO inhibitors are less protected against amines and are thus susceptible for risk situations such as hypertensive crisis when ingesting significant amounts of amines. Other amines, such as phenylethylamine, may cause migraine and an increase in the blood pressure or in a histamine that excites the smooth muscles of the uterus, the intestine, and the respiratory tract. Health risks may be reduced by use of starter cultures that are unable to produce amines and are competitive against amine-producing microorganisms. Additionally, the use of microorganisms that have amine oxidase activity and are able to degrade amines, the selection of raw materials of high quality, and good manufacturing practices assure products of high quality and reduced risks (Talon et al. 2002). Finally, the generation of nitrosamines during processing is almost negligible due to the restricted amount of nitrate and/or nitrite that can be added initially and to the low amount of residual nitrite remaining at the end of the process (Cassens 1997).

The processing conditions may favor the oxidation of cholesterol. Some oxides that can be involved in cardiovascular-related diseases (e.g., 7-ketocholesterol and 5,6-epoxycholesterol) are generated, but in general, the reported levels of all cholesterol oxides is very low, less than 0.15 mg/100g, for exerting any toxic effect (Demeyer et al. 2000).

FINISHED PRODUCT

Once the product is finished, it is packaged and distributed. Fermented sausages can be sold either whole or as thin slices (Fig. 23.8). The developed

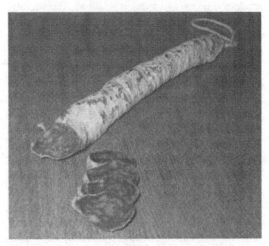

Figure 23.8. Picture of a typical small-diameter salchichón, showing its cross section.

color, texture, and flavor depend on the processing and type of product. Main sensory properties are described below.

COLOR

The color of the sausage depends on its moisture and fat content as well as its content of hemoprotein, particularly myoglobin. Color is also influenced by pH drop rate and the ultimate pH, and may be also affected by the presence of spices like red pepper. An excess of acid generation by lactobacilli may also affect color.

The characteristic color is due to the action of nitrite with myoglobin. Nitrite is reduced to nitric oxide, favored by the presence of ascorbate/erythorbate. Myoglobin and nitric oxide may then interact to form nitric oxide myoglobin, which gives the sausage its characteristic cured, pinky-red color (Pegg and Shahidi 1996). This reaction is favored at low pH. Long-processed sausages using nitrate need some time for the growth of *Micrococcaceae* before pH drops. The nitrate reductase that is present in *Micrococcaceae* reduces nitrate to nitrite, and later may further reduce it to nitric oxide, which reacts with myoglobin.

TEXTURE

The development of the consistency of fermented meats is initiated with the addition of salt and pH reduction. The water-binding ability of myofibrillar proteins decreases as the pH level approaches their isoelectric point, and water is released. The solubility of myofibrillar proteins also decreases, with a trend towards aggregation and coagulation of the proteins, forming a gel. The consistency of this gel increases with water loss during drying. So there is a continuous development of textural characteristics like firmness, hardness, and cohesiveness of meat particles during drying (Toldrá 2002). The meat:fat ratio may affect some of these textural characteristics, but in general, the final texture of the sausage mainly depends on the extent of drying.

FLAVOR

Little or no flavor is usually detected before meat fermentation, although a large number of flavor precursors is present. As fermentation and further ripening/drying progress, the combined action of endogenous muscle enzymes and microbial activity produces a high number of nonvolatile and volatile compounds with sensory impact. The accumulation of these compounds is increased and sensory perception enhanced as long as the process continues. Although not so important as in meat cooking, some compounds with sensory impact may be produced through further chemical reactions. The addition of spices also makes an intense contribution to specific flavors.

Taste

The main nonvolatile compounds contributing to the taste of fermented meats are summarized in Table 23.5. Sour taste, mainly resulting from lactic acid generation through microbial glycolysis, is the most relevant taste in fermented meats. Sourness is also correlated with other microbial metabolites such as acetic acid. Ammonia may be generated through the activity of deaminase and deamidase, usually present in yeasts and molds, reducing the intensity of the acid taste. Salty taste is usually perceived as a direct taste from salt addition. ATP-derived compounds such as inosine monophosphate and guanosine monophosphate exert some taste enhancement, while hypoxanthine contributes to bitterness. Other taste contributors are those compounds resulting from protein hydrolysis. The generation and accumulation of small peptides and free amino acids contribute to taste perception, which increases with the length of process. Some of them, for example, leucine, isoleucine, and valine, also act as aroma precursors as described below.

Aroma

The origin of the aroma mainly depends on the ingredients and processing conditions. Different pathways are responsible for the formation of volatile compounds with aroma impact (Table 23.6). As mentioned above, proteolysis creates many small peptides and free amino acids. Microorganisms can convert the amino acids leucine, isoleucine, valine, phenylalanine, and methionine to important sen-sory compounds with low threshold values. Some of the most important are branched aldehydes (2- and 3-methylbutanal and 2-methylpropanal), branched alcohols (2- and 3-methylbutanol), acids (2- and 3-methylbutanoic and 2-methylpropanoic acids), and esters (ethyl 2- and 3-methylbutanoate) (Stahnke 2002). Some of these branched-chain aldehydes may also be formed

Table 23.5. Quality Aspects: Generation or Presence of Desirable Nonvolatile Compounds Contributing to Taste in Fermented Meats

Group of Compounds	Main Representative Compounds	Routes of Generation	Presence in Final Product	Main Contribution	Expected Intensity
Peptides	Tri- and di-peptides	Proteolysis	Increases with length of process	Taste	High
Free amino acids	Glutamic acid, aspartic acid, alanine, lysine, threonine	Proteolysis	Increases with length of process	Taste	High
Nucleotides and nucleosides	Inosine monophosphate, guanosine monophosphate, inosine, hypoxanthine	ATP degradation	Around 100 mg/100g	Taste enhancement	Low
Long chain free fatty acids	Oleic acid, linoleic acid, linolenic acid, arachidonic acid, palmitic acid	Lipolysis	Increases with length of process	Taste	Low
Short chain fatty acids	Acetic acid, propionic acid	Microbial metabolism	Depends on microflora	Taste	Medium
Acids	Lactic acid	Glycolysis	Depends on initial amount of sugar and fermentation	Sour taste	High
Carbohydrates	Glucose, lactose	Remaining (nonconsumed through glycolysis)	Depends on initial amount of sugar and microflora	Sweet taste	Low
Inorganic compounds	Salt	Addition	Depends on initial amount	Salty taste	High

Table 23.6. Quality Aspects: Generation of Desirable Volatile Compounds Contributing to Aroma in Fermented Meats

Group of Compounds	Main Representative Compounds	Routes of Generation	Main Aroma	Expected Contribution
Aliphatic aldehydes	Hexanal, pentanal, octanal, . . .	Oxidation of unsaturated fatty acids	Green	High
Strecker aldehydes	2- and 3-methylbutanal, . . .	Strecker degradation of free amino acids	Roasted cocoa, cheesy-green	High
Branched-chain acids	2- and 3-methyl butanoic acid	Secondary products of previous Strecker degradation	Sweaty	Medium
Alcohols	Ethanol, butanol, . . .	Oxidative decomposition of lipids	Sweet, alcohol, . . .	Low
Ketones	2-pentanone, 2-heptanone, 2-octanone, . . .	Lipid oxidation	Ethereal, soapy	Medium
Sulfides	Dimethyldisulfide	Strecker degradation of sulfur-containing amino acids (methionine)	Dirty socks	Low
Esters	Ethyl acetate, ethyl 2-methyl-butanoate	Interaction of carboxylic acids and alcohols	Pineapple, fruity	High
Hydrocarbons	Pentane, heptane, . . .	Lipids autoxidation	Alkane	Very low
Dicarbonyl products	Diacetyl, acetoin, acetaldehyde	Pyruvate microbial metabolism	Butter	Low
Nitrogen compounds	Ammonia	Deamination, deamidation	Ammonia	Variable, depends on growth of yeasts and molds

Sources: Adapted from Flores et al. 1997, Viallon et al. 1996, Stahnke 2002, Toldrá 2002, Talon et al. 2002.

through the Strecker degradation: the reaction of amino acids with diketones. However, conditions found in sausages are far from those optimal for this kind of reaction, which needs high temperature and low water activity (Talon et al. 2002).

Methyl ketones may be formed either by β-oxidation of free fatty acids or decarboxylation of free β-keto acids. Other nonbranched aliphatic compounds generated by lipid oxidation are alkanes, alkenes, aldehydes, alcohols, and several furanic cycles.

A large number of volatile compounds are generated by chemical oxidation of the unsaturated fatty acids. These volatile compounds are mainly generated during ripening and further storage. Other low-molecular-weight volatile compounds are generated by microorganisms from carbohydrate catabolism.

The most usual compounds are diacetyl, acetoin, butanediol, acetaldehyde, ethanol, and acetic propionic and butyric acids. However, some of these compounds may be derived from pyruvate created through metabolic pathways than carbohydrate glycolysis (Demeyer and Stahnke 2002, Demeyer and Toldrá 2003). Flavor profile may have important variations depending on the type of microorganisms used as starters (Berdagué et al. 1993).

APPLICATION OF PROCESSING PRINCIPLES

See Table 23.7 for details on more references on the application of principles in the processing of fermented meat.

Table 23.7. Processing Steps and Application Principles of Fermented Meat

Processing Stage	Processing Principle(s)	References for More Information on the Principles Used
Comminution	Grinding of meat and fat and mixing with additives to form a batter for stuffing	Toldrá 2002, Demeyer and Toldrá 2004
Fermentation	Growth and development of microbial flora pH drop and acid gelation of meat proteins	Toldrá et al. 2001, Talon et al. 2002
Ripening and drying	Ripening for enzyme action and development of sensory quality Drying	Demeyer and Stahnke 2002, Toldrá 2002
Smoking	Imparting specific flavor and color Preservative effect	Ellis 2001

GLOSSARY

Aminopeptidases—exopeptidases that catalyze the release of an amino acid from the amino terminus of a peptide.

ATP—adenosine triphosphate.

Back slopping—traditional practice consisting in the addition of previous fermented sausage with successful sensory properties.

Bacteriocin—peptides of low molecular mass, produced by lactic acid bacteria, with inhibitory action against certain spoilage bacteria and food-borne pathogens.

Catalase—enzyme able to catalyze the decomposition of hydrogen peroxide into molecular oxygen and water.

Cathepsins—enzymes located in lysosomes that are able to hydrolyze myofibrillar proteins to polypeptides.

Decarboxylase—enzyme that hydrolyzes the carboxyl terminus (COOH). It is able to transform an amino acid into an amine.

DFD (dark, firm, dry)—pork meat with dark, firm, and dry characteristics due to a lack of carbohydrates in muscle and thus poor glycolysis and reduced lactic acid generation. These meats have pH values above 6.0 after 24 hours postmortem and are typical of exhausted stressed pigs before slaughtering.

Glycolysis—enzymatic breakdown of carbohydrates with the formation of pyruvic acid and lactic acid and the release of energy in the form of ATP (adenosine triphosphate).

HACCP— hazard analysis and critical control points.

Heterofermentative bacteria—bacteria that produces several end products (lactic acid, acetoin, ethanol, CO_2, etc.) from fermentation of carbohydrates.

Homofermentative bacteria—bacteria that produces a single end product (lactic acid) from fermentation of carbohydrates.

Lactate dehydrogenase—enzyme that catalyzes the oxidation of pyruvic acid to lactic acid.

Lactate racemase—enzyme that catalyzes lactic acid racemization reactions.

Lipolysis—enzymatic breakdown of lipids with the formation of free fatty acids.

Lysosomal acid lipase—enzyme that catalyzes the release of fatty acids by hydrolysis of triacylglycerols at positions 1 and 3.

MAO—monoamine oxidase.

Peroxide value—term used to measure rancidity and expressed as millimoles of peroxide taken up by 1000 g of fat.

Proteolysis—enzymatic breakdown of proteins with the formation of peptides and free amino acids.

PSE (pale, soft, exudative)—pork meat with pale, soft, and exudative characteristics resulting from an accelerated glycolysis and resulting rapid lactic acid generation. The pH drop is very fast, reaching values as low as 5.6 in just one hour postmortem.

Water activity (a_w)—indicates the availability of water in a food; defined as the ratio of the equilibrium water vapor pressure over the system and the vapor pressure of pure water at the same temperature.

REFERENCES

Baldini P, E Cantoni, F Colla, C Diaferia, L Gabba, E Spotti, R Marchelli, A Dossena, R Virgili, S Sforza, P Tenca, A Mangia, R Jordano, MC Lopez, L Medina, S Coudurier, S Oddou, G Solignat. 2000. Dry sausages ripening: Influence of thermohygrometric conditions on microbiological, chemical and physico-chemical characteristics. Food Research Int. 33:161–170.

Berdagué JL, P Monteil, MC Montel, R Talon. 1993. Effects of starter cultures on the formation of flavour compounds in dry sausages. Meat Sci. 35:275–287.

Cassens RG. 1997. Composition and safety of cured meats in the USA. Food Chem. 59:561–566.

Demeyer D. 1992. Meat fermentation as an integrated process. In: FJM Smulders, F Toldrá, J Flores, M Prieto, New Technologies for Meat and Meat Products, 21–36. Nijmegen, The Netherlands: Audet.

Demeyer D, L Stahnke. 2002. Quality control of fermented meat products. In: J Kerry, D Ledward, editors. Meat processing: Improving Quality, 359–393. Cambridge, U.K.: Woodhead Publ. Co.

Demeyer D, F Toldrá. 2004. Fermentation. In: W Jensen, CDevine, M Dikemann, editors. Encyclopedia of Meat Sciences. London: Elsevier Science. [In Press]

Demeyer DI, M Raemakers, A Rizzo, A Holck, A De Smedt, B Ten Brink, B Hagen, C Montel, E Zanardi, E Murbrek, F Leroy, F Vanderdriessche, K Lorentsen, K Venema, L Sunesen, L Stahnke, L De Vuyst, R Talon, R Chizzolini, S Eerola. 2000. Control of bioflavor and safety in fermented sausages: first results of a European project. Food Research Int. 33:171–180.

Durá A, M Flores, F Toldrá. 2002. Purification and characterization of a glutaminase from *Debaryomyces* spp. Int. J. Food Microbiol. 76:117–126.

Eerola S, R Maijala, AX Roig-Sangués, M Salminen, T Hirvi. 1996. Biogenic amines in dry sausages as affected by starter culture and contaminant amine-positive *Lactobacillus*. J. Food Sci. 61:1243–1246.

Ellis DF. 2001. Meat smoking technology. In: YH Hui, WK Nip, RW Rogers, OA Young, editors. Meat Science and Applications, 509–519. New York: Marcel Dekker Inc.

Flores J, F Toldrá. 1993. Curing: Processes and applications. In: R MacCrae, R Robinson, M Sadle, G Fullerlove, editors. Encyclopedia of Food Science, Food Technology and Nutrition, 1277–1282, London: Academic Press.

Flores M, Grimm CC, Toldrá F, Spanier AM. 1997. Correlations of sensory and volatile compounds of Spanish Serrano dry-cured ham as a function of two processing times. J. Agric. Food Chem. 45: 2178–2186.

Gray JI, MacDonald B, Pearson AM, Morton ID. 1981. Role of nitrite in cured meat flavour. A review. J. Food Prot. 44:302–312.

Hernández-Jover T, Izquierdo-Pulido M, Veciana-Nogués MT, Mariné-Font A, Vidal-Carou MC. 1997. Biogenic amines and polyamine contents in meat and meat products. J. Agric. Food Chem. 45: 2098–2102.

Hernández P, JL Navarro, F Toldrá. 1999. Effect of frozen storage on lipids and lipolytic activities in the longissium dorsi muscle of the pig. Z. Lebensm. Unters. Forsch. A 208:110–115.

Hierro E, L De la Hoz, JA Ordoñez. 1997. Contribution of microbial and meat endogenous enzymes to the lipolysis of dry fermented sausages. J. Agric. Food Chem. 45:2989–2995.

Hugas M, M Garriga, MT Aymerich, JM Monfort. 2002. Bacterial cultures and metabolites for the enhancement of safety and quality of meat products. In: F Toldrá, editor. Research advances in the quality of meat and meat products, 225–247. Trivandrum, India: Research Signpost.

Leistner L. 1992. The essentials of producing stable and safe raw fermented sausages. In: FJM Smulders, F Toldrá, J Flores, M Prieto, editors. New Technologies for Meat and Meat Products, 1–19. Nijmegen, The Netherlands: Audet.

Lücke FK. 1985. Fermented sausages. In: BJB Wood,editor. Microbiology of Fermented Foods, 41–83. London: Elsevier Applied Science.

___. 1992. Prospects for the use of bacteriocins against meat-borne pathogens. In: FJM Smulders, F Toldrá, J Flores, M Prieto,editors. New Technologies for Meat and Meat Products, 37–52. Nijmegen, The Netherlands: Audet.

Maijala R, Eerola S, Lievonen S, Hill P, Hirvi T. 1995. Formation of biogenic amines during ripening of dry sausages as affected by starter culture and thawing time of raw materials. J. Food Sci. 60: 1187–1190.

Molly K, DI Demeyer, G Johansson, M Raemaekers, M Ghistelinck, I Geenen. 1997. The importance of meat enzymes in ripening and flavor generation in dry fermented sausages. First results of a European project. Food Chem. 54:539–545.

Nissen H, AL Holck. 1998. Survival of *Escherichia coli* O157:H7, *Listeria monocytogenes* and *Salmonella kentucky* in Norwegian fermeted dry sausage. Food Microbiol. 15:273–279.

Nout, MJR. 1994. Fermented foods and food safety. Food Res. Int. 27, 291–296.

Ordoñez JA, EM Hierro, JM Bruna, L de la Hoz. 1999. Changes in the components of dry-fermented sausages during ripening. Crit. Rev. Food Sci. Nutr. 39:329–367.

Pegg BR, F Shahidi. 1996. A novel titration methodology for elucidation of the structure of preformed cooked cured-meat pigment by visible spectroscopy. Food Chem. 56:105–110.

Pegg BR, F Shahidi. 2000. Nitrite curing of meat. Trumbull, CT Food and Nutrition Press.

Roca M, K Incze. 1990. Fermented sausages. Food Reviews Int. 6:91–118.

Sanz Y, MA Sentandreu, F Toldrá. 2002. Role of muscle and bacterial exopeptidases in meat fermentation. In: F Toldrá, editor. Research advances in the quality of meat and meat products, 143–155. Trivandrum, India: Research Signpost.

Sebranek JG 2004. Semi-dry fermented sausages. In: YH Hui, LM Goddik, J Josephsen, PS Stanfield, AS Hansen, WK Nip, F Toldrá, editors. Handbook of Food and Beverage Fermentation Technology, 385–396. New York: Marcel Dekker, Inc.

Shalaby AR. 1996. Significance of biogenic amines to food safety and human health. Food Res. Int. 29:675–690.

Stahnke L. 2002. Flavour formation in fermented sausage. In: F Toldrá, editor. Research advances in the quality of meat and meat products, 193–223. Trivandrum, India: Research Signpost.

Talon R, S Leroy-Sétrin, S Fadda. 2002. Bacterial starters involved in the quality of fermented meat products. In: F Toldrá, editor. Research advances in the quality of meat and meat products, 175–191. Trivandrum, India: Research Signpost.

Toldrá F. 1992. The enzymology of dry-curing of meat products. In FJM Smulders, F Toldrá, J Flores, M Prieto, editors. New Technologies for Meat and Meat Products, 209–231. Nijmegen, The Netherlands: Audet.

Toldrá F. 1998. Proteolysis and lipolysis in flavour development of dry-cured meat products. Meat Sci. 49:s101–s110.

Toldrá F. 2002. Dry-cured Meat Products, 1–238. Trumbull, Conn.: Food and Nutrition Press.

Toldrá F, A Verplaetse. 1995. Endogenous enzyme activity and quality for raw product processing. In: K Lündstrom, I Hansson, E Winklund, editors. Composition of Meat in Relation to Processing, Nutritional and Sensory Quality, 41–55. Uppsala, Sweden: Ecceamst.

Toldrá F, Y Sanz, M Flores. 2001. Meat fermentation technology. In: YH Hui, WK Nip, RW Rogers, OA Young, editors. Meat Science and Applications, 537–561. New York: Marcel Dekker, Inc.

Toldrá F, G Gavara, JM Lagarón. 2004. Packaging and quality control. In: YH Hui, LM Goddik, J Josephsen, PS Stanfield, AS Hansen, WK Nip, F Toldrá, editors. Handbook of Food and Beverage Fermentation Technology, 445–458. New York: Marcel Dekker, Inc.

Viallon C, JL Berdagué, MC Montel, R Talon, JF Martin, N Kondjoyan, C Denoyer. 1996. The effect of stage of ripening and packaging on volatile content and flavour of dry sausage. Food Res. Int. 29:667–674.

24
Poultry: Canned Turkey Ham

E. Ponce-Alquicira

BACKGROUND INFORMATION

CANNING AS A FOOD PRESERVATION SYSTEM

Canning is the technique of preserving food in airtight containers through the use of an extensive heat treatment that inactivates enzymes and kills microorganisms that cause deterioration during storage. The airtight packaging protects the food from recontamination following sterilization, thus permitting storage at room temperature for many months without spoilage. In general canned meat products may be described as a convenience food because they offer several advantages: they contain little or no additives; retain most of the nutritive value of raw materials; and are ready to eat, shelf stable, and easy to consume and handle (Turner 1999).

This process is the basis of a large segment of the commercial food industry, a situation that probably will continue despite the development of other means of preserving food. However, the food canning industry needs to renew itself continuously, in order to keep consumers' attention, especially nowadays, when market globalization politics provides new opportunities for trading a great range of products from all over the world. Therefore, canners must not only offer safe and nutritive products, but also use attractive containers with convenient opening features, microwave heating instructions, several size serving portions, and so on. In addition, environment friendly packaging materials that provide the same stability and safety as conventional metal

The information in this chapter has been derived from a chapter in *Food Chemistry Workbook,* edited by J. S. Smith and G. L. Christen, published and copyrighted by Science Technology System, West Sacramento, California, ©2002. Used with permission.

Table 24.1. Poultry Meat Production 1999–2003

Country	1999	2000	2001	2002[a]	2003[b]
	(1000 metric tons; ready to cook or equivalent)				
Angola	8	8	8	nd[c]	nd
Argentina	885	870	870	650	600
Brazil	5,526	5,980	6,567	7,040	7,180
Canada	847	877	927	945	975
China	4,400	5,050	5,200	5,400	5,450
European Union	8,444	8,394	8,599	8,500	8,515
Hong Kong	63	65	60	60	59
Japan	1,078	1,091	1,074	1,090	1,080
Korea	390	394	413	433	440
Mexico	1,796	1,948	2,080	2,201	2,311
Russia	358	387	437	508	560
Saudi Arabia	370	390	424	445	472
South Africa	683	711	734	750	765
Thailand	980	1,070	1,230	1,320	1,380
United Arab Emirates	24	25	28	31	34
United States	15,891	16,122	16,523	17,052	17,349
Yemen	63	67	67	nd	nd

Source: United States Department of Foreign Agricultural Services 2003 (USDA 2003b).
[a]Preliminary data.
[b]Forecasted.
[c]No data available.

cans may be more attractive to consumers who demand healthy foods (Ponce-Alquicira 2002).

POULTRY MEAT

Poultry refers to any domesticated avian species, and poultry products can range from whole carcasses to cut-up carcasses, portions, boneless meat, or any further processed meat. Poultry production and consumption all over the world has increased considerably during recent years, as shown in data obtained from the Foreign Agricultural Services, U.S. Department of Agriculture (USDA 2003b) (Table 24.1). Several variables influence buying decisions and eating patterns in the short or long term, including ethnic or religious traditions, diet and health concerns, and price and availability. Recently, consumer attitudes about eating meat have been greatly influenced in relation to the saturated fat and cholesterol intake and their contribution to arteriosclerosis and heart attacks. Additionally, recent outbreaks of bovine spongiform encephalopathy (BSE) and foot-and-mouth disease have modified consumer's attitudes against red meats (Mandava and Hoogenkamp 1999).

Processed meats, including canning products, now contain less fat (under 25% or lower), where poultry, especially chicken and turkey, are popular as an alternative for manufacturing healthy low fat meat products. U.S. trade projections (Fig. 24.1) show an increasing demand for fresh poultry and processed poultry products over other meat species such as pork or beef. This may be explained by the "plain" flavor of poultry meat, which is easy to adapt to most recipes; it is an excellent choice for developing low calorie products, with the advantage that it is a low-cost protein source.

All meat and meat products, including poultry, must be subjected to inspection and declared suitable for human consumption; most countries have national laws and regulate inspections with specific requirements for the conditions in which animals are reared, transported, and slaughtered, and the products are prepared, distributed, and sold. In the United States, poultry and poultry products are subject to the Poultry Products Inspection Act, which is enforced by the FDA Food Safety and Inspection Service. In Mexico, SAGARPA (Secretary of Agriculture and Rural Development) dictates all regulations for food processing and distribution (USDA 2003a).

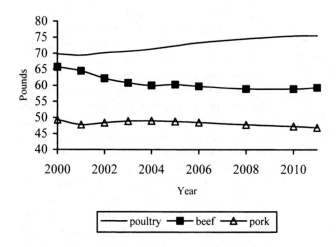

Figure 24.1. U.S. per capita boneless meat consumption project (USDA National Agricultural Statistical Service, January 2003).

TYPES OF COMMERCIAL CANNED POULTRY PRODUCTS

Canned poultry products include a wide variety of products such as reformed and emulsion-type products or purees, and soups formulated with chicken and turkey cubes. But most of these claim to be low fat products. Examples are cured breast of turkey, ham turkey, chunky chicken, chicken and vegetables, Vienna sausages, chicken soup, and chilorio (Mexican spiced poultry), shown in Figure 24. 2.

Formed Products

Formed poultry products are boneless and uniform in composition; examples are hams, loaf and restructured products. They may be produced from sectioned muscle pieces or from ground or chopped meat and shaped into a specific portion and size. Texture varies according to the initial material type; for example, hams are primarily produced from intact muscles and have more a "whole-muscle" texture, while restructured products have a smaller particle size, since they are produced from ground or chopped meat.

Formed products are prepared from defatted whole muscle pieces bound together after marinating, tumbling, and cooking. During heating, proteins form a network between meat pieces to form a continuous body. Nonmeat binders, such as soy protein, casein, or hydrocolloids, among others, can be used to enhance cohesivity between meat pieces to obtain a whole-meat-like texture (Smith 2001).

Figure 24.2. Variety of canned poultry products (includes sausages, soups, ham, pate, chilorio, etc.).

Emulsified Products

Emulsified or comminuted poultry products include frankfurters, bologna, or loaf items and are usually prepared from chilled or frozen, mechanically deboned poultry or turkey. Meat is homogenized in a cutter bowl with iced water, salt, cure, alkaline phosphates, starch, sodium erythorbate, milk or soy proteins, starch, gums, and spices to an end temperature of 15°C to avoid melting of fat, which might result in fat caps or fatting out after heating. Batter is then vacuum encased and cooked, and after peeling, sausages are canned (Smith 2001).

COMPOSITION AND PHYSICOCHEMICAL PROPERTIES OF MEAT

The process of canning begins with the selection of high quality raw materials, where skeletal muscle is the main constituent. Turkey meat contains 75% moisture, 23% protein, 1.2% lipids, and 1% minerals. Table 24.2 shows the principal skeletal muscle proteins, which are classified according to their solubility and location as sarcoplasmic, myofibrillar, and stroma fractions (Lawrie 1998). The myofibrillar (salt-soluble) fraction comprises more than 20 distinct proteins and represents about 60% of the total poultry muscle protein. Myofibrillar proteins can be divided into three groups based on their func-

tion: (1) contractile (responsible for muscle contraction), (2) regulatory (involved in regulation of muscle contraction), and (3) cytoskeletal (responsible of myofibril integrity). The contractile proteins, myosin and actin, have a large influence in muscle functionality; these proteins usually form the actomyosin complex in *postrigor*, and they contribute to functionality for comminuted and formed processed poultry products. The ratio of actin to myosin, and the ratio of free myosin to actomyosin also influence the functional properties of poultry meat (Lawrie 1998, Ponce-Alquicira et al. 2000).

Sarcoplasmic proteins play a minor role in meat protein functionality, although myoglobin and other water-soluble compounds may have a great influence on color. Myoglobin consists of a heme group bound to the histidyl (His_{93}) residue of a single polypeptide chain, as shown in Figure 24.3 (Belitz and Grosch 1999). The amount of myoglobin varies with species, age, and muscle fiber distribution; for instance, dark muscles in turkey thigh are mainly comprised of red fibers that contain more myoglobin than light breast muscles. Myoglobin in white turkey meat ranges from 0.1 to 0.4 mg/g, whereas that in dark meat ranges from 0.6 to 2 mg/g. Moreover, mechanically deboned turkey (MDT) contains some bone marrow and will have higher pigment levels than manually deboned meat (Froning and Mckee 2001, Smith 2001).

Table 24.2. Main Poultry Skeletal Muscle Proteins

Protein Fractions	Content (% of total protein)
Myofibrillar (salt soluble proteins)	
Myosin	29
Actin	13
Tropomyosin	3.2
Troponins C, I, T	3.2
Actinins	2.6
Desmin	2.1
Conectin	3.7
Sarcoplasmic (water soluble proteins)	
Myoglobin and other heme proteins	1.1
Glycolytic enzymes	12
Mitochondrial enzymes	5
Lysosomal enzymes	3.3
Stroma (insoluble proteins)	
Collagen	5.2
Elastin	0.3
Reticulin	0.5

Source: Lawrie 1998.

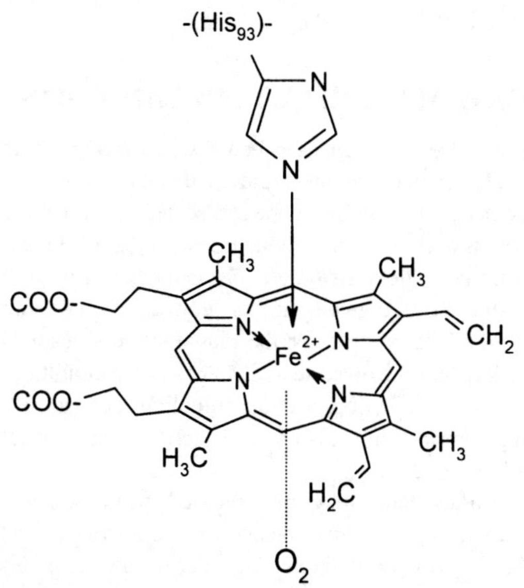

Figure 24.3. Myoglobin heme group bound to histidyl (His$_{93}$) residue of the peptide chain and oxygen. (Adapted from Belitz and Grosch 1999.)

Stroma proteins are related to meat tenderness; collagen is the major stroma protein present in flesh and in poultry skin. When collagen is present in high concentrations, it reduces the functionality of the myofibrillar proteins, diminishing binding between meat pieces in formed products. It may also reduce fat and water retention in comminuted products, especially when they are cooked at high temperatures (Lawrie 1998, Smith 2001).

Meat functionality is based on the physicochemical properties of proteins and determines its behavior during processing, storage, and consumption. Poultry meat functionality is based on three types of molecular interactions: (1) protein hydration and water-holding capacity (WHC), dependent on protein-water interactions; (2) cohesivity and gelation, based on protein-protein interactions; and (3) emulsifying, based on protein-surface-related properties. All these interactions are affected by intrinsic and extrinsic factors such as the type of protein, distribution of hydrophobic and hydrophilic groups on the protein surface, charge, and molecular flexibility. Extrinsic factors include pH value, salt concentration, phosphate salts, temperature and integrity of the meat, processing, and the addition of other nonmeat additives (Lawrie 1998).

The amount of total myofibrillar protein, the ratio of moisture to total protein, and the physicochemical condition (PSE or DFD) of the raw materials determine their functional properties. Lean poultry meat contains 19–23% protein, while mechanically deboned poultry has 14–16% protein without skin and 11–12% protein with skin. Therefore, both poultry product formulation and processing must be designed to improve protein functionality and final product quality (Damodaran 1994).

Water-holding Capacity

Water-holding capacity is the ability of meat to retain or absorb added water in the presence of an external force; this functional property is based on protein-water interactions. Water is held by muscle proteins and physically entrapped within the muscle structure in the interfilament spaces of myofibrils. Factors such as pH, salt concentration, processing, and temperature influence the protein-water binding and the quality of the poultry meat product network (Damodaran 1994, Lawrie 1998).

Protein-water interactions are highly related to the state of the meat postmortem. At the isoelectric point (pH ~5.1), myofibrillar proteins have neutral charge and tend to aggregate. However, as the pH increases during resolution of rigor mortis, proteins become more negatively charged, with an increase in repulsive forces between myofibrils that leads to swelling, allowing more water to interact with proteins; therefore, protein solubility and water-holding capacity increase as proteins become more negatively charged.

Addition of salt up to 0.6 mol/liter (2–3.5%) NaCl reduces electrostatic interactions between proteins, increasing protein extractability, solubility, and water binding in both breast and tight muscle. In addition, alkaline phosphates, in combination with salt and mechanical work, increase pH and myofibrillar protein extraction and solubilization. Additionally, chopping or tumbling disrupts the muscle, allowing the muscle fibers to absorb water and swell. However, overchopping or tumbling can lead to an excessive disintegration of muscle fibers and induce protein denaturation, as it is also associated with an increase in temperature and excessive shearing. Denatured proteins, as in PSE (pale, soft, and exudative) muscle, form aggregates that have low water affinity and reduced emulsification and foaming abilities (Lawrie 1998, Ponce-Alquicira et al. 2000).

Cohesivity

During cooking, muscle proteins denature and form a continuous, cross-linked gel network, stabilized by a series of protein-protein interactions, such as electrostatic and hydrophobic interactions, and by hydrogen and disulfide bonds. Muscle protein gelation involves a series of steps. First, when muscle proteins are heated to a critical temperature, they unfold; in a second step, unfolded molecules aggregate to form an increasingly viscous solution; then, when the gelling point is reached, molecules aggregate into a continuous gel. Myofibrillar proteins form irreversible strong gels that are responsible for the textural and sensory properties, as well as the cooking yields, of poultry products. Nevertheless, connective and sarcoplasmic proteins may interfere with the ability of myofibrillar proteins to form a gel (Damodaran 1994, Jiménez-Colmenero et al. 1994).

Emulsifying

Comminuted poultry products such as sausages may be referred to as emulsions, as the fat tissue is comminuted and dispersed in small particles into a continuous salt/protein/water matrix. The stability of this system is influenced by pH value, ionic strength, melting range of the lipid, soluble protein content, and temperature of processing.

Soluble and extracted meat proteins form a monomolecular film around the fat globules; protein polar regions orient towards the water phase, while nonpolar regions orient towards the fat droplets, to minimize free energy (Keeton 2001, Smith 2001).

Meat proteins show different emulsifying responses that decrease in the following order: myosin > actomyosin > actin > sarcoplasmic proteins. The hydrophobic heads of the myosin dip into the fat globules, while the tails interact with actomyosin in the continuous phase. Actomyosin binds water and contributes to stabilization of emulsions because of its viscous, elastic, and cohesive properties. Comminuting is necessary to extract proteins, disrupt fat, and form an emulsion; also, concentration of protein must be sufficient to form a continuous and stable film around the fat globules. During the emulsification stage, poultry batter temperature and chopping times should be monitored to avoid melting the fat globules. Stable emulsions require at least 45% myofibrillar protein in the formulation, with a maximum of 30% sarcoplasmic proteins, and connective proteins should be limited to less than 25% of total protein (Keeton 2001).

RAW MATERIALS PREPARATION

Canned turkey ham is a boneless, formed product made from cured meat pieces that are bound together into a specific shape in a sealed container and heat processed. This product retains most of the nutritional value of raw materials, is ready to eat, shelf stable, and convenient for consumers (Keeton 2001). The process for the manufacture of canned turkey ham involves several stages: meat conditioning, brine injection, vacuum tumbling, and can filling, exhaustion, closing, and sterilization (Ponce-Alquicira 2002).

Turkey ham may be prepared from boneless breast, legs, thighs, desinewed drumsticks, and MDT, with or without skin. These raw materials may be chilled or frozen, but without off color, off odor, or apparent microbial growth. The internal temperature of fresh cuts should not be above 4.4°C, and frozen materials should be below −18°C when received. Frozen cuts must be kept packaged during thawing to prevent dehydration and to avoid the risk of microbial contamination until it reaches −3.3 to −2.2°C. Turkey meat can be sliced, cubed, or ground, according to the desired final texture, but up to 33% of the meat may be finely comminuted to provide good binding for a whole-meat-like texture and good water retention. Nevertheless, temperature must be kept below 10°C during these operations to avoid the risk of microbial growth (Keeton 2001, Smith 2001, Smith and Acton 2001).

CURING AND BRINE INJECTION

Meat is usually cured by injection of brine under pressure, using a multineedle system, which facilitates and accelerates incorporation of the curing solution (see Table 24.3) (Pearson and Gillett 1999).

Table 24.3. Basic Curing Brine Formula

Component	%
Salt	2
Phosphate	0.5
Sucrose	0.5
Carrageenan	0.3
Sodium nitrite	156 ppm
Sodium erythorbate	450 ppm

Source: Ponce-Alquicira 2002.

Salt improves flavor, and in conjunction with phosphates it extracts myofibrillar proteins, producing a sticky surface that will bind meat chunks during thermal processing. Sodium chloride increases protein negative charge as well as protein repulsion, allowing more water to bind within the muscle fibers. On the other hand, alkaline phosphates increase pH and ionic strength, allowing protein to uncoil, exposing hydrophilic sites; therefore, phosphates act in a synergistic way with sodium chloride to increase WHC and protein extraction (Claus et al. 1994, Keeton 2001, Smith and Acton 2001).

Sodium nitrite is a multifunctional ingredient. It prevents the outgrowth of the spore former *Clostridium botulinum,* which grows under anaerobic conditions such as those created during canning. But some spores can survive normal heat processes and generate vegetative cells that produce lethal toxins (Claus et al. 1994, Van Laack 1994). The antibacterial properties of nitrite are based on the reaction of nitric oxide with S-H groups to form nitrosothiols (RS-NO), and on depriving anaerobic spore formers of available iron compounds with a key role in biochemical mechanisms (Ray 1996). Nitrite is also responsible for the development of the distinctive color and flavor of cured processed meats. Nitric oxide derived from sodium nitrite reacts with the heme iron of myoglobin and metmyoglobin to form nitrosylmyoglobin and the heat-stable, pink nitrosylhemochrome after cooking (Smith and Acton 2001) (see Figs 24.4 and 24.5), but color intensity

depends on the myoglobin content of the raw material (Belitz and Grosch 1999, Pearson and Gillett 1999, Fletcher 1999).

Nitrite also contributes to flavor stability, preventing warmed-over flavors by complexing the heme iron, which could promote lipid oxidation reactions (Van Laack 1994). It has been reported that nonheme iron is a potent prooxidant, released during heat processing as a result of porphyring breakdown; thus, heating accelerates the release of iron from the heme complex (Belitz and Grosch 1999). Legal limits of initial nitrite levels are 200 ppm and 156 ppm for pasteurized and sterile canned hams, respectively, with residual levels of 100 and 120 ppm (Ranken 2000). Finally, the addition of reducing agents such as sodium erythorbate accelerates curing, promotes formation of nitrosylhemochrome, and contributes to flavor and color stability (Keeton 2001, Pearson and Gillett 1999, Van Laack 1994).

VACUUM TUMBLING

Injection followed by noncontinuous tumbling cycles maximizes the quality of the product, as it permits a uniform distribution and absorption of curing ingredients, and extraction of salt-soluble proteins (Keeton 2001, Larousse and Brown 1997, Pearson and Gillett 1999). A vacuum tumbler consists of a large rotating tank with paddles inside, and jacketed walls to cool the product during tumbling. The temperature should be kept between 4 and 8°C, and the

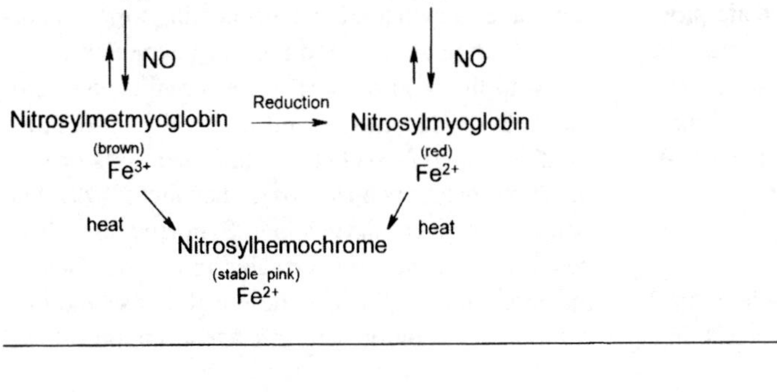

$$NaNO_2 + H_2O \xrightarrow[pH\ 5\ 4\ 6]{} \underset{\text{nitrous acid}}{HNO_2} + NaOH$$

$$3HNO_2 \xrightarrow{\text{decomposes}} \underset{\text{nitric oxide}}{2NO} + H_2O + HNO_3$$

Figure 24.4. Reaction pathway leading to the formation of nitric oxide (NO) and nitrosylhemochrome pigment. (Adapted from Pearson and Gillett 1999, Smith and Acton 2001.)

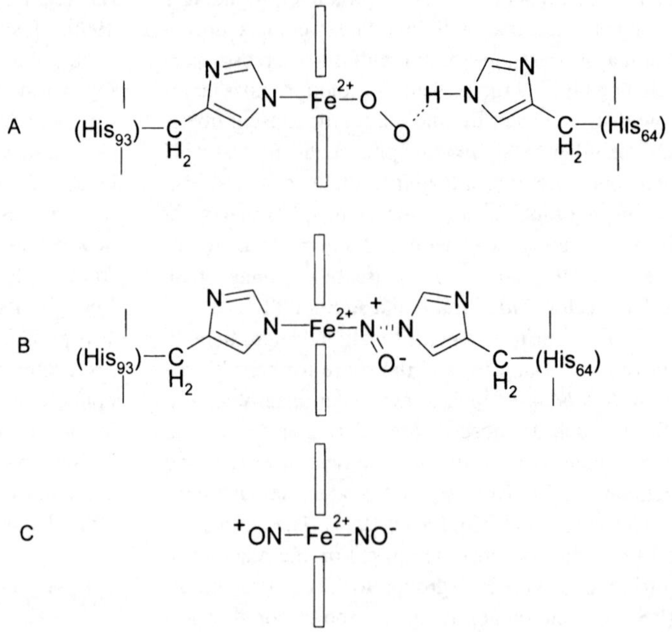

Figure 24.5. Heme ligands of
(A) myoglobin, **(B)** nitrosylmyoglobin,
and **(C)** nitrosylhemochrome. (Adapted
from Larousse and Brown 1997.)

rotation rate between 3 and 15 rpm; a higher speed can cause cell breakdown and temperature increase, reducing the quality of the final product. Vacuum tumbling has the advantage of speeding up the brine uptake, avoiding the formation of air bubbles within the product (Claus et al. 1994, Larousse and Brown 1997, Pearson and Gillett 1999).

CONTAINER FILLING

Containers should be clean and washed before filling. The cured meat mix is transferred into the appropriate containers using an automatic vacuum filler and pressed to ensure the elimination of air pockets (Guerrero-Legarreta 2001, Pearson and Gillett 1999, Turner 1999). Selection of the appropriate container is vital to maintain the stability and organoleptic properties of the product during storage, and to protect the contents against damage during transportation, storage, and distribution. Packaging materials include metal, glass, and laminated containers (Turner 1999).

METAL CANS

Containers for poultry canning can be made of steel (tinplate, tin-free steel, and nickel-plated steel), in a

variety of forms; smaller and easy-to-open cans are frequently made of aluminum. These materials are cheap and provide excellent barrier properties against gases, water vapor, light, and odors. Additionally, they can be used in on-line filling processes, have high mechanical resistance and excellent thermal conductivity, are suitable for sterile products, and can be hermetically sealed and recycled. Metal cans have some disadvantages, for instance, shipping empty cans takes up a lot of space. Also, during storage cans must be protected from moisture and/or humidity (if not lacquered or coated), and cans may be considered old-fashioned by the modern consumer because they are not suitable for use in the microwave oven (Turner 1999).

Metal containers usually have an internal, and sometimes an external, enamel coating to avoid corrosion during storage and prevent interaction of the can with the food product; the external coating provides both decorative and protective functions. Coating materials include natural oleoresins or synthetic products, such as epoxy, phenolic, acrylic, and polyester lacquers and coatings. Synthetic resins have better performance, are available in a wide range of materials, and are specially designed for use in different foods. In addition, a release agent is applied to facilitate product removal after opening (Pearson and Gillett 1999, Turner 1999). Coatings should impart no odor or flavor to the food, must be nontoxic and protect the can and contents during the required shelf life, must not flake off during can manufacture or storage, must be easy to apply and quickly cured, and must resist all temperatures encountered during processing and storage (Turner 1999).

Fabrication

Metal is fabricated into three-piece or two-piece cans. The three-piece can is composed of a body and two ends. The body is usually cylindrical (but can be rectangular, or pyramidal, etc.), and after it is formed, the two edges are brought together and sealed. The ends are made of tinplate, and one end is applied by the can manufacturer (the manufacturer's end), whereas the other is prepared by the canner (the canner's end). Round and not round two-piece containers are made from a precut aluminum alloy or steel sheet by stamping out in a cupping press or by a combination of stamping and deep drawing. Ends or covers are made from aluminum alloy, tinplate, or tin-free steel (TFS), coated on both sides (Turner 1999).

Application of ends is critical, as the end provides mechanical resistance to support the internal pressure differentials during thermal processing and cooling. It is also a barrier against all types of contamination and ensures food safety throughout storage. Sealing forms a double seam in two operations. First, the can body and the cover are brought together and clamped on a seaming chuck by a load applied vertically to the base plate; the end curl is tucked under the can flange and interlocks with it. In a second operation, the interlocked layers of metal are compressed, and the seaming compound is squeezed into voids to complete the hermetical seal (Turner 1999).

Glass Jars

Glass jars are used less often for meat products because of their fragility. They consist of a glass body, and a metal lid. The seaming panel of the metal lid has a lining of synthetic material. In households, glass jars with glass lids are used. Glass lids are fitted by means of a rubber ring.

Retort Pouches

The retort pouches are flexible, lightweight, and easily disposable laminated containers for preserv-ing foods. Heat-resistant plastic pouches are usually made of polyester (PETP) and used for ready-to-eat dishes. Laminated films made of polyester/polyethylene (PETP/PE) or polyamide/polyethylene (PA/PE) are relatively rigid containers, which are used for filling with pieces of cured ham or other kinds of prepared meat. Round containers formed out of a laminate of aluminium foil and polyethylene (PE) or polypropylene (PP) are widely used for small portions, because PE or PP permit the heat sealing of these containers, which can then even be subjected to intensive heat treatment. Retort pouches offer some advantages over other typical food containers used in canning because they are easy to handle and the thermal process is faster and therefore produces fewer flavor and texture changes. Also, the consumer can easily heat the pouch in boiling water before eating as well as save shelf space (Lin et al. 2001).

EXHAUSTION AND CLOSING

Once containers have been properly filled and pressed, they are sealed under a mechanical vacuum or by using steam to create a vacuum while the product is cooled. Air evacuation from the headspace as well as from the bulk of the turkey ham is necessary to achieve good heat penetration and to minimize alterations in color, flavor, and texture during processing and storage. Afterwards, the exterior of the container should be cleaned, and the closure must be checked prior to sterilization (Larousse and Brown 1997, Pearson and Gillett 1999, Turner 1999).

STERILIZATION

Canned foods are preserved by application of a heat treatment, which inactivates enzymes, pathogens, and microorganisms that cause deterioration during storage. Turkey canned ham is considered a commercially sterile food, because the sterilization step ensures that the product is free from viable microorganisms capable of reproducing in the food under normal, nonrefrigerated conditions of storage and commercialization. Sealed cans protect the product from recontamination after sterilization, allowing storage at room temperature for several months without spoilage (Pearson and Gillett 1999). Canning includes two steps: (1) the product is heated in a retort to high temperature for enough time to destroy spoilage and pathogenic microorganisms, and (2) the product is rapidly cooled to room temperature, where additional microbe destruction is achieved (Claus et al. 1994, Van Laack 1994). Most non-spore-forming organisms are heat labile, but some spores can survive even after heating to 120°C; however, addition of salt and nitrite decreases the thermoresistance of microorganisms (Claus et al. 1994).

THERMAL DESTRUCTION OF MICROORGANISMS

Thermal destruction of bacteria is expressed in terms of exposure to a specific temperature for a period of time; at higher temperatures, shorter periods of time are required to get the same destruction. However, the amount of microbial destruction within the food matrix depends on several factors such as the number and kind of microorganisms and the growth conditions of the microorganisms of concern. It also depends on the composition, viscosity, moisture, and pH of the food, the presence of preservatives, and the can size and shape, among others. Thus, successful sterilization requires knowledge of the rate of heat penetration at the coolest point, since heating is not homogeneous throughout the entire can due to its geometry (Ray 1996). For example, food in small containers is heated more rapidly than that in large containers; also, the center of a solid product or product near the end in a liquid canned food may be the coldest point within the food matrix. There are several mathematical relationships that describe the thermal destruction of microorganisms, the rate at which a food is heated, and the temperature of the coldest point. The number of microorganisms destroyed by the thermal process can be estimated by incorporating the destruction rate of the microorganism of concern into the heat transfer model for a food system. However, not all food systems are easily modeled. Therefore, actual time-temperature measurements can be used to establish the amount of microbial destruction during the process. In addition, microbial destruction can also be measured by inoculation of an indicator organism and measurement of the remaining population after the thermal process (Arnold et al. 2000).

The rate of inactivation of microorganisms increases in a logarithmic rate with increasing temperature and tends to follow a first-order rate reaction. The D value is the time in minutes during which the number of a specific microbial population exposed to a specific temperature is reduced by 90% or one log. It is expressed as $D_T = t$, where T is the temper-

ature and t is the time (in minutes) required for one log reduction of the microbial population. In addition, the F value represents the amount of time in minutes required to completely destroy a given number of microorganisms at a reference temperature (121.1°C for spores, 60°C for cells) (Ray 1996).

Several authors such as Larousse and Brown (1997), Pearson and Gillett (1999), and Arnold et al. (2000), among others, indicate that changes in microbial populations as a function of time can be described by Equation 24.1; while Equation 24.2 describes a first-order kinetic model, if k is the slope of the natural logarithm of survivors at any time for the microbial population, then Equation 24.2 can be integrated into Equation 24.3 to describe the reduction of microbial populations:

$$\log[N/N_0] = -t / D_T \qquad 24.1$$

where

N = microbial population at any time, t
N_0 = initial microbial population
D_T = decimal reduction time required for a one log cycle reduction in the microbial population.

$$dN / dt = -kN \qquad 24.2$$

$$\ln [N / N_0) = -kt \qquad 24.3$$

Consequently, the decimal D values and the constant k can be correlated by Equation 24.4:

$$k = 2.303 / D_T \qquad 24.4$$

Both parameters, the k and D values, describe the microbial population reduction only at a constant and specific temperature. In order to measure the influence of temperature, the thermal resistant constant (Z value) must be incorporated into the thermal death time (TDT) curve (log F vs. T) or thermal resistance curve (log D vs. T), shown in Equations 24.5 and 24.6, respectively:

$$\log [D / D_R] = -(T - T_R) / Z \qquad 24.5$$

$$\log [F / F_R] = -(T - T_R) / Z \qquad 24.6$$

The thermal death time (TDT) involves graphical integration of time-temperature data at the coolest heating point during thermal processing, and measures the microbial death rate in relation to tempera-

Table 24.4. Thermal Resistant Parameters for Some Microorganisms Having Public Health Significance

Microorganism	D^a (min)	k^b (1/min)	Z^c (°C)	Temperature (°C)
S. typhimurium	2.13–2.67	0.86–1.08		57
E. coli O157:H7	4.1–6.4	0.36–0.56		57.2
E. coli O157:H8	0.26–0.47	4.9–8.86	5.3	62.8
C. jejuni	0.62–2.25	1.0–3.72		55–56
L. monocytogenes	1.6–16.7	0.14–1.44		60
S. aureus	2.5	0.921		60
Bac. cereus	1.5–36.2	0.064–1.535	6.7–10.1	95
Clo. perfringens	6.6	0.349		104.4
Clo. botulinum B	1.19–2.0	1.152–1.935	7.7–11.3	110
Clo. botulinum E	6.8–13	0.177–0.339	9.78	74
Clo. botulinum 62A	1.79	1.287	8.5	110
Bac. subtilis	32.8	0.0702	8.74	88

Source: Arnold et al. 2000.

[a] D = decimal reduction time.

[b] k = rate constant.

[c] Z = thermal resistant constant.

ture. Thus the Z values indicate the number of degrees the temperature must be increased to decrease the microbial population by one log cycle, and measure the microbial heat resistance. Table 24.4 shows the thermal kinetic parameters for some microorganisms having public health significance; it can be seen in the table that spore-forming microorganisms have the largest D and Z values.

The rate of inactivation of microorganisms increases in a logarithmic rate with increasing temperature. Thus D and F values will decrease logarithmically with increase in temperature. Z is the change in temperature that accompanies a 10-fold change in the time for inactivation. The Z value is calculated by plotting log (D) against temperature, and increasing the temperature of a thermal process by the Z value results in a 10-fold reduction in the time required to obtain the lethality of the original process. Conversely, reducing the process temperature by the Z value necessitates a 10-fold increase in the processing time to achieve the original lethality (Arnold et al. 2000).

Commercial sterilization ensures destruction of microorganisms growing in the product under normal storage conditions. The low-acid (high pH) foods require severe heat treatment to guarantee microbial destruction to an F value of 3 (Claus et al. 1994, Pearson and Gillett 1999). The heat treatment for this type of foods is based on the total destruction of *Clostridium botulinum* type A or B spores (the

most heat-resistant spores of a pathogen) by applying the 12-*D* concept or "botulinum cook." This refers to the heat process necessary to reduce the number of surviving spores of *Cl. botulinum* from 10^{12} to 10^0, that is, to reduce the number of surviving spores by 12 log cycles. Canned turkey ham falls in the category of low-acid foods as it has a pH higher than 4.8. The reference temperature for canned low-acid foods for measuring the destruction of *Cl. botulinum* is 121°C; at this temperature the destruction time for 12 log cycles is designated as F_0. The high-acid (low pH) food (pH = 4.6) requires lower heat treatments, since *Cl. botulinum* can not germinate and outgrow at this low pH (Bratt 1999, Claus et al. 1994, Guerrero-Legarreta 2001, Larousse and Brown 1997, Smith 2001).

TEXTURE AND FLAVOR CHANGES DURING THE THERMAL PROCESS

During cooking, the extracted muscle proteins denature and form a continuous cross-linked network (stabilized by electrostatic and hydrophobic interactions and hydrogen and disulfide bonds), giving rise to the characteristic product texture. Myofibrillar proteins are mainly responsible for meat-binding and textural properties, as well as product yield (Pearson and Gillett 1999, Smith 2001). Nevertheless, stroma and sarcoplasmic proteins may interfere. In particular, collagen, a major stroma protein present in flesh and skin, diminishes WHC and binding when it is present at high concentrations, due to its shrinkage and conversion into gelatin during cooking (Claus et al. 1994, Smith 2001, Smith and Acton 2001). Heating also promotes changes in flavor, as carbonyl compounds and small quantities of hydrogen sulfide may be liberated (Cambero et al. 1992). These compounds participate in several reactions (including Millard reactions and lipid oxidation), creating a complex mixture of chemicals (i.e., aldehydes, ketones, and sulfur compounds) that provide the meaty, toasted, roasted, fatty, fruity, and sulfurous meat aroma (Adams et al. 2001, Bailey 1994, Roos 1997).

FINISHED PRODUCT

Canned poultry products can suffer several alterations due to chemical, enzymic, and microbial activity that diminish the product quality. Production of H_2, CO_2, browning, and corrosion of cans are caused by chemical reactions; liquefaction and discoloration are caused by enzymic reactions. Improper processing and handling, and storage at elevated temperatures are the main factors associated with spoilage of canned products. Alteration can take place before heat treatment due to microbial growth, chemical reactions with the container, and physical alterations when there is a delay prior to heat processing (Smith 2001). Therefore, it is recommended that heat processing be applied within 20 minutes of can closure. Spoilage is usually associated with defects and mechanical damage: improper pressure control during retorting and cooling operations may stress the seam, resulting in poor seam integrity and subsequent spoilage. Therefore, quality control of the finished product involves pH determination of the product, gas analysis of can headspace, microbiological testing, and complete external can examination for leakage, pinholes, dents, buckling, and general exterior conditions (Lin et al. 2001).

Microbial Spoilage

Depending upon the thermal treatment, microbial cells and spores can be sublethally injured or dead. The sublethally injured cells and spores are capable of repair and multiplication (Ray 1996). Canned products can have spores of thermophilic bacteria organisms (such as *Bacillus stearothermophilus, B. coagulans, Cl. thermosaccharolyticum*), but if the product is stored at 30°C or below, the spores do not germinate to cause spoilage; if the cans are stored in temperature-abused conditions to 40°C or higher, the spores germinate, multiply, and spoil the product. Microbial spoilage is generally due to germination and growth of thermophilic spore-forming bacteria, because of either inadequate cooling after heating or high storage temperatures. Growth and survival of mesophilic microorganisms is associated with inappropriate heat treatment and microbial contamination from outside (Ray 1996).

Insufficient thermal treatment or insufficient cooling allows the survival of thermophilic spores that can germinate when cans are temperature abused at 40°C for even a short period of time; once germinated, some spores can outgrow and multiply at temperatures as low as 30°C, generating acid with or without gas (Guerrero-Legarreta 2001, Larousse and Brown 1997). Germination and growth of the facultative anaerobic bacterium *B. stearothermophilus* are accompanied by acid, without gas due to fermentation of carbohydrates; the growth of the anaerobic bacterium *Cl. thermosaccharolyticum* produces H_2 and CO_2 gas and swelling of cans. Sulfide stinker spoilage is caused by the gram-negative anaerobic spore-former *Desulfotomaculum nigrificans*, which is characterized by a flat container, a darkened product, and the odor of rotten eggs, due to production of H_2S from the sulfur-containing amino acids that dissolve in the liquid and react with iron to form black-colored iron sulfide. Spoilage can be from the breakdown of either carbohydrates or proteins. *Clostridium* spp., *Cl. butyricum*, and *Cl. pasteurianum* ferment carbohydrates to produce volatile acids and H_2 and CO_2 gas, causing swelling of cans. Proteolytic microorganisms such as *Cl. sporogenes, Cl. putrefaciens,* and *Cl. botulinum* metabolize proteins and produce H_2S, mercaptans, indol, skatole, ammonia, CO_2, and H_2 (Ray 1996).

Proper sterilization processing ensures the destruction of bacteria; however, gas and other microbial metabolites can remain in the can. Furthermore, additonal microbial contamination can take place after thermal processing as a result of improper sealing, damaged and leaky containers, or the use of water of poor sanitary quality during the cooldown stage. These conditions allow different types of microorganisms to get inside cans from the environment after heating. These microorganisms can grow in the food and cause different types of spoilage, depending on the microbial type. Cans that undergo abnormally high pressure or excessive filling can suffer physical deformation; because contamination with pathogens will make the product unsafe, deformed cans must be discarded to avoid the risks of leakage (Guerrero-Legarreta 2001).

CORROSION

Corrosion of metal containers can be both internal and external, and be initiated during can manufacture or at any point during processing. External corrosion is evident in the formation of a reddish-brown ferric oxide; it may be induced by corrosive water conditions or by poor conditions during storage or shipment. Internal corrosion is not visible until it produces leakages or swelling of the container. This phenomenon is associated with elevated oxygen levels in the headspace and aggressive foods. Bubbles or loose flaps in the internal coating may occur, followed by corrosion at the point of detachment; if the base metal is exposed, the corrosion will be more extensive. The presence of nitrites also may induce internal corrosion (Larousse and Brown 1997).

APPLICATION OF PROCESSING PRINCIPLES

Table 24.5 provides recent references for more details on specific processing principles.

Table 24.5. Processing Steps and Application Principles of Canned Poultry Ham

Stages of Processing	Application and Principles	References for More Information on the Principles Used
Raw meat deboning and conditioning	Texture, water holding capacity, color, lipid oxidation, particle size	Ranken 2000, Fletcher 1999, Smith 2001
Curing and brine injection	Water holding capacity, color, lipid oxidation, flavor	Kristensen and Purslow 2001
Vacuum tumbling	Water holding capacity, protein solubility, color, lipid oxidation	Kilic and Richards 2003, Young and West 2001
Can filling	Particle size	Larousse and Brown 1997, Guerrero Legarreta 2001
Exhaustion and closing	Heat transfer	Guerrero-Legarreta, 2001, Larousse and Brown 1997, Bratt 1999
Sterilization	Protein binding, Maillard browning, lipid oxidation, water holding capacity, flavor, color, heat transfer, microbial destruction by heating	Kilic and Richards 2003, Larousse and Brown 1997, Bratt 1999, Bailey 1994, Arnold et al. 2000, Belitz and Grosch 1999

GLOSSARY

Acid foods—foods with a pH of 4.6 or lower; includes all fruits except figs; most tomatoes; fermented and pickled vegetables; jams, jellies, and marmalades.

Botulism—severe and often fatal paralytic illness caused by a nerve toxin that is produced by the bacterium *Clostridium botulinum*. Botulism is caused by eating foods that contain the botulism toxin or by consuming the spores of the botulinum bacteria, which then grow in the intestines and release toxin. Proper heat processing destroys this bacterium in canned food. Freezer temperatures inhibit its growth in frozen food. Low moisture controls its growth in dried food.

BSE—bovine spongiform encephalopathy.

Canning—packaging process in which foods are preserved in airtight vacuum-sealed containers and heat processed sufficiently to enable storing the food at normal room temperatures.

D value, decimal reduction time—the time at a given temperature, T, for a survivor curve to traverse one log cycle or, equivalently, to reduce a microbial population by 90%, $t = D_T (\log N_0 - \log N)$.

DFD (dark, firm, dry)—meat with dark, firm, and dry characteristics due to a lack of carbohydrates in muscle, and thus poor glycolysis and reduced lactic acid generation. These meats have pH values above 6.0 after 24 hours postmortem and are typical of exhausted stressed animals before slaughtering.

Enzymes—proteins with catalytic activity that accelerate many flavor, color, texture, and nutritional changes, especially when food is cut, sliced, crushed, bruised, and exposed to air. Proper blanching or hot-packing practices destroy enzymes, allowing food conservation.

Exhausting—removal of air from within and around food and from jars and canners. Exhausting or venting of pressure canners is necessary to prevent the risk of botulism in low-acid canned foods.

Headspace—unfilled space above food or liquid in containers. Allows for food expansion as containers are heated, and for forming vacuum as containers cool.

Hermetic seal—an airtight container seal that prevents reentry of air or microorganisms into packaged foods.

Low-acid foods—foods that have a pH above 4.6; acidity in these foods is insufficient to prevent the growth of the bacterium *Clostridium botulinum*. All meats, fish, seafood, and some dairy foods are low acid. To control all risks of botulism, containers of these foods must be heat processed in a pressure canner, or acidified to a pH of 4.6 or lower.

MDT—mechanically deboned turkey.

Microorganisms—organisms of microscopic size, including bacteria, yeast, and mold. Undesirable microorganisms cause disease and food spoilage.

Pasteurization—heating of a specific food enough to destroy the most heat-resistant pathogenic or disease-causing microorganism known to be associated with that food.

PA—polyamide.

PE—polyethylene.

PETP—polyester.

pH—logarithmic index for the hydrogen ion concentration in aqueous solution. Used as a measure of acidity or alkalinity. Values range from 0 to 14. A pH below 7 indicates acidity. Higher values are increasingly more alkaline.

PP—polypropylene.

Pressure canner—specifically designed metal kettle with a lockable lid; used for heat-processing low-acid food. These canners have jar racks, systems for exhausting air, and detectors to measure and control pressure and temperature.

PSE (pale, soft, exudative)— meat with pale, soft, and exudative characteristics resulting from accelerated glycolysis and rapid lactic acid generation. The pH drop is very fast, reaching values as low as 5.6 at just one hour postmortem.

SAGARPA— Secretary of Agriculture and Rural Development, Mexico.

TDT—thermal death time.

Tinplate—steel plate, plated with tin on both sides; the steel body usually is 0.22 to 0.28 mm thick, while the tin layer is from 0.385 to 3.08 μm.

TFS—tin-free steel.

USDA—U.S. Department of Agriculture.

Vacuum—state of negative pressure. Reflects how thoroughly air is removed from within a canned food. The higher the vacuum, the less air is left in the container.

Water activity—measure of unbound free water available to support biological and chemical reactions.

WHC—water-holding capacity.

Z value—number of degrees of temperature required for the thermal death time curve (log F vs. T) or thermal resistance curve (log D_T vs. T) to traverse one log cycle, $Z = (T_x - T)/ (\log F_T - \log F_{Tx})$ or $Z = (T_x - T)/ (\log D_T - \log D_{Tx})$

REFERENCES

Adams RL, DS Mottram, JK Parker, HM Brown. 2001. Flavor-protein binding: Disulfide interchange reactions between ovalbumin and volatile disulfides. J Agric Food Chem 49(9): 4333–4336.

Arnold GR, LM Crawford, RA Goldberg, M Karel, SA Miller, RM Roberts, GE Schuh, BO Schneeman, TN Urban, FF Busta, JL Kokini, IJ Pflug, MD, US Pierson. 2000. Kinetics of microbial inactivation for alternative food processing technologies. Food and Drug Administration Center for Food Safety and Applied Nutrition. J Food Sci. 65(Supplement): 6–40.

Bailey ME. 1994. Chapter 9. Maillard reactions and meat flavor development. In: F Shahidi, editor. Flavor of meat and meat products. New York: Blackie Academic Professional, Chapman and Hall.

Belitz HD, W Grosch. 1999. Food Chemistry, 2nd edition, 180–215, 257–267. New York: Springer-Verlag Berlin Heidelberg.

Bratt L. 1999. Chapter 8. Heat Treatment. In: RJ Footitt, AS Lewis, editors. The Canning of Fish and Meat. Gaithersburg, Md.: Aspen Publishers, Inc.

Cambero MI, I Seuss, KO Honikel. 1992. Flavor compounds of beef broth as affected by cooking temperature. J Food Sci 57:1285–1290.

Claus JR, JW Colby, GJ Flick. 1994. Chapter 5. Processed meats/poultry/seafood. In: DM Kinsman, AW Kotula, BC Breidenstein, editors. Muscle Foods Meat Poultry and Seafood Technology. New York: Chapman and Hall.

Damodaran S. 1994. Chapter 1. Structure-function relationship of food proteins. In: NS Hettiarachchy, GR Ziegler, editors. Protein Functionality in Food Systems. IFT Basic symposium series. New York: Marcel Dekker, Inc.

Fletcher DL. 1999. Chapter 6. Poultry meat colour. In: RI Richarson, GC Mead, editors. Poultry Meat

Science, 159–173. Abingdon, Oxford, U.K.: CAB International

Froning GW, SR McKee. 2001. Chapter 14. Mechanical separation of poultry meat and its use in products. In: Poultry Meat Processing. Washington, D.C.: CRC Press.

Guerrero-Legarreta I. 2001. Chapter 22. Meat canning technology. In: YH Hui, WK Nip, RW Rogers, OA Young, editors. Meat Science and Applications. New York: Marcel Dekker Inc.

Jiménez-Colmenero F, J Careche, J Carballo. 1994. Influence of thermal treatment on gelation of actomyosin from different myosistems. J. Food Sci., 59(1): 211–215,220.

Keeton JT. 2001. Chapter 12. Formed and emulsion products. In: AR Sams, editor. Poultry Meat Processing. Washington, D.C.: CRC Press.

Kilic B, MP Richards. 2003. Lipid oxidation in poultry döner kebab: pro-oxidative and anti-oxidative factors. J Food Sci 68(2): 686–689.

Kristensen L, P Purslow. 2001. The effect of processing temperature and addition of mono- and divalent salts on the heme- nonheme-iron ratio in meat. Food Chem 73(4): 433–439.

Larousse J, Brown BE. 1997. Food Canning Technology, 235–264, 297–332, 383–424, 489–530. New York: Wiley-VCH.

Lawrie RA. 1998. Lawrie's Meat Science, 6th edition, 11–22, 58–94, 212–254. Cambridge, England: Woodhead Publishing Ltd.

Lin RC, PH King, MR Johnston. 2001. Chapter 22A. Examination of metal containers for integrity. In: U.S. Food and Drug Administration, Center for Food Safety and Applied Nutrition Bacteriological Analytical Manual. Online. http://www.cfsan.fda.gov/~ebam/bam-toc.html (January 2003).

Mandava R, H Hoogenkamp. 1999. Chapter 19. The role of processed products in the poultry meat industry. In: RI Richardson, GC Mead, editors. Poultry Meat Science, 397–410. Abingdon, Oxford, UK.: CAB International.

Pearson AM, TA Gillett. 1999. Processed Meats, 3rd edition, 53–78, 372–413. Gaithersburg, Md.: Aspen Publishers, Inc.

Ponce-Alquicira E. 2002. Chapter 13. Canned turkey ham. In:IJS Smith, GL Christen, editors. Food Chemistry Workbook, 135–143. West Sacramento, Calif.: Science Technology System.

Ponce-Alquicira E, L Pérez Chabela, I Guerrero-Legarreta. 2000. Propiedades funcionales de la carne. In: MR Rosmini, JA Pérez-Alvarez, J Fernández-López, editors. Nuevas tendencias en la tecnología e higiene de la industria carnica, 43–50. España: Universidad Miguel Hernández.

Ray B. 1996. Fundamental Microbiology, 229–231, 381–390. Boca Raton, Fla.: CRC Press Inc.

Ranken MD. 2000. Handbook of meat product technology. Malden, Mass.: Blackwell Science, Inc.

Roos KB. 1997. How lipids influence food flavor. Food Tech 51(1): 60–62.

Smith DM. 2001. Chapter 11. Functional properties of muscle proteins in processed poultry products. In: AE Sams, editor. Poultry Meat Processing. Washington, D.C.: CRC Press.

Smith DP, JC Acton. 2001. Chapter 15. Marination, cooking and curing of poultry products. In: AR Sams, editor. Poultry Meat Processing. Washington, D.C.: CRC Press

Turner TA. 1999. Chapter 5. Cans and lids. In: RJ Footitt, AS Lewis, editors.The Canning of Fish and Meat. Gaithersburg, Md.: Aspen Publishers, Inc.

U.S. Department of Agriculture (USDA). 2003a. Poultry Outlook. Economic Research Service, USDA, Washington, D.C. http://www.ers.usda.gov/publications/outlook (January 2003).

___. 2003b. Foreign Agricultural Services, USDA, Washington, D.C. http://www.fas.usda.gov (January 2003).

Van Laack RLJM. 1994. Chapter 14. Spoilage and preservation of muscle foods. In: DM Kinsman, AW Kotula, BC Breidenstein, editors. Muscle Foods Meat Poultry and Seafood Technology. New York: Chapman and Hall.

Young OA, JWest. 2001. Chapter 3. Meat canning technology. In: YH Hui, WK Nip, RW Rogers, OA Young, editors. Meat Science and Applications. New York: Marcel Dekker Inc.

25
Poultry: Poultry Nuggets

A. Totosaus and M. de Lourdes Pérez-Chabela

BACKGROUND INFORMATION

Poultry nuggets are restructured meat products made from poultry meat or mechanically recovered meat (MRM) where the frying process will reduce product humidity and develop color and texture, but with a considerable increase in fat content. Nuggets are usually small (bite size) and are covered or breaded with a mixture of flour and spices to give characteristic flavors, color, and a crispy texture. Nuggets are preformed with a precooking step and then fried. Frying is considered one of the oldest cooking methods in existence, especially in countries where oil plays an important role (Varela 1988). The kind of oil used depends on cultural traditions and the kinds of crops available in the region (e.g., olives in Mediterranean countries, corn or soy oil in the United States, and peanut or canola oil in Asian countries). The immersion frying process is also called deep-fat frying (Singh 1995). During frying, many chemical reactions occur from heat and mass transfer. Breading nuggets with batter improves texture due to starch gelatinization, and re-sults in a unique color from Maillard reaction compounds formed during formation of the crust layer. Figure 25.1 shows the general flowchart for poultry nugget processing, and Table 25.1 (at the end of the chapter) describes the principal steps of the process.

RAW MATERIALS PREPARATION

Poultry meat must be chopped to liberate myofibrillar proteins that will act as "glue" to bind the meat pieces. The addition of a binder enhances and improves nugget texture; breading is the most common practice, but binders can also be another protein, such as egg white. A great variety of ingredients can be added for specific or ethnic flavors. Oil temperature must be hot enough to maintain constant heat transfer during the frying process. Temperatures

---

Raw Materials
↓
Meat chopping
↓
Mixing ingredients and breading
↓
Casing, precooking, and slicing
↓
Frying

Figure 25.1. Flowchart for poultry nugget production.

The information in this chapter has been derived from a chapter in *Food Chemistry Workbook*, edited by J. S. Smith and G. L. Christen, published and copyrighted by Science Technology System, West Sacramento, California, ©2002. Used with permission.

around 130–150°C are recommended. The cleanliness of oil is important because impurities may accumulate on parts of the nugget or its cover. In addition, air and water escaping from the food during frying can cause many chemical alterations.

PROCESSING STAGE 1: MEAT CHOPPING

Functional properties of meat myofibrillar proteins, mainly myosin, are responsible for many sensory characteristics of meat products. In poultry nuggets, myofibrillar proteins act as emulsifying agents, mainly to bind the meat pieces in a restructured product. Egg white proteins have the same function. The binding and gelling capacity will give form to the nuggets in the precooking stage. Use of MRM implies the use of another protein or flour to give form to the nugget.

PROCESSING STAGE 2: MIXING INGREDIENTS AND BREADING

The homogeneous mixture of the ingredients is important for the distribution of proteins that will form the matrix—entrapping water, fat, and other components. This protein matrix is different than in an emulsified meat product, where the proteins act as emulsifiers, stabilizing the fat droplets in the water-protein-salt system. The fat content in this kind of batter is low, so proteins have the function of holding the small pieces of meat together. Flavor ingredients also must be well distributed in order to have a quality product.

The breading will contribute to nugget yield and help to develop crispy texture and characteristic color. Breading is composed of wheat flour, corn flour, whole wheat flour, or a combination of two or three of these flours. Other ingredients may be used to provide the needed adhesive and other functional properties and to produce the desired appearance, color, texture, crispness, and flavor, with a great increase in product yield. Breading for chicken nuggets is recommended not to exceed 30% (by weight) of the product.

PROCESSING STAGE 3: CASING, PRECOOKING, AND SLICING

A preforming stage, where proteins form a restructured gel, is an important step before cover ingredients or breading are applied. The homogeneous batter is put into casings of the desired diameter and cooked until it reaches an internal temperature of 58°C. After cooling, the product is cut into slices 1–1.5 cm in height. The height and size of the nugget are important because of the heat necessary to reach the center of the nugget without overheating the cover ingredients, resulting in an extra golden color.

PROCESSING STAGE 4: FRYING

Deep-fat frying produces changes in food structure and properties: textural changes, attractive and tasty surface, crust, increased palatability, and browning reactions. During deep-fat frying, the fat is continuously or repeatedly used at a high temperature. Oxidative transformations usually accompany and probably precede thermal transformation of the frying medium. The fried food absorbs this heated fat and contributes considerably to the fat ingested by consumers (Guillaumin 1988).

During frying, the oil undergoes many reactions, resulting in oxidative and hydrolytic degradation. The oil is exposed to molecular tension and changes due to the energy applied to heat the oil (Stier and Blumenthal 1993). In the frying process, the high temperature employed (180°C) promotes fatty acid hydrolysis, and many oxidative reactions take place. Oxidative reactions, polymerization, and hydrolysis occur rapidly during frying, depending on process conditions such as temperature, time, and aeration (Blumenthal 1991). The complex series of reactions include oxidation, polymerization, hydrolysis, isomerization, and cyclization (Boskou 1988).

FINISHED PRODUCT

Fried foods absorb a considerable quantity of oil, and the oil contributes a great part of the flavor, odor, and color of the product. The quantity of oil absorbed depends on food composition (humidity, porosity, and surface exposed to the frying oil) and has an effect on nugget quality during storage. Although the higher fat content increases flavor, the product tends to be softer with storage time (Berry 1993). Absorption of fat during frying therefore should be kept to a minimum. Furthermore, fat-soaked foods are less palatable and carry more calories. Keeping the contact time and the surface of the food exposed to the fat at a minimum reduces absorption (Charley 1982).

APPLICATION OF PROCESSING PRINCIPLES

The main applications during processing of poultry nuggets are heat and mass transfer during frying. During immersion frying of foods, there are two distinct modes of heat transfer: conduction and convection. Conductive heat transfer under unsteady-state conditions occurs within a solid food. The rate of heat transfer is influenced by the thermal diffusivity, thermal conductivity, specific heat, and density. Convective heat transfer occurs between a solid food and the surrounding oil. The surface interactions between the oil and the food material are complicated because of the vigorous movement of water vapor bubbles escaping from the food into the oil. In addition, the water vapor bubbles entrapped on the underside of the food material prevent efficient heat transfer between the bottom side of the food and the oil. The amount of water bubbles escaping the food material decreases with longer frying times as a result of the decrease in the moisture remaining in the material. During frying, the temperatures inside a food material are restricted to values below the boiling point of liquid. Since the liquid present in foods is mostly water with some solutes, the boiling point of the liquid inside a food is slightly elevated above the boiling point of the water. As the frying process proceeds, more water evaporates from the outer regions of a food material. Consequently, the temperature of the dried regions begins to rise above the boiling point (Singh 1995).

Deep-fat frying has long been a means of food preparation for achieving desired texture and flavor attributes in a variety of food products. The breading enhances the texture, flavor, and appearance of the food. It acts as a moisture barrier to diffusion of water vapor from inside and thereby contributes to juicy meat during the holding of the product. In some products, it acts as the major carrier of the seasoning and thus the flavor system. The physicochemical phenomenon of heat-induced texturing of breading (denaturation of protein and gelatinization of starch) is the combined effect of several multiple-order chemical reactions. However, such reactions can be modeled as a pseudo first-order reaction affecting breading texture, strength, water-holding capacity, and so on. The texture of the breading in a piece of chicken will largely be dependent on the temperature history, pressure during frying, and the specific ingredients included (e.g., type of flour, browning agent, protein content, and other functional agents). It is well known that using positive pressure during frying imparts a softer texture to the breading, while using atmospheric pressure results in a crispy texture (Rao and Delaney 1995).

Figure 25.2 shows the principal process pathways during frying and the main alteration caused in oil during frying. The four sequential distinct processes that take place during frying (listed on the left side) are described as follows (Farkas 1994): During the first step, *initial heating,* which lasts a few seconds, the surface of the food heats to a temperature equal to the boiling point of oil. The mode of heat transfer between the oil and nugget in the first seconds is natural convection, and no vaporization of water occurs from the food surface. The second step, *surface boiling,* occurs when the surface boiling state due to vaporization starts. Here, the convective heat transfer changes to forced convection because of the presence of turbulence in surrounding oil. The crust, or dry region, begins to form at the surface of the food. The *falling rate* is the stage during which more internal moisture leaves the food, and the internal core temperature rises to the boiling point. Additionally, several physicochemical changes (e.g., starch gelatinization and cooking) take place in the internal core region. The thickness of the crust layer increases, and after sufficient time and more removal of moisture, the vapor transfer at the surface decreases. Finally, during the *bubble end-point* stage, after a considerable period of time, the rate of moisture removal diminishes, and no more bubbles are seen escaping from the nugget surface. As the frying process proceeds, the thickness of the crust layer increases.

As food material undergoes frying, several important changes take place in the surrounding oil. These reactions cause the formation of a great number of decomposition products, both volatile and nonvolatile. The rate of deterioration of the heated fat is greatly influenced by its degree of unsaturation, the conditions of frying, the nature of the fried food, and the presence of chemicals that minimize darkening and polymerization during prolonged heating. The significance of the unsaturation and distribution of fatty acids in the triacylglycerol molecule as well as the effect of antioxidants, silicon, and phytosterols on the stability of heated oils is discussed in terms of the mechanism proposed for antioxidant activity (Boskou 1988). The alterations that frying oil can undergo are due mainly to three factors: air, temperature, and moisture. Air and moisture are produced when the food surface begins to boil, reflecting the

Figure 25.2. Distinct stages during the frying process.

pounds (hydrocarbons, aldehydes, ketones, alcohols, acids, etc.). The compounds that can be formed during frying by thermal alteration are cyclic monomers, dimers, and polymers (Gutierrez et al. 1988).

The heat transferred from oil into a food causes several chemical and physical changes such as starch gelatinization, protein denaturation, water vaporization, and crust formation (color and flavor development). Mass transfer during frying is characterized by the movement of oil into the food and the movement of water, in the form of vapor, from the food into the oil. Frying oil becomes contaminated with components of food materials leaching into oil, water vapor condensing in oil, thermal breakdown of oil, and oxygen absorbed at the oil-air interface. These contaminants reduce oil surface tension, acting as surfactants. When the level of surfactants increases, wetting of the food surface by the oil is also increased, influencing the heat and mass transfer processes. The surfactants entering the food with the oil are suspected to influence the moisture pickup by the food during subsequent storage, hence reducing its shelf life (Singh 1995).

Table 25.1 provides recent references for more details on specific processing principles.

moisture escaping by the increase in internal core temperature. Oxygen and humidity condensation accumulate in oil during successive frying, producing oxidative and hydrolytic alterations. Oxidative alterations produce fatty acids, monoglycerides, diglycerides, and glycerol. Hydrolytic alterations produce oxidized monomers, oxidative dimers and polymers, nonpolar dimers and polymers, and volatile com-

Table 25.1. Processing Steps and Application Principles of Poultry Nuggets Elaboration

Processing Stage	Processing Principle	References for More Information on the Principle Used
Raw materials preparation	Oil preheating	Zorrila et al. 2000, Singh 1995
Meat chopping	Particle size reduction (protein liberation)	Xiong, 1997, 2000
	Homogenization	
Mixing ingredients and breading	Poultry nuggets batter homogenization	Prinyawiwatkul et al. 1997,
	Flour addition	Mukprasirt et al. 2000
Casing, precooking and slicing	Mechanical entrapment to give form	Xiong 2000, Regenstein 1989
	Moderate heat treatment	
	Slicing	
Frying	Cooking process	Resurreccion 1994, Balasubra-
	Microbial population reduction	maniam et al. 1997
	Color and texture development	

GLOSSARY

Breading—composed of wheat or corn flour or both. It contributes to poultry nuggets a crispy texture due to starch gelatinization during frying, and color development by browning reactions.

Bubble end point—final frying stage; occurs when the rate of moisture removal diminishes and no more bubbles are seen escaping from the food surface.

Conductive heat transfer—one of two modes of heat transfer during frying; takes place within the nugget batter, as a solid food.

Convective heat transfer—one of two modes of heat transfer during frying; occurs between solid food and the surrounding oil.

Crispy texture—characteristic brown or golden surface developed during frying, resulting from starch gelatinization, protein gelation, and water loss.

Falling rate—period when more internal moisture leaves the food, with the internal core temperature reaching the boiling point.

Frying—cooking with hot oil to produce a characteristic texture and color; moisture migrates from the core of the food, and heat produces a crust, or dry region, on food surfaces.

Heat transfer—see *Conductive heat transfer and Convective heat transfer.*

Hydrolytic alterations—water condensation during frying that can produce oxidized monomers, oxidative dimers and polymers, nonpolar dimers and polymers, and volatile compounds, such as hydrocarbons, aldehydes, ketones, alcohols, and acids, among others.

Initial heating—short period of time when the food surface enters into contact with oil, with no water vaporization.

Mass transfer—movement of oil into food, and water from food into oil in the form of vapor.

MRM—mechanically recovered meat.

Oxidative alterations—produced by the oxygen escaping from the food during frying, resulting in fatty acids, monoglycerides, diglycerides, and glycerol.

Surface boiling—second step during frying: the vaporization process begins to form the crust, or dry region, on food surfaces.

Thermal alterations—the compounds formed during frying by thermal alteration are cyclic monomers, dimers, and polymers.

REFERENCES

Balasubramaniam, VM, P Mallikarjunan, MS Chinnan. 1997. Heat and mass transfer during deep-fat frying of chicken nuggets coated with edible film: Influence of initial fat content. In: M Narsimhan, R Okos, S Lombardo, editors. Advances in Food Engineering, 3–6. Proceedings of COFE95.

Berry BW. 1993. Fat level and freezing temperature affect sensory, shear, cooking and compositional properties of ground beef patties. Journal of Food Science 58:34–37.

Blumenthal MM. 1991. A new look at the chemistry and physics of deep-fat frying. Food Technology 45(5): 68–71.

Boskou D. 1988. Stability of frying oils. In: G Varela, AE Bender, ID Morton, editors. Frying of Food, 174–182. Chichester: Ellis Horwood Ltd.

Charley H. 1982. Food Science, 2nd edition. New York: John Wiley and Sons.

Farkas BE. 1994. Modeling immersion frying as a moving boundary problem. Ph.D. Dissertation. University of California, Davis. Cited in RP Singh. 1995. Heat and mass transfer in food during deep-fat frying. Food Technology 49(4): 134–137.

Guillaumin R. 1988. Kinetics of fat penetration in food. In: G Varela, AE Bender, ID Morton, editors. Frying of Food, 82–90. Chichester: Ellis Horwood Ltd.

Gutierrez R, J Gonzalez-Quijano, MC Dobarganes. 1988. In: G Varela, AE Bender, ID Morton, editors. Frying of Food, 141–154. Chichester: Ellis Horwood Ltd.

Mukprasirt A, TJ Herald, DL Boyle, KD Rausch. 2000. Adhesion of rice flour-based batter to chicken drumstick evaluated by laser scanning confocal microscopy and texture analysis. Poultry Science 79:1356–1363.

Prinyawiwatkul W, KH McWatters, LR Beuchat, RD Phillips. 1997. Physicochemical and sensory properties of chicken nuggets extended with fermented cowpea and peanut flours. Journal of Agricultural and Food Chemistry 45:1891–1899.

Regenstein JM. 1989. Chapter 10. Are comminuted meat products emulsions or a gel matrix? In: JE Kinsella, WG Soucie, editors. Food Proteins, 178–194. Champaign, Ill.: The American Oil Chemists' Society.

Resurreccion AVA. 1994. Chapter 15. Cookery of muscle foods. In: DM Kinsman, AW Kotula, BC Breidenstein, editors. Muscle Foods, Meat, Poultry and Seafood Technology, 406–429. New York: Chapman and Hall, New York.

Rao VNM, RAM Delaney. 1995. An engineering perspective on deep-fat frying of breaded chicken pieces. Food Technology 49(4): 138–141.

Singh RP. 1995. Heat and mass transfer in food during deep-fat frying. Food Technology 49(4): 134–137.

Stier RF, MM Blumenthal. 1993. Quality control in deep-fat frying. Baking and Snack 15(2): 67–75.

Varela G. 1988. Current facts about the frying of foods. In: G Varela, AE Bender, ID Morton, editors. Frying of Food, 9–25. Chichester: Ellis Horwood Ltd.

Xiong YL. 1997. Chapter 12. Structure-function relationship of muscle proteins. In: S Damodaran, A Paraf, editors. Food Proteins and Their Applications, 341–392. New York: Marcel Dekker.

___. 2000. Chapter 2. Meat processing. In: S Nakai, HW Molder, editors. Food Proteins: Processing Applications, 89–145. New York: Wiley–VCH.

Zorrilla SE, CO Rovedo, RP Singh. 2000. A new approach to correlate textural and cooking parameters with operating conditions during double-sided cooking of meat patties. Journal of Texture Studies 31(5): 499–523

26
Poultry: Poultry Pâté

M. de Lourdes Pérez-Chabela and A. Totosaus

BACKGROUND INFORMATION

Liver from poultry and mammals is the most widely used organ, resulting in many styles of processed meats such as liver sausage and paste (Liu and Ockerman 2001). Pâté or liver sausage is a ready-to-eat cooked sausage, with the special feature that the meat batter can be worked at relatively high temperatures. In meat batters, the temperature is important in order to maintain the integrity of the solubilized protein-salt-water matrix. The muscle proteins (myosin, mainly) must be liberated and activated during meat chopping and the addition of ice and sodium chloride. The fat must be dispersed in the batter with the cutter to induce protein matrix gelatinization, entrapping fat and water. This results in juiciness and textural attributes in sausages. In liver pâté, the amount of lean meat is lower than in sausage formulations (15% vs. 50–60% in the average sausages). The meat added must be precooked. This reduces protein functionality and makes the product less thermostable, so that fat can be released or color and flavor affected by excessive heat. Thus, liver proteins act as the main emulsifying agent in pâté production. In the same way, fat is an important part in the formulation, but its levels are higher (47–50%) than in sausages (10–15%, depending on formulation). Also, fat in pâté has different functions than in regular sausage, where the juiciness, flavor, and texture are affected by fat. In pâté, the fat is responsible for product spreadability. That is, certain types of protein must be added to the formulation to improve emulsification. Normally, milk protein is added (1–2%) to improve emulsification and enhance flavor.

Heat treatment is the other important process in pâté production. Heat is applied to form the protein gel matrix, destroy the microbial population, extend shelf life, and make the product safe for consumption. Heat transfer is usually achieved by conduction in water or by steam in autoclaves. Two important criteria in this step are the internal temperature in the pâté and the heating effect (*F* value) necessary to destroy the microbial population.

Figure 26.1 shows a general flowchart for pâté processing, and Table 26.1 describes the principal steps of the process.

The information in this chapter has been derived from a chapter in *Food Chemistry Workbook,* edited by J. S. Smith and G. L. Christen, published and copyrighted by Science Technology System, West Sacramento, California, ©2002. Used with permission.

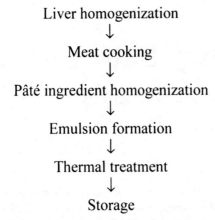

Liver homogenization
↓
Meat cooking
↓
Pâté ingredient homogenization
↓
Emulsion formation
↓
Thermal treatment
↓
Storage

Figure 26.1. Flowchart for poultry pâté production.

RAW MATERIALS PREPARATION

Fresh poultry liver is used for pâté formation, because old livers lose emulsifying capacity. Livers must be clean, with no fat or vessel. They can be chopped or cut in half-inch cubes, and frozen if not used in the next 24 hours. Lean meat cut in one-inch cubes is parboiled in salt (one-third of the total weight), not exceeding 80°C for 30–60 minutes. The broth is left to settle, concentrated by boiling, clarified, and kept at 60°C until it is added to the batter (Savic 1985).

PROCESSING STAGE 1: INGREDIENT HOMOGENIZATION

LIVER HOMOGENIZATION

Raw livers are comminuted in the cutter with half of the salt to obtain a uniform mass. The cooked meat

is added and roughly disintegrated using the lower cutter speed. The percentage of liver in the formulation has a direct influence on emulsion stability. The higher the amount of liver, the more stable are the emulsions that will be formed, with a constant fat content (Hilmes et al. 1993).

PÂTÉ INGREDIENT HOMOGENIZATION

After the meat is cooked and the fat is scalded (> 65°C), they are placed together in the cutter and homogenized at low speed. The speed is then increased, and the rest of the salt is added. If milk proteins are used, half is added at the beginning and the rest at the end of the operation (Savic 1985).

PROCESSING STAGE 2: EMULSION

EMULSION FORMATION

Hot broth is added to the batter gradually during homogenization to maintain the mixture at a constant temperature of 58–60°C, with the final temperature reaching 45°C. When the batter is thoroughly homogenized, the homogenized liver (from Processing Stage 1) is added and well distributed. The sausage emulsion is then ready. In order to improve flavor, spices such as onion can be added, and if desired, the raw pâté or liver sausage mass may be passed through an emulsifying mill (Savic 1985). Liver proteins will act as an emulsifying agent. Proteins are adsorbed at the fat/water interface, reducing interfacial tension and preventing fat globule coalescence. The parameters that influence meat batter formation and stability are chopping time (fat globule size), protein concentration (emulsifying agent concentra-

Table 26.1. Processing Steps and Application Principles of Poultry Pâté Elaboration

Stages of Processing	Application Principles
1 Raw materials preparation	Precooking
2 Ingredient homogenization	Particle size reduction (protein liberation)
	Homogenization
3 Emulsion	Fat globule dispersion
	Homogenization
4 Casing	Mechanical entrapment to give form
	Heat transfer
5 Thermal treatment	Microbial population reduction
	Fat and water entrapment in the gelled protein-water matrix
6 Packaging	Extend shelf life

tion), and mixing speed (energy input and batter heating). The last factor, that is, temperature, is the most important factor to control during emulsification (Guerrero et al. 2002).

EMULSION TEMPERATURE

The temperature of meat ingredients used in processing is a decisive factor in pâté production. Fat and lean meat must be heated to > 65°C to melt fat and denature proteins. Raw livers should be added when the meat-fat-broth mixture falls to < 60°C, in order to avoid liver protein denaturation, but is > 45°C, to ensure that the fat is melted (Savic 1985). The final fat content in the pâté has an important effect on texture, slicing, spreading, and color. Products with a high fat percentage have a fine texture and are more spreadable, but they have less stability in the emulsion (Chyr et al. 1980).

On the other hand, meat products with high fat content, for example, poultry pâté, have other problems. For example, fat not completely entrapped in the protein matrix tends to move outside the product as water. If so, liquids and melted fat occupy the empty air spaces, improving heat transfer. Furthermore, the pâté surface will be affected by the presence of fat and liquids (Waters 2002).

PROCESSING STAGE 3: PACKING (CASING)

Molded meat products can be processed in the same mold or casing. Heat and energy are transferred first to the casing and then to the batter. Air is an excellent insulator, and if air bubbles are present in the batter inside the casing, the heat transfer will be interrupted, provoking temperature differentials in the same product and extending the heat treatment, due to the low efficiency of heat transfer. If the pâté batter is in a water-permeable casing, humidity losses will cause an important reduction in final volume (Waters 2002).

PROCESSING STAGE 4: THERMAL TREATMENT

Thermal processing can be reviewed in standard literature. We are interested in its effect on poultry pâté and meat products. Thermal treatment is crucial in killing or reducing pathogens. So evaluation and recording of cooking temperatures become important parts of any food safety program, including the hazard analysis critical control point (HACCP) re-

quirements, and should be regulated (Waters 2002). Cooked meat products are heat treated to extend shelf life, by reducing the growth of, or inactivating, microorganisms. In heat or thermal treatment, the products are submerged in cooking vats or pressure cookers that contain hot water, or steam, or a mixture. Thermal treatment can be performed under pressure in retorts or autoclaves in order to reach temperatures above 100°C (sterilization). In contrast, temperatures up to 100°C can be achieved in vats (pasteurization). Some microorganisms resist moderate heat treatment, and the resulting pasteurized products must consequently be stored under controlled low temperatures to retard microbial spoilage [Food and Agriculture Organization (FAO) 1990].

The intensity of thermal treatment can be defined in physical terms. The term widely used under practical conditions is the F value, by which the lethal effect on a microbial population can be defined. The thermal death time for different microorganisms, calculated at 121°C and expressed in minutes, is used as the reference value. For example, the thermal death time for spores of *Clostridium botulinum* at 121°C is 2 minutes 45 seconds, or an F value of 2.45, is needed to inactivate the spores in the product at 121°C. This pathogenic microorganism is the most resistant and serves as reference for low pH foodstuffs (FAO 1990).

ORGANOLEPTIC, PHYSICAL, AND MICROBIAL ASPECTS OF THERMAL TREATMENT

The intensity of heat treatments has a decisive impact not only on the inactivation of microorganisms, but also on the organoleptic quality of the final product. There are products that undergo intensive temperature treatment without significant losses in quality, but some products may deteriorate considerably in taste and consistency after heat treatment. One objective of heat treatment is to destroy microorganisms. Since some resist moderate heat treatment, the resulting pasteurized products must be stored under controlled temperatures. Proper preventive measures must be taken to avoid protein denaturation or fat release due to emulsion destabilization from excessive heat application.

INTERNAL TEMPERATURE

A uniform batter is necessary to obtain the same internal temperature. When variations in composition are present in the product, final temperature will be affected. Final internal temperature only can be obtained if all the product or casing is at the same temperature. Casing surface will receive the same heat quantity, at the same rate, at different points if the heat is applied uniformly on the casing surface (Waters 2002).

F VALUE

By measuring the temperature of the product periodically during thermal treatment, the final F value can be determined. It is obvious that during thermal treatment the product temperature will rise. The temperature taken in the center of the container or casing after each minute of heat treatment corresponds to a certain F value. These partial F values are added up, and the sum is the overall F value of the product. The exact F value is of special importance to the meat producer because it ensures appropriate thermal treatment of the product, thus avoiding undercooking. It also allows the product storage time to be determined (FAO 1990).

PROCESSING STAGE 5: PACKAGING

Handling after processing is considered the primary cause of contamination in cooked meat and poultry products. Quantitative information that the processing plan meets the specific lethality performance standard for each product is necessary in order to implement a zero tolerance for the presence of pathogens in this kind of product. In poultry pâté, it is an advantage that the casing in which the batter is cooked is also the final package.

Storage time and temperature, apart from the packaging method, greatly affects the shelf life of meat products. The purpose of packaging is primarily to protect foodstuffs during the distribution process, including storage and transport, from contamination by dust, microorganisms, mold, yeast, and toxic substances, or from those influences affecting smell and taste or causing loss of moisture. Packaging should help to prevent spoilage and weight loss and to enhance consumer acceptability (FAO 1990). Therefore, the type of material for the casings must be chosen correctly to comply with the specifications in order to avoid oxygen intake and humidity loss.

FINISHED PRODUCT

Factors that limit shelf life can be intrinsic (i.e., pH value and water activity, a_w) or extrinsic (i.e., oxy-

gen, microorganisms, temperature, light, and humidity loss) (FAO 1990). Intrinsic parameters, pH and a_w, are controlled by the correct selection and managing of raw materials before and after processing, and by correct heat treatment (for both microbial reduction and gel formation). The other parameters should be controlled after the poultry pâté is finished and packed in the casing. The storage temperature for the pâté (1–4°C) is important to prevent microorganism spoilage and fat rancidity or protein denaturation.

The finished product shelf life is a reflection of the quality of the raw materials, the handling of the product during all the processing steps, and the right storage conditions to guarantee a high quality poultry pâté.

APPLICATION OF PROCESSING PRINCIPLES

Emulsion formation and heat treatment are considered the most important aspects of the manufacture of pâté or liver sausage. This means (1) formation of the protein matrix, entrapping fat and water, and (2) ensuring a physicochemical, microbiological, and sensory stable shelf life.

Heat is necessary to destroy the microbial population (denoted by F value) (Reichert 1988). The temperature in the foodstuff may be inadequate if one considers data for thermal treatments above or below 100°C, because (1) the overall heating effect cannot be predicted, due to dependence on the food dimension (diameter and length), the cooking temperature magnitude, and the cooking time; (2) a determined internal temperature cannot be maintained for a given time; (3) microorganism destruction is not lineal, but exponential; thus, the microbial count reduction depends on the applied temperature and time; and (4) changes in sensory (color, flavor, texture) and other parameters (fat separation, cooking loss, protein coagulation) also depend on time and temperature.

If so, some advantages in using F values rather than the internal temperature are the following: (1) the lethal effect on microorganisms can be significant with different food or casing sizes; (2) unnecessary alterations by cooking due to higher casing capacities or container size are avoided; and (3) different heating temperatures can be compared by referring to heat alterations (fat losses, weight loss, sensory attributes).

F value has many advantages:

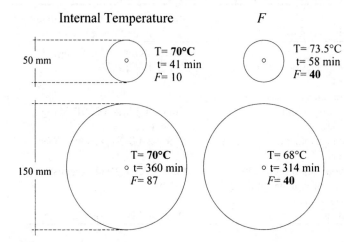

Internal Temperature   F

50 mm   T= **70°C** t= 41 min F= 10   T= 73.5°C t= 58 min F= **40**

150 mm   T= **70°C** t= 360 min F= 87   T= 68°C t= 314 min F= **40**

Figure 26.2. Comparison between heating to an internal temperature of 70°C and heating to an F value of 40 at different casing diameters.

- With F values, the lethal effect reached with different temperatures can be compared.
- In new product development, reference values are available from similar products and bibliographic references.
- Different sizes and shapes can be compared.
- Different heating procedures can be compared.

F value in the meat industry is employed only in sterilized products. Its use is recommended to improve heat treatment. Comparing casing sizes (i.e., 50 and 150 mm diameter), different F values are obtained if one uses a heating temperature of 75°C until it reaches 70°C in the geometric center than if one applies a heating effect of $F_{70°C}^{10°C} = 40$ (Reichert 1988). For 50 mm diameter casings, it takes 41 minutes to reach an internal temperature of 70°C, but only $F_{70°C}^{10°C} = 10$ can be reached. In contrast, if $F_{70°C}^{10°C} = 40$ is applied, the internal temperature reaches 73.5°C in 58 minutes, ensuring that more microorganisms are destroyed. In casings of higher diameter (150 mm), the excessive heat affects sensory attributes in the final product. It takes 360 minutes to obtain an internal temperature of 70°C, with $F = 87$. Using $F_{70°C}^{10°C} = 40$, the internal temperature will reach only 68°C in 314 minutes. Heating to the same internal temperature provokes a higher lethal effect in higher casing sizes, but it also causes more heat damage and loss of quality. A heating effect of $F_{70°C}^{10°C} = 40$ should be enough in many cases. The F value may not be adequate when attempting to reach a given internal temperature, so use of F instead of the given internal temperature may result in a short shelf life, because the microbial population has not been reduced adequately. Instead, if a given F value is followed during thermal treatment, unnecessary alteration by the heat can only be justified if the contamination of the raw materials is high (Reichert 1988) (Figure 26.2).

Table 26.2 provides recent references for more details on specific processing principles.

Table 26.2. Recent References on Application Principles of Poultry Pâté Elaboration

Processing Stage	Processing Principle	References for More Information on the Principle Used
Raw materials preparation	Precooking	Ressureccion 1994, Xiong 1997
Ingredient homogenization	Particle size reduction (protein liberation)	Xiong 1997, 2000
	Homogenization	
Emulsion	Fat globules dispersion	Xiong and Mikel 2001, Mangino 1994
	Homogenization	
Casing	Mechanical entrapment to give form	Xiong 2000, Regenstein 1989
	Heat transfer	
Thermal treatment	Microbial population reduction	Resurreccion 1994, Regenstein 1989
	Fat and water entrapment in the gelled protein-water matrix	
Packaging	Extend shelf life	Du et al. 2001, Hotchkiss 1994

GLOSSARY

Cutter—high-speed mixer with blade used to reduce particle size in order to liberate muscle proteins and disperse the fat globules in meat batters.

Emulsifying agent—compound that enhances or improves emulsion stability. It forms a film with the polar portion interacting with the nonpolar fat globule and the nonpolar portion toward the polar solvent (water). The capacity to form a high superficial area to surround fat globules is the most important characteristic.

Emulsion—system of at least two phases, one phase (liquid or solid) is dispersed in another one. In meat products the fat is dispersed in a "semi" continuous phase or matrix (water and muscle protein). The term "batter" is more appropriate because of particle size.

Functional properties (functionality)—physicochemical properties related to the development of sensory and structural properties of foods. Functional properties related to meat systems are: color, flavor, texture, solubility, emulsion and gelatinization.

FAO—United Nations Food and Agriculture Organization.

F value—number of minutes required to kill a known population of microorganisms in a given food under specified conditions. This temperature can be 121°C, and the extent of thermal treatment depends on microorganism heat resistance (z value).

Gel protein matrix—disperse phase in an ordered macromolecular network interconnected and interlinked in a three-dimensional structure that forms the continuous phase. The batter is cooked, and the gel formed is irreversible.

HACCP (hazard analysis and critical control points)—system to identify and evaluate the food safety hazards that can affect the safety of food products, institute controls necessary to prevent those hazards from occurring, monitor the performance of those controls, and routinely maintain records.

Heat conduction—heat transfer between adjacent molecules without a considerable particle displacement.

Milk protein—food additive used for emulsion stability; the most often employed are sodium caseinates.

Pasteurization—thermal treatment to destroy the vegetative cells present in food.

Shelf life—period of time that a food, raw or processed, can be stored without physical or chemical changes and without microbial spoilage that is hazardous to the consumers.

Sterilization—chemical or physical treatment to destroy viable microorganisms. Commercial sterilization may not destroy all thermal resistant spores.

Thermal treatment—heat treatment is to ensure the destruction of microorganisms and enzyme inactivation in order to avoid decomposition reactions and proliferation of pathogens. In meat, the same treatment can be used to form water-protein matrix gel and to develop color.

REFERENCES

Chyr C, J Sebranek, H Walker. 1980. Processing factors that influence the sensory quality of Brownschweiger. Journal of Food Science 45:1136–1138.

Du M, KC Nam, DU Ahn. 2001. Cholesterol and lipid oxidation products in cooked meat as affected by raw meat packaging and irradiation, and cooked meat packaging and storage time. Journal of Food Science 66(9): 1396–1401.

FAO 1990. Manual on Simple Methods of Meat Preservation. Animal Production and Health Paper 79. Rome: Food and Agriculture Organization,

Guerrero I, ML Pérez-Chabela, E Ponce-Alquicira. 2002. Curso Práctico de Tecnología de Carnes, 63–70. México City: Universidad Autónoma Metropolitana Press.

Hilmes C, S Cheon, A Fisher. 1993. Microstructure and stability of liver sausages as influenced by liver content. Fleischerei Industrieausgabe Englisch 44:3–5.

Hotchkiss JH. 1994. Chapter 18. Packaging muscle foods, In: DM Kinsman, AW Kotula, BC Breidenstein, editors. Muscle Foods, Meat, Poultry and Seafood Technology, 475–496. New York: Chapman and Hall.

Liu D-C, HW Ockerman. 2001. Chapter 25. Meat co-products. In: YH Hui, W-K Nip, RW Rogers, OA Young, editors. Meat Science and Applications, 581–603. New York: Marcel Dekker, Inc.

Mangino ME. 1994. Chapter 5. Protein interactions in emulsions: Protein-lipid interactions. In: NS Hiettiarachchy, GR Ziegler, editors. Protein Functionality in Meat Systems, 147–179. New York: Marcel Dekker, Inc.

Regenstein JM. 1989. Chapter 10. Are comminuted meat products emulsions or a gel matrix? In: JE

Kinsella, WG Soucie, editors. Food Proteins, 178–194.Champaign, Ill.: The American Oil Chemists' Society.

Reichert JE. 1988. Tratamiento Térmico de los Productos Cárnicos. Zaragoza: Editorial Acribia, S.A.

Resurreccion AVA. 1994. Chapter 15. Cookery of muscle foods. In: DM Kinsman, AW Kotula, BC Breidenstein. Muscle Foods, Meat, Poultry and Seafood Technology, 406–429. New York: Chapman and Hall.

Savic IV. 1985. Small scale sausage production. Animal Production and Health Paper 52. Rome: Food and Agriculture Organization.

Waters E. 2002. Technology of thermal processing. Carnetec 9(4): 36–39.

Xiong YL. 1997. Chapter 12. Structure-function relationship of muscle proteins. In: S Damodaran, A Paraf, editors. Food Proteins and Their Applications, 341–392. New York: Marcel Dekker, Inc.

___. 2000. Chapter 2. Meat processing. In: S Nakai, HW Molder, editors. Food Proteins: Processing Applications, 89–145. New York: Wiley-VCH.

Xiong YL, WB Mikel. 2001. Chapter 15. Meat and meat products. In: YH Hui, W-K Nip, RW Rogers, OA Young, editors. Meat Science and Applications, 351–369. New York: Marcel Dekker, Inc.

27

Seafood: Frozen Aquatic Food Products

B. A. Rasco and G. E. Bledsoe

BACKGROUND INFORMATION

Fishery products are the major source of high quality dietary protein to more than a quarter of the world's population. Aquatic foods are among the most highly valued and most perishable of all food products. Although people consume only a limited number of species of land animals as sources of muscle food, they consume hundreds of different species of aquatic animals including more than 350 species of Mollusca (e.g., clams and oysters), Arthropoda (e.g., lobsters, crabs, shrimp, and crayfish), holothurians (sea cucumbers) and Chordata (finfish) [National Academy of Sciences (NAS) 2003]. Furthermore, aquatic plants and marine mammals are also eaten.

Since 1980, seafood consumption in the United States has increased by 22%, from 12.5 to 15.3 pounds per person [U.S. General Accounting Office (GAO) 2001]. Frozen seafood products have also become an important part of the diet, including roughly 2.5 pounds of shrimp per person per year in the United States, as well as 1.7 pounds of Alaska pollack fillets, and 1.1 pound of cod (Johnson 1998). World commercial fishing fleets and aquaculture produce over 120 million metric tons of seafood annually [Food and Agriculture Organization (FAO) 2002]. None of this would be possible without the availability of economical refrigeration.

International trade in aquatic food products is critical to the balance of trade of many countries, with the United States importing 3.9 billion pounds of aquatic food products from over 160 different countries, half of this from China, Ecuador, Chile, Canada, and Thailand. Imported seafood constitutes more than half of the seafood consumed in the United States, and much of this product is frozen

---

Some information in this chapter has been derived from a chapter in *Food Chemistry Workbook,* edited by J. S. Smith and G. L. Christen, published and copyrighted by Science Technology System, West Sacramento, California, ©2002. Used with permission.

(GAO 2001), for example roughly 80% of the shrimp consumed in the United States is imported, and most of it is frozen (NAS 2003).

So these products can be made widely available, aquatic food products are often refrigerated or frozen. The widespread use of refrigeration, freezing, and cold storage has meant that aquatic food products, normally available only seasonally and within a small region, can now be sent around the world any time of the year. Until quite recently, aquatic foods were primarily harvested and consumed locally.

Improvements in the quality, availability, and price of fresh and frozen fish products, along with cheaper poultry products, have negatively impacted the canned seafood market. (With improved quality and availability of other fresh and frozen foods, a similar negative impact has been seen for other canned products, with the possible exception of tomatoes, which do not freeze well.) The exception to thermally processed products is canned tuna, which is still the most popular fish product, by volume, consumed in the United States, followed closely by frozen shrimp, and salmon in various product forms.

The primary advantages of freezing are to extend product shelf life, ensure product safety and nutritional value, and maintain product quality. Specific market advantages obtained from increasing product shelf life for aquatic food products include the ability to:

- Distribute foods over long distances far from the point of harvest;
- Distribute foods at convenient times and locations, which maximizes flexibility of shipments and simplifies logistics (at least for frozen goods when compared to fresh product);
- Store and distribute foods when a fishing season is closed for biological, stock management, or political reasons;
- Store and distribute product to markets when the animal is not in a form appropriate for harvest (e.g., spawning oysters);
- Store product on site for processing at a later date to accommodate market demand, and improve sales price for products during holidays or "off season";
- Manage the processing of whole animals, carcasses, and larger portions or cuts into units suitable for retail sale or as packaged foods;
- Store product on site to "even out" production rate at the processing plant;

- Ensure just-in-time delivery of aquatic food products to retailers, distributors, and buyers, which saves customers the cost of storage and permits delivery to customers of the highest quality product based upon market demand.

Refrigeration or *chilled* describes product temperatures above 0°C, while *freezing* describes product temperatures below 0°C. Holding and shipping live fish is *technically* a refrigeration process since product temperature is generally held at 4–10°C, depending upon the species.

RAW MATERIAL PREPARATION

Animal food products deteriorate rapidly at ambient temperatures, and aquatic food products are generally even more susceptible to deterioration. Refrigeration works by slowing metabolic processes. Reducing temperature slows the growth of pathogenic and spoilage microorganisms and reduces the rate of deteriorative biochemical and chemical reactions in the muscle and other edible tissues [Food and Drug Administration (FDA) 2001]. However, many animal and plant foods from the aquatic environment, particularly marine fish, are poikilothermic and are adapted to living at low temperatures (−1 to 10°C). For poikilotherms, refrigeration has limited effectiveness because endogeneous enzymes in these fish and the bacterial enzymes from surface microflora continue to function normally at these low temperatures. In addition, the spoilage bacteria associated with poikilothermic food sources continue to grow. This is why products from aquatic animals and plants deteriorate more quickly than foods from terrestrial sources and must be processed quickly to maintain highest quality.

In general, one can reduce deterioration and decomposition by lowering the product temperature to 4°C or less, ensuring that heat is not returned to the product, and ensuring that any heat generated by the product is promptly removed. However, to halt deterioration, the mobility of water within the food must be reduced. This is particularly significant for aquatic food products because the water content is high. Finfish contain 60–80% water on a weight basis, and some aquatic products contain over 90% water. Individuals within the industry often remark that they sell some of the most expensive water in the world. The water in any food product continues to affect the chemical activity of that product until a temperature of −40°C is achieved and maintained.

Even below −40°C, the product will still be affected by dehydration and lipid oxidation unless protected by packaging or physical barriers such as an ice glaze.

HARVEST CONDITIONS AND POSTMORTEM CHANGES

With the exception of aquaculture, where the fish can be harvested with limited stress, finfish are most commonly "stressed" when captured. As the fish pass through rigor, the ultimate pH of the fish tissue is higher than for meat, generally pH 6.4–6.6. Little glycogen is left in the muscle tissue for conversion to lactic acid during the glycolytic process that accompanies rigor. In contrast, land animals are generally rested prior to slaughter and have higher levels of glycogen and a lower ultimate pH, around 5.5 for mammalian muscle and 5.9 for chicken. The higher ultimate pH in fish is one reason why fishery products are relatively more susceptible to microbial spoilage than other muscle foods stored under the same conditions. The endogenous enzymes in the fish muscle and viscera of most commercially important species are highly active at refrigeration temperatures. Also, the microbes that grow on the external surfaces and gills and in the viscera are adapted to growing at relatively low temperatures and cause rapid spoilage.

Other factors specific to the biology of aquatic animals cause the muscle tissue to be in less than prime condition when harvested. These biological factors make proper refrigeration following capture and freezing critical for maintaining product quality. Salmon, for example, are commonly captured as they return from the ocean to spawn in a freshwater stream, often many hundreds of miles inland. In this case, the fish have stopped eating, and have also had to physiologically "readapt" to swimming in freshwater. The fish must mobilize their energy reserves (adipose fat, muscle fat, and muscle protein) for migration as well for producing roe (eggs) or milt (sperm). At a certain point during the spawning process, the salmon flesh becomes pale, soft, and flavorless. The severity of this problem is species, gender, and run dependent.

Refrigeration during storage is also required to control the production of histamine in scombroid-toxin–forming species such as tuna. Here, fish are to be held at temperatures in refrigerated brine or on ice at 40°F (4.4°C) within 12 hours of death (six hours for fish harvested in warm waters or where ambient temperature is 85°F or higher. Larger tuna are to be chilled to an internal temperature of 50°F (10°C) within six hours of death, with a recommendation that all scombroid-forming fish at receiving at a processing facility reach an internal temperature of 40°F or less (FDA 2001).

RIGOR

Rigor begins (onset of rigor) in fish within one to two hours, depending upon species and temperature. Onset of rigor is temperature dependent and occurs sooner at higher temperatures. Extremely large fish, such as bluefin tuna, weighing several hundred pounds go through rigor slowly like other large animals. As a comparison, the onset of rigor in beef muscle is within 10–24 hours postmortem at room temperature, in chicken in 2–4 hours, and in whale muscle in 50 hours.

Fish pass through rigor within hours and are generally processed postrigor. Fish should pass through rigor (resolution of rigor) before fillets are frozen to avoid toughening and shrinkage and to reduce drip loss when the product is thawed out and used (thaw rigor). One exception to processing postrigor is for certain at-sea longline processors that process high-value fish prerigor, freezing fish within two or three hours of harvest. Another exception is in aquaculture, where fish are often processed prerigor.

Fish must be carefully handled postrigor, since rough handling can tear the muscle tissue and cause the myotomes to separate. This phenomenon is called gaping. Gaping is an important quality consideration in finfish harvested from cold waters. Gaping is most prevalent in fish allowed to pass through rigor at elevated temperatures (> 17°C). Other manifestations of rough handling are discoloration and softening as a result of bruising, caused by rupturing blood vessels within the muscle tissue; and fractures of the vertebrae, which introduce blood spots into the muscle tissue.

EFFECT OF TEMPERATURE ON FISH MUSCLE CONDITION POSTMORTEM

Controlling the temperature of muscle foods is important for maintaining quality during storage. Muscle fibers contract postmortem at physiological temperatures. However, the amount of contraction decreases and is lowest around 10–20°C. At temperatures lower than 10°C, muscle contraction increases again. Contraction of muscle fibers at low temperatures causes the quality defect of cold shortening that makes muscle tissue tough. Controlled chilling is used to cool carcasses of beef and lamb to 10°C, but not lower, during the first 10 hours after slaughter because of the susceptibility of these two species to cold shortening. Pork muscle is less affected by cold shortening than beef or lamb, and chicken proceeds through rigor rapidly, so chilling has little effect. Fish muscle, with the exception of that from large pelagic species, is not highly susceptible to cold shortening. Cold shortening occurs in prerigor muscle because the sarcoplasmic reticulum cannot efficiently store calcium ions at lower temperatures. This inability to efficiently sequester calcium occurs when prerigor muscle tissue is chilled below 10°C before the pH has dropped to approximately 6.0, a point where the muscle fibers are no longer excitable and contraction no longer occurs.

A related problem is thaw shortening, which occurs when meat is frozen prerigor and then thawed rapidly. Because adenosine triphosphate (ATP) is not depleted in the muscle cells if the tissue is frozen prerigor, the muscle fibers contract rapidly during thawing, releasing large amounts of tissue fluids (drip loss), with accompanying toughening.

QUALITY CHANGES DURING COLD STORAGE

Aquatic food products deteriorate more rapidly than other muscle food products. Unlike muscle foods from terrestrial animals, aquatic muscle foods are from vertebrates (finfish) and invertebrates (crustaceans, mollusks) that are both cold blooded, and for the most part, cold adapted. Many fish, crustaceans, and mollusks are harvested from waters that are less than 10°C, and some from waters as low as −1°C! Even fish from tropical areas are harvested from waters that rarely exceed 20°C. Endogenous proteolytic enzymes and lipases in poikilothermic organisms "naturally" work at refrigeration temperatures; therefore, refrigeration does little or nothing to slow the rate of deteriorative biochemical reactions that occur in the tissues of these animals. Similarly any microbes, including spoilage flora associated with these animals, are also cold adapted and continue to grow at refrigeration temperatures. Many of the bacteria found on the surface of finfish or in the visceral cavities of finfish or mollusks have highly active proteases that cause off flavors and odors and cause the flesh to soften and discolor. There is often very little that can be done to control these deteriorative reactions.

The biochemical composition of aquatic food products also influences how quickly these foods deteriorate. Oysters, for example, can have high quantities of glycogen, which make the product taste sweet. However, bacterial or endogenous enzymes can rapidly deplete glycogen, causing the oysters to become sour during storage.

Fatty fish, particularly salmon, and pelagic fish, such as herring or mackerel, have relatively high concentrations of polyunsaturated fatty acids compared with terrestrial animals, and these fatty acids are very susceptible to oxidation. Oxidation of these fatty acids can be initiated prior to freezing by endogenous or microbial enzymes. Unfortunately, freezing does not stop lipid oxidation, and aquatic food products can become highly oxidized during frozen storage, with the development of rancid, fishy flavors and discoloration (Foegeding et al. 1996). Fish tissues have lipases and phospholipases that remain active at frozen storage temperature, yielding free fatty acids during enzymic hydrolysis. These free fatty acids are more susceptible to lipid oxidation than the native triglycerides. Although the flavor changes associated with lipid oxidation may be more pronounced with high fat fish, flavor changes also occur with lean fish as a result of oxidation of cell membrane lipids. Because of this problem with lipid oxidation, packaging fish to limit contact of the product with oxygen and ultraviolet light, even for fish with low lipid content, is critical.

Certain fish have novel ways of cycling nitrogen that can lead to quality problems during frozen storage. For example, gadoid species (cod, haddock, Alaska or walleye pollack, and hake) have high levels of trimethylamine. Trimethylamine (TMA) is enzymatically converted to trimethylamine oxide (TMAO), then to dimethylamine (DMA) and formaldehyde. The formaldehyde cross-links muscle proteins, leading to tough, dry tissue. Species of elasmobranch fish (sharks and rays) contain high levels of urea. If shark are not properly refrigerated after harvest, or if steps are not taken to remove or neutralize any ammonia that may have formed during storage, the meat can have ammonia or urine-like flavor defects.

In other fish, the decarboxylation of the amino acid histidine to histamine by bacterial enzymes presents a food safety problem (FDA 2001). Certain fish species, such as tuna and mahi-mahi, have high levels of free histidine. Histamine sensitivity can be fatal for susceptible individuals: the condition is known as scombroid poisoning.

PROCESSING STAGE 1: REFRIGERATION

Any aquatic food product should be refrigerated or cooled on ice as soon as possible after harvest. Live mollusks should be placed in refrigerated seawater, held in cold storage at 10°C or lower, or be placed in saltwater ice. Live marine mollusks can be placed *ON THE SURFACE* of freshwater ice; however, placing live marine mollusks *in* freshwater ice will kill them. Mollusks can remain alive under these conditions for five days or more.

For storage of fish at 15°C or less, the following systems are used: crushed ice, slush ice [water ice dispersed in water alone or in water containing additives (e.g., salt, organic acids, antimicrobials, sugar)], champagne ice (slush ice with gaseous carbon dioxide), and mechanical refrigeration. Landed fish are most commonly iced, and fish must be iced as soon as possible after landing. Theoretically, one pound of ice with a heat of fusion of 144 BTU/pound can reduce the temperature of several pounds of fish to near 0°F; however, factors such as air and water temperature, insulation in the fish hold, and time the fish must be held usually require that one to two pounds of ice be used for each pound of fish (Pigott and Tucker 1990). Use of refrigerated seawater (RSW) is probably the second most common method for cooling fish. This involves circulating clean seawater through refrigerated coils to temperatures near freezing. RSW has the advantage of placing less mechanical pressure on the fish and therefore causing less structural damage than stacking the fish between layers of ice. RSW can also reduce temperature more quickly and reduce the level of contamination coming onto the fish compared with that from melting ice. The problem with RSW is the increase in salt content in the fish muscle, which makes it unacceptable for some fresh markets.

Fish have a very limited refrigerated shelf life (Table 27.1). Eviscerated ("dressed") cod, other whitefish, and salmon have a shelf life of a week or less at 4°C, but fatty fish such as intact ("round") mackerel or herring should be stored no longer than a couple of days. The shelf life can be extended significantly by superchilling: tightly controlled storage conditions at lower temperatures. This technique involves holding the product at 0 to −1°C with variations of holding temperature of less than ± 0.5°C. Most fish muscle does not freeze above −2°C.

Vacuum packaging also increases the shelf life of certain products. Storage of chilled, vacuum-

Table 27.1. Refrigerated Shelf Life of Fresh and Cured Fish

Products	Approximate Days Remaining in Good Condition	
	32°F	60°F
Cod, fresh	14	1
Salmon, fresh	12	1
Halibut, fresh	14	1
Finnan haddie	28	2
Kippers	28	2
Herring, salted	1 yr	3–4 mo
Cod, dried salted	1 yr	4–6 mo

Source: Adapted from Pigott and Tucker 1990

packaged meats, including smoked fish, for up to 10 weeks is possible at 0°C. The primary concern with seafood products is the growth of *Clostridium botulinum* type E in vacuum-packaged products. This organism can grow at refrigeration temperature [38°F (3°C)] and relatively high concentrations of water-phase salt (4.5–6%) (FDA 2001).

PROCESSING STAGE 2: FREEZING

Freezing muscle foods permits storage for one year or longer at −20°C, assuming that temperature fluctuations in the storage freezer can be controlled (Table 27.2). The objective is to freeze products as rapidly as possible, forming small intracellular ice crystals.

Rapid freezing is required for aquatic food products, even more so than for muscle tissue from terrestrial animals. Muscle proteins in fish are less tolerant to changes in the ionic strength of intracellular fluids that occur during freezing than other types of muscle food. Freezing damages fish tissue. Small intracellular ice crystals form in rapidly frozen samples and create less visible tissue damage. In slowly frozen samples, large intracellular ice crystals, which rupture cell membranes, are formed.

Fish muscle myotomes are more susceptible to mechanical damage during freezing than the muscle tissue of terrestrial animals. This is due in part to the orientation of the myotomes and to the relatively weak connective structures that hold them together. Rapid freezing is also critical for maintaining the quality of other aquatic food animals that are frozen whole, including shrimp, lobster, and molluscan

Table 27.2. Practical Storage Life for Aquatic Foods (Months)

	Temperature		
	−12°C/10°F	−18°C/0°F	−24°C/−12°F
Fatty fish, glazed	3	5	> 9
Lean fish	4	9	> 12
Lean fish fillets	—	6	9
Lobster, crab, shrimp in shell	4	6	> 12
Shrimp, cooked peeled	2	5	> 9
Clams, oysters	4	6	> 9

Source: Adapted from Institut International du Froid 1986.

shellfish. These animals are often frozen without evisceration, so it is critical to freeze tissue rapidly, with as little damage as possible, so that digestive enzymes are not released into the flesh, since the visceral enzymes in these animals will remain active at low temperatures. After the food has been frozen, it must be protected with a glaze, or with packaging materials that limit surface dehydration of the product and exclude light.

FREEZING METHODS

Freezing and onboard refrigeration have made it possible to expand commercial fishing to new species that were not widely utilized until the late 1970s. The development of a factory trawler fleet and growth of whitefish fisheries around the world for surimi, fillet, and fish block production would not be possible without the ability to harvest tons of fish at a time and keep them in refrigerated seawater storage until the fish can be processed on board. Without recent developments in freezing technology, it would not be possible to hold the millions of pounds of frozen, processed product on board ship until it can be delivered hundreds of miles to shore, and from there to consumers.

Different freezing methods are employed in seafood production. Some of these are outlined in Table 27.3; also shown are common temperature and air velocity parameters for freezing different food products. Most aquatic food products are blast frozen, or frozen under conditions where the air velocity during refrigeration is very high. These freezers include air blast freezers, in which product is packaged and placed upon shelves inside a chamber. Very cold air at high velocity is blown around the chamber by powerful fans near the ceiling. After the product is frozen, it is removed from the blast freezer and placed in a storage freezer.

Sometimes large fish, such as salmon, are frozen in a blast freezer without being packaged first. In this case, the fish are frozen; removed from the freezer; glazed with water or a mixture of water, sugar, and possibly other additives; and packed for storage.

Contact-plate freezers are commonly used for freezing products that can be marketed as uniform slabs, such as blocks of fish fillets, fish mince, fish roe, and surimi. Plate freezing is used on factory processors because it is compact, efficient, and has relatively low operating costs. In a contact-plate freezer, the product is placed in a rigid pan between two large metal plates that contain circulating refrig-

Table 27.3. Freezing Methods for Fish

Product	Freezer Type	T (°F)	Air Velocity (m/s)
Fish, bulk	Air blast/batch	−30 to −40	17
	Air blast/continuous	−40	
Fish	Tunnel	−30	
	Plate	−30 to −50	
Fish	Cryogenic		
	Nitrogen	−196	
	Carbon dioxide	−78.5	

Source: Anonymous 1986.

erant. These plates are pressed down upon the product as it freezes. Plate freezing is *required* for products like fillet block, mince/block, and mince that are used for sandwich portions, fish sticks, or nuggets. In these cases, blocks must have very uniform dimensions because the secondary manufacturer cuts the block into portions of uniform size and weight.

Plate-frozen products are frozen in aluminum pans of very specific dimensions. These pans are lined with coated, paperboard, block liners that are folded to fit inside the freezer pan. The fish product is arranged inside the liner, and the lid of the liner is folded over and closed. The product is packed by weight. These pans are placed into a contact-plate freezer. Commercial freezers on ships can be 10–12 plates and contain dozens of blocks per layer. It takes approximately 2–2.5 hours to freeze a 7.7 kg block of fish in a commercial plate freezer ($-28°F$).

Cryogenic freezing, immersion freezing in liquid nitrogen or a carbon dioxide "snow," is a popular method for freezing high-value items such as shrimp and molluscan shellfish. The freezing rate is extremely rapid, and for some products, this can cause the food to crack or split. Carbon dioxide forms a snow on the food and then sublimes. Carbon dioxide is often preferred, since there is less thermal shock than with liquid nitrogen, and less physical damage to the product. For seafood, the product is placed on a conveyor and passed through a carbon dioxide snow. For nitrogen freezing systems, the product is cooled with gaseous nitrogen before the liquid nitrogen is sprayed on it. After the product is frozen, it is packaged in plastic and allowed to equilibrate to the frozen storage temperature before it is transferred to a storage freezer. These products are generally glazed. Often vacuum packaging is used.

GLAZING

To extend the shelf life of frozen whole, dressed fish, fillets, whole shrimp, or mollusks, a glaze is often applied. Glazing involves dipping or spraying water or an aqueous solution on the product after the surface has been frozen. Sometimes a cryoprotectant such as fructose, sucrose, or sorbitol; an antioxidant such as ascorbic acid; or a thickening agent (e.g., alginate) is added to the glaze. Levels of glaze on whole fish can be as high as 9% by weight. The glaze sublimes during frozen storage, protecting the product from surface dehydration, or freezer burn. The glaze also keeps oxygen from migrating into the food, thus limiting lipid oxidation. The problem with glazing is that glaze is fragile and can break if the product is bumped or dropped. Glaze fracture is more of a problem with large fish. Glaze fracture exposes product surface to dehydration and is unattractive. The presence of a good glaze on seafood is positive factor; however, to prevent economic fraud, seafood products are sold by weight after the glaze has been removed (or weight net of glaze).

PACKAGING

Proper packaging of fish products is necessary to maintain quality: it minimizes water loss or weight loss in the finished product; changes to texture; and loss or change in product flavor, color, or appearance. Furthermore, proper packaging will maintain high nutritional value, for example limiting the loss of omega-3 fatty acids through oxidation. Packaging also limits contamination during distribution and transit and protects product from mechanical damage that would affect its appearance and final market value. Examples of this include scale loss, shell breakage in mollusks or crustaceans, loss of appendages (whole shell on shrimp, crab, lobster), and product fracture (mentaiko or dyed pollack roe skeins, shrimp).

A wide variety of packaging materials is used for frozen aquatic food products. For frozen fish fillets, headed and gutted Pacific salmon, and frozen glazed crab, the product is commonly loosely wrapped in plastic and placed inside a cardboard carton for shipment to distribution centers. Cartons weigh from 40 to 800 pounds or more. Shrimp, individually quick-frozen fillets, and breaded products are commonly marketed in heat-sealed plastic bags. Some large products, such as whole tuna, are not packaged at all. Certain traditional foods such as uni (sea urchin roe brined and treated with alum) and sujiko (brined, colored, whole skeins of salmon roe) are still marketed in small wooden boxes.

Freezing packages can cause consternation and confusion on the part of regulators and the general public. Many consumers believe that all product in cans is shelf stable. However, frozen Dungeness crab meat (muscle removed from cooked crab) and razor clams are still packaged in cans with double-seamed metal ends. Although these containers are clearly labeled "keep refrigerated" or "keep frozen," thermal abuse is possible. As a result, plastic containers are replacing the cans because of food safety concerns. Similarly, salmon, lumpfish, and sturgeon caviars are packaged in glass jars with metal, lug-type closures. Products in these containers are distributed as frozen foods. People mistakenly think that these products are also shelf stable, and the "keep refrigerated" labeling is commonly ignored at retail and by the consumer.

FINISHED PRODUCT

One of the earliest food patents was issued in 1842 for refrigerated fish. However, mechanical refrigeration/freezing did not become a significant factor for the preservation of aquatic food products until the early 1950s. The development of shipboard refrigeration and freezing systems made high seas fisheries possible, by permitting vessels to harvest finfish and crustaceans from distant areas, and bring these aquatic food products to shore-based processing facilities and distribution centers. In a similar manner, the development of practical freezing technologies and refrigerated/frozen transportation systems permitted shore plants to be constructed near fishing grounds and to serve worldwide markets.

The largest volume of frozen aquatic products consists of frozen fillet blocks of whitefish such as cod, perch, and haddock that are subsequently batter coated or batter/breaded and converted to "fish sticks," "fishwiches," or other low-value products. High-value products such as fresh or frozen salmon, block-frozen or IQF (individually quick frozen) fish fillets or shrimp, frozen lobster tails, roe products such as mentaiko (pollack roe) and kozunoko (herring roe), batter-coated or batter/breaded shrimp, and a broad selection frozen shellfish are now relatively common.

The recent rapid international expansion of aquaculture worldwide now provides higher quality and less expensive aquatic foods to consumers throughout the year. Important cultured species including salmonids (Atlantic and Pacific salmons, rainbow trout), catfish, tilapia, sea bream, halibut, eels, sole/flounder, striped bass, molluscan shellfish, shrimp, and sea vegetables (e.g., nori, the common covering for sushi rolls) are now commonly available all over the world at any time. This would not be possible if it had not been for the development of practical refrigerated/frozen processing and transportation. Unfortunately *high quality* frozen or refrigerated (fresh) aquatic foods are too often unavailable because of poor handling, poor processing, or inadequate temperature control. This is still a problem that plagues the industry.

Refrigeration and freezing also made it possible to introduce new and extremely valuable products into commerce, for example caviar and fish roe products. The U.S. retail market price for Beluga is around \$200.00 per ounce. Caviar products are cured with salt, but with few exceptions, refrigeration or freezing is required to maintain product safety and quality. Other extremely valuable aquatic food products that would not be available without freezing include king crab with the shell on, giant prawns, magaro (sashimi tuna or tuna to be consumed raw), and lox [lightly salt cured, cold-smoked (effectively raw) salmon].

APPLICATIONS OF PROCESSING PRINCIPLES

PROCESSING FROZEN FISH FILLET BLOCKS

Harvest

"White fish" such as cod, pollack, or whiting, are harvested on the high seas by trawl and held on board in refrigerated seawater until they have gone through rigor. Some Pacific cod is harvested by longline, headed and gutted, and frozen shipboard within 2.5 hours at $-20°F$ in a prerigor state. Product frozen prerigor is preferred in the Japanese market.

Warm water cultured fish such as catfish or tilapia are collected from ponds and stunned by dropping the water temperature. Another way of stunning the fish is to place them into water saturated with carbon dioxide ≥ 600 ppm). After this, the fish are bled by removing the gill rakers or by cutting the vein above the heart, allowing the heart to still function and pump the blood out of the body. The fish are then placed in circulating ice water for 5–20 minutes for complete removal of the blood. These fish are then further processed and frozen.

Clearly, the postharvest stress in cultured fish is less, since there is limited struggle when they are harvested. Cultured fish are generally fasted for a couple of days before harvest, so the metabolic activity of the digestive enzymes is lower. This improves muscle quality and enhances shelf life. Seasonal variation in wild-caught fish can be a serious problem, and limits product quality. For Alaska pollack, the fish harvested during breeding season have poorer quality flesh than fish harvested later in the year. Fish are allowed to pass through rigor before they are processed into frozen products. Processing for block-frozen fillets, a common product form for pollack, is outlined in Table 27.4.

Table 27.4. Processing Frozen Fish Fillet Blocks

Process Operation	Controls to Maintain Quality and Safety
Harvest	Reduce post harvest stress, control for seasonal variation, and maintain product integrity prerigor.
Holding	Monitor progression of fish through rigor; control temperature, and maintain good sanitation to minimize adverse biochemical and microbial changes.
Eviscerating	Control deleterious enzyme reactions [endogeneous and microbial] by maintaining good sanitation, avoiding cross contamination, and keeping product cold.
Filleting	Reduce mechanical damage to cell structure through good manufacturing practices.
Freezing	Control ice crystal formation; reduce opportunities for deleterious chemical and biochemical changes by rapid freezing and proper packaging.
Frozen Storage	Reduce problems with chemical changes to product by controlling ice crystal growth; hold product at lowest practical temperature and keep temperature fluctuation in cold storage to a minimum.

Holding

Fish are held at 4–10°C until rigor has resolved. For the best quality product, fish should be processed as soon after resolution of rigor as possible. Because these animals and their accompanying microflora are adapted to cooler temperatures, deleterious biochemical reactions can occur quickly in fish. Fish generally pass through rigor "whole" and still retain visceral enzymes.

Eviscerating

Care must be taken to ensure that butchering operations are as clean and sanitary as possible to avoid cross contamination between viscera and meat. Eviscerating must be conducted under cool conditions. Often, fish processing facilities are kept at 45–50°F to maintain product quality.

Filleting

Both mechanical and hand labor are employed for filleting fish, depending upon the size of the operation and labor costs. Commercial filleting and skinning machines process hundreds of fish per hour, and this is how pollack fillets are produced on an at-sea processing vessel. Machines can be set to maximize recovery of the flesh, or to recover predominantly "light" muscle only. A deep-skinned fillet is one in which the dark muscle along the lateral line, just underneath the skin, has been removed. This tissue is darker, has a high fat content, a high concentration of heme iron, and a stronger flavor. Because the dark muscle can oxidize readily, this can result in flavor problems with the finished product.

Freezing

Rapid freezing is critical for fish fillets, to limit the formation of large intracellular ice crystals. Contact-plate freezers are used for frozen block, but a tunnel freezer (blast freezer) operated as a batch or continuous system could also be used successfully if product were first formed into blocks in frames specially designed for this purpose. Chemical changes, specifically lipid oxidation, occur in the tissue of fish such as pollack during frozen storage. Even though the lipid content of "white fish" is less than 1%, the membrane lipids are susceptible to oxidation. This oxidation can lead to stale and rancid off flavors. "Fishy" off flavors are a result of microbial decomposition occurring *before* the fish were frozen. Gadoid fish, including Atlantic and Pacific cod, hakes, and haddock, contain high levels of trimethylamine oxide (TMAO). This compound is broken down by enzymes that are active during frozen storage, causing proteins in the muscle to cross-link and cause toughening. These enzymes are more active when the tissue has been damaged, which is another reason careful freezing is important.

Frozen Storage

Frozen storage temperature must be carefully controlled to limit ice crystal growth and water migration in the frozen fish tissue. Wide fluctuations in storage temperature enhance the rate of deleterious chemical and biochemical reactions in the fish, leading to the development of off flavors. For other products, poor frozen storage conditions result in the liberation of proteolytic and lipolytic enzymes from the viscera, which causes loss of quality during storage and after the product is thawed.

Maintaining package integrity and high quality frozen storage conditions are important for maintaining product quality as well as retaining maximal economic return for the product. Improper freezing and frozen storage can lead to a loss of 5% or more in finished product weight. Besides loss of value, product labeling must reflect the proper net weight. If product is below the stated label weight, fraud is inferred, and the product is technically misbranded.

CRYOGENIC FREEZING OF SHRIMP

Harvest

Shrimp sold in the United States are from both wild harvest and culture fisheries. Aquaculture is rapidly replacing wild harvest as a consistent high quality source. The culturist delivers shrimp to the processing plant, alive but generally iced. The quality of the shrimp and the number of dead shrimp in the shipment are checked. A large number of dead shrimp in a shipment indicates harvest stress and a greater likelihood of quality problems, such as soft texture, after the product is frozen. The diet of the shrimp can affect product color, flavor, and storage quality, and this is closely monitored in good culture operations. Because of the potential risk of contamination with food-borne pathogens, unapproved aquaculture drugs, and potential environmental contaminants, tests of incoming shipments are routinely conducted by processing operations. Shrimp captured by wild harvest are held on ice or in RSW. The highest quality products, such as large Alaska spot prawns, may be held live in tanks and transported live to restaurants instead of being frozen.

A process description for cryogenic freezing of shrimp is presented in Table 27.5.

Holding

Shrimp are processed pre- and postrigor. Temperature control during the holding step prior to processing is critical. In many of the tropical areas where shrimp are cultured, there is little available refrigeration, and ice is scarce. Product quality is highly variable.

Some shrimp from wild harvest are also treated with sulfating agents after harvest to control "black spot," which results from an enzyme reaction in the shrimp tissue. Addition levels up to 10 ppm in the finished product are permissible, but use must be labeled (FDA 2001).

Peeling

Shrimp are processed either raw or cooked. Because of the amount of hand labor at some processing facilities, good manufacturing practices and sanitation are key considerations.

There are several common product forms for raw frozen shrimp. Sometimes whole head-on shrimp

Table 27.5. Cryogenic Freezing of Shrimp

Process Operation	Controls to Maintain Quality and Safety
Harvest	Reduce post harvest stress; control diet to maintain desired product flavor profile; control sources of contamination that could jeopardize product safety; maintain product integrity prerigor.
Holding	Monitor progression of animal through rigor; control temperature and maintain good sanitation to minimize adverse biochemical and microbial changes.
Peeling	Control contamination and cross contamination through good manufacturing practices to limit microbial contamination through handling; control microbial growth and deleterious biochemical reactions by keeping the product cool.
Freezing	Control ice crystal formation; reduce opportunities for deleterious chemical and biochemical changes by rapid freezing and proper packaging.
Glazing	Glaze product to limit freezer burn and oxidative changes that could occur during frozen storage.
Packaging	Chemical and biochemical changes.
Frozen storage	Reduce problems with chemical changes to product by controlling ice crystal growth; hold product at lowest practical temperature and keep temperature fluctuation in cold storage to a minimum.

are frozen. However, for the largest volume of frozen raw shrimp, the head is broken off by hand or mechanically. Visceral material near the head is removed. Product may be sold with the exoskeleton, or shell, on or off. In some cases, all of the shell except the small tail fan is removed.

Following removal of the exoskeleton, the "vein," or digestive tract, of the shrimp is commonly removed. However, it is also common to leave the vein "in," or to remove the vein by making an incision in the exoskeleton to remove the vein, yielding a deveined shell-on shrimp.

Products are often cooked, yielding a ready-to-eat product, or coated with a batter or breaded.

Freezing

Individually quick-frozen shrimp are commonly frozen in carbon dioxide snow in South American plants, and in spiral blast freezers in Asian facilities. Each type of freezing can produce an excellent product, and freezing rate is rapid.

Glazing

Shrimp are glazed with a spray of water after freezing. Shrimp may also be treated with a phosphate-containing dip, prior to or during the glazing step, to improve water retention. After the glaze sets, the shrimp are commonly packaged in plastic barrier film bags of various types, then in a cardboard master case.

Packaging

Cryogenic frozen IQF shrimp are packaged in plastic film. However, the most common product form for shrimp is a 2 or 5 kg frozen blocks. Freezing shrimp in a frozen block keeps the exposed surface of the product to a minimum and limits surface dehydration; it also tends to protect the delicate individual shrimp from physical damage.

Frozen Storage

Well-controlled cold storage is critical for maintaining the quality of high-value products such as IQF shrimp. Fluctuating frozen storage temperatures will result in loss of glaze and surface dehydration. Also, for vein-in shrimp, digestive enzymes could migrate from the vein into damaged muscle tissue and cause decomposition. Ice crystal growth results from temperature fluctuations, as follows: As the temperature in the storage freezer rises, water from small ice crystals melts; as the temperature drops again when the freezer cycles, this water will freeze onto the surface of an ice crystal, making it larger. As the freezer temperature continues to fluctuate, the smaller ice crystals gradually disappear. In their place are a smaller number of large ice crystals. These large crystals cause tissue damage. Storage freezer temperatures are preferably at $-20°C$ although, maintaining this storage temperature is not always possible.

GLOSSARY: ACRONYMS

ATP—adenosine triphosphate.
DMA—dimethylamine.
FAO—United Nations Food and Agriculture Organization.
FDA—U.S. Federal Drug Administration.
GAO—U.S. Government Accounting Office.
IIF—Intitut International du Froid.
IQF—individually quick frozen.
NAS—U.S. National Academy of Sciences.
RSW—refrigerated seawater.
TMA—trimethylamine.
TMAO—trimethylamine oxide.

REFERENCES

Bledsoe GE, CD Bledsoe, BA Rasco. 2003. Caviars and fish roe products. Critical Reviews in Food Science and Nutrition. 43(3): 317–356.

Food and Agriculture Organization (FAO) 2002. Fishery Statistics, Catches and Landings. United Nations, FAO, Rome. National Marine Fisheries Service, Fisheries Statistics and Economics Division, 2000. U.S. Department of Commerce, Silver Springs, Md. http://www.st.nmfs.gov.

Food and Drug Administration (FDA). 2001. Fish and Fishery Products Hazards and Control Guide, 3rd edition. Office of Seafood, Center for Food Safety and Applied Nutrition, FDA, Public Health Service, Department of Health and Human Services, Washington, D.C.

Foegeding EA, TC Lanier, HO Hultin. 1996. Characteristics of edible muscle tissues. In: O Fennema, editor. Food Chemistry, 3rd edition, 879–942. Marcel Dekker, Inc., New York.

General Accounting Office (GAO). 2001. Food Safety. Federal Oversight of Seafood Does Not Sufficiently Protect Consumers. U.S. GAO. Report to the Committee on Agriculture, Nutrition and Forestry, U.S. Senate, Washington, D.C. GAO-01-204.

Institut International du Froid (IIF). 1986.
Recommandations pour la Preparation et la
Distribution des Aliments Congeles, 3rd edition.
[Recommendations for the Processing and
Handling of Frozen Foods.] IIF, 177 Boulevard
Malesherbes, F-75017, Paris, France.

Johnson H. 1998. Annual Report on the United States
Seafood Industry, 6th edition. H.M. Johnson and
Associates, Bellevue, Wash.

National Academy of Sciences (NAS). 2003.
Scientific Criteria to Ensure Safe Food. National
Academy of Sciences, Washington, D.C.

Pigott GM, B Tucker. 1990. Chapter 6. Controlling
water activity. In: Seafood—Effects of Technology
on Nutrition, 151. Marcel Dekker, Inc., New York.

28
Seafood: Processing, Basic Sanitation Practices

P. Stanfield

BACKGROUND INFORMATION

The U.S. national regulatory authority for public protection and seafood regulation is vested in the Food and Drug Administration (FDA). The FDA operates an oversight compliance program for fishery products under which responsibility for the product's safety, wholesomeness, identity, and economic integrity rests with the processor or importer, who must comply with regulations promulgated by the FDA. In addition, the FDA operates a low-acid canned food (LACF) program, which is based on the hazard analysis critical control point (HACCP) concept and is focused on thermally processed, commercially sterile foods, including seafood such as canned tuna and salmon.

The seafood processing regulations, which became effective on December 18, 1997, require that a seafood processing plant (domestic and exporting foreign countries) implement a preventive system of food safety controls known as a hazard analysis crit-

---

Most data provided in this chapter have been modified from a document published and copyrighted by Science Technology System, West Sacramento, California. ©2002. Used with permission.

ical control point (HACCP) plan. A HACCP plan essentially involves (1) identifying food safety hazards that, in the absence of controls, are reasonably likely to occur in the products; and (2) having controls at "critical control points" in the processing operations to eliminate or minimize the likelihood that the identified hazard will occur. These are the kinds of measures that prudent processors already take. A HACCP plan provides a systematic way of taking those measures that demonstrates to the FDA, customers, and consumers that the firm is routinely practicing food safety by design. Seafood processors that have fully operating HACCP systems advise us that they benefit in several ways, including having a more safety-oriented workforce, less product waste, and generally, fewer problems.

Most FDA in-plant inspections consider product safety, plant/food hygiene, and economic fraud issues, while other inspections address subsets of these compliance concerns. Samples may be taken during FDA inspections in accordance with the agency's annual compliance programs and operational plans or because of concerns raised during individual inspections. The FDA has laboratories around the country to analyze samples taken by its investigators. These analyses are for a vast array of defects including chemical contaminants, decomposition, net weight, radionuclides, various microbial pathogens, food and color additives, drugs, pesticides, filth and marine toxins such as paralytic shellfish poison (PSP), and domoic acid.

In addition, the FDA has the authority to detain or temporarily hold food being imported into the United States while it determines if the product is misbranded or adulterated. The FDA receives notice of every seafood entry, and at its option, conducts wharf examinations, collects and analyzes samples, and where appropriate, detains individual shipments or invokes "Automatic Detention," requiring private or source country analysis of every shipment of product when recurring problems are found, before the product is allowed entry.

Further, the FDA has the authority to set tolerances in food for natural and man-made contaminants, except for pesticides, which are set by the Environmental Protection Agency (EPA). The FDA regulates the use of food and color additives in seafood and feed additives and drugs in aquaculture. The FDA also has the authority to promulgate regulations for food plant sanitation [i.e., good manufacturing practices (GMP) regulations], standards of identity, and common or usual names for food products.

The FDA has the authority to take legal action against adulterated and misbranded seafood and to recommend criminal prosecution or injunction of responsible firms and individuals.

The FDA conducts both mandatory surveillance and enforcement inspections of domestic seafood harvesters, growers, wholesalers, warehouses, carriers, and processors. The frequency of inspection is at the agency's discretion, and firms are required to submit to these inspections, which are backed by federal statutes containing both criminal and civil penalties.

The FDA provides financial support, by contract, to state regulatory agencies, for the inspection of food plants including those for seafood.

The FDA also operates two other specific regulatory programs directed at seafood—the Salmon Control Plan (SCP) and the National Shellfish Sanitation Program (NSSP), recently augmented by the Interstate Shellfish Sanitation Conference (ISSC). These are voluntary programs involving the individual states and the industry.

The Salmon Control Plan is a voluntary, cooperative program involving the industry, the FDA, and the National Food Processors Association (NFPA). The plan is designed to provide control over processing and plant sanitation and to address concerns about decomposition in the salmon canning industry.

Consumer concerns about molluscan shellfish are addressed through the NSSP. It is administered by the FDA and provides for the sanitary harvest and production of fresh and frozen molluscan shellfish (oysters, clams, and mussels). Participants include the 23 coastal shellfish-producing states and nine foreign countries.

The NSSP was created upon public health principles and controls formulated at the original conference on shellfish sanitation called by the Surgeon General of the U.S. Public Health Service in 1925. These fundamental components have evolved into the National Shellfish Sanitation Program Manual of Operations. A prime control is proper evaluation and control of harvest waters and a system of product identification, which enables trace-back to harvest waters.

The FDA conducts reviews of foreign and domestic molluscan shellfish safety programs. Foreign reviews are conducted under a memorandum of understanding (MOU), which the FDA negotiates with each foreign government to assure that molluscan shellfish products exported to the United States are acceptable.

The FDA's regulations on HACCP for seafood processing have been in full force since 1997. HACCP, in addition to other scientific and technical considerations, is an extension of the basics of food processing sanitation that uses the FDA's current good manufacturing practice regulations (CGMPR) and the Food Code as frames of references. The FDA considers such sanitation compliance prerequisite to HACCP planning and implementation.

This chapter discusses those prerequisites of basic sanitation for seafood processing. If you are a seafood processor and you are planning to start the HACCP program, you must first examine the current practices of your operation to ascertain that it complies with such prerequisites.

The information presented in this chapter has been modified from the CGMPR of the FDA and the USDA, the Food Code, and other documents issued by the FDA on inspection of seafood processing plants.

The format and style used in this chapter reflects the instructional process between a teacher (e.g., a training supervisor) and a student (e.g., a company personnel).

FRESH AND FROZEN FISH

SANITATION CRITICAL FACTORS

The critical factors to remember when a company officer performs a sanitation inspection of a processing plant for fresh and frozen fish are:

- Look for evidence of rodents, insects, birds, or pets within the plant.
- Observe employee practices including hygienic practices, cleanliness of clothing.
- Check to see if there are proper strength hand-dip solutions.
- Check to see if fish are inspected upon receipt and during processing for decomposition, off odor, parasites, and so on.
- Check for decomposition and parasites during an establishment inspection (EI).
- Ascertain if equipment is washed and sanitized during the day and at the beginning and end of the daily production cycle.
- Check if the fish are washed with a vigorous spray after evisceration and periodically throughout the process prior to packaging.
- Determine the method and speed of freezing for frozen fish and fish products.

- Check use of rodenticides and insecticides to assure that no contamination occurs.
- Observe handling from boats to finished package and observe any significant objectionable conditions.

Specific details on the sanitation follow.

RAW MATERIALS

- Determine what tests are conducted on incoming fish for decomposition, parasites, chemical contamination, and so on.
- Determine disposition of incoming fish that have been found to be decomposed, contain excessive parasites, or contaminated with mercury, pesticides, and so on.
- Conduct organoleptic examination of incoming fish or fish products, especially those that have been thawed for processing or held for prolonged periods of time at room temperature during processing.

  Give attention to fish arriving at the plant, as to effectiveness of elimination of decomposed fish, and check fish actually being packed. Determine percentage of decomposed units encountered, classifying each as passable (class 1), decomposed, (class 2), or advanced decomposed (class 3).
- Examine susceptible fish for parasitic infestations, (e.g., whitefish, rosefish, tullibees, ciscos, inconnus, bluefish, herring, etc.).
- Check other raw materials and storage areas for rodents, insects, filth, or other contaminating factors.
- See required specification on other raw materials for bacterial load, and so on (e.g., received under a *Salmonella*-free certificate issued by a recognized government or private agency).
- Check for misuse of dangerous chemicals including insecticides and rodenticides.
- If fish is received directly from boats, see if a hook is used for loading and unloading, or for that matter, if a hook is used for any handling of the fish.

MANUFACTURING

- Study manufacturing procedure. Include flow plan.
- Study type of equipment used as to construction, materials, ease of cleaning, and so on.

- Observe equipment cleaning and sanitizing procedures, and evaluate their adequacy.
- Observe evisceration procedure, filleting procedure, or other butchering procedures used.
- Determine source of water used in operation. Check that only potable water from an approved source is used.
- If, during processing of fish, there are long delays at room temperature, check for decomposition.
- Examine all handling steps and intermediate steps in processing that could lead to the contamination of the fish with filth and/or bacteria.
- Study holding times and temperatures during the processing operation.
- If battering and/or breading fish is involved, check the process carefully. In addition, check times and temperature, and check for other possible routes of filth and/or microbial contamination.
- Evaluate compliance with good manufacturing practices.

CONTROLS

- Check coding system. If no code marks are used, mark suspect lot packages with fluorescent crayon for later sampling.
- Review records regarding finished product assay for decomposition, parasites, microbial load, pesticides, mercury, and other quality factors.
- Study labeling used on products.
- Check use of preservatives on fish or ice.

SUMMARY AND CHECKLIST

Check on:

- Compliance with CGMPR.
- Use of adequate and proper-strength hand and equipment-sanitizing solutions.
- Proper cleanup.
- Evidence of rodents, insects, birds, domestic animals, or any other source of contamination.

Use the following list of indicators of sanitation to make a valid assessment of the operations at different stages of the process flow.

Sanitation Indicators for Filleted Fish

Stage	Assessment Items
Receive (unload fish)	• Determine condition of the fish. (Acceptable or decomposed) • Separate work area.
Store	• Suitable storage area (sanitation). • Time/temperature (icing) (quality). • Separate work area.
Wash	• Remove surface slime and dirt (sanitation). • Use of potable water.
Fillet	• Personnel sanitation. • Equipment sanitation. • Separate work area.
Skin (either hand or machine)	• Personnel sanitation. • Equipment sanitation • Separate work area. Same area as fillet operation.
Rinse	• Potable water. • Equipment sanitation. • Time/temperature (quality).
Pack (either retail or block)	• Equipment sanitation. • Personnel sanitation. • Suitable packaging materials. • Time/temperature (quality). • Separate work area.
Freeze	• Time/temperature (quality)

CANNED TUNA

SANITATION CRITICAL FACTORS

During a sanitation inspection, use the following critical factors:

- Check adequacy of firm's controls and review records covering the receipt of tuna fish. Ascertain if only tuna below the mercury guidelines and not decomposed is processed. Determine disposition of decomposed or overtolerance tuna.
- Conduct organoleptic analysis of incoming tuna and of tuna being processed.
- Check food additives to determine that only those permitted by the standards are used.
- Check usage of insecticides and rodenticides to determine that they are used properly and do not become incidental food additives.
- Study controls over the canning operation to assure that only good quality tuna is canned and that it is canned in accordance with FDA requirements.

RAW MATERIALS

- Determine adequacy of firm's controls for assuring that decomposed tuna or tuna with excessive mercury is not being canned.
- Determine disposition of lots of tuna that are rejected because of excessive mercury.
- Review firm's assay records and controls regarding mercury analysis of raw, in-process, and finished canned tuna.
- Ascertain adequacy of controls the firm utilizes to assure that the species of tuna canned are those allowed by standards.
- Conduct organoleptic analysis of incoming raw tuna, frozen tuna that has been thawed for canning, and of any tuna being held for excessively long periods at room temperature.
- Determine disposition of any tuna that is found to be decomposed (destruction, diversion, etc.).
- Check raw material storage area for presence of insects, rodents, or other possible contaminants.
- Check food additives in storage to ascertain if they are allowed in canned tuna as per 21 CFR 161.190(a)—Canned Tuna Standards.
- Check firm's storage of rodenticides and insecticides to determine that they are used in accordance with instructions and are not becoming secondary food additives.

PROCESSING

- Check firm's can seamers to determine if they are functioning properly.
- Determine adequacy of firm's check on can seaming.
- Determine if firm's retorts or continuous cookers are functioning properly.
- Review recording charts from retorts and continuous cookers to ascertain if tuna was processed at a proper time and temperature relationship.
- Determine firm's postprocessing can handling: how cans are cooled, and whether water is clean and chlorinated.
- Examine fish for organoleptic quality at critical points in the processing procedure, such as (1) in butchering state—prior to precook, (2) after precook, before being canned (no long holding time after precook), and (3) after any period the tuna has been held excessively long at room temperature.
- Evaluate firm's canning operation for compliance with the GMPR for low-acid foods (21 CFR 113).
- Check plant for proper screening and rodent proofing to eliminate insects and/or rodents.

SANITATION

Check:

- Firm's operation for compliance with GMPR for human foods [(Sanitation) 21 CFR 110].
- Firm's equipment cleaning and sanitizing operation and determine its effectiveness.
- If adequate hand-washing and sanitizing facilities have been provided and that signs are posted directing employees to use them.
- Employees' use of hand-sanitizing solutions and whether solutions are maintained at proper strength.
- Firm's usage of insecticides and rodenticides, so they do not become incidental food additives.
- Freezers for proper storage temperatures and for sanitary storage.
- Review firm's records regarding assay of finished product for mercury, decomposition, and other quality factors.
- Review firm's assay records to determine if the canned tuna complies with the Standard (21 CFR 161.190).
- Ascertain if the food additives used are permitted by the Standards and other legal requirements.

OYSTERS

Most oyster-shucking operations are handled by state inspection agencies. For procedures see FDA standard guidelines on interstate shellfish sanitation. Microbiological considerations are of prime importance in any shellfish gathering and processing plant. Time-temperature abuses enter into most problems with the products. However, the high value of these products has made economic violations even more profitable to the unethical operator. During an evaluation of sanitation, use the critical factors as follows:

- Check for evidence of contamination from the presence of cats, dogs, birds, or vermin in the plant.
- Check results of any testing conducted on incoming oysters including filth, decomposition, pesticides, or bacteria.
- Check for possible incorporation of excessive fresh water through (1) prolonged contact with water or (2) by insufficient drainage.
- Determine if employee sanitation practices preclude adding contamination (clean dress and proper use of 100 ppm chlorine equivalent hand sanitizers).
- Determine if equipment is washed and sanitized about every two hours.
- Check for time-temperature abuses that may cause rapid bacterial growth.

BLUE CRAB (FRESH AND PASTEURIZED)

SANITATION CRITICAL FACTORS

During a sanitation evaluation, use critical factors as follows:

- Check for evidence of contamination from rodents, insects, flies, birds, and domestic pets.
- Determine if employee sanitation practices preclude adding contamination (clean dress and proper use of 100 ppm chlorine equivalent hand sanitizers), particularly during pick-out of shells from crabmeat.
- Determine if equipment is washed and sanitized about every two hours.
- Check for time-temperature abuses that may cause rapid bacterial growth.
- Check testing of incoming crabs for decomposition, bacterial load, pesticides, and dead crab removal prior to processing.

- Check firm's usage of rodenticides and insecticides to determine that they do not contaminate the in-process crabs.

Let us look at the sanitation aspects of the different stages of operation.

RAW MATERIALS

- Check receiving and handling process prior to cooking.
- See if firm discards all dead crabs prior to cooking. If not, estimate percent of dead crabs utilized.
- Note any rodent or insect activity in the receiving area.
- If the firm refrigerates the live crabs prior to cooking, see if they are kept in a separate cooler from the processed crabs.
- Check results of any testing of incoming crabs including bacteriological results and pesticides.

MANUFACTURING PROCESS

To evaluate the sanitation of the manufacturing process, check on the following:

Cooking

Check product flow and determine time and temperature of cooking and type of cooker.

- Retort.
- Live steam. Check boiler compound used.
- Review recording charts for retorts.
- Determine venting procedures.

Cooling

Check time and temperature relationship and:

- How long cooked crabs are held at room temperature.
- Any processing delays between cooking, cooling, and picking.
- Whether cooled crabs are refrigerated until picked.
- Whether cooked crabs are stored in the same baskets as they are cooked in or are transferred to another container.
- If refrigerator is used for storing cooled crabs, is it used only for this purpose?

Picking

Check on the following sanitation aspects:

- If the picking table is cleaned and sanitized prior to use, at appropriate times during the day, and at the end of the day.
- If the picking table is not cleaned and sanitized between each new supply of crabs, and if all crabs on the table are picked prior to the addition of new crabs. Check handling of crab claws prior to picking.
- Pickers' hands for cuts, sores, and so on.
- That picking utensils are of proper construction: (1) See if all-metal knives, without wooden handles, are used. (2) Check to see that the workers do not wrap the handles of the knives with paper towels, cloth, or string. (3) See if all-stainless-steel or other metal shovels with steel handles and shafts are used for placing the crabs onto the picking table. Check shovel storage and see whether it is used for anything besides crabs.
- If claws are picked mechanically, obtain procedure and check operation.
- Check on how often pickers deliver the picked meat to the packing room.

Packing

- See if picked crabmeat is placed directly into the can or into holding pans. If the crab is "deboned" prior to packing, check on how long it is held.
- See if crabmeat weighed into the final can is closed and iced at frequent intervals. Determine if pickers do their own weighing and final packing.
- Check on how finished, packaged crabmeat is stored, or if it is shipped the same day it is packaged.
- See if ice used is from an approved source. Check storage of ice.

Pasteurization

- Check the can closing system and can handling prior to pasteurization.
- Check time-temperature of the pasteurization process.
- Check on how pasteurized cans are cooled and stored.
- See if the finished canned crabmeat is stored in a refrigerator prior to shipping and how long it is held prior to shipment.

- Determine shipping operation: refrigerated trucks, iced baskets, and so on.

Lighting, Ventilation, Refrigeration, Equipment

- Determine if building is adequately lighted and ventilated.
- Check if the cooling and refrigerating facilities are adequate to do the job.
- See if equipment is of proper construction.

OVERALL SANITATION

- See if the building provides for a separation of the various processes.
- See if building is so constructed as to be free from rodent or insect entry points or harborages and whether there are rodents or insects in the plant.
- Check if product contact surfaces (tables, carts, pans, knives, etc.) are of proper construction. See if seams are sealed to avoid product buildup.
- Obtain in detail the firm's plant and equipment cleaning and sanitizing procedures and check if all equipment is cleaned and sanitized as necessary.
- Determine if employee toilets and hand-washing facilities are provided, maintained, and supplied and if hand-washing facilities are located in various processing areas.
- Determine if hand-sanitizing solutions are provided at appropriate locations, maintained at proper levels at all times, and used when necessary. Check hand-sanitizing solution strength at various intervals during the inspection. Check to see if employees use hand dips when necessary.
- Evaluate the firm's operations and employee practices for compliance with the human food (sanitation) GMPR, CFR part 110, and the Food Code.
- Document any insanitary conditions noted that could lead to the contamination of the firm's crabs or crabmeat with filth and/or bacteria.
- Check storage and disposal of solid waste, for example, shells.

CHECKLIST FOR CRUSTACEA PROCESSOR

Use the following table to obtain the information necessary to make a valid assessment of the sanitation of a processor's operation.

Sanitation Assessment for the Crustacea Processors

Control Aspect	Points for Assessment
Receiving (unload)	• Determine condition (acceptable or decomposed). • Separate work area.
Sorting	• Remove miscellaneous species of incidental fish. • Further determination of condition (quality).
Age	• Sanitation. • Time/temperature.
Peeling [mechanical, types (Model A) (PCA-1.5" cook) (choice for freezing)].	• Sanitation. • Potable water. • Separate work area (for peeling, washers, separators, and, if applicable, shaker-blower)
Washers	• Sanitation. • Potable water. • Shell and debris removal (quality).
Shaker-blower (options)	• Sanitation. • Shell removal (quality).
In-house inspection	• Sanitation. • Shell removal. • Separate work area for freezing.
Size graded (machine or manual)	• Sanitation.
Package (cans or plastic)	• Sanitation. • Personnel sanitation. • Suitable packaging materials. • Time/temperature. • Separate work area.
Freeze	• Time/temperature (quality).

SCALLOPS

BACKGROUND INFORMATION

The scallop industry encompasses three primary species: (1) sea scallops, (2) bay scallops, and (3) calico scallops. The processing of sea scallops is accomplished on board the vessel actually harvesting the product. Boats that process sea scallops remain at sea for from 3 to 12 days, depending on area and catch. In most cases, the calico scallops are harvested daily and processed at shore processing plants rather than on board the vessel. The trend, however, is toward onboard processing for this species also. Bay scallops pose a unique problem in that they may be processed in a commercial plant or at home.

SANITATION CRITICAL FACTORS

During the evaluation of food plant sanitation, use the following critical factors:

• Check for evidence of contamination from rodents, insects, birds, or from domestic animals.
• Determine if equipment is washed and sanitized about every two hours.
• Check for time-temperature abuses, which could cause rapid bacterial growth and/or decomposition.
• Determine if employee practices preclude the addition of contaminants: clean dress and proper use of 100 ppm chlorine equivalent hand sanitizers.
• Determine method of icing or freezing of the scallops.

- Ascertain if incoming scallops are tested for bacterial load, decomposition, pesticides, etc. Review results of these tests.
- Check usage of pesticides and rodentcides by firm, to ascertain that they do not become incidental food additives.

RAW MATERIALS

Determine (1) geographical area where the scallops are harvested, (2) the type of scallops harvested and processed by common or species name, and (3) how scallops are handled between harvesting and processing.

PROCESSING

- Observe in detail the scallop processing operation. Make a flow plan.
- Check shucking and evisceration process, and see if this process is physically separated from the packaging and other operations.
- Determine source of water used in the scallop washing and rinsing operations. If treated by the processor, determine nature and extent of treatment.
- See if equipment used in the processing operation is of proper construction and design.
- Check firm's equipment cleaning and sanitizing operation.
- Determine time and temperature of the processing operation: (1) Check how long between harvest and shucking and determine the temperature of the scallops; (2) check how long scallops are held at ambient air temperature and determine the ambient temperature; (3) check how long between shucking and rinsing and determine the temperature of the scallops; and (4) check how long, after being iced, before scallops reach an internal temperature below 40°F.
- Check finished product packaging.
- Determine source of ice used in icing operation and, if bagged ice is used, source and type of bag, condition of bags, and conditions of storage.
- Check finished product storage facilities and condition.
- Check on the use of any food additives to determine if used at allowable levels.

OVERALL SANITATION

- See if building or vessel is free from rodent or insect activity.

- Check that toilets and hand-washing facilities provided are properly located and maintained.
- Determine strength and type of hand-sanitizing solutions used and the sanitizer's location.
- Note any employee practices that could lead to the contamination of the scallops with filth and/or bacteria.
- See if water and ice used in the process are from an approved source, and list source.
- Determine method of shell and waste material disposal.
- Evaluate the firm's operation for compliance with the human foods (sanitation) CGMPR, 21 CFR 110, and the Food Code.
- Document any insanitary conditions noted that could lead to the contamination of this firm's products with filth and/or bacteria.

SHRIMP

SANITATION CRITICAL FACTORS

Breading of shrimp has long posed a problem from an economic standpoint. In addition, the time-temperature abuses present a great potential for food poisoning organisms. The growing scarcity and consequential high value of the raw material make the breading standards even more important. Review breaded shrimp standards (21 CFR 161) prior to evaluating plant sanitation.

During a sanitation assessment, use critical factors as follows:

- Check for the presence of cats, dogs, birds, or vermin in the plant.
- Review testing of incoming shrimp. Check results of tests for decomposition, bacterial load, pesticides, and other possible adulterants.
- Evaluate operation for compliance with 21CFR 12.1–Raw Breaded Shrimp.
- Watch for any time-temperature abuses in the handling of seafood.
- Determine that employee hygienic practices are satisfactory, for example, clean dress, washing of hands, and use of 100 ppm chlorine equivalent hand sanitizers.
- Note any equipment defects that cause seafood to lodge, decompose, then dislodge into the pack.
- Observe breading operations for suspected excesses (21 CFR 161.175/6) or lack of coolant to keep batter mix below 50°F in an open system and below 40°F in closed system.

Note the misuse of pesticides, abuse of color or food additives, deviations from standards, and so on.

RAW MATERIALS: RECEIPT AND STORAGE

Determine if:

- Shrimp and other raw materials are inspected upon receipt for decomposition, microbial load, pesticides, and filth.
- Raw materials susceptible to microbial contamination are received under a supplier's guarantee. Raw material specifications exist, and only wholesome raw materials are accepted into active inventory. Determine disposition of rejected raw materials.
- Shrimp receiving and storage facilities are physically adequate.
- Frozen shrimp are stored at 0°F (-18°C) or below.
- Fresh or partially processed shrimp are iced or otherwise refrigerated to maintain a temperature of 40°F (4°C) or below until they are ready to be processed.
- Decomposed shrimp are being processed.
  - Examine shrimp as received, and again after sorting, for decomposition. Classify as passable (class 1), decomposed (class 2), or advanced decomposition (class 3). Less experienced inspectional personnel should submit some of class 2 and class 3 shrimp for confirmation by the laboratory.
  - Prompt handling and adequate sorting is necessary to prevent decomposition. Check times and temperatures.
  - Where decomposed shrimp are going into canned or cooked-peeled shrimp, collect investigational samples of the finished pack. Give attention to disposition of loads showing a high percentage of decomposition that cannot be adequately sorted, and to disposition of reject shrimp. Make certain that "bait shrimp" is denatured.
- Fresh raw shrimp are washed and chilled to ≤ 40°F (4°C) within two hours of receipt. Frozen shrimp should be held at ≤ 0°F (-18°C). Determine if they are examined organoleptically when received.
- Peeled and deveined shrimp are promptly chilled to 40°F (4°F) or below.

PLANT SANITATION

Determine if:

- The water (ice) is (1) from an approved source, (2) disinfected and contains residual chlorine, (3) sampled and analyzed for contamination, and (4) handled in a sanitary manner.
- Drainage facilities are adequate to accommodate all seepage and wash water.
- The plant has readily cleanable floors that are sloped and equipped with trap drains.
- The plant is free of the presence of vermin, dogs, cats, or birds.
- The screening and fly control is adequate.
- Offal, debris, and refuse are placed in covered containers and removed at least daily or continuously.
- Adequate hand-washing and sanitizing facilities are located in the processing area and are easily accessible to the preparation, peeling, and subsequent processing operations.
- Signs are posted directing employees handling shrimp and other raw materials to wash and sanitize their hands after each absence from the workstation.
- Employees actually wash and sanitize their hands as necessary (before starting work, after absences from the workstation, when hands become soiled, etc.)
- Hand-sanitizing solutions are maintained at 100 ppm available chlorine or the equivalent and are used.
- Persons handling food or food contact surfaces wear clean outer garments, maintain a high degree of personal cleanliness, and conform to good hygienic practices.
- Management prevents any person known to be affected with boils, sores, infected wounds, or other sources of microbiological contamination from working in any capacity in which there is a reasonable probability of contaminating the food.
- The product is processed to prevent contamination by exposure to areas involved in earlier processing steps, refuse, or other objectionable conditions or areas.
- Food contact surfaces are constructed of metal or other readily cleanable materials.
- Seams are smoothly bonded to prevent accumulation of shrimp, shrimp material, or other debris.
- Each freezer and cold storage compartment used for raw materials, in process or finished product, is fitted with required temperature indicating devices.

- Unenclosed batter application equipment is flushed and sanitized at least every four hours during plant operations, and all batter application equipment is cleaned and sanitized at the end and the beginning of the day's operation.
- Breading application equipment and utensils are thoroughly cleaned and sanitized at the end of the day's operations.
- Utensils used in processing and product contact surfaces of equipment are thoroughly cleaned and sanitized at least every four hours during operation.
- All utensils and product contact surfaces, excluding breading application equipment and utensils, are rinsed and sanitized before beginning the day's operation.
- Containers used to convey or store food are handled in a manner to preclude direct or indirect contamination of the contents.
- The nesting of containers is prohibited.

PROCESSING

Determine if:

- Raw frozen shrimp are defrosted at recommended temperatures (air defrosting at ≤ 45°F (7°C), or in running water at ≤ 70°F (21°C) in less than two hours).
- Fresh raw shrimp are washed in clean potable water and chilled to ≤ 40°F (4°C).
- Fresh shrimp are adequately washed, culled, and inspected.
- Every lot of shrimp that has been partially processed in another plant, including frozen shrimp, is inspected for wholesomeness and cleanliness.
- Shrimp entering the thaw tank are free from exterior packaging material and substantially free of liner material.
- On removal from the thaw tank, shrimp are washed with a vigorous potable water spray.
- Shrimp are removed from the thaw tank within thirty minutes after they are thawed.
- During the grading, sizing, or peeling operation, the (1) equipment is cleaned and sanitized before use, (2) water is maintained at proper chemical strength and temperature, and (3) raw materials are protected from contamination.
- Sanitary drainage is provided to remove liquid waste from peeling tables.

- Firm prohibits the practice of salvaging shrimp (i.e., repicking the accumulated hulls and shells for missed shrimp or shrimp pieces).
- Peeled and deveined shrimp are promptly chilled to ≤ 40°F (4°F).
- Peeled shrimp are transported from peeling machines or tables immediately, or if containerized, within 20 minutes.
- Peeled shrimp containers, if applicable, are cleaned and sanitized as often as necessary, but in no case less frequently than every three hours.
- When a peeler is absent from his duty post, his container is cleaned and sanitized prior to resuming peeling.
- Peeled shrimp that are transported from one building to another are properly iced or refrigerated, covered, and protected.
- Shrimp are handled minimally and protected from contamination.
- Shrimp that drop off the processing line are discarded or reclaimed.
- Shrimp are washed with a low-velocity spray or in unrecirculated flowing water at ≤ 50°F (10°C) just prior to the initial batter or breading application, whichever comes first, except in cases where a predust application is included in the process.
- Removal of batter or breading mixes or other dry ingredients from multiwalled bags is accomplished in an acceptable manner.
- Batter in enclosed equipment is assured a temperature of not more than 40°F (4°C) and disposed of at the end of each workday, but in no circumstances less often than every 12 hours.
- Batter in an unenclosed system is maintained at ≤ 50°F (10°C) and disposed of at least every four hours and at the end of the day's operation.
- Breading reused during a day's operation is sifted through a 1/2-inch or smaller mesh screen.
- Breading remaining in the breading application equipment at the end of a day's operation is reused within 20 hours and is sifted, as above, and stored in a freezer in a covered sanitary manner.
- Hand batter pans are cleaned, sanitized, and rinsed between each filling with batter or breading.

FINISHED PRODUCT PROCESS AND QUALITY ASSURANCE STANDARDS

Determine if:

- Processing and handling of finished product is (1) performed in a sanitary manner, (2) protected from contamination, and (3) arranged to facilitate rapid freezing.

- Manual manipulation of breaded shrimp is kept to a minimum.
- Aggregate processing time, excluding the time required for thawing frozen material, is less than two hours, exclusive of iced or refrigerated storage time.
- Breaded shrimp are placed into freezer within 30 minutes of packaging.
- Breaded shrimp are frozen in a plate or blast freezer at $\leq -20°F$ ($-29°C$).
- Storage freezer is maintained at $\leq 0°F$ ($-18°C$).
- In-line, environmental, and finished product samples are analyzed and evaluated at least weekly for microbial conditions. Review the analytical record, if available.
- Firm has established microbiological specifications for the final product. If so, review and report these specifications.
- Firm withholds from distribution lots that do not meet their established microbiological standards.
- Finished product is handled and stored in a manner that precludes contamination.
- Labels bear a cautionary statement to keep product frozen.

SMOKED FISH

SANITATION CRITICAL FACTORS

During an evaluation of the sanitation of a smoked fish operation, use critical factors as follows:

- Check sanitary conditions under which firm is operating, including any evidence of insanitation and contamination associated with insects, rodents, microorganisms, chemicals, or other possible sources. Check raw material and packaging material storage areas as well as other susceptible locations in the plant.
- Review raw material receiving records for DDT and other pesticides, decomposition, and bacteriological quality.
- Check food and color additives to ascertain that they are allowed for use and are being used properly.
- Observe employee practices to make sure that they are not acting as routes of contamination.
- Ascertain if the various operations including raw material receipt and storage, defrosting, brining, and so on are acceptable.

- Review recording charts to ascertain what time/temperatures of smoking have been; this may vary depending on the desired salt content the firm is trying to achieve.
- Check finished stored product (i.e., any smoked chubs in which nitrite is used) to ascertain the internal temperature based on the time since smoking (temperature within 3 hours of cooking and again within 12 hours of cooking).

PLANT SANITATION AND FACILITIES

- Check method(s) for cleaning and sanitizing utensils, conveyors, smoking racks, and other food-contact surfaces used in daily operations.
- Check the strength and adequacy of hand- and equipment-sanitizing solutions. The minimum effective chlorine concentration is 100 ppm for hand-sanitizing solutions, and 200 ppm for equipment-sanitizing solutions. Iodine solutions should be 15 ppm for hand-sanitizing solutions and 25 ppm for equipment-sanitizing solutions. Determine if maintained at proper levels.
- Determine method used to separate finished product cooling, packaging, and storage areas from the uncooked product and processing areas.
- Determine the adequacy of plant waste disposal operations.
- Check if hand-washing, toilet, and sanitizing facilities have been provided and if signs have been posted directing the employees to wash and sanitize hands following use.

RAW MATERIALS

Determine:

- Source (area and distributor) and species of fish processed by the firm including the type selected for full coverage during this inspection.
- Process condition in which bulk fish is supplied (e.g., fresh, frozen, mild cured, brined, etc.).
- Quality of fish received. Organoleptic examination should be performed and results reported.
- Raw fish handling procedures (e.g., defrosting, draining procedures encountered).
- Available chlorine or iodine concentrations in hand-dip or equipment-sanitizing solutions, if used.
- Time/temperature intervals for each step in the raw fish handling operations.
- If incoming fish are sampled and analyzed for the presence of DDT and other pesticides.

PROCESSING

Salting and Brining

Determine:

- Size of fish or pieces of fish brined, noting variations of fish size and sizing procedures.
- Form and grade of salt (NaCl) used in the brining.
- Ratio of brine to fish. Determine actual or near estimates of weight of salt, volume of water, and weight of fish being brined.
- Concentrations of brine (NaCl) solutions in degrees (salinometer) at the initiation of brining, during brining, and at the conclusion of the brining operation. A reduction in salt concentration in the brining solution after brining may indicate salt uptake by the fish during brining. (*CAUTION:* If salinometers are made of glass, the degree of salinity should be read in a plastic graduate. *Do not* put the salinometer directly into the tank with fish. It could break and contaminate the fish with glass.)
- Time/temperatures of brining solutions at different intervals during the brining process. Include total brining time.
- Method of agitation of brine solution during brining, if employed, noting number of times agitated and length of each agitation.

Heating, Cooking, and Smoking Operation

- Check equipment used during heating, cooking, and smoking operation. Include oven type, source of heat, type of smoke generators, product temperature monitoring equipment, humidity regulators, and so on. (Temperature recording devices should have an accuracy of ±2°F.)
- Determine the methods and procedures used in drying, cooking, and smoking. Include time/temperature data, results of temperature monitoring by the firm, location of their temperature probes, and product rotation practices.

Cooling

- Monitor time/temperature relationships during cooling to determine how long it takes to reach an internal temperature of 38°F.
- Determine method of cooling.
- Check observable procedures and conditions that can contribute to the microbiological contamination of the processed fish. Include observations such as extended cooling time and optimum incubation temperature, exposure to airborne contamination, improper handling, and poor in-process storage conditions.
- Determine if firm separates cooling facilities from raw processing and cooking operations.

Packaging

Determine method and types of packing including (1) time/temperature relationships during packaging, (2) any use of additives or prepackaging additive treatment (include name, quantity added, method of application, etc.), and (3) observable practices and conditions that can contribute to the microbiological contamination of the processed fish (include lack of required facilities, excessive product handling, improper storage, etc.).

STORAGE AND DISTRIBUTION

- Check type of equipment used for determining, recording, and maintaining storage temperatures.
- Determine actual storage compartment temperatures. Refrigerated storage temperatures should be 38°F or below.
- Determine method of distribution (e.g., refrigerated, iced, frozen, etc.).

LABORATORY CONTROLS

Check or determine:

- Method and frequency of sampling. Salinity testing operations: adequacy of testing procedures and frequency. Microbiological testing of processed fish, how often, methods used, adequacy of testing, and so on.
- Checks made on in-process controls and laboratory equipment.
- Use of outside laboratories, consultants, and so on. Include name, location, and tests each firm performs and how often tests are conducted.
- Results of analysis from previous lots.

OVERALL SANITATION

- Evaluate the firm's operation for compliance with 21 CFR 110—GMPR Human Foods (Sanitation).
- Evaluate the firm's cleaning and sanitizing procedures.

- Check if adequate hand-washing and sanitizing facilities are provided and if signs directing their use are provided. Evaluate the employees' use of hand dips and if they are used when necessary.
- See if hand dips and equipment-sanitizing solutions are maintained at the proper level and changed when necessary.

GLOSSARY

CGMPR—current good manufacturing practice regulations.

EPA—U.S. Environmental Protection Agency.

FDA—U.S. Food and Drug Administration.

GMP—good manufacturing practices.

HACCP (hazard analysis and critical control points)—a system to identify and evaluate the food safety hazards that can affect the safety of food products, institute controls necessary to prevent those hazards from occurring, monitor the performance of those controls, and routinely maintain records.

ISSC—Interstate Shellfish Sanitation Conference.

LACF—low-acid canned food.

MOU—memorandum of understanding.

NFPA—National Food Processors Association.

NSSP—National Shellfish Sanitation Program.

PSP—paralytic shellfish poison.

SCP—Salmon Control Plan.

USDA—U.S. Department of Agriculture.

29
Vegetables: Tomato Processing

S. A. Barringer

BACKGROUND INFORMATION

The composition of the tomato is affected by the variety, state of ripeness, year, climactic growing conditions, light, temperature, soil, fertilization, and irrigation. Tomato total solids vary from 5 to 10% (Davies and Hobson 1981), with 6% being average. Approximately half of the solids are reducing sugars, with slightly more fructose than glucose. Sucrose concentration is unimportant in tomatoes and rarely exceeds 0.1%. A quarter of the total solids consist of citric, malic and dicarboxylic amino acids, lipids, and minerals. The remaining quarter, which can be separated as alcohol-insoluble solids, contains proteins, pectic substances, cellulose, and hemicellulose.

Tomatoes are mostly water (94%), a disadvantage when condensing the product to paste. They are a reasonably good source of vitamin C and A. In 1972 tomatoes provided 12.2% of the recommended daily allowance of vitamin C, and only oranges and potatoes contribute more to the American diet (Senti and Rizek 1975). Tomatoes provided 9.5% of the vitamin A, second only to carrots. When major fruit and vegetable crops were ranked on the basis of their content of 10 vitamins and minerals, the tomato occupied sixteenth place (Rick 1978). However, when the amount that is consumed is taken into consideration, the tomato places first in its nutritional contribution to the American diet. This is because the tomato is a popular food, added to a wide variety of soup, meat, and pasta dishes.

The red carotenoid in tomatoes, lycopene, does not have any vitamin activity, but it may act as an antioxidant when consumed (Stahl and Sies 1992). A review of epidemiological studies found that evidence for tomato products was strongest for the prevention of prostate, lung, and stomach cancer, with

473

possible prevention of pancreatic, colon and rectal, esophageal, oral cavity, breast, and cervical cancer (Giovannucci 1999). The consumption of fresh tomatoes, tomato sauce, and pizza has been found to be significantly related to a lower incidence of prostate cancer, with tomato sauce having the strongest correlation (Giovannucci et al. 1995). Since anticancer correlations are typically stronger to processed tomatoes than to fresh tomatoes, several studies have looked at the effect of processing on lycopene. Tomato juice and paste have more bioavailable (absorbed into the blood) lycopene than fresh tomatoes when both are consumed with corn oil (Gartner et al. 1997, Stahl and Sies 1992). This may be because thermally induced rupture of cell walls and weakening of lycopene-protein complexes releases the lycopene, or because of improved extraction of lycopene into the lipophilic corn oil.

Fresh tomatoes are the fifth most popular vegetable consumed in the United States (16.6 pounds per capita), after potatoes (48.8), lettuce (23.3), onions (17.9), and watermelon (17.4) [U.S. Department of Agriculture (USDA) 2000]. Canned tomatoes are the most popular canned vegetable, at 74.2 pounds per capita in the United States. In the condiment category, salsa and ketchup are number one and two, respectively.

RAW MATERIALS PREPARATION

The flowchart for processing tomatoes into juice, paste, whole, sliced, or diced tomatoes is shown in Figure 29.1. After harvesting, tomatoes are transported to the processing plant as soon as possible. Once at the plant, they should be processed immediately, or at least stored in the shade. Fruit quality deteriorates rapidly while waiting to be processed. To unload, either the tomatoes are off-loaded onto an inclined belt, or the gondolas are filled with water from overhead nozzles. If water is used, gates along the sides or undersides of the gondolas are opened, allowing the tomatoes to flow out into water flumes.

GRADING

The first step the tomatoes go through is grading, to determine the price paid to the farmer. This is done at the processing facility or at a centralized station before going to the processing facility. Individual companies may set their own grading standards, use the voluntary USDA grading standards, or use locally determined standards, such as those of the

Processing Tomato Advisory Board in California. The farmer is paid based on the percentage of tomatoes in each category. Typically, companies hire USDA graders or hold an annual grading school to train their graders.

The USDA divides tomatoes for processing into categories A, B, C, and culls (USDA 1983). Grading is done on the basis of color and percentage of defects. Color can be determined visually by estimation of what percentage of the surface is red, or with an electronic colorimeter on a composite raw juice sample. Defects include worms, worm damage, freeze damage, stems, mechanical damage, anthracnose, mold, and decay. The allowable percentage of extraneous matter may also be specified. Extraneous matter includes stems, vines, dirt, stones, and trash.

Tomatoes for canning whole, sliced, or diced are graded on the basis of color, firmness, defects, and size. Solids content is unimportant, unlike in tomatoes for juice or paste. Graders must be trained to evaluate and score color and firmness. Color should be a uniform red across the entire surface of the tomato. Color is graded using USDA issued plastic color comparators, the Munsell colorimeter or the Agtron colorimeter, or the tomato is ground into juice and used in a colorimeter with a correlation equation to convert it to the Munsell scale. Firmness, or character, is important to be sure the tomato will survive canning. Soft, watery cultivars or cultivars possessing large seed cavities give an unattractive appearance and therefore receive a lower grade. Size is not a grading characteristic per se, but all tomatoes must be above a minimum agreed upon size.

The Processing Tomato Advisory Board inspects all tomatoes for processing in California. Their standards are similar to those of the USDA, but more geared for the paste industry. They inspect fruit for color, soluble solids, and damage (California Department of Food and Agriculture 2001). A load of tomatoes may be rejected for any of the following reasons: > 2% of fruit is affected by worm or insect damage, > 8% is affected by mold, > 4% is green, or > 3% contains material other than tomatoes, such as extraneous material, dirt, and detached stems.

WASHING

Washing is a critical control step in producing tomato products with a low microbial count. A thorough washing removes dirt, mold, insects, *Drosophila* eggs, and other contaminants. The efficiency

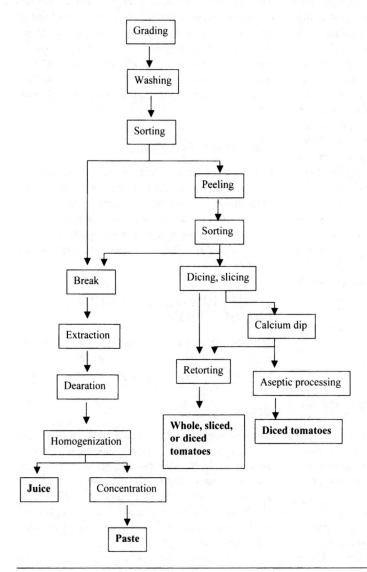

Figure 29.1. Flow diagram for tomato processing.

of the washing process will determine microbial counts in the final product (Heil et al. 1984, Zacconi et al. 1999). Several methods can be used to increase the efficiency of the washing step. Agitation increases the efficiency of soil removal. The warmer the water spray or dip, up to 90°C, the lower the microbial count (Adsule et al. 1982, Trandin et al. 1982), although warm water is not typically used because of economic concerns. Lye or surfactants may be added to the water to improve the efficiency of dirt removal; however, surfactants have been shown to promote infiltration of some bacteria into the

tomato fruit by reducing the surface tension at the pores (Bartz 1999). The washing step also serves to cool the fruit. Since tomatoes are typically harvested on hot summer days, washing removes the field heat, slowing respiration and therefore quality loss.

Tomatoes are typically transported in a water flume to minimize damage to the fruit. Therefore, tomato washing can be a separate step in a water tank or it can be built into the flume system. A water tank also serves to separate stones from the fruit, since the stones settle to the bottom. The final rinse step uses pressurized spray nozzles at the end of the soaking process. Flume water may be either recirculated or used in a counterflow system, so that the final rinse is with fresh water, while the initial wash is done with used water. In either system, the first flume frequently inoculates rather than washes the tomatoes because all of the dirt in the truck is washed into the flume water (Heil et al. 1984). When the water is reused, high microbial counts on the fruit may result if careful controls are not kept.

Chlorine is frequently added to the water. Chlorine will not significantly reduce spores on the tomato itself because the residence time is too short. However, chlorine is effective at keeping down the number of spores present in the flume water (Heil et al. 1984). When there is a large amount of organic material in the water, such as occurs in dirty water, chlorine is used up rapidly, so it must be continuously monitored.

During fluming to the next step, upright stakes may be placed at intervals within the flume. Vines and leaves that have made it this far in the process are caught on the stakes. Periodically, workers remove the trapped vines.

SORTING

A series of sorters are used in a plant. The first sorter, especially in small plants, is an inclined belt. The tomatoes are off-loaded onto the belt. The round fruit rolls down the belt and into a water flume. The leaves, sticks, stones, and rotten tomatoes are carried up by the belt and dropped into a disposal bin.

Photoelectric color sorters are used in almost every plant to remove the green and pink tomatoes. These sorters work by allowing the tomatoes to fall between conveyor belts in front of the sensor. Unacceptable tomatoes are ejected by a pneumatic finger. A small percentage of green tomatoes in tomato juice does not adversely affect the quality. Green tomatoes bring down the pH, but do not affect the color of the final product. In addition, less mature tomatoes result in a higher viscosity paste (Luh et al. 1960, Whittenberger and Nutting 1957). Pink or breaker tomatoes are a problem, however, because they decrease the redness of the juice. Both pink and green tomatoes need to be removed from the whole peel or dice line. Size sorters remove excessively small tomatoes, which would be undesirable in the can. The small tomatoes are diverted to the juice or crushed tomato line.

The final sorting step is to go past human sorters, who are more sensitive than mechanical sorters. Employees remove extraneous materials and rotten tomatoes from sorting tables. Sorting conveyors should require employees to reach no more than 20 inches, move no more than 25 feet/minute, and consist of roller conveyors that turn the tomatoes as they travel, exposing all sides to the inspectors (Denny 1997).

CORING AND TRIMMING

In the past, tomatoes were cored by machine or, more frequently, by hand, to remove the stem scar. Modern tomato varieties have been bred with very small cores so that this step is no longer needed. Trimming to remove rot or green portions is not practiced in the United States due to the high cost of labor.

JUICE, PASTE, AND SAUCE PRODUCTION

The majority of processed tomatoes are made into juice, which is condensed into paste. The paste is remanufactured into a wide variety of sauce products.

BREAK

The tomatoes are put through a break system to be chopped. Some break systems operate under vacuum to minimize oxidation. In an industrial plant operating under vacuum, no degradation of ascorbic acid occurs during the break process (Trifiro et al. 1998). When vacuum is not used, the higher the break temperature, the greater the loss of ascorbic acid (Fonseca and Luh 1976).

Tomatoes can be processed into juice by either a hot break or cold break method. Most juice is made by hot break. In the hot break method tomatoes are chopped and heated rapidly to at least 82°C to inactivate the pectolytic enzymes polygalacturonase (PG) and pectin methylesterase (PME). Inactivation of these enzymes helps to maintain the maximum viscosity. Most juice is made by the hot break method, since most juice is concentrated to paste, and high viscosity is important in tomato paste used to make other products. Most hot break processes occur at 93–99°C.

In the cold break process, tomatoes are chopped and then mildly heated to accelerate enzymatic activity and increase yield. Pectolytic enzyme activity is at a maximum at 60–66°C. Cold break juice has less destruction of color and flavor but also has a lower viscosity because of the activity of the enzymes. This juice can be made into paste, but its lower viscosity is a special advantage in tomato juice and juice-based drinks. In practice, both hot and cold break paste with excellent color and high viscosity can be purchased.

EXTRACTION

After the break system, the comminuted tomatoes are put through an extractor, pulper, or finisher to remove the seeds and skins. Juice is extracted with either a screw-type or paddle-type extractor. Screw-type extractors press the tomatoes between the screw and the screen. The screw is continually expanding along its length, forcing the tomato pulp through the screen. The expanding screw with the screen removed is shown in Figure 29.2. Screw-type extractors incorporate very little air into the juice, unlike paddle-type extractors, which beat the tomato against the screen, incorporating air. Air incorporation during extraction should be minimized because it oxidizes both lycopene and ascorbic acid. The screen size determines the finish, or particle size, which will affect viscosity and texture.

DEAERATION

Deaeration to remove dissolved air incorporated during breaking or extraction is frequently the next step. The juice is deaerated by pulling a vacuum as soon as possible, because oxidation occurs rapidly at high temperatures. Deaeration also prevents foaming during concentration. If the product is not deaerated, substantial loss of vitamin C will occur.

HOMOGENIZATION

The juice is homogenized to increase product viscosity and minimize serum separation. The homogenizer is similar to that used for milk and other dairy products. The juice is forced through a narrow orifice at high pressure, shredding the suspended solids. The creation of a large particle surface area increases product viscosity.

CONCENTRATION INTO PASTE

If the final product is not juice, the juice is next concentrated to paste. Concentration occurs in forced circulation, multiple effect, vacuum evaporators. Typically, three- or four-effect evaporators are used, and most modern equipment now uses four effects. The temperature is raised as the juice goes to each successive effect. A typical range is 48–82°C. Vapor is collected from later effects and used to heat the product in previous effects, conserving energy. The reduced pressure lowers the temperature, minimizing color and flavor loss.

The paste is concentrated to a final solids content of at least 24% NTSS (natural tomato soluble solids) to meet the USDA definition of paste. Commercial paste is available in a range of solids contents, finishes, and Bostwick consistencies. The larger the screen size, the coarser the particles and the larger the finish. Bostwick may range from 2.5 to 8 cm (tested at 12% NTSS).

ASEPTIC PROCESSING

The paste is heated in a tube-in-tube or scraped-surface heat exchanger, held for a few minutes to pasteurize the product, then cooled and filled into

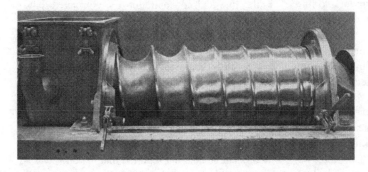

Figure 29.2. Inside of a screw-type tomato extractor.

sterile containers, in an aseptic filler. A typical process might heat to 109°C, then hold 2.25 minutes, or heat to 96°C and hold for 3 minutes. Aseptically processed products must be cooled before filling, both to maintain high quality and because many aseptic packages will not withstand temperatures above 38°C. An aseptic bag-in-drum or bag-in-crate filler is used to fill the paste into bags previously steam sterilized. Paste is typically sold in 55-gallon drums or 300-gallon bag-in-box containers.

REMANUFACTURING INTO SAUCE

Manufacturers of convenience meals buy tomato paste and remanufacture it by mixing it with water, particulates, and spices to create the desired sauce. Some sauce is made directly from fresh tomatoes during the tomato season, but this is less common. Sauce production from paste is more economical because it can be done during the off season using the equipment in tomato processing plants that would otherwise be unused. It is also cheaper to ship paste than sauce.

CANNED WHOLE OR SLICED TOMATO PRODUCTION

PEELING

Tomatoes are typically peeled before further processing. The FDA standard of identity does allow for canned, unpeeled tomatoes if the processor so desires. This is not common on the market, though there are some unpeeled salsas. This is probably because the peel is very tough and undesirable to the consumer; in additon, unpeeled tomatoes would show many blemishes that are hidden from the consumer by peeling. Some easy-peel varieties have been bred that may be suitable for canning with the peel on, since the peel is less tough. However, these varieties also have less resistance to insect and microbial attack on the plant and so are not typically used by growers.

There are two commonly used peeling methods: steam and lye. In California, most peeling is done by steam, while in the midwestern United States and Canada peeling is done with a hot lye solution. In steam peeling, the tomatoes are placed on a moving belt one layer deep and pass through a steam box in a semicontinuous process. Steam peeling is done at 24–27 psig, which equals about 127°C, for 25–40 seconds. Peel removal is possible because of rupture

of the cells just underneath the peel. Due to the high temperature and pressure, the temperature of the water inside these cells exceeds the boiling point, but remains in a liquid state. When the pressure in the chamber is released, the water changes to steam, bursting the cells. Time and temperature are the most critical factors to control to optimize the peeling process. The higher the temperature, the shorter the time required, and the more complete the peel removal. At higher temperatures, there is also less mushiness in the fruit due to cooking. The process uses relatively little water and produces little waste effluent. The waste peels that are produced can be used as fertilizer or animal feed or processed into other products, such as lycopene extract.

In lye, or caustic peeling, the tomatoes pass on a conveyor belt under jets of hot lye (sodium hydroxide) or through a lye tank in a continuous operation. The tomatoes go through a solution of 12–18% lye at 85–100°C for 30 seconds, followed by holding for 30–60 seconds to allow the lye to react. The lye dissolves the cuticular wax and hydrolyzes the pectin. The hydrolysis of the pectin in the middle lamella causes the cells to separate from each other, or rupture, causing the peel to come off. This produces wastewater that contains a high organic load and high pH. Potash, or potassium hydroxide, can be used instead of lye. The advantage of potash peeling is that the potash waste can be discarded in the fields, since it does not contain the sodium ion that is detrimental to soil quality. One processor has done this for several years with no apparent detrimental effect. In some cases, potassium hydroxide can be used at almost half the concentration of sodium hydroxide to produce the same result (Das 1997). Time in the lye, temperature of the bath, and concentration are the three major controllable factors that determine peeling efficiency. Increasing any of these factors increases the extent of peel removal. Time and temperature are linearly correlated, while time and concentration are correlated exponentially (Bayindirli 1994).

With lye peeling, various additives are frequently added to the lye bath to improve peeling. These additives work by removing the wax (Das and Barringer 1999), speeding the penetration of lye into the peel; or decreasing the surface tension of water, increasing the wettability of the cuticle. C6-C8 saturated fatty acids, especially octanoic acid, have been claimed to be very effective (Neumann et al. 1978). One processor tried octanoic acid but reported that the odor was so objectionable that the workers threatened to quit.

Wetting agents are typically used at a level of approximately 0.5 percent in the lye bath. Lye peeling typically produces a higher yield of well-peeled tomatoes than steam peeling, but disposal of the lye wastewater can be difficult (Downing 1996b). Steam gives a higher total tomato yield, but removes much less of the peel than lye (Schlimme et al. 1984). A 65% peel removal is considered good for steam peeling, while peel removal with lye is close to 100%. For this reason, lye is used exclusively in the midwestern United States, where peeled tomatoes are the most important tomato product produced.

After either steam or lye peeling, the tomatoes pass through a series of rubber disks or through a rotating drum under high-pressure water sprays to remove the adhering peel (Figure 29.3). Fruits with irregular shape and wrinkled skin are difficult to peel and result in excessive loss during the peeling step. Thus varieties prone to these characteristics are undesirable. Overpeeling is undesirable because it lowers the yield, results in higher waste, and strips the fruit of the red, lycopene-rich layer immediately underneath the peel, exposing the less attractive yellow vascular bundles.

Figure 29.3. Rubber disks used in tomato peeling.

Both fruit variety and maturity affect the efficiency of the peeling process. One study attempted to determine how well a tomato would peel based on physical structure (Mohr 1990). They found that an abrupt cell size change in the pericarp and the absence of small cells in the mesocarp correlate to better peeling.

Other proposed peeling methods include freeze-heat peeling, and hot calcium chloride. Freeze-heat peeling submerges the tomatoes in liquid nitrogen, refrigerated calcium chloride, or Freon to rupture the cells, releasing pectolytic enzymes. The tomatoes are then transferred into warm water to encourage enzyme activity (Brown et al. 1970; Leonard and Winter 1974; Thomas et al. 1976, 1978). The hot calcium chloride process is similar to peeling in boiling water, which was the standard before the discovery of lye peeling. The disadvantages of the process are that it is patented, that the tomatoes may take up more calcium than allowed in the standards of identity, and that the method requires trained operators to adjust conditions based on maturity and variety (Stephens et al. 1967, 1973). These methods have been tested in laboratories but never put into commercial practice. The other peeling method, no longer used in the United States, is to blanch the tomatoes in boiling water then hand-peel them.

MANUAL SORTING

Peeled tomatoes are inspected by hand before filling into the can. Sorters are mainly looking for rotten parts that cannot be detected by photoelectric sorters. The main defects of concern are those included in the USDA grading standards for canned product: presence of peel, extraneous vegetable material, blemished areas, discolored portions, and objectionable core material (USDA 1990). Inadequately peeled, blemished, small, or misshapen fruits are diverted to the juice line. For greatest efficiency, roller conveyors should be used to turn the tomatoes as they travel, exposing all sides to the sorters.

FILLING, ADDITIVES, AND CONTAINERS

Cans may be filled by hand; however, due to labor costs almost all manufacturers use mechanical filling. The container must be filled to not less than 90% of the container volume, and drained weight must be at least 50% of the water weight, to meet standards of identity [Code of Federal Regulations

(CFR) 2000]. The exact drained weight affects the USDA grade (USDA 1990). A headspace is left in the can to allow for expansion during retorting.

Because of the acidic nature of the fruit, enameled cans and lids are used. When unenameled cans are used, hydrogen swells may occur. These are caused by a reaction between the metal of the can and the acid in the fruit. Glass can also be used, but it is not common in the market. The tomatoes are packed into the can and filled with tomato juice. FDA standards of identity require that some form of tomato juice or puree be used as the packing medium (CFR 2000). Alternately, tomatoes may be in a "solid pack," where no packing medium is used, but this product is not currently on the market.

Heating softens the tomatoes, so calcium is typically added. Calcium can be added in the form of calcium chloride, calcium sulfate, calcium citrate, or monocalcium phosphate. The final amount of calcium cannot exceed 0.045% by weight in whole tomatoes and 0.08% in dices, slices, and wedges (CFR 2000). The calcium ion migrates into the tomato tissue, creating a salt bridge between methoxy groups on adjacent pectin chains and forming calcium pectate or pectinate. This minimizes the softening that occurs during canning. The calcium may be mixed with the cover juice or added directly to the can. Tablets may be added directly, but typically the calcium is mixed with the juice. The amount of calcium added is adjusted based on the firmness of the tomatoes. The typical range is 0–1%, with an average of 1/2%.

Most tomatoes are high-acid foods naturally; however, overly mature tomatoes and certain cultivars can result in a higher pH. The standard of identity allows organic acids to be added to lower the pH as needed. Citric acid is most common, although malic and fumaric acids are also used. Sugar may be added to offset the tartness from the added acid. Sodium chloride is frequently added for taste. The standard of identity allows calcium, organic acids, sweeteners, salt, spices, flavoring, and vegetables to be added (CFR 2000). Because of the presence of other natural components that inhibit botulinum growth, the United States allows tomatoes up to a pH of 4.7 (rather than the pH 4.6 required for other foods) to be canned as high-acid foods.

EXHAUSTING AND SEALING

Cans are typically exhausted and sealed at the same time. The old style of filling the tomatoes cold then conveying the cans through an exhaust box to be heated before sealing is seldom used. Tomatoes peeled either by steam or lye are already hot and are immediately filled, cover juice is added, and the cans are sealed. Steam is injected into the headspace of the can as the can is sealed. When the steam condenses, a partial vacuum is created, preventing "flippers," which appear spoiled to the consumer. A headspace is critical if the product is going to be retorted since the product will expand during heating. Without adequate headspace, the ends of the can will bulge out. This is referred to as a "flipper" if the end can be pushed back down, or a "hard swell" if it cannot.

CANNING/RETORTING

Because tomatoes are a high-acid food, they do not have to be sterilized. Tomato products can be hot filled and held, or can be processed in a retort as needed to minimize spoilage. Most tomato products undergo a retort process to ensure an adequate shelf life. Of the retorts, the continuous rotary retort is that most commonly used for tomato products. This retort provides agitation of the product and can handle large quantities in a continuous process. Because tomatoes are a high-acid food, the retort may operate at boiling water temperature, 100°C. Continuous rotary retorts set at 104°C for 30–40 minutes are also common. Exact processing conditions depend on the product being packed, the size of the can, and the type and brand of retort used. The key is for the internal temperature of the tomatoes to reach at least 88°C.

COOLING

After canning, the product must be cooled to 30–40°C to minimize quality loss. The product may be cooled by water or air. When cooling water is used, it should be chlorinated to 2–5 ppm free chlorine to prevent contamination of the product while the seals are soft (Downing 1996a). Even though the cans are sealed, spoilage rates increase when the water is not chlorinated. The vacuum that forms as the contents cool must draw some microorganisms into the can. A rotary water cooler may be used in a continuous process after a rotary retort. Water cooling is more efficient than air cooling; therefore, longer retort process times are recommended when water cooling is used than when air cooling is used (Downing 1996b).

DICED TOMATO PRODUCTION

Diced tomatoes have become very popular because of the increase in salsa consumption. Dices are processed in a similar manner to canned tomatoes. The major difference is that the tomatoes (peeled or unpeeled) are diced into 3/8-, 1/2-, or 1-inch cubes, inspected to remove green or blemished dices, then calcified. Calcification can occur by direct addition of calcium to the container, or by conveying the dices through a calcium bath. The dices are then packed into cans for thermal processing or aseptically packed. In the past, 80% of dices were thermally processed in no.10 cans (Floros et al. 1992). Cans are still common, but aseptic processing has increased the amount of dices sold in 55- and 300-gallon containers. Dices have an 18- to 24-month shelf life.

Calcium salts can be added as needed to increase firmness and drained weight, but the final amount of calcium cannot exceed 0.08% by weight (CFR 2000). These salts are typically in the form of calcium chloride, calcium sulfate, calcium citrate or monocalcium phosphate. For direct addition, the calcium can be added in the form of a tablet or mixed with the cover juice. For immersion, the dices are conveyed through a calcium bath, or mixed with a calcium solution that is drained off after a holding period. Immersion causes a significant loss of acid and sugar over that from addition of calcium to the can; however immersion results in significantly firmer tomatoes for the same final calcium content (Villari et al. 1997).

A number of studies have attempted to determine the best conditions for immersion of the dices. The best conditions have been determined to be dipping in 0.75% calcium for one minute (Poretta et al. 1995) or 0.43% calcium for 3.5 minutes (Floros et al. 1992). The resulting firmness is dependent on calcium concentration and time, but not temperature (Floros et al. 1992). The drained weight is dependent on the calcium concentration, time, and temperature (Poretta et al. 1995). In general, calcium concentration in the dipping solution is the most important factor. The firmness and drained weight are linearly related to the calcium content and dipping time, though the changes in firmness are much larger than the changes in drained weight (Villari et al. 1997).

Experimentally, it has been shown that pectin methylesterase (PME) further increases the firmness of the dices (Castaldo et al. 1995). The PME activity deesterifies the galacturonic acid subunits, making them available to bind to the calcium ions. The firmness of the dices can be doubled with the addition of PME. Tomato firmness can be increased more economically by processing the dices in a dip solution at a higher pH (7.5) for a longer time (five minutes) to allow the natural enzymes to act within the tomato (Castaldo et al. 1996).

Based on sensory evaluation, dices become inedible at approximately 1.5 times the legal limit of calcium in the dices (Poretta et al. 1995). It has been reported that an adverse effect can be observed at calcium contents as low as 0.045–0.050% (Villari et al. 1997). The lower the calcium content, the higher the dices score in sweetness and natural taste (Poretta et al. 1995). The higher the calcium, the higher the acidity taste and the lower the pH.

WASTE AND WASTEWATER

Wastewater disposal is a critical issue in some locations, and the high cost of disposal can put a tomato processor out of business. By volume, approximately half of the wastewater in a tomato processing plant comes from tomato washing, a third from peeling, and a fifth from canning (Napoli 1979). Most of the waste and wastewater produced during tomato processing is biodegradable and can be disposed of on fields (Pearson 1972). Lye-peeling wastewater is the major exception, if lye peeling is used. This wastewater can be disposed of in the sewer system; however, it has a high organic load and thus is expensive. Some treatment plants also object to the high pH. Some processors report that they have disposed of their potash peeling solution on their fields without any adverse effects. It is also been proposed that the lye-peeling waste be treated with HCl and reclaimed as salt for use in canning, although this is not done in practice. In most cases, lye-peeling wastewater must be disposed of in the sewer system.

Several treatment methods for reducing the organic load before disposal in the sewer system have been tried. These methods are used either to decrease the amount the plant is charged for wastewater treatment, or because local laws restrict the biochemical oxygen demand (BOD) and volume of wastewater that can be discharged into the public sewer system. Treatment methods include microbial digestion, coagulant chemicals, and membrane filtration.

MEASUREMENT OF QUALITY AND HOW IT IS AFFECTED BY GROWING CONDITIONS

COLOR AND LYCOPENE

There are several methods for measuring color. The voluntary USDA grading standards for tomatoes to be processed use the Munsell disk colorimeter (USDA 1983). The Munsell disk colorimeter consists of two spinning disks containing various percentages of red, yellow, black, and gray. As the disks spin, they visually combine to produce the same color as the tomato. USDA color comparators are plastic color standards that can be used to visually grade tomatoes. With fresh tomatoes, the Agtron colorimeter is common, especially for tomato juice and halves. The Agtron is an abridged spectrophotometer that measures the reflection at one to three wavelengths and reports the result as a color score. For processed tomato products, the Hunter colorimeter is common. The Hunter measures the L, a, and b values. The a and b values are put into a formula, dependent on the machine, to correlate to color standards provided by the University of Califonia–Davis (Marsh et al. 1980). The Agtron and Gardner can also be converted to these color scores. In the scientific literature, the L, a, and b values are converted to hue angle (arc tangent b/a).

Consumers associate a red, dark-colored tomato product with good quality. The red color of tomatoes is created by the linear carotenoid lycopene. Lycopene constitutes 80–90% of the carotenoids present. With the onset of ripening, the lycopene content increases (Davies and Hobson 1981, Rick 1978). The final lycopene concentration in the tomato depends on both the variety and the growing conditions. Some tomato varieties have been bred to be very high in lycopene, resulting in a bright red color. During growth, both light level and temperature affect the lycopene content. The effect of light on lycopene content is debated. Some authors report that shading increases lycopene content (Yamaguchi et al. 1960), while others report mixed results (McCollum 1946). The effect of temperature is much more straightforward. At high temperatures, over 30°C, lycopene does not develop (Rabinowitch et al. 1974, Tamburini et al. 1999, Yamaguchi et al. 1960).

VISCOSITY AND CONSISTENCY

For liquid tomato products, viscosity is a very important quality parameter. It is second only to color

as a measure of quality. Viscosity also has economic implications because the higher the viscosity of the tomato paste, the less needs to be added to reach the desired final product consistency. To the scientist, viscosity is determined by analytical rheometers, while consistency is an empirical measurement. To the consumer they are synonyms. Depending on the method, either the viscosity or the consistency of the product may be measured. Tomato products are non-Newtonian; therefore, many methods measure consistency rather than viscosity. The standard method for determining the consistency of most tomato products is the Bostwick consistometer. The Bostwick value indicates how far the material at 20°C flows under its own weight along a flat trough in 30 seconds. Tomato concentrates are typically measured at 12% NTSS to remove the effect of solids. Theoretically, this can be modeled as a slump flow (McCarthy and Seymour 1994). The Bostwick consistometer measures the shear stress under a fixed shear rate. Efflux viscometers such as the Libby tube (for tomato juice) and the Canon-Fenske (for serum viscosity) measure shear rate under fixed shear stress.

The viscosity of tomato products is determined by solids content, serum viscosity, and the physical characteristics of the cell wall material. The solids content is affected by the cultivar, but is primarily determined by the degree of concentration. The serum viscosity is largely determined by the pectin. Pectin is a structural cell wall polysaccharide. The primary component of pectin is polygalacturonic acid, a homopolymer of (1-4) alpha-D-galacturonic acid and rhamnogalacturonans. Some of the carboxyl groups are esterified with methyl alcohol. Pectin methylesterase (PME) removes these ester groups. This leaves the pectin vulnerable to attack by polygalacturonase (PG), which cleaves between the galacturonic acid rings in the middle of the pectin chain, greatly reducing the viscosity. During the break process, heat is used to inactivate pectolytic enzymes, but these enzymes are released during crushing and act very quickly. Genetic modification has been used to produce plants with either an antisense PME (Thakur et al. 1996b) or an antisense PG (Schuch et al. 1991) gene to inactivate the enzyme, producing juice with a significantly higher viscosity. The physical state of the cell wall fragments affects viscosity by determining how easily the particles slide past each other. Most tomato products are homogenized to create more linear particles, which increases the viscosity.

SERUM SEPARATION

Serum separation can be a significant problem in liquid tomato products. Serum separation occurs when the solids begin to settle out of solution, leaving the clear, straw-colored serum as a layer on top of the product. Preventing serum separation requires that the insoluble particles remain in a stable suspension throughout the serum. Generally, the higher the viscosity, the less serum separation occurs. Homogenization significantly reduces serum separation.

FLAVOR

The flavor of tomatoes is determined by the variety used, the stage of ripeness, and the conditions of processing. Typically, varieties have not been bred for optimal flavor, although some work has focused on breeding tomatoes with improved flavor. Processing tomatoes are picked fully ripe; therefore, the concern that tomatoes that are picked mature but unripe have less flavor is not important. Processing generally causes a loss of flavor. Processes are not optimized for the best flavor retention, but practices that maximize color usually also maximize flavor retention. When flavor is evaluated, it is done by sensory evaluation. Gas chromatography is used to determine the exact volatiles present.

Flavor is made up of taste and odor. The sweet-sour taste of tomatoes is due to their sugar and organic acid content. The most important of these are citric acid and fructose (Stevens et al. 1977). The sugar/acid ratio is frequently used to rate the taste of tomatoes, though Stevens et al. (1977) recommend against it because tomatoes with a higher concentration of both sugars and acids taste better than those with low concentrations, for the same ratio. The free amino acids, salts, and their buffers also affect the character and intensity of the taste (Petro-Turza 1987). The odor of tomatoes is created by the over 400 volatiles that have been identified in tomato fruit (Petro-Turza 1987, Thakur et al. 1996a). No single volatile is responsible for producing the characteristic tomato flavor. The volatiles that appear to be most important to fresh tomato flavor include cis-3-hexenal, 2-isobutylthiazole, beta ionone, hexenal, trans-2-hexenal, cis-3-hexenol, trans-2-trans-4-decadienal, 6-methyl-5-hepten-2-one, and 1-penten-3-one (Petro-Turza 1987, Thakur et al. 1996a).

pH AND TITRATABLE ACIDITY

The pH of tomatoes has been reported to range from 3.9 to 4.9, or in standard cultivars, 4.0 to 4.7 (Sapers et al. 1977). The critical issue with tomatoes is to ensure that they have a pH below 4.7, so that they can be processed as high-acid foods. The lower the pH, the greater the inhibition of *Bacillus coagulans,* and the less likely flat sour spoilage will occur (Rice and Pederson 1954). Within the range of mature, red ripe to overly mature tomatoes, the more mature the tomato, the higher the pH. Thus pH is more likely to be a concern at the end of the season. The USDA standards of identity allow organic acids to be added to lower the pH as needed during processing.

The acid content of tomatoes varies according to maturity, climactic conditions, and cultural method. The acid concentration is important because it affects the flavor and pH. Citric and malic are the most abundant acids. The malic acid contribution falls quickly as the fruit turns red, while the citric acid content is fairly stable (Hobson and Grierson 1993). The average acidity of processing tomatoes is about 0.35%, expressed as citric acid (Thakur et al. 1996a). The total acid content increases during ripening to the breaker stage, then decreases.

The relationship between total acidity and pH is not a simple inverse relationship. The phosphorous in the fruit acts as a buffer, regulating the pH. Of the environmental factors, the potassium content of the soil most strongly affects the total acid content of the fruit. The higher the potassium content the greater the acidity.

TOTAL SOLIDS, DEGREES BRIX, NTSS, AND SUGAR CONTENT

Tomato solids are important because they affect the yield and consistency of the finished product. Due to the time required to make total solids measurements, soluble solids are more frequently measured. Soluble solids are measured with a refractometer that measures the refractive index of the solution. The refractive index is dependent on the concentration and temperature of solutes in the solution; therefore, many refractometers are temperature controlled. The majority of the soluble solids are sugars, so refractometers are calibrated directly in percentage sugar, or degrees Brix. Natural tomato soluble solids (NTSS) are the same as degrees Brix, minus any added salt.

The sugar content reaches a peak in tomatoes

when the fruit is fully ripe (Hobson and Gierson 1993). Light probably has a more profound effect on sugar concentration in tomatoes than any other environmental factor (Davies and Hobson 1981). The seasonal trends in the sugar content of greenhouse grown tomatoes have been found to roughly follow the pattern of solar radiation (Winsor and Adams 1976). Even the minor shading that is provided by the foliage reduces the total sugar content by up to 13% (McCollum 1946).

FINISHED PRODUCT

SPOILAGE

Based on experience, spoilage of tomato products other than juice and whole tomatoes is caused by non–spore-forming aciduric bacteria (Denny 1997). These bacteria are readily destroyed by processes in which the inside of the can reaches at least 85°C. Spoilage of whole tomatoes can be caused by these same microorganisms, but whole tomatoes are also susceptible to spoilage by spore formers such as *Clostridium pasteurianum*. Juice is commonly spoiled by *Bacillus coagulans* (formerly *B. thermoacidurans*). In the past, flat sour spoilage due to *B. coagulans* was a major problem in tomato products. Flat sour spoilage causes off flavors and odors, and the pH of the juice drops to 3.5. The spores of these microbes are too resistant to heat to be destroyed by practical heat treatments at 100°C if they are present in high numbers, so they must be controlled by limiting initial levels or by processing at temperatures above the boiling point. These organisms occur in the soil and grow on some equipment (Denny 1997). The National Canners Association (NCA) recommendation for eliminating *Clostri-* *dium* spores is $F_{93°C} = 10$ minutes for pH above 4.3, and $F_{93°C} = 5$ minutes for pH below 4.3. Against spores of *B. coagulans*, the recommendation is $F_{107°C} = 0.7$ minutes at pH 4.5 (NCA 1968).

Historically, the occurrence of swelled cans is most commonly due to either hydrogen swells or growth of *C. pasteurianum*. *C. pasteurianum* produces carbon dioxide, so determination of the type of gas in the headspace is one way to determine the cause.

QUALITY CHANGES DURING PROCESSING

The type of process is important in determining how much quality loss occurs. For the same F value, significantly more vitamin C is lost during thermal processing of whole peeled tomatoes in a rotary pressure cooker than in a high-temperature, short-time (HTST) process (Leonard et al. 1986). Similarly, the texture is significantly firmer after the HTST processing (Leonard et al. 1986). During canning, the nutrient content remains fairly stable (Table 29.1). The already small lipid content decreases because of the removal of the skin. The calcium and sodium contents increase because the processors add them to improve the firmness and flavor of the tomatoes. The vitamin A content is fairly constant, while the vitamin C content is reduced by 45%. Bioavailable lycopene content increases, because processing makes the carotenoid more available to the body (Gartner et al. 1997, Stahl and Sies 1992).

Color loss is accelerated by high temperature and exposure to oxygen during processing. The red color of tomatoes is mainly determined by the carotenoid lycopene, and the main cause of lycopene degradation is oxidation. Oxidation is complex and depends

Table 29.1. Nutrition Composition of Tomatoes, Value per 100g of Edible Portion

	Raw	Canned (salt added)	Daily Values
Water (g)	93.76	93.65	
Protein (g)	0.85	0.92	50
Total lipid (g)	0.33	0.13	65
Carbohydrate, by difference (g)	4.64	4.37	300
Fiber, total dietary (g)	1.1	1.0	25
Calcium, Ca (mg)	5	30	1000
Sodium, Na (mg)	9	148	2400
Zinc, Zn (mg)	0.09	0.16	15
Vitamin C (mg)	19.1	14.2	60
Vitamin A (IU)	623	595	5000

Source: Adapted from the USDA Nutrient Database

on many factors, including processing conditions, moisture, temperature, and the presence of pro- or antioxidants. Several processing steps are known to promote oxidation of lycopene. During hot break, the hotter the break temperature, the greater the loss of color, even when operating under a vacuum (Trifiro et al. 1998). However in some varieties the break temperature affects color while in others it does not (Fonseca and Luh 1977). The use of fine screens in juice extraction enhances oxidation because of the large surface area exposed to air and metal (Kattan et al. 1956). Similarly, concentrating tomato juice to paste in the presence of oxygen degrades lycopene. It has been reported that heat concentration of tomato pulp can result in up to 57% loss of lycopene (Noble 1975). However, other authors have reported that lycopene is very heat resistant and that no changes occur during heat treatment (Khachik et al. 1992). With current evaporators it is likely very little destruction of lycopene occurs.

Processing also affects color due to the formation of brown pigments. This is not necessarily detrimental, because a small amount of thermal damage resulting in a darker serum color increases the overall red appearance of tomato paste (Leonard et al. 1986). Browning is caused by a number of reactions. Excessive heat treatments can cause browning due to caramelization of the sugars. Amadori products, representing the onset of the Maillard reaction, occur during all stages of processing, including breaking, concentrating, and canning (Eichner et al. 1996). However, during production of tomato paste the Maillard reaction is still of minor importance (Eichner et al. 1996). Degradation of ascorbic acid has been suggested to be the major cause of browning (Mudahar et al. 1986). Processing and storage at lower temperatures, decreasing the pH to 2.5, and the addition of sulfites can decrease browning (Danziger et al. 1970).

Canning significantly softens the fruit, so calcium is frequently added to increase the firmness. Varieties have been bred to be firm to withstand machine harvesting, which has also increased the firmness of canned tomatoes. Conditions during processing such as temperature, screen size, and blade speed will affect the final viscosity of the juice. Hot break juice typically has a higher viscosity than cold break juice due to inactivation of the enzymes that degrade pectin. At very high break temperatures, such as 100°C, the structure collapses and the viscosity decreases again (Trifiro et al. 1998), although this effect is not always observed (Luh and Daoud

1971). The screen size and blade speed during extraction are also important factors. The effect of screen size is not a simple relationship. A higher viscosity is produced using a screen size of 1.0 mm than either 0.5 mm or 1.5 mm (Robinson et al.1956). Other studies have found no effect of finisher size on final viscosity (Trifiro et al. 1998). The faster the blade is, the higher the viscosity. The higher the evaporation temperature is, the greater the loss of viscosity (Trifiro et al. 1998).

Factors that affect the quantity and quality of the solids determine the degree of serum separation that occurs. The higher the temperature during the break process, the less serum separation occurs (Trifiro et al. 1998). Hot break juice has less serum separation than cold break juice. This may be due to greater retention of intact pectin in the hot break juice (Luh and Daoud 1971), although Robinson et al. (1956) found that the total amount of pectin did not affect the degree of settling in tomato juice. The cellulose fiber may be more important in preventing serum separation than the pectin (Robinson et al. 1956, Shomer et al. 1984). Addition of pectinases degrades the pectin, increasing the dispersal of cellulose from the cell walls. The expansion of this cellulose minimizes serum separation (Shomer et al. 1984).

Homogenization is commonly used to shred the cells, increasing the number of particles in solution and creating cells with ragged edges that reduce serum separation. The result is particles that will not efficiently pack and settle. Of these two effects, changing the shape of the particles is more important than change in size (Shomer et al. 1984). Evaporator temperature during concentration has little effect on serum separation (Trifiro et al. 1998).

Processed tomato products have a distinctively different aroma from fresh tomato products. This is due to both the loss and the creation of volatiles. Heating drives away many of the volatiles. Oxidative decomposition of carotenoids causes the formation of terpenes and terpenelike compounds, and the Maillard reaction produces volatile carbonyl and sulfur compounds.

Many of the volatiles responsible for the fresh tomato flavor are lost during processing, especially cis-3-hexenal and hexenal (Buttery et al. 1990b). Cis-3-hexenal, an important component of fresh tomato flavor, is rapidly transformed into the more stable trans-2-hexenal; therefore, it is not present in heat-processed products (Kazeniac and Hall 1970). The amount of 2-isobutylthiazole, responsible for a tomato leaf green aroma, diminishes during manufacture of tomato puree and paste (Chung et al. 1983).

Other volatiles are created. Breakdown of sugars and carotenoids produce compounds responsible for the cooked odor. Dimethyl sulfide is a major contributor to the aroma of heated tomato products (Buttery et al. 1971, 1990b, Guadagni et al. 1968, Thakur et al. 1996a). Its contribution to the characteristic flavor of canned tomato juice is more than 50% (Guadagni et al. 1968). Linalool (Buttery et al. 1971), dimethyl trisulfide, 1-octen-3-one (Buttery et al. 1990a), acetaldehyde, and geranylacetone (Kazeniac and Hall 1970) may also contribute to the cooked aroma. Pyrrolidone carboxylic acid, which is formed during heat treatment, has been blamed for an off flavor that occasionally appears (Mahdi et al. 1961). This compound, formed by cyclization of glutamine, arises as early as the break process (Eichner et al. 1996).

Heating causes degradation of some flavor volatiles and inactivates lipoxygenase and associated enzymes that are responsible for producing some of the characteristic fresh tomato flavor (Goodman et al. 2002). However, some authors (Fonseca and Luh 1977) have found that hot break produces a better flavor, while others (Goodman et al. 2002) have found that it produces a less fresh flavor. Within one study, the flavor of one variety may be rated better as cold break juice than as hot break juice, and another variety the reverse (Fonseca and Luh 1976, 1977). This may in part be because some panelists prefer the flavor of heat-treated tomato juice to fresh juice (Guadagni et al. 1968).

Processing conditions further affect the pH and acidity of processed tomato products. During processing, the pH decreases and total acid content increases (Hamdy and Gould 1962, Miladi et al. 1969), although the citric acid content may increase (Miladi et al. 1969) or decrease (Hamdy and Gould 1962). Hot break juice has a lower titratable acidity (Gancedo and Luh 1986) and higher pH than cold break juice (Fonseca and Luh 1976, Luh and Daoud 1971). The difference is caused by breakdown of pectin by pectolytic enzymes that are still present in the cold break juice (Stadtman et al. 1977).

During heat treatment, the reducing sugar content decreases due to caramelization, Maillard reaction, and the formation of 5-hydroxymethyl furfural. The amount of sugar lost depends on the process. Studies have reported as much as a 19% loss in processed tomato juice (Miladi et al. 1969) and a 5% loss during spray drying (Alpari 1976).

QUALITY CHANGES DURING STORAGE

Changes in flavor are the most sensitive index to quality deterioration during storage, followed by color (Eckerle et al. 1984). The Maillard reaction is the major mode of deterioration during storage of canned fruit and vegetable products, in general, and leads to a bitter off flavor [Office of Technology Assessment (OTA) 1979]. A number of studies have used hedonic measurements to determine the end of shelf life for tomato products. However, many of these studies did not go on long enough to find the end of shelf life. No significant differences were found between the flavor of tomato concentrates stored for six months at 4°C and those stored at 21°C for the same period (McColloch et al. 1956). The samples at 38°C were significantly different; however, neither the fresh nor the stored sample was preferred. Canned tomatoes stored for three years at 21°C were rated fair, due to a slightly stale, bitter or tinny off flavor (Cecil and Woodruf 1963, 1962). Storage at 21°C should be limited to 24–30 months, and that at 38°C to less than a year.

There is little problem with color changes during storage. When no oxygen is present, the red pigment lycopene slowly degrades by an autocatalytic mechanism. No loss of lycopene was seen in hot break tomato puree that was stored up to a year (Tamburini et al. 1999). Cold break puree did show a loss of lycopene, likely due to enzymatic activity (Tamburini et al. 1999). In addition to degradation of lycopene, darkening occurs during storage due to nonenzymatic browning (Mudahar 1986). Typically, the color does not change during storage if the product is kept at room temperature or below (Davis and Gould 1955, Kattan et al. 1956). No difference in serum color was seen after 300 days at 20°C, for either hot or cold break tomato paste (Luh et al. 1964.) When stored at 31°C, cold break paste did darken faster than hot break paste (Luh et al. 1964). Extreme conditions of 12 months at 88°C were required to reduce the color of tomato juice to grade C (Gould 1978). Products stored at lower temperatures or shorter times were still grade A.

Vitamin C is the most labile of the nutrients, so its degradation is used as an indicator of quality. No loss in natural vitamin C was found in tomato juice after nine months of storage at up to 20°C (Gould 1978). In another study, some losses were seen at 31°C. After 1.2 years, some degradation of vitamin C was seen at storage temperatures of 6–11°C (Luh et al. 1958), but at least 80% was still present when stored at 6–20°C. At 25°C, 55% remained. When samples were fortified with vitamin C, this added vitamin C degraded at storage temperatures as low as 2°C. This occurs because the added vitamin C is not bound or protected in the juice the way the natural vitamin C is.

APPLICATION OF PROCESSING PRINCIPLES

Table 29.2. lists some examples illustrating specific processing stages and the principle(s) involved in the manufacturing of tomato products, as well as references where additional information may be found.

Table 29.2. References for Further Information on Processing Principles

Processing Stage	Processing Principle(s)	References for More Information on the Principles Used
Peeling	Raw material preparation	Downing 1996b, Gould 1992
Dicing or slicing	Size reduction	Fellows 2000, Gould 1996
Hot break	Enzyme inactivation and size reduction	Hayes et al. 1998, Thakur et al. 1996
Extraction	Mechanical separation	Hayes et al. 1998, Thakur et al. 1996
Deaeration	Mechanical separation	Downing 1996b, Smith et al. 1997
Homogenization	Size reduction	Thakur et al. 1996, Downing 1996b
Concentration to paste	Concentration	Hayes et al. 1998, Smith et al. 1997
Aseptic processing	Heat pasteurization	Hayes et al. 1998, Downing 1996b
Retorting	Heat pasteurization	Fellows 2000, Downing 1996b

GLOSSARY

Aseptic processing—process of heating the sample to pasteurize or sterilize it, followed by filling into a previously sterilized container.

BOD—biochemical oxygen demand.

Bostwick consistometer—standard device for measuring the consistency of tomato products.

Break—process step in which tomatoes are chopped and heated to inactivate enzymes.

CFR—Code of Federal Regulations.

Degrees Brix (° Brix)—percent soluble solids, typically considered to be sugar.

Extraneous matter—stems, vines, dirt, stones, and trash.

F-value—the number of minutes required to kill the desired number of microorganisms at the stated temperature, assume a z value of 10°C.

Flume—transportation system in which the product is carried by a stream of water.

Homogenization—process of shredding the particles under high pressure and shear.

HTST—high-temperature, short-time process for pasteurization or sterilization.

Lycopene—a linear carotenoid responsible for the red color in tomatoes.

NTSS—natural tomato soluble solids.

OTA—Office of Technology Assessment.

PG—polygalacturonase.

Photoelectric sorter—automated sorter that reads the color of individual products and removes the unacceptable ones.

PME—pectin methylesterase.

Polygalacturonase—enzyme responsible for rapid viscosity loss in juice, if it is active.

Retort—a sealed vessel for processing containers at high temperatures and pressures.

Serum separation—when the solids begin to settle out of solution, there is a clear, straw-colored serum as a layer on top of the product.

USDA—U.S. Department of Agriculture.

REFERENCES

Adsule PG, D Amba, H Onkarayya. 1982. Effects of hot water dipping on tomatoes. Indian Food Packer 36(5): 34–37.

Alpari A. 1976. Changes in the quality characteristics of tomato puree during spray drying. Acta Aliment, 5:303–313.

Bartz JA. 1999. Washing fresh fruits and vegetables: Lessons from treatment of tomatoes and potatoes with water. Dairy Food Environ Sanitation 19(12): 853–864.

Bayindirli L. 1994. Mathematical analysis of lye peeling in tomatoes. J Food Eng 23:225–231.

Brown HE, F Meredith, G Saldana, TS Stephens. 1970. Freeze peeling improves quality of tomatoes. J Food Sci 35:485–488.

Buttery RG, RM Seifert, DG Guadagni, LC Ling. 1971. Characterization of additional volatile components of tomato. J Agr Food Chem 19(3): 524–529.

Buttery RG, R Teranishi, RA Flath, LC Ling. 1990a. Identification of additional tomato paste volatiles. J Agric Food Chem 38:792–795.

Buttery RG, R Teranishi, LC Ling, JG Turnbaugh. 1990b. Quantitative and sensory studies on tomato paste volatiles. J Agric Food Chem 38:336–340.

California Department of Food and Agriculture. 2001. California Processing Tomato Inspection Program. West Sacramento: California Department of Food and Agriculture, Marketing Branch.

Castaldo D, L Servillo, B Laratta, G Fasanaro, G Villari, A de Giorgi, A Giovane. 1995. Preparation of high-consistency vegetable products: Tomato pulps. II. Ind Conserve 70(3): 253–258.

Castaldo D, G Villari, B Laratta, M Impembo, A Giovane, G Fasanaro, L Servillo. 1996. Preparation of high-consistency diced tomatoes by immersion in calcifying solutions. A pilot plant study. J Agric Food Chem 44:2600–2607.

Cecil SR, JG Woodruf. 1962. Long-term storage of military rations. Ga Agric Exp Stn Bull 25.

___. 1963. Stability of canned foods in long-term storage. Food Technol. 17:131–138.

Chung T-Y, F Hayase, H Kato. 1983. Volatile components of ripe tomatoes and their juices, purees and pastes. Agric Biol Chem 47(2): 343–351.

Code of Federal Regulations (CFR). 2000. 21CFR155.190 Code of Federal Regulations. Washington D.C.: U.S. General Services Administration, National Archives and Records Service, Office of the Federal Register.

Danziger MT, MP Steinberg, AI Nelson. 1970. Thermal browning of tomato solids as affected by concentration and inhibitors. J Food Sci 35:808–810.

Das DJ.1997. Factors effecting the peelability of tomatoes and methods to improve chemical peeling for tomatoes. MSc Thesis. Columbus: The Ohio State Univ.

Das DJ, SA Barringer. 1999. Use of organic solvents for improving peelability of tomatoes. J Food Process Preserv 23(4): 193–202.

Davis RB, WA Gould. 1955. The effect of processing methods on the color of tomato juice. Food Tech 9:540–547.

Davies JN, GE Hobson. 1981. The constituents of tomato fruit—The influence of environment, nutrition and genotype. Crit Reviews Food Sci Nutr 15(3): 205–280.

Denny C, editor. 1997. Tomato Products, 7th edition. Washington, D.C.: National Food Processors Association.

Downing DL, editor. 1996a. A Complete Course in Canning and Related Processes, 13th edition. Book I. Fundamental Information on Canning. Timonium, Md.: CTI Publications, Inc.

___. 1996b. A Complete Course in Canning and Related Processes, 13th edition. Book III. Processing Procedures for Canned Food Products. Timonium, Md.: CTI Publications, Inc.

Eckerle JR, CD Harvey, T-S Chen. 1984. Life cycle of canned tomato paste: Correlation between sensory and instrumental testing methods. J Food Sci 49:1188–1193.

Eichner K, I Schrader, M Lange. 1996. Early detection of changes during heat processing and storage of tomato products. In: T-C Lee, H-J Kim, editors. Chemical Markers for Processed and Stored Foods, 32–53. Washington D.C.: American Chemical Society.

Fellows PJ. 2000. Food Processing Technology Principles and Practice. Cambridge, U.K.: Woodhead Publishing Ltd.

Floros JD, A Ekanayake, GP Abide, PE Nelson. 1992. Optimization of a diced tomato calcification process. J Food Sci 57(5): 1144–1148.

Fonseca H, BS Luh. 1976. Effect of break temperature on quality of tomato juice reconstituted from frozen tomato concentrates. J Food Sci 41:1308–1311.

___. 1977. Effect of break condition on quality of canned tomato juices. Confructa 22(5/6): 176–181.

Gancedo MC, BS Luh. 1986. HPLC analysis of organic acids and sugars in tomato juice. J Food Sci 51(3): 571–573.

Gartner C, W Stahl, H Sies. 1997. Lycopene is more bioavailable from tomato paste than from fresh tomatoes. Am J Clin Nutr 66:116–122.

Giovannucci EL. 1999. Tomatoes, tomato-based products, lycopene, and cancer: Review of the epidemiologic literature. J Natl Cancer Inst 91(4): 317–331.

Giovannucci EL, A Ascherio, EB Rimm, MJ Stampfer, GA Colditz, WC Willett. 1995. Intake of carotenoids and retinal in relationship to risk of prostate cancer. J Natl Cancer Inst 87(23): 1767–1776.

Goodman C, S Fawcett, SA Barringer. 2002. Flavor, viscosity, and color analyses of hot and cold break tomato juices. J Food Sci 67(1): 404–408.

Gould WA. 1978. Quality evaluation of processed tomato juice. J Agric Food Chem 26(5): 1006–1011.

___. 1992. Tomato production, processing and technology. Baltimore, Md.: CTI Publishing Co.

___. 1996. Unit operations for the food industries. Baltimore, Md.: CTI Publications, Inc.

Guadagni DG, JC Miers, D Venstrom. 1968. Methyl sulfide concentration, odor intensity, and aroma quality in canned tomato juice. Food Technol 22:1003–1006.

Hamdy MM, WA Gould. 1962. Varietal differences in tomatoes: a study of alpha-keto acids, alpha-amino compounds, and citric acid in eight tomato varieties before and after processing. J Agric Food Chem, 10:499–503.

Hayes WA, PG Smith, AEJ Morris. 1998. The production and quality of tomato concentrates. Critical Reviews in Food Science and Nutrition 38(7): 537–564.

Heil JR, S Leonard, H Patino. 1984. Microbiological evaluation of commercial fluming of tomatoes. Food Technol 38(4): 121–126.

Hobson G, D Grierson. 1993. Tomato. In: GB Seymour, JE Taylor, GA Tucker, editors. Biochemistry of Fruit Ripening, 405–530. New York: Chapman and Hall.

Kattan AA, WL Ogle, A Kramer. 1956. Effect of processed variables on quality of canned tomato juice. Proc Am Soc Hort Sci 68:470–481.

Kazeniac SJ, RM Hall. 1970. Flavor chemistry of tomato volatiles. J Food Sci 35:519–530.

Khachik F, MB Goli, GR Beecher, J Holden, WE Lusby, MD Tenoirio, MR Barrera. 1992. Effect of food preparation on qualitative and quantitative distribution of major carotenoid constituents of tomatoes and several green vegetables. J Agric Food Chem, 40(3): 390–398.

Leonard S, F Winter. 1974. Pilot application of freeze-heat peeling of tomatoes. J Food Sci 39:162–165.

Leonard SJ, RL Merson, GL Marsh, JR Heil. 1986. Estimating thermal degradation in processed foods. J Agric Food Chem. 34:392–396.

Luh BS, HN Daoud. 1971. Effect of break temperature and holding time on pectin and pectic enzymes in tomato pulp. J Food Sci 36:1039–1043.

Luh BS, S Leonard, GL Marsh. 1958. Objective criteria for storage changes in tomato paste. Food Tech 12:347–351.

Luh BS, SJ Leonard, F Villarreal, M Yamaguchi. 1960. Effect of ripeness level on consistency of canned tomato juice. Food Technol 14:635–639.

Luh BS, CO Chichester, H Co, SJ Leonard. 1964. Factors influencing storage stability of canned tomato paste. Food Technol 18(4): 159–162.

Mahdi AA, AC Rice, KG Weckel. 1961. Effect of pyrrolidonecarboxylic acid on flavor of processed fruit and vegetable products. J Agric Food Chem 9:143–146.

Marsh GL, J Buhlert, S Leonard, T Wolcott, J Heil. 1980. Color scoring tomato products objectively. Davis: University of California.

McCarthy KL, JD Seymour. 1994. Gravity current analysis of the Bostwick consistometer for power law foods. J Texture Studies; 25(2): 207–220.

McColloch RJ, RC Rice, JC Underwood. 1956. Storage stability of canned concentrated tomato juice. Food Technol 10:568–570.

McCollum JP. 1946. Effect of sunlight exposure on the quality constituents of tomato fruits. Proc Am Soc Hortic Sci 48:413–416.

Miladi SS, WA Gould, RL Clements. 1969. Heat processing effect on starch, sugars, proteins, amino acids, and organic acids of tomato juice. Food Technol 23:691–693.

Mohr WP. 1990. The influence of fruit anatomy on ease of peeling of tomatoes for canning. Int J Food Sci Tech 25:449–457.

Mudahar GS, JS Sidhu, KS Minhas. 1986. Technical note: effect of low pH preservation on the color and consistency of tomato juice. J Food Technol 21:233–238.

Napoli MA. 1979. A study of treatment of wastewater from a peeled-tomato factory. Agric Wastes 1:143–156.

National Canners Association (NCA). 1968. Laboratory Manual for Food Canners and Processors. Vol. 1. Microbiology and Processing. Westport, Conn.: AVI Publishing Co, Inc.

Neumann HJ, S WG chultz, JP Morgan, JE Schade. 1978. Peeling aids and their application to caustic peeling of tomatoes. J Food Sci 43(5): 1626–1627.

Noble AC. 1975. Investigation of the color changes in heat concentrated tomato pulp. J Agr Food Chem 23(1): 48–49.

Noomhorn A, A Tansakul. 1992. Effect of pulper-finisher operation on quality of tomato juice and tomato puree. J Food Proc Eng 15:229–239.

Office of Technology Assessment (OTA). 1979. Open Shelf-life dating of foods. Washington, D.C.: OFA, Government Printing Office.

Pearson GA. 1972. Suitability of food processing wastewater for irrigation. J Environ Quality 1(4): 394–397.

Petro-Turza M. 1987. Flavor of tomato and tomato products. Food Rev Intl 2(3): 309–351, 1987.

Poretta S, G Poli, L Palmieri. 1995. Optimization of the addition of calcium chloride to canned diced tomatoes. Sci Aliment 15:99–112.

Rabinowitch HD, N Kedar, P Budowski. 1974. Induction of sunscald damage in tomatoes under natural and controlled conditions. Sci Hortic 2(3): 265–272.

Rice AC, CS Pederson. 1954. Factors influencing growth of Bacillus coagulans in canned tomato juice. 2. Acidic constituents of tomato juice and specific organic acids. Food Res 19:124–133.

Rick CM. 1978. The tomato. Sci Am 239(2): 66–76.

Robinson WB, LB Kimball, JR Ransford, JC Moyer, DB Hand. 1956. Factors influencing the degree of settling in tomato juice. Food Technol 10:109–112.

Sapers GM, JG Phillips, AK Stoner. 1977. Tomato acidity and the safety of home canned tomatoes. Hortscience 12:204–208.

Schlimme DV, KA Corey, BC Frey. 1984. Evaluation of lye and steam peeling using four processing tomato cultivars. J Food Sci 49:1415–1418.

Schuch W, J Kanczler, D Robertson, G Hobson, G Tucker, D Grierson, S Bright, C Bird. 1991. Fruit quality characteristics of transgenic tomato fruit with altered polygalacturonase activity. Hortscience 26:1517–1520.

Senti FR, RL Rizek. 1975. Nutrient levels in horticultural crops. HortScience 10:243–246.

Shomer I, P Lindner, R Vasiliver. 1984. Mechanism which enables the cell wall to retain homogenous appearance of tomato juice. J Food Sci 49:628–633.

Smith DS, JN Cash, WK Nip, YH Hui, editors. 1997. Processing Vegetables: Science and Technology. Lancaster, Pa.: Technomic Publishing Co.

Stadtman FH, JE Buhlert, GL Marsh. 1977. Titratable acidity of tomato juice as affected by break procedure. J Food Sci 42(2): 379–382.

Stahl W, H Sies. 1992. Uptake of lycopene and its geometrical isomers is greater from heat-processed than from unprocessed tomato juice in humans. J Nutr 122(11): 2161–2165.

Stephens TS, G Saldana, HE Brown. 1967. Peeling tomatoes by submerging in a hot solution of calcium chloride. Annu Proc J Rio Grande Valley Hort Soc 21:114–124.

___. 1973. Effect of different submergence times in hot calcium chloride on peeling efficiency of tomatoes. J Food Sci 38:512–515.

Stevens MA, AA Kader, M Albright-Holton, M Algazi. 1977. Genotypic variation for flavor and composition in fresh market tomatoes. J Am Soc Hort Sci 102(5): 680–689.

Tamburini R, L Sandei, A Aldini, F de Sio, C Leoni. 1999. Effect of storage conditions on lycopene content in tomato purees obtained with different processing techniques. Industria-Conserve 74(4): 341–357.

Thakur BR, RK Singh, PE Nelson. 1996a. Quality attributes of processed tomato products: A review. Food Rev Int 12(3): 375–401.

Thakur BR, RK Singh, DM Tieman, AK Handa. 1996b. Tomato product quality from transgenic fruits with reduced pectin methylesterase. J Food Sci 61(1): 85–87, 108.

Thomas WM, DW Stanley, DR Arnott. 1976. An evaluation of blanch, lye and freeze-heat methods for tomato peel removal. Can Inst Food Sci Technol J 9(3): 118–124.

Thomas WM, WP Mohr, DW Stanley, DR Arnott. 1978. Evaluation of conventional and freeze heat peeling methods for field tomatoes. Can Inst Food Sci Technol J 11(4): 209–215.

Trandin GG, GA Vlasov, AP Volkov, AV Kirpil. 1982. Use of hot water for washing mechanically harvested tomatoes. Konserv Ovoshch Promysh 9:22–23.

Trifiro A, S Gherardi, C Zoni, A Zanotti, M Pistocchi, G Paciello, F Sommi, PL Arelli, MAM Antequera. 1998. Quality changes in tomato concentrate production: Effects of heat treatments. Industria Conserve 73(1): 30–41.

U.S. Department of Agriculture (USDA). 1983. United States Standards for grades of tomatoes for processing. Washington D.C.: Fruit and Vegetable Division, Agricultural Marketing Service, USDA.

U.S. Department of Agriculture (USDA). 1990. United States Standards for grades of canned tomatoes. Washington DC: Fruit and Vegetable Division, Agricultural Marketing Service, USDA.

U.S. Department of Agriculture (USDA). 2000. Agricultural Statistics. Washington D.C.: U.S. Government Printing Office.

Villari G, F de Sio, R Loiudice, D Castaldo, A Giovane, L Servillo. 1997. Effect of firmness of canned peeled tomatoes dipped in calcium solution at neutral pH. Acta Alimentaria 26(3): 235–242.

Whittenberger RT, GC Nutting. 1957. Effect of tomato cell structures on consistency of tomato juice. Food Technol 11(1): 19–22.

Winsor GW, P Adams. 1976. Changes in the composition and quality of tomato fruit throughout the season, 134–142. Annu Rep Glasshouse Crops Res Inst 1975.

Yamaguchi M, FD Howard, BS Luh, SJ Leonard. 1960. Effect of ripeness and harvest date on the quality and composition of fresh canning tomatoes. Proc Am Soc Hortic Sci 76:560–567.

Zacconi C, A Causarano, P Dallavalle, A Casana. 1999. Monitoring of contaminating microflora in the production of tomato products. Industria Conserve 74(2): 133–144.

Index

LEEDS TRINITY UNIVERSITY

Lightning Source UK Ltd.
Milton Keynes UK
UKOW020711230612

194897UK00001B/23/P